FACETS OF
SYSTEMS SCIENCE

International Federation for Systems Research
International Series on Systems Science and Engineering

Series Editor: George J. Klir
State University of New York at Binghamton

Editorial Board

Gerrit Broekstra
Erasmus University, Rotterdam,
  The Netherlands
John L. Casti
Technical University of Vienna, Austria
Brian Gaines
University of Calgary, Canada

Ivan M. Havel
Charles University, Prague,
  Czechoslovakia
Manfred Peschel
Academy of Sciences, Berlin, Germany
Franz Pichler
University of Linz, Austria

Volume 1	ANTICIPATORY SYSTEMS: Philosophical, Mathematical, and Methodological Foundations Robert Rosen
Volume 2	FOUNDATIONS OF MATHEMATICAL SYSTEM DYNAMICS: The Fundamental Theory of Causal Recursion and Its Application to Social Science and Economics Arvid Aulin
Volume 3	METASYSTEMS METHODOLOGY: A New Synthesis and Unification Arthur D. Hall, III
Volume 4	PREDICTIVE SIMPLICITY: Induction Exhumed Kenneth S. Friedman
Volume 5	DYNAMICS AND THERMODYNAMICS IN HIERARCHICALLY ORGANIZED SYSTEMS Pierre Auger
Volume 6	SELF-MODIFYING SYSTEMS: A New Framework for Dynamics, Information, and Complexity Gyorgy Kampis
Volume 7	FACETS OF SYSTEMS SCIENCE George J. Klir

IFSR was established "to stimulate all activities associated with the scientific study of systems and to coordinate such activities at international level." The aim of this series is to stimulate publication of high-quality monographs and textbooks on various topics of systems science and engineering. This series complements the Federation's other publications.

A Continuation Order Plan is available for this series. A continuation order will bring delivery of each new volume immediately upon publication. Volumes are billed only upon actual shipment. For further information please contact the publisher. Volumes 1–6 were published by Pergamon Press.

FACETS OF SYSTEMS SCIENCE

GEORGE J. KLIR
State University of New York at Binghamton
Binghamton, New York

PLENUM PRESS • NEW YORK AND LONDON

Library of Congress Cataloging-in-Publication Data

Klir, George J., 1932-
    Facets of systems science / George J. Klir.
        p.   cm. -- (International Federation for Systems Research
    international series on systems science and engineering ; v. 7)
    Includes bibliographical references and index.
    ISBN 0-306-43959-X
    1. System theory.   I. Title.   II. Series: IFSR international
series on systems science and engineering ; v. 7.
Q295.K554 1991
003--dc20   91-25765
                                                                  CIP

ISBN 0-306-43959-X

© 1991 Plenum Press, New York
A Division of Plenum Publishing Corporation
233 Spring Street, New York, N.Y. 10013

All rights reserved

No part of this book may be reproduced, stored in a retrieval system, or transmitted
in any form or by any means, electronic, mechanical, photocopying, microfilming,
recording, or otherwise, without written permission from the Publisher

Printed in the United States of America

Nothing can be loved or hated unless it is first understood.
—Leonardo Da Vinci

Preface

This book has a rather strange history. It began in Spring 1989, thirteen years after our Systems Science Department at SUNY–Binghamton was established, when I was asked by a group of students in our doctoral program to have a meeting with them.

The spokesman of the group, Cliff Joslyn, opened our meeting by stating its purpose. I can closely paraphrase what he said: "We called this meeting to discuss with you, as Chairman of the Department, a fundamental problem with our systems science curriculum. In general, we consider it a good curriculum: we learn a lot of concepts, principles, and methodological tools, mathematical, computational, heuristic, which are fundamental to understanding and dealing with systems. And, yet, we learn virtually nothing about systems science itself. What is systems science? What are its historical roots? What are its aims? Where does it stand and where is it likely to go? These are pressing questions to us. After all, aren't we supposed to carry the systems science flag after we graduate from this program? We feel that a broad introductory course to systems science is urgently needed in the curriculum. Do you agree with this assessment?"

The answer was obvious and, yet, not easy to give: "I agree, of course, but I do not see how the situation could be alleviated in the foreseeable future. Systems science is still in its forming stage and there is little agreement in the professional community of what it is or what it should be. There is also no textbook for such a course and, above all, who would teach it?"

To my amazement, the question was not answered by words, but by a group of forefingers pointed, in a perfect unison, toward me. "Wishful thinking," was my immediate reaction. "Look," I continued, "for some ten years, when I was very active in the systems movement, I was asked countless number of times to deliver key-note addresses, after-dinner speeches, and the like regarding the systems movement, its history, state of the art, or its future. I eventually got so tired of these activities that I started to view them as a waste of time, diverting me from serious research. What I am really saying is that these overview lectures are not my cup of tea anymore. And, now, you want me to give such lectures for the whole semester. Although your point is well taken and I will do my best to help you, you clearly have to find someone else to teach the course."

The students did not give up. "Why not think it over and have another meeting with us," said one of them. "And, by the way, we will do our utmost to help you with the course in any way you would desire."

There were, in fact, a few more meetings. Their outcome, to make the long story short, was most unexpected: I eventually agreed to teach the course, on an experimental basis, in Fall 1989. After making this commitment, I tried to develop some enthusiasm for preparing the course, but all my efforts in this respect were in vain. This lack of enthusiasm continued until I actually started to teach the course. Then, to my surprise, my attitude toward the course changed and I actually began to enjoy preparing and teaching it. The main factor inducing this change, I suspect, was the students. I could feel their enthusiasm for the course and how grateful they were that it was offered.

The course, which was entitled *Introduction to Systems Science*, was a graduate course. It was taken by 23 highly motivated and hard-working graduate students. Lectures in the course were supplemented by heavy reading assignments each week. During the semester, the students were required to read 46 carefully selected classic papers and two classic books on systems science, *General Systems Theory* by Ludwig von Bertalanffy and *An Introduction to Cybernetics* by W. Ross Ashby. The purpose of these reading assignments was to provide the students with additional information on topics that were covered only briefly in the lectures.

The only requirement in the course was to write a term paper that would overview, in a coherent and critical fashion, the material learned from the lectures and assigned readings. In particular, a detailed critical evaluation of each assigned reading was required. Information contained in these papers turned out to be very valuable. It gave me confidence that, in general, the course was well conceived, and, at the same time, it provided me with some guidance for improving it.

One outcome of the course is this text, on which I began to work immediately after the course was completed. It has two parts. Part I, entitled "Systems Science: A Guided Tour," is based upon the class notes I prepared for the lectures; Part II, entitled "Classical Systems Literature," consists of reprints of significant papers that elaborate on some of the topics covered in Part I. The whole book, especially Part II, is heavily influenced by the feedback I received from the students. Some papers that were on the reading list of the course are not included here, while some others are.

The primary purpose of this book is to help the reader to develop an adequate general impression of what systems science is, what its main historical roots are, what its relationship is with other areas of human affairs, what its current status is, and what its role in the future is likely to be. In addition, it helps the reader to identify sources for further study of various aspects of systems science. As suggested by the title, *Facets of Systems Science*, no attempt is made in the book to cover systems science in a comprehensive way. The presentation is not technical and the use of mathematics is minimized. The few mathematical concepts that are used in the book are introduced in the Appendix to Part I. That is, the book is virtually self-contained.

The book is suitable as a text for a one-semester course similar to the one described earlier. The course should be offered as either a first-year graduate course or an upper-division undergraduate course. Although such a course is essential for programs in systems science and related areas, it could be a valuable enrichment to other programs as well. In addition to its role as a text, the book is also suitable for self-study.

For me, the book has another useful function. On numerous occasions, I was asked the question: "What is systems science?" Now, I can answer this laden question with ease: "If you are really interested in what it is, read *Facets of Systems Science.*"

Part I is written from a particular perspective. It reflects primarily my own views, as revealed more explicitly in a biographical paper, "Systems Profile: The Emergence of Systems Science," included in Part II. This can hardly be otherwise. However, the one-sided argumentation in Part I is counterbalanced, at least partially, by alternative views expressed in some of the papers in Part II.

The main difficulty in working on the book was to select a sample that would be well representative of the tremendous amount of information pertaining to systems science. Hard choices had to be made to produce a book of a reasonable size. The reports I received from the students were quite helpful in making some of the choices. In general, I tried to select papers for Part II that elaborate on or complement some specific arguments made in Part I. That is, I tried to integrate, as well as possible, the two parts of the book. To my regret, I could not include, because of the size limit, many relevant articles that I consider excellent.

I am sure that the book has numerous shortcomings. Some of them may be a result of wrong choices of material or of my ignorance about some relevant sources of information; others may be due to my writing style. These shortcomings of the book will surely emerge from its use in various classroom environments. I intend to identify them and revise the book, at some proper time in the future, to make it better and more permanent.

Binghamton, New York GEORGE J. KLIR

Note to the Reader

References to literature in Part I are designated by brackets and are identified by the author(s) and the year of publication. References to papers that are included in Part II are further identified by being printed in boldface.

After reading each individual chapter of Part I, the reader should proceed to relevant papers (if any) in Part II. The relevance of articles with respect to the individual chapters is identified in the Detailed Contents of Part II.

If confused with mathematical terminology and notation employed in Part I, the reader should consult the Appendix (pp. 191–194).

Acknowledgments

As explained above, this book would never have been written if the gentle pressure had not been applied to me by five of my doctoral students to teach a course from which it emerged. Therefore, these students should be given the same credit (or blame) as I for this literary product. They are: Cliff Joslyn, Kevin Kreitman, Mark Scarton, Ute St. Clair, and William Tastle. Two of these, Cliff Joslyn and William Tastle, played additional roles. Cliff helped me to prepare the course and, to some degree, also helped me to teach it. Bill, who took the course, read a semifinal version of the manuscript of this book and gave me many valuable comments for improvement; later, he also helped me proofread the galleys. I am very grateful to both of them for their extremely valuable support. I am also grateful to all the students who took the course, inspired me by their enthusiasm, and prepared critical analyses of the course that appeared to be very instrumental in my work on the book. Finally, I would like to express my gratitude to Bonnie Cornick, a genius who can read my handwriting, for her excellent typing of the manuscript, and to my wife, Milena, for her understanding, support, and patience during my disciplined work on the book.

Part I contains many excellent quotes and I am grateful to the copyright owners for permitting me to use the material. They are: *Behavioral Science* (James G. Miller, ed.), Gordon and Breach (New York), Hutchinson (London), Institute of Electrical and Electronic Engineers (IEEE), International Federation for Systems Research (IFSR), International Institute for Applied Systems Analysis (IIASA), John Wiley (New York), MIT Press (Cambridge, Massachusetts), *Philosophy of Science Association*, Pergamon Press (Oxford), Plenum Press (New York), Springer-Verlag (New York), and Sterling Lord Literistic, Inc. (New York).

Part II of this book consists of 35 reprinted articles, some of them appropriately shortened. The following is the list of copyright owners whose permissions to reproduce the articles in this book is gratefully acknowledged:
American Philosophical Society (Philadelphia, Pennsylvania)
Association for Cybernetics (Namur, Belgium)
Elsevier Scientific Publishers Ireland Ltd. (Limerick, Ireland)

Ervin Laszlo (Pisa, Italy)
Gordon and Breach Science Publishers (New York)
The Institute of Electrical and Electronics Engineers, Inc. (Piscataway, New Jersey)
The Institute of Management Science (Providence, Rhode Island)
International Federation for Systems Research (Vienna, Austria)
International Institute for Applied Systems Analysis (Laxenburg, Austria)
International Society for the Systems Sciences (Pomona, California)
John Wiley & Sons, Inc. (New York)
Kluwer Academic Publishers (Dordrecht, The Netherlands)
The National Council of Teachers of Mathematics, Inc. (Reston, Virginia)
Operations Research Society of America (Baltimore, Maryland)
Pergamon Press (Oxford, U.K.)
Plenum Publishing Corporation (New York)
Sage Publications, Inc. (Beverly Hills, California)
Sigma Xi Research Society (Research Triangle Park, North Carolina)
Taylor & Francis, Ltd. (London, U.K.)
The United Nations University (Tokyo, Japan)

Contents

Part I
SYSTEMS SCIENCE: A Guided Tour

Chapter 1.	What Is Systems Science?	3
Chapter 2.	More about Systems	9
Chapter 3.	Systems Movement	19
Chapter 4.	Conceptual Frameworks	41
	4.1. Deductive Approaches	42
	4.2. Inductive Approaches	46
	4.3. Epistemological Hierarchy of Systems	48
Chapter 5.	Systems Methodology	71
	5.1. General Systems Problem Solver	75
	5.2. Systems Modeling	77
	5.3. Methodological Role of the Computer	86
Chapter 6.	Systems Metamethodology	87
Chapter 7.	Systems Knowledge	101
Chapter 8.	Complexity	113
	8.1. Bremermann's Computational Limit	121
	8.2. Computational Complexity	126
Chapter 9.	Simplification Strategies	135
	9.1. Systems Simplification: A General Formulation	137
	9.2. Special Simplification Strategies	138

Chapter 10.	Goal-Oriented Systems	143
	10.1. Adaptive Systems	149
	10.2. Special Types of Goal Orientation	155
Chapter 11.	Systems Science in Retrospect and Prospect	163
	11.1. Criticism	163
	11.2. Status and Impact of Systems Science	169
	11.3. The Future of Systems Science	185
Appendix:	Mathematical Terminology and Notation	191
References	....	195

Part II
CLASSICAL SYSTEMS LITERATURE

Introduction and Comments 209

Detailed Contents 211

Author Index 651

Subject Index 657

PART I

Systems Science
A Guided Tour

CHAPTER 1

What Is Systems Science?

The aim of science is not things themselves, . . . but the relations between things; outside those relations there is no reality knowable.
—HENRI POINCARÉ

An inevitable prerequisite for this book, as implied by its title, is a presupposition that systems science is a legitimate field of scientific inquiry. It is self-evident that I, as the author of this book, consider this presupposition valid. Otherwise, clearly, I would not conceive of writing the book in the first place.

I must admit at the outset that my affirmative view regarding the legitimacy of systems science is not fully shared by everyone within the scientific community. It seems, however, that this view of legitimacy is slowly but steadily becoming predominant. It is my hope that this book, whose purpose is to characterize the essence and spirit of systems science, will have a positive influence in this regard.

What is *systems science*? This question, which I have been asked on countless occasions, can basically be answered either in terms of activities associated with systems science or in terms of the domain of its inquiry. The most natural answers to the question are, almost inevitably, the following definitions:

1. Systems science is what systems scientists do when they claim they do science.
2. Systems science is that field of scientific inquiry whose objects of study are systems.

Without further explanation, these definitions are clearly of little use.

Definition (1) is meaningful but somewhat impractical. It is meaningful since systems scientists do, indeed, exist. I, for example, claim to be one of them, and so do colleagues at my department and other departments of systems science. Hence, the meaning of systems science could, in principle, be determined by observing and analyzing our scientific activities. This strategy, however, involves some inherent practical difficulties. First, systems scientists are still a rather rare species among all scientists and, consequently, they are relatively hard to find. Second, scientific activities of scientists who are officially labeled as systems scientists vary from person to person, and, moreover, some of these activities are clearly associated with other, well-established areas of science. Third, the strategy would require a massive data collection and extensive and sophisticated data analysis.

For all the reasons mentioned, and possibly some additional ones, it is virtually impossible to utilize definition (1) in an operational way for our purpose. Therefore, let me concentrate on definition (2). To be made operational, this definition requires that some broad and generally acceptable characterization of the concept of a *system* be established.

The term "system" is unquestionably one of the most widely used terms not only in science, but in other areas of human endeavor as well. It is a highly overworked term, which enjoys different meanings under different circumstances and for different people. However, when separated from its specific connotations and uses, the term "system" is almost never explicitly defined. To elaborate on this point, let me quote from a highly relevant paper by Rosen [**1986**]:

> Let us begin by observing that the word "system" is almost never used by itself; it is generally accompanied by an adjective or other modifier: physical system; biological system; social system; economic system; axiom system; religious system; and even "general" system. This usage suggests that, when confronted by a system of any kind, certain of its properties are subsumed under the adjective, and other properties are subsumed under the "system," while still others may depend essentially on both. The adjective describes what is special or particular; i.e., it refers to the specific "thinghood" of the system; the "system" describes those properties which are independent of this specific "thinghood."
>
> This observation immediately suggests a close parallel between the concept of *system* and the development of the mathematical concept of a set. Given any specific aggregate of things; e.g., five oranges, three sticks, five fingers, there are some properties of the aggregate which depend on the specific nature of the things of which the aggregate is compared. There are others which are totally independent of this and depend only on the "setness" of the aggregate. The most prominent of these is what we call the *cardinality* of the aggregate.
>
> It should now be clear that *systemhood* is related to thinghood in much the same way as setness is related to thinghood. Likewise, what we generally call *system properties* are related to systemhood in the same way as cardinality is related to setness. But systemhood is different from both setness and from thinghood; it is an *independent category*.

To begin our search for a meaningful definition of the term "system" from a broad perspective, let us consult a standard dictionary. We are likely to find that a system is "a set or arrangement of things so related or connected as to form a unity or organic whole" (*Webster's New World Dictionary*), although different dictionaries may contain stylistic variations of this particular formulation. It follows

What Is Systems Science?

from this *common-sense definition* that the term "system" stands, in general, for a set of some things and a relation among the things. Formally, we have

$$S = (T, R), \qquad (1.1)$$

where S, T, R denote, respectively, a *system*, a *set of things* distinguished within S, and a *relation* (or, possibly, a set of relations) defined on T. Clearly, the thinghood and systemhood properties of S reside in T and R, respectively.

The common-sense definition of a system, expressed by Eq. (1.1), is rather primitive. This, paradoxically, is its weakness as well as its strength. The definition is weak because it is too general and, consequently, of little pragmatic value. It is strong because it encompasses all other, more specific definitions of systems. In this regard, this most general definition of systems provides us with a criterion by which we can determine whether any given object is a system or not: an object is a system if and only if it can be described in a form that conforms to Eq. (1.1).

For example, a collection of books is not a system, only a set. However, when we organize the books in some way, the collection becomes a system. When we order them, for instance, by authors' names, we obtain a system since any ordering of a set is a relation defined on the set. We may, of course, order the books in various other ways (by publication dates, by their size, etc.), which result in different systems. We may also partition the books by various criteria (subjects, publishers, languages, etc.) and obtain thus additional systems since every partition of a set emerges from a particular equivalence relation defined on the set. Observe now that a relation defined on a particular set of books, say the ordering by publication dates, may be applied not only to other sets of books, but also to sets whose elements are not books. For example, members of a human population may be ordered by their dates of birth.

These simple examples illustrate that the same set may play a role in different systems; each system is distinguished from the other by the different relations on the set. Similarly, the same relation, when applied to different sets, may play a role in different systems. In this case, the systems are distinguished by their sets or, in other words, by their thinghood properties.

Once we have the capability of distinguishing objects that are systems from those that are not, the proposed definition of systems science—*a science whose objects of study are systems*—becomes operational. Observe, however, that the term "system" is used in this definition without any adjective or other modifier. This indicates, according to the distinction between thinghood and systemhood, that systems science focuses on the study of systemhood properties of systems rather than their thinghood properties. Taking this essential aspect of systems science into consideration, the following, more specific definition of systems science emerges:

Systems science is a science whose domain of inquiry consists of those properties of systems and associated problems that emanate from the general notion of systemhood.

The principal purpose of this book is to elaborate on this conception of systems science. It is argued throughout the book that systems science, like any other science, contains a *body of knowledge* regarding its domain, a *methodology* for acquisition of new knowledge and for dealing with relevant problems within the domain, and a *metamethodology*, by which methods and their relationship to problems are characterized and critically examined. However, in spite of these parallels with classical areas of science, systems science is fundamentally different from science in the traditional sense. The difference can best be explained in terms of the notions of thinghood and systemhood.

It is a truism that classical science has been far more concerned with thinghood than systemhood. In fact, the many disciplines and specializations that have evolved in science during the last four centuries or so reflect predominantly the differences between the things studied rather than the differences in their ways of being organized. This evolution is still ongoing. Since at least the beginning of this century, however, it has increasingly been recognized that studying the ways in which things can be, or can become, organized is equally meaningful and may, under some circumstances, be even more significant than studying the things themselves. From this recognition, a new kind of science eventually emerged, a science that is predominantly concerned with systemhood rather than thinghood. This new science is, of course, systems science.

Since disciplines of classical science are largely thinghood-oriented, the systemhood orientation of systems science does not make it a new discipline of classical science. With its orientation so fundamentally different from the orientation of classical science, systems science transcends all the disciplinary boundaries of classical science. From the standpoint of systems science, these boundaries are totally irrelevant, superficial, and even counterproductive. Yet, they are significant in classical science, where they reflect fundamental differences, for example, differences in measuring instruments and techniques. In other words, the disciplinary boundaries of classical science are thinghood-dependent but systemhood-independent. If systems science becomes divided into special disciplines in the future, the boundaries between these disciplines will inevitably be systemhood-dependent but thinghood-independent.

Classical science, with all its disciplines, and systems science, with all its prospective disciplines, thus provide us with two distinct perspectives from which scientific inquiry can be approached. These perspectives are complementary. Either of them can be employed without the other only to some extent. In most problems of scientific inquiry, the two perspectives must be applied in concert.

It may be argued that traditional scientific inquiries are almost never totally devoid of issues involving systemhood. This is true, but these issues are handled

What Is Systems Science?

in classical science in an opportunistic, ad hoc fashion. There is no place in classical science for a comprehensive and thorough study of the various properties of systemhood. The systems perspective is thus suppressed within the confines of classical science in the sense that it cannot develop its full potential. It was liberated only through the emergence of systems science. While the systems perspective was not essential when science dealt with simple systems, its significance increases with growing complexity of systems of our current interest and challenge.

From the standpoint of the disciplinary classification of classical science, systems science is clearly cross-disciplinary. There are at least three important implications of this fact. First, systems science knowledge and methodology are directly applicable in virtually all disciplines of classical science. Second, systems science has the flexibility to study systemhood properties of systems and the associated problems that include aspects derived from any number of different disciplines and specializations of classical science. Such cross-disciplinary systems and problems can thus be studied as wholes rather than collections of the disciplinary subsystems and subproblems. Third, the cross-disciplinary orientation of systems science has a unifying influence on classical science, increasingly fractured into countless number of narrow specializations, by offering unifying principles that transcend its self-imposed boundaries.

Classical science and systems science may be viewed as complementary dimensions of modern science. As I argue elsewhere [Klir, **1985b**], the emergence and evolution of systems science and its integration with classical science into genuine two-dimensional science are perhaps the most significant features of science in the information (or postindustrial) society.

CHAPTER 2

More about Systems

What is a system? As any poet knows, a system is a way of looking at the world.
—GERALD M. WEINBERG

The common-sense definition, as expressed by Eq. (1.1), looks overly simple:

$$\mathbf{S} = (T, R)$$

- $\mathbf{S}$ — a system
- T — a set of certain things (thinghood)
- R — a relation defined on T (systemhood)

Its simplicity, however, is only on the surface. That is, the definition is simple in its form, but it contains symbols, T and R, that are extremely rich in content. Indeed, T stands for any imaginable set of things of any kind, and R stands for any conceivable relation defined on T. To appreciate the range of possible meanings of these symbols, let us explore some examples.

Symbol T may stand for a single set with arbitrary elements, finite or infinite, but can also represent, for example, a power set (the set of all subsets of another set), any subset of the power set, or an arbitrary family of distinct sets.

The content of symbol R is even richer. For each set T, with its special characteristics, the symbol stands for every relation that can be defined on the set. A relation, in general, is a subset of some Cartesian product of given sets. When T is a single set, then $R \subset T \times T$ is called a *binary relation* on T. Examples of relations of this simple form are ordering relations, equivalence relations, and compatibility relations, according to which elements of T are ordered, partitioned, or classified as desired. Relations defined on other forms of Cartesian products of T are also possible, such as $R \subset (T \times T) \times (T \times T)$, etc.

When T consists of two sets, $T = \{X, Y\}$, symbol R stands for relations not only of different types, but also of greater variety of different forms, such as

$R \subset X \times Y,$ $R \subset (X \times X \times X) \times Y,$
$R \subset (X \times X) \times Y,$ $R \subset (X \times X \times X) \times (Y \times Y),$
$R \subset X \times (Y \times Y),$ $R \subset (X \times X \times X) \times (Y \times Y \times Y),$
$R \subset (X \times Y) \times (X \times Y),$ $R \subset (X \times Y) \times (X \times Y) \times (X \times Y).$

It is now easy to see, I trust, how rapidly the number of possible relational forms increases with the increasing number of distinct sets in T, illustrating thus the tremendous richness of the systemhood symbol R. The fact that we discuss the meaning of this symbol solely in terms of mathematical relations is no shortcoming. The well-defined concept of a mathematical relation (as a subset of some Cartesian product) is sufficiently general to encompass the whole set of kindred concepts that pertain to systemhood, such as interaction, interconnection, coupling, linkage, cohesion, constraint, interdependence, function, organization, structure, association, correlation, pattern, etc.

Although the common-sense conception of systems allows us to recognize a system, when one is presented to us, it does not help us to construct it. Whence do systems arise? To address this question, let me begin with some relevant thoughts offered by Brian Gaines [1979]:

Definition: A system is what is distinguished as a system. At first sight this looks to be a nonstatement. Systems are whatever we like to distinguish as systems. Has anything been said? Is there any possible foundation here for a systems science? I want to answer both these questions affirmatively and show that this definition is full of content and rich in its interpretation. Let me first answer one obvious objection to the definition above and turn it to my advantage. You may ask, "What is peculiarly systemic about this definition? Could I not equally well apply it to all other objects I might wish to define?" i.e., A *rabbit* is what is distinguished as a rabbit." "Ah, but," I shall reply, "my definition is adequate to define a system but yours is not adequate to define a rabbit." In this lies the essence of systems theory: that to distinguish some entity as being a system is a necessary and sufficient criterion for its being a system, and this is uniquely true for systems. Whereas to distinguish some entity as being anything else is a necessary criterion to its being that something but not a sufficient one.

More poetically we may say that the concept of a system stands at the supremum of the hierarchy of being. That sounds like a very important place to be. Perhaps it is. But when we realize that getting there is achieved through the rather negative virtue of not having any further distinguishing characteristics, then it is not so impressive a qualification. I believe this definition of a system as being that which uniquely is defined by making a distinction explains many of the virtues, and the vices, of systems theory. The power of the concept is its sheer generality; and we emphasize this naked lack of qualification in the term *general systems theory*, rather than attempt to obfuscate the matter by giving it some respectable covering term such as *mathematical* systems theory. The weakness, and paradoxically the prime strength, of the concept is in its failure to require further distinctions.

It is a weakness when we fail to recognize the significance of those further distinctions to the subject matter in hand. It is a strength when those further distinctions are themselves unnecessary to the argument and only serve to obscure a general truth through a covering of specialist jargon. No wonder general systems theory is subject to extremes of vilification and

> praise. Who is to decide in a particular case whether the distinction between the baby and the bath water is relevant to the debate?
>
> What then of some of the characteristics that we do associate with the notion of a system, some form of coherence and some degree of complexity? The *Oxford English Dictionary* states that a *system* is "a group, set or aggregate of things, natural or artificial, forming a connected or complex whole." I would argue that any other such characteristics arise out of the process of which making a distinction is often a part, and are some form of post hoc rationalization of the distinction we have made. One set of things is treated as distinct from another and it is that which gives them their coherence; it is that also which increases their complexity by giving them one more characteristic than they had before—that they have now been distinguished. Distinguish the words on this page that contain an "e" from those which do not. You now have a "system" and you can study it and rationalize why you made that distinction, how you can explain it, why it is a useful one. However, none of your postdistinction rationalizations and studies of the "coherency" and "complexity" of the system you have distinguished is intrinsically necessary to it being a "system." They are just activities that naturally follow on from making a distinction when we take note that we have done it and want to "explain" to ourselves, or others, why.

The point made by Gaines in this interesting discussion is that we should not expect that systems can be discovered, ready made for us. Instead, we should recognize that systems originate with us, human beings. We construct them by making appropriate distinctions, be they made in the real world by our perceptual capabilities or conceived in the world of ideas by our mental capabilities.

These sentiments are echoed and articulated with remarkable clarity by Goguen and Varela [1979]:

> A *distinction* splits the world into two parts, "that" and "this," or "environment" and "system," or "us" and "them," etc. One of the most fundamental of all human activities is the making of distinctions. Certainly, it is the most fundamental act of system theory, the very act of defining the system presently of interest, of distinguishing it from its environment.
>
> The world does not present itself to us neatly divided into systems, subsystems, environments, and so on. These are divisions which we make ourselves, for various purposes, often subsumed under the general purpose evoked by saying "for convenience." It is evident that different people find it convenient to divide the world in different ways, and even one person will be interested in different systems at different times, for example, now a cell, with the rest of the world its environment, and later the postal system, or the economic system, or the atmospheric system.
>
> The established scientific disciplines have, of course, developed different preferred ways of dividing the world into environment and system, in line with their different purposes, and have also developed different methodologies and terminologies consistent with their motivation.

All these considerations are extremely important for proper understanding of the nature of systems, at least as I conceive them, and consequently, as they are viewed in this book. According to this view, *systems do not exist in the real world independent of the human mind.* They are created by the acts of making distinctions in the real world or, possibly, in the world of ideas. Every act must be made by some agent, and, of course, the agent is in this case the human mind, with its perceptual and mental capabilities.

The view just stated is usually referred to as *constructivist view* of reality and knowledge, or *constructivism.** The most visible contemporary proponent of this view, particularly well recognized within the systems science and cognitive science communities, is Ernst von Glasersfeld. To illuminate the constructivist view a little more, let me use a short quotation from his many writings on constructivism [Glaserfeld, 1987]:

> Quite generally, our knowledge is useful, relevant, viable, or however we want to call the positive end of the scale of evaluation, if it stands up to experience and enables us to make predictions and to bring about or avoid, as the case may be, certain phenomena (i.e., appearance, events, experiences). If knowledge does not serve that purpose, it becomes questionable, unreliable, useless, and is eventually devaluated as superstition. That is to say, from the pragmatic point of view, we consider ideas, theories, and "laws of nature" as structures which are constantly exposed to our experiential world (from which we derived them), and they either hold up or they do not. Any cognitive structure that serves its purpose in our time, therefore, proves no more and no less than just that—namely, given the circumstances we have experienced (and determined *by* experiencing them), it has done what was expected of it. Logically, that gives us no clue as to how the "objective" world might be; it merely means that we know *one* viable way to a goal that we have chosen under specific circumstances in our experiential world. It tells us nothing—and cannot tell us anything—about how many other ways there might be, or how that experience which we consider the goal might be connected to a world *beyond* our experience. The only aspect of that "real" world that actually enters into the realm of experience, are its constraints. . . .
>
> Radical constructivism, thus, is *radical* because it breaks with convention and develops a theory of knowledge in which knowledge does not reflect an "objective" ontological reality, but exclusively an ordering and organization of a world constituted by our experience.

---

*The founder of constructivism is generally considered Giambattista Vico (1668–1744), an Italian philosopher. The principal ideas from which constructivism emerged are presented in his early work, *On the Most Ancient Wisdom of the Italians* (Cornell University Press, Ithaca, New York, 1988), which was originally published in Italian in 1710. Perhaps the most visible contributor to constructivism in this century is Jean Piaget. Basic ideas of constructivism are well overviewed in some writings by Glaserfeld [1987, **1990**]. Arguments supporting constructivism, primarily of biological nature, are well presented by Maturana and Varela [**1987**].

To avoid any confusion, let me emphasize that the constructivist view does not imply that the existence of the real world independent of the human mind, is necessarily denied. This is a different issue, on which constructivism remains neutral. The constructivist view, at least from my perspective, is not an ontological view (concerned with the existence and ultimate nature of reality), but an epistemological view (concerned with the origin, structure, acquisition, and validity of knowledge).

The essence of constructivism is well captured by the following four quotes. The first is due to Giambattista Vico, the founder of constructivism:

> God is the artificer of Nature,
> man the god of artifacts.

The second quote is from a book by Stephane Leduc [1911]:

> Classes, divisions, and separations are all artificial,
> made not by nature but by man.

The third quote is due to Humberto R. Maturana, an important contributor to modern systems thinking:

> I maintain that all there is is that which the observer brings forth in his or her distinctions. We do not distinguish what is, but what we distinguish is.

The last quote is from a book by Maturana and Varela [1987], in which they bring forth convincing biological arguments in support of constructivism:

> All doing is knowing, and
> all knowing is doing.

Although the constructivist view may not be typical in classical science (at least not yet), it is undoubtedly the predominant view in contemporary systems science. Indeed, most writings on various aspects of systems science are either openly supportive of the view or, at least, compatible with it. My commitment to the constructivist view in this book reflects, therefore, not only my personal conviction, but also the mainstream of contemporary systems science.

The basic position regarding systems that I take in this book can thus be summarized as follows: Every system is a construction based upon some world of experiences, and these, in turn, are expressed in terms of purposeful distinctions made either in the real world or in the world of ideas.

Given some world of experiences, systems may be constructed in many different ways. Each construction employs some distinctions as primitives and others for characterizing a relation among the primitives. The former distinctions represent thinghood, while the latter represent systemhood of the system constructed.

The chosen primitives may be not only simple distinctions, but also sets of distinctions or even systems constructed previously. To illustrate this point, let all words printed on this page be taken as primitives and those that are verbs be distinguished from those that are not. Clearly, the primitives themselves are in this case rather complex systems, based upon many distinctions (visual, grammatical, semantic), but all these finer distinctions clustered around each word are taken for granted and left in the background when we choose to employ the words as primitives in a larger system. We consider the words as things whose recognition is assumed, and focus on the extra distinction by which verbs are distinguished from other words. This distinction imposes an equivalence relation on the set of words, which partitions the set into two equivalence classes, the class of all verbs and the class of all nonverbs on this page.

Although much more could be said about the common-sense conception of systems (see, e.g., a thorough discussion by Marchal [1975]), I believe that enough has already been said here for our purpose. To make the concept of system useful, the common-sense definition must be refined in the sense that specific classes of ordered pairs (T, R), relevant to recognized problems, must be introduced. This can be done in one of two fundamentally different ways:

a. By restricting T to certain kinds of things;
b. By restricting R to certain kinds of relations.

Although the two types of restrictions are independent of each other, they can be combined.

Restrictions of type (a) are exemplified by the traditional classification of science into disciplines and specializations, each focusing on the study of certain kinds of things (physical, chemical, biological, economic, social, etc.) without committing to any particular kind of relations. Since different kinds of things are based on different types of distinctions, they require the use of different senses or measuring instruments and techniques. Hence, this classification is essentially experimentally based.

Restrictions of type (b) lead to fundamentally different classes of systems, each characterized by special kinds of relations, with no commitment to any particular kind of things on which the relations are defined. Since systems characterized by different types of relations require different theoretical treatment, this classification, which is fundamental to systems science, is predominantly theoretically based.

A prerequisite for classifying systems by their systemhood properties is a conceptual framework within which these properties can properly be codified. Each framework determines the scope of systems conceived. It captures some basic categories of systems, each of which characterizes a certain type of knowledge representation, and provides a basis for further classification of systems within each category. To establish firm foundations of systems science, a comprehensive framework is needed to capture the full scope of systemhood properties.

More about Systems

The issue of how to form conceptual frameworks for codifying systemhood properties, and the associated issue of how to classify systems by their systemhood properties are addressed in Chap. 4. Without waiting for the full discussion of these issues, however, we are able to introduce one rather important systemhood-dependent and thinghood-independent classification of systems right not, using only the common-sense definition of systems. This classification is based upon an important equivalence relation defined on the set of all systems of interest, say all systems captured by the common-sense definition. According to this relation, which is called an *isomorphic relation*, two systems are considered equivalent if their systemhood properties are totally preserved under some suitable transformation from the set of things of one system into the set of things of the other system.

To illustrate the notion of isomorphic systems, let us consider two systems, $S_1 = (T_1, R_1)$ and $S_2 = (T_2, R_2)$. Assume, for the sake of simplicity, that T_1, T_2 are single sets and $R_1 \subset T_1 \times T_1, R_2 \subset T_2 \times T_2$. Then S_1 and S_2 are called *isomorphic systems* if and only if there exists a transformation from T_1 to T_2 expressed in this case by a bijective function $h: T_1 \to T_2$, under which things that are related in R_1 are also related in R_2 and, conversely, things that are related in R_2 are also related in R_1. Formally, systems S_1 and S_2 are isomorphic if and only if, for all $(x_1, x_2) \in T_1 \times T_1$,

$$(x_1, x_2) \in R_1 \text{ implies } [h(x_1), h(x_2)] \in R_2$$

and, for all $(y_1, y_2) \in R_2$,

$$(y_1, y_2) \in R_2 \text{ implies } [h^{-1}(y_1), h^{-1}(y_2)] \in R_1.$$

This definition must be properly extended when T_1, T_2 are families of sets and R_1, R_2 are not binary but n-dimensional relations with $n > 2$.

The notion of isomorphic systems imposes a binary relation on $\mathcal{S} \times \mathcal{S}$, where $\mathcal{S}$ denotes the set of all systems. Two systems, S_1 and S_2, are related by this relation (the pairs S_1, S_2 and S_2, S_1 are contained in the relation) if and only if they are isomorphic.

It is easy to verify that the *isomorphic relation* (or *isomorphism*) is reflexive, symmetric, and transitive. Consequently, it is an equivalence relation defined on the set of all systems (or on its arbitrary subset), which partitions the set into equivalence classes. These are the smallest classes of systems that can be distinguished from the standpoint of systemhood. In fact, we may view each of these equivalent classes as being characterized by a unique relation, defined on some particular set of things, which is freely interpreted in terms of other sets of things within the class. This unique relation may be taken as a canonical representative of the class.

Which set of things should be chosen for these canonical representations? Although the choice is arbitrary, in principle it is essential that the same selection criteria be used for all isomorphic classes. Otherwise, the representatives would not be compatible and, consequently, it would be methodologically difficult to deal with them. Therefore, it is advisable to define the representatives as systems whose sets of things consist of some standard symbols that are abstract (interpretation-free), such as integers or real numbers, and whose relations are described in some convenient standard form.

Since the canonical representatives of isomorphic equivalence classes of systems are devoid of any interpretation, it is reasonable to call them general systems. Hence, *a general system is a standard and interpretation-free system chosen to represent a particular equivalence class of isomorphic systems.*

Observe that, according to this definition, each isomorphic equivalence class may potentially contain an infinite number of systems, owing to the unlimited variety of thinghood, but it contains only one general system. The number of isomorphic equivalence classes and, thus, the number of general systems, may also be potentially infinite owing to the unlimited variety of systemhood.

Each isomorphic equivalence class of systems contains not only a general

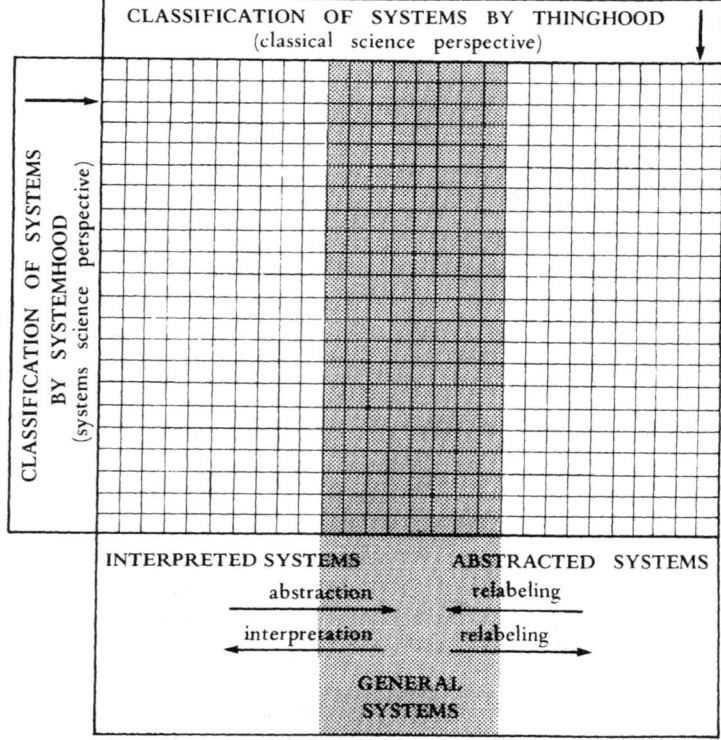

Figure 2.1. Two ways of classifying systems and the role of general systems.

More about Systems

system and its various interpreted systems, but also other abstract systems, different from the general system. The isomorphic transformations between the general system and the other abstract systems (described by the bijective function h) are rather trivial. They are just arbitrary replacements of one set of abstract symbols with another set. It seems appropriate to call these transformations *relabelings*.

The transformations between a general system and its various interpreted systems are by far not trivial since they involve different types of distinctions made in the real world, and these, in turn, are subject to different constraints of the real world. The isomorphic transformation from an interpreted system into the corresponding general system, which may be called an *abstraction*, is always possible. The inverse transformation, which may be called an *interpretation*, is not guaranteed and must be properly justified in each case. Indeed, relations among things based upon distinctions made in the real world cannot be arbitrary, but must reflect genuine constraints of the real world, as represented in our world of experiences. Hence, each interpreted system determines uniquely its representative general systems by the isomorphic transformation, but not the other way around.

The independence (or orthogonality) of the two ways of classifying systems, by thinghood and by systemhood, is visually expressed in Fig. 2.1. This figure also illustrates the role of general systems and their connection to other abstracted systems and to interpreted systems.

The two dimensions of science, which reflect the two-dimensional classification of systems symbolized by Fig. 2.1., are complementary. When combined in scientific inquiries, they are more powerful than either of them alone. The traditional perspective of classical science provides a meaning and context to each inquiry. The perspective of systems science, on the other hand, provides a means for dealing with any desirable system, regardless of whether or not it is restricted to a particular discipline of classical science.

CHAPTER 3

Systems Movement

The more science becomes divided into specialized disciplines, the more important it becomes to find unifying principles.
—HERMAN HAKEN

Systems science is a phenomenon of the second half of this century. It developed within a movement that is usually referred to as *systems movement*. In general, systems movement may be characterized as a loose association of people from different disciplines of science, engineering, philosophy, and other areas, who share a common interest in ideas (concepts, principles, methods, etc.) that are applicable to all systems and that, consequently, transcend the boundaries between traditional disciplines.

Systems movement emerged from three principal roots: *mathematics, computer technology*, and a host of ideas that are well captured by the general term *systems thinking*.

Since at least the publication of Newton's *Principia* in 1687, mathematics has played a key role in describing and dealing with systems in various areas of science and engineering. Prior to the Twentieth Century, however, mathematics was capable of dealing only with rather simple systems, consisting of a mere handful of variables (usually two or three) related in a functional way. This was adequate for typical problems in science and engineering until physics became concerned with processes at the molecular level in the late Nineteenth Century. It was obvious that the available mathematical methods were completely useless for investigating processes of this sort. They were not useless in principle, but owing to the enormous number of entities involved: a gas in a small closed space would typically contain in the order of 10^{23} molecules. The need for fundamentally new mathematical tools resulted eventually in statistical methods, which turned out to be applicable not only to the study of molecular processes (statistical mechanics), but to a host of other areas such as the actuarial profession, design of large telephone exchanges, and the like.

While the classical mathematical tools, exemplified by the calculus and differential equations, are applicable to problems that involve only a very small number of components that are related in perfectly predictable ways, the applicability of statistical tools is exactly the opposite: they require a very large number of components and a high degree of unpredictability (randomness). These two types of mathematical tools are thus complementary. When one of them excels, the other totally fails. Despite their complementarity, these mathematical tools can deal only with problems that are clustered around the two extremes of complexity and randomness scales. In his classic paper, Warren Weaver [1948] refers to them as problems of *organized simplicity* (simple, deterministic) and problems of *disorganized complexity* (complex, random). He describes them with unmatched clarity:

> The classical dynamics of the nineteenth century was well suited for analyzing and predicting the motion of a single ivory ball as it moves about on a billiard table. In fact, the relationship between positions of the ball and the times at which it reaches these positions forms a typical nineteenth-century problem of simplicity. One can, but with a surprising increase in difficulty, analyze the motion of two or even of three balls on a billiard table. There has been, in fact, considerable study of the mechanics of the standard game of billiards. But, as soon as one tries to analyze the motion of ten or fifteen balls on the table at once, as in pool, the problem becomes unmanageable, not because there is any theoretical difficulty, but just because the actual labor of dealing in specific detail with so many variables turns out to be impracticable.
>
> Imagine, however, a large billiard table with millions of balls rolling over its surface, colliding with one another and with the side rails. The great surprise is that the problem now becomes easier, for the methods of statistical mechanics are applicable. To be sure the detailed history of one special ball cannot be traced, but certain important questions can be answered with useful precision, such as: On the average how many balls per second hit a given stretch of rail? On the average how far does a ball move before it is hit by some other ball? On the average how many impacts per second does a ball experience?
>
> Earlier it was stated that the new statistical methods were applicable to problems of disorganized complexity. How does the word "disorganized" apply to the large billiard table with the many balls? It applies because the methods of statistical mechanics are valid only when the balls are distributed, in their positions and motions, in a helter-skelter, that is to say a disorganized, way. For example, the statistical methods would not apply if someone were to arrange the balls in a row parallel to one side rail of the table, and then start them all moving in precisely parallel paths perpendicular to the row in which they stand. Then the balls would never collide with each other nor with two of the rails, and one would not have a situation of disorganized complexity.
>
> From this illustration it is clear what is meant by a problem of disorganized complexity. It is a problem in which the number of variables is very large, and one in which each of the many variables has a behavior which is individually erratic, or perhaps totally unknown. However, in spite of this helter-skelter, or unknown, behavior of all the individual variables, the system as a whole possesses certain orderly and analyzable average properties.

Systems Movement

Weaver further argues that problems of organized simplicity and disorganized complexity cover only a tiny fraction of all systems problems. Most problems are located somewhere between the two extremes of complexity and randomness; Weaver calls them problems of *organized complexity* and explains the reason for coining this name:

> This new method of dealing with disorganized complexity, so powerful an advance over the earlier two-variable methods, leaves a great field untouched. One is tempted to oversimplify, and say that scientific methodology went from one extreme to the other—from two variables to an astronomical number—and left untouched a great middle region. The importance of this middle region, moreover, does not depend primarily on the fact that the number of variables involved is moderate—large compared to two, but small compared to the number of atoms in a pinch of salt. The problems in this middle region, in fact, will often involve a considerable number of variables. The really important characteristic of the problems of this middle region, which science has as yet little explored or conquered, lies in the fact that these problems, as contrasted with the disorganized situations with which statistics can cope, show the essential feature of organization. In fact, one can refer to this group of problems as those of organized complexity.
>
> What makes an evening primrose open when it does? Why does salt water fail to satisfy thirst? Why can one particular genetic strain of microorganisms synthesize within its minute body certain organic compounds that another strain of the same organism cannot manufacture? Why is one chemical substance a poison when another, whose molecules have just the same atoms but assembled into a mirror-image pattern, is completely harmless? Why does the amount of manganese in the diet affect the maternal instinct of an animal? What is the description of aging in biochemical terms? What meaning is to be assigned to the question: Is a virus a living organism? What is a gene, and how does the original genetic constitution of a living organism express itself in the developed characteristics of the adult? Do complex protein molecules "know how" to reduplicate their pattern, and is this an essential clue to the problem of reproduction of living creatures? All these are certainly complex problems, but they are not problems of disorganized complexity, to which statistical methods hold the key. They are all problems which involve dealing simultaneously with a *sizable number of factors which are interrelated into an organic whole*. They are all, in the language here proposed, problems of organized complexity.
>
> On what does the price of wheat depend? This too is a problem of organized complexity. A very substantial number of relevant variables is involved here, and they are all interrelated in a complicated, but nevertheless not in helter-skelter, fashion.
>
> How can currency be wisely and effectively stabilized? To what extent is it safe to depend on the free interplay of such economic forces as supply and demand? To what extent must systems of economic control be employed to prevent the wide swings from prosperity to depression? These are also

obviously complex problems, and they too involve analyzing systems which are organic wholes, with their parts in close interrelation.

How can one explain the behavior pattern of an organized group of persons such as a labor union, or a group of manufacturers, or a racial minority? There are clearly many factors involved here, but it is equally obvious that here also something more is needed than the mathematics of averages. With a given total of national resources that can be brought to bear, what tactics and strategy will most promptly win a war, or better: what sacrifices of present selfish interest will most effectively contribute to a stable, decent, and peaceful world?

These problems—and a wide range of similar problems in the biological, medical, psychological, economic, and political sciences—are just too complicated to yield to the old nineteenth century techniques which were so dramatically successful on two-, three-, or four-variable problems of simplicity. These new problems, moreover, cannot be handled with the statistical techniques so effective in describing average behavior in problems of disorganized complexity.

These new problems, and the future of the world depends on many of them, require science to make a third great advance, an advance that must be even greater than the nineteenth-century conquest of problems of simplicity or the twentieth-century victory over problems of disorganized complexity. Science must, over the next 50 years, learn to deal with these problems of organized complexity.

Instances of problems with the characteristics of organized complexity are abundant, particularly in the life, behavioral, social, and environmental sciences, as well as in applied fields such as modern technology or medicine. Some of the problem areas that involve organized complexity are especially profound, such as cancer research, the study of aging, or the rich area of difficult and diverse problems associated with modern technology. This last area is well characterized by George B. Dantzig in his 1979 Distinguished Lecture at the International Institute for Applied Systems Analysis in Laxenburg, Austria, an institute that has played an important role in this new thrust of science into the domain of organized complexity:

It is not easy to paint a picture of just how complex modern technology is. One way to start is to list the activities of a small town. By using the classified section of the telephone directory, I can list a few activities of the town of Richmond, California. Here are those that begin with the letters Br: Bridge Builders, Bridge Tables, Broadcasting Stations, Brochures, Brokers, Bronze, Brushes, Brooches, Brakes, Brandies, Brazing, Bricks, Brick Stain, Bric-a-Brac. I counted over 6,000 activities in all.

Another way to see the diversity of the material side of life is to look at a catalog of electronic supply items that are for sale. These are thousands upon thousands of different kinds of resistors, condensers, vacuum tubes, transis-

Systems Movement

tors, cables, sockets, knobs, switches, dials, circuit boards, cabinets. Look up the number of different items listed in a chemical supply catalog or a Sears, Roebuck catalog, and again the number of different items runs into many thousands. A modern university can have a hundred different departments. The United States Government has nearly 2,000 different kinds of offices in San Francisco alone, each presumably carrying out a different function for the public good. So far we have spoken only of diversity, but complexity has other dimensions.

The Leontief input–output model of the national economy of the United States classifies industries into about 400 major types and requires data for each of these industries about how much it shipped (or received) from every other industry. The resulting 400×400 table contains 160,000 numbers. Each region of the country has such an input–output table, and there are many regions. Each number in an input–output table expresses a dependency of one industry upon another; the transactions between regions and industries represent further dependencies; there are a great number of cross combinations. Countries depend on each other in the same way.

There are also time dependencies: facilities are built and maintained for future use; material is stockpiled for future use; people are trained for future jobs. There are locational dependencies as well: men, material, and facilities are moved to new locations, not only on the surface of the globe but below and above.

While we may easily understand the ins and outs of each small part of this vast web of activities, the problem is how to track all the interactions at once. We know that the powerful forces of population growth, shortages of raw materials, food, energy, growing affluence, and so on, are rapidly reshaping this complexity. There is a fear that the structure that interconnects these activities may not hold up very well under these stresses. We see the possibility of all kinds of system failure if we let the changes go on uncontrolled.

There is no doubt that both analytical and statistical methods are relevant to systems science. However, the primary orientation of systems science, at least as I view it, is to study systems and associated problems that possess the characteristics of organized complexity. Studying the domain of organized complexity requires, in comparison with the domains of organized simplicity and disorganized complexity, a much higher level of systemhood expertise. To achieve such a level of expertise for a traditional scientist, in addition to the ever-increasing thinghood-oriented knowledge pertaining to his or her specialization, is virtually impossible owing to the fundamental limits of the human mind. Hence, the role of developing and applying the systemhood expertise must be undertaken by a scientist of a different kind, *a systems scientist*, whose specialization is this very expertise.

One of the major difficulties with problems of organized complexity is that one cannot often avoid massive combinatorial searches of various kinds without losing relevance. This is a weakness of the human mind, while, at the same time, it is a strength of the computer. This explains why attempts to investigate systems with the characteristics of organized complexity had not been successful prior to the emergence of computer technology. This also explains why the evolution of systems science has been so closely correlated with the evolution of computer technology.

Systems science is strongly dependent upon the computer, which is its laboratory (Chap. 7) as well as its most important methodological tool (Chap. 5). It is thus not surprising that the visibility of modern systems ideas, which have been emerging in science and engineering since the beginning of this century, significantly increased shortly after the fully automatic digital computers were built in the late 1940s and early 1950s. It is also not surprising that the beginnings of systems movement are clearly associated with this period. Since these beginnings, systems movement and computer technology have been developing side by side and have been influencing each other.

Systems science, an offspring of systems movement, has a particularly strong linkage with computer technology. The steadily increasing computing power opens new methodological possibilities, which stimulate novel developments in mathematics, and these, in turn, make advances in systems science. These three areas—mathematics, computer technology, and systems science—are thus intimately interrelated. Computer technology has a key role in this relationship; it is a sort of catalyzer that influences the conversion of methodological possibilities into methodological actualities.

I hope it is now clear that mathematics and computer technology, which are fundamental for describing and dealing with systemhood phenomena, played an important role in the emergence of systems movement. The crucial factor in this emergence, however, was a host of systemhood-oriented ideas that have arisen in philosophy, science, and engineering since the beginning of this century. Let me overview some of these ideas, which, in my opinion, were decisive in the initial, formative stage of systems movement.

Perhaps the most important of these ideas that contributed to the emergence of systems movement were ideas associated with a cluster of related views that are referred to as *holism* (from Greek *holos*, which means a whole). The basic idea of holism is well captured by the famous statement, "The whole is more than the sum of its parts," which has a long history and whose author is, presumably, Aristotle.

It is well known that holism was already well entrenched in classical Greek philosophy (as well as, for example, in the ancient Chinese philosophy of Taoism) and, consequently, it is not a product of this century. However, it reappeared, rather forcefully, at the turn of the century, as an antithesis of *reductionism*, a methodological view predominant in science since about the sixteenth century. The latter claims that properties of a whole are explicable in terms of properties of the constituent elements. Holism rejects this claim and maintains that a whole cannot be analyzed, in general, in terms of its parts without some residuum.

The rejuvenation of holism in this century was a result of the recognition (or frustration) in some disciplines of science that reductionism was not able to explain certain phenomena belonging to these disciplines. Let me look at just two disciplines, in which the need for holism was most visible: psychology and biology.

The holistic movement in psychology, which is referred to as *Gestalt theory* (or *Gestalt psychology*) originated in Germany in the 1910s and 1920s. The word *Gestalt* means in German *an organized whole whose parts belong together*. To characterize the spirit of Gestalt theory, let me use words of one of its founders, Max Wertheimer:*

> It has long seemed obvious—that is, in fact, the characteristic tone of European science—that "science" means breaking up complexes into their component elements. Isolate the elements, discover their laws, then reassemble them, and the problem is solved. All wholes are reduced to pieces and piecewise relations between pieces.
>
> The fundamental "formula" of Gestalt theory might be expressed in this way: There are wholes, the behavior of which is not determined by that of their individual elements, but where the part-processes are themselves determined by the intrinsic nature of the whole. It is the hope of Gestalt theory to determine the nature of such wholes . . .
>
> We hear a melody and then, upon hearing it again, memory enables us to recognize it. But what is it that enables us to recognize the melody when it is played in a new key? The sum of the elements is different, yet the melody is the same; indeed, one is often not even aware that a transposition has been made. . . . Is it really true that when I hear a melody I have a sum of individual tones (pieces) which constitute the primary foundation of my experience? Is not perhaps the reverse of this true? What I really have, what I hear of each individual note, what I experience at each place in the melody is a *part* which is itself determined by the character of the whole.

In biology, the limits of reductionism began to be felt in the late Nineteenth Century. As a reaction, various biologists argued that holistic thinking is essential for adequate understanding of some biological phenomena. This was paralleled with similar developments in psychology, which led eventually to the formulation of the program of gestalt psychology. It was only in the late 1920s when holistic thinking in biology became formulated in a coherent and programmatic manner. The most instrumental in this formulation, which is usually referred to as

---

*Three German psychologists, Max Wertheimer, Wolfgang Köhler, and Kurt Koffka, are usually recognized as founders of *Gestalt theory*. Our quotation is from a conference address by Wertheimer in Berlin in 1924; English translation of this address is included with other classical writing on Gestalt theory in a useful source book [Ellis, 1938]. Classic monographs on Gestalt theory were written by Köhler [1929] and Koffka [1935]; textbooks by Hartmann [1935] and Katz [1950] are recommended as excellent introductions to Gestalt theory.

organismic biology, were two biologists, Paul Weiss and Ludwig von Bertalanffy. While Bertalanffy played the dominant role in the actual formulation of the program of organismic biology [Bertalanffy, 1952], the main contribution of Weiss in this regard was his thorough and convincing characterization of holistic thinking within the realm of biology. The following quotation is a sample from his writings [Weiss, 1969]:

> When people use the phrase *"The whole is more than the sum of its parts,"* the term *"more"* is often interpreted as an algebraic term referring to numbers. However, a living cell certainly does not have more content, mass or volume than is constituted by the aggregate mass of molecules which it comprises. . . . the *"more"* (than the sum of parts) in the above tenet does not at all refer to any measurable quantity in the observed systems themselves; it refers solely to the necessity for the observer to supplement the sum of statements that can be made about the separate parts by any such additional statements as will be needed to describe the *collective behaviour* of the parts, when in an organized group. In carrying out this upgrading process, he is in effect doing no more than *restoring information content* that has been lost on the way down in the progressive analysis of the unitary universe into abstracted elements.
>
> In this neutral version appears to lie the reconciliation between reductionism and holism. The reductionist likes to move from the top down, gaining precision of information about fragments as he descends, but losing information content about the larger orders he leaves behind; the other proceeds in the opposite direction, from below, trying to retrieve the lost information content by reconstruction, but recognizes early in the ascent that that information is not forthcoming unless he has already had it on record in the first place. The difference between the two processes, determined partly by personal predilections, but largely also by historical traditions, is not unlike that between two individuals looking at the same object through a telescope from opposite ends.

To illustrate that a whole may indeed contain more information than the total information contained in its parts, a simple example is in order. Consider that three attributes were monitored on a large computer installation to determine conditions under which the utilization of the central processing unit (CPU) is high (H) and under which it is low (L). Assume that the utilization is studied in connection to the utilization of two other units of the computer, channel 1 (CH1) and channel 2 (CH2), whose utilizations for our purpose are also distinguished as either high (H) or low (L). Assume further that the three attributes were monitored simultaneously for two hours during which the computer handled a typical workload and in each period of 5 sec the utilizations of all the three observed units were recorded. For each unit, if its utilization was smaller than a certain percentage of time (defined by the investigator), it was considered low (L) during the interval; otherwise, it was considered high (H). Assume now that the following triples were observed repeatedly with approximately the same frequencies:

Systems Movement

CH1	CH2	CPU
L	L	L
L	H	H
H	L	H
H	H	L

This experimental result is a ternary relation among three things (utilizations of the three computing units). The things and this particular relation among them form a simple system (according to the common-sense conception of systems discussed in Chap. 2), which represents the whole (or the overall system) in the context of the given investigation.

The ternary relation obtained by the monitoring indicates that utilizations of the three computing units are strongly related: the relation contains only four possible triples out of eight potential triples. It provides the investigator with an important insight: to keep the utilization of the CPU high, make arrangements under which the utilizations of the two channels are mixed (L and H or H and L) as frequently as possible.

Assume now that we have only knowledge associated with three parts of our overall system, each consisting of utilizations of two of the three units and a binary relation between them. This knowledge may have been obtained by three simpler monitors, each of which allows us to monitor only two units simultaneously. Clearly, each of these monitors must give us the same observation regarding the two observed units as the larger monitor, but it does not give us any observations regarding the third unit. That is, observations of the three monitors can be obtained from the table of observations of the larger monitor by selecting from it the three possible pairs of columns:

CH1	CH2	CH2	CPU	CH1	CPU
L	L	L	L	L	L
L	H	H	H	L	H
H	L	L	H	H	H
H	H	H	L	H	L

How much can we infer from these parts (binary relations) about the whole (the ternary relation)? We can easily see that we can infer very little. Since each of the binary relations contains all possible pairs L and H, we may conclude that the attributes (utilizations of the three computing units) are pairwise independent. Yet, they are far from being independent when considered all together as a whole: they are not triplewise independent. It is clear that the ternary relation cannot be understood in terms of the binary relations in this case. Without knowing the ternary relation, we have no way to reconstruct it from the binary relations. In fact, we can find by simple combinatorial considerations that our ternary relation is only one of 35 different ternary relations that result in the same binary relations. Hence, knowing only the binary relations, we cannot decide which of the 35 ternary relations is the actual one.

Perhaps the most comprehensive and thorough exposition of holism is covered in a classic book *Holism and Evolution* by Jan Christiaan Smuts [1926]. In this book, Smuts elevates holistic thinking into a general self-organizing principle that is fundamental to evolution of nature at all levels, from inorganic matter through the many forms of life to the human mind and the human society. The book is important not only from an historical perspective, as it certainly contributed to the eventual emergence of systems movement, but also from the standpoint of current developments in systems science. It is particularly relevant to recent interests in studying self-organizing processes through which order develops from chaos. The coherent holistic view carefully developed and clearly expressed by Smuts seems still the best framework for pursuing these studies in a unified fashion.

To get a better feeling for this important book, let me present here a few quotations from it:

> The creation of wholes, and ever more highly organized wholes, and of wholeness generally as characteristic of existence, is an inherent character of the universe. There is not a mere vague indefinite creative energy or tendency at work in the world. This energy or tendency has specific characters; the most fundamental of which is whole-making. And the progressive development of the resulting wholes at all stages—from the most inchoate, imperfect, inorganic wholes to the most highly developed and organized—is what we call Evolution. The whole-making, holistic tendency, or Holism, operating in and through particular wholes, is seen at all stages of existence, and is by no means confined to the biological domain to which science has hitherto restricted it. With its roots in the inorganic, this universal tendency attains clear expression in the organic biological world, and reaches its highest expressions and results on the mental and spiritual planes of existence. Wholes of various grades are the real units of Nature. Wholeness is the most characteristic expression of the nature of the universe in its forward movement in time. It marks the line of evolutionary progress. And Holism is the inner driving force behind that progress.
>
> It is evident that if this view is correct, very important results must follow for our conceptions of knowledge and life. Wholes are not mere artificial constructions of thought, they point to something real in the universe; and Holism as the creative principle behind them is a real *vera causa*. It is the motive force behind Evolution. We thus have behind Evolution not a mere vague and indefinable creative impulse or *elan vital*, the bare idea of passage or duration without any quality or character, and to which no value or character could be attached, but something quite definite. Holism is a specific tendency, with a definite character, and creative of all characters in the universe, and thus

fruitful of results and explanations in regard to the whole course of cosmic development.

It is possible that some may think I have pressed the claims of Holism and the whole too far; that they are not real operative factors, but only useful methodological concepts or categories of research and explanation. There is no doubt that the whole is a useful and powerful concept under which to range the phenomena of life especially. But to my mind there is clearly something more in the idea. The whole as a real character is writ large on the face of Nature. It is dominant in biology; it is everywhere noticeable in the higher mental and spiritual developments; and science, if it had not been so largely analytical and mechanical, would long ago have seen and read it in inorganic nature also. The whole as an operative factor requires careful exploration. That there are wholes in Nature seems to me incontestable. That they cover a very much wider field than is generally thought and are of fundamental significance is the view here presented. But the idea of the whole is one of the neglected matters of science and to a large extent of philosophy also. It is curious that, while the general viewpoint of philosophy is necessarily largely holistic, it has never made real use of the idea of the whole. The idea runs indeed as a thread all through philosophy, but mostly in a vague intangible way. . . .

It is very important to recognize that the whole is not something additional to the parts: it is the parts in a definite structural arrangement and with mutual activities that constitute the whole. The structure and the activities differ in character according to the stage of development of the whole; but the whole is just this specific structure of parts with their appropriate activities and functions. . . . The concept of Holism and the whole is as nearly as possible a replica of Nature's observed process, and its application will prevent us from appearing to run the stuff of reality into a mould alien to Nature. It will, therefore, enable us to explain Nature from herself, so to say, and by her own standards. In this way justice can be done to the concrete character of natural phenomena. . . .

A whole is a synthesis or unity of parts, so close that it affects the activities and interactions of those parts, impresses on them a special character, and makes them different from what they would have been in a combination devoid of such unity or synthesis. That is the fundamental element in the concept of the whole. It is a complex of parts, but so close and intimate, so unified that the characters and relations and activities of the parts are affected and changed.

Holism is also creative of all values. Take the case of organic Beauty. It is undeniable that Beauty rests on a holistic basis. Beauty is essentially a product of Holism and is inexplicable apart from it. Beauty is of the whole; Beauty is a relation of parts in a whole, a blending of elements of form and colour, of foreground and background of expression and suggestion, of structure and function, of structure and field, which is perceived and appreciated as harmonious and satisfying, according to laws which it is for Aesthetics to determine. . . .

> For me the great problem of knowledge, indeed the great mystery of reality, is just this: How do elements or factors a and b come together, combine and coalesce to form a new unity or entity x different from both of them? To my mind this simple formula of synthesis sums up all the fundamental problems of matter and life and mind. The answer to this question will in some measure supply the key to all or most of our great problems. My answer has already been given; it is in the word Holism.

The status of a system as either a whole (an overall system) or a part (a subsystem) is, of course, not absolute. The same system may be viewed in one context as a whole and in another context as a part. We may say, more poetically, that a part is a whole in a role (in one context), and a whole is a part in a role (in another context). This duality makes it possible to represent systems hierarchically in the sense that a system conceived as a whole may consist of interconnected parts that themselves are systems, and each of these parts may again consist of interconnected parts that are systems, etc., until some primitive parts are reached that do not quality as systems. The nature of this hierarchy is concisely captured by Goguen and Varela [**1979**]:

> At a given level of the hierarchy, a particular system can be seen as an outside to systems below it, and as an inside to systems above it; thus, the status (i.e., the mark of distinction) of a given system changes as one passes through its level, in either the upward or the downward direction. The choice of considering the level above or below corresponds to a choice of treating the given system as autonomous or controlled (constrained).

To emphasize the dual role of a system at some level of the hierarchy, either as a part or a whole, Arthur Koestler suggested a new term, *holon* [Koestler and Smythies, 1969]:

> A part, as we generally use the word, means something fragmentary and incomplete, which by itself would have no legitimate existence. On the other hand, there is a tendency among holists to use the word "whole" or "Gestalt" as something complete in itself which needs no further explanation. But wholes and parts in this absolute sense do not exist anywhere, either in the domain of living organisms or of social organizations. What we find are intermediary structures on a series of levels in ascending order of complexity, each of which has two faces looking in opposite directions: the face turned towards the lower levels is that of an autonomous whole, the one turned upward that of a dependent part. I have elsewhere [Koestler, 1967] proposed the word "holon" for these Janus-faced sub-assemblies—from the Greek *holos*—whole, with the suffix *on* (cf. neutr*on*, prot*on*) suggesting a particle or part.

Holism, in opposition to reductionism, was undoubtedly one of the main roots from which systems movement sprang. Initially, a tendency toward a full commitment to holism and a total rejection of reductionism was quite visible in systems movement. Over the years, this extreme position became slowly moderated. Now, the two doctrines are viewed, by and large, as complementary. While holism is accepted as a thesis that is correct on logical grounds (as documented by our simple example of computer monitoring and endless other examples) and, consequently, desirable to follow as an ideal guiding principle, it is recognized that its applicability is often limited on pragmatic grounds. For example, simultaneous monitoring of a large number of variables may be technically impossible or impractical, computational demands for dealing with a desirable overall system may exceed our computational limits, the overall system may be incomprehensible to the human mind, etc. The complementarity of holism and reductionism is well described by Goguen and Varela [**1979**]:

> Most discussions place holism/reductionism in polar opposition. This seems to stem from the historical split between empirical sciences, viewed as mainly reductionist or analytic, and the (European) schools of philosophy and social science that grope toward a dynamics of totalities.
>
> Both attitudes are possible for a given descriptive level, and in fact they are complementary. On the one hand, one can move down a level and study the properties of the components, disregarding their mutual interconnection as a system. On the other hand, one can disregard the detailed structure of the components, treating their behavior only as contributing to that of a larger unit. It seems that both these directions of analysis always coexist, either implicitly or explicitly, because these descriptive levels are mutually interdependent for the observer. We cannot conceive of components if there is no system from which they are abstracted; and there cannot be a whole unless there are constitutive elements. . . .
>
> These descriptive levels haven't been generally realized as complementary largely because there is a difference between publicly announced methodology and actual practice, in most fields of research in modern science. A reductionist attitude is strongly promoted, yet the analysis of a system cannot begin without acknowledging a degree of coherence in the system to be investigated; the analyst has to have an intuition that he is actually dealing with a coherent phenomenon. Although science has publicly taken a reductionist attitude, in *practice* both approaches have always been active. It is not that one has to have a holistic view as opposed to a reductionist view, or vice versa, but rather that the two views of systems are complementary. . . . Reductionism implies attention to a lower level, while holism implies attention to a higher level. They are intertwined in any satisfactory description; and each entails some loss relative to our cognitive preferences, as well as some gain.

By studying the relationship between wholes and parts from the methodological point of view (as overviewed later), current systems thinking goes far beyond thinking molded from either reductionism or holism. We may say that is represents a synthesis of the reductionist thesis and the holistic antithesis, as poetically expressed by Patrick Suppes [1983]:

> I am for the delicate dance from parts to wholes
> and back again. We should not be captured at
> either end. The dance should go forever.

Enough of holism and reductionism. Let me turn to other developments during the first half of this century that influenced the emergence of systems movement. One of them was the increasing awareness that there were many phenomena and problems that could not be studied within the boundaries of individual disciplines of science. This eventually led to the emergence of *interdisciplinary areas*, such as biophysics, biochemistry, physiological psychology, and social psychology. The existence of these interdisciplinary areas was probably the first step leading to the recognition that systems may be defined across disciplinary boundaries. We may say that it was the first step in recognizing the notion of systemhood. Another step was the recognition of *isomorphies* (often called *analogies*) between systems describing different physical phenomena, such as mechanical, electrical, acoustic, and thermal. Once an isomorphic (relation preserving) correspondence was established between two or more areas of physics, any method developed in one area became readily applicable to corresponding problems in the other areas.

The discovery of isomorphies between different areas of physics contributed to a new way of thinking about systems. Systemhood similarities became more and more recognized as at least equally important as thinghood differences. Moreover, the discovered isomorphies were not only of theoretical significance, but also of great practical value. They made it possible, for example, to transfer methods from a methodologically well-developed area to areas methodologically less developed. A visible result of this possibility was the notion of *generalized circuits*, a framework within which well-developed methods for analyzing electric circuits were transferred through established isomorphies to less developed areas of mechanical, acoustic, magnetic, and thermal systems [Thorn, 1963].

The established isomorphies made it also possible to study systems indirectly, in terms of other systems, isomorphic to them. For example, experimental investigations of dynamic properties of new designs of automobiles, air crafts, helicopters, or rockets may be done in wind tunnels on models of these man-made objects, appropriately scaled, rather than on the objects themselves. This is often more convenient, cheaper, and safer. The experiments may also be performed on objects that belong to a different area of physics, say on electric circuits whose dynamic properties are known to be isomorphic with those of the investigated objects. This may even be more convenient and less costly than the construction and use of scale models.

Basic concepts, principles, and methods regarding the utilization of isomorphies for dealing with practical problems, mostly in engineering, were eventually incorporated into a new subject area referred to as the *theory of similarity* or *similitude* [Skoglund, 1967; Szucz, 1980]. This theory also deals with the construction and use of *analog computers*. These are devices whose basic units are physical models of some mathematical operations (addition, multiplication, integration, etc.) and functions (exponential, trigonometric, etc.). They are used for solving algebraic or differential equations. When appropriate units of an analog computer are connected according to the form of a mathematical equation, the solution is obtained by measuring relevant physical quantities.

The ideas of holism, the emergence of interdisciplinary areas in science, and the increasing recognition of the existence and utility of isomorphies between disciplines of science created a growing awareness among some scholars that certain concepts, ideas, principles, and methods were applicable to systems in general, regardless of their disciplinary categorization. This eventually led to the notions of general systems, general theory of systems, general systems research, and the like.

There seems to be no doubt that the terms *general systems* and *general systems research* (or *general systems theory*) are due to Ludwig von Bertalanffy. Although he introduced them orally in the 1930s, the first written presentation appeared only after World War II; they are included, together with some of his later writings in one of his books [Bertalanffy, 1968]. According to Bertalanffy, general systems research is a discipline whose subject matter is "the formulation and derivation of those principles which are valid for 'systems' in general."

Bertalanffy was not only the originator of the idea of general systems research, but also one of the principal organizers of systems movement that sprang from this idea. In 1954, he and three other distinguished scholars with similar systems ideas, Kenneth Boulding (an economist), Ralph Gerard (a physiologist), and Anatol Rapoport (a mathematical biologist), happened to spend some time together as Fellows at the just founded *Center for Advanced Study in the Behavioral Sciences* in Palo Alto, California. These four scholars, who are often referred to as the founding fathers of systems movement, apparently influenced each other in a highly synergetic fashion. This synergy led to the formation of the first organization fully devoted to the promotion of systems thinking, the *Society for General Systems Research* (SGSR), in December 1954. The Society was founded with the following four objectives:

1. To investigate the isomorphy of concepts, laws, and models from various fields, and to help in useful transfers from one field to another;
2. To encourage development of adequate theoretical models in fields which lack them;
3. To minimize the duplication of theoretical effort in different fields; and
4. To promote the unity of science through improving communication among specialists.

These objectives are as meaningful now as they were when the Society was founded. Their spirit is well captured in the following extract from a classic paper by Boulding [**1956**]:

> General Systems Theory is a name which has come into use to describe a level of theoretical model-building which lies somewhere between the highly generalized constructions of pure mathematics and the specific theories of the specialized disciplines. . . .
>
> Because in a sense mathematics contains all theories it contains none; it is the language of theory, but it does not give us the content. At the other extreme, we have the separate disciplines and sciences with their separate bodies of theory. Each discipline corresponds to a certain segment of the empirical world, and each develops theories which have particular applicability to its own empirical segment. Physics, Chemistry, Biology, Psychology, Economics and so on all carve out for themselves certain elements of the experience of men and develop theories and patterns of activity (research) which yield satisfaction in understanding, and which are appropriate to their special segments.
>
> In recent years increasing need has been for a body of systematic theoretical construction which will discuss the general relationships of the empirical world. This is the quest of General Systems Theory. It does not seek, of course, to establish a single, self-contained "general theory of practically everything" which will replace all special theories of particular disciplines. Such a theory would be almost without content, for we always pay for generality by sacrificing content, and all we can say about practically everything is almost nothing. Somewhere however between the specific that has no meaning and the general that has no content there must be, for each purpose and at each level of abstraction, an optimum degree of generality.

Over the years, the Society for General Systems Research has become the main professional organization supporting all facets of the emerging systems movement. Each year, the Society organizes an Annual Meeting and publishes a Yearbook entitled General Systems. In 1988, the name of the Society was changed to *International Society for the Systems Sciences* (ISSS), reflecting perhaps the maturity of the field.

Ideas quite similar to those associated with general systems research, although more focusing on information processes in systems such as communication and control, were proposed in the late 1940s under the name *cybernetics*. This name was coined by the promoter of these ideas, Norbert Wiener; it is based on the Greek word *kybernetes*, which means *steerman*. In his seminal book, Wiener [1948] defines cybernetics as the study of "control and communication in the animal and in the machine."

To capture the essence of Wiener's motivation to introduce this new field, I can hardly do better than use his own words [Wiener, 1984]:

Since Leibniz there has perhaps been no man who has had a full command of all the intellectual activity of his day. Since that time, science has been increasingly the task of specialists, in fields which show a tendency to grow progressively narrower. . . . Today there are few scholars who can call themselves mathematicians or physicists or biologists without restriction. A man may be a topologist or an acoustician or a coleopterist. He will be filled with jargon of his field, and will know all its literature and all its ramifications, but, more frequently than not, he will regard the next subject as something belonging to his colleague three doors down the corridor, and will consider any interest in it on his own part as an unwarrantable breach of privacy. . . . There are fields of scientific work, which have been explored from the different sides of pure mathematics, statistics, electrical engineering, and neurophysiology; in which every single notion receives a separate name from each group, and in which important work has been triplicated or quadruplicated, while still other important work is delayed by the unavailability in one field of results that may have already become classical in the next field.

It is these boundary regions of science which offer the richest opportunities to the qualified investigator. They are at the same time the most refractory to the accepted techniques of mass attack and the division of labor. If the difficulty of a physiological problem is mathematical in essence, ten physiologists ignorant of mathematics will get precisely as far as one physiologist ignorant of mathematics, and no further. If a physiologist who knows no mathematics works together with a mathematician who knows no physiology, the one will be unable to state his problem in terms that the other can manipulate, and the second will be unable to put the answers in any form that the first can understand. . . . A proper exploration of these blank spaces on the map of science could only be made by a team of scientists, each a specialist in his own field but each possessing a thoroughly sound and trained acquaintance with the fields of his neighbors; all in the habit of working together, of knowing one another's intellectual customs, and of recognizing the significance of a colleague's new formal expression. The mathematician need not have the skill to conduct a physiological experiment, but he must have the skill to understand one, to criticize one, and to suggest one. The physiologist need not be able to prove a certain mathematical theorem, but he must be able to grasp its physiological significance and to tell the mathematician for what he should look. We have dreamed for years of an institution of independent scientists, working together in one of these backwoods of science, not as subordinates of some great executive officer, but joined by the desire, indeed by the spiritual necessity, to understand the region as a whole, and to lend one another the strength of that understanding.

Cybernetics is based upon the recognition that certain information-related problems, such as some problems of communication and control, can be meaningfully and beneficially studied, to some extent, independently of any specific context. When Wiener proposed cybernetics, the circumstances were favorable.

Communication theory and control theory made tremendous progress during World War II, and a new field of information-processing machines (computers) was just emerging. These developments resulted in a rich body of knowledge bound together by the notion of information. Claude Shannon [1948] showed how to measure information for the purposes of telecommunication and control, and established some basic laws of information that govern systems; control theory provided practitioners with a respectable inventory of rigorously formulated concepts, such as stability, positive and negative feedback, observability, controllability, and feedforward and feedback control, as well as with principles and methods for analyzing and designing control systems of various types; and the emerging area of information-processing machines began to develop basic ideas regarding the design of general purpose computing machines, which involved, for example, the issues of physical encoding of information, processing of higher types of information in terms of elementary logical operations, controlling the computation process, and the like.

During World War II, Wiener worked, as a mathematician, on various war-related engineering projects, in which aspects of communication, control, and information processing were predominant. At the same time, he was attracted to certain issues of neurophysiology and, eventually, he became involved in some neurological projects jointly with a Mexican neurophysiologist Arthuro Rosenblueth. He discovered that communication, control, and information processing were also fundamental in this area. This apparently helped him to recognize the cross-disciplinary nature of problems connected with information and led eventually to his formulation of cybernetics.

General systems research and cybernetics have developed side by side since their emergence, and a considerable cross-fertilization has occurred between them. Perhaps the most important person in this cross-fertilization was W. Ross Ashby, whose profound contributions to various issues of cybernetics were consistently formulated and dealt with as systems problems.* There are different opinions about the relationship between general systems research and cybernetics. In one opinion, which seems to have become predominant within systems

---

*According to a survey regarding influences among systems researchers, whose results are published as Appendix B in [Klir, 1978], W. Ross Ashby was by far the most influential person in systems movement. The survey showed that Ashby had a major influence on almost twice as many systems researchers as the second most influential person, Ludwig von Bertalanffy, and almost three times as many systems researchers as the third most influential persons, Norbert Wiener and Anatol Rapoport. The great impact of Ashby's work on systems movement can be explained, at least partially, by the superior clarity of his writings, his unusual capability to recognize important principles where others see only trivialities, his great gift for essence-preserving simplification, his broad interests, encompassing both cybernetics and general systems research, and his meticulous scholarship. He wrote two book masterpieces [Ashby, 1952, 1959] and many influential articles, most of which are included in a book [Ashby, 1981] edited by Conant.

movement, cybernetics is a subarea of general systems research that focuses on the study of information processes in systems, particularly communication and control. I fully share this opinion [Klir, 1970] since I consider all properties and problems connected with the notion of information as fundamentally systemhood properties and problems. Indeed, I cannot conceive of the possibility to conceptualize information without any reference to a system of some sort.

Another major factor in the formation of systems movement can best be captured by the term *mathematical* systems theory. This term, in fact, represents a broad variety of mathematical theories (emerging mostly in the 1960s) which attempt to formalize fundamental systems concepts and develop a formal framework for formalizing and dealing with systems problems. These general theories of systems evolved, by and large, from various special systems theories, which in the late 1950s and early 1960s were already mathematically well developed for specific purposes within the engineering milieu. Most visible of these special theories were the mathematical theories of control, electric and generalized circuits, switching circuits, and finite-state automata.*

One of the most comprehensive mathematical systems theories was initiated and its foundations developed by Mihajlo Mesarovic. The significance of this theory is that it is based on the common-sense conception of systems. That is, the theory begins with only one axiom by which the concept of a system is formulated at the most general level: a system is a family of sets and a relation defined on the sets. Additional axioms are then added to formalize pragmatically significant special classes of systems. The most comprehensive exposition of the theory is given in two books by Mesarovic and Takahara [1975, 1988].

The emergence of systems movement was also connected with some developments in the areas of engineering and management in this century. Throughout the century, it became increasingly important in both of these areas to think in terms of systems. Engineering was challenged with designing systems of rapidly increas-

---

*One major predecessor of modern mathematical systems theories was the theory of feedback control. Although various devices based upon feedback control have presumably been built since the Third Century B.C. [Mayr, 1970], conscious efforts to develop a theory of systems with feedback control began only in the 19th century. These efforts were simulated primarily by the success of the centrifugal governor invented around 1790 by James Watt to control the speed of the steam engine [Bennett, 1979]. The theory was already well developed in the late 1930s and further extended and perfected during World War II. Methods of analysis and design of electric circuits, which were essential in the development of modern control theory, also played an important role in the formation of more general mathematical systems theories [Zadeh, **1962**].

Another major predecessor of modern mathematical systems theories was the theory of finite state automata. This theory, which emerged after World War II, was primarily motivated by problems associated with the design of digital computers and some questions connected with the idea of artificial intelligence. A comprehensive coverage of automata theory was prepared by Booth [1967]. An offspring of automata theory, which is more intimately connected with the design issues, is the theory of switching circuits [Klir, 1972b].

ing complexity, from telephone networks through production automation to the design of computers and computer-based systems. The situation in management was quite similar; the challenge came from the increasingly more complex organizations to be managed and from the increasing complexity of the associated decision making.

The developments in engineering and management were somewhat connected. Clearly, the increasing complexity of engineering tasks made increasing demands on management. These demands increased drastically during World War II, when complex military problems entered into the realm of management. Since critical strategic and tactical decisions had to be made quickly, interdisciplinary teams of scientists, mathematicians, engineers, and managers were formed to analyze the issues involved, such as optimal scheduling and resource allocation, cost–benefit and risk analysis, planning, budgeting, and the like. An outgrowth of these activities was a new discipline for which the name *operations research* was coined. In general, operations research can be characterized as the study of possible activities or operations within a particular institutional and organizational framework (e.g., a firm, a military organization, or a government) for the purpose of determining an optimum plan for reaching a given goal.

The main methodological resources of operations research are the various optimization methods for single as well as multiple objective criteria, decision-making methods, and methods derived from game theory.

After World War II, some new disciplines evolved from operations research. Two of them, with the strongest connection to systems movement are referred to as *systems analysis* and *systems engineering*. The aim of systems analysis is to use systems thinking and methodology (including methodological tools inherited from operations research) for analyzing complex problem situations that arise in private and public enterprises and organizations as a support for policy and decision making [Miser and Quade, 1985]. Systems engineering, on the other hand, is oriented to planning, design, construction, evaluation, and maintenance of large-scale systems that may involve both machines and human beings [Flagle *et al.*, 1960].

This concludes a summary of the most visible developments that contributed to the emergence and evolution of systems movement. Although numerous other developments could be justifiably mentioned in this regard, it is not the purpose of this text to trace the history of systems ideas as completely as possible.*

---

*Let me only mention one additional development of considerable historical significance. It is now increasingly recognized that an important early precursor of general systems theory was a theory developed by A. Bogdanov (1873–1928), a prominent Russian thinker, at the beginning of this century under the names *tektology* (inspired by the Greek word "tekton," which means "a builder") or *general organizational science*. Ideas pertaining to tektology were published in different forms between 1912 and 1928. Unfortunately, they sank into oblivion, primarily to their rejection by Lenin and their suppression by Soviet authorities. As a consequence, one of Bogdanov's books on tektology became

What is the current status of systems movement? After the *Society for General Systems Research* was formed in 1954, other professional societies oriented to systems research or cybernetics were formed in different countries. Perhaps the most visible, active, and stable have been the *American Society for Cybernetics*, the *Austrian Society for Cybernetic Studies*, and the *Netherlands Society for Systems Research*. Since 1980, systems movement has been united under the auspices of the *International Federation for Systems Research* (IFSR). The aims of the Federation, which was officially incorporated in Austria on March 12, 1980, are "to stimulate all activities associated with the scientific study of systems and to coordinate such activities at the international level."

Systems movement is now supported not only organizationally through the IFSR and its member societies, but also by a respectable number of scholarly journals and other publications. Furthermore, there are some academic programs in systems science, systems engineering, cybernetics, and related areas that are already well established and stable, and additional programs seem to emerge at a steady pace.

---

available to the English speaking world only some 60 years after its original publication [Bogdanov, 1980]. An excellent exposition of the development of systems thinking in the Soviet Union, which elaborates on the significance of Bogdanov's ideas, was prepared by Susiluoto [1982].

CHAPTER 4

Conceptual Frameworks

The only justification for our concepts is that they serve to represent the complex of our experiences; beyond this, they have no legitimacy.
—ALBERT EINSTEIN

To characterize the domain of systems science more specifically requires a conceptual framework within which systems are characterized. Each framework determines a scope of systemhood properties that can be described within it and leads to some specific systemhood-based taxonomy of systems. To capture the full scope of systemhood phenomena we are currently able to envision, a comprehensive framework is needed.

There are many systems conceptual frameworks, but only a few of them qualify as comprehensive frameworks, suitable for characterizing fully the domain of systems science. These comprehensive frameworks were developed in different ways, based on different motivations and presumptions. As a consequence, the differences in terminology and mathematical formulation among them are considerable. Little work has been done to rigorously compare the systemhood categories that emerge from these seemingly different frameworks. We can only make a sensible judgement, based on intuitive grounds and some limited evidence [Islam, 1974], that differences in systemhood categories emerging from these broad frameworks are relatively minor and can be reconciled.

There are essentially two fundamentally different approaches to developing a conceptual framework for systems science. I refer to them as a deductive approach and an inductive approach. In the *deductive approach*, one begins with an axiomatic characterization of the notion of systemhood that attempts to be as general as possible. Special systemhood categories are then introduced by additional axiomatic requirements. In the *inductive approach*, a conceptual framework is developed by collecting examples of systems employed in various disciplines of science, engineering, and other areas, abstracting them from their specific thinghood interpretations, categorizing them, integrating them into a coherent whole, and, finally, filling any obvious gaps in this emerging whole.

In this section, both of the approaches are illustrated, but no attempt is made to fully describe the various existing frameworks. For our purpose in this text, it is sufficient to use just one (any one) of the frameworks. For two reasons, I decided to use my own framework, which is usually referred to in the literature as the GSPS (*General Systems Problem Solving*) framework. The first reason is that this framework, which is based upon the inductive approach, allows me to explain the basic categories of systems by simple examples to which, I am sure, the reader can relate; the framework is such that no heavy mathematical treatment is needed to explain the basic ideas. The second reason is that my own familiarity with this framework is obviously better than with any of the other frameworks.

4.1. Deductive Approaches

Perhaps the most general and best developed conceptual framework based upon the deductive approach is due to Mihajlo Mesarovic and his research associates. This framework is described in detail in two books by Mesarovic and Takahara [1975, 1988]; it is also well overviewed in a paper by Mesarovic [1972].

The starting point in formulating the framework is the notion of *general system* as a relation on abstract sets.

$$\mathbf{S} \subset \times \{V_i | i \in I\}, \quad (4.1)$$

where I is an index set and $\times$ denotes the Cartesian product of sets V_i. When I is a finite set, (4.1) can be written as

$$\mathbf{S} \subset V_1 \times V_2 \times \cdots \times V_n. \quad (4.2)$$

The following is Mesarovic's own explanation of the meaning of this definition and the rationale for choosing it as a nucleus from which his framework is developed [Mesarovic, 1972]:

> The components of the relation, V_i, are referred to as (the system's) objects. An object stands for a feature or the characteristic in terms of which the system is described; the set V_i is the totality of alternative ways in which the respective feature is observed or experienced. The system, then, is the totality of proper combinations of the appearances of the system's objects.
>
> The following remarks will help to clarify some of the reasons for adopting the concept of a system as a relation. A system is defined in terms of observed features or, more precisely, in terms of the relationship between these features rather than what they actually are (physical, biological, social or other phenomena). This is in accord with the nature of the system field and its concern with the organization and the interrelationships of components into an (overall) system, rather than with the specific mechanisms within a given phenomenological framework.
>
> The notion of a system as given in Eq. 4.1 is perfectly general. On one hand, if the system is described by more specific mathematical constructs, say a set of equations, it is obvious that these indeed serve to define or specify a

Conceptual Frameworks

relation as given in Eq. 4.1. Different systems, of course, have different methods of specification, but they all are but relations as given in Eq. 4.1. On the other hand, in the case of the most incomplete information, when the system can be described only in terms of a set of verbal statements, they still, by their linguistic function as statements, define a relation as in Eq. 4.1. Indeed, every statement contains two basic linguistic categories; nouns and functors—nouns denoting objects, functors denoting the relationship between them. For any proper set of verbal statements there exists a (mathematical) relation (technically referred to as a model for these statements). The adjective "proper" refers here, of course, to the conditions for the axioms of a set theory. In short, then, a system is always a relation, as given in Eq. 4.1, and various types of systems are more precisely defined by the appropriate methods—linguistic statements, mathematical constructs, computer programs, and so on.

A system is defined as a set (of a particular kind, i.e., a relation). It stands for the collection of all appearances of the object of study rather than for the object of study as such. This is necessitated by the use of mathematics as the language for the theory in which a "mechanism," a function or a relation is defined as a set, that is, by means of all proper combination of components. Such a characterization of a system ought not to create any difficulty since the set-relation, with additional specifications, contains all information about the actual "mechanism" that we can legitimately use in further development of the theory starting from a given framework, as defined by the objects in terms of which the system is defined.

The specification of a given system is often expressed in terms of equations defined on appropriate variables. To every variable corresponds a systems object that represents the range of the respective variable. In stating that a system is defined by a set of equations on a set of variables one says that the system is a relation on the respective systems objects specified by the variables, each one with a corresponding object as a range, and such that for any combination of elements from the objects, that is, the values for the variables, the given set of equations is satisfied.

As we proceed along the path of formalization, that is, from more general to better structured and more specific, the next question is, What are the methods of systems specification, of defining a given relation as being distinct from others defined on the same objects? There are two basic approaches here: the *input–output approach* (referred to also as terminal causal, stimuli–response, and the like) and the *goal-seeking approach* (referred to also as decision-making, problem-solving, teleological, etc.).

In the input–output approach, objects of two types are distinguished, *inputs* (stimuli),

$$X = \times\{V_i | i \in I_x\},$$

and *outputs* (responses),

$$Y = \times\{V_i | i \in I_y\},$$

where $\{I_x, I_y\}$ is a partition of I. The *input–output system* is then defined as a relation on inputs and outputs,

$$S \subset X \times Y. \qquad (4.3)$$

Input–output systems are further classified according to types of constructive procedures by which outputs are determined for given inputs.

In the goal-seeking approach, two objects, a decision object, D, and a value object, V, are distinguished in addition to the inputs X and outputs Y. Furthermore, two functions are defined,

$$O: X \times D \to Y,$$
$$P: D \times Y \to V,$$

which are called an *outcome function* and a *performance function*, respectively. It is also assumed that a preference ordering is defined on V and that every subset of V has a minimum. Using all these concepts, the *goal-seeking system* is then defined as a relation of the form (4.3) whose elements are determined by the following optimization problem: for any $x \in X$ and any $y \in Y$, $(x, y) \in S$ if and only if there exists $d_x \in D$ such that

$$P(d_x, O(x, d_x)) \leq P(d, O(x, d))$$

for all $d \in D$ and $y = O(x, d_x)$. That is, for every input $x \in X$, the output $y \in Y$ is such that the performance function P is minimized subject to the constraints specified by the output function O. Hence, the goal of the system is to minimize P.

Again, the rationale for formulating the goal-seeking system is best described by Mesarovic [1972] himself:

> For proper understanding of the goal-seeking approach the following remarks will be helpful. We have defined the goal-seeking activity of the system in the example as minimization. This, of course, is just a special case. Many other approaches (satisfaction, general problem solving, etc.) can be used to define the goal-seeking activity.
>
> The goal-seeking procedure, such as minimization in the given example, is introduced solely in order to provide a specification of the given system, that is, a relation on inputs and outputs. In general, the only objects actually observed are X and Y; D and V are additional objects assumed for convenience of an appropriate and efficient specification of the system. There is little point (within the systems theory as such) to arguing whether the system is actually pursuing such a goal or not. All it matters is that the systems functioning can be described and is most appropriately described in such a manner. In general there may be more than one way to describe the system as goal-seeking. Also, there might be a case in which the systems functioning can be described only as goal seeking, while an input–output transformation specification is not given. That does not mean that the system fails to satisfy some kind of causality requirements and has some intrinsically different character from,

Conceptual Frameworks

say, a system described by a set of differential equations. It means only that within the family of constructive procedures which we are currently using to describe input–output transformation there is none which corresponds to the observed input–output relationship. After all, there is no reason to believe that all input–output transformations ought to be describable by the transformation procedures currently used. The availability of a goal-seeking description, then, can be simply considered as a convenience and indeed a necessity for an efficient specification of systems.

The next step in formulating the Mesarovic framework is an introduction of dynamic aspects into the input–output system (defined, possibly, by the goal-seeking approach). One way of doing that is to require that $X = A^T$ and $Y = B^T$, where A and B are arbitrary sets, called input and output alphabets, and T is a linearly ordered set, called a time set. The resulting system

$$\mathbf{S} \subset A^T \times B^T$$

is called a *general time system*.

Another way of introducing dynamic aspects into the input–output system is to introduce a time set, T, a state set, Z, and two functions,

$$\rho: Z \times X \times T \to Y \times T,$$
$$\phi: Z \times X \times T^2 \to Z,$$

referred to as a response function and a state-transition function, respectively. When the functions ρ and ϕ are such that for each pair $(x, y) \in \mathbf{S}$, there exists a state $z \in Z$ for which

$$\rho(z, x, t) = (y, t) \text{ if and only if } (x, y) \in \mathbf{S},$$
$$\rho(\phi(z, x, t, t'), x, t') = (y, t'),$$

they are said to define a *general dynamic system*.

Mesarovic also introduces the notion of a complex system, which is viewed as a system whose objects are systems in their own right. That is, a complex system is defined by a relation that describes interactions among a set of simple systems.

This brief overview of some basic categories of systems that emerge from the Mesarovic framework is sufficient for our purpose. I trust the reader has now a general impression of the framework. The advantage of this framework is its sheer generality and mathematical rigor. The disadvantage is its distance from typical systems that occur in praxis. That is, it takes a rather long path of qualifications in this framework to obtain practical systems from the general system expressed by Eq. (4.1).

Several other broad systems frameworks based upon the deductive approach are described in the literature. The common feature of these frameworks is that they attempt to capture by one set of axioms both the theory of finite-state machines (automata) and the theory of continuous systems described by differential equations.

Perhaps the most visible framework in this category, which has been well tested on many practical applications, was developed by Wayne Wymore. It is described in detail in two books by Wymore [1969,1976], and well overviewed in one of his papers [Wymore 1972]. Frameworks similar in spirit, but quite different in terminology and notation, were also developed by Michael Arbib [1965,1966] and Lotfi Zadeh [1969]. The motivation for developing these frameworks is well described by Arbib [1965]:

> In the abstract theory of automata, our preoccupation is with machines which carry out computations, or logical manipulations, on their inputs to produce their outputs. Automata are digital, in that the inputs and outputs are always assumed to come from some finite "alphabet" or symbols; and the operation of the machines proceeds in discrete steps, i.e., at times $t = 0, 1, 2, 3 \ldots$ on some suitable scale.
>
> In control theory, however, we consider the inputs of our machines as variables which we may alter in such a way as to control the output or "states" of our machine. Our preoccupation is with executing the control in an economical manner, so as to minimize some "cost function." The inputs and outputs are usually assumed to be continuously variable, and in fact, to take the values in some Euclidean space.
>
> Despite the disparity in goals and assumptions, much of the basic apparatus is common to the two theories.

Although these frameworks, based on the reconciliation of automata theory and control theory, are important, I do not deem it necessary to cover them here for at least two reasons. First, they are subsumed under the Mesarovic framework, as shown by Islam [1974] for the framework developed by Wymore. Second, they require some knowledge of both automata theory and control theory, which, I assume, not all readers of this book will have.

4.2. Inductive Approaches

Contrary to deductive approaches, in which systems frameworks are developed in terms of intuitively justifiable axiomatic requirements, inductive approaches proceed basically by forming generalizations from examples of systems observed in different disciplines of science, engineering, and other areas of human endeavor. That is, these various disciplines are considered the only indigenous source of systems ideas from which a systems conceptual framework should emerge.

Collecting examples of systems and making generalizations is, of course,

Conceptual Frameworks

only a part of the overall process by which a systems framework is developed in an inductive way. Categories of systems discovered from examples must also be properly ordered and integrated into a coherent whole. When this whole begins to emerge, its nature may suggest some additional categories of systems, not observed as yet, but fitting into specific nonoccupied places in it.

The most primitive notion in systems derived by the inductive approach is the notion of a *variable*, viewed basically in the same way as in science, engineering, or mathematics. A variable is a thing that has a name (label), which distinguishes it from other variables under consideration, and a particular set of entities through which it manifests itself. These entities are usually referred to as *states* (or *values*) of the variable; the whole set is called a *state set*.

A variable is always an abstract (mathematical) thing. It may or may not represent some observable or measurable attribute of the real world. In the former case, states of the variable correspond, according to some observation or measurement rule, to specific manifestations of the attribute.

Using the notion of a variable as a primitive, systems are then conceived as sets of variables together with a relation recognized among their state sets. For any given purpose, the first item in conceptualizing a system is the selection of relevant variables, as well characterized by Ross Ashby [1956]:

> At this point we must be clear about how a "system" is to be defined. Our first impulse is to point at the pendulum and to say "the system is that thing there". This method however, has a fundamental disadvantage: every material object contains no less than an infinity of variables and therefore of possible systems. The real pendulum, for instance, has not only length and position; it has also mass, temperature, electric conductivity, crystalline structure, chemical impurities, some radio-activity, velocity, reflecting power, tensile strength, a surface film of moisture, bacterial contamination, and optical absorption, elasticity, shape, specific gravity, and so on and on. Any suggestion that we should study "all" the facts is unrealistic and actually the attempt is never made. What is necessary is that we should pick out and study the facts that are relevant to some main interest that is already given. . . .
> . . . The system now means, not a thing, but a list of variables.

When a set of variables is established, by which relevant distinctions on an object of interest are characterized, and a relationship among the variables is expressed in some form, we say that a system is defined on the object. Which category this system belongs to, according to a conceptual framework, depends on the type of the relation involved. That is, a conceptual framework classify systems by the types of relations among their variables. In the inductive approach, the types of relations employed for classifying systems are not chosen arbitrarily, but, by and large, they are abstracted from examples of systems that are actually used in praxis. This inherent connection between theory and praxis is the principal advantage of the inductive approach.

Although there are numerous systems frameworks based upon the inductive approach, only two of them appear to be sufficiently broad to qualify as a basis for systems science. One of them is my own framework, the GSPS framework mentioned earlier. Some aspects of this framework are overviewed in this section and employed as a vehicle for further discussions in this book. The second framework is due to Bernard Zeigler [1974, 1976a,b]. These two frameworks are very similar in spirit, even though they differ in some, relatively minor aspects. Their initial developments seem independent of each other, but later they influenced one another in a constructive way.

According to the GSPS framework, which is described in detail in one of my books [Klir, 1985a], basic categories of systems are distinguished from each other by epistemological characteristics; they are called *epistemological categories* (or types) of systems.

The term "epistemological" refers to distinctions regarding knowledge: epistemology is the study of the origin, nature, methods, and limits of knowledge. That is, epistemological categories of systems differ from one another on various aspects of knowledge regarding a given set of variables. They are basic forms for knowledge representation, and are partially ordered by their knowledge content. A system at a higher level, according to this ordering, contains all knowledge available in the corresponding systems at any lower level, but it also contains some additional knowledge. This ordering forms a semilattice, which is usually called an *epistemological hierarchy of systems*. Zeigler [1974] refers to similar ordering of categories of systems in his conceptual framework as *knowledge hierarchy* or *hierarchy of system specifications* [Zeigler, 1976a].

Further distinctions between systems, which are methodological in nature, can be distinguished in each epistemological category. Although important for methodological research, these methodological distinctions are not essential for discussing the foundations of systems science. Consequently, they are only of secondary interest in this book.

Some of the more important methodological distinctions are introduced in Chap. 5. In the rest of this chapter, however, I focus solely on the epistemological categories of systems.

4.3. Epistemological Hierarchy of Systems

The GSPS epistemological hierarchy of systems is derived from some primitive notions: an *investigator* (a specialist in a subject area) and his environment, an *investigated object* (a part of the world) and its *environment*, and an *interaction* between the investigator and the object. These notions are left undefined; they are used solely in their common-sense connotation. The term "sys-

Conceptual Frameworks

tem," on the other hand, is used for well-defined abstractions (on various epistemological levels) representing some features of the investigated object.

On the lowest level in the epistemological hierarchy, *a system is what is distinguished as a system by the investigator* [Gaines, 1979]. That is to say, the investigator makes a choice regarding the manner in which he wants to interact with the investigated object. The choice is not completely arbitrary in most instances; it is at least partially determined by the purpose of the investigation, investigative constraints (such as availability of specific measuring instruments, financial and time limitations or legal restrictions), and available knowledge relevant to the investigation.

A specific investigator–object interaction, which results in a system at the lowest epistemological level, can be described in various ways. In the GSPS framework, it is defined by a set of variables, a set of potential states (values) recognized for each variable, and (if the system is not purely mathematical) some operational way of describing the meaning of the variables and their states in terms of the associated real-world attributes and their manifestations.

The set of variables is always partitioned into two subsets. Variables in these subsets are called *basic variables* and *supporting variables* or, using a simpler terminology, *variables* and *supports*. Aggregate states of all supporting variables (or supports) form a *support set* within which changes in states of the basic variables occur. The most frequent examples of supporting variables are those representing time, space, or various populations of individuals of the same kind.

A system defined by a set of variables and a set of supports, their state sets, and, possibly, their real-world interpretations is usually called a *source system*. This name reflects the fact that such a system is, at least potentially, a source of empirical data. Other names used in the literature include *experimental frame* (coined by Zeigler [1976b]), *primitive system*, or *dataless system*.

Basic variables of a source system may further be partitioned into *input variables* and *output variables*. Under such a partition, states of input variables are viewed as conditions which affect the output variables. Input variables are not subject of inquiry within a given system. They are viewed as being determined by some agent that is not part of the system under consideration. Such an agent is referred to as an *environment* of the system; it includes in many cases the investigator.

Systems whose variables are classified into input and output variables are called *directed systems*; those for which no such classification is given are called *neutral systems*.

Various special, methodologically significant properties of state sets of individual variables of a source system may be recognized. These are called methodological distinctions. They include, for instance, distinctions between discrete and continuous variables, crisp and fuzzy variables, and variables of different scales (nominal, ordinal, interval, ratio, etc.).

Properties of individual variables of a source system can be viewed as thinghood properties and properties associated with their relationship as systemhood properties. In this sense, the source system is devoid of any systemhood properties. It is solely a frame that allows us to conceptualize presumed systemhood properties, but it does not contain any such conceptualization.

The relationship among variables of a source system is conceptualized only on epistemological levels higher than the source system. Systems on different higher epistemological levels are distinguished from each other by the level of knowledge regarding this relationship. A system on a higher level entails all knowledge of the corresponding systems on any lower levels and, at the same time, contains some additional knowledge that is not available on lower levels. The source system, clearly, must be contained in the definition of any higher level system.

As an example of a source system, let me describe a seven-variable ecological system that was actually defined on Lake Ontario for the purpose of data gathering during a period of twelve months in 1972 and 1973. The seven variables are as follows:

Variable 1, *Temperature* (in °C);
Variable 2, *Soluble reactive phosphorus* (SRP, in μg/liter);
Variable 3, *Soluble ammonia* (NH_3, in μg/liter);
Variable 4, *Total filtered nitrite and nitrate* (NO_2/NO_3, in μg/liter);
Variable 5, *Chlorophyll a* (in μg/liter);
Variable 6, *Zooplankton biomass* (in μg/liter);
Variable 7, *Solar radiation* (langleys/day).

Five states are defined in Table 4.1 for each of these variables in terms of specific intervals of values of possible measurements. Measurements of these variables were taken in time (on a daily basis) and in space (at 32 different locations that

Table 4.1. Ecological System Defined on Lake Ontario: Variables and State Sets

Variables	State identifiers[a]				
	1	2	3	4	5
1. Temperature (°C)	2–4.9	5–7.9	8–11.9	12–14.9	15–18
2. SRP (μg/liter)	0–2.9	3–5.9	6–8.9	9–11.9	12–15
3. NH_3 (μg/liter)	2–5.9	6–9.9	10–13.9	14–17.9	18–20
4. CO_3/NO_2-N (μg/liter)	40–89.9	90–139.9	140–189.9	190–229.9	230–260
5. Chlorophyll (μg/liter)	0–2.9	3–3.9	4–4.9	5–5.9	6–8
6. Zooplankton (μg/liter)	0–14.9	15–50.9	51–119.9	120–184.9	185–250
7. Light (langleys/day)	70–139.9	140–179.9	180–219.9	220–259.9	260–300

[a]Measurements for each variable have been classified into five states.

Conceptual Frameworks 51

encompass all of the area of Lake Ontario); hence time and space are supports in this example. The system is viewed as directed, with the climatological variables 1 (temperature) and 7 (light) declared as input variables and the ecological variables 2–6 as output variables.

As another example of a source system, let the support set be the population of all residents of a township that live in rented homes, and let the following six variables and their states be defined with respect to this population at some particular period of time:

1. Age (in years);
2. Sex (male, female);
3. Housing type (apartment, atrium house, tower block, terraced house);
4. Influence on management (low, medium, high);
5. Contact with other residents (low, medium, high);
6. Satisfaction with housing conditions (low, medium, high).

When a source system is supplemented with data, i.e., by actual states of the variables within the defined support set, we obtain a system on the next higher epistemological level. Systems on this level are called *data systems*. In general, any set of data is defined by a function whose domain and range are, respectively, the support set and the set of overall states of the basic variables.

An example of a data set, which supplements the ecological source system defined previously, is given in Table 4.2. That is, the source system and this data set, which has the form of a time series, constitute a data system. In general, every time series qualifies as a data set.

Other examples of data, illustrating differences in support sets, are given in Figs. 4.1 and 4.2. The data set, shown only partially in Fig. 4.1, consists of

Table 4.2. Data Set: Average Values of 32 Locations on Lake Ontario and of Daily Observations in Each Month

Variables	Month[a]	1972								1973		
		A	M	J	J	A*	S	O	N	D	J	F*	M
1. Temperature		1	1	2	4	5	5	4	3	3	2	2	1
2. SRP		5	4	2	1	1	1	2	3	4	5	5	5
3. NH_3		2	1	2	5	4	4	4	2	2	2	2	2
4. NO_3/NO_2		5	5	3	1	1	1	3	4	4	5	5	5
5. Chlorophyll		2	2	4	5	5	4	2	1	1	1	1	1
6. Zooplankton		1	1	2	5	5	4	3	3	3	2	2	2
7. Light		1	2	2	2	3	4	5	5	3	1	1	1

[a]Months marked with asterisks indicate interpolated values; light values are classified up to the saturation light level of 300 langleys/day.

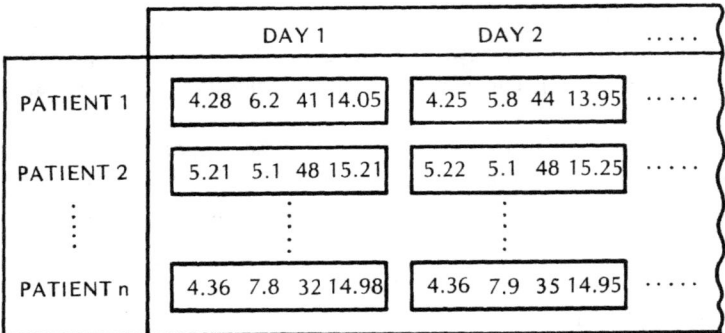

Figure 4.1. Example of a medical data system.

(a) Pacific sector of the Southern Ocean.

TIME	\multicolumn{4}{c}{SPACE ⟶ l}									
	\multicolumn{13}{c}{Degrees of latitude}
	50	52	54	56	58	60	62	64	66	68	70	72	74	76
Jan.	0	0	0	1	1	1	1	1	2	3	3	4	4	4
Feb.	0	0	0	0	1	1	1	1	2	2	3	3	3	4
March	0	0	0	0	0	1	1	1	2	2	3	4	4	4
April	0	0	0	0	0	1	1	1	2	3	4	4	5	5
May	0	0	0	0	1	1	1	2	3	4	4	5	5	5
June	0	0	0	1	1	2	2	3	4	4	5	5	5	5
July	0	0	1	1	1	2	3	4	4	4	5	5	5	5
Aug.	0	1	1	1	2	3	4	4	4	5	5	5	5	5
Sept.	0	1	1	1	2	3	4	4	5	5	5	5	5	5
Oct.	0	1	1	1	2	3	3	4	4	5	5	5	5	5
Nov.	0	1	1	1	2	2	2	3	4	4	5	5	5	5
Dec.	0	0	1	1	1	2	2	2	3	4	4	4	5	5

(b) Atlantic sector of the Southern Ocean.

	\multicolumn{13}{c}{Degrees of latitude}
	50	52	54	56	58	60	62	64	66	68	70	72	74	76
Jan.	0	0	0	0	0	1	1	1	1	2	4	4	4	4
Feb.	0	0	0	0	0	0	0	1	1	1	3	4	4	3
March	0	0	0	0	0	0	0	0	1	1	3	4	4	4
April	0	0	0	0	0	0	0	1	1	2	4	5	5	5
May	0	0	0	0	0	1	1	1	2	4	4	5	5	5
June	0	0	0	0	0	1	1	2	3	4	5	5	5	5
July	0	0	0	0	0	1	1	3	4	5	5	5	5	5
Aug.	0	0	0	0	0	1	2	3	4	4	5	5	5	5
Sept.	0	0	0	0	1	1	1	3	4	5	5	5	5	5
Oct.	0	0	0	0	1	1	1	3	4	5	5	5	5	5
Nov.	0	0	0	0	0	1	1	3	4	4	5	5	5	5
Dec.	0	0	0	0	0	1	1	2	2	3	4	5	5	5

Figure 4.2. Example of a climatological data system: satellite-derived five-year monthly averages (1973–1977) of sea-ice cover in the Southern Ocean.

Conceptual Frameworks 53

measurements of four physiological variables (red blood cell count, white blood cell count, hematocrit, and hemoglobin) on a population of patients suffering from anemia and in time. For example, patient 1, on day 1, had a red blood cell count of 4.28, white blood cell count of 6.2, hematocrit of 41, and a hemoglobin of 14.05.

The data shown in Fig. 4.2 (obtained by satellite imagery) specify the percentage of ice cover at different degrees of latitude in the Pacific and Atlantic sectors of the Southern Ocean during individual months over a period of one year. Here the supports are time and space and the only basic variable involved (percentage of ice cover, c) has six states:

0, no ice cover ($c = 0$);
1, low ice cover ($c \in (0, 25]$);
2, medium ice cover ($c \in (25, 50]$);
3, high ice cover ($c \in (50, 75]$);
4, very high ice cover ($c \in (75, 100)$);
5, full ice cover ($c = 100$).

Higher epistemological categories of systems involve some *support-invariant characterization of a relation* (constraint, dependence, etc.) among the basic variables. In general, the relation can be utilized for generating states of the basic variables within the support set.

On the next level higher than the level of data systems, each system is represented by one overall support-invariant characterization (time-invariant, space-invariant, population-invariant, etc.) of the relation among the basic variables of the associated source system and, possibly, some additional variables. Each of the additional variables is defined in terms of a specific *translation rule* in the support set. A translation rule is basically a bijective function by which each element of the support set is assigned another (unique) element of the same support set. For example, each particular discrete time t is assigned the time $t - a$ for some particular integer a; we say that time t is translated into time $t - a$. Since the overall relation among basic variables and those derived from them by various translation rules (e.g., lagged variables) can be utilized for describing a process by which states of the basic variables are generated within the delimited support set, systems on this level are called *generative systems*.

Finite-state machines (deterministic or probabilistic), Markov chains, and differential equations with constant coefficients are examples of generative systems. Each of them characterizes a relation among the variables involved, which does not change within the relevant support set (time or space set). This relation makes it possible to generate specific data sets, one for each initial or boundary condition and for a given support set. Observe, for example, that the solution of a set of differential equations for specific initial or boundary conditions represents a data set (a function of time or space).

As a simple example, consider a single variable, v, with two states, 0 and 1, which alternate in discrete time, t:

$$t = 0\ 1\ 2\ 3\ 4\ \ldots$$
$$v(t) = 0\ 1\ 0\ 1\ 0\ \ldots$$

How can this alternating sequence of zeroes and ones (useful, for example, as a computer clock) be generated? It is easy to see that, for any t, $v(t)$ and $v(t + 1)$ are related in such a way that $v(t + 1)$ is uniquely determined by the knowledge of $v(t)$. The relation consists of two pairs:

$v(t)$	$v(t + 1)$
0	1
1	0

It can also be expressed as

$$v(t + 1) = 1 - v(t).$$

For any given t, values $v(t)$ and $v(t + 1)$ are always observed together. We may thus view them as values of two variables, s_1 and s_2, defined by the equations

$$s_1(t) = v(t),$$
$$s_2(t) = v(t + 1).$$

Here, s_2 is defined in terms of v by a simple translation rule, $t + 1$, applied to the time set; s_1 is identical with v (i.e., it is defined by the identity translation rule). The relation now becomes a relation between two variables observed at the same time:

$$s_2(t) = 1 - s_1(t).$$

While $s_1(t)$ and $s_2(t)$ are time varying, the relation between them is time invariant in the sense that it holds for every value of t. Given now the initial condition $s_1(0) = 0$, the given data sequence $v(t)$ can be readily generated by the previous equation and the definition of the variables s_1 and s_2. These variables, which are defined in terms of basic variables by specific translation rules, are called in the GSPS terminology *sampling variables*. Furthermore, s_2 is called a *generated variable* and s_1 is called a *generating variable*.

When the support of a generative system is discrete time, sampling variables are defined, in general, by equations of the form

$$s_k(t) = v_i(t + \alpha), \tag{4.4}$$

where s_k, v_i, α denote, respectively, a sampling variable, a basic variable, and an integer that characterizes the time translation involved. For each discrete time t (represented by an integer), the state of s_k is defined as the state of v_i at time $t + \alpha$. When $\alpha = 0$, s_k is identical to v_i; this trivial translation is called an identity translation rule. When $\alpha < 0$ or $\alpha > 0$, then s_k represents past or future states of v_i, respectively.

Conceptual Frameworks

Given a set of basic variables, a larger set of sampling variables can be defined by equations of type (4.4). In each equation, a sampling variable s_k is uniquely assigned to a specific pair (v_i, α). A set of these pairs, by which desirable sampling variables are defined, is usually called a *mask* [Klir, 1985a]. It may be viewed as a "window" through which appropriate samples of data are obtained, as illustrated in Fig. 4.3. In this example, the sample of data obtained for the given mask is the seventuple $(s_1(8), s_2(8), \ldots, s_7(8)) = (2.7, 6, 0.8, 3.2, 0.8, 4, 0.9)$ when it is applied to the data matrix for $t = 8$.

When given data are exhaustively sampled by a chosen mask, the resulting set of overall states (n-tuples) represents a time-invariant relation among the sampling variables that is implied by the data. Conversely, a time-invariant relation can be used for generating future states (predictions) of some basic variables (output variables) on the basis of past and present states of these variables and, possibly, present and past states of input variables. The generated variables are in this case the rightmost sampling variables in the mask, but some of these may be input variables (i.e., variables generated by the environment of the system rather than the system itself). The remaining sampling variables in the mask are generating variables.

The time-invariant relation can also be used for generating past states (retrodictions). In this case, the generated variables are the leftmost sampling variables that are defined in terms of the output variables.

Let G and $\overline{G}$ denote, respectively, the set of generated variables (viewed as

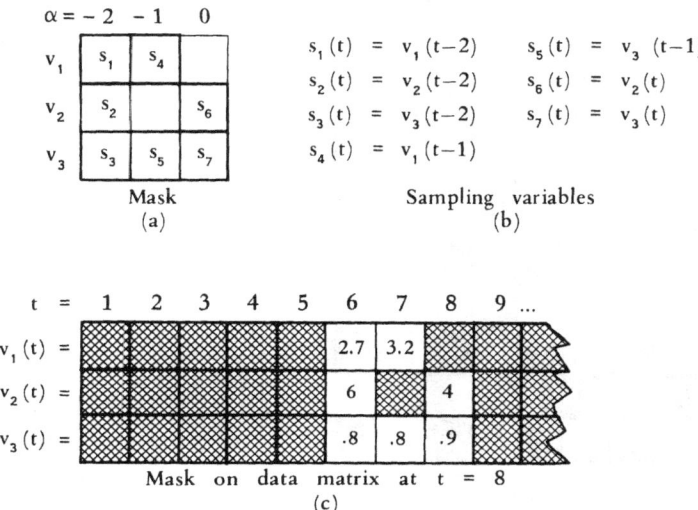

Figure 4.3. An example of a mask and sampling variables.

either predictive or retrodictive variables) and the set of generating variables (including the input variables) of a mask, and let $\mathbf{g} \in G$ and $\bar{\mathbf{g}} \in \bar{G}$. Then, ideally, states $\mathbf{g}$ are determined by a function

$$f: \bar{G} \to G,$$

which for each state $\bar{\mathbf{g}}$ gives us a unique prediction (or retrodiction) $\mathbf{g} = f(\bar{\mathbf{g}})$. This function is called a *behavior function*.

If a given time-invariant relation among sampling variables can be expressed in terms of a behavior function, the system is called *deterministic*; otherwise, it is called *nondeterministic*. States of generated variables of indeterministic systems are not determined uniquely by the states of the generating variables. Instead, they are generated with some uncertainty, which may, for example, be expressed in terms of conditional probabilities of $\mathbf{g}$ given $\bar{\mathbf{g}}$, $p(\mathbf{g}|\bar{\mathbf{g}})$, defined on the time-invariant relation $R \subset G \times \bar{G}$. These probabilities are called *behavior probabilities*.

To illustrate that the concepts introduced for generative systems based upon time support are similar for systems based upon space support, consider a single variable, v, with two states, black and white, which alternate in a two-dimensional space according to the usual chessboard pattern (Fig. 4.4a). The support consists here of 64 squares of the spatial grid. It is described by coordinates x and y.

How can the chessboard pattern be generated from a simple initial (boundary) condition? If we choose, for example, the color of the left top square as the initial condition, it is easy to see that the first mask shown in Fig. 4.4b can be used for this purpose, where s_2 and s_3 are generated variables and s_1 is a generating variable. These sampling variables are defined by the equations

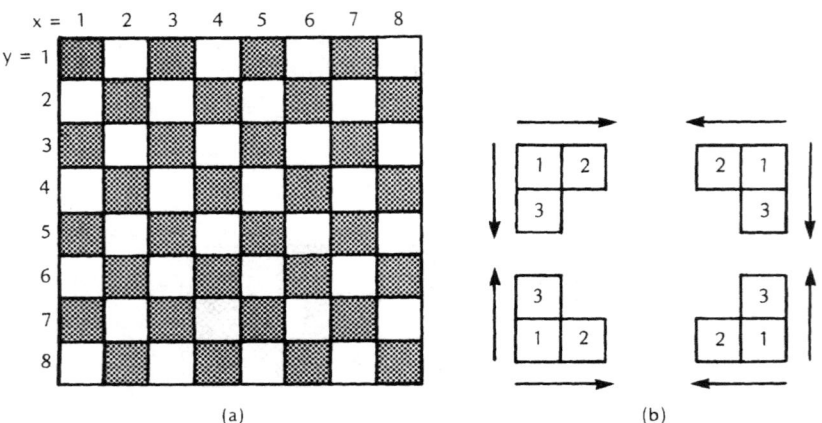

Figure 4.4. Illustration of spatial invariance.

Conceptual Frameworks

$$s_1(x, y) = v(x, y),$$
$$s_2(x, y) = v(x + 1, y),$$
$$s_3(x, y) = v(x, y + 1).$$

The space-invariant relation consists of two triples:

$s_1(t)$	$s_2(t)$	$s_3(t)$
B	W	W
W	B	B

This relation can be expressed in terms of the behavior function f defined as follows:

$$f(B) = W, W,$$
$$f(W) = B, B.$$

The whole chessboard pattern can be generated by moving the mask from the left top square in either of the two directions shown in Fig. 4.4b. Observe that this mask is only one of four different masks, shown in Fig. 4.4b, by which the chessboard pattern can be generated. Each of them is associated with specific generating orders and requires a different initial condition (regarding the color in one of the four corners of the chessboard).

For systems with continuous variables and supports, support-invariant relations among variables are characterized, in general, by differential equations with constant coefficients. In these systems, sampling variables cannot be defined directly in terms of basic variables and translation rules. Instead, they are defined indirectly, in terms of derivatives (of different orders) of the basic variables. The solution of a set of differential equations for specific initial or boundary conditions is a set of functions (one for each basic variable) defined on the support set. These functions represent in this case data that are generated by the differential equations.

As a simple example, let function $v(t) = \sin t$ be considered as data, where v denotes a basic variable and t stands for time whose values are taken from the set of nonnegative real numbers. How can these data be generated? An obvious answer is that they can be generated by a suitable differential equation, if one can be found, whose solution under appropriate initial condition is the given function.

To find a differential equation suitable for our purpose, we have to explore derivatives of v. Taking the first derivative,

$$\dot{v}(t) = \cos t,$$

and the second derivative,

$$\ddot{v}(t) = -\sin t,$$

we can see that

$$v(t) = -\ddot{v}(t).$$

This differential equation is clearly time invariant. Its solution, under the initial conditions $\dot{v}(0) = 1$ and $v(0) = 0$, is the given function $v(t) = \sin t$. Viewing v as the generated variable and $\ddot{v}$ as the generating variable, v can be generated, for example, on an analog computer by the scheme shown in Fig. 4.5.

Further climbing up the epistemological hierarchy involves two principles of integrating systems as components in larger systems. According to one of these principles, several systems that share some variables or interact in some other way are viewed as subsystems integrated into one overall system. Systems of this sort are called *structure systems*. The subsystems forming a structure system are often called its *elements*. When elements of structure systems are themselves structure systems, we call the overall system a *second-order structure system*. *Higher-order structure systems* are defined recursively in the same way. That is, elements of a structure system of order n are structure systems of order $n - 1$, their elements are structure systems of order $n - 2$, etc., until structure systems of order 1 are reached, whose ultimate elements are either source systems, data systems, or generative systems. Different categories of structure systems are thus distinguished by their orders and by the type of their ultimate elements.

A common form of representing structure systems are block diagrams. As an example, let me use a structure system that is based on the seven ecological variables defined in Table 4.1. It was well established in this case, both by ecological background knowledge and by systems analysis of relevant data, that the seven-dimensional relation among the variables can be adequately represented by five ternary relations among appropriate subsets of the seven variables. The five subsets of variables, which define elements of a structure system, are represented by the blocks in Fig. 4.6. Connections between blocks represent variables; arrows distinguish input and output variables of each element. Variables 1 and 7 are input

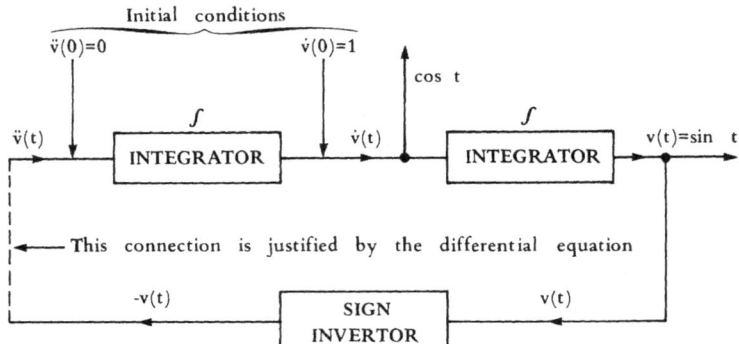

Figure 4.5. The scheme of gernerating the function $v(t) = \sin t$ by the time-invariant differential equation $v(t) = -\ddot{v}(t)$ on an analog computer.

Conceptual Frameworks 59

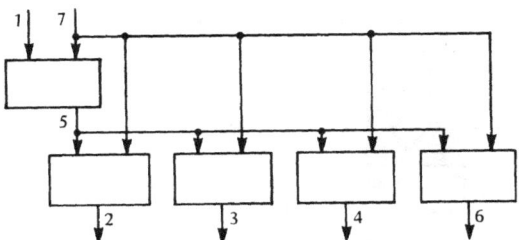

Figure 4.6. A structure system defined in terms of the ecological variables specified in Table 4.1.

variables of the whole system, which means that they are determined by the environment of the system.

The block diagram in Fig. 4.6 may represent different types of structure systems, depending on the types of their elements. The elements may be source systems, data systems, generative systems, or, possibly, structure systems of some order. In the last case, the block diagram would represent a structure system of one order higher than the order of its elements.

The connections between elements of a structure system cannot be totally arbitrary. To avoid inconsistencies, each variable must be determined exactly by one of the elements or by the environment of the whole system. That is, no variable is allowed to be an output variable of more than one element. Each variable shared by several elements must also have the same characterization in each of these elements.

As another example, let me describe a structure system defined for the purpose of studying the probation services offered by the State of New York for cases originating from complaints that are processed by criminal courts. The system, whose block diagram is given in Fig. 4.7, characterizes the flow of the case workload through the criminal court and probation institutions. It represents a framework for data gathering and processing. The support in this system is time. Depending on the specific questions to be addressed, observations are made on a monthly, weekly, or even daily basis starting on some fixed date. The variables of the system are defined as follows:

v_1, The total number of complaints received by the criminal court (during each individual period of observation—month, week, or day);
v_2, The number of complaints that are carried toward the arraignment;
v_3, The number of complaints that are dismissed;
v_4, The number of cases that are held over for sentencing;
v_5, The number of cases that are acquitted or discharged;
v_6, The number of cases that are assigned for probation;
v_7, The number of cases that are not assigned to probation (this includes

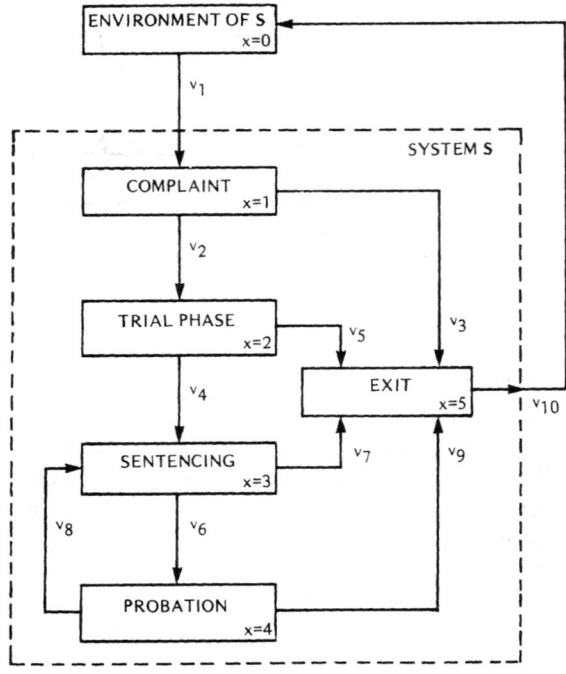

Figure 4.7. Block diagram of the structure system defined for criminal courts of the State of New York.

cases where a fine or restitution is the only punishment, those where imprisonment is assigned, and those where an unconditional or conditional discharge is used);

v_8, The number of cases that violate the conditions of probation;
v_9, The number of cases that are discharged from probation;
v_{10}, The number of cases that are discharged from the criminal court institutions.

The system consists of five elements shown by the blocks in Fig. 4.7, which are given some suggestive names describing their roles in the overall system. Also shown in Fig. 4.7 is the environment as a separate block outside the system of concern. Elements of the system are initially source systems, which later become data systems.

The purpose of the next example is to illustrate the concept of structure systems of higher orders. The example describes a simple structure system of second order, in terms of which a logic design of a serial binary adder is conceptualized.

Conceptual Frameworks 61

A serial binary adder is a device by which two binary numbers, properly encoded in some physical signals, are added. The adder receives at its inputs encoded digits of the two numbers, starting with the least significant digits, and produces at its output a series of encoded digits that represents the sum of the two numbers.

There are alternative ways of conceptualizing the adder at an abstract level (ignoring the physical encoding) as a generative system. One way is to utilize the notion of *carry* in describing the operation of addition. In this way, the adder is a generative system with discrete time support, t, and four basic variables, each assuming two states, 0 and 1. Let x_1, x_2 denote input variables whose values $x_1(t)$, $x_2(t)$ represent digits of the two binary numbers to be added, and let y, c denote output variables whose digits $y(t)$, $c(t)$ represent digits of the sum and the carry, respectively. To describe the operation of addition as a time-invariant relation, we need, in addition to the values of these four variables, also the value $c(t-1)$. Let c', defined by the equation

$$c'(t) = c(t-1),$$

denote this additional sampling variable (previous carry). The relation among the five sampling variables can now be expressed in terms of the eight quintuples given in Table 4.3. We can easily see by inspection that both y and c are functions of x_1, x_2, c' and, consequently, the system is deterministic. Assume now that we want to implement these functions in terms of the following simple logic functions:

a	b	$\vee$	$\vee$	$\not\vee$	$\equiv$
0	0	0	0	1	1
0	1	0	1	0	0
1	0	0	1	0	0
1	1	1	1	0	1

Table 4.3. Behavior Functions Characterizing the Operation of Binary Addition

x_1	x_2	c'	y	c
0	0	0	0	0
0	0	1	0	1
0	1	0	0	1
0	1	1	1	0
1	0	0	0	1
1	0	1	1	0
1	1	0	1	0
1	1	1	1	1

A particular implementation is shown in Fig. 4.8. It is presented as a structure system of second order, which consists of three elements: SUM, CARRY, and MEMORY. If these elements are viewed as generative systems, the resulting structure system is of first order. Elements SUM and CARRY represent the two behavior functions defined in Table 4.3, while MEMORY implements the translation rule by which the sampling variable c' (previous carry) is defined (it requires a memory device that can keep the input state for one discrete time and, then, release it as output state). If, however, the elements themselves are viewed as specific structure systems, as shown for elements SUM and CARRY in Fig. 4.8, the overall structure systems is of order 2.

It should now be clear how structure systems are constructed by integrating other systems (including structure systems of lower orders) into larger wholes. This is one of two possible integrating principles by which larger systems are constructed.

According to the second integrating principle, an overall systems is viewed as varying within the relevant support (time, space, population, etc.). The change from one system to another in the delimited class is described by a replacement procedure that is invariant with respect to the support employed. Overall systems of this type are called *metasystems*. In principle, the replacement procedure of a metasystem may also change. Then, an invariant higher-level procedure is needed to describe the change. Systems of this sort are called *metasystems of second order*. *Higher-order metasystems* are then defined recursively in the same way.

Examples of metasystems are time-varying finite state machines, tessellation

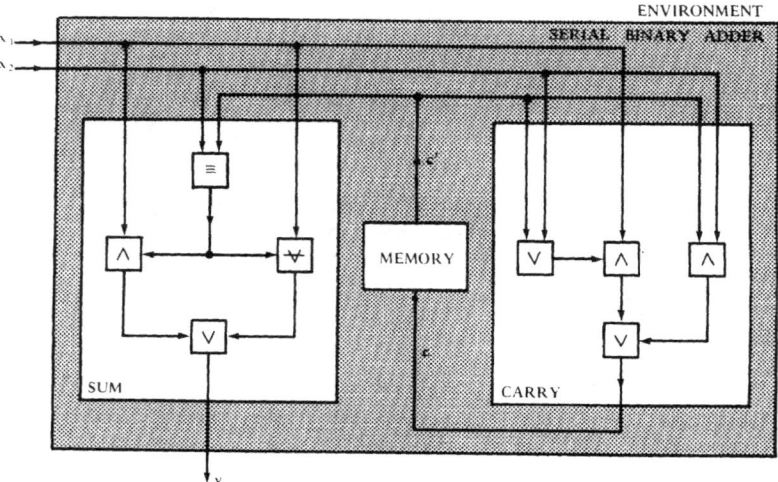

Figure 4.8. Serial binary adder conceptualized as a structure system of second order.

Conceptual Frameworks

automata, or differential equations whose coefficients are not constant. Metasystems (of the various orders) are important for capturing systems phenomena that involve change, such as adaptation, self-organization, morphogenesis, autopoiesis, evolution, etc. Evolution, for example, is conceptualized in terms of transitions from lower- to higher-order metasystems by Turchin [1977].

Replacement procedures of metasystems of different orders can be defined in many different ways. In fact, any procedures that describe changes in states of variables in systems of any category may readily be used at the metasystem level for describing changes in systems instead. The replacement procedures may even include random decisions. No attempt is made here to categorize all possible replacement procedures. Instead, some typical replacement procedures are illustrated by a few simple examples.

The first example describes the functioning of a set of traffic lights at an intersection for a 24-hr period as a homogeneous metasystem that consists of three elements defined as data systems. Each of the three elements contains the same variables and state sets. Variables describing the lights for traffic bound north–south, south–north, east–west, and west–east are denoted by NS, SN, EW, and WE, respectively, and those describing the lights for left turns for traffic bound north–east, south–west, east–south, and west–north are denoted by NE, SW, ES, and WN, respectively. The support is time; 1 sec is the smallest recognizable interval of time in terms of which all other relevant time intervals are defined.

Data matrices d_1, d_2, d_3 of the three elements D_1, D_2, D_3 are defined in Fig. 4.9a; their time sets are specified directly by the relevant time intervals. The data matrices are periodical and are defined by their first periods. As indicated in the figure, the individual systems D_1, D_2, D_3 represent the traffic control at night, at periods of normal traffic during day, and during rush hours, respectively. The systems, viewed as elements of a metasystem, replace each other at specific times during each period of 24 hr. A convenient manner of defining the replacement function is in this case the labeled diagram in Fig. 4.9b. Its nodes represent the three elements of the metasystem, each arrow from D_i to D_j ($i, j = 1, 2, 3$) indicates that D_i is replaced by D_j, and the label attached to the arrow specifies the time at which the replacement is made.

As the second example, consider a patient whose kidneys do not function properly at times. His condition is monitored in terms of several variables. When necessary, the functioning of his kidneys is replaced by the so-called hemodialysis machine. The same monitoring continues even when the machine is used, but some additional variables must be observed during such periods. Two source systems (experimental frames), say E1 and E2, can thus be recognized for the purpose of monitoring the patient. One of them is associated with the periods during which the natural kidneys function adequately, while the other one represents periods during which the hemodialysis machine is employed. System E1 contains the following four variables:

v_1, Water in urine (measures to an accuracy of 0.1 liters in the range of 0–1 liter);

v_2, Glucose in urine (measured to an accuracy of 20 g in the range of 0–200 g;

v_3, Urea in urine (measured to an accuracy of 5 g in the range of 0–50 g);

v_4, Blood urea nitrogen (only two states are defined by the observation channel, say states 1 and 0, depending on whether or not the actual value reaches at least 150 mg per 100 ml of blood).

ELEMENT D_1: night traffic control.

$t_i \in$	[0, 20)	[20, 30)	[30, 50)	[50, 60)
NE = SW	g	y	r	r
NS = SN	g	y	r	r
ES = WN	r	r	g	y
EW = WE	r	r	g	y

ELEMENT D_2: normal traffic control.

$t_i \in$	[0, 15)	[15, 25)	[25, 55)	[55, 65)	[65, 80)	[80, 90)	[90, 110)	[110, 120)
NE = SW	g	y	r	r	r	r	r	r
NS = SN	r	r	g	y	r	r	r	r
ES = WN	r	r	r	r	g	y	r	r
EW = WE	r	r	r	r	r	r	g	y

ELEMENT D_3: traffic control during rush hours.

$t_i \in$	[0, 30)	[30, 40)	[40, 50)	[50, 60)
NE = SW	r	r	r	r
NS = SN	g	y	r	r
ES = WN	r	r	r	r
EW = WE	r	r	g	y

(a)

6 a.m.   7 a.m., 4 p.m.

11 p.m.   9 a.m., 6 p.m.

REPLACEMENT PROCEDURE r

(b)

Figure 4.9. Three-mode traffic light system as an example of metasystem.

Conceptual Frameworks

System **E2** contains all of these variables plus two additional variables:

v_5, Temperature of blood (measured to an accuracy of 0.2°F in the range of 97–100°F);

v_6, Blood pressure (measured to an accuracy of 2 mm of mercury column in the range of 110–130 mm).

These variables are essential for the hemodialysis machine, which must maintain each of them in a narrow range. All of the introduced variables are observed in time. The actual time set (frequency of observation) depends on the seriousness of patient's condition as well as other factors and there is no need to define it for the purpose of this example.

The two source systems can be viewed as a metasystem under the following replacement procedure r: if $v_4 = 1$, replace $\mathbf{E}_1$ by $\mathbf{E}_2$; if $v_4 = 0$, replace $\mathbf{E}_2$ by $\mathbf{E}_1$.

As another example, consider a structure system whose elements are arranged in an $n \times n$ array. Assume that each of the elements, which are often called cells of the array, is coupled only to cells that are adjacent to it in the array. For example, a 5×5 array is shown in Fig. 4.10. Individual cells in the array can be identified conveniently by two integers $i, j \in \{0, 1, \ldots, n - 1\}$ that label rows and columns of the array, respectively. As indicated in Fig. 4.10, they can be also identified by a single integer

$$c = ni + j.$$

Let c be called a cell *identifier*.

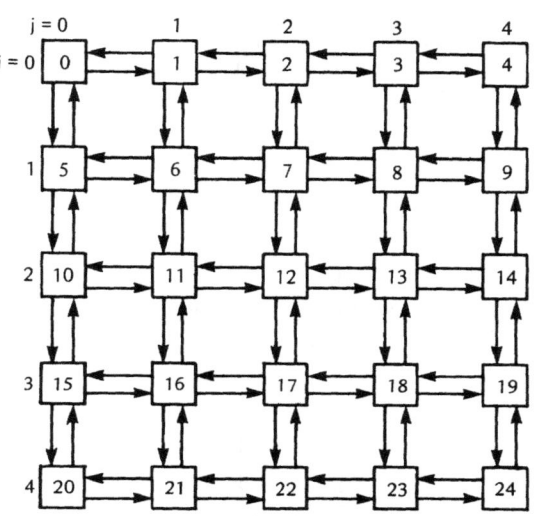

Figure 4.10. 5×5 cellular array.

Assume that the *environment* of each cell c (except the boundary cells) consists of its four adjacent cells, as shown in Fig. 4.11. The cell has four input variables v_{c-n}, v_{c-1}, v_{c+1}, v_{c+n}, one from each of the adjacent cells. It has one output variable, which is coupled to all the adjacent cells. The internal environment of each of the boundary cells (cells in rows $i = 0, n - 1$ and columns $j = 0, n - 1$) is degenerated in an obvious way.

Let all cells in a cellular array, say the one in Fig. 4.10, be deterministic generative systems defined by the same totally ordered time set and a behavior function

$$v'_c = f_c(v_{c-n}, v_{c-1}, v_c, v_{c+1}, v_{c+n})$$

where v'_c represents the next state of v_c and $c \in \{0, 1, \ldots, 24\}$; it must of course, be specified how this function is interpreted for the boundary cells, where some of the input variables are not present. Assume further that only two states, 0 and 1, are distinguished for each of the variables. When $v_c = 1$ ($v_c = 0$), let cell c be called *active* (*inactive*, respectively).

Given a cellular array, a set of structure systems can be defined on it, each characterized by a subset of its cells. For example, there are 2^{25} (more than 3.3×10^7) structure systems for the cellular array in Fig. 4.10. It is sometimes desirable to integrate structure systems in this set into a metasystem by a suitable replacement procedure. As a simple example, let the replacement procedure be defined as follows: cell c is included in the structure system if and only if it is active or at least one cell in its internal environment is active, i.e., if and only if

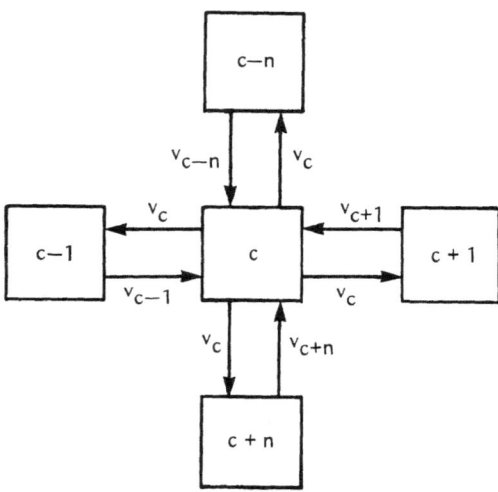

Figure 4.11. Environment of a cell in a cellular array of the type shown in Fig. 4.10.

Conceptual Frameworks

$$v_{c-n} + v_{c-1} + v_c + v_{c+1} + v_{c+n} \geq 1.$$

To show a specific metasystem of this kind, let us use the proposed replacement procedure and let

$$v'_c = [(v_{c-5} + v_{c-1} + v_{c+1})(\text{mod } 2) + v_{c+5}](\text{mod } 2)$$

be the behavior function of cells in the 5 × 5 cellular array. Variables that are not available in the environment of a cell are simply excluded from the formula. This metasystem generates sequences of structure systems, one for each initial structure system. Short segments of three such sequences are shown in Fig. 4.12, where the black and gray squares identify cells that are included in the individual structure systems, and distinguish active cells (black squares) from inactive cells (gray squares); white squares identify cells that are not included in the various structure systems.

Variations of the metasystems introduced in this example are possible by using different behavior functions for the cells, or different replacement procedures. More radical variations can be produced by using different arrays, generally k-dimensional, where $k \geq 1$. The members of this class of metasystems are usually referred to in the literature as tessellation automata.

The next example is a metasystem whose elements, $\mathbf{D}_1, \mathbf{D}_2, \ldots, \mathbf{D}_n$, are data systems. Each of these data systems consists of one variable, v, its state set, V, and a single support, t, which is viewed, in general, as an index identifying the location in a sequence of states from V. The data systems differ in their data, $\mathbf{d}_i$,

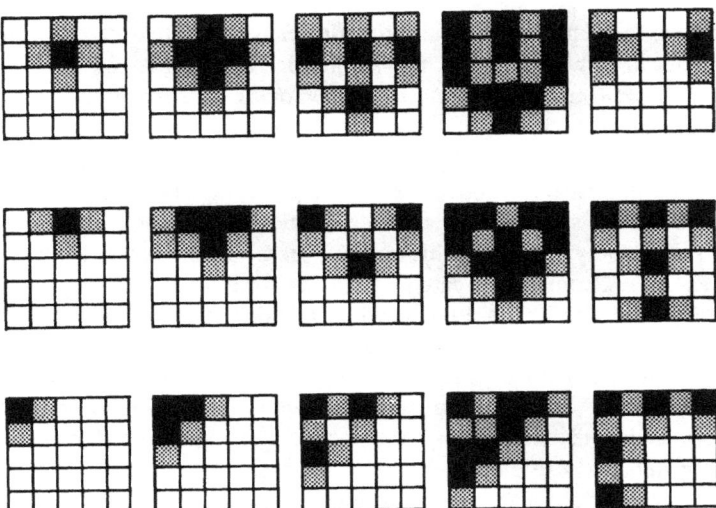

Figure 4.12. Segments of three possible sequences of structure systems.

and the associated support sets, T_i. The first data system, $\mathbf{D}_1$, is defined explicitly, while the remaining data systems are defined implicitly, by a replacement function,

$$r: \mathbf{D}_i \to \mathbf{D}_{i+1}$$

applied for $i = 1, 2, \ldots n - 1$. One way of implementing this function is to define a replacement of each data entry, α, in $\mathbf{D}_n$ with a string of states, $p(\alpha)$, from V. These replacements, $\alpha \to p(\alpha)$, are usually called production rules, and the metasystems based upon them are called Lindenmayer systems, L-systems, or developmental systems [Herman and Rozenberg, 1975].

As a specific example, let $V \in \{0, 1, \ldots, 9\}$ and let the production rules be defined by the table

α	0	1	2	3	4	5	6	7	8	9
$p(\alpha)$	12	93	49	61	25	87	78	34	9	9

Then, for example, if the initial data system has the data array [0], the metasystem generates a sequence of data systems with the following data arrays:

[0]
[12]
[9349]
[961259]
[9789349879]
[93499612599349]
[961259978934987996125 9]
[978934987993499612599 3499789349879]

The described epistemological categories of systems are ordered in a semilattice form. As already mentioned, this semilattice ordering is usually referred to under the name *epistemological hierarchy of systems*. Mathematically, it is a meet semilattice, whose maximum element is the category of source systems. A portion of the semilattice is described by an inverted Hasse diagram in Fig. 4.13. The arrows in the diagram symbolize the ordering $\leq$. Letters E, D, G denote the categories of source systems (experimental frames), data systems, and generative systems, respectively, Letter S, which always precedes some other letter or letters, is used as an operator for a category of structure systems. The category is delimited by the letters that are preceded by this operator S; these letters specify the type of systems that are used as elements in the structure system. For example SG denotes the category of structure systems whose elements are generative systems. Symbol S^2 denotes a category of structure systems of second order. For example, S^2D denotes the category of structure systems whose elements are also structure systems whose elements, in turn, are data systems. Clearly, symbols $S^3, S^4, \ldots$ would designate higher categories of structure systems (not shown in Fig. 4.13). Symbols M and M^2 are used in a similar way as symbols S and S^2, but

Conceptual Frameworks

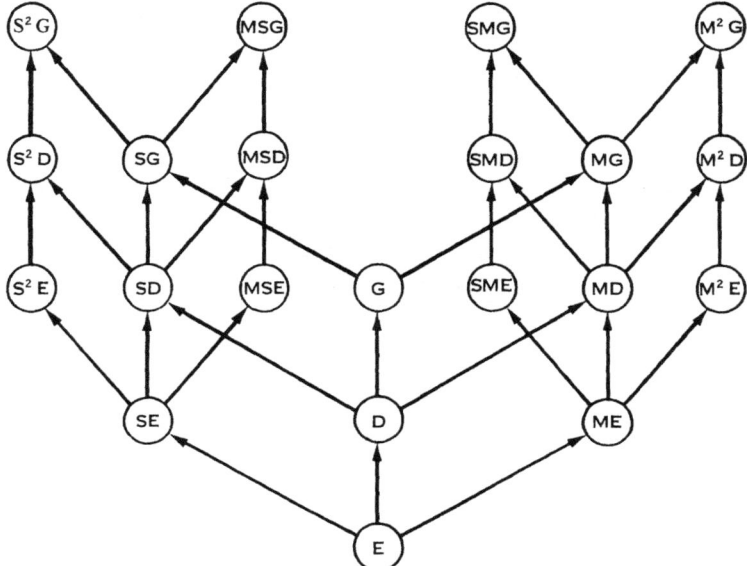

Figure 4.13. A portion of the semilattice describing the ordering of epistemological systems categories.

they denote categories of metasystems of first and second orders. Again, symbols M^3, M^4, ... would designate higher categories of metasystems. Combinations SM and MS denote, respectively, categories of structure systems whose elements are metasystems and categories of metasystems whose elements are structure systems.

The diagram in Fig. 4.13 describes only a part of the epistemological hierarchy of systems. It can be extended in an obvious way to combinations such as S^3G, S^2MG, SMSD, M^2SE, SMSMG, etc. Clearly, the last letter in each of these descriptors, which designates the type of the ultimate elements in a structure system or a metasystem of some order, must be either E, or D, or G: the ultimate elements are either source systems, or data systems, or generative systems.

CHAPTER 5

Systems Methodology

As our ability to solve problems expands, the scale of the problems attached themselves seem to expand at a similar rate. As a result there always exist over the horizon new categories of problems of greater size to tackle.
—DAVID M. HIMMELBLAU

Systems methodology is understood in this book as a family of coherent collections of methods for dealing with the various systems problems that emanate from the conceptual framework employed.* Thus, for example, one systems methodology is based upon the GSPS framework outlined in Chap. 4. Systems methodologies based upon different but equally general conceptual frameworks are capable of covering, by and large, the same class of problems. With some adjustment, methods developed under one framework can usually be converted into methods for dealing with comparable systems problems under another framework.

A problem is viewed as a systems problem if (i) it involves systems of some sort, and (ii) it is concerned only with the systemhood properties and not with the thinghood properties. This means that systems methodology is based on the assumption that problems regarding systemhood properties can be extracted from overall problems that involve not only systemhood properties, but thinghood

---

*Two kinds of systems methodology are sometimes distinguished in the literature: hard and soft systems methodologies. This distinction, which is well explicated by Flood and Carson [1988], has emerged from the efforts of some systems researchers to develop a methodology that can deal with ill-structured problems, in which objectives or purposes are themselves problematic. Such a methodology is now usually called soft, while methodology that is based on the assumption that problems can be structured, through a problem formulation stage, in terms of well-defined objectives and constraints is called hard. Perhaps the most important proponent of soft systems methodology is Peter Checkland [1981].

The issue of soft methodology versus hard systems methodology is somewhat controversial. Although hard methodology, if applicable, is generally recognized as superior to soft methodology, there is no doubt that many practical problems are not easily amenable to it. The question is whether this is due to some intrinsic properties of the problems involved or due to our lack of understanding of the full spectrum of our own cognitive capabilities. For example, can the expertise in using soft systems methodology be elicited from someone like Peter Checkland and utilized in the knowledge base of an expert system? These questions, which are by no means trivial, are beyond the scope of this book.

properties as well. Is it meaningful and useful to divide overall problems in this way? I believe it is. In fact, we employ this division in solving simple everyday problems when we use arithmetic, as is well expressed by Zeigler [1976b]:

> Nobody questions the role of arithmetic in the sciences, engineering, and management. Arithmetic is all pervasive, yet it is a mathematical discipline having its own axioms and logical structure. Its content is not specific to any other disciplines but is directly applicable to them all. Thus students of biology and engineering are not taught how to add differently—the different training comes in what to add, when to do it, and why.
>
> The practice of modelling and simulation too is all pervasive. However it has its own concepts of model description, simplification, validation, simulation, and exploration, which are not specific to any particular discipline. These statements would be agreed to by all. Not everyone, however, would say that the concepts named can be isolated and abstracted in a generally useful form.

Systemhood aspects of problems can be learned and dealt with independently of the thinghood aspects in basically the same way as we learn and use arithmetic independently of the things to which it is applied. In both cases, the abstraction from thinghood properties is useful. It leads to the development of methods that have broad, cross-disciplinary applicability. They allow us to answer questions concerning systemhood properties regardless of the nature of things in terms of which they are formulated. The implications of a well-developed systems methodology are far reaching, but not as easy to grasp as the role of arithmetic in our daily life.

The role of systems methodology as an aid to scientific inquiry is illustrated in Fig. 5.1. Two levels are distinguished, one expressed by the inner rectangles and one by the outer rectangles:

1. (Inner rectangles) To use a particular systems methodology in some field of inquiry, the investigator must not only be an expert in his own field, but he must also be sufficiently familiar with the conceptual framework upon which the methodology is based (e.g., the GSPS framework outlined in Chap. 4). That is, he must be able to formulate specific (interpreted) systems and associated problems emerging in his field in terms of the conceptual framework involved. Once a specific system and associated problem are properly formulated, it is easy to convert them to their isomorphic general systems counterparts and, then, the available systems methodology (developed for general systems) can readily be utilized. When a solution is obtained, it is converted back to the specific (interpreted) system.

2. (Outer rectangles) Many systems investigations are of sufficient complexity that the investigator could make meaningful use of more information than that provided solely by the solution to a particular systems problem. This additional information about the system involved can also be obtained, when requested, by converting it from the general system into the corresponding specific system.

Systems Methodology

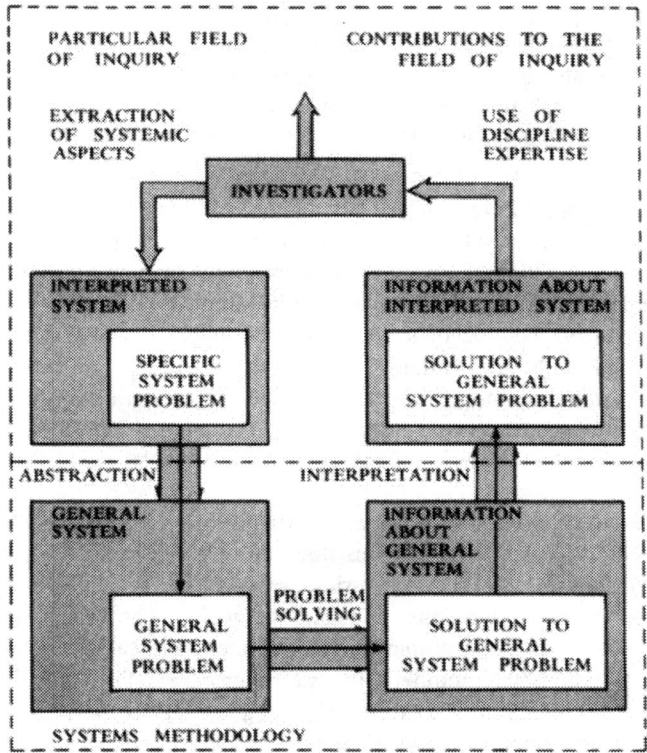

Figure 5.1. The role of systems methodology in science.

We can see that the utilization of a thinghood-independent systems methodology (developed in terms of general systems) for the study of specific systems requires an interface with the disciplines of science involved. Such an interface consists of two dual and alternately used procedures: *abstraction* and *interpretation*. In a scientific investigation, the application of these processes has, in general, an on-going and nonterminating character. This vital characteristic of science is well depicted by G. Spencer Brown [1957]:

> Science is a continuous living process; it is made up of activities rather than records; and if the activities cease it dies. Science differs from mere records in much the same way as a teacher differs from a library . . . Scientific knowledge, like negative entropy, tends constantly to diminish. It is prevented from dwindling completely into anecdote only by the attitude which seeks to repeat experiments and confirm results without end . . . Science is a significant game: one player tries to reduce significance to insignificance by asking more questions, while another seeks to counter his activities by doing more

experiments. The scientist, like the chess-enthusiast, often plays both sides himself . . . Repetitions of scientific results serve two purposes. First, they inhibit alternative questions which would tend to reduce their significance; and secondly, each successful repetition tends to increase the significance which such question might reduce. We thus have a race between the questions and the results.

The abstraction-interpretation scheme depicted in Fig. 5.1 is not restricted to science. It is applicable equally well to other areas such as engineering, management, or medicine. Although problems in these areas (systems design, testing, diagnosis, decision making, and the like) are different from the problems involved in scientific inquiry, the role of systems methodology in assisting the various specialists to deal with their systems problems is essentially the same.

How do systems problems emerge from a conceptual framework? How can they be operationally described to form a basis for the development of a comprehensive systems methodology? Let me answer these questions in terms of the GSPS framework.

Since systems problems must be, by definition, formulated in terms of systems, let me first inspect all systems that can be described within the GSPS framework. While they enjoy infinite variety, we need to capture them by a finite set of methodologically significant systems categories. The most fundamental categories of systems are those employed in the epistemological hierarchy outlined in Chap. 4. Although the number of systems categories in the hierarchy is potentially infinite (but countable), their ordering allows us to restrict the development of systems methodology at any given time to a finite number of categories at some lower levels of the hierarchy (e.g., those depicted in Fig. 4.13).

Each epistemological category of systems may further be classified by relevant methodological distinctions. The aim of methodological distinctions is to distinguish systems that require different methods when involved in systems problems. These distinctions do not affect the epistemological ranks of systems and, consequently, they are criteria for a secondary classification of systems. In general, they are introduced opportunistically, for the purpose of making systems methodology efficient and responsive to special needs.

Some important methodological distinctions refer to variables (both basic and supporting) and their state sets. They include, for instance, the distinctions between crisp and fuzzy variables, between discrete and continuous variables, and between variables based upon different measurement scales (nominal, ordinal, interval, ratio, metric, etc.). Examples of other important methodological distinctions, applicable only to some higher epistemological categories, are distinctions between deterministic and nondeterministic systems, between consistent and inconsistent systems, and between linear and nonlinear systems.

The epistemological and methodological criteria for classifying systems form a basis for defining methodologically significant types of systems, which, in

Systems Methodology

turn, form a basis for defining types of systems problems. In general, a system problem is either a question or some other request regarding one or more given systems, or it is a transition from a given initial system to some terminal system that satisfies given requirements. In the latter case, the transition may involve some intermediary systems between the initial system and the terminal system.

Problem requirements are basically objectives and constraints. For methodological purposes, their infinite variety must be properly categorized into a finite number of types, in a similar way as explained for systems. Types of systems and types of requirements result then in types of systems problems, which, in turn, form a basis for the development of systems methodology. We need a methodological tool (a coherent collection of methods) for dealing with each of the recognized problem types.

5.1. General Systems Problem Solver

The systems problem-solving expertise (or systemhood expertise) may, in principle, be implemented on a computer in the form of an expert system. Such an expert system is complementary to the usual expert systems, which are predominantly oriented to thinghood expertise in various special areas of science, engineering, and some other professions. The two types of expert systems together should form a far better computer support for dealing with overall problems than either of them alone.

A sketch of a prospective expert system oriented to systemhood expertise is given in Fig. 5.2. Let this expert system, which is described in more detail in one of my books [Klir, 1985a], be called a *General Systems Problem Solver* (or GSPS). It consists of three parts: a part implementing the GSPS conceptual framework, an operational part, and two interface units.

The *conceptual framework*, outlined previously, characterizes systems types (based on epistemological and methodological criteria), requirement types, and problem types. It is a linguistic medium in terms of which the user communicates with the GSPS to formulate systems problems emerging from various problem-solving situations. The user represents in this interaction the thinghood expertise. There is also a possibility, at least in principle, of replacing the user with a thinghood-oriented expert system capable of interacting with the GSPS.

The *operational part* consists of four functional units: a methodological unit, a metamethodological support unit, a knowledge base unit, and a control unit. It is assumed that each of these units is equipped with appropriate reasoning strategies.

The *methodological unit* consists of a set of methodological tools for dealing with systems problems of at least some of the types that emanate from the conceptual framework. The term "methodological tool" is used here for a package

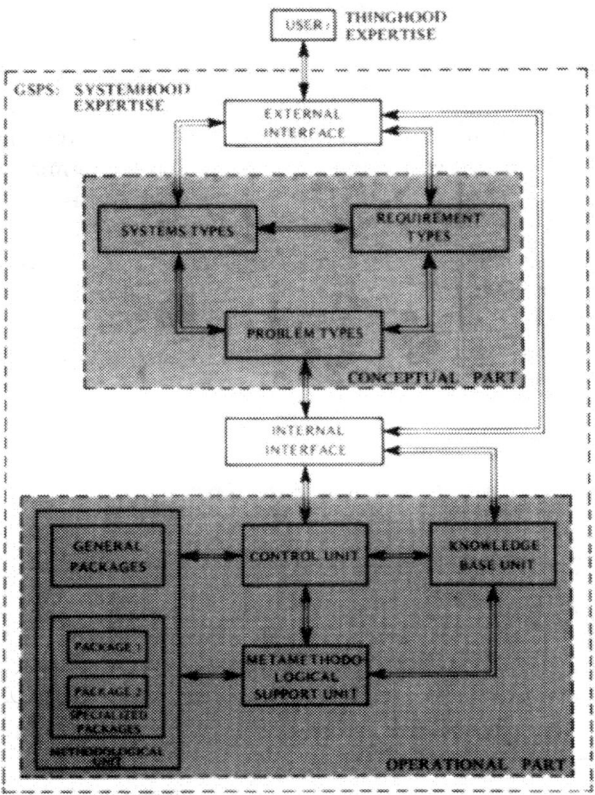

Figure 5.2. A sketch of the General Systems Problem Solver.

of methods (and the corresponding computer programs) addressing one of the problem types. The methodological tools are divided into general packages and specialized packages. The general packages are designed for problem types based upon the most general methodological distinctions (e.g., nondeterministic systems with fuzzy and nominal-scale variables); all other packages, which can be ordered by their generality, are called specialized.

The *metamethodological support unit* contains information about the various characteristics of the methodological tools available. This information is primarily used for matching problems with methodological tools, but it is also available to the user upon specific request. The various metamethodological issues are discussed in Chap. 6.

The *knowledge-base unit* contains useful knowledge regarding the various types of systems and problems such as theoretical or experimental laws, principles, or rules of thumb. This aspect of systems science is discussed in Chap. 7.

Systems Methodology   77

The necessary coordination of the three described units of the operational part is performed by a *control unit*. It basically makes decisions, according to the user's requests and other conditions, about which unit to activate and how.

The user–GSPS communication is facilitated by two interface units, called an *external interface* and an *internal interface*. The former is involved in the formulation of problems; the latter is associated with the various metamethodological considerations and the utilization of the knowledge-base unit.

The following broad classes of problem-solving activities seem to be the primary sources of problems for systems methodology, in general, and for a systemhood-oriented expert system (such as the GSPS), in particular:

   a. *Systems inquiry*. The full scope of activities by which we attempt to construct systems that are adequate models of some aspect of reality.
   b. *Systems design*. The full scope of activities by which we attempt to construct systems that are adequate models of desirable man-made objects.

The purpose of systems inquiry is to understand some phenomenon of reality, to make adequate predictions or retrodictions, to learn how to control the phenomenon in any desirable way, and to utilize all these capabilities for various ends. The purpose of systems design is to prescribe operations by which a conceived and desirable artificial object can be constructed in such a way that relevant objective criteria are satisfied within given constraints.

We can see that in both systems inquiry and systems design, we "construct systems that are adequate models of" something. That is, the construction of systems models, or *systems modeling*, permeates in a very fundamental way all disciplines of science, be they oriented to the natural or to the artificial.

5.2. Systems Modeling

What does it mean to construct a system that is a model of something else? First, it means that a system, when considered on its own, is not a model. It becomes a *model* only in a relationship to another system, which is usually referred to as an *original*. Furthermore, this relation is not arbitrary, but it must be a *homomorphic relation*.

A homomorphic relation (or homomorphism) between two systems is contingent upon a function from relevant entities of one system (the original) onto the corresponding entities of the other system (the modeling system) under which the relation among the entities is preserved. If the function, which is called a *homomorphic function*, is bijective, the relation is preserved completely (we say that the two systems are *isomorphic*); otherwise, it is preserved only in a simplified form. We say, in general, that the modeling system is a homomorphic (or isomorphic) image of the original.

The entities to which the homomorphic function is applied and the relation that is preserved under the function depend on the type of the systems involved. That is, the homomorphism among systems assumes different forms for different types of systems. In fact, one way of defining systems of some type is to define them indirectly (as mathematical categories), by characterizing the nature of the homomorphic relation among them.

To illustrate the notion of homomorphic systems, let us consider two simple systems, $S_1 = (X, R)$ and $S_2 = (Y, Q)$, where X, Y denote sets of overall states of some variables and $R \subset X \times X$, $Q \subset Y \times Y$ are binary relations that describe possible transitions from present states to next states. Such systems could be described, more precisely, as special generative systems, possibly nondeterministic. We say that S_2 is a *weak homomorphic image* of S_1 if and only if there exists an onto function, $h: X \to Y$, such that for all $x_1, x_2 \in X$,

$$(x_1, x_2) \in R \text{ implies } [h(x_1), h(x_2)] \in Q.$$

Sometimes, the following additional condition is required for h, under which the homomorphism is called a *strong homomorphism*: for all $y_1, y_2 \in Y$,

$$(y_1, y_2) \in Q \text{ implies } (x_1, x_2) \in R$$
$$\text{for some } x_1 \in h^{-1}(y_1) \text{ and } x_2 \in h^{-1}(y_2).$$

If S_2 is a homomorphic image of S_1 under h, then S_2 together with h is called a *model* of S_1. Clearly, the same system may be combined with different homomorphic functions (assuming they exist) to form different models of the same original; it may also be used for modeling different originals. The opposite is also true: the same original may be modeled by different systems under appropriate homomorphisms.

The modeling relation is illustrated for the simple state-transition system in Fig. 5.3a; it is exemplified in Figs. 5.3b and 5.3c by the weak (regular) homomorphism and the strong homomorphism, respectively. Observe that the relation of system S_2 in Fig. 5.3b contains two pairs (γ, α) and (γ, γ), which violate the additional condition required by the strong homomorphism.

It is now clear under which conditions a system may be employed for modeling another system. Since systems are mathematical objects, homomorphisms between systems are defined in terms of mathematical functions. This raises a question: how can a homomorphism be defined when we want to employ a system for modeling some aspect of reality. Clearly, the homomorphism cannot be defined mathematically since reality, or any aspect of it, is not a mathematical object. The only possibility is to define the homomorphism physically, by the construction and use of appropriate measuring instruments or, in some cases, by our own sensors.

A system that is employed in this sort of modeling may be called an *interpreted system*. It is a system (a mathematical object) whose entities are given

Systems Methodology

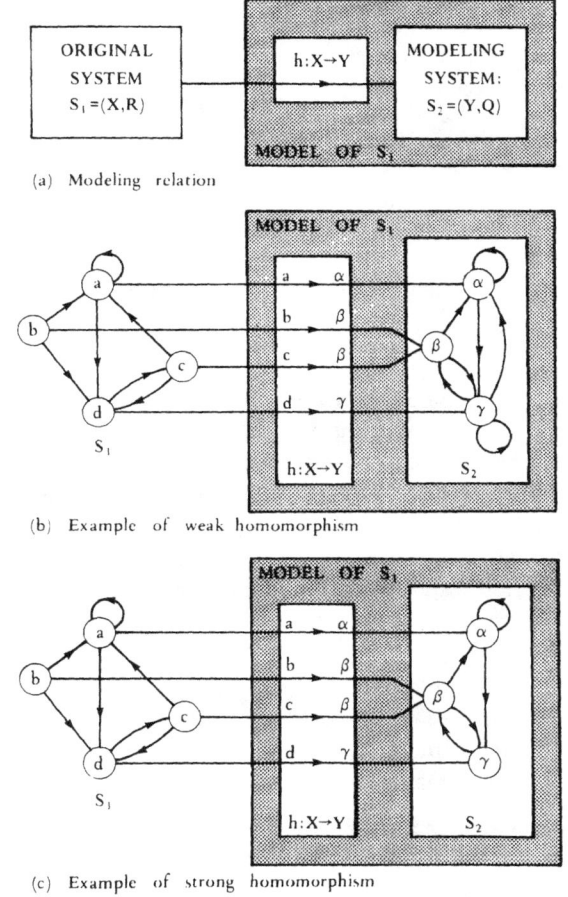

Figure 5.3. Modeling relation based upon weak or strong homomorphism.

a real-world interpretation by the employed measuring instruments or by our own sensors. That is, the interpreted system is actually a model whose homomorphism is defined physically rather than mathematically; it is a system together with the measuring instruments involved, by which some presumed states of nature (mathematically not tractable) are mapped into well-defined states of variables of the system (Fig. 5.4).

It is important to realize that the modeling relation can be defined only within each particular epistemological category of systems. As already hinted, the epistemological categories of systems are true categories in the sense of category

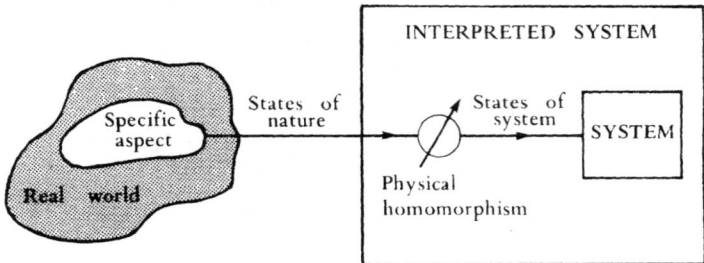

Figure 5.4. Illustration of the notion of an interpreted system.

theory. They can be distinguished from one another (within category theory) by distinctions in their homomorphisms. Consequently, the epistemological hierarchy of systems represents a natural base for a classification of systems models.

Another useful way of classifying systems models is to consider whether the two systems involved in the modeling relation, the original system and the modeling system, are interpreted or noninterpreted (purely mathematical). This classification criterion results in the following four classes of systems models:

	Original system	Modeling system
I.	Interpreted	Mathematical
II.	Mathematical	Interpreted
III.	Interpreted	Interpreted
IV.	Mathematical	Mathematical

Let me characterize each of these classes of models and illustrate them by simple examples. For the sake of simplicity, I use examples of models that fully preserve relevant relations of their originals (i.e., they are all isomorphic models). I should emphasize, however, that this feature of systems models is desirable only in some instances. It is essential in any circumstances in which it is advisable to replace an original with a suitable model, but all details of the original are required to be preserved in the model. The use of a suitable model instead of a given original may have some advantages or may even be necessary. It may, for example, be cheaper, faster, less dangerous, easier to understand or control, legally accepted, less controversial, better adjusted to the human scale, or more convenient in some other respects.

In many other instances, we search for models that are simplifications of their originals. The role of these models is to suppress certain details in the originals, which are undesirable for specific purposes. In fact, every interpreted system (the system and its real-world interpretation) is precisely a model of this kind.

Class I of systems models consists of mathematical models of all kinds.

Systems Methodology

These models are based on accepted physical and other laws of nature. They make it possible to answer questions regarding interpreted systems by mathematical reasoning rather than by experimenting with real-world objects. For example, they allow us to answer questions concerning man-made objects before they are actually constructed. In general, these models allow us to perform *gedanken experiments* (thought experiments) on mathematical models of hypothetical interpreted systems. For example, the dynamic behavior of electric currents and voltages in an hypothetical electric circuit may be studied in terms of appropriate differential equations rather than by constructing the actual circuit and making relevant measurements on it.

Class II of systems models is best exemplified by computers of all kind, whether analog, digital, or hybrid. Also included are various special interpreted systems, each of which is designed as a universal model for mathematical systems of some particular class. Examples are linear analyzers (to solve linear algebraic equations), polynomial analyzers (to investigate properties of polynomial functions), or electrolytic tanks (to deal with partial differential equations). The use of models in this class consists in manipulating (physically) some variables of a suitable interpreted system and making measurements of other variables of the system by which answers to various mathematical questions are obtained.

One of the simplest examples of models in class II is the slide rule, by which products (or quotients) of two numbers are determined by adding appropriate distances on two rules. Isomorphic functions for input and output variables of this well-known example are specified in Fig. 5.5, which is self-explanatory. Observe that the model produces the same results as the original, but it is applicable only to numbers expressed with a limited precision that reflects the inherent limitations of observation or measurement accuracy.

Another example of models in class II is the simple electric circuit with two semiconductor diodes whose diagram is shown in Fig. 5.6a. This example illustrates how logic functions can be modeled by an interpreted system, and, consequently, it illustrates one of the basic principles underlying the operation of digital computers.

Semiconductor diodes are nonlinear devices whose conductivity depends on the polarity of the voltage applied to them. This is specified in Fig. 5.6b, where letters H, L denote, respectively, high or low levels of voltage.

The circuit has two inputs, v_1 and v_2, and one output v_3, which represent voltage levels with respect to a constant reference level, V. Selecting appropriate values of voltage representing H and L, the circuit behaves according to the function given in Fig. 5.6c. In our terminology, this is the behavior function of a generative system interpreted in terms of voltages in the semiconductor circuit.

Consider now two logic functions, AND and OR, which are defined by the following tables, where x_1, x_2 denote arbitrary propositions and F, T denote the falsity and the truth of these propositions, respectively:

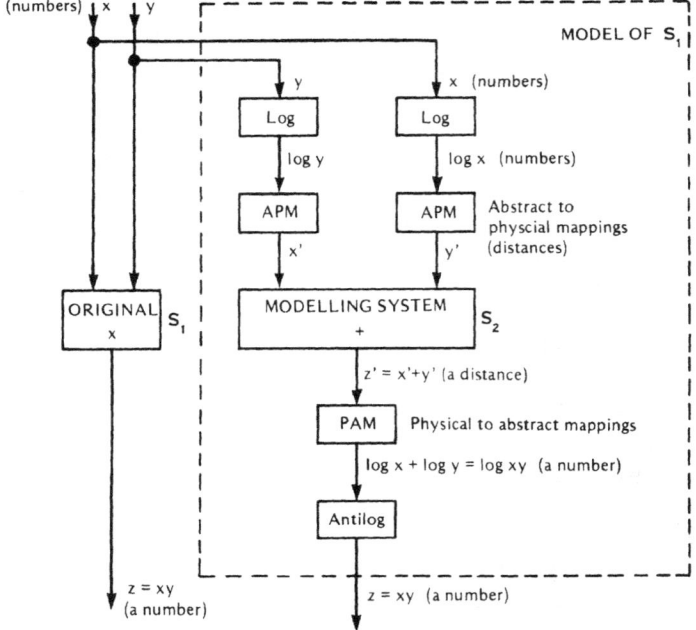

Figure 5.5. Slide rule as a model.

x_1	x_2	$x_3 = x_1$ AND x_2	x_1	x_2	$x_3 = x_1$ OR x_2
F	F	F	F	F	F
F	T	F	F	T	T
T	F	F	T	F	T
T	T	T	T	T	T

Either of these logic functions can be modeled by our interpreted system depending on the chosen isomorphic function, h, by which the truth values of the logic variables x_1, x_2, x_3 are connected with the states of the interpreted variables v_1, v_2, v_3. To model the AND function, we have to choose $h(F) = H$ and $h(T) = L$; to model the OR function, we must choose $h(F) = L$ and $h(T) = H$. This example thus illustrates that the same physical device can be employed, through an interpreted system defined on it, for modeling different mathematical systems.

Systems models in class III play an important role in engineering. In this class, originals as well as modeling systems are interpreted. Thus, for example, the dynamic behavior of a mechanical system may be modeled by an electrical system provided that both of these interpreted systems can be modeled by the same set of differential equations. Systems modeling of this sort is the subject of the

Systems Methodology

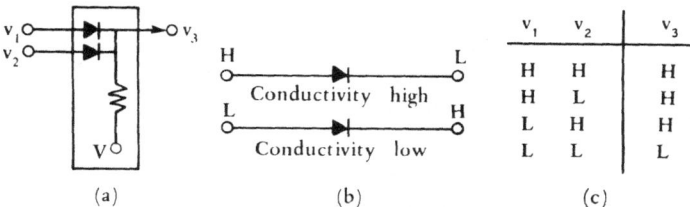

Figure 5.6. Physical modeling of logic functions.

theory of similarity or *similitude* [Skoglund, 1967; Szucz, 1980]. Other examples of models in this class include computers simulated on other computers, aircraft simulators for training pilots, and medical simulators such as artificial hearts or kidneys.

Class IV consists of models that are of great importance to applied mathematics. They are associated with various kinds of mathematical transformations (such as Laplace or Fourier transformations) with the aid of which mathematical systems of one kind (e.g., differential equations) are modeled by other mathematical systems (e.g., algebraic equations). Instead of dealing with the original, we can use a modeling system, which is usually considerably simpler, and then transform the results obtained back to the original. In some cases (e.g., the Laplace transformation), detailed vocabularies have been compiled which specify the isomorphic correspondence between the originals and the modeling systems.

An important classification of systems models of various aspects of reality and methodological issues involved in systems modeling is based on the purposes for which the models are constructed. In science, systems models are constructed for the purposes of explanation, prediction, and, sometimes, retrodiction. They are referred to as *explanatory models*, *predictive models*, and *retrodictive models*, respectively. Possible approaches to constructing models of these kinds extend over a wide spectrum that is bounded by two extreme approaches, which are usually called a postulational approach and a discovery approach. To characterize them, let me employ the GSPS epistemological hierarchy of systems categories outlined in Chap. 4.

In the *postulational approach*, a hypothetical generative system or an epistemologically higher system is postulated. This postulated system is a result of the scientists's background knowledge, experience, insight, intuition, and the like. It is a frame for a specific type of *deductive reasoning*: given a particular set of premises regarding conditions of the system's environment as well as initial or boundary conditions of the system itself, the system allows us to derive specified conclusions pertaining to the explanation, prediction, or retrodiction of the phenomenon of inquiry. The *validity* of the postulated system as a model of some specific aspect of reality depends on the degree of agreement of the derived conclusions with relevant empirical evidence.

The *discovery approach* is data-driven. That is, models are derived by processes that discover patterns in data and utilize them for making *inductive inferences*. This approach is characterized by climbing up, from a given data system, the epistemological hierarchy of systems. The criterion of validity of models derived in this way does not involve the issue of the agreement between the empirical evidence and deductive inferences made from the model: the model was derived from given data and, hence, the agreement is perfect by definition; it involves rather the issue of the justification of inductive reasoning.

Neither of the two extreme (and idealized) approaches is actually ever used in its pure form in praxis. In the postulational approach, the hypothetical system is not chosen arbitrarily from among an infinite number of competitors, but on the basis of relevant knowledge accumulated throughout the history of science as well as the individual experience of the investigator. Induction is thus involved in the postulational approach as well, but it is hidden in the general notions of "accumulated knowledge" and "experience." In the discovery approach, we do not consider all systems that conform to the given data (their set is usually infinite), but only those satisfying also a variety of relevant extraevidential considerations. Consequently, background knowledge and experience play as important roles in the discovery approach as they do in the postulational approach.

Let me turn now to engineering, where the purpose of constructing systems models is not to explain, predict, or retrodict, but to prescribe operations by which a conceived man-made object can be built. These models are called *prescriptive models*.

The construction of prescriptive models, which is usually referred to as *systems design*, involves two principal stages. The aim of the first stage is to formulate the design problem on the basis of the conceived object to be designed. This involves a selection of relevant variables and a conceptualization of a support-invariant relation among the variables that adequately characterizes the conceived object. The support-invariant relation may be conceptualized in terms of a generative system or any epistemologically higher systems type (a structure system or a metasystem of some order). In addition, suitable *technological primitives* must be chosen by which the relation can be realized. Each of these primitives is a man-made object, which we view, for the purpose of systems design, as a specific interpreted system. That is, we distinguish only certain variables on the object, which are relevant to our problem, and the known relationship among them. This relationship is, of course, a result of some previous modeling (explanatory, predictive); in systems design, we take its validity for granted. The formulation of the design problem must also include a list of *objective criteria* and *constraints* which may be connected with cost, reliability, safety, maintainability, usability, time limitations, legal constraints, and many other aspects.

The second stage in the design problem consists in constructing a system that implements the required relation in terms of the interpreted systems representing

Systems Methodology

the chosen technological primitives and which satisfies the required objective criteria and constraints. This stage involves a climbing up the epistemological hierarchy of systems, from a generative system (or, sometimes, a data system) to a structure system, a metasystem, or a combination of the two. The basic issue is to find a suitable *decomposition* of the required relation into the relations available in terms of the technological primitives. Since there are usually a tremendous number of such decompositions, the real challenge is to determine one of them that also satisfies the given objectives and constraints. To reduce the complexity of this task, the decomposition is usually conceptualized in terms of structure systems or metasystems of higher orders. This important issue, which is touched upon in Chap. 8, is well explicated by Herbert Simon [**1962**].

Prescriptive models are constructed not only for the purpose of building new objects, but also for various purposes of *policy or decision-making*, as explained by Simon [1988]:

> Generally, modeling serves policy. We construct and run models because we want to understand the consequences of taking one decision or another. Predictive models are only a special case, where we seek to predict events we cannot control in order to adapt to them better. We do not expect to change the weather, but we can take steps to moderate its effects. We predict populations so that we can plan to meet their needs for food and education. We predict business cycles in order to plan our investments and production levels.
>
> Of course, our models may, and frequently do, also contain policy variables. Our demographic models may contain variables that represent birth control policies or the effects of public health measures. (Usually, we simply incorporate these in the estimates of birth and death rates.) Our economic models may contain variables representing government fiscal and monetary policies. Our model of the eutrophication of a lake may include the quantities of phosphates we drain into it. When we include such variables in models, our intent in modeling changes from simple prediction to prescription.
>
> When our goal is prescription rather than prediction, then we can no longer take it for granted that what we want to compute are time series. We do not care what the water quality will be in the lake next year or the year after that. What we care about is how much the phosphate input needs to be reduced to eliminate the unwanted eutrophication. Moreover, given the crudity of our data and the approximate nature of our model of the eutrophication process, we are interested mainly in orders of magnitude. This means that we can greatly simplify our calculations, possibly with the result that the back of an envelope will suffice.

One additional type of systems modeling should be mentioned for the sake of completeness: *performance modeling*. The aim of performance modeling is to develop models by which various performance characteristics of manufactured products (utilization, reliability, quality, human factors) can be evaluated under various conditions. The primary utilization of performance models is to identify factors that influence performance of the investigated products and determine ways in which their performance can be improved. These models are thus both explanatory and prescriptive.

5.3. Methodological Role of the Computer

The computer is perhaps the most important methodological tool in systems science.* Yet, its role is not to replace the human mind, but rather to supplement it, in a *symbiotic* fashion. It is recognized that, in spite of its weaknesses, the human mind has certain faculties that are superior even to the most powerful computers. The current understanding of these faculties is rather rudimentary. In spite of the progress made by artificial intelligence, cognitive science, physiology, and other relevant areas, it is reasonable to expect that there will always be a large residuum in the abilities of the human mind that will not be understood.

Intuition, insight, and the ability of global comprehension are possibly the most valuable assets of the human mind, particularly one that is appropriately trained. However, complex systems frequently possess properties that are counterintuitive and resistant to global comprehension. As such, they represent traps for the human mind in the sense that they may guide it into illusory insights. To discover such traps, it is usually unavoidable to perform the tedious work of detailed analysis of the system on hand. While the human mind is weak and severely limited in this respect, it is exactly this domain of detailed analysis where the computer, equipped with appropriate methodology, excels. This ability gives the computer an important role as an *intuition safeguard* and *intuition amplifier*.

The symbiosis of the human being (scientist, decision maker, designer, and the like) with the methodologically equipped computer, such as the GSPS, makes it possible to invent and utilize new approaches to various intellectual tasks, far superior than those applicable by either of them alone. The strength of the human being is his experience in the area of study, understanding and taking advantage of the context of investigation, intuition, global comprehension, feeling for the right solution, visual and auditory capabilities, creativity, and the like. The strength of the computer lies in its computational power, the ease with which it can handle a tremendous number of operations, far exceeding the human capacity in this respect. It is the computational power which, when properly utilized, can significantly enhance the human intellectual qualities by providing the human being with desirable detailed analyses and, as already mentioned, help him to avoid the many counterintuitive traps associated with complex systems.

---

*The increasingly important connection between systems science (especially systems methodology) and computer science has recently become a subject of growing interest, as illustrated by the emergence of a new area called "Computer Aided Systems Theory" (CAST); it was initiated by Franz Pichler in the late 1980s [Pichler & Moreno-Diaz, 1990].

CHAPTER 6

Systems Metamethodology

> *Nothing is best, nothing is worst.*
> *Each thing, seen in its own light,*
> *Stands out in its own way.*
> *It can seem to be "better'*
> *Than what is compared with it*
> *On its own terms.*
> *But seen in terms of the whole,*
> *No one thing stands out as "better".*
> —CHUANG TZU

As argued previously, the principal aim of systems science is to understand the phenomenon of systemhood as completely as possible. The first step in achieving this aim is to divide the whole spectrum of conceivable systems into significant categories. The second step is to study the individual categories of systems and their relationship, and to organize the categories into a coherent whole. The third step is to study systems problems that emerge from the underlying set of organized systems categories. Finally, we address methodological issues regarding the various types of systems problems.

This attitude toward systems problem solving is considerably different from that of mathematics. Indeed, pure mathematics is basically oriented to the development of various axiomatic theories and applied mathematics attempts to develop some of these theories as potentially useful methodological tools in various problem areas. Each mathematical theory is derived from some specific assumptions (axioms) and, consequently, the use of any methodological tool based upon the theory is restricted to problems that conform to these assumptions. If a problem does not conform to them and an applied mathematician trained in the methodology still wants to use it, he has to adjust (reformulate) the problem to make it fit the assumptions. This means, however, that a different problem is now solved. The problem adjustment is often not stated explicitly and, as a consequence, an impression is created that the original problem was solved while, in fact, it was not.

Applied mathematics provides thus the various users (scientists, engineers, etc.) with a set of methodological tools, each derived from a mathematical theory which, in turn, is based on a specific set of assumptions. Mathematical theories are most frequently developed for assumptions that are interesting or convenient from the mathematical point of view. As a consequence, they produce methods that cover rather small and scattered parts of the whole spectrum of systems problems.

In contrast with applied mathematics, systems methodology is committed to the investigation of the domain of systems problems as a coherent whole. In particular, it attempts to identify pragmatically rich subproblems, i.e., subproblems that occur in as many genuine systems problems as possible. This emphasis on comprehensiveness and pragmatic significance in pursuing methodological research is quite different from the usual emphasis in mathematics to pursue research of methodological areas based on convenient (and often arbitrary) mathematical properties.

Hence, the *primacy of problems* in systems methodology is in sharp contrast with the *primacy of methods* in applied mathematics. It is the most fundamental commitment of systems methodology to develop methods or solving genuine systems problems in their natural formulation. Simplifying assumptions, if unavoidable, are introduced carefully, for the purpose of making the problem manageable and, yet, distort it as little as possible. The methodological tools for dealing with a problem are of secondary importance; they are chosen in such a way as to best fit the problem rather than the other way around. Moreover, the tools need not be only mathematical in nature. They may consist of a combination of mathematical, computational, heuristic, experimental, or any other desirable methodological traits.

In order to choose an appropriate method for a specific problem, relevant characteristics of prospective methods must be determined. This is a subject of *systems metamethodology*—the study of systems methods as well as methodologies (integrated collections of methods). Let any particular study whose aim is to determine some specific characteristics of a method (or methodology) be called a *metamethodological inquiry*. Examples of the most fundamental characteristics of methods, which are relevant to virtually all problems, are computational complexity, performance, and generality of the methods involved.

Computational complexity is a characterization of the time or space (memory) requirements for solving a problem by a particular method. Either of these requirements is usually expressed in terms of a single parameter that represents the size of the problem, e.g., the number variables in the given system. The dependence of the required time or memory space on the problem size is usually called a time complexity function or space complexity function, respectively. Either of these functions can be used for comparing different methods for dealing with the same problem type. Further aspects of computational complexity are discussed in Chap. 8.

Performance of a method is characterized by the degree of its success in dealing with the class of problems to which it is applicable. It can be expressed in various ways, typically by the percentage of cases in which the desirable solution is reached, by the average closeness to the desirable solution, or by a characterization of the worst case solution. Methods whose performance is not perfect are usually called heuristic methods. They are employed as a means for reducing computational complexity.

Generality of a method is determined by the assumptions under which it operates, e.g., by the axioms of a mathematical theory upon which the method is based. A particular set of assumptions, upon which several different methods may be based, is often referred to as a *methodological paradigm*. Each assumption contained in a methodological paradigm restricts the applicability of the associated methods and, consequently, restricts the set of possible solutions in some specific way.

In some instances, characteristics of methods or methodologies can be obtained mathematically. For example, worst-case complexity functions have been determined for many methods involved in systems problem solving [Garey and Johnson, 1979]. In many cases, however, mathematical treatment is not feasible. For example, it is often impossible to determine mathematically the performance of a heuristic method or the complexity function of a method for typical (average) problems of a given type. One way of obtaining the desired characteristics in these cases is to perform experimental investigations of the methods involved. That is, the application of the investigated method or methodology is simulated on a computer for a set of typical problems of a given type. Results obtained for these problems are then summarized in a desirable form to characterize the method or methodology and, possibly, compare it with its various competitors. This and other kinds of experimental investigations on computers are further discussed in Chap. 7.

Let me now return to the notion of methodological paradigms. By definition, methods based upon the same paradigm are equivalent in the sense that they share the same set of possible solutions. This set consists of all solutions to the problem except those that violate any of the assumptions that constitute the paradigm. That is, each methodological paradigm excludes some of all possible solutions to the problem.

When the set of solutions excluded by one methodological paradigm is a proper subset of solutions excluded by another one, then the former paradigm is considered more general than the latter. Any generalization of a methodological paradigm extends the set of possible solutions to a given problem and makes it often possible to reach a better solution. At the same time, however, it usually requires methods with greater computational complexities. Generality and computational complexity of methods thus conflict with each other.

The most general methodological paradigm for each problem type in unique: it is the *assumption-free paradigm*. However, this extremum of generality is virtually meaningless in praxis owing to the extreme computational demands associated with it.

In principle, each problem type is associated with a set of methodological paradigms. These paradigms can be ordered by their generality. Each of them may be associated with a set of methods. These methods are equally general (by definition), but they may differ from each other in computational complexity, performance, and other characteristics.

The principal role of systems metamethodology is to evaluate and compare methods available for the various types of problems, eliminate those that are inferior, and help to select methods best suited for dealing with a problem in a given context. Let this metamethodological problem be called a *method identification problem*. This problem can be formulated as follows.

Given a specific problem type t, let M_t denote the *set of all available methods* for dealing with t. Let $\overset{g}{\leq}$, $\overset{c}{\leq}$, and $\overset{p}{\leq}$ denote *preference orderings* on M_t that are based on the generality, computational complexity, and performance of the methods involved, and let $\overset{\alpha}{\leq}$, $\overset{\beta}{\leq}$, ... denote some additional (optional) preference orderings on M_t. For the sake of simplicity, let $x \leq y$ mean that x is preferred to y. For example, $x \overset{g}{\leq} y$ means that x is more general than y, and $x \overset{\beta}{\leq} y$ means that x performs better than y. All preference orderings can be combined in a *joint preference ordering*, $\overset{*}{\leq}$, by the following definition: for all $x, y \in M_t$, $x \overset{*}{\leq} y$ if and only if $x \overset{g}{\leq} y$ and $x \overset{p}{\leq} y$ and $x \overset{\alpha}{\leq} y$ and $x \overset{c}{\leq} y$ and $x \overset{\beta}{\leq} y$, etc. The *solution set*, S_t, of the method identification problem consists of those methods in M_t that are either equivalent or incomparable in terms of the joint preference ordering $\overset{*}{\leq}$. Formally, we have

$$S_t = \{x \in M_t | \text{for all } y \in M_t, y \overset{*}{\leq} x \text{ implies } x \overset{*}{\leq} y\}.$$

When the solution set consists of more than one method, the user may employ some additional criteria to make a unique choice.

In order to illustrate the meaning of methodological paradigms, let me discuss their role in one category of systems problems, most typical in systems design. This problem type can be described in terms of the GSPS epistemological hierarchy of systems as follows: given a desirable generative system (determined by previous stages of a design problem), a set of primitive elements (generative systems that describe relevant traits of available technological components), and a set of objective criteria and constraints, determine a structure system that implements the required overall generative system solely by the primitive elements and satisfies the given objective criteria and constraints. To facilitate further discussion, let this phase of systems design be referred to as *structure design*.

Assume, for the sake of simplicity, that only one objective criterion is employed in structure design: to minimize the cost of the final product. Assume further that the cost is directly proportional to the number of elements in the structure system. A method for structure design under this objective criterion and

Systems Metamethodology

within the most general methodological paradigm, the assumption-free paradigm, is rather obvious. First, we inspect each of the primitives. If any of them is identical with the desirable behavior system (which is extremely rare in praxis), the design is trivially completed. Otherwise, we consider various pairs of the given primitives (of the same or different types) as elements of the prospective structure system. For each pair, we examine all meaningful connections of output variables of the elements to their input variables. If any of the resulting structure systems implements the desirable behavior system, the design terminates. Otherwise, we have to repeat the exercise with three elements drawn from the given pool of primitives. If no structure system with three elements implements the required behavior system, we must explore structure systems with four elements, etc.

It is easy to see that the number of structure systems that must be constructed and examined within the assumption-free paradigms grows extremely rapidly with the number of primitives, their numbers of input and output variables, and the minimum number of elements for which the required behavior system is obtained. In fact, this number is usually beyond any realistic computing power and, consequently, the assumption-free paradigms have virtually no practical value. Hence, we must resort to various special paradigms. The challenge of systems metamethodology is to select a paradigm that is not too restrictive while, at the same time, it is computationally tractable and has an acceptable performance.

An important difference, for example, is between paradigms that do not allow *feedbacks* in the designed structure system and paradigms that do allow them. Clearly, the paradigm that allows feedbacks is more general. It may result in better solutions while, at the same time, it is computationally more demanding. To illustrate these issues, let me discuss a simple example.

Consider the design of a digital decoder that has three input variables, x_1, x_2, x_3, and three output variables, z_1, z_2, z_3, and whose behavior function is specified in Table 6.1a. Assume that it is required to implement the decoder in terms of only one type of primitive elements, whose behavior function is given in Table 6.1b. These elements, which represent the logical NOR function, are known to be universal in the sense that every logic function can be expressed as a suitable composition of the NOR functions. Assume further that the only objective criterion is to minimize the cost, which is expressed by the number of the NOR elements in the design.

When feedbacks are not considered, the minimum cost design can be obtained easily by various algebraic minimization methods [Klir, 1972b]. One solution is shown in Fig. 6.1a.

When feedbacks are considered, we can obtain a better solution, as shown in Fig. 6.1b. This solution saves three NOR elements, or one third of the cost of the previous solution. However, the design process is considerably more complex in this case since feedbacks may introduce inconsistencies into the system. Let me explain this issue in terms of our example.

Table 6.1. Functions Involved in the Systems Given in Fig. 6.1

(a)

x_1	x_2	x_3	z_1	z_2	z_3
0	0	0	1	1	1
0	0	1	0	1	0
0	1	0	1	0	0
0	1	1	1	0	0
1	0	0	0	0	1
1	0	1	0	1	0
1	1	0	0	0	1
1	1	1	0	0	0

(b)
NOR element

a	b	c
0	0	1
0	1	0
1	0	0
1	1	0

(c)

x_1	x_2	x_3	y_1	y_2	y_3
0	0	0	0	0	0
0	0	1	0	0	1
0	1	0	0	1	0
0	1	1	0	1	0
1	0	0	1	0	0
1	0	1	0	0	1
1	1	0	1	0	0
1	1	1	?	?	?

(d)

Present state			Next state		
y_1	y_2	y_3	y_1	y_2	y_3
0	0	0	1	1	1
0	0	1	0	1	1
0	1	0	1	1	0
0	1	1	0	1	0
1	0	0	1	0	1
1	0	1	0	0	1
1	1	0	1	0	0
1	1	1	0	0	0

Systems Metamethodology

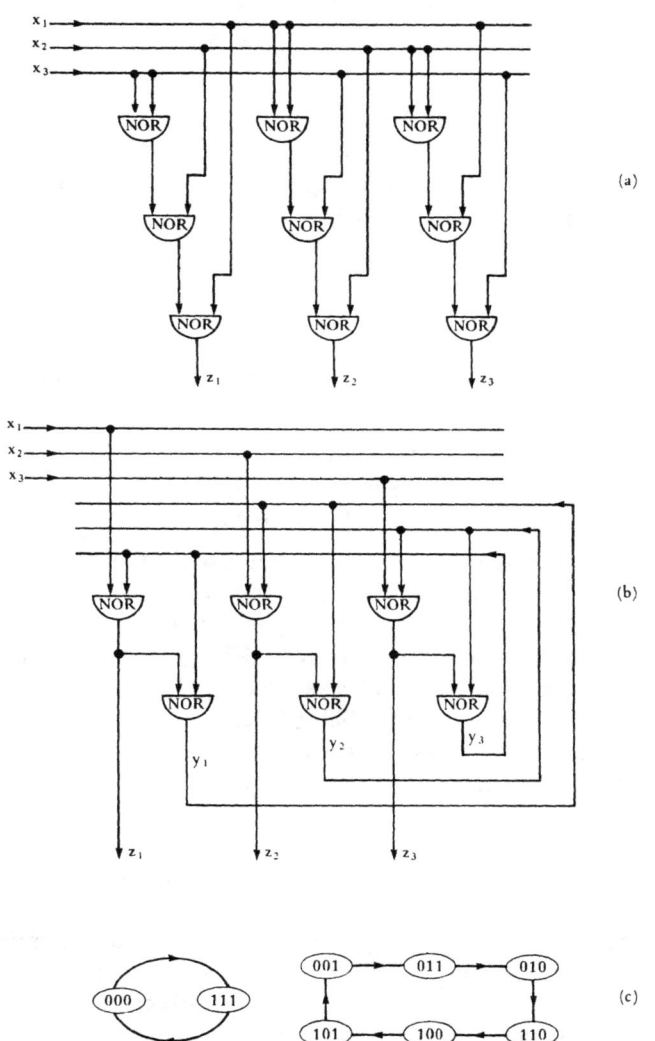

Figure 6.1. An example of systems design based on two distinct paradigms. (a) A paradigm without feedbacks. (b) A paradigm with feedbacks. (c) Inconsistencies of the feedback variables y_1, y_2, y_3 in (b), which do not affect the proper functioning of the system.

The six NOR elements in our system are described by the following six equations, where the symbol $\wedge$ denotes the logical AND function (conjunction) and the bars over some of the letters denote negations of the respective variables:

$$z_1 = \bar{x}_1 \wedge \bar{y}_3, \quad y_1 = \bar{z}_1 \wedge \bar{y}_3,$$
$$z_2 = \bar{x}_2 \wedge \bar{y}_1, \quad y_2 = \bar{z}_2 \wedge \bar{y}_1,$$
$$z_3 = \bar{x}_3 \wedge \bar{y}_2, \quad y_3 = \bar{z}_3 \wedge \bar{y}_2.$$

Variables y_1, y_2, y_3 are in these equations the feedback variables whose consistency must be verified. After substituting for z_1, z_2, z_3 into the last three equations from the first three equations, we obtain

$$y_1 = x_1 \wedge \bar{y}_3,$$
$$y_2 = x_2 \wedge \bar{y}_1,$$
$$y_3 = x_3 \wedge \bar{y}_2.$$

These are functional equations for the feedback variables y_1, y_2, y_3. Any three functions,

$$y_1 = f_1(x_1, x_2, x_3),$$
$$y_2 = f_2(x_1, x_2, x_3),$$
$$y_3 = f_3(x_1, x_2, x_3),$$

for which the equations are satisfied is a solution of the functional equations. The system is consistent if the solution exists and is unique. By inspection of the functional equations (with no need to employ one of the existing methods for solving functional equations of this sort [Klir, 1972b]), we can find that the solution exists and is unique for all states of the variables x_1, x_2, x_3 except the state $x_1 = x_2 = x_3 = 1$. The solution is given in Table 6.1c.

What happens in the system when $x_1 = x_2 = x_3 = 1$? In this case, the functional equations assume the form

$$y_1 = \bar{y}_3,$$
$$y_2 = \bar{y}_1,$$
$$y_3 = \bar{y}_2.$$

Applying each state of variables y_1, y_2, y_3 to the right-hand side of the equations, we obtain another state on the left-hand side of the equations. This results in the state-transition function expressed in Table 6.1d. We can see that this function involves two cycles of transitions, which are shown in Fig. 6.1c. Depending on the initial state of the variables, one of the two cycles applies. In either case, the feedback variables are inconsistent since they do not have a unique (stable) state. Fortunately, these inconsistencies do not affect the output variables: observe that the output variables z_1, z_2, z_3 are independent of the feedback variables y_1, y_2, y_3 when $x_1 = x_2 = x_3 = 1$. Hence, the system produces states of the output variables

Systems Metamethodology

in a consistent manner in spite of the internal inconsistencies involving the feedback variables for one of the input states.

As another example of the role of methodological paradigms, let us consider the process of designing *finite-state dynamic systems*. In particular, let us consider four paradigms, each based on a particular mathematical theory, that have been employed for this purpose: finite-memory machine, Moore machine, Mealy machine, and combined machine [Klir, 1972b]. All of these paradigms are based on the assumption that one subsystem of the designed structure system is a temporary storage of states of some variables, while the remaining subsystems represent function dependencies among appropriate variables. The paradigms differ in the nature of the function dependencies, which affects the constraints imposed upon the structure of the system to be designed. These constraints are shown in terms of the block diagrams in Fig. 6.2, where **S** denotes the storage subsystem and **F**, $\mathbf{F}_1$, $\mathbf{F}_2$ denote the required function dependencies.

In the finite-memory machine, states $\mathbf{y}(t)$ of output variables at time t depend on states $\mathbf{x}(t)$ of input variables at time t and on a relevant sequence of previous

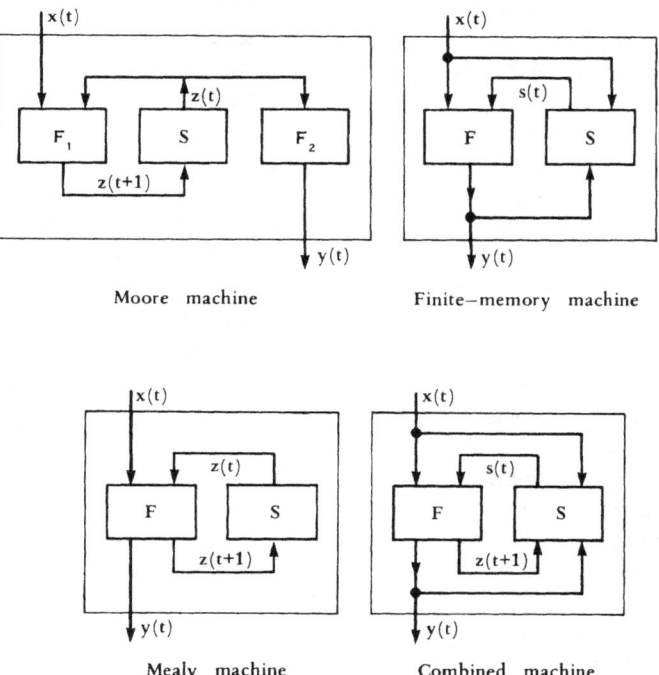

Figure 6.2. Methodological paradigms for finite-state dynamical systems.

input and output states that is stored in **S** (denoted in Fig. 6.2 as s(t)—the content of **S** at time t).

The remaining three paradigms also contain, in addition to input and output variables, so-called internal variables; their state at time t (or $t + 1$) is denoted in Fig. 6.2 by **z**(t) [or **z**($t + 1$), respectively]. In the Moore machine, the internal next state **z**($t + 1$) is a function of both the present input state **x**(t) and internal state **z**(t), while the present output state is a function of only the present internal state **z**(t). In the Mealy machine, both **y**(t) and **z**($t + 1$) are functions of **x**(t) and **z**(t). The combined machine is an integration of the finite-memory and Mealy machines: **y**(t) and **z**($t + 1$) depend not only on **x**(t) and **z**(t), but also on an appropriate sequence on input, output, and internal states [expressed again by the symbol s(t)].

It is well known that the class of behavior functions that can be captured in terms of the finite-memory machine paradigm is a proper subset of those captured by any of its three competitors. We say that the finite-memory machine is a less powerful paradigm than the other three. Moore, Mealy, and combined machines, on the other hand, are equally powerful in this sense. This equivalence (mathematically well established) has often been a source of the following fallacy: since the three paradigms are equivalent and, consequently, it does not matter which of them is used as a base for designing dynamic systems, we may as well use the one that is the easiest to deal with mathematically—the Moore machine. That this is indeed a fallacy can be seen by examining the block diagrams in Fig. 6.2. Each of them clearly represents a global constraint for the structure system to be designed. While the three paradigms are equivalent in terms of the classes of generative systems they can capture, they are not equivalent in terms of the classes of structure systems they represent. For example, the Mealy machine allows the dependence of the output state on the input state, which is not directly allowed by the Moore machine. Hence, there are structure systems that may emerge from the Mealy machine, but not from the Moore machine. On the other hand, every structure system that can be expressed in terms of the Moore machine can be described in terms of the Mealy machine as well. The Mealy machine is thus a more general paradigm than the Moore machine in the context of designing dynamic systems.

We can see that the combined machine includes all that is contained in the Mealy machine, and, in addition, it allows the output and next states to depend on appropriate sequences of previous input, output, and internal states. Hence, the combined machine is the most general paradigm of the three paradigms under consideration.

It follows from these observations that the ordering of paradigms by their generality may be different at different epistemological levels. These orderings, one at each level, are based upon the subset relations between sets of solutions to relevant problems that are captured by the individual paradigms. These subset

Systems Metamethodology

relations and, hence, the orderings of the four paradigms of the finite-state dynamical systems at the levels of generative systems and structure systems are depicted in Fig. 6.3.

To illustrate more specifically the fundamental difference between the paradigms discussed, a comparison between the Mealy paradigm and the finite-memory paradigm at the level of structure systems is shown in Fig. 6.4 for the serial binary adder introduced earlier (see Fig. 4.8). We can see that the two paradigms are based on totally different masks. Moreover, they are based on totally different behavior functions. The Mealy paradigm involves the two functions of three variables given in Table 4.3; the finite-memory paradigm involves one function of five variables,

$$z = \{x + y + [x' + y' + (z' - x' - y')(\text{mod } 2) - z']/10\}(\text{mod } 10),$$

where mod 2 (or mod 10) denotes the remainder obtained when the relevant integer is divided by 2 (or 10, respectively). It is obvious that the two paradigms will lead to very different designs, regardless of the objective criteria, constraints, and primitive elements employed.

Paradigms of finite-state dynamical systems can be studied even more comprehensively in terms of the nine couplings (between two elements and an environment) shown in Fig. 6.5. Symbols $x(t)$, $y(t)$, $z(t)$, and $s(t)$ have the same

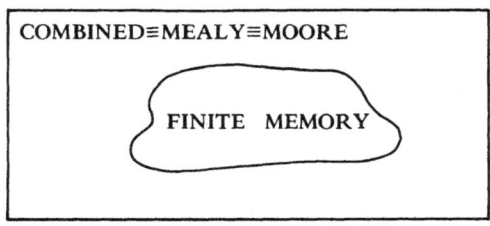

GENERATIVE SYSTEMS

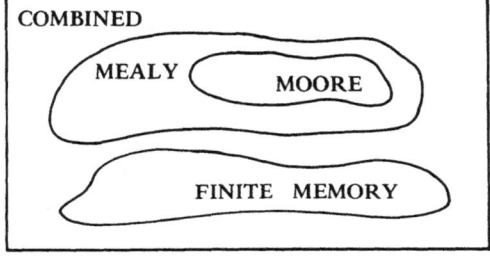

STRUCTURE SYSTEMS

Figure 6.3. Ordering of paradigms of finite-state dynamical systems.

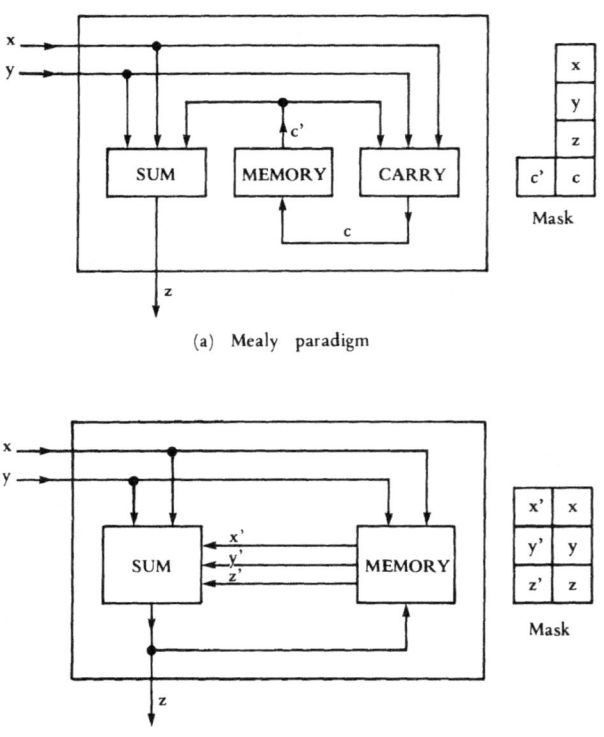

Figure 6.4. Binary adder conceptualized in terms of two distinct paradigms.

Systems Metamethodology

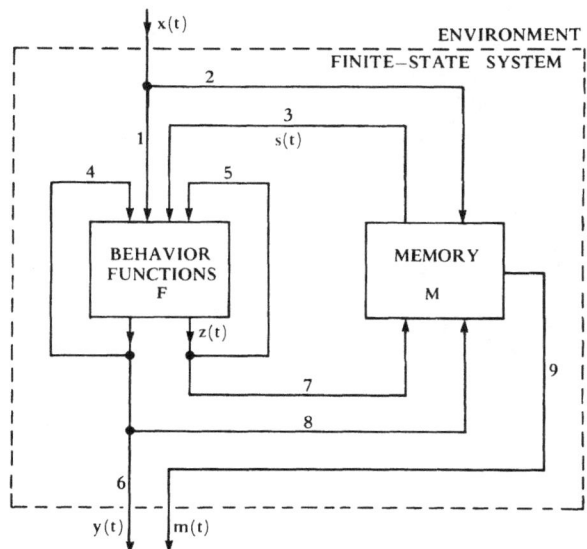

Figure 6.5. A framework for studying paradigms of finite-state dynamical systems.

meaning as before; symbol $m(t)$ denotes some portion of the memory content. Different paradigms are distinguished from each other by the exclusion or inclusion of these various couplings and, in some cases, by the nature of included couplings. I leave it to the reader to identify meaningful paradigms within this framework and determine their ordering in terms of generality at the level of structure systems.

The meaning of systems metamethodology should be sufficiently clear now. However, some additional metamethodological issues are touched upon at a few other places in the subsequent chapters.

CHAPTER 7

Systems Knowledge

Great knowledge sees all in one. Small knowledge breaks down into the many.
—CHUANG TZU

In every traditional discipline of science, we develop systems models of various phenomena of the real world. Each of these models, when properly validated, represents some specific knowledge regarding the relevant domain of inquiry. In systems science, the domain of inquiry consists of knowledge structures themselves—the various categories of systems that emerge from the conceptual framework employed. That is, the objects of investigation in systems science are not objects of reality, but systems of certain specific types.

Knowledge pertaining to systems science, or *systems knowledge*, is thus different from knowledge in traditional science. It is not knowledge regarding various aspects of reality, as in traditional science, but rather knowledge regarding the various types of systems in terms of which knowledge in traditional science is organized. That is, it is *knowledge concerning knowledge structures*. This knowledge is, of course, applicable to the processes of acquisition, management, and utilization of knowledge in every discipline of traditional science.

Systems knowledge can be obtained either mathematically or experimentally. Mathematically derived systems knowledge is the subject of the various mathematical systems theories, each applicable to some class of systems. It consists of theorems regarding issues such as controllability, stability, state equivalence, information transmission, decomposition, homomorphism, self-organization, self-reproduction, and many others. This area, which is well developed for some classes of systems, is beyond the scope of this introductory text.* Some systems

---

*The literature devoted to the various issues of mathematical systems theory is extensive. Let me mention only a few references that seem to represent well the nature of mathematically derived systems knowledge. The most important in this regard are perhaps two monographs by Mesarovic and Takahara [1975, 1988], primarily due to the high generality of some of their results. Another book comparable with these monographs, but more suitable as a textbook, was written by Windeknecht [1971]. An excellent textbook exposition of classical mathematical systems theory, including both linear and nonlinear systems, is given in two books by Casti [1977, 1985]. These books cover virtually

theorems with broad applicability are also mentioned on a few occasions in the subsequent chapters.

Systems knowledge can also be obtained experimentally. Although systems (knowledge structures) are not object of reality, they can be simulated on computers and in this sense made real. We can then experiment with the simulated systems for the purpose of discovering or validating various hypotheses in the same way as other scientists do with objects of their interests in their laboratories. In this sense, computers may be viewed as laboratories of systems science. Experimentation with systems on computers is not merely possible, but it may give us knowledge that is otherwise unobtainable.

The computer has, in fact, a dual role in systems science. In one of the roles, it is a methodological tool for dealing with systems problems. In the other role, it serves as a laboratory for experimenting with systems. It is the second role that is of interest to us in the chapter. Let me overview five distinct purposes for which the computer can be used as a laboratory.

The best known use of the computer for experimenting with systems is characterized as follows: a system that is established as a model of some aspect of reality is simulated on a computer for the purpose of generating scenarios under various assumptions regarding the environment of the system as well as various parameters of the system itself. This sphere of activity, which is usually referred to as *computer simulation*, is of particular significance to policy and decision making. The user (a policy or decision maker) can explore, through the generated scenarios, consequences of the various possible policies or decisions. Notwithstanding its significance, this way of experimenting with systems does not produce new systems knowledge. Consequently, it is not pursued here further.

Another purpose for experimenting with systems on the computer is to explore various systems as prospective models of a real-world phenomenon. The explored systems, which are based on an hypothetical explanation of the phenomenon, are of the same type and differ from each other only in values of some adjustable parameters. Systems based on different values of the parameters are simulated on the computer and their behaviors are compared with the behavior of the real thing. The aim is to determine the right values of the parameters, for which the behavior of the model is closest to the observed behavior. Again, this experimentation does not contribute to systems knowledge.

The third purpose for which we experiment with systems on the computer is

---

all basic theorems regarding controllability, reachability, observability, realizability, stability, and other properties of systems. In addition, they contain extensive bibliographical notes. A mathematical theory of hierarchically structured systems was developed by Mesarovic, Macko, and Takahara [1970]. There are many laws of information that have a universal applicability to all systems. Some of these laws were formulated by Ashby [1956, **1958a**, 1969], Conant [1969, 1974, **1976**], and Aulin [1982, 1989]. Most theorems regarding finite-state machines can be found in books by Gill [1962], Booth [1967], Hartmanis and Stearns [1966], and Klir [1972b].

Systems Knowledge

connected with *abduction* or *forming mathematical conjectures*. In this case, many relevant examples of systems that are of interest are examined on the computer, looking for regularities and developing thus empirical support for a mathematical conjecture. When the support becomes convincing, it is appropriate to try to prove the conjecture formally. A construction of the formal proof may also be inspired by regularities observed in the examples. In fact, this only describes how mathematicians typically work. By allowing a mathematician to examine many more examples (and even more complex examples) than otherwise possible, the computer plays the role of an *intuition amplifier*. Clearly, mathematical results relevant to systems science can be obtained in this way. One may, for example, study the relationship between structure systems of some type and data produced by these structure systems. The aim in such a study is to classify the structure systems and the data in such a way that each class of structure systems would correspond to exactly one class of data. The type of data generated by a structure system may then be inferred from the type of the structure system without resorting to complete analysis. Systems knowledge of this sort is useful as a heuristic aid.

The computer can also be used for *discovering or validating laws of systems science*. In contrast to laws of nature, laws of systems science characterize properties of various categories of systems rather than categories of real-world objects. We perform experiments of some kind on the computer with many different systems of the same category. The aim of this experimentation is to discover useful properties characterizing the category of systems or, alternatively, to validate some conjectures regarding the category.

This use of the computer conforms perhaps best to the notion of systems science laboratory. It was first conceived in the late 1960s by Gardner and Ashby [1970], when they performed some of the most exemplary experiments of this kind. Their experiments are simple and, yet, highly illustrative. Let me describe them as an example of this important role of the computer in systems science.

Systems that are investigated in these experiments are linear dynamic systems described by sets of differential equations of the form

$$\frac{dx_i}{dt} = a_{i1}x_1 + a_{i2}x_2 + \cdots + a_{in}x_n, \qquad (7.1)$$

where x_i ($i = 1, 2, \ldots, n$) denote variables, t denotes time, and $a_{i1}, a_{i2}, \ldots, a_{in}$ are constant coefficients in the ith equation. Each set of equations of this form may also be described, more concisely, in the matrix form

$$\dot{\mathbf{x}} = \mathbf{A}\mathbf{x}, \qquad (7.2)$$

where $\mathbf{x}$, $\dot{\mathbf{x}}$ denote vectors of the variables and their derivatives, respectively, and $\mathbf{A}$ is a matrix of the coefficients.

Questions that Gardner and Ashby attempted to answer for this category of systems are: "What is the chance that a large system will be stable? If a large system is assembled (connected) at random, or has grown haphazardly, should we expect it to be stable or unstable? And how does the expectation change as n, the number of variables, tends to infinity?" [Gardner and Ashby, 1970].

The equilibrium state (a state that does not change with time) of any system described by Eq. (7.1) is obtained when $x_1 = x_2 = \cdots = x_n = 0$. A particular system is considered stable when, given a sufficiently large amount of time, it converges to this equilibrium state, from any initial state. Some systems in this category are stable and some are not stable. There are several methods by which it can be determined from the nature of matrix $\mathbf{A}$ in Eq. (7.2) whether the system described by the matrix is stable or not. Although the issue of how stability of investigated systems was actually tested is not essential for understanding the experiments, let me mention that Gardner and Ashby employed the so-called Hurwitz criterion. According to this criterion, a system is stable if and only if the real parts of roots of $\mathbf{A}$'s characteristic equation are all negative. The characteristic equation of $\mathbf{A}$ has the form

$$\det(\mathbf{A} - \lambda \mathbf{I}) = 0,$$

where λ is the unknown, $\mathbf{I}$ is the identity matrix, and "det" indicates that we take a determinant of the respective matrix.

For each experiment, values of three variables were recorded:

n, The number of variables in the system, by which the size of the system is expressed;
c, Connectance of the system, defined by the percentage of direct connections (influences) between variables and expressed in terms of the percentage of nonzero entries in A outside the main diagonal;
s, Stability of the system (stable, unstable).

For each pair of values of n and c within the experimental range, the probability of stability, $p(n, c)$, was estimated by the formula

$$p(n, c) = \frac{N_s(n, c)}{N(n, c)},$$

where $N_s(n, c)$ and $N(n, c)$ denote, respectively, the number of stable systems and the number of all systems examined (with n variables and connectance c).

A selection of results is shown in Fig. 7.1. We can see that the function $p(n, c)$ decreases with c for each value of n. As n increases, $p(n, c)$ converges to a step function of c. When $n = 10$, $p(n, c)$ may already be regarded as the step function whose critical value is $c = 13\%$. This result is important since it enables us to answer questions of stability regarding sufficiently large systems (with at least ten variables) in the investigated category in a simple way: when the connectance is below 13%, the system is almost certainly stable; when it is above 13%, the system is almost certainly unstable.

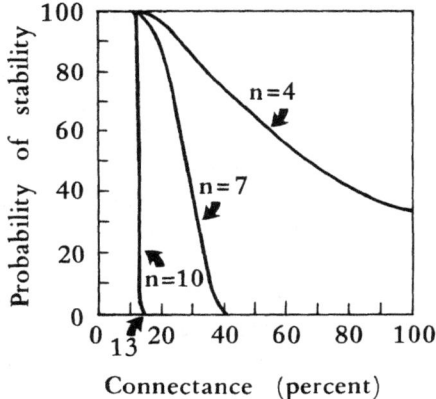

Figure 7.1. The effect of connectance on stability.

It is certainly appropriate to view the described result regarding the critical connectance as a law of systems science. It is significant that this law was established by experiments of basically the same nature as those in traditional science, with only one exception. In traditional science, we experiment with objects of the real world; in systems science, we experiment with systems (abstractions) that are simulated on the computer and made real in this sense.

It is also significant that some systems laws can be obtained only experimentally. Although mathematical analysis can always be employed, in principle, it often provides us with results pertaining only to special, limiting cases (e.g., $n \to \infty$) rather than cases of practical utility.

The potential of the computer as a laboratory for discovery or validation of systems laws has not been fully realized as yet. Although the pioneering experiments by Gardner and Ashby influenced other researchers to use the computer in a similar way,* reported experiments and results are rather scattered. No concerted

---

*The experiments (first described in 1968 in a report at the University of Illinois in Urbana) were extended to a more general class of dynamic systems described by nonlinear and time-varying differential equations by Makridakis and Faucheux [1973]. Somewhat similar experimental studies, biologically motivated, were conducted by Kaufmann [1969] for networks of simple discrete systems. Different experiments were performed by Walker [1971] and Gelfand and Walker [1977] for networks of simple finite-state machines. The aim of these experiments was to determine the dependence of cycle length and other behavioral characteristics on the size of the overall system for various types of finite-state machines. A representative sample of the many recent experiments performed with systems on the computer is described in a paper by Pilette, Sigal and Blamire [1990].

This role of the computer as a laboratory of systems science is discussed, in the context of computer technology of the 1990s, by Cartwright [1991]. The discussion is illustrated by several examples of experimental studies of chaotic behavior of certain deterministic nonlinear systems.

effort has been made thus far to codify the existing systems knowledge and to investigate the various categories of systems in a more organized fashion.

Let me discuss now one additional use of the computer as a laboratory, its use for *metamethodological inquiries*. In each such inquiry, a particular systems methodology (a coherent collection of methods for dealing with systems problems of some type) is applied to a set of problems chosen from the delimited class. Relevant results for all these problems, each of which is viewed as an experiment, are then summarized in a desirable form to characterize the methodology and, possibly, compare it with its competitors.

This sort of experimentation provides us with knowledge regarding systems methodologies rather than knowledge regarding systems. These two kinds of knowledge obviously play very different, but equally important, roles in systems science.

In some cases, a metamethodological inquiry can also be employed for validating various methodological principles that are based upon the investigated methodologies. To illustrate this possibility, let me describe, in conceptual terms, an experimental inquiry by which a novel principle of inductive reasoning was established. This principle, which is called a *reconstruction principle of inductive reasoning*, is based upon a methodology that is referred to as *reconstructability analysis* [Klir, 1985a]. Only some aspects of this methodology, which are directly relevant to the principle in question, are introduced here.

One of the problems addressed by reconstructability analysis is usually called the *reconstruction problem*. This problem is concerned with the question of how a given generative system, assumed to be employed in modeling some aspect of reality, can be broken down into appropriate subsystems. A collection of subsystems, which in our terminology is a structure system, is appropriate when it is sufficient for reconstructing the original overall system to an acceptable degree of approximation.

Two preference orderings are essential for comparing competing structure systems in the reconstruction problem. One is expressed in terms of the amount of information that is lost when the overall system is replaced with a structure system. Let this preference ordering be called *information ordering*. How to actually measure information and information loss in this context depends on the way in which systems are formalized. Although some aspects of this measurement are discussed in Chap. 8, this background is not essential for a conceptual understanding of the experiments I intend to describe here.

The second preference ordering involved in the reconstruction problem is

---

Computer experimentation of this sort has recently become almost routine for the study of certain classes of systems, such as *dynamic cellular automata* [Wolfram, 1986; Toffoli and Margolus, 1987], *developmental systems* (also called L-systems) [Goel and Rozehnal, 1991], and systems based on *fractal geometry* [Becker and Dorfler, 1989].

Systems Knowledge

connected with the size of subsystems contained in the structure system: smaller subsystems are preferred. This latter preference ordering may be called a *complexity ordering* since a reduction in the size of a system tends to reduce the size of its description. It is only a partial ordering.

The *solution set of the reconstruction problem* applied to a given overall system is the set of all its structure systems that are either equivalent or noncomparable in terms of the information and complexity orderings. This means typically that the solution set consists of all structure systems whose information distance is minimal at each level of complexity.

A general scheme of the experiments designed to validate the reconstruction principle of inductive inference is given in Fig. 7.2. Let me use this figure to describe the experiments.

Each experiment begins with a selection of a generative system of some type. Assume that this system is, in general, a nondeterministic system characterized by appropriate conditional probabilities. It is denoted in Fig. 7.2 by **G** (given generative system). This system is simulated on a computer and employed for generating a data matrix (a time series). Generative systems of the same type as **G** are then inferred from different segments of the data matrix by a suitable modeling methodology. For each data segment containing some specific number of observations, we obtain a system denoted in Fig. 7.2 by **D** (inferred from data).

The next step in the experiment is to solve the reconstruction problem for system **D**. Let S_c denote a structure system of complexity c that is in the solution set, let R_c denote the overall generative system reconstructed solely from informa-

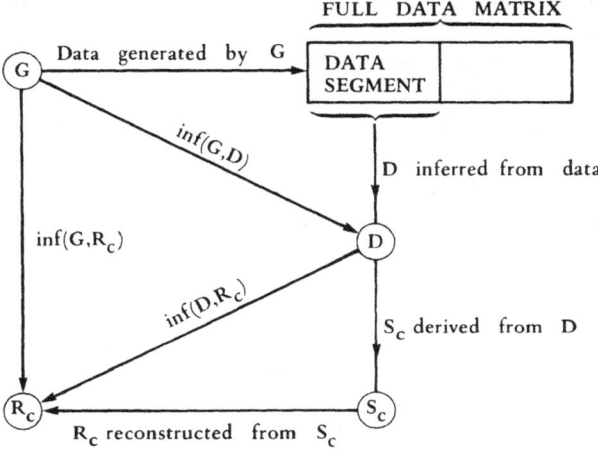

Figure 7.2. General scheme of the experiments designed to validate the reconstruction principle (rule) of inductive reasoning.

tion contained in elements of $\mathbf{S}_c$, and let $\inf(\mathbf{D}, \mathbf{R}_c)$ denote the amount of information that is lost when $\mathbf{D}$ is replaced with $\mathbf{R}_c$ (reconstructed from $\mathbf{S}_c$). Since $\mathbf{S}_c$ is a member of the solution set, $\inf(\mathbf{D}, \mathbf{R}_c)$ is (by definition) the smallest information loss at complexity level c.

Since system $\mathbf{D}$ was inferred from some specific data, its conformation to the data is perfect. As a consequence, any predictions made by $\mathbf{D}$ reflect the very properties of the given data that are captured by the system. This way of making predictions (or, generally, inductive inferences) is usually called a *straight rule*. This rule is well characterized by the following precept offered by Rescher [1980]:

> When a certain percentage of population P have in fact been observed to have a particular trait T, then adopt *this very value* as your answer to the question: "What proportion of the entire population have the trait T?"

In contrast to $\mathbf{D}$, the reconstructed system $\mathbf{R}_c$ does not conform to the given data completely [unless $\inf(\mathbf{D}, \mathbf{R}_c) = 0$]. That is, $\mathbf{R}_c$ contains some *novelty* with respect to the data and, consequently, predictions made by $\mathbf{R}_c$ violate the straight rule. This raises an interesting and important question: which of the two systems, $\mathbf{D}$ or $\mathbf{R}_c$, does contain more information about $\mathbf{G}$? The question can also be reformulated: which amount of lost information is smaller, $\inf(\mathbf{G}, \mathbf{D})$ or $\inf(\mathbf{G}, \mathbf{R}_c)$? Clearly, if

$$\inf(\mathbf{G}, \mathbf{D}) < \inf(\mathbf{G}, \mathbf{R}_c),$$

$\mathbf{D}$ is preferable (more credible) in making predictions. If the inequality is inverted, however, $\mathbf{R}_c$ is preferable, and in case of equality, $\mathbf{D}$ and $\mathbf{R}_c$ are equally good.

Although the question cannot be answered if $\mathbf{G}$ is not known, which is the case in the usual data-driven systems modeling, empirical support for an answer can be obtained by simulation experiments of the type just described. Since $\mathbf{G}$ is known in each experiment, values of $\inf(\mathbf{G}, \mathbf{D})$ and $\inf(\mathbf{G}, \mathbf{R}_c)$ can be determined and compared. Such experiments were actually performed on a large scale for probabilistic systems with small number of variables [Klir and Parviz, 1986]. They allowed us to answer the question in a conditional way: under some specific conditions within the experimental range, $\mathbf{D}$ is preferable (more credible) as a predictive model; under other conditions, $\mathbf{R}_c$ is preferable. In the former case, the straight rule is appropriate in making predictions or other inductive inferences. In the latter case, a different rule is preferable, which may be called a *reconstruction rule*: use the reconstructed system $\mathbf{R}_c$ as a guide for making inductive inferences.

The conditions that qualify the preference of either the straight rule or the reconstruction rule are expressed in terms of four parameters:

n, The number of variables in the systems;
s, The number of states each variable may assume (assumed, for the sake of simplicity, to be the same for all variables);

Systems Knowledge

N, The number of observations in data employed in the experiments;
c, The level of complexity of the structure system involved.

The studies were restricted (due to tremendous computational demands) to systems with five variables or less, five states per variable or less, and data with up to 2,000 observations. All variables in each experiment were assumed to exhibit the same number of states (s). Thirty experiments were performed for each selected combination of values of the four parameters, and values of inf ($\mathbf{G}$, $\mathbf{D}$) and inf ($\mathbf{G}$, $\mathbf{R}_c$) were averaged over the thirty experiments.

The study showed that, on average, the reconstruction rule performs better than the straight rule within a restricted domain of values of the four parameters. The domain depends on the risk we are willing to accept that the rule may occasionally fail. The risk is expressed in terms of the percentage of relevant experiments in which the rule actually failed. Given values of n, s, c, and the acceptable percentage of potential failure, f, the reconstruction rule performs better than the straight rule when data contain no more than some specific number of observations $x(n, s, c, f)$. Although the values of $x(n, s, c, f)$ cannot be specified precisely, they can be loosely estimated from the experimental outcomes with sufficient precision for practical applications. Considering, for example, $n = 4$, $s = 3$, $c = 1,2,3,4,5$, and either $f = 25\%$ or $f = 15\%$, we obtain the following approximate values:

$x(4, 3, 1, 25) = 150,$ $x(4, 3, 1, 15) = 40$
$x(4, 3, 2, 25) = 600,$ $x(4, 3, 2, 15) = 50$
$x(4, 3, 3, 25) = 550,$ $x(4, 3, 3, 15) = 100$
$x(4, 3, 4, 25) = 250,$ $x(4, 3, 4, 15) = 80$
$x(4, 3, 5, 25) = 0,$ $x(4, 3, 5, 15) = 0$

Another way of comparing systems $\mathbf{D}$ and $\mathbf{R}_c$ in their relationship to system $\mathbf{G}$ is to compare them in terms of the overall states they generate. An example of experimental results of this kind is shown for $n = 4$, $s = 2$, and $c = 1, 2, 3, 4, 5$ in Fig. 7.3. Each point of the plots in this figure represents the average value of experimental outcomes of 30 experiments. The interrupted horizontal line in each set of plots indicates the percentage of all possible states that are actually generated by $\mathbf{G}$. The plots labeled with $\mathbf{D}$ and $\mathbf{R}_c$ specify, for each value of N and c, the percentage of all possible overall states that are generated by the respective systems. The middle plot, labeled with $\mathbf{GR}_c$, specifies (for each N, c) the percentage of all possible overall states that are generated by both $\mathbf{R}_c$ and $\mathbf{G}$.

The plots in Fig. 7.3 can be employed for estimating the number of correct states that are not in the data, but can be recovered by solving the reconstruction problem on $\mathbf{D}$. For each particular values of c and N, the distance between the corresponding points on plots $\mathbf{GR}_s$ and $\mathbf{D}$ indicates the percentage of recovered correct states, while the distance between the points on plots $\mathbf{R}_s$ and $\mathbf{GR}_s$ indicates

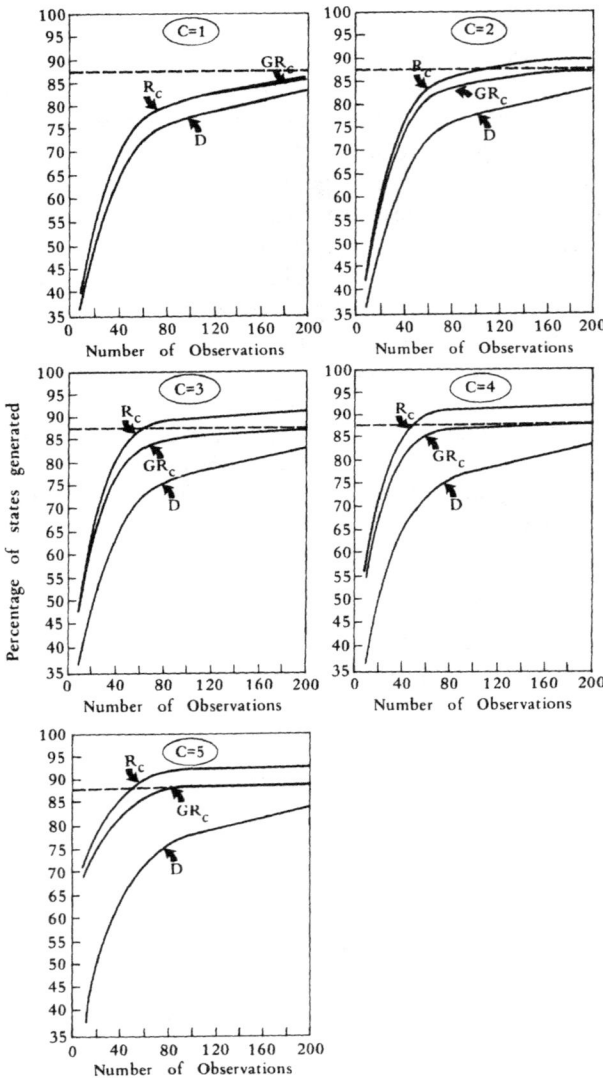

Figure 7.3. Recovery of states by solving the reconstruction problem for $n = 4$ and $n = 2$.

Systems Knowledge

the percentage of incorrect states generated by $\mathbf{R}_s$. Thus, for example, for $c = 2$ and $N = 30$, there are about 8% of recovered correct states and only about 1% of incorrect states on average. Since the total number of possible states is 16 in this case, 8% represents 1.28 states, and 1% corresponds to 0.16 states on the average. Hence, we can expect that at least one correct state would be recovered by solving the reconstruction problem in a typical situation when $n = 4$, $s = 2$, $c = 2$, and $N = 30$; an incorrect state may also be produced occasionally, but only once in about six or seven systems inquiries.

The reconstruction rule of inductive reasoning states that, under certain conditions, some justifiable novelty (something not explicitly contained in the available data) can be produced by a reconstruction method at the expense of pure conformation to data. The significance of the described simulation experiments is that they allow us to delimit the conditions under which the reconstruction rule is superior to the usual straight rule. I hope they illustrate the great potential of the computer for metamethodological inquiries in systems science.

CHAPTER 8

Complexity

Complexity is a paradoxical newcomer to the history of science. By a twist of semantic perversity, in proposing an intelligence of complexity, we look first for support from the complexity of intelligence.
—JEAN-LOUIS LE MOIGNE

Complexity is perhaps as important a concept for systems science as the concept of a system. It is a difficult concept, primarily because it has many possible meanings. While various specific meanings of complexity have been proposed and discussed on many occasions, there is virtually no sufficiently comprehensive study that attempts to capture its general characteristics. The reason for this situation is well expressed by John Casti [1986]:

> The notion of system complexity is much like St. Augustine's description of time: "What then is time [complexity]? If no one asks me, I know; if I wish to explain it to one that asks, I know not." There seems to be fairly well-developed intuitive ideas about what constitutes a complex system, but attempts to axiomatize and formalize this sense of the complex all leave a vague, uneasy feeling of basic incompleteness, and a sense of failure to grasp important aspects of the essential nature of the problem.

In line with the purpose of this book, an attempt is made in this chapter to capture the essence of complexity, rather than to discuss its various narrow, technical meanings. My primary aim is to show that the concept of complexity has many faces while, at the same time, it is associated with some general properties that remain invariant when one face is replaced with another.

To begin with a broad perspective, let us consult a common dictionary first; we find that complexity is "the quality or state of being complex," i.e.,

- "Having many varied interrelated parts, patterns, or elements and consequently hard to understand fully"; or
- "marked by an involvement of many parts, aspects, details, notions, and necessitating earnest study or examination to understand or cope with" (*Webster's Third International Dictionary*).

This common-sense characterization of complexity does not contain any qualification regarding the kind of entities to which it is applicable. As such, it can be applied to virtually any kinds of entities, material or abstract, natural or manmade, products of art or science. Regardless of what it is that is actually considered as being complex or simple, its degree of complexity is, according to the common-sense characterization, associated with the number of recognized parts as well as the extent of their interrelationship; in addition, complexity is given a somewhat subjective connotation since it is related to the ability to understand or cope with the thing under consideration.

We can see that the common-sense characterization of complexity assumes an interaction between the object (a part of the world that may have "many varied interrelated parts . . .") and a human being (or, perhaps, a computer) for whom it may be difficult "to understand or cope with" the object. This means that the complexity of an object for a particular human being depends on the way he interacts with it (i.e., on his interests and capabilities). More poetically, we may say that the *complexity of an object is in the eyes of the observer*.

In most cases, there is virtually an unlimited number of ways one can interact with an object. As a consequence, the interaction is almost never complete. It is based on a limited (and, usually, rather small) number of attributes that the observer is capable of distinguishing on the object and that are relevant to his interests. These attributes are not available to the observer directly, but only in terms of their abstract images, which are results of perception or some specific measurement procedures. These abstract images are usually called variables. When a set of variables is established as a result of our interaction with an object of interest, we say that a system (or, more precisely, a source system) is distinguished on the object.

Since we deal with systems distinguished on objects and not with the objects themselves (in their totality), it is not operationally meaningful to view complexity as an intrinsic property of objects. This does not mean, however, that I deny the existence of complexity of objects in the ontological sense. I only recognize that the notion of object complexity is epistemologically and methodologically vacuous, in contrast to the notion of systems complexity. This point is well expressed by Ashby in one of his last writings [Ashby, 1973]:

> The word "complex," as it may be applied to systems, has many possible meanings, and I must first make my use of it clear. There is no obvious or preeminent meaning, for although all would agree that the brain is complex and a bicycle simple, one has also to remember that to a butcher the brain of a sheep is simple while a bicycle, if studied exhaustively (as the only clue to a crime) may present a very great quantity of significant detail.
>
> Without further justification, I shall follow, in this paper, an interpretation of "complexity" that I have used and found suitable for about ten years. I shall measure the degree of *"complexity"* by *the quantity of information*

required to describe the vital system. To the neurophysiologist the brain, as a feltwork of fibers and a soup of enzymes, is certainly complex; and equally the transmission of a detailed description of it would require much time. To a butcher the brain is simple, for he has to distinguish it from only about thirty other "meats," so not more than $\log_2 30$, i.e., about 5 bits, are involved. This method admittedly makes a system's complexity purely *relative to a given observer*; it rejects the attempt to measure an absolute, or intrinsic, complexity; but this acceptance of complexity as something in the eye of the beholder is, in my opinion, the only workable way of measuring complexity.

Others have expressed their views on this important issue differently, though in the same spirit. The following two imaginative quotes should reinforce the point I want to make here:

We can only hope for explicit models of the world and not for reality itself or even a small part of it [Suppes, 1977].

One of the functions of the experimental method is to substitute simple artificial systems for the complex systems that Nature presents to us [Simon, 1977a].

Complexity (in the epistemological and methodological sense) is thus associated with systems, that is, some abstractions distinguished on objects that reflect the way in which the objects are interacted with. Systems, however, have many different faces, each represented by one of the epistemological categories of systems and, possibly, by some methodological distinctions within the category (Chap. 4). As a consequence, the concept of complexity, when applied to these various systems categories, has many different faces as well. That is, different types of systems give the concept of complexity different meanings, each of which requires a special treatment.

Notwithstanding the differences in complexities of the various systems types, *two general principles of systems complexity* can be recognized; they are applicable to any of the systems types and can thus be utilized as guidelines for a comprehensive study of systems complexity.

According to the *first general principle*, the complexity of a system (of any type) should be proportional to the amount of information required to describe the system. Here, the term "information" is used solely in a syntactic sense; no semantic or pragmatic aspects of information are employed. One way of expressing this *descriptive complexity*, perhaps the simplest one, is to measure it by the number of entities involved in the system (variables, states, components) and the variety of relationship among the entities. Indeed, everything else being the same, our ability to understand or cope with a system tends to decrease when the number of entities involved or the variety of their relationship increase. There are, of course, many different ways in which descriptive complexity can be measured even within a particular category of systems. However, each way of measuring complexity must satisfy at least the following general requirements, which are rather obvious on intuitive grounds:

1. The complexity should be expressed in terms of nonnegative real numbers.
2. If system **A** is a strong homomorphic image of system **B**, then the complexity of **A** should not be greater than the complexity of **B**.
3. If systems **A** and **B** are isomorphic, then their complexities should be equal.
4. If system **C** consists of two subsystems, **A** and **B**, that do not interact with each other and neither is a homomorphic image of the other one, then the complexity of **C** should be equal to the sum of the complexities of **A** and **B**.

It seems that any additional requirements can be introduced only in the context of specific categories of systems.

Descriptive complexity can also be characterized in a universal way, independent of the nature of systems to which it is applied. In this sense, descriptive complexity of a system (of any type) is defined to be *the size of the shortest description of the system in some standard language* or, alternatively, the size of the smallest program in a standard language by which the system can be simulated on a canonical universal computer. The primary advantage of this definition of descriptive complexity is that it is theoretically sound and applicable to all systems, regardless of their classification. Its primary weakness is methodological: it is rather difficult to determine in many cases the shortest description of a system.

According to the *second general principle*, systems complexity should be proportional to the amount of information needed to resolve any uncertainty associated with the system involved (predictive, retrodictive, prescriptive). Here, again, syntactic information is used, but information that is based on a *measure of uncertainty*.

Uncertainty, which is an inherent property of every nondeterministic system, is now well understood in several mathematical frameworks. It was first conceived in terms of set theory by Hartley [1928]. He derived a simple class of functions,

$$I(A) = K \log_b |A|,$$

as the only meaningful functions by which to measure the amount of information needed to resolve the uncertainty associated with $|A|$ alternatives that are left undecided (or to characterize one particular element of the set A); $|A|$ denotes the cardinality of a finite set A and $K > 0$, $b > 1$ are constants that distinguish individual functions in this class. Later, the uniqueness of this class of functions to measure uncertainty in set-theoretic terms was proven axiomatically by Rényi [1970]. A choice of one of these functions, by choosing some particular values of K and b, determines basically the unit by which uncertainty is measured. An

Complexity

intuitively appealing unit of uncertainty (and the associated information) results in $I(A) = 1$ when the set A contains 2 elements. In this case, the *unit of uncertainty* expresses the total ignorance regarding the truth or falsity of *one proposition*; it was given the name *bit*, which is an abbreviation of "*binary digit*" (values 0 and 1 of a binary digit are often used for encoding truth values of a proposition). The Hartley measure of uncertainty in bits is thus expressed by the unique function

$$I(A) = \log_2 |A|. \tag{8.1}$$

Another classical mathematical framework for conceptualizing uncertainty is probability theory. A measure of *probabilistic uncertainty* (and the associated information) was established by Shannon [1948]. This measure, whose basic form is

$$H(p(x)|x \in X) = -\sum_{x \in X} p(x)\log_2 p(x), \tag{8.2}$$

where $(p(x)|x \in X)$ denotes a probability distribution on a finite set X, is usually called the *Shannon entropy*. It is well justified, in a number of alternative ways, as a unique measure of uncertainty conceptualized in terms of probability theory under the assumption that the units of measurement are bits [Klir and Folger, 1988; Rényi, 1971; Shannon, 1948]. In practical applications, the Shannon entropy is usually used in a conditional form (based on appropriate conditional probabilities) or in a generalized form that involves two probability distributions.

Since the Shannon entropy plays an important role in systems science, let me overview Shannon entropies of joint, marginal, and conditional probability distributions defined on two sets, X and Y. In agreement with a common practice in the literature, let me simplify the notation by using $H(X)$ instead of $H(p(x)|x \in X)$ to denote the simple entropy defined by Eq. (8.2). On two sets, X and Y, we can recognize three types of entropies:

1. Two *simple entropies* based on the marginal probability distributions,

$$H(X) = -\sum_{x \in X} p(x)\log_2 p(x), \tag{8.2'}$$

$$H(Y) = -\sum_{y \in Y} p(y)\log_2 p(y). \tag{8.3}$$

2. A joint entropy defined in terms of the joint probability distribution, $p(x, y)$, on $X \times Y$,

$$H(X,Y) = -\sum_{x \in X}\sum_{y \in Y} p(x, y)\log_2 p(x,y). \tag{8.4}$$

3. Two *conditional entropies* defined in terms of weighted averages of local conditional entropies as

$$H(X|Y) = \sum_{y \in Y} p(y) \sum_{x \in X} p(x|y)\log_2 p(x|y), \tag{8.5}$$

$$H(Y|X) = -\sum_{x \in X} p(x) \sum_{y \in Y} p(y|x) \log_2 p(y|x). \qquad (8.6)$$

In addition to the simple, joint, and conditional entropies, the function

$$T(X, Y) = H(X) + H(Y) - H(X, Y) \qquad (8.7)$$

is frequently used in the literature as a measure of the strength of relationship (in the probabilistic sense) between elements of sets X and Y. This function is called *information transmission*.

The following are some fundamental properties of the various entropies and the information transmission:

$$H(X|Y) = H(X,Y) - H(Y), \qquad (8.8)$$
$$H(Y|X) = H(X,Y) - H(X), \qquad (8.9)$$
$$T(X,Y) = H(X) - H(X|Y), \qquad (8.10)$$
$$T(X,Y) = H(Y) - H(Y|X), \qquad (8.11)$$
$$H(X) - H(Y) = H(X|Y) - H(Y|X), \qquad (8.12)$$
$$H(X,Y) \leq H(X) + H(Y), \qquad (8.13)$$
$$H(X) \geq H(X|Y), \qquad (8.14)$$
$$H(Y) \geq H(Y|X). \qquad (8.15)$$

These properties, as well as their generalizations to more than two sets [Ashby, 1969] can be easily derived from the basic definitions [Klir and Folger, 1985].

Since the mid-1960s, the two classical theories capable of conceptualizing uncertainty, classical set theory and probability theory, have been generalized into fuzzy set theory and fuzzy measure theory. *Fuzzy set theory* deals with sets whose boundaries are not sharp. That is, the change from nonmembership to membership in a fuzzy set is gradual rather than abrupt. *Fuzzy measure theory* deals with measures that are not necessarily additive (as probability measures are), but only monotonic with respect to the ordering based upon the relation "to be a subset of." Each of these general theories encompasses various special theories (including the two classical theories). Well-justified measures of uncertainty in some of these new theories are now available [Klir and Folger, 1988], but their coverage is beyond the scope of this book.

Studying uncertainty from the broader perspectives of fuzzy set theory and fuzzy measure theory made us aware that more than one type of uncertainty must be distinguished. At this time, we recognize three types of uncertainty, which are distinguished from one another by the suggestive names *nonspecificity*, *dissonance* (conflict), and *fuzziness* (vagueness). Nonspecificity is exemplified by the Hartley measure in classical set theory: the greater the set of alternatives that are left undecided in a situation (e.g., predictions or prescriptions), the less specific the situation is; when only one alternative is possible, the situation is fully specific. Dissonance is exemplified by the Shannon entropy in probability theory: it is required that probabilities be assigned to mutually exclusive alternatives and, hence, they conflict with one another and create dissonance; the greater the lack of discrimination among the probabilities, the greater the dissonance. Fuzziness (or vagueness) of a fuzzy set expresses the lack of distinctions between its members and nonmembers.

While only one type of uncertainty is captured by either of the classical theories (nonspecificity in classical set theory and dissonance in probability theory), some of the broader theories capture more than one type of uncertainty. Fuzzy sets, for example, have not only degrees of fuzziness, but also degrees of nonspecificity: the fuzzy set of all numbers *close* to zero (defined by an appropriate membership grade function) is certainly less specific than the set of all numbers *very close* to zero (defined by a comparable membership grade function). Two or three types of uncertainty also coexist in some special theories of fuzzy measures. Both nonspecificity and dissonance appear, for example, in the so-called *evidence theory*, in which the additivity requirement of probability theory is replaced with two weaker requirements of subadditivity and superadditivity [Klir and Folger, 1988].

Both descriptive complexity and uncertainty-based complexity are connected with information: information needed to describe a system and information needed to resolve uncertainty embedded in it. These two complexities (and the associated kinds of information) conflict with each other. When we want to reduce one of them, the other is likely to increase or, at best, remain the same. This trade-off is one of the most fundamental methodological issues in systems science.

To capture adequately phenomena within the realm of organized complexity (biological, medical, economic, social, etc.), we need models of great descriptive complexity. When the required complexity for obtaining realistic models becomes unmanageable, we must simplify. In every simplification, unfortunately, we are doomed to lose something. When we insist that the simplified models give us predictions with no uncertainty, we often lose relevance of these predictions to the real world. We can preserve some of this relevance only by allowing some uncertainty in the models. That is, we can trade certainty for relevance under a given limit of acceptable descriptive complexity. Furthermore, as previously explained, there are various types of uncertainty we can utilize in this trade.

The relationship among relevance (or credibility) and the two kinds of complexities of systems models, which is of utmost importance to systems modeling, is not as yet well understood. In general, we try to construct highly relevant (credible) models that are simple (in the descriptive sense) and, if possible, we want to avoid uncertainty. Unfortunately, these objectives conflict with one another in a rather complicated way. Although uncertainty is undesirable when considered alone, it becomes very valuable when considered in connection with descriptive complexity and relevance. It is the only commodity that can be traded for a reduction of complexity of a model, an increase in its relevance, or both. By investigating the broad theories of fuzzy sets and fuzzy measures, we try to extend the scope of this important commodity. This purpose of the broad theories of uncertainty is still not generally understood [Klir, 1989]. It is appropriate to quote in this context Lotfi Zadeh [1973], the founder of the theory of fuzzy sets:

> Given the deeply entrenched tradition of scientific thinking which equates the understanding of a phenomenon with the ability to analyze it in quantitative terms, one is certain to strike a dissonant note by questioning the growing tendency to analyze the behavior of humanistic systems as if they were mechanistic systems governed by difference, differential, or integral equations.
>
> Essentially, our contention is that the conventional quantitative techniques of system analysis are intrinsically unsuited for dealing with humanistic systems or, for that matter, any system whose complexity is comparable to that of humanistic systems. The basis for this contention rests on what might be called the *principle of incompatibility*. Stated informally, the essence of this principle is that as the complexity of a system increases, our ability to make precise and yet significant statements about its behavior diminishes until a threshold is reached beyond which precision and significance (or relevance) become almost mutually exclusive characteristics. It is in this sense that precise quantitative analyses of the behavior of humanistic systems are not likely to have much relevance to the real-world societal, political, economic, and other types of problems which involve humans either as individuals or in groups.
>
> An alternative approach . . . is based on the premise that the key elements in human thinking are not numbers, but labels of fuzzy sets, that is, classes of objects in which the transition from membership to nonmembership is gradual rather than abrupt. Indeed, the pervasiveness of fuzziness in human thought processes suggests that much of the logic behind human reasoning is not the traditional two-valued or even multivalued logic, but a logic with fuzzy truths, fuzzy connectives, and fuzzy rules of inference. In our view, it is this fuzzy, and as yet not well-understood, logic that plays a basic role in what may well be one of the most important facets of human thinking, namely, the ability to *summarize* information—to extract from the collections of masses of data impinging upon the human brain those and only those subcollections which are relevant to the performance of the task at hand.
>
> By its nature, a summary is an approximation to what it summarizes. For many purposes, a very approximate characterization of a collection of data is sufficient because most of the basic tasks performed by humans do not require a high degree of precision in their execution. The human brain takes advantage of this tolerance for imprecision by encoding the "task-relevant" (or "decision-relevant") information into labels of fuzzy sets which bear an approximate relation to the primary data. In this way, the stream of information reaching the brain via the visual, auditory, tactile, and other senses is eventually reduced to the trickle that is needed to perform a specific task with a minimal degree of precision. Thus, the ability to manipulate fuzzy sets and the consequent summarizing capability constitute one of the most important

Complexity

assets of the human mind as well as a fundamental characteristic that distinguishes human intelligence from the type of machine intelligence that is embodied in present-day digital computers.

Viewed in this perspective, the traditional techniques of system analysis are not well suited for dealing with humanistic systems because they fail to come to grips with the reality of the fuzziness of human thinking and behavior. Thus to deal with systems radically, we need approaches which do not make a fetish of precision, rigor, and mathematical formalism, and which employ instead a methodological framework which is tolerant of imprecision and partial truths.

The two types of complexity introduced thus far, the descriptive complexity and the uncertainty-based complexity, pertain to systems. Yet another face of complexity exists, a complexity that pertains to systems problems. This complexity, which is usually referred to as *computational complexity*, is a characterization of the time or space (memory) requirements for solving a problem by a particular algorithm. Either of these requirements is usually expressed in terms of a single parameter that represents the size of the problem.

To appreciate the significance of computational complexity for systems methodology, let me first address the question of fundamental computational limits.

8.1. Bremermann's Computational Limit

The following conjecture is the central theme of a paper by Hans Bremermann [1962]:

> No data processing system, whether artificial or living, can process more than 2×10^{47} bits per second per gram of its mass.

To process a certain number of bits means in this statement to transmit that many bits over one or several channels within the computing system. Let me overview the arguments by which Bremermann derived this conjecture.

It is obvious that information that is to be acted upon by a machine must be physically encoded in some manner. Assume that it is encoded in terms of energy levels within the interval $[0,E]$ of energy of some sort; E is viewed as the total energy available for this purpose. Assume further that energy levels can be measured with an accuracy of only ΔE. Then, the most refined encoding is defined in terms of markers by which the whole interval is divided into $N = E/\Delta E$ equal subintervals, each associated with the energy amount ΔE. If at each instant no more than one of the levels (represented by the markers) is occupied, then

$$\log_2 (N + 1)$$

is the maximum number of bits that are representable by energy E; $N + 1$ is used here to account for the case in which none of the levels is occupied. If, instead of one marker with energy levels in $[0,E]$, K markers $(2 \leq K \leq N)$ are used simultaneously, then

$$K \log_2 (1 + N/K)$$

bits become representable. The optimal utilization of the available amount of energy E is obtained when N markers with levels in the interval $[0,\Delta E]$ are used. In this optimal case, N bits of information can be represented.

In order to represent more information by the same amount of energy, it is desirable to reduce ΔE. This is possible only to a certain extent since the resulting levels must be distinguished by some measurement process which, regardless of its nature, always has some limited precision. The extreme case is expressed by the Heisenberg principle of uncertainty: energy can be measured to the accuracy of ΔE if the inequality

$$\Delta E \Delta t \geq h$$

is satisfied, where Δt denotes the time duration of the measurement, $h = 6.625 \times 10^{-27}$ ergs/sec is Plank's constant, and ΔE is defined as the mean deviation from the expected value of the energy involved. This means that

$$N \leq \frac{E \Delta t}{h}.$$

Now, the available energy E can be expressed in terms of the equivalent amount of mass m by Einstein's formula

$$E = mc^2,$$

where $c = 3 \times 10^{10}$ cm/sec is the velocity of light in a vacuum. If we take the upper (most optimistic) bound of N in the inequality, we get

$$N = \frac{mc^2 \Delta t}{h}.$$

Substituting the numerical values for c and h, we obtain

$$N = 1.36 m \Delta t \times 10^{47}.$$

For a mass of 1 g ($m = 1$) and time of 1 sec ($\Delta t = 1$), we obtain the value

$$N = 1.36 \times 10^{47},$$

which implies the conjecture.

Using the limit of information processing obtained for one gram of mass and one second of processing time, Bremermann then calculates the total number of bits processed by a hypothetical computer the size of the Earth within a time

Complexity

period equal to the estimated age of the Earth. Since the mass and age of the Earth are estimated to be less than 6×10^{27} grams and 10^{10} years, respectively, and each year contains approximately 3.14×10^7 seconds, this imaginary computer would not be able to process more than 2.56×10^{92} bits or, when rounding up to the nearest power of ten, 10^{93} bits. The last number—10^{93}—is usually referred to as *Bremermann's limit* and problems that require processing more than 10^{93} bits of information are called *transcomputational problems*.

Bremermann's limit seems at first sight rather discouraging for system problem solving, even though it is quite conservative (more reasonable assumptions would lead to a number smaller than 10^{93}). Indeed, many problems dealing with systems of even modest size exceed it in their information-processing demands. Consider, for example, a system of n variables, each of which can take k different states. The set of all overall states of the variables consists clearly of k^n states. In each particular system, however, the actual overall states are restricted to a subset of this set. There are 2^{k^n} such subsets. Suppose we need to select, identify, distinguish, or classify one system from the set of all possible systems of this sort. Then, under the assumption that the most efficient method of searching is used, in which each bit of information (the answer to a dichotomous question) allows us to cut the remaining choices in half,

$$\log_2 2^{k^n} = k^n$$

bits of information have to be processed. The problem becomes transcomputational when

$$k^n > 10^{93}.$$

That happens, e.g., for the following values of k and n:

k	2	3	4	5	6	7	8	9	10
n	308	194	154	133	119	110	102	97	93

The problem of transcomputationality arises in various contexts. One of them is pattern recognition. Consider, for example, a $q \times q$ spatial array of the chessboard type, each square of which can have one of k colors. There are clearly k^n color patterns, where $n = q^2$. Suppose we want to determine the best classification (according to certain criteria) of these patterns. This requires a search through all possible classifications of the patterns. In the case of only two classes, the problem becomes isomorphic to the previous one. For two colors, for example, the problem becomes transcomputational when the array is 18×18; for a 10×10 array, the problem becomes transcomputational when nine colors are used. This pattern recognition problem is directly relevant to physiological studies of the retina, but its complexity is tremendous. The retina contains about a million light-sensitive cells. Even if we consider (for simplicity) that each of the cells have only two states (active and inactive), the attempt to study the retina as a whole would require the processing of

$$2^{1,000,000} \equiv 10^{300,000}$$

bits of information. This is far beyond Bremermann's limit.

Another context in which the same problem occurs is the area of testing large-scale integrated digital circuits. These are tiny electronic chips with considerable complexity and a large number of inputs and outputs. For properly defined electric signals (each, usually, with two ideal states), the individual outputs should represent some specific logic functions of the logic variables associated with the inputs. To test a particular integrated circuit means to analyze it as a "black box": to determine the actual logic functions it implements, solely by manipulating the input variables and observing the output variables. For each output variable, the testing problem is thus basically the same as the problem previously discussed for $k = 2$ (unless a multiple-valued logic is used). It follows that testing of circuits, for example, with 308 inputs and one output is a transcomputational problem. However, it is well known that the practical complexity limits of this testing problem are considerably lower. Some currently manufactured large-scale integrated circuits cannot be in fact completely tested. The focus is thus on developing testing methods that can be practically implemented and guarantee only that the testing be *almost complete*, that, say, well over 90% of all possibilities be tested.

A more detailed characterization of the complexity of this problem, from the practical domain to Bremermann's limit, is expressed by Fig. 8.1. The figure shows the dependence of the time (in years) required to select (identify, classify, distinguish, etc.) one logic function of n variables under the consideration of different information-processing rates in the range from 10 through 10^{100} bits per second. Two significant values of time are also shown in the figure: L indicates the approximate age of the oldest fossil records of life on the Earth; M shows the approximate time since men first appeared on the Earth.

The testing example is in no way exceptional. Genuine systems problems are notorious for their huge demands on information-processing capabilities. This point is illustrated by specific examples on various occasions elsewhere in this book. It is also well depicted by Bremermann [1962] in the conclusion of his paper:

> The experience of various groups who work on problem solving, theorem proving and pattern recognition all seem to point in the same direction: These problems are tough. There does not seem to be a royal road or a simple method which at one stroke will solve all our problems. My discussion of ultimate limitations on the speed and amount of data processing may be summarized like this: Problems involving vast numbers of possibilities will not be solved by sheer data processing quantity. We must look for quality, for refinements, for tricks, for every ingenuity that we can think of. Computers faster than

Complexity

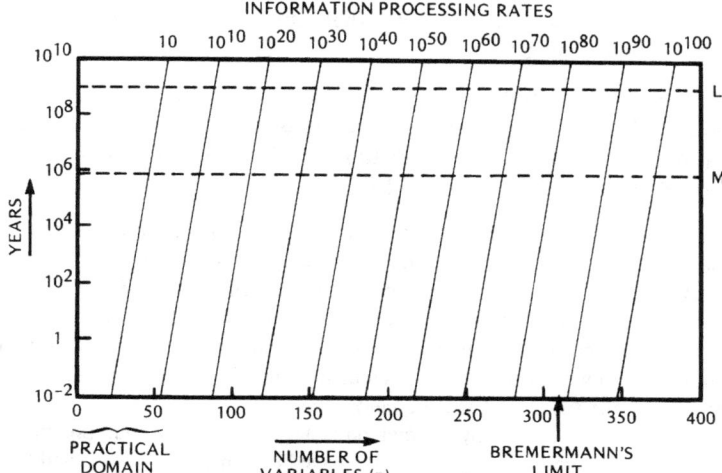

Figure 8.1. Time required to select or identify one logic function of n variables for information processing rates of $10, 10^{10}, \ldots, 10^{100}$ bits per second.

those of today will be a great help. We will need them. However, when we are concerned with problems in principle, present-day computers are about as fast as they ever will be.

If a problem is transcomputational, it is obvious that it can be dealt with only in some modified form. It is desirable to modify it no more than is necessary to make it manageable. The most natural way of modifying a problem is to soften its requirements. For instance, a requirement of getting the best solution may be replaced with a requirement of getting a good solution, instead of requiring a precise solution we may accept an approximate solution, and so on. Such softening of requirements permits the use of heuristic methods, in which vast numbers of unpromising possibilities are ignored, or approximate (fuzzy) methods, in which substantial aggregation takes place.

Bremermann's limit allows us to make only the most rudimentary categorization of systems problems by their complexities. It does not say anything about the actual practical computational limits. Nevertheless, it is a useful benchmark for a preliminary evaluation of each problem situation, as emphasized by Ashby [1973]:

> One of its most obvious consequences, yet one almost universally neglected today, is that, before the study of a complex system is undertaken, at least a rough estimate of its informational demands should be made. Should the estimate be 2000 bits we have little to worry about, but should it prove to be

10^{300} bits we would know that our whole strategic approach to the system needs re-formulating.

For purposes of practical problem solving, this simple benchmark—10^{93}—must be supplemented, of course, by sharper bounds on problem complexity. Its principal significance is epistemological: it is an indicator of fundamental limits to our knowledge, as explained by Ashby [1968]:

> The most obvious fact is that we, and our brains, are themselves made of matter, and are thus absolutely subject to the limit. Not only are we subject as individuals, but the whole cooperative organization of World Science is also made of matter, and is therefore subject to it. Thus both the total information that I can use personally, and the information that World Science can use, are limited, on any ordinary scale, to about 10^{80} bits. Whatever our science will become in the future, all will lie below this ceiling.
>
> We cannot claim any special advantage because of our preeminent position in the world of organisms. We have been shaped, and selected to be what we are, by the process of natural selection. As a selection, this process can be measured by an information-measure; it is therefore subject to its limits. In any type of selection, under any planetary conditions, a planetary surface made of matter cannot produce adaptation faster than the rate of the limit. However good we may think we are, 10^{80} measures something that we do not exceed. The science of the future will be built by brains that cannot have had more than 10^{80} bits used in their preparations, and they themselves will advance only by something short of 10^{80}. This is our informational universe: what lies beyond is unknowable.

[Note: Ashby derives the value of 10^{80} from the Bremermann limit for one second and one gram by considering "centuries of time and tons of computers" (e.g., about ten thousand centuries and 10^{15} tons of mass). It is obviously not important for the argument whether we take 10^{80} or 10^{93}].

Bremermann's limit works well, as a simple benchmark, for problems whose information-processing demands exceed it, but it does not say much about the remaining problems. Even if a problem is not rejected by Bremermann's limit, it may still be practically intractable. A more refined understanding of the notion of computational complexity is thus needed.

8.2. Computational Complexity

Computational properties of problems are studied under the general theory of algorithms. This general theory includes three large subject areas: the theory of computability, design of algorithms, and the theory of computational complexity. It is beyond the scope of this book to cover these areas in any depth. It is desirable, however, to provide the reader with a brief overview, focusing primarily on computational complexity, of those results and issues that are of particular significance to systems methodology. No proofs of the summarized results are presented here.*

An *algorithm* is understood intuitively as a set of instructions, expressed in some language, for executing a sequence of operations for solving a problem of some specific type. Algorithms are required to be *finite*, i.e., each algorithm must terminate after a finite number of steps (operations) have been executed.

The intuitive notion of an algorithm was formalized in several ways, including formalizations based on the concepts of Turing machines, Markov algorithms, and recursive functions, which were all proven to be equivalent. One of the concepts—that of a *Turing machine*—is envisioned as a simple device that consists of a finite-state control unit and a tape. The control unit has a *memory*, which makes it capable of being in any one of a finite set of states, say set $Z = \{z_1, z_2, \ldots, z_n\}$. The tape is potentially infinite in both directions, and is marked off along its length into spaces of equal size. Each of these spaces, referred to as cells, has written on it a symbol from a finite set of symbols, say set $X = \{x_0, x_1, \ldots, x_m\}$. One of the symbols, say symbol x_0, is always interpreted as a *blank space* (empty cell). Communication between the control unit and tape is provided by a *read-write head*, which is capable of reading symbols from the tape and writing over the symbols that are written on it. Only one cell of the tape is accessible to the head at any time.

The control unit of a Turing machine operates in discrete steps. In each step it replaces the current state with a new one, and performs a single operation of one of the following three types:

1. It replaces the current symbol on the tape with a new one;
2. It moves the tape by one cell to the left or right;
3. It stops the computation (the so-called halt operation).

The new state as well as the operation performed are uniquely determined by the current state and the symbol read on the tape.

Let z_c, z_n denote the current and next state of a Turing machine, respectively, let x_r denote the symbol that is read on the tape, and let y_p denote the operation performed. Then, given an initial string of symbols on the tape (any cell for which a symbol is not given is assumed to be blank) and a particular initial state a computation on the Turing machine is defined by an ordered set of quadruples

$$(z_c, x_r, z_n, y_p)$$

---

*Computational complexity has been extensively investigated since the early 1970s. A good overview of the main issues regarding computational complexity and results available in the late 1970s was prepared by Garey and Johnson [1979]. An excellent coverage of computational complexity is in the book of Harel [1987], which is self-contained and contains extensive bibliographical notes on the subject.

If no two quadruples in the set are allowed to begin with the same pair z_c, x_r, the Turing machine is said to be *deterministic*; otherwise, it is said to be *nondeterministic*.

A hypothesis that has become known as *Church's thesis* (or the Church–Turing thesis), and which has been generally accepted, states that any function regarded naturally as computable can be computed on a deterministic Turing machine. According to this hypothesis, a Turing machine is taken to be a precise formal equivalent of the intuitive notion of an algorithm. The hypothesis cannot be proven mathematically, but it is well justified by informal arguments and empirical evidence. It can be overthrown only by proposing an alternative formalization of computation, generally acceptable on intuitive grounds and capable of describing computations that are beyond the capabilities of Turing machines. The existence of such a formalization is considered highly unlikely.

In general, a problem is considered unsolvable if no algorithm exists by means of which a solution can be obtained. The notion of deterministic Turing machines together with Church's thesis have made possible the study of the existence of algorithms for various problems in a formal manner. To prove that a problem is unsolvable, it is sufficient to prove that it cannot be solved by a Turing machine. Such proofs of unsolvability have been obtained for a number of problems.

Unsolvable problems form one of three primary classes of problems. The second class consists of problems that have not been proven unsolvable, but for which no algorithms are known for solving them. These are thus problems whose solvability status has not been resolved as yet.

All remaining problems are solvable. That is, they are solvable in principle. In practice, however, many of them cannot be solved due to their excessive demands on computing resources such as computing time and memory size. Since the required computing time is usually the single factor that determines whether or not a problem is practically solvable, computational complexity has been predominantly studied in terms of this single resource.

The practical solvability of a problem depends on

1. The algorithm employed for solving the problem;
2. The size of the particular systems involved in the problem;
3. The computational power of the computing resources available.

Given a particular algorithm for solving a problem, it is convenient to express its time requirements in terms of a single variable that represents the size of the systems involved in the problem. This variable, which is often called the *size of a problem instance*, is supposed to express the amount of input data needed to describe the particular systems.

Given a particular systems problem instance, let n denote its size. Then, the time requirements of a specific algorithm for solving the problem are expressed by a function

$$f: \mathbb{R} \to \mathbb{R}$$

such that $f(n)$ is the largest amount of time needed by the algorithm to solve a problem instance of size n. Function f is usually called a *time complexity function*.

It has been recognized that it is useful to distinguish two classes of algorithms by the rate of growth of their time complexity functions. One class consists of algorithms whose time complexity functions can be expressed in terms of a polynomial. They are called *polynomial time algorithms*. Since the degree of each polynomial is considerably more significant, especially for large values of n, than its coefficients and lower-order terms, it is useful to classify polynomial time complexity functions by their order. A function f is said to be of complexity $O(n^k)$, where k is a positive integer, if and only if there is a constant $c > 0$ such that

$$f(n) \leq cn^k$$

for all $n \geq n_0$, where n_0 is a positive integer that usually represents the smallest size of the problem instances involved. For example, function

$$f(n) = 25n^2 + 18n + 31$$

is of complexity $O(n^2)$ since

$$f(n) \leq 74n^2$$

when $n_0 = 1$, or

$$f(n) \leq 42n^2$$

when $n_0 = 2$, etc.

The second class of algorithms consists of those whose time complexity functions are not bounded by complexity $O(n^k)$ for some k. They are usually referred to as *exponential time algorithms*.

The distinction between the polynomial and exponential time algorithms is significant, especially when considering large problem instances. This is illustrated in Table 8.1 by showing differences in growth rates for several time complexity functions. The computing times in this table are based on the assumption that the computing is performed at a rate of one million operations per second. When comparing, for instance, n^2 with n^{10}, we can see that the degree of a polynomial time complexity function has a considerable effect on practical limitations of the corresponding algorithms. However, polynomial time algorithms are substantially more responsive than exponential time algorithms to increases in computing power (except for small values of n). This can be seen by comparing plots of some polynomial and exponential time complexity functions in Fig. 8.2 and, even more explicitly, by examining the actual increases in the ranges of applicability due to increases in computing speed, as illustrated by the formulas in Table 8.2.

Because of the essential differences between polynomial and exponential

Table 8.1. Illustration of Growth Rates of Several Polynomial and Exponential Time Complexity Functions

Time complexity function	1	10	20	30	40	50	100
n	0.000001 sec	0.00001 sec	0.00002 sec	0.00003 sec	0.00004 sec	0.00005 sec	0.0001 sec
n^2	0.000001 sec	0.0001 sec	0.0004 sec	0.0009 sec	0.0016 sec	0.0025 sec	0.01 sec
n^5	0.000001 sec	0.1 sec	3.2 sec	24.3 sec	1.7 min	5.2 min	2.8 hr
n^{10}	0.000001 sec	2.8 h	118.5 days	18.7 yr	3.3 centuries	31.0 centuries	3.2×10^4 centuries
2^n	0.000002 sec	0.001 sec	1.0 sec	17.9 min	12.7 days	35.7 yr	4×10^{14} centuries
3^n	0.000003 sec	0.059 sec	58 min	6.5 yr	3,855 centuries	2×10^8 centuries	1.6×10^{32} centuries
10^n	0.00001 sec	2.8 hr	3.2×10^4 centuries	3.2×10^{14} centuries	3.2×10^{24} centuries	3.2×10^{34} centuries	3.2×10^{84} centuries
2^{2^n}	0.000004 sec	5.7×10^{292} centuries	$10^{3 \cdot 10^5}$ centuries	$10^{3 \cdot 10^8}$ centuries	$10^{3 \cdot 10^{11}}$ centuries	$10^{3 \cdot 10^{14}}$ centuries	$\sim 10^{3 \cdot 10^{29}}$ centuries
n^n	0.000001 sec	2.8 h	3.3×10^{10} centuries	6.5×10^{28} centuries	3.8×10^{48} centuries	$\sim 2.8 \times 10^{69}$ centuries	$\sim 3.2 \times 10^{184}$ centuries
$n!$	0.000001 sec	3.6 sec	771.5 centuries	8.4×10^{16} centuries	2.6×10^{32} centuries	$\sim 9.6 \times 10^{48}$ centuries	$\sim 2.9 \times 10^{142}$ centuries

Complexity

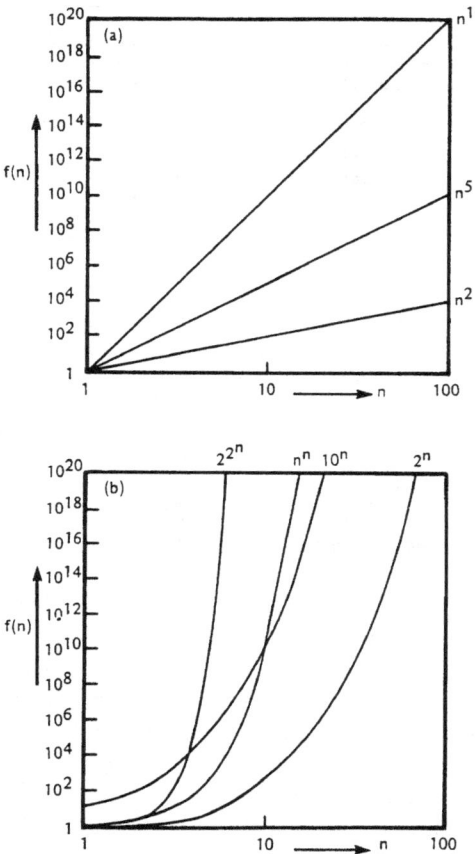

Figure 8.2. Plots of some typical complexity functions: (a) polynomial; (b) exponential.

time complexity functions, polynomial time algorithms are considered *efficient*, while exponential time algorithms are considered *inefficient*. As a consequence, problems for which it can be proven that they are not solvable by polynomial time algorithms are viewed as *intractable*, while problems for which polynomial time algorithms are known are viewed as *tractable*. The latter problems are usually called *P-problems* (i.e., solvable in polynomial time); the set of all such problems is called the *problem class P*.

It is known that differences among standard schemes used in practice for encoding problems as well as differences in the computer types used do not affect the classification of problems into tractable and intractable. Standard encoding schemes and computer types are known to differ from each other at most

Table 8.2. Effects of Increases in Computing Speed
on Problem-Solving Capabilities for Some Time Complexity Problems

Time complexity function	Current computer technology	Size of largest problem instance solvable in some unit of time with			
		Technology hundred times faster	Technology thousand times faster	Technology million times faster	Technology X times faster
n	n_1	$100n_1$	$1{,}000n_1$	$1{,}000{,}000n_1$	Xn_1
n^2	n_2	$10n_2$	$31.6n_2$	$1{,}000n_2$	$\sqrt{X}n_2$
n^5	n_3	$2.5n_3$	$3.98n_3$	$15.8n_3$	$\sqrt[5]{X}n_3$
n^{10}	n_4	$1.58n_4$	$2n_4$	$3.98n_4$	$\sqrt[10]{X}n_4$
2^n	n_5	$n_5 + 6.64$	$n_5 + 9.97$	$n_5 + 19.93$	$n_5 + \log X/\log 2$
3^n	n_6	$n_6 + 4.19$	$n_6 + 6.29$	$n_6 + 12.58$	$n_6 + \log X/\log 3$
10^n	n_7	$n_7 + 2$	$n_7 + 3$	$n_7 + 6$	$n_7 + \log X/\log 10$

polynomially. Alternative encoding schemes or computer types may thus influence the practical range of solvability of a problem, but they do not affect its tractability status.

It turns out that for most of the problems encountered in practice, neither is a polynomial time algorithm known to solve them, nor have they been proven intractable. A common trait of such problems is that they can be "solved" in polynomial time by nondeterministic computers such as nondeterministic Turing machines. Such problems are called *NP*-problems (nondeterministic polynomial time problems) and form a set called the *problem class NP*. The term "solve" is used here in the sense that if the machine guesses the solution, it can verify its correctness in polynomial time. The notion of a nondeterministic polynomial time algorithm is thus used solely as a convenient definitional device for capturing the notion of polynomial time verifiability of a proposed (guessed) solution of the actual problem. It is known that any NP problem can be solved by a deterministic algorithm with time complexity $O(2^{p(n)})$, where p is a polynomial function.

The class NP contains the class P because any problem that is solvable in polynomial time on a deterministic Turing machine is also solvable (i.e., verifiable) in polynomial time on a nondeterministic Turing machine. A considerable number of NP-problems have been proven to have the property that every other NP-problem can be converted to them in polynomial time. Such problems are distinguished as *NP-complete problems*.

Since the class NP consists of many practically important problems, it is highly desirable to resolve its status. The question of whether or not NP-problems are intractable is therefore one of the most important questions in mathematics, computer science, and systems science. Its implications for systems problem solving are quite profound. The question is often stated in the form "is NP = P?".

Complexity

It can be answered by proving for any of the NP-complete problems that it is either a P-problem (i.e., solvable in polynomial time) or a problem inherently intractable (i.e., solvable only in exponential time). If any one of the NP-complete problems is proven intractable, then NP ≠ P. If, on the other hand, such a problem is proven tractable, then NP = P. Since there are strong indications that NP ≠ P under the usual rules of inference, the question becomes primarily one of discovering some unorthodox rules of inference under which any one of the NP-complete problems could be proven tractable.

The classification of problems from the standpoint of their solvability and computational complexity is summarized in Fig. 8.3. The class denoted as coNP consists of problems that are complementary to the NP-problems in the sense that their answers are complements of the answers obtained for the corresponding NP-problems. It is not known whether NP = coNP, but it is known that the intersection NP ∩ coNP is not empty and contains all P-problems as well as some other problems.

Although computational complexity has been predominantly studied in terms of the time it takes to perform a computation, the amount of computer memory required is frequently just as important. This requirement is usually referred to as

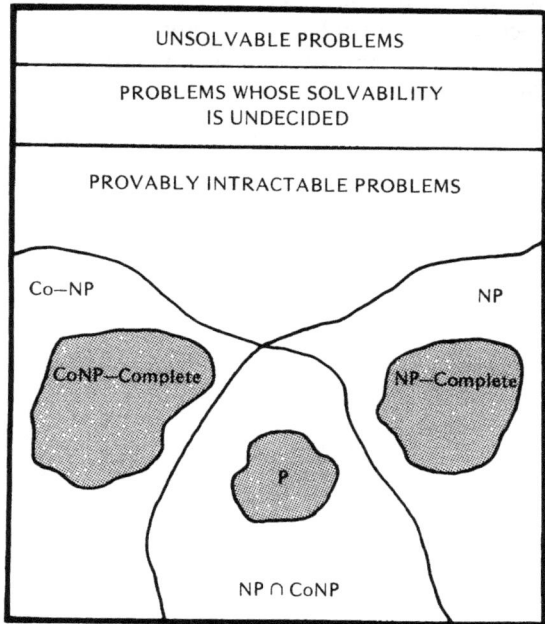

Figure 8.3. Classification of problems from the standpoint of their solvability and tractability.

the *space requirement*. It can be studied in terms of a *space complexity function*, analogous to the time complexity function. It is known, however, that any problem solvable in polynomial time can be solved in polynomial space as well. Indeed, the number of cells operated on by the read-write head of a Turing machine in a particular computation (which represents the space requirement) cannot exceed the number of steps involved in the computation (which represents the time requirement). It is not certain, however, whether all problems that are solvable in polynomial space are solvable in polynomial time. It is for this reason that the time complexity is used to classify problems as either tractable or intractable. In practice, however, both of these requirements are equally important.

From a broader, more realistic perspective, the size of a problem instance is not the only determinant of its computational complexity. That is, problem instances of the same type and size may have very different computational demands. Most studies in the area of computational complexity are oriented primarily to the characterization of the worst-case problem instances. Although this orientation is theoretically sound, it usually results in estimates that are rarely reached in practice and are therefore too pessimistic. To ameliorate this situation, the worse-case estimates are sometimes supplemented with average-case estimates. However, such estimates are based on the assumption that all problem instances are equally likely, which does not necessarily reflect the actual probability distribution of problem instances encountered in practice. The problem of determining the actual distribution for various problem types is predominantly an empirical problem. This problem can be studied, in principle, by monitoring and analyzing problem instances requested by users of the various systems problem packages. Any such study is an example of a metamethodological inquiry (Chap. 6).

CHAPTER 9

Simplification Strategies

The Scientist, like the Pilgrim, must wend a straight and narrow path between the Pitfalls of Oversimplification and the Morass of Overcomplication
—RICHARD BELLMAN

In some contexts, complexity is a desirable property, i.e., we search, within given constraints, for systems with a high degree of complexity. Cryptography and the design of random number generators are two typical examples of such contexts. In some situations, a certain degree of complexity is a necessary condition for obtaining some specific systems properties, usually referred to as *emergent properties*. Self-reproduction, learning, and evolution are examples of such properties.

In other contexts, which seem to predominate in systems problem solving, we search for simple systems or attempt to simplify existing systems. The importance of systems simplicity and simplification methods is well depicted by Herbert Simon [1977]:

> The human species has survived and thrived in the world, simple or complex as it may be, not so much through the speed and power of its computational capacities, as by exploiting the fact that the systems of interest to it represent highly special cases that can often be analyzed by relatively simple means, provided their underlying structure is detected. This argues for a strategy of searching for that structure, of pattern induction—a skill that is rather highly developed in the animal kingdom—followed by special analysis and heuristic problem-solving search, rather than brute-force analysis of very general classes of highly interconnected complex systems.

Similar sentiments are expressed by Edward Teller [1980], a physicist:

> Simplicity, for me, is best characterized in a story from the art traditionally the favorite of mathematicians and scientists: music. When Mozart was fourteen years old, he listened to a secret mass in Rome, Allegri's *Miserere*. The composition had been guarded as a mystery; the singers were not allowed to transcribe it on pain of excommunication. Mozart heard it only once. He was then able to reproduce the entire score.

Let no one think that this was exclusively a feat of prodigious memory. The mass was a piece of art and, as such, had threads of simplicity. The structure is the essence of art. The child who was to become one of the world's greatest composers may not have been able to remember the details of this complicated work, but he could identify the threads, remember them and reinvent the details having listened once with consummate attention. These threads are not easily discovered in music or in science. Indeed, they usually can be discerned only with effort and training. Yet the underlying simplicity exists and once found makes new and more powerful relationships possible.

Gerald Weinberg [1972] goes even further when he suggests defining systems science as a science of simplification. After describing the remarkable simplification which Newton successfully developed for mechanics, Weinberg summarized the importance of simplification as follows:

Newton was a genius, but not because of superior computational power of his brain. Newton's genius was on the contrary his ability to simplify, idealize, and streamline the world so that it became, in some measure, tractable to the brains of perfectly ordinary men. By studying the methods of simplification which have succeeded and failed in the past, the general systems theorist hopes to make the progress of human knowledge a little less dependent on genius.

Weinberg is not alone in giving methods of simplification high prominence in systems science. Ashby [1964], when discussing the various implications of Bremermann's computational limit, arrives at similar conclusions:

The systems theorist may be defined as a man, with resources not possibly exceeding 10^{100}, who faces problems and processes that go vastly beyond this size. What is he to do?

At this point, it seems to me, he must make up his mind whether to accept this limit or not. If he does not, let him attack it and attempt to find a way of defeating it. If he does accept it, let him accept it wholeheartedly and consistently. My own opinion is that this limit is much less likely to yield than, say, the law of conservation of energy. The energy law is essentially empirical, and may vanish overnight, as the law of conservation of mass did, but the restriction that prevents a man with resources of 10^{100} from carrying out a process that genuinely calls for more than this quantity rests on our basic ways of thinking about cause and effect, and is entirely independent of the particular material on which it shows itself.

If this view is right, systems theory must become based on methods of simplification, and will be founded, essentially, on the science of simplification. . . .

The science of simplification has, I think, been well started by the mathematicians in their studies of homomorphisms, but much remains to be done, and m my questions essential for applications have to be answered. . . . Can the me: ds of simplification be classified and studied systematically?

… The science of simplification clearly has its own techniques and its own sophistication. The systems theorist of the future, I suggest, must be an expert in how to simplify.

It follows from these observations that systems complexity is primarily studied for the purpose of developing sound methods by which systems that are incomprehensible or unmanageable can be simplified to an acceptable level of complexity. Such methods are crucial for dealing with phenomena of organized complexity [Weaver, **1948**].

By definition, systems with the characteristics of organized complexity are rich in factors that cannot easily be justified as negligible. And, by the same token, they are not sufficiently complex and random to yield meaningful statistical averages. That means that they are susceptible to neither of the two simplification strategies invented by science. And, yet, simplification is unavoidable in most instances. Even if a problem regarding a highly complex system can be successfully handled without any simplification by a computer, the solution must be eventually reduced to a level of complexity that is acceptable to the mind of, say, a decision maker who is in a position to utilize it. Since neither the Newtonian nor statistical simplification strategies are applicable, new avenues to the simplification of systems are needed.

When we simplify a system, we want to reduce both the complexity based on descriptive information and the complexity based on the uncertainty information. Unfortunately, these two complexities conflict with each other. In general, when we reduce one, the other increases or, at best, remains unchanged. Based on these considerations, a general problem of simplification can be formulated as follows.

9.1. Systems Simplification: A General Formulation

Given a system **S** of some type, let $\mathcal{A}_s$ denote the set of all its simplifications that are considered *admissible* in a given context. For example, $\mathcal{A}_s$ may be the set of simplifications of **S** that can be obtained by a particular simplification method. Let $\stackrel{d}{\leq}$ and $\stackrel{u}{\leq}$ denote, respectively, the two *preference orderings* on $\mathcal{A}_s$ that are based upon the descriptive complexity and the uncertainty-based complexity. These orderings are, in general, only weak orderings (i.e., reflexive and transitive relations on $\mathcal{A}_s$). The uncertainty ordering $\stackrel{u}{\leq}$ may, in fact, stand for several orderings based on different types of uncertainty. Let $\stackrel{\alpha}{\leq}, \stackrel{\beta}{\leq}, \ldots$ denote addi-additional (optional) orderings on $\mathcal{A}_s$, which express special preferences specified by the user of the system. For all the individual preference orderings, we define a *joint preference ordering*, $\stackrel{*}{\leq}$, as follows: for all $\mathbf{X}, \mathbf{Y} \in \mathcal{A}_s$, $\mathbf{X} \stackrel{*}{\leq} \mathbf{Y}$ if and only if $\mathbf{X} \stackrel{d}{\leq} \mathbf{Y}, \mathbf{X} \stackrel{u}{\leq} \mathbf{Y}, \mathbf{X} \stackrel{\alpha}{\leq} \mathbf{Y}, \mathbf{X} \stackrel{\beta}{\leq} \mathbf{Y}$, etc. When the simplification problem is solved, we obtain a *solution set*, $\mathcal{S}_s$, which consists of those admissible simplifications in

$\mathcal{A}_s$ that are either equivalent or incomparable in terms of the joint preference ordering $\overset{*}{\leq}$. Formally, we have

$$\mathcal{S}_s = \{\mathbf{X} \in \mathcal{A}_s \mid \text{for all } \mathbf{Y} \in \mathcal{A}_s, \mathbf{Y} \overset{*}{\leq} \mathbf{S} \text{ implies } \mathbf{X} \overset{*}{\leq} \mathbf{Y}\}.$$

Observe that this general formulation of the simplification problem is analogous to the method identification problem formulated in Chap. 6.

There are two fundamental variations of the simplification problem. In one of them, it is required that the descriptive complexity be reduced to a specified level. Let $\mathcal{A}_{s,d}$ denote the subset of $\mathcal{A}_s$ that contains only systems whose descriptive complexity is d, and let $\overset{*}{\leq}_d$ denote a joint preference ordering on $\mathcal{A}_{s,d}$, which involves only the uncertainty ordering $\overset{u}{\leq}$ and, possibly, some optional orderings $\overset{\alpha}{\leq}$, $\overset{\beta}{\leq}$, etc. The solution set, $\mathcal{S}_{s,d}$, consists of those systems in $\mathcal{A}_{s,d}$ that are either equivalent or incomparable in terms of $\overset{*}{\leq}_d$. Formally, we have

$$\mathcal{S}_{s,d} = \{\mathbf{X}_d \in \mathcal{A}_{s,d} | \text{for all } \mathbf{Y}_d \in \mathcal{A}_{s,d}, \mathbf{Y}_d \overset{*}{\leq}_d \mathbf{X}_d \text{ implies } \mathbf{X}_d \overset{*}{\leq}_d \mathbf{Y}_d\}.$$

In the second variation of the simplification problem, the requirement is inverted: it is required that the uncertainty-based complexity be reduced to a specified level. Let $\mathcal{A}_{s,u}$ denote in this case the subset of $\mathcal{A}_s$ that contains only systems whose total uncertainty (expressed, if necessary, by some aggregation of uncertainties of distinct types) is u, and let $\overset{*}{\leq}_u$ denote a joint preference ordering on $\mathcal{A}_{s,u}$, which involves only the descriptive complexity ordering $\overset{d}{\leq}$ and, possibly, some optional orderings $\overset{\alpha}{\leq}$, $\overset{\beta}{\leq}$, etc. The solution set, $\mathcal{S}_{s,u}$, is formally defined as

$$\mathcal{S}_{s,u} = \{\mathbf{X}_u \in \mathcal{A}_{s,u} | \text{for all } \mathbf{Y}_u \in \mathcal{A}_{s,u}, \mathbf{Y}_u \overset{*}{\leq}_u \mathbf{X}_u \text{ implies } \mathbf{X}_u \overset{*}{\leq} \mathbf{Y}_u\}.$$

All simplification strategies can be formulated as special cases of these general formulations. They differ from one another by:

- The type of the system to be simplified;
- The set of simplifications of the given system that are declared as admissible;
- The implied meanings of the two complexity orderings;
- The nature of additional (optional) preference orderings.

It is clear from these considerations that the general formulation subsumes a large class of special simplification problems. Let us look at some important categories of these problems.

9.2. Special Simplification Strategies

One way of reducing the descriptive complexity of a system of any type is to exclude some variables from the system. *Excluding variables* from any relation reduces the relation in two ways. First, its dimension is reduced since some sets in

Simplification Strategies

its Cartesian product are excluded. Second, its cardinality is reduced since overall states that were distinguished only by the excluded variables become equivalent.

When our aim is to reduce uncertainty-based complexity, we have to involve an inverse procedure. That is, we add some variables to the system, usually some input or generating variables (Chap. 4). Each added input or generating variable contributes, at least potentially, some information that, in turn, reduces the uncertainty regarding the output or generated variables.

Another way of reducing descriptive complexity of a system is to partition states of some variables of the system into equivalence classes and replace each equivalence class with one state. This simplification strategy is usually referred to as *coarsening* or *quantizing* of variables; it reduces cardinalities of relations involved, but it leaves their dimensions intact. While descriptive complexity is always reduced by coarsening of variables, uncertainty-based complexity may be affected by coarsening of variables in either way.

An important strategy for making very complex systems manageable is to break them down into appropriate subsystems. One aspect of systems manageability is expressed in terms of the size of computer memory required to store the system; this may in some cases be adopted as the measure of descriptive complexity. Consider, for example, n variables, each of which has k states. When dealing with the overall system of these variables, nk^n memory cells, each of which can store any one of k states, must be made available for storing states of the system. On the other hand, when a structure system consisting of all subsystems with two variables is used, the number of memory cells that are needed for the same purpose is $k^2 n(n-1)$. This number grows with increasing values of k and n at a considerably lower rate than that of nk^n, as illustrated in Fig. 9.1 for $k = 10$. If the structure system contained only some of the two-variable subsystems, the comparison would be even more favorable. Consider, for example, the overall ecological system with seven five-state variables specified in Table 4.1 and a seven-dimensional relation among the variables, which can adequately be represented by a structure system with five three-dimensional relations specified by the block diagram in Fig. 4.6. In this case, $7 \times 5^7 = 546{,}875$ memory cells, each of which can store any one of the five states, is needed for the overall system, but only $5 \times 3 \times 5^3 = 1{,}875$ memory cells are needed for the structure system. Although for some small values of n and k, structure systems may require more memory space than the corresponding overall system, it is clear that their memory requirements are far less demanding in most cases of practical significance, especially for large values of k and n.

Another aspect of systems manageability is connected with the number of possible systems that must be distinguished in some problems. Consider again n variables, each with k states and assume that the constraint among the variables is expressed solely by listing the actual overall states. Then, there are 2^{k^n} distinct overall systems of this kind. If the constraint among the same variables can be

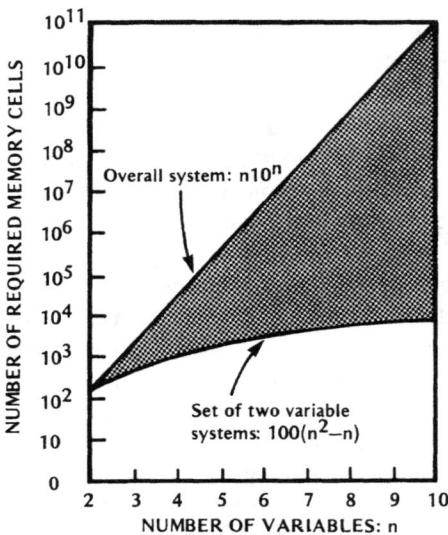

Figure 9.1. Comparison of memory requirements for an overall system and the associated structure system with two variable subsystems ($k = 10$).

expressed in terms of a structure system that consists of all two-variable subsystems, the number of all such structure systems is

$$2^{k^2-1} n(n-1).$$

To identify one particular element of a finite set X requires $\log_2 |X|$ bits of information, where $|X|$ denotes the number of elements in X. Hence, to identify one overall system requires in our case k^n bits, while to identify one structure system requires $(k^2-1)+\log_2 (n^2-n)$ bits of information. These information demands are compared for $k = 2$ and $2 \leq n \leq 12$ in Table 9.1. It is obvious that the rapid growth of the difference between these two information demands with increasing n is even more pronounced for larger values of k.

The descriptive complexity of a system can thus be reduced by breaking the system into its appropriate subsystems; this is a general principle, which is

Table 9.1. Information Demands in Identifying a Single System with n Variables, Each with Two States

n	2	3	4	5	6	7	8	9	10	11	12
Overall system	4	8	16	32	64	128	256	512	1024	2048	4096
Structure system	4	5.6	6.6	7.3	7.9	8.4	8.8	9.2	9.5	9.8	10

independent of the adopted measure of descriptive complexity. On the other hand, the uncertainty complexity increases or, at best, remains the same when the system is replaced by its subsystems. This means that simplifying systems by breaking them into subsystems is a special case of the general simplification problem formulated earlier. This special, but very important, simplification strategy has been extensively investigated since the mid-1970s within the context of *reconstructability analysis*—a set of computer-aided methodological tools for dealing with the relationship between wholes and parts, i.e., between overall systems and their various subsystems [Klir 1985a, **1986**].

Conceptualizing systems as structure systems, possibly of higher orders, is certainly an efficient way of managing complexity. Such systems are organized hierarchically: each system consists of a network of interconnected subsystems, each of which consists again of a network of its own subsystems, and so on, until some ultimate subsystems are reached that are not further divided into more primitive subsystems.

For a long time, the power of organizing systems hierarchically has been recognized and utilized with great success in the sciences of the artificial. This organizing principle has undoubtedly been one of the basic tools of good designers, artists, and managers. It has also played a key role in the process of developing efficient mass production by allowing the division of labor in manufacturing complex products.

The significance of hierarchically organized systems has also been recognized for a long time in the sciences of the natural. It has repeatedly been observed that virtually all complex systems that we recognize in the real world (that is models of the real world) have the tendency to organize hierarchically. Thus, for example, biological cells seem to group naturally into organs, while organs group into organisms, organisms group into populations of animals, and the latter group into ecosystems. The fact that we tend to perceive the world as hierarchically organized might have some ontological significance, but it may as well be solely of epistemological nature, reflecting the way in which the human brain and mind have evolved to deal with the complexity of the real world. Regardless of its ontological significance, it is undeniable that hierarchically organized systems play an important pragmatic role in our comprehension and management of reality, be it natural or man-made. The same point is well expressed by Simon [**1962**]:

> If there are important systems in the world that are complex without being hierarchic, they may to a considerable extent escape our observation and our understanding. Analysis of their behavior would involve such detailed knowledge and calculation of the interactions of their elementary parts that it would be beyond our capacities of memory or computation.

Although the importance of multilevel hierarchical systems (structure systems of higher orders) has been recognized for a long time by philosophers, scientists, engineers, managers, and artists, the first attempt to develop mathematical frameworks for conceptualizing such systems appeared only in the early 1970s [Mesarovic, Macko, and Takahara, 1970]. These frameworks emerged from mathematical system theory and were largely motivated by problems from engineering and management. It took more than a decade before these mathematical frameworks were applied to various specific contexts in the sciences of the natural and appropriate theories of multilevel hierarchical systems in these contexts were developed [Auger, 1989].

Hierarchical organizations are important not only for reducing descriptive complexity of systems, but they play also a significant role in reducing computational complexity of some systems problems. For example, many systems problems require a search through many hypotheses that are ordered by relevant preference orderings. One class of such problems are, in fact, the simplification problems formulated earlier in this chapter. If we can find a meaningful property by which the set of hypotheses can be partitioned in such a way that the preference ordering is preserved, than the search can be performed in two stages. In the first stage, we search only through equivalence classes of hypotheses, using an appropriate canonical representation of each equivalence class. After a desirable equivalence class is determined, we then search through the class to determine the actual members of the solution set. Since the number of equivalence classes is usually substantially smaller than the total number of hypotheses, this strategy of two-stage search appreciably reduces computational demands. If the partitioning of hypotheses can be arranged hierarchically on several levels, the reduction of computational complexity may be phenomenal.

Another way of dealing with very complex systems that possess the characteristics of organized complexity, perhaps the most significant one, is to allow imprecision in describing properly aggregated data. Here, the imprecision is not of a statistical nature, but rather of a more general modality, even though the possibility of imprecise statistical descriptions is included as well. The mathematical frameworks for this new modality are, as already mentioned, the theories of fuzzy sets and fuzzy measures. Again, this simplification strategy is not applicable only to reducing descriptive complexity of systems, but also to reducing computational complexity of systems problems.

CHAPTER 10

Goal-Oriented Systems

> . . . all systems are adaptive, and the real question is what they are
> adaptive to and to what extent.
> —LOTFI A. ZADEH

Literature dealing with various issues that emanate from recognized categories of goal-oriented systems is voluminous and growing rapidly. The subject of *goal-orientation* does not always appear in the literature under this general and neutral term. More frequently, it is discussed under other names, which designate special types of goal orientation. Typical examples are: regulation, control, self-organization, learning, autopoiesis, self-reproduction, self-correction, adaptation, evolution. No attempt is made in this chapter to cover this broad subject comprehensively since each of the special types of goal-orientation alone could easily occupy a whole book. Instead, the focus here is on a few key concepts and issues pertaining to goal-oriented systems.

The most fundamental concepts for conceptualizing goal-oriented systems are the concepts of a goal and a performance. Defining these concepts is a prerequisite for defining goal-oriented systems.

The concept of a *goal* of a system can be defined in many different ways. According to a general view I take, which encompasses virtually all special views, the goal of a system is "in the eyes of a cognitive agent" (observer, investigator, user, designer). That is, given a system of some type, a goal is defined in terms of some specific restriction of the systemhood properties that a cognitive agent dealing with the system considers desirable under given circumstances. Some examples of desirable goals are: keeping an output variable of a system within a specific and usually small range of values (point regulation); restricting the state transitions of a system to a specific cycle of states (path regulation); keeping a specific external behavior of a structure system invariant under some changes (malfunctions) in its elements (self-correction); acquiring in an autonomous way (through the regular operation of a system, with no specific interferences from outside) a particular spatial, temporal, or functional relationship (self-organization).

A system may thus be viewed from the standpoint of different goals. It satisfies each of them to some degree. This degree, which is called the *performance of the system with respect to the goal*, should measure (in some reasonable manner) the closeness between the actual and desirable manifestations of those systemhood properties that are involved in the goal. It is often expressed in terms of an appropriate function, which is called a *performance function*.

As a simple example, consider a generative system with one input variable, x, and one output variable, y, both of which assume two states, 0 and 1. Assume that the system is supposed to serve as a random generator with the following desirable conditional probabilities:

$$p(0|0) = 0.5, p(1|0) = 0.5,$$
$$p(0|1) = 0.9, p(1|1) = 0.1.$$

From the standpoint of a designer or a user of the system, these probabilities represent the goal. The behavior of the actual system, when designed and constructed, may be observed and analyzed. Assume that the estimated actual probabilities are

$$p_a(0|0) = 0.4, \quad p_a(1|0) = 0.6,$$
$$p_a(0|1) = 0.85, \quad p_a(1|1) = 0.15.$$

As a reasonable performance function, Perf, normalized to the interval [0,1], we may use the function

$$\text{Perf}(p, p_a) = 1 - \frac{d(p, p_a)}{D},$$

where

$$d(p, p_a) = \sum_{x,y} |p(y|x) - p_a(y|x)|$$

is a distance between the desirable probabilities, expressing the goal, and the actual probabilities, and D is the largest possible value of the distance. In our example, $d(p, p_a) = 0.3$, $D = 4$ (for an arbitrary goal), and

$$\text{Perf}(p, p_a) = 1 - \frac{0.3}{4} = 0.925.$$

Hence, the actual system satisfies the goal with the degree of 0.925 (or to 92.5%). If our goal were, for example, a deterministic behavior by which the input state is complemented at the output [$p(0|0) = p(1|1) = 0$ and $p(0|1) = p(1|0) = 1$], we would obtain $d(p, p_a) = 1.1$, $D = 4$, and $\text{Perf}(p, p_a) = 0.725$.

Observe that a given system satisfies different goals of the same type to different degrees and, similarly, a given goal is satisfied by different systems of the same type to different degrees. This indicates that goal-oriented systems may be defined in two distinct ways. One way is to consider goal-orientation as a fuzzy concept: given a goal and an associated performance function, every system

Goal-Oriented Systems

compatible with the goal is goal-oriented to a degree that is equal to the value of the performance function. Another way of defining goal-oriented systems is to choose some threshold value of performance (typically greater than 0.5) and consider a system as goal-oriented only if its performance with respect to the relevant goal exceeds the threshold.

A substantially different way of defining goal-oriented systems, which seems operationally more meaningful, is to view the notion of goal-orientation in relative terms. That is, one system is viewed as goal-oriented with respect to another system of the same type and a specified goal if it performs better (according to some performance function) with respect to the goal. Furthermore, the amount by which the performance of one system exceeds the performance of another system may be viewed as the degree of goal-orientation of the former system with respect to the latter one.

A system that has a positive degree of goal-orientation with respect to another system must contain some traits, other than those included in the latter system, that are responsible for its improved performance. Let us call these *goal-seeking traits*. Such traits are, for instance, some additional variables or states of variables, additional elements or couplings in structure systems, and the like.

To illustrate the general notion of goal-seeking traits, let me use an example. Consider a computer system in which the utilization of three expensive units is of particular interest. Three variables, v_1, v_2, v_3, are defined, one for each unit, by which activities of the units are described in time. Each of the variables has two states: 0, which indicates that the unit is not active at the time of observation, and 1, which indicates that the unit is active. The goal, g, is to keep all the units active all the time.

Suppose that an extensive hardware monitoring of the variables was performed and the probability distribution p given in Table 10.1a was estimated from the observed frequencies. How can we determine the performance of the system characterized by this probability distribution with respect to the goal? The most natural way is to use the expected number of active units (normalized to the unit interval) as a performance function:

$$\text{Perf}(g, p) = \tfrac{1}{3}[3p(111) + 2p(011) + 2p(101) + 2p(110) \\ + p(001) + p(010) + p(100)].$$

In our case, we obtain $\text{Perf}(g, p) = 0.62$.

Assume now that a new unit, say a communication channel, is added to the computer system in such a manner that it affects the activities of the three units under consideration. Assume further that the unit is relatively inexpensive, when compared with the other units, so that its own utilization is not important. The unit is introduced only for the purpose of enhancing the utilization of the other three units. The goal thus remains the same.

Let variable v_4 be defined for the new unit in the same way as the other

Table 10.1. Illustration of a Goal-Seeking Variable

(a)

v_1	v_2	v_3	p
0	0	1	0.15
0	1	0	0.20
1	0	0	0.10
1	1	0	0.25
1	1	1	0.30

(b)

v_1	v_2	v_3	v_4	p'
0	1	1	1	0.10
1	0	0	0	0.02
1	0	1	0	0.03
1	1	0	0	0.04
1	1	0	1	0.01
1	1	1	0	0.25
1	1	1	1	0.55

variables are defined for their units. Assume that hardware monitoring was performed again for the new, extended system, which includes variable v_4; it resulted in the probability distribution p' given in Table 10.1b. Using our performance function, we obtain Perf$(g, p') = 0.93$. We can see that the performance increased by 31%. This increase was caused by the new unit, which affects our system through variable v_4. This variable represents the goal-seeking trait of the extended system; it may be called a *goal-seeking variable*. The degree of goal-orientation of the extended system with respect to the original system is 0.31.

This example illustrates that a goal-seeking variable is a variable whose inclusion in the system contributes positively toward achieving the considered goal. An exploration of possible ways in which states of goal-seeking variables can be generated is important for comprehending the notion of goal-oriented systems and, especially, for developing methods by which desirable goal-oriented systems can be designed. Such an exploration leads necessarily to some specific types of structure systems. Each of them can be viewed as a paradigm that describes a principle (scheme, form) in terms of which states of goal-seeking variables are generated. Let these paradigms be called *paradigms of goal-oriented systems*.

The most primitive paradigms of goal-oriented systems are usually conceived as structure systems with two elements. One of the elements is a system in terms of which the goal is defined. This system models some phenomenon of the real world; it is usually called a *goal-implementing element*. The other element, which is called a *goal-seeking element*, is a system that generates states of a goal-seeking variable. By applying this variable as an additional input to the goal-implementing element, its performance with respect to the goal increases.

Block diagrams of four paradigms of goal-oriented systems based upon these two elements are displayed in Fig. 10.1, where **A** and **B** denote the goal-implementing and goal-seeking elements, respectively; **x** and **y** denote the input and output variables, respectively, in terms of which the goal is defined, and **z** denotes the goal-seeking variable. In general, each of the symbols **x**, **y**, **z** may stand for more than one variable.

Goal-Oriented Systems

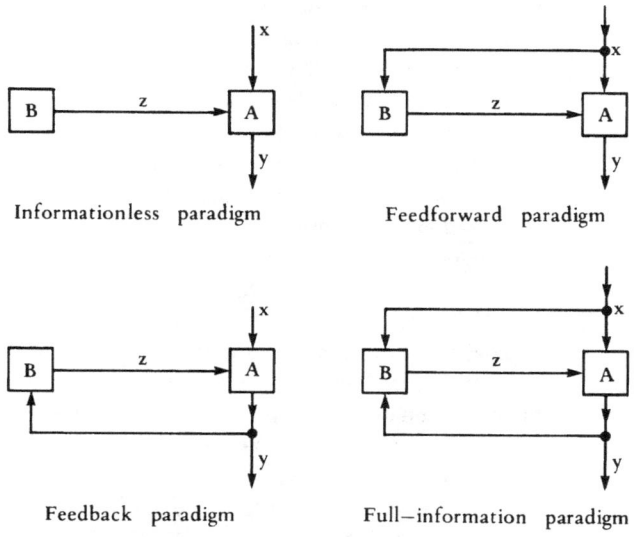

Figure 10.1. Simple paradigms of goal-oriented systems.

In systems inquiries, the paradigms of goal-oriented systems provide the investigator with a useful description, according to which some of the investigated variables are viewed as contributing toward a desired goal. The basic problem is to determine which of the variables under investigation exhibit a high degree of this goal-seeking capability.

In designing goal-oriented systems, on the other hand, each paradigm specifies a frame within which the designer is required to operate. In other words, each paradigm represents a set of assumptions (restrictions) regarding the system to be designed that the designer must not violate. The paradigms can be partially ordered by the severity of their restrictions. The more severe the restriction, the less freedom is left to the designer to perform his task and, consequently, the less general is the paradigm.

A typical problem of designing a goal-oriented system within one of the paradigms can be characterized as follows. Given a relationship between states of **x** and **y** (usually called an *essential relationship*), and given a desirable goal defined in terms of the variables involved, we want to design a system (element **B**) that is capable of influencing the relationship toward the goal. This can be accomplished only by identifying an appropriate variable whose states exert some influence upon the essential relationship. Our task is to design a system that produces states of this variable in such a way that they actually influence the relationship toward the desirable goal. This requires that an adequate model of the ternary relationship among **x**, **y**, **z** be determined first, in which **x** and **z** are input variables and **y** is an output variable. This model is then employed for specifying behavior of the goal-seeking system (element **B**) in such a way that the performance of achieving the goal (defined in some appropriate way) is maximized. In this specification, **x** and **y** are input variables and **z** denotes output variables.

We can see that each of the paradigms expressed in Fig. 10.1 restricts the design of the goal-seeking system in some way. The *informationless paradigm* is clearly the most restrictive and, consequently, the least powerful in achieving a high performance. The *feedback paradigm* is less restrictive since it allows the designer to utilize information about the output variable. Design methods based upon this paradigm are well developed, but they can achieve only a limited performance. This limitation is basically caused by the fact that the feedback paradigm allows the goal-seeking element **B** only to react to any undesirable changes in states of the output variables **y**, but it does not provide a mechanism for anticipating them. Such a mechanism is available in the *feedforward paradigm*, which allows the designer to utilize anticipatory information about the input variable. If **A** is a good and deterministic model of the corresponding real phenomenon, the feedforward paradigm may allow the designer to achieve a higher performance than the feedback paradigm. If **A** is not a good model, however, the feedback paradigm may lead to a higher performance.

The feedback and feedforward paradigms are thus not comparable. Each of them is superior under certain circumstances and inferior under other circumstances. However, they are both inferior to the *full-information paradigm*, which allows the designer to utilize information about both **x** and **y**.

The following are the main issues involved in the design of goal-oriented systems within any of the four paradigms shown in Fig. 10.1:

1. The goal and performance function must be made explicit. If only the goal is given in the problem statement, it is up to the designer to choose an appropriate performance function. If only the performance function is given, it is normally assumed that the goal is represented by a behavior function (defined in terms of **x** and **y**) for which the performance function reaches its maximum.

2. A suitable goal-seeking variable (or, in general, a set of variables) must be selected. This variable cannot be arbitrary; it must exert some influence upon the essential relation (in the interpreted, physical sense) due to which the performance increases. The main difficulty in selecting a proper goal-seeking variable is that its influence upon the goal variables is usually not fully known. The designer must thus investigate this influence and develop appropriate models of the extended goal-implementing system **A** for different goal-seeking variables.

3. Once a goal-seeking variable is accepted and the extended goal-oriented system properly modeled, the next step is to determine some particular way in which the goal-seeking variable is generated. The objective is to determine a behavior function of the goal-seeking system **B** for which the performance function reaches its maximum within the constraints of the given paradigm. Various optimization methods can be employed for solving this problem. Their

Goal-Oriented Systems

choice depends, primarily, on the methodological type of the system involved as well as on the nature of the performance function.

4. The goal-seeking system **B** must be designed in terms of available elements. This is a standard problem of systems design.

10.1. Adaptive Systems

A system designed to be goal-oriented performs as expected only under the assumption that neither the essential relation of the goal-implementing element nor the goal changes. If this assumption is not satisfied, the performance may deteriorate or even fall below an acceptable threshold. In order to maintain a high level of performance despite the changes, the goal-seeking element must be capable of adaptation.

There are various reasons why the essential relation or the goal may change. In general, a change in the essential relation means that the goal-implementing system was not properly conceptualized during the process of design. For example, some essential input variables were not considered or some states of the considered variables were not conceived. Once the goal-seeking element **B** is designed and implemented for a particular goal-implementing element **A**, changes in **A** are not under the control of the user. Changes in the goal, on the other hand, are fully determined by the user. In general, he may consider it desirable to change the goal because the circumstances for which the goal-oriented system was designed changed.

In order to enable a goal-oriented system to adapt to changes in the goal, its goal-seeking element must be designed for a set of alternative goals and supplemented with a special input variable, **g**, whose states designate the goals. Let us call this kind of goal-oriented system, in which the adaptation is restricted to a specific set of goals, a *multigoal-oriented system*. A block diagram of its paradigm with full information is shown in Fig. 10.2, where g denotes the variable that specifies the goal. This *goal-designating variable* is controlled either by the user or by another system to which it is coupled.

Changes in the goal may be determined by a goal-generating system that is included as an element in the multigoal-oriented system itself. This possibility is illustrated by the block diagram in Fig. 10.3a. Elements **A** and **B** form a basic multigoal-oriented system (as in Fig. 10.2); element **C** is a *goal-generating element*. An overall goal is expressed in this case by a sequence of subgoals, which are specified by states of variable **g**. This sequence is determined by the goal-generating elements in terms of variables **x**, **y**, **z**. Let systems of this sort be called *autonomous multigoal-oriented systems*.

One obvious motivation for considering autonomous multigoal-oriented systems is to simplify the overall optimization problem involved in the design of a

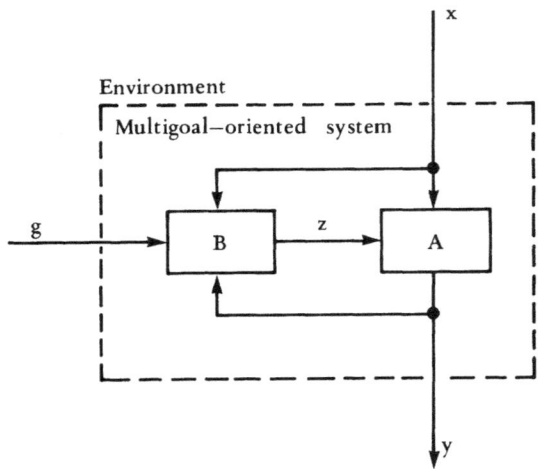

Figure 10.2. Full-information paradigm of multigoal-oriented system.

goal-seeking element. The optimization problem is decomposed into several simpler optimization problems, each of which is demarcated by specific conditions expressed in terms of the variables involved. From this point of view, it is more appropriate to consider the system as an ordinary goal-oriented system with a two-level goal-seeking element, as illustrated in Fig. 10.3b.

Observe that both block diagrams in Fig. 10.3 are the same as far as couplings between the three elements and environment are concerned. They differ solely in the way in which the elements are conceptually combined into larger elements. This subtle difference is an illustration of a basic phenomenon associated with systems of all kinds: the same system can be considered from different viewpoints when subjected to some operation of coarsening or refinement.

Goal-oriented systems with a *k-level goal-seeking element* ($k > 2$) can be defined recursively on the basis of the block diagram in Fig. 10.3b. Any such system represents a hierarchical decomposition of the overall goal into subgoals, and consists of k levels of decomposition.

Let us consider now goal-oriented systems that adapt to changes in the goal-implementing element **A**. The goal-seeking element of any such system must be able to perform the following two tasks:

1. To process data involving time series of states of variables **x**, **y**, **z** and update periodically the model of the goal-implementing relation;
2. To react to changes in the goal-implementing element by modifying the goal-seeking element in such a way that the performance is maximized within the constraints embedded in the system.

Goal-Oriented Systems

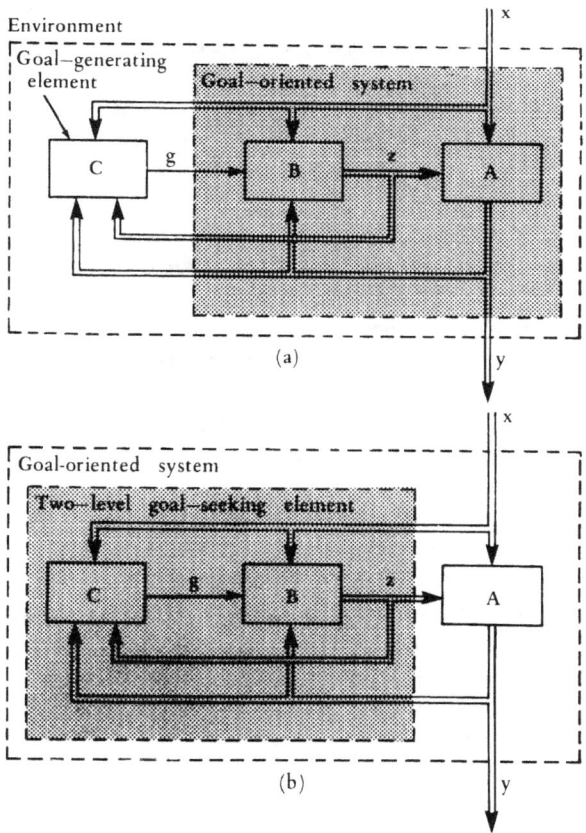

Figure 10.3. General structure paradigm of a goal-oriented system viewed either as (a) an autonomous multigoal-oriented system, or (b) a goal-oriented system with two-level goal-seeking element.

That is, the goal-seeking element must be capable of performing data-driven systems modeling of appropriate type and of dealing with a relevant type of optimization problem.

To illustrate the notion of goal-oriented systems that are adaptive to changes in the goal-implementing system, let me describe a particular adaptive system of this kind. The object on which the system is defined is a *computer* equipped with a mechanism that allows it to move within some *operating area*. For the sake of simplicity, let the operating area be a square area that is divided into smaller squares, called *cells*, by n rows and n columns in chessboard fashion. Let rows and columns be labeled by identifiers i and j, respectively ($i, j = 0, 1, \ldots n - 1$), and let the individual cells be labeled by a single identifier

$$k = ni + j.$$

Clearly, $k = 0, 1, \ldots n^2 - 1$.

The computer also has extensive *error-correcting capabilities*. During its execution of computing jobs, it is able to identify malfunctions in hardware, regardless of the kind of disturbances in the environment that caused them. It can also eliminate, within certain limits, the effect of malfunctions on the proper execution of the computing jobs. When the number of hardware malfunctions becomes too large, beyond the error-correcting capabilities of the computer, normal operation of the computer is threatened. It is thus desirable to counter the unknown disturbances by embedding the computer in an adaptive goal-oriented system whose goal is to minimize the number of hardware malfunctions in the long run. Since the computer can move, a natural way of countering external disturbances is to move selectively within the operating area according to some strategy designed to seek this goal.

A general block diagram of the proposed goal-oriented system is shown in Fig. 10.4. Two variables, which are observed in time, are involved at this level: m, a goal-implementing variable, which represents the number of malfunctions that occurred in the computer during each defined period of time; c, a goal-seeking variable, which specifies (controls) the location of the computer. Since the nature of disturbances and their variability is now known, the goal-implementing element **A**, in itself, can be viewed only as a source system. It accepts at each time a state of variable c and generates a state of variable m. The manner in which m (representing malfunctions) is generated is not known and may change.

The goal-seeking element **B** monitors variable m and produces a sequence of states of the goal-seeking variable c in a stochastic manner described later. States

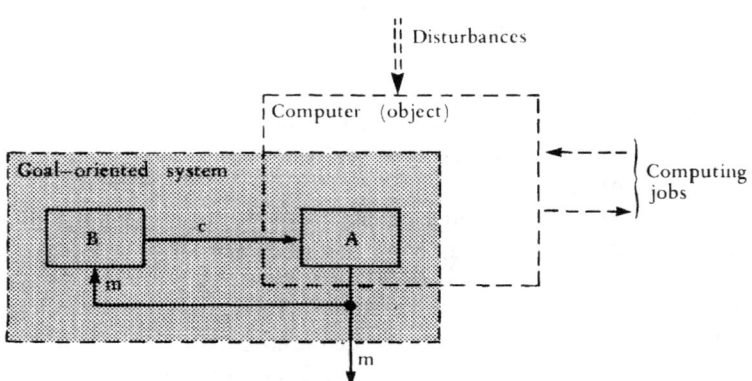

Figure 10.4. General block diagram of the adaptive goal-oriented system whose goal is to minimize defects that affect its normal functioning.

Goal-Oriented Systems

of variable c, which control the movement of the computer, represent the index k by which the individual cells of the operating area are identified; the initial state of c must identify the cell in which the computer is placed initially. At each discrete time $\Delta t, 2\Delta t, 3\Delta t, \ldots$, the computer may move from cell k only to one of the four cells that are adjoined to cell k, cells $k-1, k+1, k-n, k+n$, or may remain in cell k. That is, the computer does not move during each open time interval $(i\,\Delta t, (i+1)\Delta t)$, $i = 1, 2, \ldots$. It is assumed that the time needed to move from one cell to another is negligible with respect to the time interval Δt.

For any given present state of variable c, let c' denote the next state of c. When $c = k$ for some particular $k \in \{0, 1, \ldots, n^2-1\}$, then c' may assume any state k' in the neighborhood set

$$N_k = \{k, k-1, k+1, k-n, k+n\}$$

of cell k, with the exception of boundary cells, where one or two of the states in N_k do not exist. The actual next state, k', is chosen from the set N_k randomly, on the basis of conditional probabilities $p(k'|k)$, as illustrated in Fig. 10.5. These probabilities are determined by some procedure based upon the mean number of malfunctions observed in the relevant cells. For this purpose, a pair of values is stored in the goal-seeking element for each cell k: the number of previous visits of the computer to the cell, v_k, and the mean number of malfunctions observed in the cell, m_k. After each new visit to the cell, these values are replaced with new values, v'_k and m'_k, calculated by the formulas

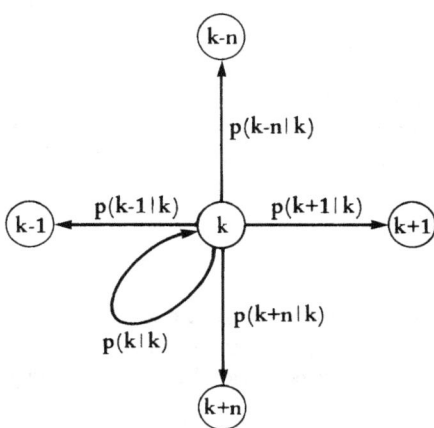

Figure 10.5. Conditional probabilities associated with a control decision in cell k of the discussed self-preserving system.

$$v'_k = v_k + 1,$$
$$m'_k = \frac{v_k m_k + a}{v_k + 1},$$

where a denotes the number of malfunctions that occurred during the visit (current value of variable m). Assume that m'_k is calculated to an accuracy, say 0.01.

Since the goal is to avoid disturbances that cause malfunctions in the computer, the probabilities $p(k'|k)$ should be made indirectly proportional, according to some rule, to the mean number of disturbances in the respective cells. However, to make the system adaptive to unexpected changes in the spatial distributions of disturbances, all of the states should be assigned nonzero probabilities. Hence, if m_k is too large so that $1/m_k < 0.01$, then $1/m_k$ is set to α, where $\alpha > 0$ is chosen constant; let $\alpha = 0.01$.

According to these requirements and conventions, probabilities $p(k'|k)$ may be determined in the following way. First, we determine

$$q_k = \begin{cases} 1/m_k & \text{if } m_k \geq 0.01 \\ 0.01 & \text{if } m_k < 0.01 \end{cases}$$

Then, we have

$$p(k'|k) = \frac{q_{k'}}{q_k + q_{k-1} + q_{k+1} + q_{k-n} + q_{k+n}}$$

for all $k' \in N_k$ and each particular $k \in \{0, 1, \ldots, n^2 - 1\}$. Obvious adjustments must be made in calculating the probabilities for the various boundary cells.

Before the described goal-oriented system is put into operation, its memory may be filled with any information about the spatial distribution of malfunctions that is available at that time. If no information is available, then states of all variables in the memory are set to zero. Once put into operation, the goal-oriented system controls the movements of the computer according to its model of the computer environment (operating area). The model is represented by the content of the memory of the goal-seeking element. It is periodically updated by the monitored variables m and c. The system acts according to its anticipation of the effect of the environment on the proper operation of the computer. Systems of this kind, which are able to develop a model of their environment and use it in an anticipatory manner, are the most sophisticated adaptive goal-oriented systems. They are usually referred to in the literature as *anticipatory systems*

Let us reflect now on three aspects of the described adaptive system. First, the hardware malfunctions can be divided into three categories:

1. The malfunctions that originate from within the computer hardware itself (hardware defects);

Goal-Oriented Systems

2. The malfunctions that are caused by disturbances distributed evenly within the operating area;
3. The malfunctions that are caused by local disturbances within the operating area.

It is obvious that the goal-seeking variable (location in the operating area) can influence only malfunctions in category 3.

Second, it should be emphasized that the system described reacts to any event that endangers its normal activity (correct execution of requested computing jobs). It does not require that the nature of such events be predetermined: everything that threatens its ability to operate normally evokes a reaction tending to preserve this ability. It is thus reasonable to call the system a *self-preserving system* (i.e., a system that attempts to preserve its ability to operate normally).

Third, the minimal value α of q_k, which is fixed for a particular system, has an important effect on the way in which the system adapts to changes in the environment. If α is too small compared to the normal values of q_k, as determined by the malfunctions in categories 2 and 3, the system is slow in recognizing changes in the environment. If α is too large, the system is fast in recognizing changes, but its model of the environment is underutilized when the changes are substantially slower than the rate at which the computer can move.

10.2. Special Types of Goal Orientation

The concept of goal-oriented systems, in general, and their structure paradigms, in particular, enjoy a broad range of applicability under various special types of goal orientation. One type, which has been studied quite extensively, is *regulation*. The goal of a regulator is usually to keep the output variables of the system constant in spite of disturbances, which are represented by the input variables. Many theoretical results regarding regulation, particularly for the feedback paradigm, can be found in a typical textbook on control theory. Let me only mention a general law pertaining to regulation, which is called a *law of requisite variety*. This law, which is due to Ashby (1956, **1958a**), states that the variety of the regulated (output) variable can be reduced to an acceptable level (which is the purpose of the regulator) only by a sufficient variety of the regulating (goal-seeking) variable. It is often paraphrased by the simple assertion: "only variety can destroy variety." Since variety, when conceived in probabilistic terms, can be measured by the Shannon entropy, a probabilistic version of the law of requisite variety is the inequality

$$H(Y) \geq H(X) - H(Z) + H(Z|X) - K \qquad (10.1)$$

where X, Y, Z are, respectively, state sets of variables x, y, z of a regulator. In the inequality, $H(Y)$ is the variety (uncertainty) of the output (regulated) variable y of the regulator. The purpose of the regulator is to reduce this variety as much as possible (ideally to zero) and, consequently, $H(Y)$ may be employed to measure the regulating performance. The other terms in the inequality have the following meaning: $H(X)$ is the variety of the input variable x, which is viewed in the context of regulation as a source of disturbances; $H(Z)$ is the variety of the regulating variable z; $H(Z|X)$ is the average uncertainty in selecting an optimal state of z, given a state of x (a disturbance); K expresses an inherent effect of z on the reduction of $H(Y)$ independently of information about x, as in the information-free paradigm [if there is no information transmission from x to z, then $H(Z|X) = H(Z)$ by Eq. (8.10) and the right-hand side in (10.1) becomes $H(X) - K$]. We can see that, according to the law of requisite variety (10.1), the variety of disturbances can be countered effectively by selecting a regulating variable whose variety, the K-factor, and information transmission with the input variable are as high as possible.

Another type of goal orientation is *decision making*. Elements **A** and **B** of the goal-oriented structure paradigms become decision-implementing and decision-making elements, respectively. States of input variables represent so-called "states of nature" (e.g., relevant external circumstances, possible moves of an opponent, recognized characteristics of some sort, and the like). States of the output variables represent outcomes, on which a *utility function* is defined. The goal of the system is to maximize the utility function. The role of the goal-seeking variables is to make selections from a set of decision alternatives that affect the outcomes in a positive way with respect to the goal. Given this role, they may be called, e.g., decision-making, selection-making, or utility-seeking variables. In this way, a theory of decision making may be conceived in terms of the various paradigms of goal-oriented systems.

More sophisticated goal-oriented systems, whose importance have been steadily growing since the early 1960s, are systems that are capable, with no explicit outside help, of improving their performance while pursuing their goals. These systems are usually called *self-organizing systems*. As the name suggests, a self-organizing system is a system that tends to improve its performance in the course of time by making its elements better organized for achieving the goal. This formulation includes the special case in which the goal is to achieve a high degree of organization (order) of relevant entities from low degree of their organization (disorder, chaos). In this case, the "goodness" of organization is viewed as its *degree of orderliness* defined in some pertinent manner. Consider, for example, a generative system with an input state set, X, an output state set, Y, and a relationship between these sets characterized by conditional probabilities $p(y|x)$ for all $x \in X$ and all $y \in Y$. In this simple example, it seems most natural to associate

Goal-Oriented Systems

the degree of orderliness with the degree of predictability, and define it, in terms of the conditional Shannon entropy $H(Y|X)$, by the performance function

$$\text{Perf}(H) = 1 - \frac{H(Y|X)}{\log_2 |Y|}$$

where $|Y|$ denotes the number of elements in set Y and $\log_2 |Y|$ is the largest possible value of $H(Y|X)$.

Haken [1977] coined the name *synergetics* (from Greek *synergos*, which means *working together*) for the study of processes by which organizations on various levels (physical, chemical, biological, social) develop spontaneously through synergetic (cooperative) interactions of many elements, usually of the same kind or of a few different kinds. In justifying this new field of study, he says:

> The spontaneous formation of well organized structures out of germs or even out of chaos is one of the most fascinating phenomena and most challenging problems scientists are confronted with. Such phenomena are an experience of our daily life when we observe the growth of plants and animals. Thinking of much larger time scales, scientists are led into the problems of evolution, and, ultimately, of the origin of living matter. When we try to explain or understand in some sense these extremely complex biological phenomena it is a natural question, whether processes of self-organization may be found in much simpler systems of the unanimated world.
>
> In recent years it has become more and more evident that there exist numerous examples in physical and chemical systems where well organized spatial, temporal, or spatio-temporal structures arise out of chaotic states. Furthermore, as in living organisms, the functioning of these systems can be maintained only by a flux of energy (and matter) through them. In contrast to man-made machines, which are devised to exhibit special structures and functionings, these structures develop spontaneously—they are self-organizing. It came as a surprise to many scientists that numerous such systems show striking similarities in their behavior when passing from the disordered to the ordered state. This strongly indicates that the functioning of such systems obeys the same basic principles.

Although self-organizing processes involved in some phenomena observed in physics, chemistry, and biology are now well understood, other phenomena (for example the fascinating osmotic growths produced at the turn of this century by Leduc [1911]) represent still a considerable challenge to systems science [Klir, Hufford, and Zeleny, 1988].

There are obviously many different types of self-organization, depending on the category of systems within which it is conceptualized, the goal, the performance function, the principle by which the pertinent organization is properly altered in the process of pursuing the goal, and, perhaps, other factors. Rather than to try to overview these various types, I consider it more meaningful to illustrate the basic idea of self-organization by describing one type of self-organizing systems in more detail.

Self-organizing systems chosen here to exemplify the idea of self-organization are called *autopoietic systems*. The term "autopoiesis" is of Greek origin and literally means "self-production." However, the self-production involved in autopoietic systems is not arbitrary, but must satisfy certain requirements.

In general, autopoietic systems function within a finite and discrete space-time support set. The space is usually two dimensional or three dimensional, but a k-dimensional space ($k \geq 1$, finite) can also be used. An overall characterization of autopoietic systems is that they form and maintain in time a spatially distinguished unit by a set of movement and production rules defined for entities of several different types distributed within the space.

Autopoietic systems are thus goal oriented. The goal is some kind of a boundary, usually called a *topological boundary*, that allows the observer to recognize a part of the space as a unit. Some of the movement or production rules that are essential for achieving this goal in a particular autopoietic system may thus be viewed as goal-seeking traits of the system.

The idea of autopoietic systems originated in biology, where it is exemplified by a great variety of instances. Some of the simplest biological objects whose formation and maintenance as spatial units can be well characterized in terms of autopoietic systems are biological cells, as well depicted by Milan Zeleny (1981):

> We observe self-production phenomena intuitively in living systems. The cell, for example, is a complex production system, producing and synthesizing macromolecules of proteins, lipids, and enzymes, among others; it consists of about 10^5 macromolecules on the average. The entire macromolecular population of a given cell is renewed about 10^4 times during its lifetime. Throughout this staggering turnover of matter, the cell maintains its distinctiveness, cohesiveness, and relative autonomy. It produces myriads of components, yet it does not produce only something else—*it produces itself*. A cell maintains its identity and distinctiveness even though it incorporates at least 10^9 different constitutive molecules during its life span. This maintenance of unity and wholeness while the components themselves are being continuously or periodically disassembled and rebuilt, created and decimated, produced and consumed, is called "autopoiesis."

An autopoietic system is usually described in terms of components of certain types which are often given such suggestive names as "substrates," "catalysts," "holes," "links," and the like. At every instant of the defined time set, each recognized location of the defined space is occupied by exactly one component of a particular type. The components undergo spatial transformations in time according to specific rules (movement and production rules). If these rules are chosen

Goal-Oriented Systems

properly, the system is capable of forming and maintaining a topological boundary of some kind (its goal) and may be thus considered as an autopoietic system.

One way of conceptualizing an autopoietic system is to view it as a metasystem that produces, for each initial data system, a series of data systems according to some replacement procedure. The data systems are all defined in terms of the same spatial support and the same variable whose states may have some real-world interpretation. What matters, however, is not what the states mean, but how they participate in the replacement procedure.

To be more specific, let me describe a simple autopoietic system that is discussed in more detail in a paper by Varela, Maturana, and Uribe (**1974**), where the idea of autopoietic systems was first conceived. The variable of the system consists of five states, which are given the following labels and suggestive names: 0, hole; 1, catalyst; 2, substrate; 3, link; 4, bonded link. The support of the variable is a two-dimensional spatial grid with ten rows and ten columns. In each of the 100 cells of the grid, the state of the variable is interpreted as a label of the element that is located in the cell (with the exception of state 0, which indicates that no element is placed in the cell). The replacement procedure is defined by the following rules:

1. Two neighboring substrates either of which is in the neighborhood of a catalyst are joined to form a link.
2. Neighboring links are joined to form bonded links. (A closed bonded link constitutes a boundary the maintenance of which is the goal of the autopoietic metasystem.)
3. Randomly selected links, whether free or bonded, disintegrate, yielding two substrates or two links, respectively. (The resulting components may later rebond.)
4. Substrates may move randomly into any neighboring empty space, passing through a single chain of bonded links if necessary. Links may move into empty spaces and may also displace substrates, either pushing them into adjacent holes or trading positions with them. Catalysts have all the freedom of movement of links, and may displace them. However, unlike substrates, neither links nor catalysts may pass through bonded-link segments. Bonded links do not move.

The procedure begins with an initial data matrix, whose entries are required to contain either substrates or catalysts (states 1 or 2). Applying the rules of the replacement procedure, this initial matrix is replaced with a new matrix and that, in turn, is again replaced with another matrix, and so on. The result is a series of data matrices. The system is considered autopoietic if a closed space defined by bonded links (states 4), which is called a topological boundary, is eventually formed in the spatial grid.

A particular example of forming the topological boundary is shown in Fig.

2	2	2	2	2	2	2	2	2	2
2	2	2	2	2	2	2	2	2	2
2	2	2	2	2	2	2	2	2	2
2	2	2	2	2	2	2	2	2	2
2	2	2	2	2	2	2	2	2	2
2	2	2	2	1	2	2	2	2	2
2	2	2	2	2	2	2	2	2	2
2	2	2	2	2	2	2	2	2	2
2	2	2	2	2	2	2	2	2	2
2	2	2	2	2	2	2	2	2	2

t = 0

2	2	2	2	2	2	2	2	2	2
2	2	2	2	2	2	2	2	2	2
2	2	3	3	3	2	2	2	2	2
2	2	3	0	0	0	3	2	2	2
2	2	3	0	1	0	2	2	2	2
2	2	2	0	0	0	2	2	2	2
2	2	2	2	2	2	0	2	2	2
2	2	2	2	2	2	2	2	2	2
2	2	2	2	2	2	2	2	2	2
2	2	2	2	2	2	2	2	2	2

t = 1

2	2	3	2	2	2	2	2	2	2
2	0	2	2	4	2	2	2	2	2
2	2	3	2	4	4	4	2	2	2
2	2	4	0	3	0	3	2	2	2
2	2	4	0	1	0	4	2	2	2
2	2	4	4	4	4	4	2	2	2
2	2	2	2	2	2	3	2	2	2
2	2	2	2	2	2	2	2	2	2
2	2	2	2	2	2	2	2	2	2
2	2	2	2	2	2	2	2	2	2

t = 2

2	2	3	2	2	2	2	2	2	2
2	0	2	2	4	2	2	2	2	2
2	2	3	2	4	4	4	2	2	2
2	2	4	0	3	0	3	2	2	2
2	2	4	0	1	0	4	2	2	2
2	2	4	4	4	4	4	2	2	2
2	2	2	2	2	2	3	2	2	2
2	2	2	2	2	2	2	2	2	2
2	2	2	2	2	2	2	2	2	2
2	2	2	2	2	2	2	2	2	2

t = 3

2	2	3	2	2	2	2	2	2	0
2	2	2	2	4	2	2	2	2	2
2	2	4	3	4	4	4	2	2	2
2	2	4	0	3	0	4	2	2	2
2	2	4	2	1	3	4	2	2	2
2	2	4	4	4	4	4	2	2	2
2	2	2	2	2	2	0	2	2	2
2	2	2	2	2	2	2	2	2	2
2	2	2	2	2	2	2	2	2	2
2	2	2	2	2	2	2	2	2	2

t = 4

2	2	2	2	0	2	2	2	2	2
2	2	3	4	4	2	2	2	2	2
2	2	4	0	4	4	4	2	2	2
2	2	4	0	3	2	4	2	2	2
2	2	4	3	1	3	4	2	2	2
2	2	4	4	4	4	4	2	2	2
2	2	2	2	2	2	3	2	2	2
2	2	2	2	2	2	2	2	2	2
2	2	2	2	2	2	2	2	2	2
2	2	2	2	2	2	2	2	2	2

t = 5

2	2	2	2	2	0	2	2	2	2
2	2	4	4	4	2	2	2	2	2
2	2	4	2	4	4	4	2	2	2
2	2	4	3	3	2	4	2	2	2
2	2	4	0	1	3	4	2	2	2
2	2	4	4	4	4	4	2	2	2
2	2	2	2	2	2	3	2	2	2
2	2	2	2	2	2	2	2	2	2
2	2	2	2	2	2	2	2	2	2
2	2	0	2	2	2	2	2	2	2

t = 6

Figure 10.6. Illustration of the formation of a topological boundary within an autopoietic system.

Goal-Oriented Systems 161

10.6. We can see that the boundary begins to emerge in this case at $t = 2$ and is completed at $t = 6$.

The literature regarding the various types of goal-oriented systems is very extensive, and its comprehensive overview is beyond the scope of this book.*

---

*A classic work on adaptive systems is a monograph by Holland [1975]. Properties of adaptive systems are thoroughly examined by Gaines [1972]. Methods for designing adaptive control systems are well overviewed in a book by Landau [1979]. Conrad [1983] shows that adaptability is essential for the life process at all levels, from the molecular and cellular levels to the ecological levels, and develops a general framework for dealing with adaptation at these various levels. Ideas regarding anticipatory systems have been pursued primarily by Rosen [**1979b**, 1985b]. Although regulation has been studied extensively, general principles of regulators are primarily due to Ashby and Conant [Ashby, 1956; Conant, 1969; Conant and Ashby, **1970**]. Although it is quite natural to view decision-making situations as goal-oriented systems, this view has rarely been pursued in the literature on decision making. One of these rare exceptions is a book by Kickert [1980]. The possibility of self-organizing systems was first demonstrated by Farley and Clark [1954], when they published their successful results of computer simulation of a self-organizing system. After this publication, interest in self-organizing systems increased rapidly, resulting in three edited volumes devoted to the subject in the early 1960s: Yovits and Cameron [1960], Foerster and Zopf [1962], Yovits, Jacobi and Goldstein [1962]. As well summarized by Ashby [**1962a**], basic principles of self-organizing systems were already known at that time. Except for some engineering applications [Ivakhnenko, 1970], research on self-organizing systems became considerably less visible after the active period in the early 1960s. It reappeared with full strength only in the late 1970s and especially in the 1980s. The following books are representative of this new wave of interest in self-organizing systems: Nicolis and Prigogine [1977], Prigogine [1980], Jantsch [1980], Farlow [1984], Krinsky [1984], Yates [1987], Haken [1988], and Dalenoort [1989]. Autopoietic systems, which belong to the general class of self-organizing systems, have lately become an independent subject of research. The idea of autopoiesis was proposed in a classic paper by Varela, Maturana, and Uribe [**1974**]. Various aspects of autopoietic systems are well covered in four books: Varela [1979], Zeleny [1980, 1981], and Maturana and Varela [1980]. The idea of self-reproducing systems, which also belong to the broader category of goal-oriented systems, is due to John Von Neumann. He presumably formulated the idea in the early 1950s, but it was published only posthumously [Neumann, 1966]. A good discussion of the notion of self-reproduction was published by Ashby [**1962b**].

CHAPTER 11

Systems Science in Retrospect and Prospect

Knowledge grows by accretion, but we gain truth by pruning the tree of knowledge.
—KENNETH E. BOULDING

No historical reflection upon systems science and its impact on other areas of human endeavor can be definitive at this time since systems science is currently still in the process of forming. It is by far not established as yet to a degree comparable with traditional disciplines of science such as, e.g., physics, chemistry, psychology, or economics. One of the difficulties in examining systems science in this formative stage is the lack of unified terminology. Thus, the very notion of *systems science*, as conceived in this book, is often discussed in the literature under the names *systems research* or *systems theory*, sometimes with the adjective *general*.

Since the emergence of systems movement in the 1940s and 1950s, the progress in forming systems science as a new field of inquiry has been quite impressive. During this period, various aspects of the developing systems science has been subjected to criticism. Some of this criticism was justified, focusing mainly on exaggerated claims of certain systems ideas, and influenced the formation of systems science in a positive way. Other criticism was ill-conceived, sometimes quite vicious, and has been refuted. Let me overview some of the main arguments involved.

11.1. Criticism

One of the earliest criticisms of general systems theory, pursued vigorously by R. C. Buck, a mathematician, was concerned with the concept of a system. It focused on the following definition proposed by James Miller [1953]:

> A system is a bounded region in space-time, in which the component parts are associated in functional relationship.

The essence of Buck's criticism is summarized in this quote [Buck, 1956]:

> One is unable to think of anything, or any combination of things, which could not be regarded as a system. And, of course, a concept that applies to everything is logically empty.

This claim of logical emptiness of general systems theory was later echoed by other critics. Although it seemed to present general systems theory with a serious challenge, the claim is not substainable upon closer scrutiny. First, it applies only to a common-sense conception of systems. That is, it does not apply, for example, to the GSPS framework (outlined in Chap. 4), within which a potentially infinite number of epistemologically distinct categories of systems are defined. Moreover, Miller's definition, by its reference to "a bounded region in space-time," is not in line with the modern, constructivist view, which is becoming predominant in systems science. It seems that it characterizes an object of the real world, in which a functional relationship among components is assumed on ontological grounds, rather than a system constructed on the object on pragmatic grounds.

Regardless of these broader considerations, Buck's claim can be refuted even in the specific context of the common-sense definition of systems. This was achieved, for example, by J. H. Marchal through his careful analysis of the common-sense definition of the concept of a system [Marchal, 1974]. Let me use a relevant quote from his paper, where the term "the definition" stands for "S is a system only if S is a set of elements and relations between the elements":

> Turning to Buck's objection we see that it is really twofold: his claim is that we are unable to think of (a) anything or (b) any combination of things which cannot be regarded as a system. With the definition in hand and the twofold nature of Buck's objection made explicit, it will be seen that the objection is neither accurate nor decisive.
>
> First, what we are (un)able to think of in a situation such as this is relevant only if it produces a counterexample. That is, what might be of interest is an example that satisfies the explication but fails to satisfy our preanalytic intuitions about systems, or, alternately, something that we took pre-analytically to be a system which fails to satisfy the explication. Short of this, alluding to what we are (un)able to think of is simply quite unhelpful.
>
> Second, I think that we can describe things that fail to satisfy the explication: thus showing that the definition does not lead to the kind of vacuity indicated in Buck's objection. For example, the piece of paper on which I am writing is in no obvious way either a thing (or combination of things) which is a set of elements and relations among the elements. So both (a) and (b) of Buck's objection miss their mark. [However, the retort might be, the piece of paper *can be so regarded* (represented, construed, viewed, seen)! And just what does this mean? Possibly it means that under some reasonable description the piece of paper can be *seen to* satisfy (a) or (b) and thus the explication. Well and good: let us have the description, or a method for

routinely generating such descriptions, and then decide if we have a counterexample.]

Third, it should be pointed out that the presupposition that great generality or even "total universality" results in vacuity is mistaken; it is not at all clear that a very general concept such as that of a system is, because of its generality, logically empty (whatever that might mean here). The claim, were it made, that everything is countable would not lead us to say that counting (or the concept of countability) is vacuous. Or, again, the claim that everything can be *construed* as a set or as a member of some set would not lead us to reject the concept of a set as vacuous. Quite to the contrary, sometimes the ubiquitous applicability of such procedures and concepts is a mark of their usefulness, if not of their indispensability.

I think that the analogy with set theory is illuminating. It is worth noting that the concept of a set is itself intelligible, by and large, *only in the context of some set theory*. Within such a theory questions about sets can be raised and answered, while the employment of the concept outside the theory, in some "ordinary" fashion, leads, as we know, to infamous paradoxes. And, of course, the general question of what can be construed as a set is not answered by the theory (any more than the question what English sentences can be construed as formulae is answered by some predicate logic). I think that the concept of a system would also benefit from a more formal, theoretical, explication. The move to a formal theory of systems is not, as it was in set theory, generated out of "contradictory" intuitions about systems; rather, it comes from the seeming dead end reached in nontheoretical discussion and the expressed dissatisfaction with informal analyses. If a formal theory could be developed (and just as with set theory, there might well be more than one viable alternative), then questions about systems could be raised and answered in a way that might avoid the difficulties found in the informal employment of the concept. At the same time, the ordinary-discourse ubiquity of the concept as characterized is to be preserved, for it is just this which makes its theoretical development inviting, not vacuoos.

Except for the paper by Buck, systems movement has been virtually free of criticism throughout the 1950s and 1960s. Then, in the 1970s, it suddenly became a fashionable target of critics of all trades. The most visible critiques of various aspects of systems movement from this period are contained in three books written by Ida R. Hoos [1972], David Berlinski [1976], and Robert Lilienfeld [1978]. These books, by and large, focus their criticisms on some early claims and expectations of systems research, often overly enthusiastic and exaggerated, and on some ill-conceived applications of systems methodology, particularly in the social sciences.

The critique by Hoos is the most specific of the three. Her target is not systems science, as conceived in this book, but systems analysis or systems approach (she uses the two terms interchangeably), which she views as offsprings of operations research. The following is her own characterization of the book:

The research reported in this book represents a critical investigation of the state-of-the-art of systems analysis. The technique is examined in theory and practice, in its own circumscribed, structured, and simulated world and in the real world, where solutions must face pragmatic test and not merely satisfy an abstract set of conditions. The assumptions implicit in the approach as it developed over time and their validity as a basis for social planning are analyzed. They are examined in the context of the specific areas in which systems analyses are being applied, viz., education, crime, health, welfare, land use, transportation, and pollution of land, sea, and air. The experience with information systems as components of these areas and as entities in themselves are examined in detail. The process, procedures, and products of systems analysis are analyzed for the social, cultural, political, and economic factors that influence the adoption of this problem-solving technique in its various forms, notably cost–benefit ratios and planning–programming–budgeting.

Using numerous case studies, Hoos criticizes some exaggerated and highly oversold claims of systems analysis as a tool for making public policy. Her critique is, with some exceptions, sound and appeared at the right time.

Berlinksi's critique is aimed at a larger target, including not only systems analysis, but also general systems theory and mathematical system theory, particularly the area of dynamical systems formulated in terms of differential equations. His criticism is more vigorous, but, unfortunately, far less substantiated than the one by Hoos. Moreover, it contains a fair number of obvious errors or misunderstandings. Berlinski focuses primarily on the misuse of mathematics by some systems researchers or, in his own words, on "the use of mathematical methods for largely ceremonial reasons." The main point of his criticism is that it is not warranted to transfer mathematical methods (he almost exclusively refers to methods based on ordinary differential equations) from physics, where they work well, into the realm of social sciences or even biological sciences. In his own words:

> Ultimately, I see my essay as a fragment of a larger and philosophically more stimulating mass dealing with the question of the extent to which minds constituted roughly as ours are constituted may hope to achieve an adequate understanding of social, political, and biological life. There is no obvious reason why the interiors of certain distant stars should be the subject of deep physical theories while the gross features of ordinary experience remain, by the standards of theoretical physics or pure mathematics, inaccessible to sophisticated speculation. Systems analysis I take to be a bouncy and outrageous attempt to pretend that such speculation, and the understanding that it sometimes prompts, is in fact within our grasp; to show that this is not so is not a very difficult task, but it is one that ought to be performed.

The criticism regarding the issue of transferring mathematical method from physics to biological or social sciences, which Berlinski illustrates primarily by the industrial, urban, and world dynamics of Jay W. Forrester [1961, 1968, 1969, 1971, 1975], is well taken. However, this criticism is applicable only to some activities within the systems movement (as well as some activities outside it) and it is by no means justified to extend it to the whole movement, as Berlinski seems to imply.

The book by Lilienfeld [1978], which is the most comprehensive of the three, was written as a critique of systems theory (or systems thinking—the author uses these terms interchangeably), particularly its societal claims. It consists of three parts. In the first part, which consumes approximately one half of the book, the author overviews historical roots and major characteristics of systems theory. He describes relevant views of Ludwig von Bertalanffy, Norbert Wiener's formulation of cybernetics, ideas developed by W. Ross Ashby, Shannon's information theory, operations research, systems analysis, artificial intelligence, and aspects of systems theory that developed within economics (input–output theory, game theory, decision theory).

The second part of the book is devoted to the societal claims of some systems thinkers. The author focuses mainly on the writings by Ervin Laszlo, Walter Buckley, Karl Deutsch, and David Easton.

In the third part, Lilienfeld tries to portray systems theory (or systems thinking) as an ideology of new scientific and technocratic elites. Similarly as in the book by Berlinski, considerable attention is given in this context to the work of Jay Forrester and to his association with the Club of Rome.

Although the overview part of the book is excellent (even though its scope is limited), the critical part is less commendable. For one thing, the criticism is restricted only to the ideas and claims of some representative pioneers of the systems movement, while it virtually ignores the mainstream of systems research at that time. Moreover, the criticism has a distinctly emotional flavor, exhibiting strong, almost religious, antisystem and, indeed, even antiscientific sentiments.

In spite of its shortcomings, the book by Lilienfeld contains some points of critique that are well taken. They include the criticism of "the passion for quantification" and exaggerated societal and philosophical claims of some systems theorists, empty importation of systems terminology into various traditional disciplines, and the lack of efforts to justify various simplifying assumptions by most systems analysts.

The best part of the book is, in my opinion, the critique of systems analysis, which is partially based on the previous critique of Ida Hoos. I fully agree with the criticism of so-called systems analysts who claim mastery in any discipline and ability to solve any problem (depending on the availability of funds). However, I consider it absurd to put these people in one bag with serious systems researchers and to claim, as the author does in the last part of the book, that all these people form a new social elite through their enormous influence on politicians, administrators and businessmen.

Although published in the 1970s, the books by Hoos, Berlinski, and Lilienfeld are still the most challenging critiques of various aspects of the systems movement. To see them in a proper historical perspective, let me quote from a short paper by Ervin Laszlo [1980]:

> Publications such as those discussed here may represent points along a curve that marks the trajectory of a theory innovation. When a new theory or conceptual mode of thought appears, it is almost totally ignored by the adherents of the established paradigm (or paradigms). As it wins adherents here and there, than more eyebrows are raised in the circles of the establishment: some definitions or findings attributed to the challenger become known (falsely, as often as not). There are occasional references to it in books and journals. In time, it becomes acceptable to include a criticism of it in one's writings, and may even become fashionable to do so. The day is not far away when it becomes good academic politics to show one's sophistication by producing a full-fledged critique of the challenger. At first, these are somewhat condescending—after all, one does not want to commit the mistake of taking a new theory too seriously, lest one be misunderstood by one's colleagues. The purpose is to expose the sins and crudities of the challenger, making fun of it in an erudite manner. Whether the critic has truly understood and properly represented the criticized theory is of small moment, since most of his colleagues are certain neither to know nor to care.
>
> If the challenger continues to gain ground in the intellectual community (or in society as a whole), the critics tend to become more expert. Now they read up a little more on the challenger and try to understand it before shooting at it. The shots, however, still come (for a time) from the home base of another mode of thought which, for the critic, grasps the real truth and uses the proper logic.
>
> This is how far we appear to have come today, with Berlinski–Hoos representing the penultimate, and Lilienfeld (sofar) ultimate phase. That we did come this far is a remarkable achievement of systems thinking. It has become an innovation that is legitimate to criticize, and indeed good business to do so. Books on it sell, and are used even by one's establishment colleagues.
>
> There could—and in this case I believe will—be more advanced points along the trajectory aptly described as the "rise of systems theory." A logical next stage indicating its rise would be the appearance of books and studies which undertake a consistent meta-paradigmatic exploration of the merits and faults of the new theory vis-à-vis the older schools, without using some of the latter as an axiomatic basis for criticizing it. Subsequent to this we shall witness the publication of an increasing number of critical essays which already move within the conceptual universe of the new field. These constitute internal critiques, exploring inconsistencies, correcting biases, and suggesting further applications and developments. For systems theory this stage is yet to come, but its coming is prepared by the previous stages, including those that have just been attained.

After reading the three critiques, an uninitiated reader is likely to get the impression that systems thinking and methodology are fully embraced and well supported by the government and business alike. This, unfortunately, is not the case, with the exception of some methods that evolved from operations research and are subsumed now under the fashionable term "systems analysis." These methods, as properly argued by the critics, have often been grossly oversold. Genuine systems science, on the other hand, has received only modest recognition and support thus far, and its impact on other areas, which has a great potential, has by far not been fully realized as yet. Let me elaborate a little more on these issues.

11.2. Status and Impact of Systems Science

Systems science, by its very nature, contributes a new perspective to science that is complementary to the perspective of traditional science. There are some important factors that favor the development of systems science, but there are also some factors that tend to block it. Let me quote from a paper by Laszlo [1975], in which these positive and negative factors are well overviewed (although Laszlo refers to general systems theory, using the terminology common in the 1970s, his observations are equally relevant, with a few exceptions, under current situation to systems science):

> Factors favoring the development of general systems theory operate both internally and externally to science. There is an intrinsic trend within science itself to maximize the scope of theories consistently with their precision. There are also extrinsic pressures on science to overcome traditional boundaries in producing multidisciplinary theories applicable to societal problems.
>
> Modern science has made great progress by adopting the analytical method of identifying and, if possible, isolating the phenomena to be investigated. If effective isolation is not feasible, e.g., in the life and social sciences, it is replaced by the theoretical device of averaging the values of inputs and outputs to the investigated object, and varying the quantities with the needs of the experiment. Thus influences from what has often been disparagingly called the rest of the world can be disregarded. It appears, however, that the rest of the world is an important factor in many areas of investigation. The consequences of disregarding it are not immediately evident for a good detailed knowledge of the immediate phenomena in a short time-range can nevertheless be won. But the spin-offs, or side effects, of the phenomena will be incalculable, and such effects are not the secondary phenomena they were taken to be in the past. They are the results of the complex strands of interdependence which traverse all realms of empirical investigation but which science's analytical method selectively filters out. Hence, we get much detailed knowledge of local phenomena, and a great deal of ignorance of the interconnections between such phenomena. The analytical method produced the explosion of contemporary scientific information, and the dearth of applicable scientific knowledge. It has also engendered wasteful parallelisms in research due to failures in the transfer of models and data between

disciplines. . . . Scientists value theory refinement as well as theory extension, although they do so to differing degrees. The routine experimentalist, mainly involved with puzzles that can be solved through a suitable application of existing theories and techniques, tends to disparage the philosophizing of colleagues bent on the revision and refinement of the theories themselves. But scientists who perceive internal inconsistencies in their frameworks of explanation are greatly concerned with overcoming them through the creation of new, more general postulates, embracing existing theories as special cases, or reinterpreting them in the light of new axioms. Although the emphasis changes from person to person, from scientific community to scientific community, and from period to period, depending on the problems encountered in the given field, it remains true that, on the whole, the progress of science involves the integration of loosely joined, lower level concepts and hypotheses in mathematically formulated general theories. . . .

The historical trend in modern science is to counterbalance segmentation and specialization in patterns of research and experimentation. . . . In almost every case concrete societal problems call for interdisciplinary research and the integration of hitherto separately investigated variables. . . . Disciplinary compartmentalization is useful only if it is coupled with transdisciplinary integration. . . .

Intrinsic trends to balance fragmentation in mathematically elegant general theories, and extrinsic trends to overcome the limitations in the application of fragmented knowledge are factors which favor the evolution of any theory of integrative potential. They favor the development and acceptance of general system theory inasmuch as that theory is specifically designed to integrate theories of different fields of science, and to make possible the societal application of the integrated scientific knowledge. But general system theory is not thereby automatically elected as the paradigm of contemporary general theory in science. There is a great deal of scepticism that focuses on general theories as such, and on a general system theory especially.

The factors that block the progress on general system theory are due partly to intellectual and organizational inertia, and partly to confusion and suspicions centering on the general system concept itself.

Every theory innovation faces resistance due to intellectual inertia; the tendency of persons trained to work with earlier theories to fail to perceive, or perceive and fail to take seriously or perceive, take seriously, but feel threatened by, the innovation. Such factors of intellectual inertia have created resistance for the acceptance of general theories within individual disciplines—they were experienced by Maxwell, Lavoisier, Pareto, Parsons, Skinner, to mention but a few. . . .

Resistance to theories moving across disciplinary boundaries is stronger than resistance within the disciplines. It is due to several additional factors, including indifference and fear. A scientist confronted with a theory that did not originate in his field and is not confined to it can shrug off its meaning: it does not concern him in his professional capacity. Thus an ecologist taking a systems approach to information and energy flows in the ecologies he studies may disavow interest in and responsibility for a general theory of systems which would apply, in addition to ecologies, to economics or politics. Mutatis mutandis, with other scientists working with systems concepts in any of the sciences of complexity.

The other generic blocking factor is fear. Specialists rely on knowing more about their specialty than anyone else. They may not know, or even care, about theories and phenomena not directly connected with their specialties, as long as they feel assured that they are masters in their own corner of the scientific edifice. It is unsettling to them to find that some general theorists claim to know their field, and indeed offer interpretations of their findings with which they themselves are not familiar. Instead of welcoming such interests from other scientists and seeking to strengthen the linkages of their specialty with other fields, they feel threatened and tend to block overtures for collaboration. These are psychosociological factors which operate in science no less than in other organizations. General system theory is fully exposed to them. Persistent neglect in some quarters and isolation in others manifest their effects. . . .

Western science, while decentralized in its administrative structure, is also more difficult to move from its present tracks. These are deeply disciplinary in nature. Monies and prestige are vested in academic departments, and the departmental structure of colleges and universities is almost exclusively disciplinary. There have been experiments with multidisciplinary departments, but outside of colleges of general studies, few have managed to survive. . . .

Left to themselves, departments in general show unwillingness to stake their precious resources on new ventures leading beyond the known disciplinary boundaries. Administrators charged with curriculum tasks likewise show unease and unwillingness when faced with multidisciplinary proposals. While agreeing on the need for such programs, they are, for the most part, unfamiliar with the conceptual content of the required offerings. What, for example, are general systems? How much can one say about them? It would seem to many that, as soon as one goes into detail, one is constrained to speak about and conduct research on some special kind of system. What then is the point of institutionalizing a program based on general systems?

Furthermore, who is qualified to teach or investigate general systems? Those who claim such qualification include engineers, life scientists, social scientists, and philosophers. But are the engineers not merely talking about artificial systems, the life scientists about the living systems, the social scientists about social systems, and the philosophers about conceptual systems? If so, they could well pursue their investigations and teaching programs within their existing departments. . . .

The term general system theory is subject to basic misunderstandings. These often originate with the careless use of language but have a tendency to harden into metaphysical doctrines.

The practitioners of general system theory have an unfortunate tendency to speak of general systems as a subject that has predicates. . . . Current usage associates general with system instead of associating it with theory—and taking system as a predicate of theory. Thus we have general-system theory instead of general system-theory. This plays havoc with the legitimacy of the field. Elementary reflection discloses that there is no such real world entity as a general system. There is a kind of theory known as system theory, and there is a general form of this theory: *general* system theory. Assuming the contrary is nonsense. It is to assume that there is a theory of general systems. In fact, there is only a general theory of systems. . . .

The semantic confusion discussed here is a more than academic interest. It impedes the development and acceptance of general system theory. Those who, seduced by the careless linguistic habits of its practitioners believe that it is a theory of a curious entity called general system, view it with understandable scepticism. They are prevented from appreciating that general system theory is not the investigation of a mythological beast labelled general system but the investigation of the full scope of the phenomena conceptualized as various kinds of systems. . . .

A common fallacy is to hold that all general system theorists are metatheorists. In that event, they would not be system theorists, but theorists of systemtheories. What they would be concerned with would not be systems, but theories of systems—and such theories themselves are not systems. Those who investigate existing general theories of systems are not system theorists, but historians or philosophers of science, specialized to the field of general system theory. . . . The designation of the field as one of metatheory is false. As a source of confusion, the designation is dangerous, for it subsumes general system theory within another field and thus removes its individual raison d'être. . . .

Some of the skepticism confronting general system theory is due to a (mistaken) belief that it is another terrible generalization. The most frequently heard opinions accuse it of being either a generalization of a theory of organism, or of a theory of automata depending on whether the accuser has heard more of von Bertalanffy or Ross Ashby, for example. Yet general system theory is innocent of such charges. It is not a generalized theory, but a general theory. We must not confuse the historical origins of a theory with its actual status and orientation.

In spite of the various unfavorable factors mentioned by Laszlo, systems science has made significant progress since the emergence of systems movement in the 1940s and 1950s. Perhaps the most fundamental role in forming systems science as a legitimate field of inquiry was played by the various conceptual frameworks (as overviewed in Chap. 4) through which the notion of systemhood is properly characterized and codified. Categories of systems that emerge from these frameworks delineate fairly precisely the domain of systems science.

Major progress has been made in using the computer as a laboratory. In fact, numerous laws of the various categories of systems have already been determined by properly designed experiments on computers, contributing thus to knowledge base of systems science. Although this knowledge base is still rather small when compared with other, well-established areas of science, it is slowly but steadily growing.

The progress in systems methodology over the last few decades has also been quite encouraging. On the conceptual level, it is important that systems problems are now well characterized and codified in terms of the various categories of systems. On the pragmatic level, it is clear that the methodologies for some important classes of problems have advanced considerably.

One such class consists of problems concerned with the relationship between overall systems and their various subsystems. These problems are modern formulations of some aspects of the age-old philosophical problem of wholes and parts [Lerner, 1963]. They are also closely connected with the polemic between the doctrines of reductionism and holism. Perhaps the most important outcome of a serious methodological work on these problems, as exemplified by the area referred to as *reconstructability analysis* [Klir, 1985a, **1986**], is that it provides us with new insights regarding the nature of the relationship between wholes and parts. The methodology resulting from this work goes far beyond the thinking emerging from both reductionism and holism. From the viewpoint of reconstructability analysis, these doctrines are only two extreme positions in a broad spectrum of methodological possibilities. Reconstructability analysis recognizes that it is often essential, and sometimes unavoidable, to reduce a complex system into appropriate subsystems in order to make it manageable. It makes us aware, however, that the choice of the subsystems is critical; some may represent the overall system quite well while others may be highly inadequate. We should not look for subsystems that look "natural," but rather for those that allow us to reconstruct the overall system with as high accuracy as possible.

As a rule, genuine systems problems are computationally extremely difficult. They are often made tractable by overly strong simplifying assumptions that are usually not explicitly stated. The resulting methods can then deal with sizable systems emerging from practical applications and produce "impressive" results, but the significance of these results is questionable at best. Such methodological dishonesty, which is one of the primary sources of the previously overviewed critiques by Hoos, Berlinski, and Lilienfeld, is contrary to the spirit of systems science. The latter is not interested in producing immediately marketable methodological tools at the cost of convenient simplifying assumptions whose validity is dubious, but rather in pursuing basic methodological research involving genuine systems problems. To overcome the exploding computational complexity, which is characteristic of the latter problems, is never easy and it usually requires years of concentrated research for each problem category. In his fight with computational complexity, the systems scientist has to explore various methodological strategies such as, for example, the structuring of the computation hierarchically in terms of appropriate equivalence classes of the alternatives involved, the use of various heuristic procedures, the utilization of relevant background knowledge, the utilization of uncertainty of various types as a commodity that is traded for reduction of computational complexity, or the use of specialized computer technology based on massive parallel processing.

In spite of the difficulties and little support, systems methodology progresses in some areas rather well. Let me use the problems subsumed under reconstructability analysis as an example. This is a particularly difficult problem area since the number of collections of subsystems of a given overall system grows doubly exponential with the number of variables in the overall system. When reconstructability analysis was conceived in the mid 1970s [Klir, 1976], we were able to deal with systems of no more than seven or eight variables. Now, after more than a decade of research that involved virtually all the mentioned methodological strategies, we can handle systems with hundreds or even thousands of variables [Conant, 1988].

In parallel with the progress in systems methodology, we can also observe a progress at the metamethodological level. For one thing, the role of the computer in metamethodological studies (by which, for example, the performance of heuristic methods can be evaluated) has now become almost routine. However, one major task of systems metamethodology has not been accomplished as yet: to establish a rigorous relationship among the various conceptual frameworks overviewed in Chap. 4.

The progress in systems science can also be examined in terms of the growth of relevant publications, academic programs, and activities. In the mid 1970s, it was determined that the literature pertaining to systems science had been doubled approximately every four years since 1950 [Klir and Rogers, 1977]. This trend was still observed in the early 1980s [Trappl, Horn, and Klir, 1985], as expressed by the plot in Fig. 11.1. This is a good indicator of the progress in systems science. There is also some evidence, although less documented, that the number of academic programs and professional activities bearing upon systems science has been slowly but steadily growing since the 1950s.

How has the steady progress in systems science impacted other areas of science? This question is important but rather premature since systems science is still in an early stage of its evolution. Moreover, the question is too broad to be adequately answered by any one person. To try to answer it anyhow, I can, at best, only direct the reader to the relevant literature.

The main impact of systems science on traditional science is, undoubtedly, its cross-disciplinary orientation. As a result of arguments pursued by systems science for decades, scientists are now becoming, in general, more sensitive to the limitations of their own disciplines. They tend to be considerably more aware now than a few decades ago that significant real-world problems involve almost always

Systems Science in Retrospect and Prospect

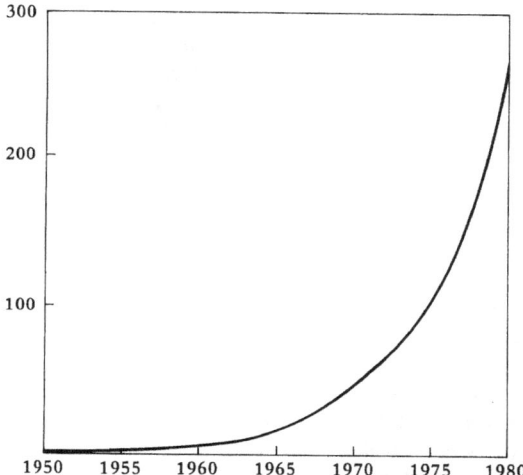

Figure 11.1. The increase of significant new publications (books and papers in refereed journals) pertaining to systems science in the period 1950–1980.

aspects that transcend disciplinary boundaries. This impact is, of course, only indirect and, consequently, it is virtually impossible to characterize it more specifically. Another indirect impact of systems science on the traditional science is the increasingly habitual use of systems thinking in the latter. For example, it is now quite common in most areas of traditional science to think in terms of systems concepts such as hierarchy, homomorphism, regulation, feedback, stability, adaptivity, complexity, information, and many others.

Some areas of traditional science have also been impacted by systems science more directly. In general, those most developed have been impacted least. *Physics*, for example, has been almost untouched by systems science. This does not mean, however, that systems science has no potential role in physics. In fact, the discussion of this role, which is potentially profound, has already started (see *Special Issue of the International Journal of General Systems on Systems Thinking in Physics*: Vol. 11, No. 4, 1985, pp. 279–345). The following are some ideas expressed by one of the main proponents of systems thinking in physics, Paul A. LaViolette [1985]:

> The concept of the open system has proven to be applicable to a wide variety of fields. The functioning of biological organisms, business organizations, social systems, human personalities, and frameworks of knowledge may in each case be understood as systems whose structure is continuously sustained through the operation of import and export transactions and transformation processes. The open system concept is also found to be applicable to

understanding the behavior of nonequilibrium phenomena studied in the physical sciences. However, there is one discipline which has thus far evaded the inroads of this general systems concept. This is microphysics, a field which is currently still framed within the mechanistic paradigm. Subatomic particles are today viewed in much the same way that atoms were once viewed in the time of Democritos, namely as closed, inert systems. Except, now it is understood that from time to time these "billiard balls" undergo abrupt changes of their internal structure (and identity) either through spontaneous decay or as a result of mutual collision. It is often claimed that modern science has "dematerialized" matter, particles now being represented in terms of probability density functions, rather than as solid bodies. However, the mechanistic fingerprint has left a deep impression on physics and its conceptualizations are still very much with us.

The question which naturally comes to the mind of the general system theorist is whether microphysical phenomena are indeed so different from other natural phenomena that they are the only ones for which open systems concepts are unsuitable. Or, is it that physics is still in its infancy and has not yet emerged from the mechanistic, closed system paradigm which also at one time characterized these other sciences. There are many who would agree that the latter may be the case. Indeed, of the sciences, microphysics deals with phenomena which take place on a scale that is quite far removed from direct experience. In studying a biological organism or a business organization it is quite easy to demonstrate, either through experiment or observation, that the continued existence of such systems depends on the import, transformation, and export of currencies such as chemical substances, energy, capital, and human labor. Detecting a comparable "currency" which might be actively sustaining a subatomic particle is quite a different story. In probing such a microscopic level of nature, one inevitably encounters an observational barrier which prevents the direct detection of processes at the subquantum level. Since all of our measuring probes are necessarily composed of matter or energy (subatomic particles or energy quanta), the most microscopic level we are able to directly sense by necessity is the quantum level. Even then we are prevented from simultaneously determining the precise position and energy of the particle under investigation, a restriction commonly known as the Heisenberg Uncertainty Principle. Therefore our understanding about the ultimate nature of particles and fields and the possible existence or nonexistence of an underlying dynamic, form-sustaining substrate must necessarily be arrived at through a process of inference.

The "unconquered" territory of physics, therefore, presents the following challenge to the general system theorist: Namely, is it possible to devise an open system model of space and time which provides for the emergence of quantum structures and fields, and which may perhaps be more suitable as a description of microphysical phenomena than present theoretical frameworks? Or more broadly speaking, is it possible to devise a new *methodology* for representing microphysical phenomena, one that is framed within the open system paradigm? It is the purpose of this paper to present such a new approach.

When the time is ripe, systems thinking, exemplified by these ideas, is likely to make a paradigm shift in physics. Thus far, however, the interest of physicists in ideas like these is lukewarm at best.

When we proceed from physics to *biology*, we find a considerably stronger connection with systems science. This is not surprising since biology belongs by its very nature to the realm of organized complexity. The first serious discussions regarding the role of systems thinking, theory, and methodology in biology started in the late 1960s [Mesarovic, 1968]. These discussions involved systems scientists and biologists of both theoretical and experimental orientations. To illustrate the spirit of these discussions, let me use two quotes. In the first quote, D.F. Bradley, a submolecular biologist, illuminates an important role of systems thinking in biology by arguing the need of modeling biological phenomena as multilevel systems:

> Most biological phenomena can be examined at many different magnifications. At each magnification interesting and even useful observations can be made and often prediction of future behavior or even correction of malfunctions can also be made. It is rather surprising that this is the case because there are very compelling reasons for believing that macroscopic biological behavior represents the summation of countless numbers of intermolecular reactions and interreactions, each of which follows physical laws that operate at the angstrom level. Recognizing this, a being from outer space might logically assume that the proper course of study of earth life should begin at the submolecular level and proceed stepwise with larger and larger groups of individual reactions until the meter level was reached. Our sciences did not, however, develop historically in such a logical sequence. Sociology, economics and political science, psychology and psychiatry, anatomy, physiology, biochemistry, chemistry and physics have developed in parallel rather than sequential ways and often with little intercommunication. In the course of this development we have learned to treat illnesses of the whole being, such as mental disease, without referring to, or even knowing much about, the physiology, biochemistry, chemistry or physics underlying either the normal or the diseased condition. Mendel learned a great deal about heredity and could predict behavior in unborn organisms without knowing of the existence of deoxyribonucleic acid (DNA), much less that it carries the genetic information. . . .
>
> Analysis at any given level may "work" for solving some problems and not for others. Psychiatry may prove useful in treating neurotics but not psychotics. The tremendous difficulties inherent in attempting to analyze social or emotional behavior at the molecular level makes attempts to analyze them at the sociological and psychiatric level well worthwhile. It is far too early to make valid predictions of the behavior of such large molecular

systems organized into a living system and under genetic and environmental constraints. The time to begin thinking about the possibility of analyzing at levels of higher magnification is when analysis at the chosen level, the level which *a priori* seems most appropriate, fails. . . .

Success at problem solving at levels of low magnification is therefore to be applauded since it obviates the necessity of working a great deal harder at higher magnification levels. The best single criterion for going to higher magnification levels is persistent failure to solve specific or general problems at the *a priori* most appropriate level.

There are currently working in biology investigators who have moved from lower to higher magnification levels and those who have moved in the opposite direction. Doctors of medicine have become biochemists, or even molecular physicists, and physicists have left physics for biochemistry. The former are usually dissatisfied with the slowness with which progress is being made in solving problems at the medical level and turn to smaller systems whose problems they hope to solve more completely, precisely and rapidly. The physicist-turned-biochemist usually wishes to apply his proven problem-solving abilities on more complex systems which appear to hold more human interest. Collaboration between those who have moved in opposite directions to meet at the same level can be very productive, an outstanding example being the collaboration between the physicist, F.H.C. Crick and the biologist, J.D. Watson, which resulted in the determination of the three-dimensional structure of DNA.

In the second quote, R.E. Kalman, a mathematical systems theorist explains his view of systems theory in biology:

Today systems theory is making an increasingly important impact on systems technology in providing new solutions to old problems or suggesting new kinds of systems. I have an (unproved) theorem which claims that "Systems Technology" = "Artificial Biology"; after all, the aim of the systems theorist is to create systems which approach or perhaps even surpass capabilities normally observed only in the living world. The aims of the systems theorists are not unlike those of the biologists, though we must remember that the two groups work under very different kinds of constraints. The systems theorist does not claim that computers will provide ready-made models for explaining the brain, but he is optimistic that the *methods* devised to gain a deep understanding of inanimate computers will have some relevance to the understanding of living computers. The systems theorist is not an engineer—he wants to know the capabilities and utilization of computers, but he does not worry about the fine constructional details. By analogy, the systems theorist will add little to the experimental phases of brain research, but he should be very useful (if he really is a good theoretician) in the scrutiny and evolution of models which embody and explain the experimental results.

As anticipated here by Kalman, systems theory has indeed played a major role in the formation of modern theoretical biology. This is well explained by

Robert Rosen [1978], perhaps the most important liaison between systems science and theoretical biology:

> Biology is the science of life and of the living. The term "life" encompasses a unique range of phenomena, and biology is correspondingly a unique science in many ways. Among these is the fact that biology alone seems to sit at the intersection of all other sciences, so that no advance in any science can be without impact on biology; and conversely, any advance in biology ultimately propagates into every corner of science.
>
> The basic object of study in biology is the individual organism. The basic question is: how does it work? What are the basic forces which shaped it and maintain it? In attempting to come to grips with this basic question, we soon find that it is not one question, but many different questions, the answer to each of which gives rise to a different branch of biology.
>
> For instance, we may ask how an organism maintains its structure and organization against the disordering forces of its ambient environment. By its very nature, the dominant ideas in physiology will be those of homeostasis, stability and control.
>
> We may ask how an organism comes to have the particular structure and organization which characterize it. There are two kinds of answers to this question. One is in terms of development; the elaboration of this structure and organization from its simplest direct antecedent, usually a zygote or fertilized egg. Another kind of answer is found in evolutionary biology, the study of the continuity of life over geological time back to, and including, the origin of life on the planet. The dominant feature in both these areas is not so much homeostasis but what Waddington has termed homeorrhesis: the stability of dynamical processes. Characteristically we have to study here the emergence of complex structures from simpler ones, and hence both of these areas are intimately involved with ideas of self-organization.
>
> We may ask how it is that organisms tend to resemble their ancestors, and why they resemble their proximal ancestors more than their remote ancestors. The study of this question resides in the science of genetics. The prevailing idea here is that of a code or program, which is successively "read out" or expressed in each individual organism, and is transmitted from ancestor to descendants modulated by a sexual mechanism. The idea of a code carries with it a corresponding duality between genotype (the program) and phenotype (an expression of the program), a duality which is so characteristic of organisms.
>
> We can ask how it is that an individual organism adapts or modifies its behavior in response to environmental fluctuations. Here the objects of study are the transducers between environmental stimulus and organismic response; the most important such transducer is the central nervous system of higher animals. The dominant ideas here are those of computation and communication.
>
> We can ask how a number of organisms will interact with each other and with their environment. Here again, a number of answers can be given to such a question. One kind of answer will lead off into population biology and

ecology. Another kind of answer will lead in the direction of the study of communities and social structures (including human societies in all their aspects). Still another kind of answer will take us into psychology, with its ideas of learning, conditioning, intelligence and consciousness.

Resting on all of these sciences which collectively comprise biology, we find a number of unique technologies which seek to regulate or control biological processes: agriculture; husbandry; and medicine. . . .

Biology has provided for system theory a unique source of challenge, of inspiration, and a field for applications. Let us note a few. The field of cybernetics is largely an outgrowth of the physiological concept of homeostasis, or the "constancy of the *milieu interior,*" enunciated by Claude Bernard in the 19th century. This was combined with the recognition that we could explicitly design devices which would similarly exhibit adaptive behavior in order to maintain constancy of some associated quantity. Out of this came the integrative notion of *feedback control*; it is significant to note that Wiener subtitled his definitive work on cybernetics, "Control and Communication in the Animal and the Machine." The dissection of an organism, or a machine, into a set of interacting feedback loops, provides a specific analysis of dissimilar systems behaving similarly.

Another example: what we now call the theory of automata arose in large part through explicit and tacit attempts to understand the behavior of the brain. Turing's initial approach to the machines with bear his name was an abstraction, not from a mechanical device, but from how a human being would go about solving a mathematical problem, or computing a number. The neural nets of McCulloch & Pitts [1943] were an explicit attempt to model the brain in terms of a network of threshold elements. It was quickly recognized that such ideas were intimately related to the switching networks of communication theory; to digital computation in its widest sense; to the mathematical theory of effective or recursive processes (and hence to the foundation of mathematics itself); to linguistics; and to psychology. An entire research area, generally called "artificial intelligence," has grown up to exploit the possibilities of generating brainlike behavior in technological systems through the employment of automata-theoretic concepts.

Still a third example may be mentioned, the theory of "self-organization" or adaptive systems whose paradigms are most often drawn from the phenomena of developmental biology, but which are also found in learning systems, social systems, linguistic systems in their widest sense, in technological artifacts, and in purely physical systems exhibiting co-operative effects.

From all of these diverse sources, the concept of a *system* has been distilled; both as an object of study in its own right, and as a vehicle for the study of behaviors in the abstract and the concrete. It thus comprises (a) the study of particular classes of systems, defined in terms of some kind of common property shared by all the systems in the class, and (b) the study of what is common to different classes of systems, *qua* systems. . . .

We pointed out that an organism can simultaneously be considered under many different headings. We could view an organism in a physiological context, a developmental context, an evolutionary context, a genetic context, a physical context, and so on. . . . On the other hand, we also saw that many systems which are not usually regarded as organisms can exhibit behaviors which are also shared, or realized, by organisms. Thus, the class of systems which are in some sense self-organizing, exhibit such behavior; as do classes of cybernetic systems, computation and communication systems, social systems, linguistic systems, hierarchical systems, and so on.

From these simple heuristic considerations, we see immediately why it is that (a) biology and general system theory have been so intimately related, and (b) biology plays such a unique and pivotal role in the sciences. For there is hardly any class of systems which does not either represent a possible class of models for organic behavior, or which may not be modelled by organisms, or both.

Examples of the cross-fertilization between biology and systems science are plentiful. Some of the most visible examples of systems notions emerging from biology are the various types of goal-oriented systems (Chap. 10), such as self-organizing, self-reproducing, antopoietic, adaptive, anticipatory, or morphogenetic systems. On the other hand, the *general theory of living systems* developed by James G. Miller [1978] is perhaps the most visible example of using systems thinking for structuring biological and even social knowledge.

In spite of the tremendous potential of utilizing systems thinking, knowledge, and methodology in biology, this potential has not been fully realized as yet. It seems that this underutilization of systems science in biology is caused by a considerable gap that currently exists between theoretical biology and experimental biology. Consequently, advances in theoretical biology, closely correlated with advances in systems science, have little influence on experimental biologists. The latter, whose views are predominantly reductionistic, are currently dominant in biology, both in numbers and influence. As a consequence, theoretical biology has, paradoxically, much greater impact on some areas outside biology (artificial intelligence, artificial life, systems science) than on biology itself.

Let me turn now to social sciences. Since the emergence of cybernetics and general systems research in the late 1940s and early 1950s, there have been growing expectations that these new cross-disciplinary areas will play an increasingly important role in our quest for understanding social phenomena. These expectations have been fulfilled in different social sciences to various degrees. In economics, for example, this degree is rather low, as explained in detail by Boulding [1972]. This is unfortunate for at least two reasons. First, economics could greatly benefit from some systems theories that are now well developed, most notably the theory of feedback control. In spite of its rather obvious applicability to the study of economic phenomena, the theory of feedback control has virtually been ignored by most economists. Although the connection between the theory and economics was developed by Oskar Lange [1970], the influence of his work on mainstream economics is not noticeable. Second, under the influence of the cross-disciplinary perspective of systems science, economics could have a chance to liberate itself from its notorious isolationism, which takes it for granted that the economic process can meaningfully be studied as a closed system. That is, orthodox economic models do not involve social, political, ecological, legal, and other relevant aspects, which are considered outside the self-imposed, rigid boundaries of economics. While the study of a phenomenon in isolation is often acceptable in natural sciences, this investigative strategy is totally unrealistic in social sciences, where phenomena that we artificially classify as economic, political, social, ecological, etc., are strongly interconnected. As a consequence, the history of economics is a history of a persistent discrepancy between economic predictions and economic reality. In order to develop more realistic economic models, systems open to relevant phenomena outside economics will have to be considered.

In other areas of social sciences, systems science has been more influential. Systems thinking, for example, is now quite common in political science, management science, and other areas.* At the same time, methodological research in systems science has responded to special methodological needs, often cross-disciplinary, of social sciences.†

To illustrate the great potential of systems thinking in social sciences, let me overview the work of Arvin Aulin [1982, 1989], which focuses on the use of the law of requisite variety and the law of requisite hierarchy for global studies of regulating capabilities of human societies.

As previously explained (Chap. 10), the *law of requisite variety* states that a

---

*The purpose of this note is to provide the reader with representative references that describe connections between the various facets of systems science and some other areas. Neither the list of areas nor the references listed under each area are exhaustive.

Philosophy: Laszlo [1971], Bunge [**1977**], Blauberg, Sadovsky, and Yudin [1977].
Biology: Rosen [1970, **1972**, 1978, 1991], Miller [1978].
Ecology: Margaleff [1968], Caswell *et al.* [1972], Holling [1973], Patten [1978], Halfon [1979], Patten and Auble [1981], Patten and Odum [1981], Allen and Starr [1982].
Management: Beer [1966, 1979, 1981], Lee [1970], Beckett [1971], Cleland and King [1972].
Political Science: Easton [1965], Deutch [1966], Steinbruner [1974], La Porte [1975], Coming [1983], Ziegenhagen [1986].
Social Sciences: Berrien [1968], Buckley [1967], Chadwick [1971], Phillips [1976], Cavallo [1979, 1982], Hanken [1981], Hanken and Reuver [1981].
Cognitive Sciences: Gray, Duhl, and Rizzo [1969], Bateson [1972, 1979], Pask [1975, 1976], Sayre [1976], Lowen [1982].

†Methods that emerged from the needs of social sciences are well exemplified by the *interpretive structural modeling*, developed by Warfield [1976], by the so-called *Q-analysis*, which was developed by Atkin [1976, 1977], and by the methodology developed by Forrester [1961, 1968, 1969, 1971, 1975], which is usually referred to as *Forrester's systems dynamics*.

necessary condition for a successful regulation is that the variety of the regulating system must be at least as large as the variety of the regulated system. The *law of requisite hierarchy* is due to Aulin himself. It states that the lack of regulatory capability of a simple regulator can be compensated for, to some degree, by conceptualizing the regulator as a hierarchical multilevel structure system. In general, the weaker the regulatory abilities of available regulators, the more hierarchy is needed to achieve the same regulatory performance, if it is possible at all.

Survival of a human society depends on keeping certain essential variables (food, energy sources, medical supplies, etc.) within their survival regions in spite of disturbances. This can be accomplished by organized human production. Production forces of the human society (i.e., labor and production means) can thus be viewed as its regulator. The weaker the production forces, the less capable is this regulator and, using the principle of requisite hierarchy, the higher the degree of hierarchy that is necessary for effective regulation and survival of the society. This means that, given a specific production level of the society, some specific minimum degree of hierarchy is necessary for its survival. When the degree of hierarchy is increased, the regulatory capability of production forces of the society increases until an upper limit of the degree of hierarchy is reached, beyond which further increase has no effect on the regulation. That is, the law of requisite hierarchy establishes, in the context of social systems, a relationship between the production capacity of a human society and the degree of hierarchy associated with the organization of the society that is needed for efficient regulation. For each production level, there is a specific degree of hierarchy for which the regulation is most efficient. In general, the degree of requisite hierarchy increases with decreasing level of production capacity.

Another important concept in Aulin's global studies of human societies is the concept of *steering* of human actions (individual or social). Aulin makes a clear distinction between steering of human actions from outside and their *self-steering*. He views human actions as interactions between subjects and objects, i.e., between conscious actors and parts of the real world. Actions are steered from outside by conditioning the consciousness of the actor, i.e., his cognitive beliefs, values, and norms. On the other hand, actions based on self-steering result from belief, values and norms that are developed by observations and generalizations.

In general, Aulin characterizes a self-steering system as a goal-oriented system that always expands its states as well as goals, i.e., it never returns to the same state or goal. This means that a *self-steering system* is constantly producing something new, both in terms of goals and means of achieving them, insofar as it is not forced into a state that is outside the domain of self-steering.

Aulin argues that self-steering in a system increases its regulatory capability. This implies, using the principle of requisite hierarchy, that any increase in the degree of self-steering in a system reduces the degree of hierarhy in its organization demanded for effective regulation. Self-steering and hierarchy are thus reciprocal notions and, consequently, the degree of self-steering can be expressed in terms of the degree of hierarchy.

The lower limit of the degree of hierarchy in each particular society is absolute in the sense that it cannot be exceeded without risking the survival of the population of the society. The upper limit of the degree of hierarchy is not absolute. It can be exceeded without jeopardizing the survival of the population. When exceeded, however, no improvement in the regulation is achieved, while, at the same time, the level of self-steering in the society is proportionally reduced. Since self-steering is associated with freedom, creativity, and progress, its reduction beyond the desirable minimum (corresponding to the upper limit of the degree of hierarchy) restricts unnecessarily the overall progress of the society. Among other things, it restricts the improvement of productive forces, while, at the same time, it evokes social movements resisting severe restriction of freedom and highly visible inequality caused by the unnecessarily high degree of hierarchical organization of the society.

In general, an increase of the degree of self-steering in a human society means progress in human emancipation. The potential for self-steering depends on the level of production. The actual degree of self-steering, however, is not solely the result of the production level, but it depends on other factors (political, administrative, etc.) as well. Unless self-steering is totally suppressed, production level in every human society tends to increase in time. This implies that the maximum and minimum degrees of hierarchy of the society steadily decrease and, consequently, the maximum and minimum degrees of self-steering increase.

The increase of the potential degree of self-steering (i.e., its maximum and minimum values) of a human society in the course of time, due to the increase in production level, and the actual degree of self-steering in the society over the same period of time (determined by political and other factors) is a fundamental macroscopic characteristic of the society. Aulin uses this characteristic for introducing and comparing various types of human societies. He also uses it for describing some historical as well as current societies.

Aulin's work is important since it clearly demonstrates that the use of modern systems thinking for studying social phenomena is very powerful. Without resorting to any ideological arguments, he can show, for example, that the Marxist theory leads to conclusions that violate the laws of requisite variety and requisite hierarchy, laws that must be satisfied by all systems.

I trust that these few remarks, together with the relevant articles in Part II, will provide the reader with a general but adequate impression of the current status of systems science and its real or potential impact on other areas of science. To conclude the book, let me make a few additional remarks expressing my opinion regarding the future of systems science.

11.3. The Future of Systems Science

There is little doubt that, in the foreseeable future, systems science will continue to be the principal intellectual base for making advances into the territory of organized complexity, a territory that still remains, by and large, virtually unexplored. The challenge offered by organized complexity narrows down fundamentally to one question: how to deal with systems and associated problems whose complexities are beyond our information-processing limits? We are well aware now that the answer to this question lies primarily in the relationship among four key characteristics of systems models: *complexity, uncertainty, credibility,* and *usefulness*. However, we are still far from understanding this intricate relationship, in which all the characteristics are themselves multidimensional entities with their own structures. Although some results relevant to this issue have recently been obtained, most notably results regarding the notion of uncertainty [Klir and Folger, 1988], this only scratches the surface of the whole issue. I believe that basic research concerning this relationship will be the main preoccupation of systems science in the foreseeable future, extending likely well into the Twenty-First Century.

Since the computer is the principal laboratory in systems science, further progress of systems science will undoubtedly be closely correlated with advances in computer technology and science. As far as computer hardware is concerned, its overall progress is usually expressed in terms of several factors such as the number of ordinary instructions that are executed by a single central processing unit (CPU) of the high-end type, the number of transistors on a single CPU chip, the number of bits per chip of a dynamic random access memory, the number of bits per square inch of a magnetic disk storage, and the speed and resolution of input and output devices. Each of these factors has shown a steady exponential growth, with an overall hardware capability increase of approximately 20% each year. This trend is likely to continue until physical limits are reached or the currently predominant electronic hardware is replaced with hardware based on radically different principles such as optical, chemical, or superconductive hardware.

The progress in computer hardware is only one factor contributing to increases in the overall capabilities of computing systems. Another factor is the underlying computer architecture, in particular architecture based on massive parallel processing. Computer systems with perhaps as many as several million simple processors are currently under consideration. Problems in designing such systems are formidable, but there is no doubt that they will be given great attention in the years to come.

In its relationship with computer technology, systems science is not only a beneficiary, but also a benefactor. For example, one of the main issues of current research on parallel computers is to determine the right architectures for the various classes of systems problems. Systems science, one of whose main concerns has been a comprehensive study of systems problems, should be an important resource in this respect. For some problem areas involving computer technology, such as modeling and performance evaluation of large computing systems, software engineering, and knowledge engineering, systems science is also an important methodological resource.

The increasing computing power also influences new developments in mathematics, and these, in turn, produce new methodological types of systems, enlarging thus the inventory of systems categories. The most visible recent trend in mathematics is to generalize. As a rule, each generalization opens new methodological possibilities while, at the same time, tending to increase the computational complexity involved. To counterbalance this increase in computational complexity, we need an adequate increase in computing power. This explains why some of these generalizations have come to the fore only lately: they became practical only when the available computing power became sufficient to support them.

It is interesting to observe that a generalization of a mathematical theory results usually in a conceptually simpler theory. This is a consequence of the fact that some properties of the former theory are not required in the latter. At the same time, the more general theory has always a greater expressive power, which, however, is achieved only at the cost of greater computational demands. This conceptual simplicity of some mathematical theories whose expressive power is very high allows us to generate patterns of great complexity by simple rules that are repeated many times. Here, the high complexity is manifested on the level of data systems; it is produced, with the help of powerful computer technology, by simple systems on some epistemologically higher level. This phenomenon, which has lately emerged in systems science, is increasingly referred to by the catchy phrase *simple models of complex systems* [Goel and Rozehnal, 1991].

The current trend of generalizing mathematical theories is manifested by the following representative examples of generalizations:

- From quantitative to *qualitative theories*;
- From functions to *relations*;
- From graphs to *hypergraphs*;
- From ordinary geometry (Euclidean as well as non-Euclidean) to *fractal geometry*;
- From ordinary automata to *dynamic cellular automata*;
- From linear to *nonlinear theories*;
- From regular to *singular* (*catastrophy theory*);
- From precise analysis to *interval analysis*;
- From probability to *fuzzy measures*;
- From classical set theory and logic to *fuzzy set theory and logic*;

Systems Science in Retrospect and Prospect

- From regular languages to *developmental languages*;
- From intolerance to inconsistencies of all kinds to the *logic of inconsistency*;
- From single objective to *multiple objective criteria optimization*.

These generalizations have enriched not only our insights but, together with computer technology, extended also our capabilities for modeling the intricacies of the real world. It is not within the scope of this introductory book to cover these recent developments in mathematics, in spite of their importance for systems science.*

The study of relatively new categories of systems—such as cellular automata, developmental systems, neural networks, systems based on fractal geometry, fuzzy systems, and nonlinear chaotic systems—will likely be one of the most

---

*The following are bibliographical remarks regarding the recent developments in mathematics that are of significance to systems science. The growing interest in *qualitative mathematics* is well documented by a recent book on systems modeling and simulation [Fishwick and Luker, 1991]. Although thinking in terms of *mathematical relations* rather than functions is fundamental to systems science, there is no publication oriented to systems science that covers mathematical relations in a comprehensive way. An exception are *binary relations* on the Cartesian product of one set; these are usually called directed graphs and are covered in the literature under the name *graph theory* [Harary et al., 1965; Berge, 1973]. Generalizations of graphs are *hypergraphs* [Berge, 1973], which are important for dealing with structure systems (a hypergraph is a finite set with a family of its nonempty subsets that cover the set). A good overview of issues involved in nonlinear systems was prepared by Casti [1985]. Special areas emerging from the theory of nonlinear systems include *catastrophe theory* [Thom, 1975; Zeeman, 1977; Poston and Stewart, 1978; Casti, 1979; Arnold, 1984], which focuses on the study of singularities of mathematical functions; *theory of chaos* [Barnsley and Demko, 1986; Holden, 1986; Gleick, 1987; Devaney, 1987; Becker and Dorfler, 1989; Rasband, 1990], under which unpredictable bifurcations in behavior of dynamic deterministic systems, due to sensitive dependence upon initial conditions, are studied; and *fractal geometry* [Mandelbrot, 1977; Barnsley, 1988; Feder, 1988; Fleischmann et al., 1989], whose descriptive power enables us (with the help of the computer) to generate geometrical objects of most peculiar shapes (e.g., shapes described by functions that are continuous and yet nondifferentiable at every point) by transformations (often quite simple) of similarity and scaling on metric spaces, resulting in patterns that repeat themselves at smaller and smaller scales. Generalizations that emerged from classical automata and formal language theories include *Petri nets* [Peterson, 1981], *dynamic cellular automata* [Wolfram, 1986; Toffoli and Margolus, 1987] and *developmental systems* (or *L-systems*) [Herman and Rozenberg, 1975; Prusinkiewicz and Lindenmayer, 1990; Goel and Rozehnal, 1991]. Generalizations in mathematics that attempt to deal with the various types of uncertainty include the following areas: *interval analysis* [Moore, 1979], *fuzzy set theory and logic* [Dubois and Prade, 1980; Zimmermann, 1985; Kandel, 1986; Yager et al., 1987; Klir and Folger, 1988], *fuzzy measure theory* [Wang and Klir, 1992] and its various special subtheories, such as *evidence theory* [Shaffer, 1976] and *possibility theory* [Dubois and Prade, 1988]. Especially important for systems science is the notion of *fuzzy systems* [Negoita and Ralescu, 1975; Negoita, 1981; Pedrycz, 1989; Klir, 1990].

Basic ideas regarding the management of inconsistencies (usually local) in mathematical descriptions of systems were developed by Rescher [1976] and Rescher and Brandom [1980].

Another visible trend in the role of mathematics in systems science is the use of category theory [Padulo and Arbib, 1974; Mesarovic and Takahara, 1975, 1988; Takahara and Takai, 1985].

active areas of research in systems science in the coming years. Another active area will be, in my opinion, the study of the various types of goal orientation, in particular self-organization, self-preservation, self-reproduction, and autopoiesis.

On a broader scene, the genuine cross-disciplinary and holistic perspective of systems science will undoubtedly continue to counterbalance, by and large indirectly, the one-sided reductionist and discipline-oriented attitude that still prevails in contemporary science. The profound long-term implications of this ongoing influence upon science and even culture are well described by Norman D. Cook [1980]:

> The value of general systems theory lies not only in what it can bring to the individual academic disciplines, but also—and more importantly, I believe—in what it can bring in terms of a unified world-view and a fundamental, scientific conception of nature and human life—a world-view which is both open to precise, reductionist analysis and comprehensible to our mundane and poetic common sense. Thus far, Western science has excelled in reductionist analysis and has thereby led us into a world of mass communications, global transportation, computer technology and molecular medicine. But is has also led us into a psychological–philosophical wasteland where, in spite of and often because of stupendous material progress, our sense of the wholeness and unity of human existence and our daily appreciation of it have disappeared. Despite the hopes of the dreaming romantic, however, there is no retreat from modern science—either in practical, technological terms or in terms of the scientific world-view. And yet, there is the real hope of a more unified scientific world-view than that which is currently scientific "common sense"—a world-view which does not leave us with ridiculous, semi-scientific odds and ends . . . aggression centers, death instincts, genes for drug addiction, and magnetic monopoles . . . without a rational synthesis. Of course, the technological developments of reductionist science are a well-recognized mixed blessing, but the mixed nature of the reductionist's world-view is less widely appreciated. On the one hand, reductionist science has given us the ability to search for (and often find) single-element causes for the malfunction of larger systems, but, on the other hand, we apparently lose the ability to consider whole systems. Disease is traced back to the microorganism or even molecule which "causes" it, despite the fact that life and its irregularities are unquestionably systemic phenomena. The organisms' fundamental "attitude" to the pathogens may indeed determine the effectiveness of the attack by the smaller molecule, cell or organism on the larger system. In psychology, the outstanding questions concerned with the nature of consciousness, will, and individuality have been all but abandoned as the search for "aggression centers," "satiation centers," etc. proceeds. Without condemning such obviously relevant research, it is nonetheless apparent that the systemic phenomenon of man is not clarified in this type of reductionist psychology. At the social level, again the reductionist approach singles out a

few key elements—the man or men central to a social change—and fails to see the development of the social system itself.

Certainly the wholist's and the reductionist's approaches to reality differ fundamentally from one another, but there is nothing inherently contradictory between them. As the reductionist continues to define with ever greater precision the elements and few-body interactions of the elements within natural systems, the wholist can—and, indeed, must be able to—put those same elements within ever more comprehensive wholistic, many-bodied frameworks.

Although meaningful wholistic work necessarily must be built upon (and, chronologically, follow) the reductionist elucidation of the elements involved, the wholistic approach is not merely a rephrasing of the questions already answered by the reductionist. To the contrary, the wholist's questions as well as answers are apt to differ radically from the reductionist's. By putting the reductionist's remarkably well-defined bits and pieces within a rational context, we may be able to re-open the wholistic approach to natural phenomena which has long been neglected and even ridiculed by a scientific community devoted to rigorous reductionism.

Most important among the current wholistic modes of thought is "general systems theory." It offers a potentially more wholistic conception of natural phenomena than is now presented by the scientific world and yet without abandoning traditional scientific precision and verifiability. By dealing with man, society, or any unit of natural organization in terms of its basic properties as an abstract "system," the fundamentals of the given systems of nature may be discerned not only as the isolated, disconnected units of reductionist science, but also as units coordinated within a system which in turn functions within an environment. That is, individual objects can be rationally studied within their natural contexts.

What is the likely role of systems science in the emerging information society? There seems to be a general agreement that information society will be a fundamentally different environment for organizations than was industrial society. It is expected that the amount of available knowledge, the level of complexity, and the degree of turbulence will be significantly greater in the information society than they were in the industrial society. In addition, it is also expected that even the absolute growth rates of these three factors will be significantly greater than in the past.

In information society, organizations will have no choice but to deal with this radically different environment. To survive, they will have to adjust their structures, processes, and technologies to be able to cope with the new environment. This implies that designs of organizations in information society will be qualitatively different from those in industrial society.

Confronted with an environment that is characterized by more and increasing knowledge, complexity, and turbulence, organizations will be faced with challenging demands on decision making. Not only will decision making be considerably more complex, but decisions will also be required more frequently and will have to be made faster.

To keep organizations compatible with the environment, substantial portions of decision making in information society will be concerned with organizational innovations, i.e., radical changes in produced goods or services, as well as in the technologies, processes, and structures of the organizations themselves. In general, demands on organizational innovations will be more frequent, more extensive, and will have to be implemented faster than in the past.

All these demands on organizations in information society indicate that organizations will be required to function as *anticipatory systems*, i.e., systems that possess on-going capabilities of building relevant systems models of their environments and are able to use these models for making decisions and actions that optimize specific goals. This means that an on-going systems modeling of relevant aspects of the environment will be an essential feature of the decision-making infrastructure of organizations. This implies that expertise in systems science will be in increasing demand by organizations in the information society.*

---

*The following are the main journals that publish regularly articles relevant to systems science:
  Behavioral Sciences (Journal of the International Society for the Systems Sciences);
  Complex Systems (Complex Systems Publications, Champaign, Illinois);
  Cybernetics and Systems (Hemisphere, New York);
  Fuzzy Sets and Systems (North-Holland, Amsterdam);
  IEEE Transactions on Systems, Man, and Cybernetics (published by the IEEE Systems, Man, and Cybernetics Society);
  International Journal of General Systems (Gordon and Breach, New York);
  International Journal of Systems Science (Francis & Taylor, London);
  Kybernetes (MCB University Press, Bradford, U.K.);
  Kybernetika (Journal of the Czechoslovak Association for Cybernetics);
  Revue Internationale de Systemique (AFCET, Paris);
  Systems Practice (Plenum Press, New York and London);
  Systems Research (Journal of the International Federation for Systems Research).
Other valuable sources of information are:
  International Book Series on Systems Science and Engineering (sponsored by the International Federation for Systems Research and published by Plenum Press, New York);
  General Systems Yearbook (published by the International Society for the Systems Sciences);
  Bibliographies [Klir and Rogers, 1977; Trappl, Horn, and Klir, 1985].

Appendix

Mathematical Terminology and Notation

Throughout this book, *sets* are denoted by capital letters and their members by lower-case letters. To indicate that an individual object x is a *member* (or *element*) of a set X, we write $x \in X$; otherwise, if x is not a member of X, we write $x \notin X$.

A finite set can be defined by naming all its members. The set X whose members are $x_1, x_2, \ldots, x_n$ is usually written as

$$X = \{x_1, x_2, \ldots, x_n\}.$$

A set can also be defined by specifying properties satisfied by the members of the set. This type of definition is usually written as

$$X = \{x | x \text{ has properties } p_1, p_2, \ldots\},$$

where the symbol $|$ denotes the phrase "such that."

A set whose elements are themselves sets is often called a *family of sets*. It can be defined in the form

$$\{X_i | i \in I\},$$

where I is an *index set*; sets X_i in this family are distinguished from one another by the *index i*.

If every member of set X is also a member of set Y, then X is called a *subset* of Y, and this is written as $X \subset Y$. If $X \subset Y$ and $Y \subset X$, then the sets contain the same members. We call them *equal sets* and write $X = Y$. To indicate that sets X and Y are not equal, we write $X \neq Y$.

The set that contains all the possible elements of concern in each particular context or application is called the *universe of discourse* in that context. The set that contains no elements is called the *empty set* and is denoted by $\emptyset$. The family of all the subsets of a particular set X is called the *power set* of X and it is denoted by $\mathcal{P}(X)$.

The *relative complement of a set X* with respect to another set Y, which is denoted by $Y - X$, is the set of all the members of Y that are not also members of X. If the set X is the universe of discourse, the complement is absolute and is usually denoted by $\overline{X}$.

The *union of sets* X and Y is the set containing all the elements that belong either to set X alone, to set Y alone, or to both set X and set Y. This is denoted by $X \cup Y$. For a family of sets $\{X_i | i \in I\}$, the union is defined as

$$\bigcup_{i \in I} X_i = \{x | x \in X_i \text{ for some } i \in I\}.$$

The *intersection of sets* X and Y is the set containing all the elements belonging to both set X and Y; it is denoted by $X \cap Y$. For a family of sets $\{X_i | i \in I\}$, the intersection is defined by

$$\bigcap_{i \in I} X_i = \{x | X \in X_i \text{ for all } i \in I\}.$$

Any two sets X and Y are called *disjoint* if they have no elements in common, that is, if $X \cap Y = \emptyset$.

A family of pairwise disjoint nonempty subsets of a given set X is called a *partition* of X if the union of these subsets yields the original set X. The subsets that form a partition are usually called its *blocks*. Given a partition of X, each element of X belongs to one and only one of its blocks.

The Cartesian product of sets X and Y, denoted by $X \times Y$, is the set of all ordered pairs in which the first element is a member of X and the second element is a member of Y. The two sets may be equal, or they themselves may be Cartesian products. Thus, for example, $X \times X$ is the set of all ordered pairs that can be formed from elements of set X, and $(X \times Y) \times (X \times Y)$ is the set of all ordered pairs that can be formed from elements (pairs) of the Cartesian product $X \times Y$. Cartesian products may also be defined on more than two sets. For example, $X \times Y \times Z$ is the set of all ordered triples in which the first, second, and third elements are members of sets X, Y, Z, respectively. A convenient shorthand notation is often employed for Cartesian products in which X^2 denotes $X \times X$, X^3 denotes $X \times X \times X$, etc.

Given a Cartesian product of some sort, any subset of it is a relation. The fact that some elements of the Cartesian product (ordered pairs, triples, pairs of pairs, etc.) are not allowed in a relation means that the sets involved in the Cartesian product can manifest themselves only in a constrained way, i.e., the choice made from some of them restricts the possible choices from others.

A relation on $X \times Y$ is called a *function* if each element of X occurs at most once as the first element in the ordered pairs contained in the relation. To designate that a relation on $X \times Y$ is a function, f, we write $f: X \rightarrow Y$. The symbol $f(x)$ denotes the unique element of Y that is paired with the element x of X by function f; it is called the image of x. Set X is called a *domain* of f; set Y is called a *range* of f. A function is called *bijective* if each element of X occurs in exactly one pair of the relation and, similarly, each element of Y occurs in exactly one pair of the relation. Clearly, every bijective function on $X \times Y$ requires that sets X and Y contain the same number of elements.

Symbol f^{-1}, where f stands for a function, denotes the set of all pairs contained in f with the inverted order of their elements; $f^{-1}(y)$ denotes the set of all elements in X that are paired with element y by function f. When f is bijective, $f^{-1}(y)$ is a single element of X.

Various significant types of binary relations on X (i.e., relations of the type $R \subset X \times X$) are distinguished on the basis of three different characteristic properties: reflexivity, symmetry, and transitivity.

A relation $R \subset X \times X$ is *reflexive* if and only if $(x, x) \in R$ for every $x \in X$, that is, if every element of X is related to itself. Otherwise, R is called *irreflexive*. If $(x, x) \notin R$ for every $x \in X$, the relation is called *antireflexive*.

A relation $R \subset X \times X$ is *symmetric* if and only if for every pair $(x, y) \in R$, it is also the case that $(y, x) \in R$. That is, whenever x is related to y through a symmetric relation, then y is also related to x through the same relation. If this condition is not satisfied for one or more pairs, the relation is called *asymmetric*. If $(x, y) \in R$ and $(y, x) \in R$ implies $x = y$, the relation is called *antisymmetric*. If either $(x, y) \in R$ or $(y, x) \in R$ whenever $x \neq y$, then the relation is called *strictly antisymmetric*.

A relation $R \subset X \times X$ is *transitive* if and only if $(x, z) \in R$ whenever both $(x, y) \in R$ and $(y, z) \in R$ for at least one $y \in X$. That is, if x is related to y and y is related to z in a transitive relation, then x is also related to z. A relation that violates this property for at least one triple x, y, z is called *nontransitive*. If $(x, z) \notin R$ whenever both $(x, y) \in R$ and $(y, z) \in R$, then the relation is called *antitransitive*.

A relation $R \subset X \times X$ that is reflexive, symmetric, and transitive is called an *equivalence relation*. For each element x in X, we can define the set

$$E_x = \{y | (x, y) \in R\}$$

which contains all the elements of X that are related to x by the equivalence relation R. The element x is itself contained in E_x owing to the reflexivity of R; because R is symmetric and transitive, each member of E_x is related to all the other members of E_x. Furthermore, no member of E_x is related to any element of X that is not in E_x. This set E_x is referred to as an equivalence class of R with respect to x. The members of each equivalence class are considered equivalent to each other and distinct from elements in other equivalence classes. The family of all equivalence classes of X defined by an equivalence relation $R \subset X \times X$, which is usually denoted by X/R, forms a partition of X.

A relation $R \subset X \times X$ that is reflexive and symmetric is usually called a *compatibility relation*. An important concept associated with compatibility relations are compatibility classes. A *compatibility class* imposed by R is a subset A of X such that $(x, y) \in A$ for all $x, y \in A$. A *maximal compatible* is a compatibility class that is not contained within any other compatibility class. The family consisting of all the maximal compatibles induced by R on X is called a *complete cover* of X with respect to R.

It is obvious that equivalence relations are special cases of compatibility relations such that the complete covers are partitions of X. This is a consequence of the property of transitivity that equivalence relations possess. Compatibility relations classify elements of X into subsets of compatible elements (where each element relates to any other), but some of these subsets overlap.

Another important class of relations are *ordering relations*. While both equivalence and compatibility relations require symmetry, ordering relation require asymmetry (or antisymmetry) and transitivity. There are several types of ordering relations.

A relation $R \subset X \times X$ that is reflexive, antisymmetric, and transitive is called a *partial ordering*. The common symbol $\leq$ is suggestive of the properties of partial orderings. Thus, $x \leq y$ denotes that $(x, y) \in R$ and signifies that x *precedes* y (or y *succeeds* x); x is called a *predecessor* of y (and y is called a successor of x). If we need to distinguish several partial orderings, such as P, Q, R, we use the symbols $\leq_P$, $\leq_Q$, $\leq_R$, respectively.

A partial ordering $\leq$ on X does not guarantee that all pairs of elements x, y in X are comparable in the sense that either $x \leq y$ or $y \leq x$. If all pairs of elements are comparable, the partial ordering becomes *total ordering* (or *linear ordering*). Such an ordering is characterized by—in addition to reflexivity and antisymmetry—a property of *connectivity*: for all $x, y \in X$, $x \neq y$ implies either $x \leq y$ or $y \leq x$.

Let X be a set on which a partial ordering is defined and let A be a subset of X. If $x \in X$ and $x \leq y$ for every $y \in A$, then x is called a *lower bound* of A on X with respect to the partial ordering. If $x \in X$ and $y \leq x$ for every $y \in A$, then x is called an *upper bound* of A on X with respect to the partial ordering. If a particular lower bound of A succeeds (is greater than) any other lower bound of A, then it is called the *greatest lower bound*, or *infimum*, of A. If a particular upper bound precedes (is smaller than) every other upper bound of A, then it is called the *least upper bound*, or *supremum*, of A.

A partially ordered set X any two elements of which have a greatest lower bound (also referred to as a *meet*) and a least upper bound (also referred to as a *join*) is called a *lattice*. The meet and join of elements x and y in X are often denoted by $x \wedge y$ and $x \vee y$, respectively.

A partially ordered set any two elements of which have only a greatest lower bound is called a *lower semilattice* or *meet semilattice*. A partially ordered set any two elements of which have only a least upper bound is called an *upper semilattice* or join semilattice.

References

Ackoff, R. L. [1973], Science in the systems age: Beyond IE, OR, and MS, *Operations Research*, **21**(3), 661–671.
Allen, T. F. H., and Starr, T. B. [1982], *Hierarchy*. University of Chicago Press, Chicago.
Arbib, M. A. [1965], A common framework for automata theory and control theory, *SIAM Journal of Control, Ser. A*, **3**(2), 206–222.
Arbib, M. A. [1966], Automata theory and control theory: A rapprochement, *Automatica*, **3**, 161–189.
Arnold, V. I. [1984], *Catastrophe Theory*. Springer-Verlag, New York.
Ashby, W. R. [1952], *Design for a Brain*. John Wiley, New York.
Ashby, W. R. [1956], *An Introduction to Cybernetics*. John Wiley, New York.
Ashby, W. R. [1958a], Requisite variety and its implications for the control of complex systems, *Cybernetica*, **1**, 83–99.
Ashby, W. R. [1958b], General systems theory as a new discipline, *General Systems Yearbook*, **3**, 1–6.
Ashby, W. R. [1962a], Principles of the self-organizing system. In Foerster and Zopf [1962].
Ashby, W. R. [1962b], The self-reproducing system. In C. A. Muses (ed.), *Aspects of the Theory of Artificial Intelligence*. Plenum Press, New York, pp. 9–18.
Ashby, W. R. [1964], Introducory remarks at panel discussion. In Mesarovic [1964].
Ashby, W. R. [1968], Some consequences of Bremermann's limit for information processing systems. In H. Oestreicher and D. Moore (eds.), *Cybernetic Problems in Bionics*. Gordon and Breach, New York, pp. 69–76.
Ashby, W. R. [1969], Two tables of identities governing information flows within large systems, *ASC Communications*, **1**, 3–8.
Ashby, W. R. [1973], Some peculiarities of complex systems, *Cybernetic Medicine*, **9**(2), 1–6.
Ashby, W. R. [1981], *Mechanisms of Intelligence: Ross Ashby's Writings on Cybernetics* (R. Conant, ed.). Intersystems, Seaside, California.
Atkin, R. H. [1976], *Mathematical Structures in Human Affairs*. Heineman, London.
Atkin, R. H. [1977], *Combinatorial Connectives in Social Systems*. Birkhäuser, Basel and Stuttgart.
Auger, P. [1989], *Dynamics and Thermodynamics in Hierarchically Organized Systems: Applications in Physics, Biology and Economics*. Pergamon Press, Oxford.
Aulin, A. [1982], *The Cybernetic Laws of Social Progress: Towards a Critical Social Philosophy and a Criticism of Marxism*. Pergamon Press, Oxford.
Aulin, A. [1989], *Foundations of Mathematical System Dynamics*. Pergamon Press, Oxford.
Barnsley, M. [1988], *Fractals Everywhere*. Academic Press, San Diego.

ns, pp. 163–177.

Barnsley, M. F. and **Demko, S. G.** (eds.) [1986], *Chaotic Dynamics and Fractals*. Academic Press, Orlando, Florida.
Barto, A. G. [**1978**], Discrete and continuous models, *International Journal of General Systems*, **4**(3), 163–177.
Bateson, G. [**1967**], Cybernetic explanation, *American Behavioral Scientist*, **10**(8), 29–32.
Bateson, G. [1972], *Steps to an Ecology of Mind*. Chandler, San Francisco.
Bateson, G. [1979], *Mind and Nature*. E.P. Dutton, New York.
Becker, K.-H., and **Dorfler, M.** [1989], *Dynamical Systems and Fractals: Computer Graphics Experiments in Pascal*. Cambridge University Press, Cambridge.
Beckett, J. A. [1971], *Management Dynamics: The New Synthesis*. McGraw-Hill, New York.
Beer, S. [1966], *Decision and Control*. John Wiley, New York.
Beer, S. [1979], *The Heart of Enterprise*. John Wiley, New York.
Beer, S. [1981], *Brain of the Firm*. John Wiley, New York.
Bellman, R. [1972], *Adaptive Control Processes: A Guided Tour*. Princeton University Press, Princeton, New Jersey.
Bennett, S. [1979], *A History of Control Engineering: 1800–1930*. IEE, New York.
Berge, C. [1973], *Graphs and Hypergraphs*. North-Holland, New York.
Berlinski, D. [1976], *On Systems Analysis: An Essay Concerning the Limitations of Some Mathematical Methods in the Social, Political, and Biological Sciences*. MIT Press, Cambridge, Massachusetts.
Berrien, F.K. [1968], *General and Social Systems*. Rutgers University Press, New Brunswick, New Jersey.
Bertalanffy, L. von [1952], *Problems of Life: An Evolution of Modern Biological and Scientific Thought*. C.A. Watts, London.
Bertalanffy, L. von [1968], *General Systems Theory: Foundations, Development, Applications*. George Braziller, New York.
Bertalanffy, L. von [1980], *Systems View of Man*. West View Press, Boulder, Colorado.
Blauberg, I.V., **Sadovsky, V.N.**, and **Yudin, E.G.** [1977], *Systems Theory: Philosophical and Methodological Problems*. Progress, Moscow.
Bogdanov, A. [1980], *Essays in Tektology: The General Science of Organization*. Intersystems, Seaside, California (originally published in Russian in 1921).
Booth, T. L. [1967], *Sequential Machines and Automata Theory*. John Wiley, New York.
Boulding, K. L. [**1956**], General systems theory—The skeleton of science, *Management Science*, **2**, 197–208.
Boulding, K. E. [**1972**], Economics and general systems. In Laszlo [1972].
Bowler, T. D. [1981], *General Systems Thinking: Its Scope and Applicability*. North-Holland, New York.
Bremermann, H. J. [1962], Optimization through evolution and recombination. In M. C. Yovits *et al.* (eds.), *Self-Organizing Systems*. Spartan Books, Washington, D.C., pp. 93–106.
Bremermann, H. J. [1967], Quantifiable aspects of goal-seeking self-organizing systems. In M. Snell (ed.), *Progress in Theoretical Biology*. Academic Press, New York, pp. 59–77.
Brown, G. S. [1957], *Probability and Scientific Inference*. Longmans, Green and Co., London.
Buck, R. C. [1956], On the logic of general behavioral systems. In H. Feigl and M. Scriven (eds.), *Minnesota Studies in the Philosophy of Science*, Vol. 1. University of Minnesota Press, Minneapolis, pp. 223–238.

References

Buckley, W. [1967], *Sociology and Modern Systems Theory*. Prentice-Hall, Englewood Cliffs, New Jersey.
Bunge, M. [**1977**], The GST challenge to the classical philosophies of science. *International Journal of General Systems*, 4(1), 29–37.
Cartwright, T. J. [1991], Experimental systems research: Towards a laboratory for the general theory, *Cybernetics and Systems*, 22(1), 135–149.
Casti, J. L. [1977], *Dynamical Systems and Their Applications: Linear Theory*. Academic Press, New York.
Casti, J. L. [1979], *Connectivity, Complexity, and Catastrophe in Large-Scale Systems*. John Wiley, New York.
Casti, J. L. [1985], *Nonlinear System Theory*. Academic Press, Orlando, Florida.
Casti, J. L. [1986], On system complexity: Identification, measurement, and management. In Casti and Karlqvist [1986].
Casti, J. L. [1989], *Alternate Realities: Mathematical Models of Nature and Man*. John Wiley, New York.
Casti, J. L., and **Karlqvist, A.** (eds.) [1986], *Complexity, Language, and Life: Mathematical Approaches*. Springer-Verlag, New York.
Caswell, H., *et al.* [1972], A introduction to systems science for ecologists. In B. C. Patten (ed.), *Systems, Analysis and Simulation in Ecology*, Vol. II. Academic Press, New York.
Cavallo, R. E. [1979], *The Role of Systems Methodology in Social Science Research*. Martinus Nijhoff, Boston.
Cavallo, R. E. (ed.) [1982], *Systems Methodology in Social Science Research: Recent Developments*. Kluwer-Nijhoff, Boston.
Chadwick, G. [1971], *A Systems View of Planning*. Pergamon Press, Oxford.
Checkland, P. B. [**1976**], Science and systems paradigm. *International Journal of General Systems*, 3(2), 127–134.
Checkland, P. [1981], *Systems Thinking, Systems Practice*. John Wiley, New York.
Churchman, C. W. [1979], *The Systems Approach and Its Enemies*. Basic Books, New York.
Cleland, C. I., and **King, W. R.** [1972], *Management: A Systems Approach*. McGraw-Hill, New York.
Coming, P. A. [1983], *The Synergism Hypothesis: A Theory of Progressive Evolution*. McGraw-Hill, New York.
Conant, R. C. [1969], The information transfer required in regulatory processes, *IEEE Transactions on Systems Science and Cybernetics*, **SSC-5**(4), 334–338.
Conant, R. C. [1974], Information flows in hierarchical systems, *International Journal of General Systems*, 1(1), 9–18.
Conant, R. C. [**1976**], Laws of information which govern systems, *IEEE Transaction on Systems, Man, and Cybernetics*, **SMC-6**(4), 240–255.
Conant, R. C. [1988], Extended dependency analysis of large systems. Part I: Dynamic analysis; Part II: Static analysis. *International Journal of General Systems*, 14(2), 97–141.
Conant, R. C., and **Ashby, W. R.** [**1970**], Every good regulator of a system must be a model of that system, *International Journal of Systems Science*, 1(2), 89–97.
Conrad, M. [1983], *Adaptability: The Significance of Variability from Molecule to Ecosystem*. Plenum Press, New York.
Cook, N. D. [1980], *Stability and Flexibility: An Analysis of Natural Systems*. Pergamon Press, Oxford.

Dalenoort, G. J. (ed.) [1989], *The Paradigm of Self-Organization: Current Trends in Self-Organization.* Gordon and Breach, New York.
Deutch, K. W. [1966], *The Nerves of Government: Models of Political Communication and Control.* Free Press, New York.
Devaney, R. L. [1987], *An Introduction to Chaotic Dynamical Systems.* Addison-Wesley, Reading, Massachusetts.
Dubois, D., and **Prade, H.** [1980], *Fuzzy Sets and Systems: Theory and Applications.* Academic Press, New York.
Dubois, D., and **Prade, H.** [1988], *Possibility Theory.* Plenum Press, New York.
Easton, D. [1965], *Systems Analysis of Political Life.* Prentice-Hall, Englewood Cliffs, New Jersey.
Ellis, W. D. (ed.) [1938], *A Source Book of Gestalt Psychology.* Routledge & Kegan Paul, London.
Farley, B. G., and **Clark, W. A.** [1954], Simulation of self-organizing systems by digital computer, *IRE Transacations*, **IT-4**(3), 76–84.
Farlow, S. J. [1984], *Self-Organizing Methods in Modeling GMDH Type Algorithms.* Marcel Dekker, New York.
Feder, J. [1988], *Fractals.* Plenum Press, New York.
Fishwick, P. A., and **Luker, P. A.** (eds.) [1991], *Qualitative Simulation Modeling and Analysis.* Springer-Verlag, New York.
Flagle, C. D., *et al.* (eds.) [1960], *Operations Research and Systems Engineering.* The John Hopkins Press, Baltimore.
Fleischmann, M., **Tildesley, D. J.**, and **Ball, R. C.** [1989], *Fractals in the Natural Sciences.* Princeton University Press, Princeton, New Jersey.
Flood, R. L., and **Carson, E. R.** [1988], *Dealing With Complexity: An Introduction to the Theory and Application of Systems Science.* Plenum Press, New York.
Foerster H. von, and **Zopf, G. W.** (eds.) [1962], *Principles of Self-Organization.* Pergamon Press, New York.
Foerster, H. von [1981], *Observing Systems.* Intersystems, Seaside, California.
Forrester, J. W. [1961], *Industrial Dynamics.* MIT Press, Cambridge, Massachusetts.
Forrester, J. W. [1968], *Principles of Systems.* Wright-Allen Press, Cambridge, Massachusetts.
Forrester, J. W. [1969], *Urban Dynamics.* MIT Press, Cambridge, Massachusetts.
Forrester, J. W. [1971], *World Dynamics.* Wright-Allen, Cambridge, Massachusetts.
Forrester, J. W. [1975], *Collected Papers of Jay W. Forrester.* Wright-Allen Press, Cambridge, Massachusetts.
Gaines, B. R. [1972], Axioms for adaptive behaviour, *International Journal of Man-Machine Studies*, **4**, 169–199.
Gaines, B. R. [1977], System identification, approximation and complexity, *International Journal of General Systems*, **3**(3), 145–174.
Gaines, B. R. [1979], General systems research: quo vadis? *General Systems Yearbook*, **24**, 1–9.
Gaines, B. R. [**1984**], Methodology in the large: Modeling all there is, *Systems Research*, **1**(2), 91–103.
Gallopín, G. C. [1981], The abstract concept of environment, *International Journal of General Systems*, **7**(2), 139–149.
Gardner, M. R., and **Ashby, W. R.** [1970], Connectance of large dynamic (cybernetic) systems: Critical values of stability. *Nature*, **228**(5273), 784.
Garey, M. R., and **Johnson, D. S.** [1979], *Computers and Intractability: A Guide to the Theory of NP-Completeness.* W.H. Freeman, San Francisco.

Gelfand, A. E., and **Walker, C. C.** [1977], The distribution of cycle lengths in a class of abstract systems, *International Journal of General Systems*, **4**(1), 39–45.
Gill, A. [1962], *Introduction to the Theory of Finite-State Machines*. McGraw-Hill, New York.
Glasersfeld, E. von [1987], *The Construction of Knowledge: Contribution to Conceptual Semantics*. Intersystems, Seaside, California.
Glasersfeld, E. von [**1990**], An exposition of constructivism: Why some like it radical. In R. B. Davis *et al.* (eds.), *Constructivist Views on the Teaching and Learning of Mathematics*. JRME Monographs, Reston, Virginia.
Gleick, J. [1987], *Chaos: Making a New Science*. Viking, New York.
Goel, N., and **Rozehnal, I.** [1991], Some non-biological applications of L-systems. *International Journal of General Systems*, **18**(4).
Goguen, J. A., and **Varela, F. J.** [**1979**], Systems and distinctions: Duality and complementarity, *International Journal of General Systems*, **5**(1), 31–43.
Gray, W., **Duhl, F. J.**, and **Rizzo, N. D.** (eds.) [1969], *General Systems Theory and Psychiatry*. Little, Brown and Company, Boston.
Haken, H. [1977], *Synergetics*. Springer-Verlag, New York.
Haken, H. [1988], *Information and Self-Organization: A Macroscopic Approach to Complex Systems*. Springer-Verlag, New York.
Halfon, E. (ed.) [1979], *Theoretical Systems Ecology*. Academic Press, New York.
Hall, A. D., III [1989], *Metasystems Methodology: A New Synthesis and Unification*. Pergamon Press, Oxford.
Hanken, A. F. G. [1981], *Cybernetics and Society: An Analysis of Social Systems*. Heyden, Philadelphia, Pennsylvania.
Hanken, A. F. G., and **Reuver, H. A.** [1981], *Social Systems and Learning Systems*. Martinus Nijhoff, Boston.
Harary, F., *et al.* [1965], *Structural Models: An Introduction to the Theory of Directed Graphs*. John Wiley, New York.
Harel, D. [1987], *Algorithmics: The Spirit of Computing*. Addison-Wesley, Reading, Massachusetts.
Hartley, R. V. L. [1928], Transmission of information, *The Bell System Technical Journal*, **7**, 535–563.
Hartmanis, J., and **Stearns, R. E.** [1966], *Algebraic Structure Theory of Sequential Machines*. Prentice-Hall, Englewood Cliffs, New Jersey.
Hartmann, G. W. [1935], *Gestalt Psychology: A Survey of Facts and Principles*. Ronald Press, New York.
Hartnett, W. E. (ed.) [1977], *Systems: Approaches, Theories, Applications*. D. Reidel, Boston.
Herman, G. T., and **Rozenberg, G.** [1975], *Developmental Systems and Languages*. North-Holland, New York.
Holden, A. V. (ed.) [1986], *Chaos*. Princeton University Press, Princeton, New Jersey.
Holland, J. H. [1975], *Adaptation in Natural and Artificial Systems*. University of Michigan Press, Ann Arbor.
Holling, C. S. [1973], Resilience and stability of ecological systems, *Annual Review of Ecology and Systematics*, **4**, 1–23.
Hoos, I. R. [1972], *Systems Analysis in Public Policy: A Critique*. University of California Press, Berkeley and Los Angeles.
Islam, S. [1974], Toward integration of two system theories of Mesarovic and Wymore, *International Journal of General Systems*, **1**(1), 35–40.
Ivakhnenko, A. G. [1970], Heuristic self-organization in problems of engineering cybernetics, *Automatica*, **6**(2), 207–219.

Jantsch, E. [1980], *The Self-Organizing Universe: Scientific and Human Implications of the Emerging Paradigm of Evolution.* Pergamon Press, Oxford.
Kandel, A. [1986], *Fuzzy Mathematical Techniques With Applications.* Addison-Wesley, Reading, Massachusetts.
Katz, D. [1950], *Gestalt Psychology: Its Nature and Significance.* Ronald Press, New York.
Kauffman, S. A. [1969], Metabolic stability and epigenesis in randomly constructed genetic nets, *Journal of Theoretical Biology*, **22**, 437–467.
Kickert, W. J. M. [1980], *Organization of Decision-Making: A Systems-Theoretical Approach.* North-Holland, Amsterdam and New York.
Klir, G. J. [1969], *An Approach to General Systems Theory.* Van Nostrand Reinhold, New York.
Klir, G. J. [1970], On the relation between cybernetics and general systems theory. In J. Rose (ed.), *Progress in Cybernetics.* Gordon and Breach, London, pp. 155–165.
Klir, G. J. [1972a], *Trends in General Systems Theory.* Wiley-Interscience, New York.
Klir, G. J. [1972b], *Introduction to the Methodology of Switching Circuits.* Van Nostrand Reinhold, New York.
Klir, G. J. [1976], Identification of generative structures in empirical data, *International Journal of General Systems*, **3**(2), 89–104.
Klir, G. J. (ed.) [1978], *Applied General Systems Research: Recent Developments and Trends.* Plenum Press, New York.
Klir, G. J. [1985a], *Architecture of Systems Problem Solving.* Plenum Press, New York.
Klir, G. J. [1985b], The emergence of two-dimensional science in the information society, *Systems Research*, **2**(1), 33–41.
Klir, G. J. [1986], Reconstructability analysis: An offspring of Ashby's constraint theory, *Systems Research*, **3**(4), 267–271.
Klir, G. J. [1988], Systems profile: The emergence of systems science, *Systems Research*, **5**(2), 145–156.
Klir, G. J. [1989], Is their more to uncertainty than some probability theorists might have us believe? *International Journal of General Systems*, **15**(4), 347–378.
Klir, G. J. (ed.) [1990], *Special Issue of the International Journal of General Systems on a Quarter Century of Fuzzy Systems*, **17**(2-3), 89–294.
Klir, G. J., and **Folger, T. A.** [1988], *Fuzzy Sets, Uncertainty, and Information.* Prentice-Hall, Englewood Cliffs, New Jersey.
Klir, G. J., **Hufford, K. D.**, and **Zeleny, M.** [1988], Osmotic growths: A challenge to systems science, *International Journal of General Systems*, **14**(1), 1–17.
Klir, G. J., and **Parviz, B.** [1986], General reconstruction characteristics of probabilistic and possibilistic systems, *International Journal of Man-Machine Systems*, **25**, 367–397.
Klir, G. J., and **Rogers, G.** (eds.) [1977], *Basic and Applied Systems Research: A Bibliography.* SUNY, Binghamton, New York.
Koestler, A. [1967], *The Ghost in the Machine.* Macmillan, New York.
Koestler, A., and **Smythies, J. R.** (eds.) [1969], *Beyond Reductionism: New Perspectives in the Life Sciences.* Hutchinson, London.
Koffka, K. [1935], *Principles of Gestalt Psychology.* Kegan Paul, Trench, Trubner & Co., London.
Köhler, W. [1929], *Gestalt Psychology.* Horace Liveright, New York.
Krinsky, V. I. (ed.) [1984], *Self-Organization: Autowaves and Structure far from Equilibrium.* Springer-Verlag, New York.
Krohn, K. B., and **Rhodes, J. L.** [1968], Complexity of finite semigroups, *Annals of Mathematics*, **88**, 128–160.

References

Landau, Y. D. [1979], *Adaptive Control: The Model Reference Approach*. Marcel Dekker, New York.
Lange, O. [1970], *Introduction to Economic Cybernetics*. Pergamon Press, Oxford (Polish original published in 1965).
La Porte, T. R. (ed.) [1975], *Organized Social Complexity: Challenge to Politics and Policy*. Princeton University Press, Princeton, New Jersey.
Laszlo, E. [1971], *Introduction to Systems Philosophy*. Gordon and Breach, New York.
Laszlo, E. (ed.) [1972], *The Relevance of General Systems Theory*. George Braziller, New York.
Laszlo, E. [1975], The meaning and significance of general systems theory, *Behavioral Science*, **20**(1), 9–24.
Laszlo, E. [1980], Some reflections on systems theory's critics, *Nature and Systems*, **2**(1), 49–53.
Laviolette, P. A. [1985], An introduction to subquantum kinetics: I. An overview of the methodology, *International Journal of General Systems*, **11**(4), 281–293.
Leduc, S. [1911], *The Mechanism of Life*. Rebman, London.
Lee, A. M. [1970] *Systems Analysis Frameworks*. John Wiley, New York.
Lerner, D. (ed.) [1963], *Parts and Wholes*. Free Press, New York.
Lilienfeld, R. [1978], *The Rise of Systems Theory: An Ideological Analysis*. Wiley-Interscience, New York.
Lofgren, L. [1977], Complexity of descriptions of systems: A foundational study, *International Journal of General Systems*, **3**(4), 197–214.
Lowen, W. [1982], *Dichotomies of the Mind: A Systems Science Model of the Mind and Personality*. John Wiley, New York.
Makridakis, S., and **Faucheux, C.** [1973], Stability properties of general systems, *General Systems Yearbook*, **18**, 3–12.
Mandelbrot, B. B. [1977], *The Fractal Geometry of Nature*. W.H. Freeman, New York.
Marchal, J. H. [1975], The concept of a system, *Philosophy of Science*, **42**(4), 448–467.
Margaleff, R. [1968], *Perspectives in Ecological Theory*. University of Chicago Press, Chicago.
Maturana, H. R., and **Varela, F. J.** [1980], *Autopoiesis and Cognition: The Realization of the Living*. D. Reidel, Boston.
Maturana, H. R., and **Varela, F.** [1987], *The Tree of Knowledge: The Biological Roots of Human Understanding*. New Science Library, Shambhala.
Mayr, O. [1970], *The Origins of Feedback Control*. MIT Press, Cambridge Massachusetts.
McCulloch, W. A., and **Pitts, W.** [1943], A logical calculus of the ideas immanent in nervous activity, *Bulletin of Mathematical Biology*, **5**, 115–133.
Mesarovic, M. D. (ed.) [1964], *Views of Genreal Systems Theory*. John Wiley, New York.
Mesarovic, M. D. (ed.) [1968], *Systems Theory and Biology*. Springer-Verlag, New York.
Mesarovic, M. D. [1972], A mathematical theory of general systems. In Klir [1972a].
Mesarovic, M. D., **Macko, D.**, and **Takahara, Y.** [1970], *Theory of Hierarchical Multilevel Systems*. Academic Press, New York.
Mesarovic, M. D., and **Takahara, Y.** [1975], *General Systems Theory: Mathematical Foundations*. Academic Press, New York.
Mesarovic, M. D., and **Takahara, Y.** [1988], *Abstract Systems Theory*. Springer-Verlag, New York.
Miller, J. G. [1953], Profits and problems of homeostatic models in the behavioral sciences. *Chicago Behavioral Science Publications*, No. 1, University of Chicago Press, Chicago.
Miller, J. G. [1978], *Living Systems*. McGraw-Hill, New York.

Miller, J. G. [**1986**], Can systems theory generate testable hypotheses?: From Talcott Parsons to living systems theory, *Systems Research*, **3**(2), 73–84.
Miser, H. J., and Quade, E. S. (eds.) [1985], *Handbook of Systems Analysis*. North-Holland, New York.
Moore, R. E. [1979], *Methods and Applications of Interval Analysis*. SIAM, Philadelphia.
Negoita, C. V. [1981], *Fuzzy Systems*. Abacus Press, Tunbridge Wells, U.K.
Negoita, C. V., and Ralescu, D. A. [1975], *Applications of Fuzzy Sets to Systems Analysis*. Birkhäuser, Basel and Stuttgart.
Neumann, J. von [1966], *Theory of Self-Reproducing Automata*. University of Illinois Press, Urbana, Illinois.
Nicolis, G., and Prigogine, I. [1977], *Self-Organization in Nonequilibrium Systems: From Dissipative Structures to Order through Fluctuations*. Wiley-Interscience, New York.
Padulo, L., and Arbib, M. A. [1974], *System Theory: A Unified State-Space Approach to Continuous and Discrete Systems*. W.B. Saunders, Philadelphia.
Pask, G. [1975], *The Cybernetic of Human Learning and Performance*. Hutchinson, London.
Pask, G. [1976], *Conversation Theory: Applications in Education and Epistemology*. Elsevier, Amsterdam.
Pattee, H. H. [**1986**], Universal principles of measurement and language functions in evolving systems. In Casti and Karlqvist [1986].
Patten, B. C. [1978], Systems approach to the concept of environment, *Ohio Journal of Science*, **78**(4), 206–222.
Patten, B. C., and Auble, G. T. [1981], System theory and the ecological niche, *The American Naturalist*, **117**(6), 893–922.
Patten, B. C., and Odum, E. P. [1981], The cybernetic nature of ecosystem, *The American Naturalist*, **118**, 886–895.
Pedrycz, W. [1989], *Fuzzy Control and Fuzzy Systems*, John Wiley, New York.
Peterson, J. L. [1981], *Petri Net Theory ad the Modeling of Systems*. Prentice-Hall, Englewood Cliffs, New Jersey.
Philips, P. C. [1976], *Holistic Thought in Social Science*. Stanford University Press, Stanford, California.
Pichler, F., and Moreno-Diaz, R. (eds.) [1990], *Computer Aided Systems Theory*. Springer-Verlag, Berlin and Heidelberg.
Pilette, R., Sigal, R., and Blamire, J. [1990], Stability–complexity relationship within models of natural systems, *Biosystems*, **23**(4), 359–370.
Poston, T., and Stewart, I. [1978], *Catastrophe Theory and Its Applications*. Pitman, New York.
Prigogine, I. [1980], *From Being to Becoming*. W.H. Freeman, San Francisco.
Prigogine, I. [**1985**], New perspectives on complexity. In *The Sciences and Praxis of Complexity*. The United Nations University, Tokyo, pp. 107–118.
Prusinkiewicz, P., and Hanan, J. [1989], *Lindenmayer Systems, Fractals, and Plants*. Springer-Verlag, New York.
Prusinkiewicz, P., and Lindenmayer, A. [1990], *Algorithmic Beauty of Plants*. Springer-Verlag, New York.
Rapoport, A. [1984], *General Systems Theory: Essential Concepts and Applications*. Abacus, Tunbridge Wells, U.K.
Rasband, S. N. [1990], *Chaotic Dynamics of Nonlinear Systems*. John Wiley, New York.
Rényi, A. [1970], *Probability Theory*. North-Holland, Amsterdam (Chap. IX, Introduction to Information Theory, pp. 540–616).
Rescher, N. [1976], *Plausible Reasoning*. Van Gorcum, Amsterdam.

Rescher, N. [1977], *Methodological Pragmatism: A Systems-Theoretic Approach to the Theory of Knowledge.* New York University Press, New York.
Rescher, N. [1980], *Induction.* Basil Blackwell, Oxford.
Rescher, N., and **Brandom, R.** [1980], *The Logic of Inconsistency.* Basil Blackwell, Oxford.
Rosen, R. [1970], *Dynamical System Theory in Biology.* Wiley-Interscience, New York.
Rosen, R. [**1972**], Some system theoretical problems in biology. In Laszlo [1972].
Rosen, R. [**1977**], Complexity and system description. In Hartnett [1977].
Rosen, R. [1978], Biology and systems research: An overview. In Klir [1978].
Rosen, R. [**1979a**], Old trends and new trends in general systems research, *International Journal of General Systems*, **5**(3), 173–184.
Rosen, R. [**1979b**], Anticipatory systems in retrospect and prospect, *General Systems Yearbook*, **24**, 11–23.
Rosen, R. [**1981**], The challenges of system theory, *General Systems Bulletin*, **11**, 7–10.
Rosen, R. [**1985a**], The physics of complexity, *Systems Research*, **2**(2), 171–175.
Rosen, R. [**1985b**], *Anticipatory Systems: Philosophical, Mathematical, and Methodological Foundations.* Pergamon Press, Oxford.
Rosen, R. [**1986**], Some comments on systems and system theory, *International Journal of General Systems*, **13**(1), 1–3.
Rosen, R. [1991], *Life Itself: A Comprehensive Inquiry Into the Nature, Origin and Fabrication of Life. Part I: Epistemology.* Columbia University Press, New York.
Rugh, W. J. [1981], *Nonlinear System Theory: The Volterra/Wiener Approach.* John Hopkins University Press, Baltimore.
Sage, A. P. [1977], *Methodology for Large Scale Systems.* McGraw-Hill, New York.
Sandquist, G. M. [1985], *Introduction to Systems Science.* Prentice-Hall, Englewood Cliffs, New Jersey.
Sayre, K. [1976], *Cybernetics and Philosophy of Mind.* Humanities Press, Atlantic Highlands, New Jersey.
Shafer, G. [1976], *A Mathematical Theory of Evidence.* Princeton University Press, Princeton, New Jersey.
Shannon, C. E. [1948], The mathematical theory of communication, *The Bell System Technical Journal*, **27**, 379–423, 623–656.
Simon, H. A. [**1962**], The architecture of complexity, *Proceedings of the American Philosophical Society*, **106**, 467–482.
Simon, H. A. [1969], *The Sciences of the Artificial.* MIT Press, Cambridge, Massachusetts.
Simon, H. A. [1977a], How complex are complex systems? In F. Suppe and P. D. Asquith (eds.) *PSA 1976*, Vol. 2, Philosophy of Science Association, East Lansing, Michigan, pp. 507–522.
Simon, H. A. [1977b], *Models of Discovery.* D. Reidel, Boston.
Simon, H. A. [1988], Prediction and prescription in systems modeling, *IIASA Manuscript*, Laxenburg (Austria).
Skoglund, V. [1967], *Similitude: Theory and Applications.* International Textbook Co., Scranton, Pennsylvania.
Smuts, J. C. [1926], *Holism and Evolution.* Macmillan, London. (Reprinted in 1973 by Greenwood Press, Westport, Connecticut.)
Steinbruner, J. D. [1974], *The Cybernetic Theory of Decision.* Princeton University Press, Princeton, New Jersey.
Suppes, P. [1983], Reductionism, atomism and holism, In *Parts and Wholes* (Proceedings of the International Workshop in Lund, Sweden), Vol. 1, Forskningsradsnamnden, Stockholm, pp. 134–140.

Suppes, P. [1977], Some remarks about complexity. In F. Suppes and P. D. Asquith (eds.), *PSA 1976*, Vol. 2, Philosophy of Science Assoc., East Lansing, Michigan, pp. 543–547.
Susiluoto, I. [1982], *The Origins and Development of Systems Thinking in the Soviet Union.* Annales Academia Scientiarum Fennicae (Finnish Academy of Sciences), Helsinki.
Szucz, E. [1980], *Similitude and Modelling*. Elsevier, New York.
Takahara, Y., and **Takai, T.** [1985], Category theoretical framework of general systems, *International Journal of General Systems*, **11**(1), 1–33.
Teller, E. [1980], *The Pursuit of Simplicity*. Pepperdine University Press, Malibu, California.
Thom, R. [1975], *Structural Stability and Morphogenesis*. Addison-Wesley, Reading, Massachusetts.
Thorn, D. C. [1963], *An Introduction to Generalized Circuits*. Wadsworth, Belmont, California.
Toffoli, T., and **Margolus, N.** [1987], *Cellular Automata Machines: A New Environment for Modeling*. MIT Press, Cambridge, Massachusetts.
Trappl, R. (ed.) [1983], *Cybernetics: Theory and Applications*. Hemisphere, Washington, D.C.
Trappl, R., **Horn, W.**, and **Klir, G. J.** (eds.) [1985], *Basic and Applied General Research: A Bibliography 1977–1984*. Supplement of *Cybernetics and Systems*, Vol. 16. Hemisphere, New York.
Turchin, V. F. [1977], *The Phenomenon of Science*. Columbia University Press, New York.
Van Gigch, J. P. [1974], *Applied General Systems Theory*. Harper & Row, New York.
Varela, F. J. [1979], *Principles of Biological Autonomy*. North-Holland, New York.
Varela, F. J., **Maturana, H. R.**, and **Uribe, R. B.** [**1974**], Autopoiesis: The organization of living systems, its characterization and a model, *Biosystems*, **5**(4), 187–196.
Walker, C. C. [1971], Behavior of a class of complex systems: The effect of system size on properties of terminal cycles, *Journal of Cybernetics* **1**(4), 55–67.
Wang, Z., and **Klir, G. J.** [1992], *Fuzzy Measure Theory*. Plenum Press, New York.
Warfield, J. N. [1976], *Societal Systems*. Wiley-Interscience, New York.
Weaver, W. [**1948**], Science and complexity, *American Scientist*, **36**, 536–544.
Weinberg, G. M. [**1972**], A computer approach to general systems theory. In Klir [1972a], pp. 98–141.
Weinberg, G. M. [1975], *An Introduction to General Systems Thinking*. John Wiley, New York.
Weir, M. [1984], *Goal-Directed Behavior*. Gordon and Breach, New York.
Weiss, P. A. [1969], The living system: Determinism stratified. In Koestler and Smythies [1969].
Wiener, N. [1948], *Cybernetics or Control and Communication in the Animal and the Machine*. MIT Press, Cambridge, Massachusetts.
Windeknecht, J. G. [1971], *General Dynamical Processes: A Mathematical Introduction*. Academic Press, New York.
Wolfram, S. (ed.) [1986], *Theory and Applications of Cellular Automata*. World Scientific, Singapore.
Wymore, A. W. [1969], *A Mathematical Theory of Systems Engineering: The Elements*. John Wiley, New York.
Wymore, A. W. [1972], A wattled theory of systems. In Klir [1972a].
Wymore, A. W. [1976], *Systems Engineering Methodology for Interdisciplinary Teams*. John Wiley, New York.
Yager, R. R., *et al.* [1987], *Fuzzy Sets and Applications: Selected Papers by L. A. Zadeh*. John Wiley, New York.

References

Yates, F. E. (ed.) [1987], *Self-Organizing Systems: The Emergence of Order*. Plenum Press, New York.
Yovits, M. C., and **Cameron, S.** (eds.) [1960], *Self-Organizing Systems*. Pergamon Press, Oxford.
Yovits, M. C., **Jacobi, G. T.**, and **Goldstein, G. D.** (eds.) [1962], *Self-Organizing Systems—1962*. Spartan Books, Washington, D.C.
Zadeh, L. A. [**1962**], From circuit theory to system theory, *IRE Proceedings*, **50**(5), 856–865.
Zadeh, L. A. [1969], The concept of system, aggregate, and state in system theory. In Zadeh and Polak [1969].
Zadeh, L. A. [1973], Outline of a new approach to the analysis of complex systems and decision processes, *IEEE Transactions on Systems, Man, and Cybernetics*, **SMC-1**(1), 28–44.
Zadeh, L. A., and **Desoer, C. A.** [1963], *Linear System Theory*, McGraw-Hill, New York.
Zadeh, L. A., and **Polak, E.** (eds.) [1969], *System Theory*. McGraw-Hill, New York.
Zeeman, E. C. [1977], *Catastrophe Theory: Selected Papers 1972–1977*. Addison-Wesley, Reading, Massachusetts.
Zeigler, B. P. [1974], A conceptual basis for modelling and simulation, *International Journal of General Systems*, **1**(4), 213–228.
Zeigler, B. P. [1976a], The hierarchy of system specifications and the problem of structural inference. F. Suppe and P. D. Asquith (eds.), *PSA 1976*, Vol. 1, Philosophy of Science Association, East Lansing, Michigan, pp. 227–239.
Zeigler, B. P. [1976b], *Theory of Modelling and Simulation*. John Wiley, New York.
Zeleny, M. (ed.) [1980], *Autopoiesis, Dissipative Structures, and Spontaneous Social Orders*. Westview Press, Boulder, Colorado.
Zeleny, M. (ed.) [1981], *Autopoiesis: A Theory of Living Organization*. North-Holland, New York.
Ziegenhagen, E. A. [1986], *The Regulation of Political Conflict*. Praeger, New York.
Zimmermann, H. J. [1985], *Fuzzy Set Theory—And Its Applications*. Kluwer–Nijhoff, Boston.

PART II

Classical Systems Literature

Introduction and Comments

The purpose of this part of the book is to provide the reader with some representative original papers pertaining to the various facets of systems science discussed in Part I, "Systems Science: A Guided Tour." These papers were carefully selected to reinforce Part I, and their order of presentation mirrors the order of presentation of topics there. Hence, they should be read in parallel with the individual chapters of Part I that they bear upon (as indicated in the Detailed Contents).

The papers were chosen for their authoritativeness, originality, clarity, or historical significance. Highly technical papers, whose reading would require special background knowledge, were avoided. In a few cases, the papers chosen were shortened by excluding their less relevant parts.

When reading the two parts of the book in parallel, the reader should be aware of occasional dissonance. One source of this dissonance is connected with terminology. While the presentation in Part I is terminologically consistent, terminologies used in different papers in Part II vary significantly. To alleviate the dissonance, the reader should resort to common sense to interpret, where appropriate, the papers in terms of the terminology introduced in Part I. Thus, for example, some authors use the terms "systems theory," "general systems theory," "systems research," and the like, while, in fact, they mean something very similar, if not identical, to what is called "systems science" in Part I.

Another source of dissonance is connected with the philosophical views that underlie the various presentations. While Part I is based upon the constructivist view, some papers in Part II do not necessarily adhere to this view. The philosophical views are usually not stated explicitly in the papers, but can be inferred from the line of reasoning in each of them. Papers based on views different from constructivism should be considered as complementary to the material in Part I. They enrich the whole book and allow the reader himself or herself to reflect on this fundamental issue.

Let me briefly overview the papers selected for Part II. Papers 1 and 2 are basically concerned with the *meaning of systems science*. The latter paper is historically a sort of *systems science manifesto*: it was the first paper that argued the legitimacy of systems science and explained similarities and differences between systems science and science in the traditional sense.

Paper 3 is an excellent exposition of *constructivism* by one of its most visible current advocates. Although the author does not use systems terminology in this paper, the implications of the expressed view for systems science are easy to see.

Papers 4–13 cover various aspects of the *history of systems movement*. Their purpose is to characterize the evolution of systems movement and to elaborate on some particular developments within it that are only touched upon in Part I. The reader might be surprised that no writings by Ludwig von Bertalanffy, an important figure in systems movement, are included here. The reason is solely pragmatic. His key papers on general systems theory are contained in a carefully edited book [Bertalanffy, 1968], whose paperback edition is available at a reasonable price. Rather than reprinting some of the papers here, it seems more sensible to recommend the book mentioned as a supplementary reading to this book.

Papers 14–16 discuss three broad *methodological issues*: the *role of computers* in large-scale systems modeling, the similarities and differences between *discrete and continuous systems*, and the issue of the relationship between wholes and parts, as addressed in an operational way by a methodology referred to as *reconstructability analysis*.

Papers 17 and 18 are included here to exemplify laws of systems that are derived mathematically. Included are the famous *law of requisite variety* (due to Ashby) and the various *laws that emerge from information theory*.

Papers 19–25 cover a broad range of issues regarding *systems complexity, computational complexity*, and the role of *simplification in science*.

Papers 26–31 describe characteristics of some important types of goal-oriented systems in more detail than they are described in Part I. Covered in the papers are regulators, self-organizing systems, anticipatory systems, autopoietic systems, self-reproducing systems, and evolutionary systems.

Papers 32–34 describe the relationship between systems science and three other areas: philosophy of science, biology, and economics. They discuss how these areas have influenced systems science and vice versa.

The last paper, Paper 35, is a broad overview. It attempts to capture the whole spectrum of views, principles, theories, etc., that pertain to systems science.

Detailed Contents

Paper 1.	Some Comments on Systems and System Theory [Rosen, **1986**]: Chap. 1	213
Paper 2.	The Emergence of Two-Dimensional Science in the Information Society [Klir, **1985b**]: Chap. 1	217
Paper 3.	An Exposition of Constructivism: Why Some Like It Radical [Glasersfeld, **1990**]: Chap. 2	229
Paper 4.	General Systems Theory—The Skeleton of Science [Boulding, **1956**]: Chap. 3	239
Paper 5.	General Systems Theory as a New Discipline [Ashby, **1958b**]: Chap. 3	249
Paper 6.	Science and the Systems Paradigm [Checkland, **1976**]: Chap. 3	259
Paper 7.	Old Trends and New Trends in General Systems Research [Rosen, **1979a**]: Chap. 3	269
Paper 8.	Cybernetic Explanation [Bateson, **1967**]: Chap. 3	285
Paper 9.	Systems and Distinctions; Duality and Complementarity [Goguen and Varela, **1979**]: Chap. 3	293
Paper 10.	The Challenges of System Theory [Rosen, **1981**]: Chap. 3	303
Paper 11.	From Circuit Theory to System Theory [Zadeh, **1962**]: Chap. 3	309
Paper 12.	Science in the Systems Age: Beyond IE, OR, and MS [Ackoff, **1973**]: Chap. 3	325
Paper 13.	Systems Profile: The Emergence of Systems Science [Klir, **1988**]: Chap. 3	337
Paper 14.	Methodology in the Large: Modeling All There Is [Gaines, **1984**]: Chap. 5	355
Paper 15.	Discrete and Continuous Models [Barto, **1978**]: Chap. 5	377
Paper 16.	Reconstructability Analysis: An Offspring of Ashby's Constraint Analysis [Klir, **1986**]: Chap. 5	397

Paper 17.	Requisite Variety and Its Implications for the Control of Complex Systems [Ashby, **1958a**]: Chap. 7	405
Paper 18.	Laws of Information Which Govern Systems [Conant, **1976**]: Chap. 7	419
Paper 19.	Science and Complexity [Weaver, **1948**]: Chap. 8	449
Paper 20.	The Architecture of Complexity [Simon, **1962**]: Chap. 8	457
Paper 21.	Complexity and System Descriptions [Rosen, **1977**]: Chap. 8	477
Paper 22.	New Perspectives on Complexity [Prigogine, **1985**]: Chap. 8	483
Paper 23.	The Physics of Complexity [Rosen, **1985a**]: Chap. 8	493
Paper 24.	The Simplification of Science and the Science of Simplification [excerpts from Weinberg, **1972**]: Chap. 9	501
Paper 25.	Introductory Remarks at Panel Discussion [Ashby, **1964**]: Chap. 9	507
Paper 26.	Every Good Regulator of a System Must Be a Model of That System [Conant and Ashby, **1970**]: Chap. 10	511
Paper 27.	Principles of the Self-Organizing System [Ashby, **1962a**]: Chap. 10	521
Paper 28.	Anticipatory Systems in Retrospect and Prospect [Rosen, **1979b**]: Chap. 10	537
Paper 29.	Autopoiesis: The Organization of Living Systems, Its Characterization and a Model [Varela, Maturana, and Uribe, **1974**]: Chap. 10	559
Paper 30.	The Self-Reproducing System [Ashby, **1962b**]: Chap. 10	571
Paper 31.	Universal Principles of Measurement and Language Functions in Evolving Systems [Pattee, **1986**]: Chap. 10	579
Paper 32.	The GST Challenge to the Classical Philosophies of Science [Bunge, **1977**]: Chap. 11	593
Paper 33.	Some Systems Theoretical Problems in Biology [Rosen, **1972**]: Chap. 11	607
Paper 34.	Economics and General Systems [Boulding, **1972**]: Chap. 11	621
Paper 35.	Can Systems Theory Generate Testable Hypotheses? From Talcott Parsons to Living Systems Theory [Miller, **1986**]: Chap. 11	631

Some Comments on Systems and System Theory

Robert Rosen

For a long time, people have been trying to characterize or define the notion of *system*. After all, "systems" are supposed to be what System Theory is about. The results so far have been contradictory and unsatisfactory. This confusion at the foundations has led many to conclude that there is no such thing as a "system" and hence to deny that System Theory is about anything.[1,3] Even those most sympathetic to the notion have difficulties at this level. The very founders of System Theory did not try to say what a system was; and as for System Theory, they characterized it only obliquely, by saying it comprised all studies of interest to more than one discipline.[5] They thereby begged the entire question.

I propose to approach this question in another way. Let us begin by observing that the word "system" is almost never used by itself; it is generally accompanied by an adjective or other modifier: physical system; biological system; social system; economic system; axiom system; religious system; and even "general" system. This usage suggests that, when confronted by a system of any kind, certain of its properties are to be subsumed under the adjective, and other properties are subsumed under the "system," while still others may depend essentially on both. The adjective describes what is special or particular; i.e., it refers to the specific "thinghood" of the system; the "system" describes those properties which are independent of this specific "thinghood."

This observation immediately suggests a close parallel between the concept of a *system* and the development of the mathematical concept of a *set*. Given any specific aggregate of things; e.g., five oranges, three sticks, five fingers, there are some properties of the aggregate which depend on the specific nature of the things of which the aggregate is composed. There are others which are totally independent of this and depend only on the "set-ness" of the aggregate. The most prominent of these is what we can call the *cardinality* of the aggregate.

The perception that the cardinality of a specific aggregate is independent of its "thinghood" is very old. The Greeks, for example, were well aware that a proposition like "two sticks plus three sticks equals five sticks" is of a completely different character from

---

From *Int. J. Gen. Syst.* **13**, 1. Copyright © 1986 Gordon and Breach, Science Publishers, Inc.

"sticks float on water," or "sticks are flammable." In fact, they were well aware that the first of these propositions is not about sticks at all. When they asked themselves what this proposition actually was about, they were led to postulate the existence of a separate universe, out of time and space; a universe of eternal verity uncorrupted by "thinghood," in which one could navigate by unaided reason alone. The truths of this universe were perceived as the highest kind of truths and the search for them comprised the study of pure mathematics. We also see in these developments the beginning of Platonic Idealism, in which the Idea of a thing is related to that thing as the cardinality of a specific aggregate is related to that aggregate.

However, as, e.g., van der Waerden points out,[4] the Greeks themselves only imperfectly made the separation between thinghood and cardinality. For the Greeks, a number was a composite entity involving both aspects; a number was always associated with a unit or counter. That is, there was for the Greeks no "pure" number five; there were only things like five feet; five square feet; five cubic feet. This enormously restricted their ability to do arithmetic, because arithmetic rules did not apply to the counters or units (the consistent elaboration of these ideas led, much later, to the Dimensional Analysis of Fourier). Even today, many languages (e.g., Japanese) persist in attaching numbers to complicated systems of counters. Indeed, the idea of a pure number had to evolve separately, and much later, in India and the Middle East.

We should also observe that a pure number is hard to *define* in logical terms. We learn very early what a number is, by a process of ostension and abstraction; for most of us, a number is an immediately apprehensible primitive and any attempt to *define* it in terms of things more primitive still seems both pointless and excessively pedantic. Indeed, "common sense" tells us that any primitive concept which allows us to define a pure number must be infinitely more alien and unfamiliar than the concept being defined; thus, we lose something, rather than gaining something, from such misguided efforts.

And yet the concept of number or cardinality does depend on a still more primitive aspect of any aggregate; the aspect which we may call its *set-ness*. This aspect is, in fact so obvious, so apparent to all of us that it was rendered totally invisible for two thousand years; it was not even noticed explicitly until it was pointed out by Cantor, only a little over a hundred years ago. The rest, as they say, is history.

It should now be clear that *systemhood* is related to thinghood in much the same way as set-ness is related to thinghood. Likewise, what we generally call *system properties* are related to systemhood in the same way as cardinality is related to set-ness. But systemhood is different from both set-ness and from thinghood; it is an *independent category*. Most of the difficulties of "General System Theory" have arisen from a confusion between systemhood and set-ness; the attempt to treat those properties of a particular system which depend on its systemhood as if it were simply a part of the Theory of Sets.

The allure of reducing systemhood to set-ness is evident. Sets possess a definite calculus; the Theory of Sets can be (or looks like it can be) *axiomatized*. However, both these facts rest ultimately on our experience with the most prominent set-theoretic property; i.e., with cardinality. Once we leave these relatively safe havens, the waters of Set Theory become murky indeed. As relatively recent experience with the Axiom of Choice (i.e., the proposition that a cartesian product of an infinite family of non-empty sets is itself non-empty) or the Continuum Hypothesis (the proposition that there is no cardinal strictly between the integers and the reals) has shown[2] there are many inequivalent ways to axiomatize our restricted experience with set-ness, just as there are many ways to axiomatize our restricted experience with little pieces of planes and plane figures. In fact, since the axiom systems in terms of which set-ness is characterized are themselves *systems*, it may well be that the attempt to define systemhood in terms of set-ness is cart before horse.

Be this as it may, the many parallels between Theory of Systems and Theory of Sets remain. Just as cardinality allows us to compare aggregates independent of their thinghood, so do system-theoretic properties allow us to compare particular systems independent of their thinghood. Thus, the founders of System Theory were perhaps very wise in characterizing it as they did, and letting it go at that.

As we noted earlier, any particular system exhibits properties which depend entirely on its thinghood and it exhibits properties which depend entirely on its systemhood. And, of course, it cannot be stressed too strongly that it will manifest properties which depend on both. The failure to recognize this last obvious fact also leads us to much trouble and confusion. For instance, there are those (and they are many) who claim that there is nothing in the world except thinghood and the properties which arise exclusively from it. The general study of thinghood is essentially *physics*; hence it then follows that the study of any system is subsumed in its physics. This is the essence of *reductionism*. On the other hand, we find a complementary claim that all apparent properties of thinghood are already subsumed under systemhood; this essentially is *holism*. Both appear to me equally in error, as our parallel discussion of thinghood and set-ness makes immediately clear.

The task of General System Theory is a large one. It is not only to characterize and study the notion of systemhood, and the properties which depend only upon it. More than this, it is up to General System Theory to divide the pie between what we have called thinghood, set-ness and systemhood, and study the hybrid properties which depend on more than one of these primitives.

References

1. D. Berlinsky, *On Systems Analysis*, MIT Press, Cambridge, MA, 1976.
2. P. J. Cohen, *Set Theory and the Continuum Hypothesis*, W. A. Benjamin, New York, 1966.
3. J. Monod, *Chance and Necessity*, Knopf, New York, 1971.
4. B. L. Van der Waerden, *Science Awakening*, P. Noordhoff, Groningen, 1954.
5. L. Von Bertalanffy, *General Systems Theory*, George Braziller, New York, 1968.

2

The Emergence of Two-Dimensional Science in the Information Society

George J. Klir

1. Three Levels of Societies

It has increasingly been recognized that a number of countries, primarily the United States and other countries in the West, are at some unique historical crossroad of great significance. This crossroad is usually described as a transition from an industrial into post-industrial phase of society. It is compared in its significance with the previous major societal transition—the change from the pre-industrial into industrial society. Although the transition into the industrial society occurred in most Western countries in the nineteenth and early twentieth centuries, most of the world today is still characterized by the pre-industrial society.

Some of the main characteristics of the *pre-industrial society* are: the labor force is overwhelmingly involved in the extractive industries such as agriculture, mining, forestry or fishing; the main resources are raw materials and natural power such as human or animal muscles, wind or water; life is basically a game against nature. Characteristics of the *industrial society* are radically different: the major part of the labor force is involved in goods-production based on machine technology; the main resources are the financial capital and created energy such as electricity, coal, oil, gas or nuclear energy; life is primarily a game against fabricated nature.

The term '*post-industrial society*' was originally used for a society which is primarily involved in services such as transportation, utilities, trade, finance, health-care, education, arts, research, government, recreation and others.[6,14,27] This view is based on the observation of the societal trends in the United States after World War II, where for the first time more than 50% of the labor force became engaged in the production of various services, and the expectation that these trends will continue. (At present, over 70% of the labor force in the United States is engaged in services, compared with less than 30% in industry, and only about 3% in agriculture.) Whether these trends will continue or not is a matter of current discussion among various forecasters. For example, Jonathan Gershuny and some others argue that the service society will eventually evolve into a different kind of society, referred to as a self-service society.[20]

While I do not intend to take any position on this particular issue, I want to express my views regarding other characteristics of the post-industrial society. They are quite similar to those expressed by Daniel Bell[6] and, more recently, John Naisbitt[30]: the main difference between an industrial and a post-industrial society is that the sources of innovation are derived increasingly from the codification of theoretical knowledge, rather than from 'random' inventions; the strategic resources of the post-industrial society become information and theoretical knowledge, just as the strategic resources of the industrial society are energy and financial capital; the post-industrial society rests on a knowledge theory of value rather than a labor theory of value; life is primarily a game between persons such as research team members, educators and students, doctors and patients, etc.

In summary, the post-industrial society is basically an *information society*. Indeed, the real increase in the so-called service sector has been in the various information related occupations (programmers, systems analysts, educators, accountants, managers, secretaries, etc.). In fact, more than 60% of the labor force in the United States has now information-related jobs.[30] It is thus not surprising that the emergence and development of this new society has been strongly correlated with the emergence and development of computer (or information processing) technology as well as a number of associated intellectual developments such as cybernetics, general systems theory, information theory, decision analysis and artificial intelligence. All these intellectual developments have one thing in common; they deal with such systems problems in which informational or structural aspects are highly predominant while, on the other hand, the kind of entities which form the systems is less significant. It is increasingly recognized that it is useful to view these interrelated intellectual developments as parts of a larger field of inquiry. Since this emerging field of inquiry is oriented to various general phenomena of systems, such as communication, control, regulation, adaptation, recognition, self-organization, self-production, etc., it is reasonable to call it *systems science*. The aim of this article is to characterize this field—systems science, describe its relationship to other areas of science, and discuss its significance and prospects in the post-industrial information society.

2. Systems Science

When consulting a common dictionary, we find that science is defined as 'systematized knowledge derived from observation, study, and experimentation carried on in order to determine the nature or principles of what is being studied' (*Webster's Third International Dictionary*). We may easily convert this definition of science into a definition of systems science by declaring that 'what is being studied' are systems. To make this definition meaningful requires, of course, to elaborate properly upon the term 'systems'. This, however, is not an extraordinary requirement, which is exempt from definitions of other, well-established sciences. Physics, for instance, which is defined by the same dictionary as 'the science dealing with properties, changes, interactions, etc. of matter and energy', requires to elaborate upon the terms 'mass' and 'energy'; this, indeed, is not a small task.

When looking again into the same dictionary, we find that the term '*system*' is defined as 'a set or arrangement of things so related or connected as to form a unity or organic whole'. It follows immediately from this common-sense definition that any given entity must satisfy two requirements to qualify as a system: it must consist of a set of things of some sort and these things must be related or connected in some recognizable manner. Beyond these two requirements, the definition remains neutral. It does not impose any restrictions on either the kind of things or the kind of relation among them. This is desirable for our purpose, since systems science should capture the full scope of the phenomena recognized as various kinds of systems. However, such a broad conception of a system is of little pragmatic value since it does not possess enough distinguishing characteristics. To make it pragmatically more useful, various additional distinguishing characteristics must be employed, by which pragmatically significant classes of systems are introduced.

Considering the two basic distinguishing characteristics of systems contained in the common-sense definition, it is clear that all additional distinctions are basically of two types:

(i) those applicable to the things involved in the system, and
(ii) those applicable to the relations recognized among the things.

Clearly, distinctions of type (i) introduce such classes of systems that each of them is based on a specific kind of things. Distinctions of type (ii), on the other hand, introduce totally different classes of systems, each characterized by a specific type of relations. The term '*relation*' is used here in a broad sense to encompass not only the well-defined concept of a mathematical relation, but the whole set of kindred concepts such as an interaction, interconnection, coupling, linkage, cohesion, constraint, interdependence, organization, structure, etc.

The *thing-oriented classification of systems* is closely associated with the traditional classification of science into disciplines and specializations, each focusing on the study of certain kinds of things without committing to any particular kind of relations. Since our interactions with different kinds of things require different experimental procedures and instruments, this classification is essentially *experimentally based*.

The *relation-oriented classification* of systems is of primary concern to systems science, which focuses on those phenomena of systems which are independent of the kind of things involved in the systems. Since systems characterized by different types of relations require different theoretical treatment, this classification is predominantly *theoretically based*.

Systems science may thus be viewed as a science whose domain of inquiry consists of those properties of systems and associated problems which emanate from the relation-oriented classification of systems. As any other science, systems science has a body of knowledge regarding its domain, a methodology (a coherent collection of methods) for the acquisition of new knowledge as well as the utilization of the knowledge for dealing with relevant problems, and a metamethodology (a characterization of the methodology).

Systems science knowledge can be obtained either mathematically or experimentally. Examples of *mathematically derived knowledge* are: Ashby's law of requisite variety,[1,2] Aulin's law of requisite hierarchy,[4,5,33] the principles of maximum and minimum entropy,[8,23,34] the various laws of information that govern systems,[13] and the many results obtained within a mathematical systems theory.[29,31,37]

As far as the *experimentally derived systems knowledge* is concerned, it is obtained by performing and analyzing experiments with systems simulated on a computer or, possibly, in some other way. The computer plays undoubtedly the most important role in this respect and it is perfectly proper to view it as the *systems science laboratory*. Indeed, the computer allows the systems scientist to perform experiments in exactly the same way other scientists do in their laboratories, although the experimental entities he deals with are abstract constructs (simulated on the computer) rather than specific phenomena of the real world.

3. Taxonomy of Systems

As argued previously, little can be said about all possible systems that conform to the common-sense definition. This class is too large and, consequently, it is logically almost empty. It is thus essential that one of the first tasks of systems science be the development of an appropriate taxonomy of systems. This is not peculiar to systems science. In fact, the highly complex hierarchy of currently recognized disciplines and specializations in traditional science is a result of the same effort: divide the domain of inquiry in order to enrich the potential knowledge content.

There is clearly an unlimited number of ways of how to classify systems. A pragmatically sound taxonomy of systems, however, should take into account such classes of systems that have proved useful in the various traditional disciplines of science, engineering and other areas of human endeavor. This explains why the various taxonomies that are now available show considerable similarities.

As an example, let me briefly describe my own taxonomy of systems, which I began to develop in the late 1960s.[24] It is now utilized in a computer expert system, referred to as *General Systems Problem Solver* or *GSPS*, which is designed to deal with systems problems.[25] This expert system is currently under implementation.

The skeleton of the GSPS taxonomy of systems is a *hierarchy of epistemological* types of systems. It consists of the most fundamental types of systems, which seem vital, in one form or another, to the development of systems science.[25,38-40] The hierarchy is derived from the following three primitive notions: an *investigator* (observer) and his environment, and investigated (observed) *object* and its environment, and an *interaction* between the investigator and object.

On the lowest level of the epistemological hierarchy, referred to as level 0, a system is determined by the manner in which the investigator interacts with the object of investigation. The choice of this interaction is not completely arbitrary in most instances; it is at least partially determined by the capabilities and interests of the investigator. There are various ways of conceptualizing this interaction as a system. In the GSPS framework, it is done by defining a set of *variables*, a set of potential *states* (values) recognized for each variable, and some operational way of describing the *meaning* of the variables and their states in terms of the associated real world attributes and their manifestations. In addition, the set of variables is partitioned into two subsets, referred to as *basic and supporting variables*. Aggregate states of all supporting variables form a *support set* within which changes in states of the basic variables occur. The most frequent examples of supporting variables are time, space and various populations of individuals of the same kind (social groups, sets of countries, manufactured products of the same kind, etc.). The term '*source system*' has been usually used for systems of this type to indicate that it is appropriate to view such systems as sources of empirical data regarding specific attributes of the investigated object.

Source systems can usefully be classified by various criteria through which methodologically significant special properties of the variables or state sets are distinguished. According to one such criterion, the basic variables may be partitioned into *input and output variables*. Under such partition, states of input variables are viewed as conditions which affect the output variables. Input variables are not subject of inquiry but are viewed as being determined by some agent which is not part of the system under consideration. Such an agent is referred to as an *environment* of the system[16]; it includes, in many cases, the investigator. It is important that the notion of input variables not be confused with the notion of independent variables.

Systems whose variables are classified into input and output variables are called *directed systems*; those for which no such classification is given are called *neutral systems*. A number of additional distinctions are recognized for state sets associated with the involved variables (basic or supporting) and provide a basis for further methodological classification of source systems. They include, for instance, the distinction of: (i) crisp and fuzzy state sets; (ii) discrete and continuous state sets; (iii) state sets with no special properties, ordered state sets (partially or totally), state sets which form a metric space, etc.

Systems on different higher epistemological levels are distinguished from each other by the level of knowledge regarding the variables of the associated source system. A higher level system entails all knowledge of the corresponding systems on any lower levels and contains some additional knowledge which is not available on the lower levels.

When the source system is supplemented by *data*, i.e., by actual states of the basic variables within the defined support set, we view the new system (a source system with data) as a system defined on epistemological level 1. Systems at this level are called *data systems*. Depending on the problem, data may be obtained by observation or measurement (as in the problem of systems modelling) or are defined as desirable states (as in the problem of systems design).

Higher epistemological levels involve knowledge of some *support-invariant relational characteristics* of the variables involved through which the data can be generated (in a deterministic, stochastic or fuzzy fashion) for appropriate boundary conditions. On level 2, these characteristics are represented by an overall support-invariant relation among the basic variables of the corresponding source system and, possibly, some additional variables. Each of the additional variables is defined in terms of a specific *translation rule* in the parameter set, applied either to a basic variable of the source system or to a hypothetical (unobserved) variable, introduced by the user (modeller, designer) and usually referred to as an *internal variable*. As the overall support-invariant relation (time-invariant, space-invariant, population-invariant) describes an overall process by which states of the basic variables are generated within the support set, systems defined on level 2 are called *generative systems*.

On epistemological level 3, the system is defined in terms of a set of generative systems (or, sometimes, lower level systems), referred to as *subsystems* of the overall system. The subsystems may be coupled in the sense that they share some variables or may interact in some other way.[35] Systems on this level are called *structure systems*.

On epistemological level 4 and higher levels, the system is allowed to change within the parameter set defined by the associated source system. On level 4, the characterization of the changes (a rule, relation, procedure) is support-invariant; such systems are called *metasystems*. On level 5, the characterization is allowed to change within the support set according to a support-invariant higher level characterization; such systems are called *meta-metasystems* (or metasystems of second order). In a similar fashion, *metasystems of higher orders* are defined.

A simplified overview of the epistemological systems hierarchy is given in Fig. 1.

Since a source system is included in each of the higher level systems, all methodological distinctions which are applicable to source systems are applicable to all higher level systems as well. In addition, however, each of the higher level systems offers further methodological distinctions such as memoryless/memory-dependent, linear/non-linear, deterministic/probabilistic/fuzzy, loopless/with loops etc.

The whole taxonomy of systems can now be summarized as follows. The largest and most fundamental classes of systems are those associated with the described epistemological types. They are further classified by the various methodological distinctions. The more of these distinctions are introduced, the smaller classes of system are obtained. The smallest classes of systems are reached when systems become totally equivalent in terms of their relations, i.e., *isomorphic* with each other. Since systems in each particular isomorphic equivalence class may be based on quite different kinds of things, but these differences are irrelevant to systems science, it is desirable to define convenient representative systems for these classes. Such systems should be interpretation-free (i.e., insensitive to the kind of things involved) and expressed in some convenient standard form; they are usually called *general systems*.

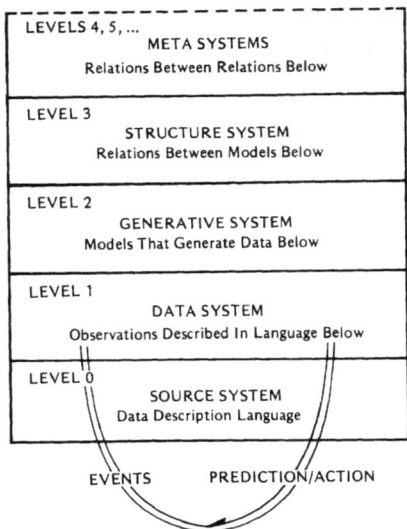

Figure 1. Epistemological systems hierarchy: a simplified overview. Interaction with the world is mediated through the source system to give a data system that is modelled through the levels above.

4. Systems Methodology

The taxonomy of systems and the notion of general systems, as introduced in the last section, are instrumental in characterizing my conception of systems *methodology*. I view it as a coherent collection of methods for studying classes of general systems of the various types and for dealing with problems associated with these systems types.

As mentioned previously, we can study any class of systems by performing appropriate experiments with selected members of the class on a computer. At least three kinds of methods are involved in such experimental studies:

(i) methods by which representative systems are chosen from the investigated class;
(ii) methods by which the chosen systems are simulated on the computer and the required experiments performed on them;
(iii) methods by which the obtained experimental data are processed in order to obtain answers to the investigative questions.

While computer simulation is now methodologically well developed, its use for systems inquiries of the described kind is still rather rudimentary.

A comprehensive study of systems problems that can be formulated within an accepted taxonomy of systems is another role of systems methodology. It consists of characterizing the full scope of systems problems and finding or developing adequate methods for dealing with them. Methods are often borrowed from other areas, most notably from mathematics, but only if they fit well the problems involved.

Some problems involve single systems. They have the form of a question or request regarding some property of the given system. Computer simulation is perhaps the most convenient way of dealing with these problems, especially when the systems involved are highly complex.

Most systems problems involve two or more systems. Typically, one of the systems is given (an initial system) and one or more systems, which satisfy given requirements (objective, constraints), are determined as a solution to the problem (terminal systems). In some problems, however, the aim is not to determine systems that satisfy given requirements, but rather to answer specific questions regarding a relationship between two given systems.

An important class of systems problems are problems of *systems identification*. They appear in many variations on different levels of the epistemological hierarchy of systems. In spite of their diversity, all systems identification problems can be adequately captured by the following general formulation, which was originally proposed by Gaines for the identification of generative systems from data systems.[15] Given a system s of some type, let X denote a set of admissible systems, either of the same type as s (e.g., a set of considered simplifications of s) or of a different type (e.g., a set of systems compatible with s on a higher level of the epistemological hierarchy). Let $\stackrel{a}{\leq} \stackrel{b}{\leq} \stackrel{c}{\leq}, \ldots$ denote preference orderings defined by the user (a scientist, designer, decision maker) on X. These orderings may express, for example, a degree of misfit with respect to some property of s, complexity, uncertainty, cost, etc.; they may be weak, partial, total or any other kinds of orderings. In general, 'smaller than' in each of the orderings has the meaning of 'preferable to.' The orderings usually conflict with each other. For example, when reducing complexity, the misfit and uncertainty have tendencies to increase. All preference orderings involved can be combined in a *joint preference ordering* $\overset{*}{\leq}$ by the following definition

$$(\forall x, y \in X)(x \overset{*}{\leq} y \Leftrightarrow x \overset{a}{\leq} y \text{ and } x \overset{b}{\leq} y \text{ and } x \overset{c}{\leq} y \text{ and } \ldots) \tag{1}$$

The *solution set* X_s of any of the many types of identification problems is then expressed by exactly the same form:

$$X_s = \{x \in X | (\forall y \in X)(y \overset{*}{\leq} x \Rightarrow x \overset{*}{\leq} y)\} \tag{2}$$

Clearly, X_s consists of those systems in X which are either equivalent or incomparable in terms of the joint preference ordering $\overset{*}{\leq}$. Observe that, regardless of the nature of the individual preference orderings, the joint preference ordering is always weak (i.e., reflexive and transitive).

The class of identification problems was chosen to illustrate that the infinite variety of systems problems can be organized in terms of a manageable number of problem categories. Although each category consists of problems of great diversity, which may involve different types of systems and requirements, these problems are in fact isomorphic. Such a categorization of systems problems has profound methodological implications and makes it feasible to develop and build powerful expert systems for dealing with broad spectra of systems problems.[25]

5. Systems Metamethodology

In systems methodology, at least as I view it (as expressed by the GSPS[25]), methods are always considered in the context of specific problems. This attitude toward problem solving is considerably different from that of mathematics. Indeed, pure mathematics is basically oriented to the development of various axiomatic theories and applied mathematics attempts to develop some of these theories as potentially useful methodological tools in various problem areas. Hence, the primacy of problems in systems methodology is in sharp contrast with the primacy of methods in mathematics.

In order to choose an appropriate method for a specific problem, relevant characteristics of prospective methods must be determined. This is a subject of *systems metamethodology*—the study of systems methods as well as methodologies (integrated collections of methods).

In some instances, characteristics of methods or methodologies can be obtained mathematically. For example, exact time complexity function or their lower and upper bounds have been determined for many methods involved in systems problem solving.[18] In most cases, however, mathematical treatment is not feasible. For example, it is usually impossible to determine mathematically the performance of a heuristic method. In such cases, the only way to obtain the desired characteristics is to perform empirical investigations of the methods involved. Metamethodological inquiries can thus be performed mathematically or empirically.

In each empirical metamethodological inquiry, the application of the investigated method or methodology is simulated on a computer for a set of problem instances chosen from a delimited class. Results obtained for all these problem instances are then summarized in a desirable form to characterize the method or methodology and, possibly, compare it with its various competitors. This use of the computer for obtaining quantitative metamethodological results has been explored only recently, primarily for the purpose of evaluating some aspects of reconstructability analysis—a methodology for dealing with the relationship between systems and their various subsystems[9-12,25]; the results obtained are very encouraging.[21,26]

6. Two-Dimensional Science

One of the major characteristics of each of the three levels of societies introduced in Section 1—the pre-industrial, industrial and information societies—is its intellectual base of technology. The pre-industrial society is characterized by common sense, the method of trial and error, craft skills and the emphasis on tradition. The industrial society is primarily characterized by machine technology based on advanced disciplines of science and their engineering counterparts. It is important to realize that science in industrial society is basically one-dimensional in the sense that its various disciplines and specializations emerge primarily due to differences in experimental (instrumentation) procedures rather than differences in the relational properties of the investigated systems.

The information society is clearly characterized by the emergence of the computer (information) technology and a new dimension in science. According to this new dimension, which is referred to in this paper as systems science, systems are recognized and classified by their relational properties rather than the kind of entities that form the relations. Such alternative point of view transcends the artificial boundaries between the experimentally based sciences and makes it possible to develop a genuine cross-disciplinary methodology, more adequate for dealing with the large-scale societal problems inherent in the information society.

Hence, my opinion about the role of science in the three types of societies can be summarized as follows: the pre-industrial society is basically *prescientific*; the industrial society is associated with *one-dimensional science*, which is essentially experimentally based; the information society is characterized by the emergence of a new dimension in science, the theoretically-based science or systems science, and its integration with the experimentally-based science. Science in the information society can thus be described as a *two-dimensional science*. The significance of this radically new paradigm of science is not fully realized as yet, but its implications for the future are, in my opinion, quite profound.

At this time, systems science is in its infant age, comparable with the classical experimental science in the seventeenth century. Although considerable progress has been made in systems science during the last two or three decades, this only scratches the surface. Adequate taxonomies of systems are now available, but we are only starting to develop knowledge and methodology regarding the various classes of systems.

The computer has become well established as our principal laboratory, but it is still rather primitive for our purposes. Further progress in systems science will undoubtedly be closely correlated with advances in computer technology. We begin to understand, however, that there are definite limits in this regard. One such limit, which was determined by simple considerations based on quantum theory by Bremermann, is expressed by the following proposition: 'No data processing system, whether artificial or living, can process more than 2×10^{47} bits per second per gram of its mass'.[7] Hence, a hypothetical computer the size of Earth could not process during the whole period of the estimated age of Earth more than about 10^{93} bits of information. The last number—10^{93}—is usually referred to as Bremermann's limit. It implies that there are definite limits to our knowledge.

Although Bremermann's limit seems large at first sight, there are many systems problems dealing with systems of even modest size that exceed it in their information processing demands.[25] Furthermore, practical limits representing current computer technology are considerably lower, say 10^{10}–10^{20} bits. If a problem exceeds the limit of an available computer, it is obvious that it can be dealt with only in some modified form. The most natural way of modifying a problem is to soften its requirements. For instance, a requirement of getting the best solution may be replaced with a requirement of getting a good solution, instead of requiring a precise solution we may accept an approximate solution, and so on. Such softening of requirements permits the use of heuristic methods, in which vast numbers of unpromising possibilities are ignored, or approximate (fuzzy) methods, in which substantial aggregation takes place.

It is reasonable to expect that systems knowledge, methodology, and metamethodology will become so extensive that some division of labor in the second dimension of science—systems science—will become unavoidable. This will eventually lead to a division of systems science into different areas of expertise, in a similar way as the disciplines of traditional science have evolved. At this time, however, it is premature to speculate about the nature of these areas.

References

1. W. R. Ashby, *An Introduction to Cybernetics*, John Wiley, New York (1956).
2. W. R. Ashby, Requisite variety and its implications for the control of complex systems. *Cybernetica* **1** (1958), 83–99.
3. W. R. Ashby, Some consequences of Bremermann's limit for information-processing systems. In H. Oestreicher and D. Moore (eds.), *Cybernetic Problems in Bionics*, pp. 69–76. Gordon & Breach, New York (1968).
4. A. Aulin, *The Cybernetic Laws of Social Progress*. Pergamon Press, New York (1982).
5. A. Y. Aulin Ahmavaara, The law of requisite hierarchy. *Kybernetes* **8** (1979), 259–266.
6. D. Bell, *The Coming of Post-Industrial Society*. Basic Books, New York (1973).
7. H. J. Bremermann, Optimization through evolution and recombination. In M. C. Yovits *et al.* (eds.), *Self-Organizing Systems*, pp. 93–106. Spartan Books, Washington, D.C. (1962).
8. R. Christensen, *Entropy Minimax Sourcebook, Vol. I: General Description*. Entropy, Lincoln, Mass. (1981).
9. R. E. Cavallo and G. J. Klir, Reconstructability analysis: overview and bibliography. *Int. J. Gen. Syst.* **7** (1981), 1–6.
10. R. E. Cavallo and G. J. Klir, Reconstructability analysis of multi-dimensional relations: a theoretical basis for computer-aided determination of acceptable systems models. *Int. J. Gen. Syst.* **5** (1979), 143–171.
11. R. E. Cavallo and G. J. Klir, Reconstructability analysis: evaluation of reconstruction hypotheses. *Int. J. Gen. Syst.* **7** (1981), 7–32.

12. R. E. Cavallo and G. J. Klir, Reconstruction of possibilistic behavior systems. *Fuzzy Sets Syst.* **8** (1982), 175–197.
13. R. C. Conant, Laws of information which govern systems. *IEEE Trans. Syst., Man, Cybernet.* **SMC-6** (1976), 240–255.
14. V. R. Fuchs, *The Service Economy*. National Bureau of Economic Research, New York (1968).
15. B. R. Gaines, System identification, approximation and complexity. *Int. J. Gen. Syst.* **3** (1977), 145–174.
16. G. C. Gallopin, The abstract concept of environment. *Int. J. Gen. Syst.* **7**, 139–149.
17. M. R. Gardner and W. R. Ashby, Connectance of large dynamic (cybernetic) systems: critical values for stability. *Nature* **228** (1970), 784.
18. M. R. Garey and D. S. Johnson, *Computers and Intractability: A Guide to the Theory of N P-Completeness*. W. H. Freeman, San Francisco (1979).
19. A. E. Gelfand and C. C. Walker, The distribution of cycle lengths in a class of abstract systems. *Int. J. Gen. Syst.* **4** (1977), 39–45.
20. J. Gershuny, *After Industrial Society?* Humanities Press, Atlantic Highlands, New Jersey (1978).
21. A. Hai, An empirical investigation of reconstructability analysis. Ph.D. dissertation, Thomas J. Watson School of Engineering, Applied Science and Technology, SUNY-Binghamton (1984).
22. M. Higashi, G. J. Klir and M. A. Pittarelli, Reconstruction families of possibilistic structure systems. *Fuzzy Sets Syst.* **12** (1984), 37–60.
23. E. T. Jaynes, Where do we stand on maximum entropy? In R. L. Levine and M. Tribus (eds.), *The Maximum Entropy Formalism*, pp. 15–118. MIT Press, Cambridge, Mass. (1979).
24. G. J. Klir, *An Approach to General Systems Theory*. Van Nostrand, New York (1969).
25. G. J. Klir, *Architecture of Systems Problem Solving*. Plenum Press, New York (1985).
26. G. J. Klir and A. Hai, The computer as a metamethodological tool: a case of reconstructability analysis. *Proc. Int. SGSR Conf.*, New York, pp. 237–242 (1984).
27. R. Lewis, *The New Service Society*. Longman, London (1973).
28. S. Makridakis and C. Faucheux, Stability properties of general systems. *Gen. Syst. Yearbook* **18** (1973), 3–12.
29. M. D. Mesarovic and Y. Takahara, *General Systems Theory: Mathematical Foundations*. Academic Press, New York (1975).
30. J. Naisbitt, *Megatrends*. Warner Books, New York (1982).
31. L. Padulo and M. A. Arbib, *Systems Theory*. W. B. Saunders, Philadelphia (1974).
32. L. Polya, An empirical investigation of the relations between structure and behavior for nonlinear second-order systems. Ph.D. dissertation, School of Advanced Technology, SUNY-Binghamton (1981).
33. B. Porter, Requisite variety in the systems and control sciences. *Int. J. Gen. Syst.* **2** (1976), 225–229.
34. J. E. Shore and R. W. Johnson, Axiomatic derivation of the principle of maximum entropy and the principle of minimum cross-entropy. *IEEE Trans. Inform. Theory* **IT-26** (1980), 26–37.
35. Y. Takahara and B. Nakao, A characterization of interactions. *Int. J. Gen. Syst.* **7** (1981), 109–122.
36. C. C. Walker, Behavior of a class of complex systems: the effect of system size on properties of terminal cycles. *J. Cybernet.* **1** (1971), 55–67.
37. T. G. Windeknecht, *General Dynamical Processes: A Mathematical Introduction*. Academic Press, New York (1971).
38. B. P. Zeigler, A conceptual basis for modelling and simulation. *Int. J. Gen. Syst.* **1** (1974), 213–228.
39. B. P. Zeigler, The hierarchy of system specifications and the problem of structural inference. In P. Suppe and P. P. Asquith (eds.), *PSA 1976*, pp. 227–239. Philosophy of Science Ass., East Lansing, Michigan (1976).
40. B. P. Zeigler, *Multifacetted Modelling and Discrete Event Simulation*. Academic Press, London (1984).

3

An Exposition of Constructivism: Why Some Like it Radical

Ernst von Glasersfeld

> *Man, having within himself an imagined World of lines and numbers, operates in it with abstractions, just as God, in the universe, did with reality.*
> —GIAMBATTISTA VICO*

When the Neapolitan philosopher Giambattista Vico published his treatise on the construction of knowledge,† it triggered quite a controversy in the *Giornale de'Letterati d'Italia*, one of the most prestigious scholarly journals at the time. This was in the years 1710–1712. The first reviewer, who remained anonymous, had carefully read the treatise and was obviously shocked by the implications it had for traditional epistemology—all the more so because, as he conceded, the arguments showed great learning and were presented with elegance. He was therefore impelled to question Vico's position, and he very politely suggested that one thing was lacking in the treatise: the proof that what it asserted was true.‡

Today, those constructivists who are "radical" because they take their theory of knowing seriously, frequently meet the same objection—except that it is sometimes expressed less politely than at the beginning of the 18th century. Now, no less than then, it is difficult to show the critics that what they demand is the very thing constructivism must do without. To claim that one's theory of knowing is *true*, in the traditional sense of representing a state or feature of an experiencer-independent world, would be perjury for a *radical* constructivist. One of the central points of the theory is precisely that this kind of "truth" can never be claimed for the knowledge (or any piece of it) that human reason produces.

---

*Vico's reply to his critics, included in the 2nd edition of *De Antiquissima Italorum Sapientia*, 1711; reprinted in Vico (1858), p. 143.
†*De Antiquissima Italorum Sapientia*, Naples, 1710; reprinted with Italian translation, 1858.
‡Giornale de'Letterati d'Italia, 1711, vol. V, article VI; reprinted in Vico (1858), p. 137.

From *Constructivist Views on the Teaching and Learning of Mathematics*, R. B. Davis, C. A. Maher, and N. Noddings, eds. Copyright © 1990 by the National Council of Teachers of Mathematics, Inc., Reston, Virginia.

To mark this radical departure, I have in the last few years taken to calling my orientation a *theory of knowing* rather than a "theory of knowledge." I agree wholeheartedly with Noddings (1990) when she says that radical constructivism should be "offered as a *postepistemological* perspective." One of the consequences of such an appraisal, however, must be that one does not persist in arguing against it as though it *were* or purported to be a traditional theory of knowledge. Another consequence—for me the more important one—is that constructivism *needs* to be radical and must explain that one can, indeed, manage *without* the traditional notion of Truth. That this task is possible may become more plausible if I trace the sources of some of the ideas that made the enterprise seem desirable.

In retrospect, the path along which I picked up relevant ideas (somewhat abbreviated and idealized) led from the early doubts of the Pre-Socratics, via Montaigne, Berkeley, Vico, and Kant, to thinkers who developed instrumentalism and pragmatism at the turn of this century, and eventually to the Italian Operational School and Piaget's genetic epistemology.

The Way of the Sceptics

To Xenophanes (6th century B.C.) we may credit the insight that even if someone succeeded in describing exactly how the world really is, he or she would have no way of knowing that it was the "true" description.* This is the major argument the sceptics have repeated for two thousand five hundred years. It is based on the assumption that whatever ideas or knowledge we have must have been derived in some way from our experience, which includes sensing, acting, and thinking. If this is the case, we have no way of checking the truth of our knowledge with the world presumed to be lying beyond our experiential interface, because to do this, we would need an access to such a world that does *not* involve our experiencing it.

Plato tried to get around this by claiming that some god had placed the pure ideas inside us and that experience with the fuzzy, imperfect world of the senses could only serve to make us "remember" what was really true. Thus, there would be no need (and no way) to check our knowledge against an independent external reality. Consequently, in Plato's famous metaphor, the man who is led out of the cave of his commonplace experience is blinded by a splendid vision. But his vision is the pure realm of an interpersonal soul and not the fuzzy world perceived by the senses.† From my point of view, Plato created an ingenious poetic or "metaphysical" myth, but not a rational theory of knowing.

The sceptics' position, developed into a school under Pyrrho at the end of the next century, was diligently compiled and documented by Sextus Empiricus about 200 A.D. It smoldered under the theological debates of the middle ages and burst into full flame in the 16th century when the works of Sextus Empiricus were rediscovered. Descartes set out to put an end to it, but succeeded only in strengthening the side he was opposing (cf. Popkin, 1979). The British Empiricists then helped to harden the sceptical doctrine by their detailed analyses. First, Locke discarded the secondary (sensory) properties of things as sources of "true" information about the real world. Then, Berkeley showed that Locke's arguments

---

*Cf. Hermann Diels (1957); Xenophanes, fragment 34.
†Cf. Plato's *The Republic* in *Great Dialogues of Plato* (1956), p. 312ff.

Exposition of Constructivism

applied equally to the primary properties (spatial extension, motion, number, etc.), and finally Hume delivered an even more serious blow by attributing the notion of causality (and other relations that serve to organize experience) to the conceptual habits of the human knower.

The final demolition of realism was brought about when Kant suggested that the concepts of space and time were the necessary forms of human experience, rather than characteristics of the universe. This meant that we cannot even imagine what the structure of the real world might be like, because whatever we call structure is necessarily an arrangement in space, time, or both.

These are extremely uncomfortable arguments. Philosophers have forever tried to dismantle them, but they have had little success. The arguments are uncomfortable because they threaten a concept that we feel we cannot do without. "Knowledge" is something of which we are quite sure that we have a certain amount, and we are not prepared to relinquish it.

The trouble is that throughout the occidental history of ideas and right down to our own days, two requisites have been considered fundamental in any *epistemological* discussion of knowledge. The first of these requisites demands that whatever we would like to call "true knowledge" has to be independent of the knowing subject. The second requisite is that knowledge is to be taken seriously only if it claims to *represent* a world of "things-in-themselves" in a more or less veridical fashion. In other words, it is tacitly taken for granted that a fully structured and knowable world "exists" and that it is the business of the cognizing human subject to discover what that structure is.

The weakness of the sceptics' position lies in its polemical formulation. It always sounds as though the traditional epistemologists' definition of knowledge were the only possible one. Hence, when Montaigne says "*la peste de l'homme c'est l'opinion de savoir*" (mankind's plague is the conceit of knowing)* it sounds as though we ought to give up all knowing. But he was referring to absolutistic claims of experiential knowledge and was discussing them in the context of the traditional dogmatic belief that religious revelation is unquestionable. He had in mind *absolute truth*, and he was castigating those who claimed that a rational interpretation of experience (of which "scientific observation" is, after all, a sophisticated form) would lead to such truth. He certainly did not intend to discredit the kind of *know-how* that enabled his peasants to make a good wine.

In short, what the sceptics failed to stress was that, though no truths about a "real" world could be derived from experience, experience nevertheless supplied a great deal of useful knowledge.

The Changed Concept of Knowledge

Unbeknownst to Kant, who in the 1780s hammered this limitation in with his *Critiques* of pure and practical reason, Giambattista Vico had come to a very similar conclusion in 1710. The human mind can *know* only *what the human mind has made*, was his slogan and, more like Piaget than Kant, he did not assume that space and time were necessarily *a priori* categories, but suggested that they, too, were human constructs (Vico, 1858).

Pursuing this way of thinking, one is led to what I have called "a reconstruction of the

---

*Montaigne wrote this in his *Apologie de Raymond Sebond* (1575–1576); cf. *Essais*, 1972, Vol. 2, p. 139.

concept of knowledge" (von Glasersfeld, 1985). Some reconstruction is needed because, on the one hand, one can no longer maintain that the cognizing activity should or could produce a true representation of an objective world, and on the other, one does not want to end up with a solipsistic form of idealism. The only way out, then, would seem to be a drastic modification of the relation between the cognitive structures we build up and that "real" world which we are inclined to assume as "existing" beyond our perceptual interface.* Instead of the illusory relation of "representation," one has to find a way of relating knowledge to reality that does not imply anything like *match* or *correspondence*.

Neither Vico nor Kant explicitly mentioned such a conceptual alternative. It was supplied, however, in Darwin's theory of evolution by the concept of *fit*. Once this relational concept has been stripped of its erroneous formulation in the slogan "survival of the fittest" (cf. Pittendrigh, 1958; von Glasersfeld, 1980), it offers a way around the paradox of the traditional theory of knowledge. As far as I know, this was first suggested by William James (1880).† George Simmel (1885) elaborated it, and Aleksandr Bogdanov (1909) developed it into a comprehensive *instrumentalist* epistemology. Hans Vaihinger (1913), who had been working at his "Philosophy of As If" since the 1870s and who probably was quite unaware of Vico, reintroduced the idea of *conceptual construction*.

Piaget's Contribution

Today, in retrospect, these and other authors can be cited as "sources" of constructivism. However, the great pioneer of the constructivist theory of knowing today, Jean Piaget, started from Kant and arrived at his view of cognition as a biologist who looked at intelligence and knowledge as biological functions whose development had to be explained and mapped in the ontogeny of organisms.

In interpreting Piaget, it is important to remember that his publications range over an astounding variety of topics and are spread over more than half a century.‡ As with any versatile and original thinker, his ideas did not cease to develop and change (Vuik, 1981). It is, therefore, not surprising that one can spot contradictions in his work. An obvious instance is his theory of stages, which was gradually superseded by his theory of equilibration. Thus it is not too difficult to dismiss Piaget on the strength of one or two quotations; or, what is even more frequent, on the strength of what superficial summarizers have said about him. It is also likely that arguments about what Piaget *actually believed* will continue and that different scholars will provide different interpretations. In my view, the following basic principles of radical constructivism emerge quite clearly if one tries to comprise as much as possible of Piaget's writings in *one* coherent theory—but I would argue for these principles even if they could be shown to diverge from Piaget's thinking.

---

*Though most philosophers, today, would agree that the ontological realm is *perceptually* inaccessible, they balk at Kant's suggestion that it is also *conceptually* inaccessible to us. They are therefore still stuck with the paradox that they have no way of *showing* the truth of the ontological claims they make.
†This reference was brought to my attention by a personal communication from Jacques Vonèche (Geneva, 1985).
‡See, for instance, Kitchener's recent article (1989) on Piaget's early work on the role of social interaction and exchange.

1. (a) Knowledge is not passively received either through the senses or by way of communication.
   (b) Knowledge is actively built up by the cognizing subject.
2. (a) The function of cognition is adaptive, in the biological sense of the term, tending toward *fit* or *viability*;
   (b) Cognition serves the subject's organization of the experiential world, not the discovery of an objective ontological reality.

One cannot adopt these principles casually. If taken seriously, they are incompatible with the traditional notions of knowledge, truth, and objectivity, and they require a *radical* reconstruction of one's concept of *reality*. Instead of an inaccessible realm beyond perception and cognition, it now becomes the experiential world we actually live in. This world is not an unchanging independent structure, but the result of distinctions that generate a physical and a social environment to which, in turn, we adapt as best we can.

Consequently, one cannot adopt the constructivist principles as an absolute truth, but only as a working hypothesis that may or may not turn out to be viable. This is the main reason why the constructivist orientation is unequivocally *postepistemological* (Noddings, 1990).

The Concept of Viability

To relinquish the inveterate belief that knowledge must eventually *represent* something that lies beyond our experience is, indeed, a frightening step to take. It constitutes a feat of *decentering* that is even more demanding than the one accomplished by a few outstanding thinkers in the 16th century who realized that the earth was not the center of the universe. Because it goes against an age-old habit, it is immensely difficult to accept that, no matter how well we can predict the results of certain actions we take or the "effects" of certain "causes" we observe, this must never be interpreted as a *proof* that we have discovered how the "real" world works.*

The key to this insight lies in what Piaget formulated in the phrase *"l'objet se laisse faire"* ("the object allows itself to be treated"; 1970; p. 35). At the symposium on the occasion of his 80th birthday he repeated the phrase and explained it further: "When one comes to have a true theory, this is because the object permitted it; which amounts to saying that it contained something analogous to my actions" (Inhelder *et al.* 1977; p. 64). In this context—as in so many in Piaget's works—it is important to remember that an "object" is never a *thing-in-itself* for Piaget, but something that the cognizing subject has constructed by making distinctions and coordinations in his or her perceptual field (Piaget, 1937).

That is all very well, one might say, but how does it come about that the reality we construct is in many ways remarkably stable? And, one might also ask why, if we ourselves construct our experiential reality, can we not construct any reality we might like?

---

*Paul Feyerabend's recent comment (1987) on the famous letter Cardinal Bellarmino wrote in the context of Galileo's trial, makes this point in exemplary fashion: "To use modern terms: astronomers are entirely safe when saying that a model has predictive advantages over another model, but they get into trouble when asserting that it is therefore a faithful image of reality. Or, more generally: the fact that a model works does not by itself show that reality is structured like the model" (p. 250).

The first question was answered in a categorical way by George Kelly: "To the living creature, then, the universe is real, but it is not inexorable unless he chooses to construe it that way" (1955; p. 8). The living creature, be it fish, fowl, or human, thrives by abstracting regularities and rules from experience that enable it to avoid disagreeable situations and, to some extent, to generate agreeable ones. This "abstracting of regularities" is always the result of *assimilation*. No experience is ever the same as another in the absolute sense. Repetition and, consequently, regularity can be obtained only by disregarding certain differences. This notion of assimilation is at the core of Piaget's scheme theory. No schemes could be developed if the organism could not isolate situations in which a certain action leads to a desirable result. It is the focus on the result that distinguishes a scheme from a reflex and makes possible the form of learning that Piaget called *accommodation*. It takes place when a scheme does *not* lead to the expected result. This produces a perturbation, and the perturbation may lead either to a modification of the pattern that was abstracted as the "triggering situation" or to a modification of the action. All this, I want to emphasize, concerns the *experiential* world of the acting organism, not any "external" reality. And the patterns a cognizing organism can and does abstract from experience depend on the operations of distinction and coordination the organism can and does carry out.* This was brilliantly demonstrated for a variety of organisms more than fifty years ago by Jakob von Uexküll (1933/1970).

The second question—why we cannot construct any reality we like—can be raised only if the concept of *viability* is misunderstood or ignored. The absurdity of solipsism stems from the denial of *any* relation between knowledge and an experiencer-independent world. Radical constructivism has been careful to stress that all action, be it physical or conceptual, is subject to constraints. I can no more walk *through* the desk in front of me than I can argue that black is white at one and the same time. What constrains me, however, is not quite the same in the two cases. That the desk constitutes an obstacle to my physical movement is due to the particular distinctions my sensory system enables me to make and to the particular way in which I have come to coordinate them. Indeed, if I now *could* walk through the desk, it would no longer fit the abstraction I have made in prior experience. This, I think, is simple enough. What is not so simple is the realization that the fact that I am able to make the particular distinctions and coordinations *and establish their permanence in my experiential world*, does not tell me anything other than the fact that it is one of the things my *experiential reality* allows me to do. Using a spatial metaphor, I have at times expressed this by saying that the viability of an action shows no more than that the "real" world leaves us room to act in that way. Conversely, when my actions fail and I am compelled to make a physical or conceptual accommodation, this does not warrant the assumption that my failure reveals something that "exists" beyond my experience. Whatever obstacle I might conjecture, can be described only in terms of my own actions. (In this context, it is important to remember that the constructivist theory holds that perception is not passive, but under all circumstances the result of action; cf. Piaget, 1969.)

The constraints that preclude my saying that black is white are, of course, not physical but conceptual. The way we use symbols to handle abstractions we have made from experience, *requires* among other things that we exclude contradiction (cf. von Glasersfeld,

---

*The focus on "operations of distinction" is a major contribution of Humberto Maturana's biological approach to cognition (1980); the notion as such, however, is implicit in much of Piaget's work, e.g., his *Mechanisms of Perception* (1969).

1989). Consistency, in maintaining semantic links and in avoiding contradictions, is an indispensable condition of what I would call our "rational game."

The Question of Certainty

The domain of mathematics is in some sense the epitome of the rational game. The certainty of mathematical results has often been brought up as an argument against constructivism. To indicate that the theoretical infallibility of mathematical operations (in practice, mistakes may, of course, occur) cannot be claimed as proof that these operations give access to an ontological reality, I have compared this generation of certainty to the game of chess. At the painful moment when you discover that your opponent can put you into a "checkmate" position, you have no way of doubting it and your shock is as real as any shock can be. Yet, it is obvious that the certainty you are experiencing springs from nothing but the conceptual relations that constitute the rules of the game; and it is equally obvious that these conceptual relations are *absolute* in the sense that if I broke them and thus destroyed the certainty they generate, I would no longer be playing that particular game.

The comparison with chess has caused remonstrations, and I would like to clarify my position. I still believe that the certainty in mathematics springs from the same conceptual source, but this does not mean that I hold mathematics to be like chess in other ways. The biggest difference is that the elements to which the rules of chess apply are all specific to the game. Flesh and blood kings cannot be put into "mate" positions, equestrian knights move unlike their chess namesakes, and living queens show their power in ways that are inconceivable on the chess board. In contrast, the elements to which the rules of mathematics are applied are not free inventions. In counting, for example, the elements start out as ordinary things that have been abstracted from ordinary experience, and the basic abstract concepts, such as "oneness" and "plurality," have a life of their own *before* they are incorporated in the realm of mathematics. It is precisely this connection with everyday experience and conceptual practice that leads to the contention that mathematics "reflects" the real world.

The "imagined world of lines and numbers" of which Vico speaks in the quotation I have put at the beginning of this essay, is in no sense an arbitrary world. At the roots of the vast network of mathematical abstractions are the simple operations that allow us to perceive discrete items in the field of our experience, and simple relational concepts that allow us to unite them as "units of units." On subsequent levels of abstraction, the re-presentations of sensory-motor material of everyday experience (Piaget's "figurative" elements) drop out, and what remains is the purely "operative," i.e., *abstractions from operations.*

None of this is developed in a free, wholly arbitrary fashion. Every individual's abstraction of experiential items is constrained (and thus guided) by social interaction and the need of collaboration and communication with other members of the group in which he or she grows up. No individual can afford *not* to establish a relative fit with the consensual domain of the social environment.*

---

*Lest this be interpreted as a concession to realism, let me point out that, in the constructivist view, the term "environment" always refers to the environment *as experientially constructed by the particular subject*, not to an "objective" external world.

An analogous development takes place with regard to mathematics, but here the social interaction specifically involves those who are active in that field. The consensual domain into which the individual must learn to fit is that of mathematicians, teachers, and other adults insofar as they practice mathematics. The process of adaptation is the same as in other social domains, but there is an important difference in the way the degree of adaptation can be assessed. In the domain of everyday living, *fit* can be demonstrated by sensory-motor evidence of successful interaction (e.g., when an individual is asked to buy apples, he returns with items that the other recognizes as apples). The only observable manifestation of the demand as well as of the response, in the abstract reaches of the domain of mathematics, are symbols of operations. The operations themselves remain unobservable. *Understanding* can therefore never be demonstrated by the presentation of results that may have been acquired by rote learning.* This is one of the reasons why mathematics teachers often insist (to the immense boredom of the students) on the exact documentation of the algorithm by means of which the result was obtained. The flaw in this procedure is that any documentation of an algorithm is again a sequence of symbols that in themselves do not demonstrate the speaker's or writer's understanding of the symbolized operations. Hence, the production of such a sequence, too, may be the result of rote learning.

Other contributions to this volume will illustrate how a constructivist approach to instruction deals with this problem. They will also show that the constructivist teacher does not give up his or her role as a guide—but this leadership takes the form of encouraging and orienting the students' constructive effort rather than curtailing their autonomy by presenting ready-made results as the only permitted path.

Here, I would merely stress the sharp distinction which, in my view, has to be made between *teaching* and *training*. The first aims at the students' conceptual fit with the consensual domain of the particular field, a fit which, from the teacher's perspective, constitutes *understanding*. The second aims at the students' behavioral fit which, from the teacher's perspective, constitutes acceptable *performance*.

This is not to say that rote learning and the focus on adequate performance should have no place in constructively oriented instruction. But it *does* mean that, where the domain of mathematics is concerned, instruction that focuses on performance alone can be no better than trivial.

Concluding Remarks

If one seriously wants to adopt the radical constructivist orientation, the changes of thinking and of attitudes one has to make are formidable. It is also far from easy to maintain them consequentially. Much like physical habits, old ways of thinking are slow to die out and tend to return surreptitiously.

In everyday living we do not risk much if we continue to speak of lovely sun*sets* and say that tomorrow the sun will *rise* at such and such a time—even though we now hold that it

---

*Thinking, conceptual development, understanding, and meaning are located in someone's head and are never directly observable A formidable confusion was generated by the behaviorist program that tried to equate meaning with observable response.

is the earth that moves and not the sun. Similarly, there is no harm in speaking of knowledge, mathematical and other as though it had ontological status and could be "objective" in that sense; as a way of *speaking* this is virtually inevitable in the social interactions of everyday life. But when we let scientific knowledge turn into belief and begin to think of it as unquestionable dogma, we are on a dangerous slope.

The critics of Copernicus who argued that his system must be "wrong" because it denied that the earth is the center of the universe, could not claim to be "scientific"—they argued in that way for political and religious reasons. Science, as Bellarmino pointed out, produces hypotheses, and as such, they may or may not be useful. Their use may also be temporary. The science we have today holds that neither the earth nor the sun has a privileged position in the universe. Like the contemporary philosophers of science, constructivists have tried to learn from that development and to give up the traditional conception of knowledge as a "true" representation of an experiencer-independent state of affairs. That is why radical constructivism does not claim to have found an ontological truth but merely proposes a hypothetical model that may turn out to be a useful one.

Let me conclude with a remark that is not particularly relevant to the teaching of mathematics but might be considered by educators in general. Throughout the two thousand five hundred years of Western epistemology, the accepted view has been a *realist* view. According to it, the human knower can attain some knowledge of a *really existing* world and can use this knowledge to modify it. People tended to think of the world as governed by a God who would not let it go under. Then faith shifted from God to science and the world that science was mapping was called "Nature" and believed to be ultimately understandable and controllable. Yet, it was also believed to be so immense that mankind could do no significant harm to it. Today, one does not have to look far to see that this attitude has endangered the *world we are actually experiencing*.

If the view is adopted that "knowledge" is the conceptual means to make sense of experience, rather than a "representation" of something that is supposed to lie beyond it, this shift of perspective brings with it an important corollary: the concepts and relations in terms of which we perceive and conceive the experiential world we live in are necessarily generated by ourselves. In this sense it is *we* who are responsible for the world we are experiencing. As I have reiterated many times, radical constructivism does not suggest that we can construct anything we like, but it *does* claim that within the constraints that limit our construction there is room for an infinity of alternatives. It therefore does not seem untimely to suggest a theory of knowing that draws attention to the knower's responsibility for what the knower constructs.

References

Anonymous (1711). Osservazioni, *Giornale de'Letterati d'Italia*, 5(6) (reprinted in Vico, 1858; pp. 137–140).

Bogdanov, A. (1909). Science and philosophy. In (anonymous editor), *Essays on the Philosophy of Collectivism*, Vol. 1. St. Petersburg.

Diels, H. (1957). *Die Vorsokratiker*. Hamburg: Rowohlt.

Feyerabend, P. (1987). *Farewell to Reason*. London/New York: Verso.

Inhelder, B., Garcia, R., and Vonèche, J. (1977). *Epistemologie génétique et équilibration*. Neuchâtel/Paris: Delachaux et Niestlé.
James, W. (1880). Great men, great thoughts, and the environment, *Atlantic Monthly*, **46**, 441–459.
Kelly, G.A. (1955). *A Theory of Personality—The Psychology of Personal Constructs*. New York: Norton.
Kitchener, R. (1989). Genetic epistemology and the prospects for a cognitive sociology of science: A critical synthesis, *Social Epistemology*, 3(2), 153–169.
Maturana, H. (1980). Biology and cognition. In H. Maturana and F. Varela, *Autopoiesis: The Organization of the Living*. Dordrecht: Reidel.
Montaigne, Michel de (1972). *Essais*, Vol. 2. Paris: Librairie Générale Française.
Noddings, N. (1990), Constructivism in mathematics education. In R. B. Davis *et al.* (Eds.), *Constructivist Views on the Teaching and Learning of Mathematics*. JRME Monographs, Reston, Virginia.
Piaget, J. (1937). *La construction du réel chez l'enfant*. Neuchâtel: Delachaux et Niestlé.
Piaget, J. (1969). *Mechanisms of Perception*. (Translation by G.N. Seagrim) New York: Basic Books.
Piaget, J. (1970). *Le structuralisme*. Paris: Presses Universitaires de France.
Pittendrigh, C.S. (1958). Adaptation, natural selection, and behavior. In A. Roe and G.G. Simpson (Eds.), *Behavior and Evolution*. New Haven, CT: Yale University Press.
Plato, *Great Dialogues of Plato* (1956). New York: New American Library.
Popkin, R. (1979). *The History of Scepticism from Erasmus to Spinoza*. Berkeley/Los Angeles: University of California Press.
Simmel, G. (1895). Ueber eine Beziehung der Selectionslehre zur Erkenntnistheorie, *Archiv für systematische Philosophie*, **1**, 34–45.
Vaihinger, H. (1913). *Die Philosophie des Als Ob*. Berlin: Reuther & Reichard.
Vico, G.-B. (1858). *De antiquissima Italorum sapientia* (1710) Naples: Stamperia de'Classici Latini.
von Glasersfeld, E. (1980). Adaptation and viability, *American Psychologist*, 35(11), 970–974.
von Glasersfeld, E. (1985). Reconstructing the concept of knowledge, *Archives de Psychologie*, **53**, 91–101.
von Glasersfeld, E. (1989). Abstraction, re-presentation, and reflection. In L.P. Steffe (Ed.), *Epistemological Foundations of Mathematical Experience*. New York: Springer.
von Uexküll, J. (1970). *Streifzüge durch die Umwelten von Tieren und Menschen* (with George Kriszat). Frankfurt am Main: Fischer (originally published in 1933).
Vuyk, R. (1981). *Overview and Critique of Piaget's Genetic Epistemology*, Vols. 1 & 2. New York: Academic Press.

4

General Systems Theory—The Skeleton of Science

Kenneth E. Boulding

General Systems Theory is a name which has come into use to describe a level of theoretical model-building which lies somewhere between the highly generalized constructions of pure mathematics and the specific theories of the specialized disciplines. Mathematics attempts to organize highly general relationships into a coherent system, a system however which does not have any necessary connections with the "real" world around us. It studies all thinkable relationships abstracted from any concrete situation or body of empirical knowledge. It is not even confined to "quantitative" relationships narrowly defined—indeed, the developments of a mathematics of quality and structure is already on the way, even though it is not as far advanced as the "classical" mathematics of quantity and number. Nevertheless because in a sense mathematics contains all theories it contains none; it is the language of theory, but it does not give us the content. At the other extreme we have the separate disciplines and sciences, with their separate bodies of theory. Each discipline corresponds to a certain segment of the empirical world, and each develops theories which have particular applicability to its own empirical segment. Physics, Chemistry, Biology, Psychology, Sociology, Economics and so on all carve out for themselves certain elements of the experience of man and develop theories and patterns of activity (research) which yield satisfaction in understanding, and which are appropriate to their special segments.

In recent years increasing need has been felt for a body of systematic theoretical constructs which will discuss the general relationships of the empirical world. This is the quest of General Systems Theory. It does not seek, of course, to establish a single, self-contained "general theory of practically everything" which will replace all the special theories of particular disciplines. Such a theory would be almost without content, for we always pay for generality by sacrificing content, and all we can say about practically everything is almost nothing. Somewhere however between the specific that has no meaning and the general that has no content there must be, for each purpose and at each level of abstraction, an optimum degree of generality. It is the contention of the General Systems

---

From *Manage. Sci.* **2**, 197. Copyright © 1956 by Institute of Management Science, Providence, Rhode Island.

Theorists that this optimum degree of generality in theory is not always reached by the particular sciences. The objectives of General Systems Theory then can be set out with varying degrees of ambition and confidence. At a low level of ambition but with a high degree of confidence it aims to point out similarities in the theoretical constructions of different disciplines, where these exist, and to develop theoretical models having applicability to at least two different fields of study. At a higher level of ambition, but with perhaps a lower degree of confidence it hopes to develop something like a "spectrum" of theories—a system of systems which may perform the function of a "gestalt" in theoretical construction. Such "gestalts" in special fields have been of great value in directing research towards the gaps which they reveal. Thus the periodic table of elements in chemistry directed research for many decades towards the discovery of unknown elements to fill gaps in the table until the table was completely filled. Similarly a "system of systems" might be of value in directing the attention of theorists towards gaps in theoretical models, and might even be of value in pointing towards methods of filling them.

The need for general systems theory is accentuated by the present sociological situation in science. Knowledge is not something which exists and grows in the abstract. It is a function of human organisms and of social organization. Knowledge, that is to say, is always what somebody knows: the most perfect transcript of knowledge in writing is not knowledge if nobody knows it. Knowledge however grows by the receipt of meaningful information—that is, by the intake of messages by a knower which are capable of reorganizing his knowledge. We will quietly duck the question as to what reorganizations constitute "growth" of knowledge by defining "semantic growth" of knowledge as those reorganizations which can profitably be talked about, in writing or speech, by the Right People. Science, that is to say, is what can be talked about profitably by scientists in their role as scientists. The crisis of science today arises because of the increasing difficulty of such profitable talk among scientists as a whole. Specialization has outrun Trade, communication between the disciples becomes increasingly difficult, and the Republic of Learning is breaking up into isolated subcultures with only tenuous lines of communication between them—a situation which threatens intellectual civil war. The reason for this breakup in the body of knowledge is that in the course of specialization the receptors of information themselves become specialized. Hence physicists only talk to physicists, economists to economists—worse still, nuclear physicists only talk to nuclear physicists and econometricians to econometricians. One wonders sometimes if science will not grind to a stop in an assemblage of walled-in hermits, each mumbling to himself words in a private language that only he can understand. In these days the arts may have beaten the sciences to this desert of mutual unintelligibility, but that may be merely because the swift intuitions of art reach the future faster than the plodding leg work of the scientist. The more science breaks into sub-groups, and the less communication is possible among the disciplines, however, the greater chance there is that the total growth of knowledge is being slowed down by the loss of relevant communications. The spread of specialized deafness means that someone who ought to know something that someone else knows isn't able to find it out for lack of generalized ears.

It is one of the main objectives of General Systems Theory to develop these generalized ears, and by developing a framework of general theory to enable one specialist to catch relevant communications from others. Thus the economist who realizes the strong formal similarity between utility theory in economics and field theory in physics is probably in a better position to learn from the physicists than one who does not. Similarly, a specialist who works with the growth concept—whether the crystallographer, the virologist, the cytologist, the physiologist, the psychologist, the sociologist or the economist—will be more sensitive to the contributions of other fields if he is aware of the many similarities of the growth process in widely different empirical fields.

There is not much doubt about the demand for general systems theory under one brand name or another. It is a little more embarrassing to inquire into the supply. Does any of it exist, and if so where? What is the chance of getting more of it, and if so, how? The situation might be described as promising and in ferment, though it is not wholly clear what is being promised or brewed. Something which might be called an "interdisciplinary movement" has been abroad for some time. The first signs of this are usually the development of hybrid disciplines. Thus physical chemistry emerged in the third quarter of the nineteenth century, social psychology in the second quarter of the twentieth. In the physical and biological sciences the list of hybrid disciplines is now quite long—biophysics, biochemistry, astrophysics are all well established. In the social sciences social anthropology is fairly well established, economic psychology and economic sociology are just beginning. There are signs, even, that Political Economy, which died in infancy some hundred years ago, may have a re-birth.

In recent years there has been an additional development of great interest in the form of "multisexual" interdisciplines. The hybrid disciplines, as their hyphenated names indicate, come from two respectable and honest academic parents. The newer interdisciplines have a much more varied and occasionally even obscure ancestry, and result from the reorganization of material from many different fields of study. Cybernetics, for instance, comes out of electrical engineering, neurophysiology, physics, biology, and even a dash of economics. Information theory, which originated in communications engineering, has important applications in many fields stretching from biology to the social sciences. Organization theory comes out of economics, sociology, engineering, physiology, and Management Science itself is an equally multidisciplinary product.

On the more empirical and practical side the interdisciplinary movement is reflected in the development of interdepartmental institutes of many kinds. Some of these find their basis of unity in the empirical field which they study, such as institutes of industrial relations, of public administration, of international affairs, and so on. Others are organized around the application of a common methodology to many different fields and problems, such as the Survey Research Center and the Group Dynamics Center at the University of Michigan. Even more important than these visible developments, perhaps, though harder to perceive and identify, is a growing dissatisfaction in many departments, especially at the level of graduate study, with the existing traditional theoretical backgrounds for the empirical studies which form the major part of the output of Ph.D. theses. To take but a single example from the field with which I am most familiar. It is traditional for studies of labor relations, money and banking, and foreign investment to come out of departments of economics. Many of the needed theoretical models and frameworks in these fields, however, do not come out of "economic theory" as this is usually taught, but from sociology, social psychology, and cultural anthropology. Students in the department of economics however rarely get a chance to become acquainted with these theoretical models, which may be relevant to their studies, and they become impatient with economic theory, much of which may not be relevant.

It is clear that there is a good deal of interdisciplinary excitement abroad. If this excitement is to be productive, however, it must operate within a certain framework of coherence. It is all too easy for the interdisciplinary to degenerate into the undisciplined. If the interdisciplinary movement, therefore, is not to lose that sense of form and structure which is the "discipline" involved in the various separate disciplines, it should develop a structure of its own. This I conceive to be the great task of general systems theory. For the rest of this paper, therefore, I propose to look at some possible ways in which general systems theory might be structured.

Two possible approaches to the organization of general systems theory suggest themselves, which are to be thought of as complementary rather than competitive, or at least as two roads each of which is worth exploring. The first approach is to look over the empirical universe and to pick out certain general *phenomena* which are found in many different disciplines, and to seek to build up general theoretical models relevant to these phenomena. The second approach is to arrange the empirical fields in a hierarchy of complexity of organization of their basic "individual" or unit of behavior, and to try to develop a level of abstraction appropriate to each.

Some examples of the first approach will serve to clarify it, without pretending to be exhaustive. In almost all disciplines, for instance, we find examples of populations—aggregates of individuals conforming to a common definition, to which individuals are added (born) and subtracted (die) and in which the age of the individual is a relevant and identifiable variable. These populations exhibit dynamic movements of their own, which can frequently be described by fairly simple systems of difference equations. The populations of different species also exhibit dynamic interactions among themselves, as in the theory of Volterra. Models of population change and interaction cut across a great many different fields—ecological systems in biology, capital theory in economics which deals with populations of "goods," social ecology, and even certain problems of statistical mechanics. In all these fields population change, both in absolute numbers and in structure, can be discussed in terms of birth and survival functions relating numbers of births and of deaths in specific age groups to various aspects of the system. In all these fields the interaction of population can be discussed in terms of competitive, complementary, or parasitic relationships among populations of different species, whether the species consist of animals, commodities, social classes or molecules.

Another phenomenon of almost universal significance for all disciplines is that of the interaction of an "individual" of some kind with its environment. Every discipline studies some kind of "individual"—electron, atom, molecule, crystal, virus, cell, plant, animal, man, family, tribe, state, church, firm, corporation, university, and so on. Each of these individuals exhibits "behavior," action, or change, and this behavior is considered to be related in some way to the environment of the individual—that is, with other individuals with which it comes into contact or into some relationship. Each individual is thought of as consisting of a structure or complex of individuals of the order immediately below it—atoms are an arrangement of protons and electrons, molecules of atoms, cells of molecules, plants, animals and men of cells, social organizations of men. The "behavior" of each individual is "explained" by the structure and arrangement of the lower individuals of which it is composed, or by certain principles of equilibrium or homeostasis according to which certain "states" of the individual are "preferred." Behavior is described in terms of the restoration of these preferred states when they are disturbed by changes in the environment.

General Systems Theory

Another phenomenon of universal significance is growth. Growth theory is in a sense a subdivision of the theory of individual "behavior," growth being one important aspect of behavior. Nevertheless there are important differences between equilibrium theory and growth theory, which perhaps warrant giving growth theory a special category. There is hardly a science in which the growth phenomenon does not have some importance, and though there is a great difference in complexity between the growth of crystals, embryos, and societies, many of the principles and concepts which are important at the lower levels are also illuminating at higher levels. Some growth phenomena can be dealt with in terms of relatively simple population models, the solution of which yields growth curves of single variables. At the more complex levels structural problems become dominant and the complex interrelationships between growth and form are the focus of interest. All growth phenomena are sufficiently alike however to suggest that a general theory of growth is by no means an impossibility.

Another aspect of the theory of the individual and also of interrelationships among individuals which might be singled out for special treatment is the theory of information and communication. The information concept as developed by Shannon has had interesting applications outside its original field of electrical engineering. It is not adequate, of course, to deal with problems involving the semantic level of communication. At the biological level however the information concept may serve to develop general notions of structuredness and abstract measures of organization which give us, as it were, a third basic dimension beyond mass and energy. Communication and information processes are found in a wide variety of empirical situations, and are unquestionably essential in the development of organization, both in the biological and the social world.

These various approaches to general systems through various aspects of the empirical world may lead ultimately to something like a general field theory of the dynamics of action and interaction. This, however, is a long way ahead.

A second possible approach to general systems theory is through the arrangement of theoretical systems and constructs in a hierarchy of complexity, roughly corresponding to the complexity of the "individuals" of the various empirical fields. This approach is more systematic than the first, leading towards a "system of systems." It may not replace the first entirely, however, as there may always be important theoretical concepts and constructs lying outside the systematic framework. I suggest below a possible arrangement of "levels" of theoretical discourse.

(i) The first level is that of the static structure. It might be called the level of *frameworks*. This is the geography and anatomy of the universe—the patterns of electrons around a nucleus, the pattern of atoms in a molecular formula, the arrangement of atoms in a crystal, the anatomy of the gene, the cell, the plant, the animal, the mapping of the earth, the solar system, the astronomical universe. The accurate description of these frameworks is the beginning of organized theoretical knowledge in almost any field, for without accuracy in this description of static relationships no accurate functional or dynamic theory is possible. Thus the Copernican revolution was really the discovery of a new static framework for the solar system which permitted a simpler description of its dynamics.

(ii) The next level of systematic analysis is that of the simple dynamic system with predetermined, necessary motions. This might be called the level of *clockworks*. The solar system itself is of course the great clock of the universe from man's point of view, and the deliciously exact predictions of the astronomers are a testimony to the excellence of the clock which they study. Simple machines such as the lever and the pulley, even quite complicated machines like steam engines and dynamos fall mostly under this category. The greater part of the theoretical structure of physics, chemistry, and even of economics falls into this category. Two special cases might be noted. Simple equilibrium systems really fall into the dynamic category, as every equilibrium system must be considered as a limiting case of a dynamic system, and its stability cannot be determined except from the properties of its parent dynamic system. Stochastic dynamic systems leading to equilibria, for all their complexity, also fall into this group of systems; such is the modern view of the atom and even of the molecule, each position or part of the system being given with a certain degree of probability, the whole nevertheless exhibiting a determinate structure. Two types of analytical method are important here, which we may call, with the usage of the economists, comparative statics and true dynamics. In comparative statics we compare two equilibrium positions of the system under different values for the basic parameters. These equilibrium positions are usually expressed as the solution of a set of simultaneous equations. The method of comparative statics is to compare the solutions when the parameters of the equations are changed. Most simple mechanical problems are solved in this way. In true dynamics on the other hand we exhibit the system as a set of difference or differential equations, which are then solved in the form of an explicit function of each variable with time. Such a system may reach a position of stationary equilibrium, or it may not—there are plenty of examples of explosive dynamic systems, a very simple one being the growth of a sum at compound interest! Most physical and chemical reactions and most social systems do in fact exhibit a tendency to equilibrium—otherwise the world would have exploded or imploded long ago.

(iii) The next level is that of the control mechanism or cybernetic system, which might be nicknamed the level of the *thermostat*. This differs from the simple stable equilibrium system mainly in the fact that the transmission and interpretation of information is an essential part of the system. As a result of this the equilibrium position is not merely determined by the equations of the system, but the system will move to the maintenance of any *given* equilibrium, within limits. Thus the thermostat will maintain *any* temperature at which it can be set; the equilibrium temperature of the system is not determined solely by its equations. The trick here of course is that the essential variable of the dynamic system is the *difference* between an "observed" or "recorded" value of the maintained variable and its "ideal" value. If this difference is not zero the system moves so as to diminish it; thus the furnace sends up heat when the temperature as recorded is "too cold" and is turned off when the recorded temperature is "too hot." The homeostasis model, which is of such importance in physiology, is an example of a cybernetic mechanism, and such mechanisms exist through the whole empirical world of the biologist and the social scientist.

(iv) The fourth level is that of the "open system," or self-maintaining structure. This is the level at which life begins to differentiate itself from not-life: it might be called the level of the *cell*. Something like an open system exists, of course, even in physico-chemical equilibrium systems; atomic structures maintain themselves in the midst of a throughput of electrons, molecular structures maintain themselves in the midst of a throughput of atoms. Flames and rivers likewise are essentially open systems of a very simple kind. As we pass up the scale of complexity of organization towards living systems, however, the property of self-maintenance of structure in the midst of a throughput of material becomes of dominant importance. An atom or a molecule can presumably exist without throughput: the existence of even the simplest living organism is inconceivable without ingestion, excretion and metabolic exchange. Closely connected with the property of self-maintenance is the property of self-reproduction. It may be, indeed, that self-reproduction is a more primitive or "lower level" system than the open system, and that the gene and the virus, for instance, may be able to reproduce themselves without being open systems. It is not perhaps an important question at what point in the scale of increasing complexity "life" begins. What is clear, however, is that by the time we have got to systems which both reproduce themselves and maintain themselves in the midst of a throughput of material and energy, we have something to which it would be hard to deny the title of "life."

(v) The fifth level might be called the genetic-societal level; it is typified by the *plant*, and it dominates the empirical world of the botanist. The outstanding characteristics of these systems are first, a division of labor among cells to form a cell-society with differentiated and mutually dependent parts (roots, leaves, seeds, etc.), and second, a sharp differentiation between the genotype and the phenotype, associated with the phenomenon of equifinal or "blueprinted" growth. At this level there are no highly specialized sense organs and information receptors are diffuse and incapable of much throughput of information—it is doubtful whether a tree can distinguish much more than light from dark, long days from short days, cold from hot.

(vi) As we move upward from the plant world towards the animal kingdom we gradually pass over into a new level, the "animal" level, characterized by increased mobility, teleological behavior, and self-awareness. Here we have the development of specialized information-receptors (eyes, ears, etc.) leading to an enormous increase in the intake of information; we have also a great development of nervous systems, leading ultimately to the brain, as an organizer of the information intake into a knowledge structure or "image." Increasingly as we ascend the scale of animal life, behavior is response not to a specific stimulus but to an "image" or knowledge structure or view of the environment as a whole. This image is of course determined ultimately by information received into the organism; the relation between the receipt of information and the building up of an image however is exceedingly complex. It is not a simple piling up or accumulation of information received, although this frequently happens, but a structuring of information into something essentially different from the information itself. After the image structure is well established most information received produces very little change in the image—it goes through the loose structure, as it were, without hitting it, much as a sub-atomic particle might go through an atom without hitting anything. Sometimes, however, the information is "captured" by the image and added to it, and sometimes the information hits some kind of a "nucleus" of the image and a reorganization takes place, with far reaching and radical changes in behavior in apparent response to what seems like a very small stimulus. The difficulties in the prediction of the behavior of these systems arises largely because of this intervention of the image between the stimulus and the response.

(vii) The next level is the "human" level, that is of the individual human being considered as a system. In addition to all, or nearly all, of the characteristics of animal systems man possesses self consciousness, which is something different from mere awareness. His image, besides being much more complex than that even of the higher animals, has a self-reflexive quality—he not only knows, but knows that he knows. This property is probably bound up with the phenomenon of language and symbolism. It is the capacity for speech—the ability to produce, absorb, and interpret *symbols*, as opposed to mere signs like the warning cry of an animal—which most clearly marks man off from his humbler brethren. Man is distinguished from the animals also by a much more elaborate image of time and relationship; man is probably the only organization that knows that it dies, that contemplates in its behavior a whole life span, and more than a life span. Man exists not only in time and space but in history, and his behavior is profoundly affected by his view of the time process in which he stands.

(viii) Because of the vital importance for the individual man of symbolic images and behavior based on them it is not easy to separate clearly the level of the individual human organism from the next level, that of social organizations. In spite of the occasional stories of feral children raised by animals, man isolated from his fellows is practically unknown. So essential is the symbolic image in human behavior that one suspects that a truly isolated man would not be "human" in the usually accepted sense, though he would be potentially human. Nevertheless it is convenient for some purposes to distinguish the individual human as a system from the social systems which surround him, and in this sense social organizations may be said to constitute another level of organization. The unit of such systems is not perhaps the person—the individual human as such—but the "role"—that part of the person which is concerned with the organization or situation in question, and it is tempting to define social organizations, or almost any social system, as a set of roles tied together with channels of communication. The interrelations of the role and the person however can never be completely neglected—a square person in a round role may become a little rounder, but he also makes the role squarer, and the perception of a role is affected by the personalities of those who have occupied it in the past. At this level we must concern ourselves with the content and meaning of messages, the nature and dimensions of value systems, the transcription of images into a historical record, the subtle symbolizations of art, music, and poetry, and the complex gamut of human emotion. The empirical universe here is human life and society in all its complexity and richness.

(ix) To complete the structure of systems we should add a final turret for transcendental systems, even if we may be accused at this point of having built Babel to the clouds. There are however the ultimates and absolutes and the inescapable unknowables, and they also exhibit systematic structure and relationship. It will be a sad day for man when nobody is allowed to ask questions that do not have any answers.

One advantage of exhibiting a hierarchy of systems in this way is that it gives us some idea of the present gaps in both theoretical and empirical knowledge. Adequate theoretical models extend up to about the fourth level, and not much beyond. Empirical knowledge is deficient at practically all levels. Thus at the level of the static structure, fairly adequate descriptive models are available for geography, chemistry, geology, anatomy, and descriptive social science. Even at this simplest level, however, the problem of the adequate description of complex structures is still far from solved. The theory of indexing and cataloging, for instance, is only in its infancy. Librarians are fairly good at cataloguing books, chemists have begun to catalogue structural formulae, and anthropologists have begun to catalogue culture trails. The cataloguing of events, ideas, theories, statistics, and empirical data has hardly begun. The very multiplication of records however as time goes on will force us into much more adequate cataloguing and reference systems than we now have. This is perhaps the major unsolved theoretical problem at the level of the static structure. In the empirical field there are still great areas where static structures are very imperfectly known, although knowledge is advancing rapidly, thanks to new probing devices such as the electron microscope. The anatomy of that part of the empirical world which lies between the large molecule and the cell, however, is still obscure at many points. It is precisely this area however—which includes, for instance, the gene and the virus—that holds the secret of life, and until its anatomy is made clear the nature of the functional systems which are involved will inevitably be obscure.

The level of the "clockwork" is the level of "classical" natural science, especially physics and astronomy, and is probably the most completely developed level in the present state of knowledge, especially if we extend the concept to include the field theory and stochastic models of modern physics. Even here however there are important gaps, especially at the higher empirical levels. There is much yet to be known about the sheer mechanics of cells and nervous systems, of brains and of societies.

Beyond the second level adequate theoretical models get scarcer. The last few years have seen great developments at the third and fourth levels. The theory of control mechanisms ("thermostats") has established itself as the new discipline or cybernetics, and the theory of self-maintaining systems or "open systems" likewise has made rapid strides. We could hardly maintain however that much more than a beginning had been made in these fields. We know very little about the cybernetics of genes and genetic systems, for instance, and still less about the control mechanisms involved in the mental and social world. Similarly the processes of self-maintenance remain essentially mysterious at many points, and although the theoretical possibility of constructing a self-maintaining machine which would be a true open system has been suggested, we seem to be a long way from the actual construction of such a mechanical similitude of life.

Beyond the fourth level it may be doubted whether we have as yet even the rudiments of theoretical systems. The intricate machinery of growth by which the genetic complex organizes the matter around it is almost a complete mystery. Up to now, whatever the future may hold, only God can make a tree. In the face of living systems we are almost helpless; we can occasionally cooperate with systems which we do not understand: we cannot even begin to reproduce them. The ambiguous status of medicine, hovering as it does uneasily between magic and science, is a testimony to the state of systematic knowledge in this area. As we move up the scale the absence of the appropriate theoretical systems becomes ever more noticeable. We can hardly conceive ourselves constructing a system which would be in any recognizable sense "aware," much less self conscious. Nevertheless as we move towards the human and societal level a curious thing happens: the fact that we have, as it were, an inside track, and that we ourselves *are* the systems which we are studying, enables us to utilize systems which we do not really understand. It is almost inconceivable that we should make a machine that would make a poem: nevertheless, poems *are* made by fools like us by processes which are largely hidden from us. The kind of knowledge and skill that we have at the symbolic level is very different from that which we have at lower levels—it is like, shall we say, the "knowhow" of the gene as compared with the knowhow of the biologist. Nevertheless it is a real kind of knowledge and it is the source of the creative achievements of man as artist, writer, architect, and composer.

Perhaps one of the most valuable uses of the above scheme is to prevent us from accepting as final a level of theoretical analysis which is below the level of the empirical world which we are investigating. Because, in a sense, each level incorporates all those below it, much valuable information and insights can be obtained by applying low-level systems to high-level subject matter. Thus most of the theoretical schemes of the social sciences are still at level (ii), just rising now to (iii), although the subject matter clearly involves level (viii). Economics, for instance, is still largely a "mechanics of utility and self interest," in Jevons' masterly phrase. Its theoretical and mathematical base is drawn largely from the level of simple equilibrium theory and dynamic mechanisms. It has hardly begun to use concepts such as information which are appropriate at level (iii), and makes no use of higher level systems. Furthermore, with this crude apparatus it has achieved a modicum of success, in the sense that anybody trying to manipulate an economic system is almost certain to be better off if he knows some economics than if he doesn't. Nevertheless at some point progress in economics is going to depend on its ability to break out of these low-level systems, useful as they are as first approximations, and utilize systems which are more directly appropriate to its universe—when, of course, these systems are discovered. Many other examples could be given—the wholly inappropriate use in psychoanalytic theory, for instance, of the concept of energy, and the long inability of psychology to break loose from a sterile stimulus-response model.

Finally, the above scheme might serve as a mild word of warning even to Management Science. This new discipline represents an important breakaway from overly simple mechanical models in the theory of organization and control. Its emphasis on communication systems and organizational structure, on principles of homeostasis and growth, on decision processes under uncertainty, is carrying us far beyond the simple models of maximizing behavior of even ten years ago. This advance in the level of theoretical analysis is bound to lead to more powerful and fruitful systems. Nevertheless we must never quite forget that even these advances do not carry us much beyond the third and fourth levels, and that in dealing with human personalities and organizations we are dealing with systems in the empirical world far beyond our ability to formulate. We should not be wholly surprised, therefore, if our simpler systems, for all their importance and validity, occasionally let us down.

I chose the subtitle of my paper with some eye to its possible overtones of meaning. General Systems Theory is the skeleton of science in the sense that it aims to provide a framework or structure of systems on which to hang the flesh and blood of particular disciplines and particular subject matters in an orderly and coherent corpus of knowledge. It is also, however, something of a skeleton in a cupboard—the cupboard in this case being the unwillingness of science to admit the very low level of its successes in systematization, and its tendency to shut the door on problems and subject matters which do not fit easily into simple mechanical schemes. Science, for all its successes, still has a very long way to go. General Systems Theory may at times be an embarrassment in pointing out how very far we still have to go, and in deflating excessive philosophical claims for overly simple systems. It also may be helpful however in pointing out to some extent *where* we have to go. The skeleton must come out of the cupboard before its dry bones can live.

5

General Systems Theory as a New Discipline

W. Ross Ashby

The emergence of general system theory is symptomatic of a new movement that has been developing in science during the past decade: Science is at last giving serious attention to systems that are intrinsically complex. This statement may seem somewhat surprising. Are not chemical molecules complex? Is not the living organism complex? And has not science studied them from its earliest days? Let me explain what I mean.

Science has, of course, long been interested in the living organism; but for two hundred years it has tried primarily to find, within the organism, whatever is *simple*. Thus, from the whole complexity of spinal action, Sherrington isolated the stretch reflex, a small portion of the whole, simple within itself and capable of being studied in functional isolation. From the whole complexity of digestion, the biochemist distinguished the action of pepsin or protein, which could be studied in isolation. And avoiding the whole complexity of cerebral action, Pavlov investigated the salivary conditioned reflex—an essentially simple function, only a fragment of the whole, that could be studied in isolation.

The same strategy—of looking for the simple part—has been used incessantly in physics and chemistry. Their triumphs have been chiefly those of identifying the *units* out of which the complex structures are made. The triumph has been in analysis, not in synthesis. Thus today the biochemist knows more about the amino-acids of which egg-protein is composed than he does about the white of egg from which they have been obtained. And the physiologist knows more about the individual nerve cell in the brain than he does about the action of the great mass of them in integration.

Thus until recently the strategy of the sciences has been largely that of analysis. The units have been found, their properties studied, and then, somewhat as an after-thought, some attempt has been made to study them in combined action. But this study of synthesis has often made little progress and does not usually occupy a prominent place in scientific knowledge.

Even when a study of synthesis seems to be made, the synthesis is often found, on closer examination, to be that in which the interaction between the parts is as slight as

---

From *Gen. Syst. Yearb.* 3, 1. Copyright © 1958 by Society for General Systems Research (now International Society for the Systems Sciences).

possible. We notice for instance how often the combinations that are treated in physics and chemistry occur under the operation of simple addition. Thus two masses in the pan of a balance have a mass that is the simple sum of the separate masses. Similarly two wave forms in an electrical network are usually studied in the *linear* case—the case in which the two patterns combine by simple addition.

Now combination by simple addition is the very next thing to no combination at all. Thus one penny combines with one penny to give just two, precisely because pennies do not in fact interact to any appreciable extent. Contrast this merely nominal combination with what happens when, say, acid is brought together with alkali, or rabbit is brought together with rabbit. Here there is real interaction, and the outcome cannot be represented as a simple sum. Thus, for a century or more, science has advanced chiefly by analysing complex wholes into simple parts. Synthesis has, on the whole, been neglected.

The rule "analyse into parts, and study them one at a time" was so widely followed that there was some danger of its degenerating into a dogma; and the rule was often regarded as the touchstone of what was properly scientific.

Perhaps the first worker to face squarely up to the fact that not all systems allow this analysis into single parts was Sir Ronald Fisher. His problem was to get information about how the complex system of soil and plants would react to fertilizers by giving crops. One method of study is to analyse plant and soil into a host of little physical and chemical subsystems, get to know each subsystem individually, and then predict how the combined whole would respond. He decided that this method would be far too slow, and that the information he wanted could be obtained by treating soil and plant as a complex whole. So he proceeded to conduct experiments in which the variables were not altered one at a time.

At first, scientists were shocked; but second thoughts have convinced us that his methods were sound. Thus Fisher initiated a new scientific strategy. Faced with a system of great complexity, he accepted the complexity as an essential, a non-ignorable property; and showed how worth while information could be obtained from it. He also showed that this could be done only if the worker accepted the need for a new scientific strategy.

What I have said is, of course, equivalent to saying that whereas physics and chemistry, given a system, promptly breaks it to pieces in order to study the parts, there is arising a new discipline that studies the system without breaking it to pieces. The internal interactions are left intact, and the system is, in the well known words, studied as a whole. What methods are there for the study of such *intact* systems? What general methods, in other words, can general system theory follow?

Two main lines are readily distinguished. One, already well developed in the hands of von Bertalanffy and his co-workers, takes the world as we find it, examines the various systems that occur in it—zoological, physiological, and so on—and then draws up statements about the regularities that have been observed to hold. This method is essentially empirical.

The second method is to start at the other end. Instead of studying first one system, then a second, then a third, and so on, it goes to the other extreme, considers the set of "all conceivable systems" and then reduces the set to a more reasonable size. This is the method I have recently followed. Since it may seem at first sight to be somewhat recklessly speculative, I would like to consider briefly its justification.

The method of considering *all* possible systems, regardless of whether they actually exist in the real world, has already been used, and shown its value, in many well established sciences. Crystallography, for instance, studies on the one hand those crystals that actually occur in Nature; and it also studies, in its mathematical branch, all forms that are conceptually possible. It has been found that the set of all conceivable crystals must still obey certain laws; and mathematical crystallography can make confident predictions about what will be found in certain cases.

Whence come these laws? Dare one dogmatize about what Nature may do? In this case one can, and the reason is that we have the option of saying what *we* mean by a "crystal." When we define it as something that shows certain properties of symmetry, we can go on to say that it must also show certain other properties of symmetry because the latter are necessarily implied by the former—they are, as it were, the same properties seen from another viewpoint.

Mathematical crystallography thus forms a background or framework, more comprehensive than the empirical material, on which the empirical material—the real crystals— can find their natural places and be appropriately related to one another. Few will deny the value of the mathematical theory, for without it the study would be a chaos of special cases.

The method of considering more than the actual has also long been used, to advantage, in physics. Much of its theory is concerned with objects that do not exist and never have existed: particles with mass but no volume, pulleys with no friction, springs with no mass, and so on. But to say that these entities do not exist does not mean that mathematical physics is mere fantasy. The massless spring, though it does not exist, in the real world, is a most important concept; and a physicist who understands its theory is better equipped to deal with, say, the balance of a watch than one who has not mastered the theory.

I would suggest that a similar logical framework would be desirable as a part of general system theory. The forms occurring in the real world are seldom an orderly or a complete set. If they are to be related to one another, and higher relations and laws investigated, a rigorous logic of systems must be developed, forming a structure on which all the real forms may find their natural places and their natural relations.

Can such a structure be developed? Can one reasonably start by considering the class of "all conceivable systems?" I suggest one can.

The first objection to be met is that the class is ridiculously wide. It includes for instance the "system" that consists of the three variables: the temperature of this room, its humidity, and the price of dollars in Singapore. Most people will agree that this set of variables, as a "system," is not reasonable, though it certainly exists. How then do we proceed?

Consideration of many typical examples shows that the scientist is, in fact, highly selective in his choice of systems for study. Large numbers of possible aggregations of variables are dismissed by him as "not suitable for study." The criteria he uses are often well known to him intuitively, though seldom stated explicitly. What is often also not recognized explicitly is the intensity of the selection used. Eddington tells the story of the empirical scientist who threw a net into the sea, examined the catch, and then announced the empirical law "all sea creatures are more than two inches long." In studying systems we do not, one hopes, proceed quite as naively as this; but that there *are* subtle laws that have an epistemological, rather than an empirical, basis, can hardly be doubted. Thus while little can be said about "all possible" systems, a good deal can be said about the very special subset of those systems that are accepted by the scientist as being "suitable for study." I shall give some examples a little later.

What is the criterion that the scientist applies when he decides whether a proposed set of variables does or does not form a "natural" system? We can see something of what is necessary by first thinking of the parallel case in energetics. For a system to be suitable for study by the physicist no energy must enter or leave it except as the experimenter directs. Such a system is usually described as "closed" to energy, but the adjective is not well chosen, as often an important part of the investigation is the addition of, say, a measured quantity of heat to it to provoke changes. I shall refer to such a system as "energy-tight."

In the same way, the systems suitable for study in the biological world, while freely open to energy, must be closed to all sources of disturbance, or variation, or entropy (in Shannon's sense) except as directed by the experimenter. They must be, in the technical sense, "information-, or noise-tight." This is the net that catches the systems that come to the empirical scientist for study. It imposes a considerable degree of selection from the set of all conceivable systems. And the selection imposes a number of special properties on the systems that conform to it. Some of the properties are obvious—we need not bother with them; but some are subtle and appear only in disguised form: they have to be discovered, and their true origin identified. Thus we can now ask: how can we identify those properties of a system that are direct consequences of the scientist's insistence that it shall be information-tight?

The Black Box

To answer this question there is, in my opinion, no finer approach than that given by the so-called Problem of the Black Box. It arises in electrical engineering, but its range is really far greater—perhaps as great as the range of science itself.

We imagine that the Investigator has before him a Black Box that, for any reason, cannot be opened. It has various inputs—switches that he may move up or down, terminals to which he may apply potentials, photoelectric cells on to which he may shine lights, and so on. Also available are various outputs—terminals on which a potential may be measured, lights that may flash, pointers that may move over a graduated scale, and so on. The Investigator's problem is do what he pleases to the inputs, and to make such observations on the outputs as he pleases, and to *deduce* what he can of the Box's contents.

In its original, specifically electrical, form, the problem was to deduce the contents in terms of known elementary components. Our problem however is somewhat wider. The questions we are interested in, in general system theory, are such matters as:

What *general* rules of strategy should guide the exploration, when the Box is not limited to the electrical but may be of any nature whatever?

When the raw data have been obtained from the outputs, what operations should *in general* be applied to the data if the deductions made are to be logically permissible?

Finally, the most basic question of all:

What can in principle be deduced from the Box's behaviour, and what is fundamentally not deducible? That is: given that the Investigator has certain finite resources for exploration and observation, what limitations does this finiteness impose on his knowledge of the Black Box?

At first, the questions may seem to be too general to be answerable, but it is now clear that this is not so. The modern development of communication theory can give substantial guidance in this matter, for what *we* are considering can be viewed as a compound system,

General Systems Theory as a New Discipline

composed of Box and Investigator. He acts on the Box when he stimulates it, and the Box acts on him when it gives him a dial-reading as observation. Thus each acts on the other. The interactions that occur between them are as subject to the laws of communication as any other interaction between two sub-systems. (I should make it clear here that the communication theory involved is not the theory which is restricted to the ergodic case.)

In practice, there are no absolute bounds given in relation to any particular Black Box. By however many ways we have tested it there are always further ways at least conceivable. By however many senses or instruments we have observed it, there are always further ways. For however long we have observed it, we could always go on longer. Eventually however the time will always come, for practical reasons, when the exploration and observation must stop—at least for the time being, when the scientist stops to *think* about the Box, and to draw deductions from his data. I shall assume that from now on that the investigation has reached this stage. Thus certain definite inputs have been used, certain variables observed, and a protocol of finite length recorded.

It is now axiomatic that whatever the interaction, it will eventually appear as a protocol of events, stating in general the succession of states taken by each part as they occurred in time. This protocol can now be regarded as a message that contains information about the Box's nature. It is axiomatic that: the Investigator's knowledge about the Box, in any of its aspects, must be essentially a re-coding of what is in the protocol; he may not claim more.

From this point of view, to discover something about the Black Box—something that has permanence—is to discover a constraint in the protocol. The study of a system can thus be summed up in a few words: to discover the constraints, the statistical structure, in the protocol. Should the Investigator find none—should the protocol, as an information source, show maximal entropy, and therefore no redundancy—then he will say, simply, "I can make *nothing* of its behavior; it is totally chaotic." Thus, *any deduction about the nature of the Black Box must be essentially a re-coding of the redundancies (or constraints) in its protocol.*

Suppose now that the Investigator announces that he has discovered a "property" of the Box—some characteristic of its behavior that holds all the way through the protocol. If he describes the property in a suitably compact statement, we can see that he is carrying out precisely the process so important in communication: he is re-coding the protocol so as to pass on a simpler message that still contains the important information but without redundancy; for the redundant part is passed on once and for all by a compact statement about this permanent "property."

Perhaps the most important property that is testable on the protocol is that of whether the system is suitable for study at all! That is, whether it is information-tight. What this means, in essence, is that the protocol should be invariant in time, in regard to the constraints it shows. If this is so, the Investigator may legitimately claim that the system, at least so far as this property is concerned, is information-tight, that is, not subject to unpredictable vagaries. In fact, at the level of fundamental concepts, such invariance may be regarded as the operational definition of what is meant by "information-tight."

The next fundamental property that can be deduced from the protocol is whether the system is or is not behaving in a "machine-like" way. By that I mean whether knowledge of its present state (as shown at the output) and of the conditions within which it is working (that is, the state of its input) is *sufficient* to determine what it will do next. Whether the Box is behaving in a machine-like way does not require study of its internal details; by a straightforward, operationally-defined, process the question can be answered from the protocol. Space does not allow me to describe the method at the moment. I will merely remark that the method also allows the Investigator to establish whether the system, though not strictly determinate, is determinate in the statistical sense of behaving with an unvarying probability.

An important question that often arises in the investigation of any particular system is: what functional connections exist between the parts? (Here I refer exclusively to the functional aspects, that is, of what affects what.) When we know the connections between the parts we often draw a diagram of the immediate, or direct effects. So we get the endocrinologist's diagram showing how the various glands, tissues, and nervous centers act on each other; and the business administrator's diagram showing how the various departments are connected. Can such a diagram be obtained from a Black Box by deduction, from the protocol?

It can, up to an isomorphism.

Let me ignore the qualification for a moment. The fact then is that *functional connections within a Black Box can be deduced from observations made from without.* Information in this respect is to be found in the protocol. To find something of the connections does not demand the opening of the Box.

But not everything of the internal connections can be so deduced. The protocol contains information about the connections that will enable us to specify the connections only up to an isomorphism. No re-coding of the protocol can go past this limit: the information necessary just does not exist in the protocol.

To see what this limitation means, let me make clear what is meant by two Black Boxes being "isomorphic."

Suppose we have before us two Black Boxes. We are privileged to look inside.

The first contains a heavy wheel, which can rotate, and a dial outside showing its position. Attached to the wheel is a spring, and the input is a lever attached to the spring's other end. So moving the lever distorts the spring and applies a force to the wheel, making it turn, or perhaps to stop turning. If the lever is given a complicated sequence of positions, the wheel will respond with some complicated sequence of turning, which will show on the dial.

The second Box is electrical, and contains an inductance and a capacitance in series. The input is a lever that controls a variable potential; it is under the experimenter's control and can be varied in any arbitrary way. Recorded as output on a dial is the total amount of current that passes round the circuit. If now the experimenter moves the lever in some arbitrary way, the potential will affect the components, causing a varying current, which will show as a complex sequence of changes at the output.

Let us further suppose that the various constants—stiffness of spring, mass of wheel, inductance, and capacitance—have been set once and for all so as to impose a certain relationship between the two systems. (As this example is for illustration only, I need not specify further.)

It will then be found that when the first Box is taken, and a particular input applied to the spring, and the consequent movement of the wheel observed, application to the second Box of the *same* values to *its* input will evoke the *same* pattern of change at the output. In other words, equality of the two inputs is always followed by equality at the two outputs. In fact, through an infinity of possibilities, whenever the two Boxes are given the same

General Systems Theory as a New Discipline

trajectory of input, no matter how long or complex, the outputs, however long or complex, will also be equal. Thus if the actual mechanisms are covered up, leaving only inputs and outputs visible, the two Boxes become indistinguishable, for they will respond similarly to all possible tests that can be applied. The two machines are then said to be "isomorphic."

Suppose now that a differential analyser has been programmed to predict the behavior of one of the Boxes for all possible inputs. Since the analyser, when given an input, gives the same output as the Box, the analyser is by definition isomorphic with the Box. Thus an analogue computer might correctly be defined as a machine that can easily be made isomorphic with any of a wide class of dynamic systems.

It has been shown, as I said, that the internal connections can be deduced from the protocol *up to an isomorphism*. It is also readily provable that no deduction, on a given protocol, can go further. We thus encounter here one of the fundamental limitations that I spoke of earlier.

It must not be thought that this limitation is merely technical, to be swept away by the invention of some new gimmick. What it means is that any finite protocol can give only a certain amount of information about any particular question of connection; and that for further questions the information in the protocol *does not exist*.

Knowing that this limitation exists may sometimes be of value in saving us from attempting the impossible. Thus suppose we have before us not only the two Black Boxes just mentioned, *and* the differential analyser suitably programmed to copy them, but also an engineer, so well trained that, if told the input he can predict what the output will be, perhaps by drawing a graph of the actual changes. If we regard *him* as a neuronic mechanism, then we have said that this mechanism is itself isomorphic with the other three; for, given the same input to all four, all four will produce identical outputs. If now we become neurophysiologists, and start to think about the engineer's brain and the neuronic connections within it, we are warned before we start that the pattern of connections is not uniquely defined by the engineer's behavior. Anything we say can only be about a class of mechanisms. And this fact should be reflected in what we try to say about the mechanism. Some of our difficulties in treating the theory of these neuronic mechanisms may be due to our tending to forget this fact.

Degrees of Freedom

Yet another characteristic of the Black Box that can be deduced rigorously from the protocol is the number of its degrees of freedom. By this is meant the number of variables that must be observed or specified if its behavior is to become determinate, that is, unique, single-valued, not subject to random variations. As example, take the simple pendulum, of fixed length. It has a determinate trajectory only if both its position and its velocity are specified; so it has two degrees of freedom. And a desk calculator that multiples eight figures by eight has sixteen degrees of freedom, for only when sixteen figures have been specified does the number that will appear as product become determined.

To return to the Black Box. We assume that its output is shown on a row of dials. Now this row may show what is occurring internally either completely or only partially. Thus, if the Box really contained a simple pendulum, the output might tell us only its *position* at each moment. Study of this Box would soon show that knowledge of its output was not sufficient to make the output's behavior predictable.

Now the number of degrees of freedom is an intrinsic property of a system and can be deduced by finding how many observations have to be made if the behavior is to become predictable. Thus suppose that some new Box has actually 20 degrees of freedom internally, and that 5 dials are reporting on the events within. With fifteen degrees of freedom unobserved, the Investigator will find that the behavior of the Box, as shown on the dials, is not determinate. (The apparent indeterminacy comes from the fact that fifteen variables are not being taken account of.) What is important is that the Investigator can restore determinacy by taking account of earlier values of the variables he *can* see, on the dials. And as fifteen degrees of freedom are not directly observable, the necessary information can be obtained by making an extra fifteen observations on the same five dials (three on each, say).

Thus Black Box theory leads us naturally into the theory—most important for those who study the brain—of the mechanism that, for whatever reason, is not wholly accessible to observation.

Thus we are led to a statement that can be proved rigorously (though for simplicity I shall omit here the qualifications that are strictly necessary):—When a system is really determinate, but cannot be observed at every significant point, determinacy can be restored by the use of supplementary observations, at the same points, that is, on the same dials, about what happened earlier. And the total number of observations to be made must always equal the number of the system's degrees of freedom. In other words, we can find how many degrees of freedom the Black Box has by finding how many observations on the dials are necessary to make correct prediction possible.

Memory

I have just said that when the Box is not completely observable, the Investigator may restore predictability *by taking account of what happened earlier*. Now this process of appealing to earlier events is also well known under another name. Suppose, for instance, that I am at a friend's house and, as a car goes past outside, his dog runs to a corner of the room and cringes. To me the behavior is causeless and inexplicable. Then my friend says "He was run over by a car a month ago." The behavior is now accounted for *by my taking account of what happened earlier*.

The psychologist would say I was appealing to the concept of "memory," as shown by the dog. What we can now see is that the concept of "memory" arises most naturally in the Investigator's mind when not all of the system is accessible to observation, so that he must use information of what happened earlier to take the place of what he cannot observe now. "Memory," from this point of view, is not an objective and intrinsic property of a system but a reflection of the Investigator's limited powers of observation. Recognition of this fact may help us to remove some of the paradoxes that have tended to collect around the subject.

I referred earlier to the fact that when the scientist decides to include only "natural" or "reasonable" systems in the set he studies, he selects somewhat intensively, and may well impose peculiarities that are later discovered empirically, just as the ichthyologist discovered that all sea creatures exceed two inches in length.

We are not surprised, then, when study of the Black Box shows that certain properties, long known to be common in the real world, are necessary consequences of *our* act in only accepting for study such systems as are information- and noise-tight.

Space is running out, so I must pass over these properties somewhat briefly.

The first depends on the fact that every noise-tight system, if subjected to no disturbance at its input, that is if "isolated," cannot *gain* information. Any change in the quantity of the information can then only be a decrease. Every isolated system shows this decrease when it goes to a state of equilibrium; for when many trajectories, from many distinct initial states, converge to one state of equilibrium, the system, when at the equilibrium, has lost the information about which initial state it came from. This is a first example of the general principle that information about what happened earlier in the system tends always to decay.

A more elaborate instance of what is essentially the same principle occurs when a noise-tight system is subjected to a long sequence of events as input. Let us regard the system's state now as showing various traces of what has happened to it in the past. If, as is usually the case, the system's capacity for information is finite, information about what has happened to it in the remoter past tends to be swamped and destroyed by information about what has happened recently. More formally: if a noise-tight system is subjected to a long sequence of events as input, then the longer the sequence, the more will its final state depend on which sequence was applied rather than upon which state the system happened to start from.

In psychology the phenomenon has long been known in various forms. We know it from everyday experience when we notice that a group of boys of varied characteristics, if put through a uniform experience, such as being sent to sea, becomes later more characterized by the fact that they are sailors than by their previous idiosyncracies. The same phenomenon has also been encountered in the laboratory as retroactive inhibition, which names the fact that later learning tends to destroy earlier learning.

Various more or less complex mechanisms have been invented to explain these well known phenomena. The possibility however exists that they may in some cases be due to the fact that the Scientist will only investigate such systems as are information-tight. He thereby unwittingly selects such systems as must show the phenomenon to some degree.

To conclude, let me offer some justification for the title of this paper, which suggests that general systems theory should be regarded as a new discipline.

You will have noticed that a good deal of what I have had to say has not been concerned directly with the Black Box but rather with what the Investigator can or cannot achieve when faced with one. *We*, the system theorists, have in fact been studying, not a Black Box, but a larger system composed of two parts, the Black Box and the Investigator, each acting on the other. We have used communication theory in its non-ergodic form to deduce something of the laws of their interacting. Thus if the Investigator is a scientist studying the Box, *we* are metascientists, for we are studying both; we are working at an essentially different level.

What I have been able to say cannot do more than to introduce the general idea of a mathematical aspect of general system theory. I hope I have made clear that it is possible and reasonable to work not upwards from the empirical but downward from the abstract and general. I hope I have shown that such a study promises worthwhile results, and that it may help to provide us with what is urgently needed in our studies of such complex systems as the brain and society, namely, a logic of mechanism.

6

Science and the Systems Paradigm

Peter B. Checkland

Introduction

We live in a largely artificial world, one made by man as a result of the most powerful activity man has discovered: the activity of science. The intellectual and practical adventure which began in Western Europe in the 16th and 17th Centuries with Copernicus, Kepler, Galileo, Newton, has made our world. Only 100 years ago there was little doubt that the application of science, leading to the creation of wealth and the elimination of much disease had shown the way to a happier future. Today, noting the manifest inability of the most scientifically advanced countries to solve the problems of the real world (as opposed to the self-defined, artificial problems of the laboratory) we wonder whether the fragmentation of science into its many separate disciplines is not a significant weakness.

Problems in the real world (of the kind: What shall we do about urban decay? Shall I marry this girl? How shall we design our schools?) are characterised by their many facets, and their "solution"—the process of forming a judgement about them—calls for integrative thinking which fragmented science seems not to provide.

The call for "interdisciplinary teams" to tackle social problems is a popular one, but 30 years' experience in Management Science has emphasised not that this is a successful way to tackle such problems, rather the fact that it is quite remarkably difficult for specialists from one discipline to understand the concepts and language of another.

What we need is not interdisciplinary teams but transdisciplinary concepts, concepts which serve to unify knowledge by being applicable in areas which cut across the trenches which mark traditional academic boundaries. "Structuralism" is one such concept, its methods being applied in areas which include linguistics, biology, ethnology and mathematics. This paper is concerned with another: the paradigm "system," and the use of it in real-world problem solving which is known as "a systems approach." Although the latter phrase is modish, and is frequently used as a mere talisman, it will be argued that systems concepts, emerging from science but in fact complementary to that tradition, can both help in problem solving and make a contribution to integrating the presently scattered sciences.

From *Int. J. Gen. Syst.* **3**, 127. Copyright © 1976 by Gordon and Breach Science Publishers Ltd.

Science and Reductionism

Modern science is a unique contribution of Western civilisation. Religion, art, crafts are found in all civilisations, and practical technological discoveries, such as the compass or paper making were made in ancient civilisations which did not have science in the modern sense.[1] But science as we know it emerged as a specific and unique activity in Western Europe around 350 years ago. Although recent scholarship has emphasised "the essential continuity of the Western scientific tradition from Greek times to the seventeenth century and, therefore, to our own day,"[2] it is generally agreed that in the 16th and 17th centuries there was a scientific *revolution*, one which, as Butterfield puts it

> outshines everything since the rise of Christianity and reduces the Renaissance and Reformation to the rank of mere episodes.[3]

The revolution was the result of many factors, which include: a questioning of authority; a determination to treat nature as a source of phenomena which could be investigated rather than as symbolic of spiritual truths (as Augustine had done); and the development of a means of doing the investigating, namely the experimental method backed by rational thought. The revolution succeeded because, as Crombie puts it, its protagonists asked

> questions within the range of an experimental answer, by limiting their inquiries to physical rather than metaphysical problems, concentrating their attention on accurate observation of the kinds of things there are in the natural world and the correlation of the behaviour of one with another rather than on their intrinsic natures, on proximate causes rather than substantial forms, and in particular on those aspects of the physical world which could be expressed in terms of mathematics.[4]

Summarising savagely for brevity we may say that the core of the revolution was a belief in rational modes of thought applied to the design and subsequent analysis of experiments. This combination of rational thinking and experimentation is outstanding in that *it works*! It provides factual information which is different in kind from that derived from consulting the Oracle at Delphi, reading the entrails of slaughtered goats, casting horoscopes or turning up the Tarot cards. Within well-defined limits, scientific evidence is trustworthy, and that is why it is the cult of science rather than, say, astrology or necromancy which has created our world.

How then may we best describe this most potent human activity? There is no agreed version of such a description and this is in part due to a frequent confusion between the *sociology* of science, including how scientists behave in going about their work, and the *logic* of the activity itself. In the last century science was said to be "organised common sense," and many definitions emphasize the systematic collection and classification of facts; Ziman's version[5] is that science is "public knowledge," implying a readiness and a commitment to the testing of assertion so that what accumulates is "a consensus of rational opinion."

I suggest that science may best be described as a particular kind of *learning system*, and that an account of the nature of that system can cover some of the crucial aspects of both the sociology and the logic of science. Such a description emphasises the ongoing nature of

Science and the Systems Paradigm

the process of gathering scientific knowledge and hence the temporary nature of the knowledge gained. Today's knowledge is supplanted as better experiments reveal more.

For a "hard" experimental science such as physics or chemistry—the ideal type to which the other "softer" sciences may be referred—the learning system may be characterised by three R's: reductionism, repeatability and refutation. We may *reduce* the complexity of the variety of the real world in experiments whose results are validated by their *repeatability*, and we may build knowledge by the *refutation* of hypotheses.[6] The reductionist approach of science is a major source of its power. By isolating only a part of nature as the object of concern, and systematically investigating a few variables in the artificially simple world of the laboratory, we help to ensure that the results will be intelligible and, hopefully, unequivocal. And we use reductionism in another sense in explanation, explaining the results using the smallest possible number of concepts, only importing more elaborate concepts when defeated in this. It is thus highly satisfactory to most scientists that chemistry is increasingly explicable in terms of the concepts of physics—mass, force, energy etc. And perhaps, thinks the reductionist, there will eventually be a reduction of biology to chemistry, and so to physics . . . and of psychology to biology . . . and so on?[7]

That the facts gained from the reductionist experiments are repeatable is what makes them public knowledge, and hence different in kind from assertions which cannot be tested. Although you may dispute the *interpretations* I put upon, say, my experiments with magnets and iron filings, if I hypothesise that magnets attract iron filings because they are made of iron rather than because of their shape, you may repeat my experiments with iron and with plastic filings and see for yourself! We may continue to argue about interpretation but we shall be forced to agree about *the happenings in the experiments*, which are repeatable. This example is trivial but the principle is not. It is the repeatability of the experimental facts which places this knowledge in a different category to opinion, preference and speculation.

Finally the learning system of science is characterised by the ability to refute hypotheses, and it is by this ability that scientific knowledge can be said to grow. This way of viewing the progress of science is, of course, Popper's,[8] and it may be compressed in his words: "The method of science is the method of bold conjectures and ingenious and severe attempts to refute them."[9] This rather backhanded approach to the accumulation of knowledge means that the current content of science is always a set of temporary hypotheses which have not yet been demolished, a point beautifully made by novelist Vladimir Nabokov:

> When a hypothesis enters a scientist's mind, he checks it by calculation and experiment, that is, by the mimicry and pantomime of truth. Its plausibility affects others, and the hypothesis is accepted as the true explanation for the given phenomenon, until someone finds its faults. I believe the whole of science consists of such exiled or retired ideas; and yet at one time each of them boasted high rank; now only a name or a pension is left.[10]

This view is of course needed because of our inability to justify in logic the method of inductive reasoning; but that is not a question to concern us here. Nor need we be concerned that actual scientists, having human weaknesses, may often be observed seeking evidence which supports rather than refutes a hypothesis; the principle of conjecture and refutation is required by logic, and is unaffected by the behaviour of mortals. (In fact, a number of the most distinguished scientists including Medawar, Monod, Eccles and Bondi have attested to its importance in their thinking.[11])

Science, then, is a successful learning system. What has its application led to in its own structure? The most dominating single characteristic of science is the reductionism of its approach, in both of the senses discussed above. This, the unifying principle in any application of the method of science, has had the ironic effect of fragmenting science into a hierarchy of separate disciplines. The weltanschauung implied by the existence of science is that the universe may be taken to consist of sets of phenomena together with the outcome of their operation, the phenomena being the subject of investigation by the different disciplines of science. Most basic is physics, concerned with the most fundamental properties and their interactions: mass, momentum, force, energy etc. Chemistry presents more complex phenomena, which at first seem different in kind from those of physics, as when for example ammonia and hydrochloric acid gas mix and a white solid settles out. This and similar findings suggest that we have here phenomena worthy of investigation in their own right—although explanations reduced to physics will always be welcome! When a seed is planted and subsequently produces a plant, obviously both physics and much chemistry are involved, but there seems to be a level of complexity involved beyond both physics and chemistry, and we draw another arbitrary boundary to create the science of biology. . . . The process continues through medical science and psychology to the immeasurably complex phenomena which are the concern of the so-called social sciences.

Pantin makes an illuminating distinction between the "restricted" sciences (relatively simple phenomena, well designed reductionist experiments possible) and the "unrestricted" sciences which study such complex effects that they are necessarily more descriptive, and experiments with controls are not always possible. He writes

> The physical sciences are restricted because they discard from the outset a very great deal of the rich variety of natural phenomena. By selecting simple systems for examination they introduce a systematic bias into any picture of the natural world which is based on them. On the other hand, that selection enables them to make rapid and spectacular progress with the help of mathematical models of high intellectual quality. In contrast, advance in the unrestricted sciences, as in geology and biology, is necessarily slower. There are so many variables that their hypotheses need continual revision.[12]

These difficulties become even greater in economics, anthropology and sociology, all "unrestricted" sciences. Although Durkheim and his fellow pioneers of sociology were clear that it was to be, in Durkheim's own words, "a distinct and autonomous science" whose concern was to elucidate the laws governing societal behaviour, progress has been slow while questions of methodology and epistemology unheeded in physics are still unresolved. These questions are not here my concern, and in any case do not affect the overall picture of a hierarchy of sciences concerned to discover the relations of things and events at difference levels of complexity by following a method which has been shown to be, in a material sense at least, spectacularly successful. The implication of scientific activity has been that the successes of the restricted sciences would, given time, be extended to the unrestricted sciences including social science.

It is now necessary to challenge that view, to examine the need for other, complementary approaches and to examine briefly the systems movement as a source of an alternative approach.

A Systems Approach and Holism

A remarkable feature of science has been the lack of concern shown by professional scientists about its fragmentation. This must stem from a weltanschauung (by definition, unquestioned) that the world contains groups of phenomena and that science is a means of establishing the laws which operate for a given group. A particular focus of interest may create a new discipline—for example "biochemistry," placed at the border between biology and chemistry but the divisions of this kind, even if recognised as arbitrary, are not taken to be a problem, puzzle or paradox. Social science, on this view, is simply concerned with more complex phenomena than physics, and it will no doubt take a little longer for its patterns and regularities to be teased out. There has been relatively little concern for the question: What is the nature of this scale of complexity which, intuitively, places physics, chemistry, biology, and the social sciences in sequence?[13]

This unconcern is not surprising, in that when issues touching on the complexity of the phenomena have arisen within particular disciplines reductionist solutions have often been found; it is comforting to imagine that this will happen again. A striking example from the natural sciences is the question of vitalism. In the last century when chemists first isolated and purified the so-called "organic" chemicals known to occur in living things, there was debate about whether or not they were characterised by some mysterious "vital" component which made them intrinsically different from substances like caustic soda and sulphuric acid. In this case the work of Wöhler and Kolbe eventually established that known organic compounds could be synthesised in the laboratory from wholly inorganic starting materials, and the controversy died. Chemists accepted that the difference between organic and inorganic compounds was simply a matter of structure. But in biology the vitalist debate was active (and heated) as recently as the 1920s. Did living matter contain a vital constituent enabling it to express purpose—Driesch's "entelechy"? Here again a reductionist explanation is now accepted, although it will need the creation of living from non-living material in the laboratory to clinch the argument. Biologists now believe that this is possible in principle,[14] and they accept that the difference between living and non-living material lies simply in the greater degree of organisation of the former.

But in spite of the triumphs of reductionist molecular biology in recent years, especially the elucidation of the genetic mechanisms, there has been in biology a school of thought not satisfied with a comfortable reductionism.[15] This school, as represented by its pioneer Ludwig von Bertalanffy, was based firstly on a belief that for biologists the concern is for *the organism as a whole*, and secondly on an intuitive feeling that descriptions of organisms as functions and partial processes, with observed behaviour the result of conditioned responses to stimuli, were too sparse for the richness to be explained. Bertalanffy developed from the 1930s theories of biological organisms as "open systems" able creatively to maintain a certain degree of organisation even though the organism's constituent parts and its environment continually changed.[16] This is possible through a continuous exchange of materials, energy and information between organism and environment, enabling the organism to exist in a state of high statistical improbability. Obviously our bodies may be treated as open systems, and the concept is at least an evocative metaphor for social groupings like organisations and societies.[17]

This thinking in biology is one of the main sources of "the systems movement," the name of those from many different disciplines who are interested in working out the consequences of holistic thinking, and using the concept "system" to increase understanding or to help solve problems. There are systems thinkers in science, technology, economics, management science, psychology, sociology, anthropology, geography, political science, history, philosophy, art. . . . It is the argument of this paper that the systems movement is usefully complementary to the science movement, and can have a unifying influence on many disciplines which in the science view are separate.

The systems movement is identified by a conscious use of the concept "system" and by the holistic thinking which this implies. The movement's holism is best understood with reference to its opposite: reductionism. The reductionist is committed to explanation in terms of the smallest number of the most fundamental entities, which in science means those of physics. The reductionist is the most enthusiastic wielder of Ockham's razor. But even in physics and physical chemistry there are phenomena—such as those connected with heat flow—which have *no meaning at all* in terms of individual atoms and molecules but which are repeatably measurable and which lead to theory able to explain the observations. Such *emergent properties* are characteristic of a given level of complexity. Now of course there may eventually be reductionist explanations of what are initially thought to be emergent properties—just as Driesch's entelechy is no longer invoked—and they should be welcomed. But the fact that, for example, dynamics is not able to explain even thermodynamics, for which the emergent properties described as "temperature" and "entropy" are needed, suggests that the anticipation that the universe will always seem to contain *wholes* which have properties not entirely explicable in terms of their parts seems not unreasonable! Popper expresses this idea as follows:

> Now I want to make it clear that as a rationalist I wish and hope to understand the world and that I wish and hope for a reduction. At the same time, I think it quite likely that there may be no reduction possible; it is conceivable that life is an *emergent* property of physical bodies.[18]

And later in the same article adds:

> So I would say: let us work out in every case the arguments for emergence in detail, *at any rate before attempting reduction.*

This is close to being the stance in the systems movement, and is the more satisfactory coming as it does from a severe critic of naive holism.[19]

Thus while the weltanschauung of science is that the world consists of groups of phenomena which may be investigated by the method of science, the counter-weltanschauung of the systems movement is that the world consists of a complex of wholes which we term "systems." The systems thinker assumes that the world will exhibit emergent properties at virtually all levels of complexity and that it will be useful to examine the world in terms of the wholes which exhibit those properties, and to develop principles of "wholeness." The long term programme of the systems movement may be taken to be the search for the conditions governing the existence of emergent properties and a spelling out of the relation between such properties and the wholes which exhibit them.

Of course, given the 30-odd years' history of conscious systems thinking, compared with the 300+ years' history of science, we should not be surprised that systems ideas are in a primitive state and that use of the phrase "a systems approach," deplorably, frequently amounts to little more than flag-waving.

Science and the Systems Paradigm

The systems work which I have been engaged in for several years may serve to illustrate one of the many approaches being taken within the systems movement. I have been concerned with attempts to incorporate specifically systemic concepts into a methodology for tackling problems which we perceive as "real-world problems." This is an area with which science manifestly cannot cope, in spite of the near-heroic attempts of management science to bring to it the methods of science.

The work of my colleagues and I[20] starts from an intuitive description of the world in terms of four system types: natural systems (atoms, molecules, rocks, plants, animals, ecological systems) designed physical systems (tools, machines, etc.) designed abstract systems (mathematics, philosophies) and human activity systems.[21] We may *learn* from natural systems; we may *use* both types of designed system; and we may *seek to design and implement* ("engineer") human activity systems. Natural systems have much to teach us concerning "wholeness," the ones which exist being those which have been shown to be viable through the course of evolution; designed systems are available to help us, in the form of both physical artefacts and formal abstractions helpful to the thinking process; human activity systems constitute the social arena in which real world problems are perceived, or appear, or disappear.

After more than 100 systems studies consisting of attempts to tackle something which a client perceives to be "a problem," systems methodology has been developed.[22,23,24]

My own work has been directed towards "soft," ill-defined problems, ones which cannot be expressed in precise terms. Their crucial characteristic is taken to be that they are *unstructured*, and the aim has been to find a way of using systems ideas to tackle them without having first to define them sharply, without forcing them into a structured form. This is deemed necessary because if a problem is stated in a well-defined form its solution is usually implicit in the definition, and such "solutions," in real-world problems, tend to pass the problems by. In the management science area, for example, we all learn to recognise the queueing theory enthusiast who readily sees problems as queueing problems, thus enabling him comfortably to use his favourite technique.

The tentative methodological solution so far is to use systems concepts as a means of problem analysis, these being a source of specific guidance for action, while being at a sufficiently high level of abstraction to avoid "distortion" of the problem. But the analysis is not of "the problem," it is a more open analysis of the *situation within which a problem is perceived*. The initial analysis enables credible viewpoints with respect to the problem to be postulated. Systems ideas then enable notional conceptual models of human activity systems which embody those viewpoints to be constructed. These models are then brought alongside what is perceived to be the present situation and debate ensues which aims to define changes which meet two criteria: Are they arguably *desirable*? Are they, in the situation we have, *feasible*? If desirable and feasible changes emerge then action in the original problem situation is possible. If changes meeting those two criteria do not emerge, then we may explore more radical, or more conservative viewpoints via further conceptual models—and in so doing *learn*. The methodology is in fact a learning system (in contrast to those methodologies for "hard" systems engineering which deal in terms of "design" and "optimisation" and "implementation"), one rather different from the learning system of science discussed above.

This very condensed account of it will now be expanded only briefly, since it is described elsewhere.[23,24]

The initial analysis looks at the problem situation and, in essence, notes two kinds of element in it, elements of *structure*, the relatively-slow-to-change features (organisation, reporting structures, attitudes which are institutionalised in these, etc.), and elements of *process*, the constantly-changing "doing" elements. Finally an attempt is made to characterise the relation between structure and process, the *"climate"* of the situation.

On the basis of this initial analysis a number of possible human activity systems believed to be *relevant* to the problem situation and its improvements are named. These names are built up into "root definitions" which encapsulate a certain viewpoint. Guidelines to ensure that the root definitions are well-formed have recently been developed.[25] In one systems study, carried out for the British charity organisation Oxfam, my colleagues selected as insightful a root definition which took Oxfam to be a "persuading-by-transferring-information" system.[26] In this view Oxfam transfers information about the needs of the "third" (under-developed) world to the richer world in order to stimulate the developed countries to devote more of their resources to helping the poorer parts of the world. Other views which might be adopted in this particular case might be to see Oxfam as a "fire brigade" operation, concerned to move help quickly to areas struck by natural disasters; or to view it in more explicitly political terms. Consider a public issue: the UK Government's problem of what to do about the uneconomic Anglo/French project to develop the supersonic passenger aircraft Concorde. This might be taken to be simply a huge *task*, creating efficiently a specified aircraft while meeting various constraints including economic ones. This would take as given the other aspects of the project. Or *they* could form the basis of the view: Concorde as "an Anglo/French collaboration" system, or as a system to prevent the Americans taking the lead in at least one advanced technology, or as an environment-threat system. The point is that the Concorde project *is* all of these things according to particular viewpoints. The methodology defines a number of them and builds systems models of human activity systems which embody the viewpoints defined.

The modelling language used is a subtle and powerful one: all the verbs in the English language! A structured set of verbs is assembled which describes the minimum necessary set of activities required in a human activity system which *is* that described by the root definition. Guidelines for checking such models for basic inadequacy have been developed[23] and the basic model obtained may now be enriched from many sources. If, for example, the systems thinker responds to the Tavistock Institute concept of the "sociotechnical" system,[27] Ackoff's "system of systems concepts,"[28] or Beer's analogy between organisations and the central nervous system,[29] then the models may be made "valid" in these terms. (I personally have found it particularly useful at this stage to examine the models in terms of Vickers' concept of "an appreciative system."[30])

The notional models are now brought back to the problem situation and "compared" with the initial structure/process/climate analysis. Doing that provides the structure for the necessary debate about changes, the debate being with *concerned actors* from the problem situation. The debate may lead to agreement on changes or perhaps to exploration of new viewpoints additional to those considered in the first iteration: the ongoing learning system continues to learn. . . .

It may be noted here that, in the terms of the deepest systems analysis of learning systems yet made, that by Churchman,[31] this methodology is "Singerian." Churchman, in "The Design of Inquiring Systems" examines historical texts concerned with the theory of knowledge (those of Leibniz, Locke, Kant, Hegel and Singer) in modern systems terms, as if those authors were seeking to design optimal inquiring systems. On the foundation of a basic Leibnizian model of a system which builds nets of contingent facts, Churchman examines: the Lockean inquirer, which seeks consensus on the facts among the community of inquirers; the Kantian/Hegelian inquirer which opposes every world view to its deadly enemy in the search for higher level truth; and the Singerian inquirer which accepts inquiry as continuous process without end, a process which summons an heroic mood to go on both defending the status quo and rocking the boat.

In these terms, the debate phase in my methodology is hopefully Lockean; it seeks the consensus of concerned actors. The structure of the debate, through rival conceptual models based on opposed root definitions is Kantian/Hegelian. Overall, the acceptance that no systems study is ever terminated, but only redefines new problems is, again hopefully, a practical manifestation of a Singerian inquirer.

This, then, summarises ongoing work in one corner of the system movement. It tries to look at problems in social systems as manifold, rather than as phenomenological problems, and it aims to throw light on a particular kind of "whole"—the human activity system. Beyond that it may contribute something to our understanding of the principles of "wholeness." The work is in its infancy but if successful will surely contribute towards solving the methodological problems of the "unrestricted sciences" where progress by the method of science has become so slow and dubious.

Science and the Systems Paradigm

The first part of this paper has described a particular view of the activity of science and its elegant tool, reductionism. The second has considered a counter approach based upon the belief that there will be emergent properties at the levels of complexity observable in the real world. This second, "systems approach," searches for relations between emergent properties and the wholes of which they are characteristic.

Neither is "right" and the other "wrong." We perceive a complex world outside ourselves and if we are to understand it we must reduce its variety; for that we need tools of *analysis* such as science has provided. But the world is a whole—everything is connected to everything else—and to restore the whole we need means of *integration* such as systems thinking may provide. So the 300-year-old Science Movement and the 30-year-old Systems Movement are not foes; they complement each other. Recognition of that fact might help towards a greater unity of knowledge.

References and Notes

1. A point made by E. H. Hutten, *The Origins of Science*. Allen and Unwin, London, 1962.
2. A. C. Crombie, *Augustine to Galileo: The History of Science AD 400–1650*. Falcon Educational Books, London, 1952.
3. H. Butterfield, *Origins of Modern Science*. Bell and Sons, London, 1950.
4. A. C. Crombie, Reference 2 above, quoted in P. P. Wiener and A. Noland (Eds.), *Roots of Scientific Thought*. Basic Books, New York, 1957.
5. J. Ziman, *Public Knowledge: The Social Dimension of Science*. Cambridge University Press, London, 1968.

6. The point is made in a review of Bertalanffy's *General System Theory*: P. B. Checkland "Systemic, not Systematic," *Journal of Systems Engineering*, 3, No. 2, 1972, pp. 149–150.
7. A lively satirical account of reductionism is given in "The Failure of Reductionism," one of the Gifford Lectures, 1971/72, by H. C. Longuet-Higgins, published in *The Nature of Mind*. Edinburgh University Press, Edinburgh, 1972.
8. K. R. Popper, *The Logic of Scientific Discovery*. Hutchinson, London, 1959.
9. The quotation is from "Two Faces of Common Sense," a chapter in *Objective Knowledge*. Oxford University Press, London, 1972.
10. The quotation is from "Ultima Thule," in *A Russian Beauty and Other Stories*. Weidenfeld and Nicolson, London, 1973.
11. The accolades are quoted in Bryan Magee's *Popper*. Fontana, London, 1973.
12. C. F. A. Pantin, *The Relations between the Sciences*. Cambridge University Press, London, 1968.
13. This issue is posed in these terms in an essay by A. Koestler "The Yogi and the Commissar II" in the collection of that title, Cape, London, 1945.
14. See, for example, recent books by the Nobel prize winners J. Monod and F. Jacob, respectively, *Chance and Necessity*. Collins, London, 1972, and *The Logic of Living Systems*. Allen Lane, London, 1974.
15. The concerns of this school are well represented in the proceedings of the Alpbach Symposium: *Beyond Reductionism*. Edited by A. Koestler and J. R. Smythies, Hutchinson, London, 1969.
16. L. von Bertalanffy, *General System Theory*. George Braziller, New York, 1968.
17. There is a considerable literature of systems-oriented work in the behavioural and organisational sciences. See, for example, J. A. Litterer, *Organisations:* Vol. 1., *Structure and Behaviour*, Vol. 2., *Systems, Control and Adaption*. John Wiley, New York, 1969.
18. Both quotations are from the chapter of Popper's *Objective Knowledge* mentioned in Reference 9.
19. Noticeably in *The Poverty of Historicism*. Routledge and Kegan Paul, London, 1957.
20. Its context is a Masters Course described in: P. B. Checkland, G. M. Jenkins, "Learning by Doing: Systems Education at Lancaster University," *Journal of Systems Engineering*, 4, No. 1, 1974, pp. 40–51.
21. P. B. Checkland, "A Systems Map of the Universe," *Journal of Systems Engineering*, 2, No. 2, 1971, pp. 107–114.
22. G. M. Jenkins, "The Systems Approach," *Journal of Systems Engineering*, 1, No. 1, 1969, pp. 3–49.
23. P. B. Checkland, "Towards a Systems-based Methodology for Real-World Problem Solving," *Journal of Systems Engineering*, 3, No. 2, 1972, pp. 87–116.
24. P. B. Checkland, "The Development of Systems Thinking by Systems Practice—A Methodology from an Action Research Programme." In: *Progress in Cybernetics and Systems Research* (Vol. 11) edited by R. Trappl and F. de P. Hanika, Hemisphere, Washington D.C., 1975, pp. 278–283.
25. P. B. Checkland and D. S. Smyth, "The Structure of Root Definitions," *Discussion Paper* 2/75, Department of Systems, University of Lancaster, 1975.
26. The study was carried out by T. R. Barnett, F. Schwarz and J. A. Gilbert.
27. F. E. Emery and E. L. Trist, "Socio-technical Systems." In: *Management Science, Models and Techniques* (Vol. 2), edited by C. W. Churchman and M. Verhulst, Pergamon, London, 1960.
28. R. L. Ackoff, "Towards a System of Systems Concepts," *Management Science*, 17, No. 11, 1971, pp. 661–671.
29. S. Beer, *Brain of the Firm*. Allen Lane, London, 1972.
30. The concept is developed in a series of books by Vickers from 1965; see especially *The Art of Judgement*. Chapman and Hall, London, 1965; *Value Systems and Social Progress*. Tavistock, London, 1971; *Freedom in a Rocking Boat*. Allen Lane, London, 1970; *Making Institutions Work*. Associated Business Programmes, London, 1973.
31. C. W. Churchman, *The Design of Inquiring Systems*. Basic Books, New York, 1971.

7

Old Trends and New Trends in General Systems Research

Robert Rosen

It is a great honor for me to have been invited to present this year's Ludwig von Bertalanffy Memorial Lecture. I was privileged to have known Ludwig von Bertalanffy personally in the last few years of his life, when he came to the State University of New York at Buffalo in the late 1960s. Although his appointment was in the Faculty of Social Sciences, his offices were located in the Center for Theoretical Biology, of which I was then Assistant Director. I cannot imagine a more fitting arrangement; von Bertalanffy's roots were solidly anchored in the theory of biological systems, of which he initiated some of the deepest pioneering studies; from these roots grew unique insights into the character of social and behavioral systems, as well as the creation of a General Systems Theory which bound all these activities together into an integrated intellectual edifice of unique amplitude and power.

I myself am primarily a theoretical biologist. My experience convinces me that the study of organic phenomena forces one down the path that von Bertalanffy was among the first to illuminate, and it is therefore no accident that the pioneers of systems theory were also deeply concerned with theoretical biology. For biology occupies a unique position, at the confluence of all sciences and technologies. In its microscopic realms it looms on the horizon of the physicist and chemist; at macroscopic levels organisms provide examples of control of complex behaviors which have inspired technologists, scientists and philosophers from Leonardo to Norbert Wiener and beyond; and organisms are both examples of and constituents of higher-order social, political and economic structures. As a result, there is no advance in any science which is without impact on biology; conversely, any development in biology tends to radiate out into every corner of natural philosophy. To understand the phenomena of organic nature *per se*, and to understand how they are related to other scientific realms, is one of the main forces which led to the creation of a theory of general systems, and determined its shape and thrust.

Ironically, it is only during the past few hundred years, since the rise of empiricism and reductionism, that we have come to recognize that life requires explanation. Before then, back to the dawn of thought, life was regarded as itself an explanatory principle, or as a primitive, irreducible feature of the world. The diagnostic of life was the capacity for autonomous movement; therefore any entity exhibiting autonomous movement was to that extent alive, and the spirit which animated it explained its movement. Consequently the stars and planets were alive; the sun, the moon and the clouds were alive. Moreover, in earliest times as well as in the present, man has striven to explain what he does not understand in terms of what he does (or thinks he does). Initially, man had no understanding of what we would now call inanimate nature; what he really understood was his society. He knew what it was for; he knew his place in it; he knew how it worked. Therefore, it was not surprising that he would attempt to explain natural phenomena in terms of a social or familial organization of vital units, not unlike the one he knew so well. Thus were born the various pantheons which were the General System Theories of their day, in which natural phenomena were considered to be understood when they were appropriately personified.

All of this was exorcised with astonishing rapidity at the beginning of the 18th century, although of course its basis had been eroding for a long time before that. The ultimate event in the banishment of life as an explanatory principle was probably the publication of Newton's *Principia*; the continuing influence of this book on the thought of the world is impossible to overestimate (although, like most mathematical books, it has not been widely read). For Newton was able to show that, on the basis of a small number of very simple universal laws, it was possible to explain, and to predict, all the celestial motions which had previously been regarded as autonomous or volitional. Newton also provided, in his development of the differential and integral calculus, the technical tools through which such predictions could be made, thereby inevitably tying natural phenomena to mathematical representation through appropriate symbolic encoding.

Newton's analysis was universally valid for any system consisting of material particles. Therefore, insofar as any natural systems can be regarded as consisting of material particles, the laws of Newtonian dynamics must be satisfied. The consequences of this are profound and immediate; in the words of Laplace.[1]

> An intelligence knowing, at a given instant of time, all forces acting in nature, as well as the momentary position of all things of which the universe consists, would be able to comprehend the motions of the largest bodies of the world as well as the lightest atoms in one single formula, provided his intellect were sufficiently powerful to subject all data to analysis; to him nothing would be uncertain; both past and future would be present in his eyes.

As noted above, the liberating and galvanizing effect of these developments was, for good or ill, perhaps the decisive influence in creating the form and character of our present civilization; its consequences continue to unfold at the present time and will doubtless continue to do so throughout whatever future is left to us. To pursue these consequences as they have propagated through history is of consuming interest, but this is not my present concern. I wish to note here only two conceptual consequences of the Newtonian picture, which are most important to understand. These are:

1) On the basis of the Newtonian picture and its subsequent ramifications, we may say that we have come to have some understanding of the phenomena of inanimate nature. We have accomplished this without the postulation of any vital principle. But we must pay a heavy price for this, and the price is the following: *life now becomes something which itself must be explained.* As noted earlier, we always explain what we do not understand in terms

of what we do; therefore life must be explained in the same way that the motion of the stars and planets was explained. Indeed, we are left no other possible mode of explanation of life than that exemplified by Newtonian analysis; nothing else can constitute explanation, almost by definition.

2) We have seen that the universality of the Newtonian world-view rests on the extent to which we can decompose any natural system into a set of material particles. The belief that any natural system can be so decomposed, and that the laws governing the motions of these particles can be determined, is the essence of *reductionism*. Reductionism thus serves to bring all natural systems under the purview of the Laplacian Spirit mentioned above. Concomitantly, the problem of actually analyzing a specific system to determine the character of the particles of which it is composed, and the nature of the forces acting on these particles, is an empirical problem, and not a conceptual or theoretical one. Thus, along with the growth of reductionism we find a corresponding rise in *empiricism*.

Empiricism initially is an extension of the commonsense view that in order to determine the properties of a system of interest, one must observe that system directly. The empirical approach dovetailed neatly with reductionism not only in the conceptual context we outlined above, but also in connection with technological developments; clearly one could only understand a machine (e.g., a clock) by taking it apart, examining each of its constituents in detail, and determining how the operation of each constituent plays a particular role in the overall functioning of the machine. But the empiricist-reductionist view can be pursued independently, separated from the conceptual framework which gives it meaning; it then becomes fraught with dangers and pitfalls. For instance: measurement and observation are primarily ways for discriminating between systems; that is, they necessarily emphasize the idiosyncratic features of systems, rather than what makes them similar. Worse still, an exclusively empirical approach to natural systems has a way of degenerating into such things as positivism; the belief that the only qualities of systems which are real or meaningful are those which can be directly measured. Invariably such a belief is coupled with an extremely narrow view as to what in fact constitutes a measurement or observation, and leads inexorably to an assertion that whatever falls outside this narrow framework does not exist. Worst of all, an exclusively empirical approach gives rise to an utterly false dichotomy between "experiment" and "theory," which has done boundless mischief in the world, and about which we shall have much more to say subsequently.

Now the phenomena of biology, i.e., the behaviors exhibited by living systems, have always posed a critical challenge to both the Newtonian synthesis, and to the reductionist-empiricist approach to which it indirectly gave rise. This was clearly perceived by Einstein, who wrote once to Leo Szilard.[2] "One can best feel in dealing with living things how primitive physics still is. . . ." We can sense the nature of the challenge of biology in that fact that, despite centuries of effort, no satisfactory definition of life or the living state has been forthcoming; no specific criterion or test which may be applied to a given system to effectively decide whether it is living or not. We all feel, I think, that there is some essential, directly perceivable distinction between the organic and the inorganic realms, but it has proved impossible to quantify this distinction. This failure is itself an important empirical fact, which as we shall see arises from a variety of circumstances. Prime among these, I believe, is that our perception of a system as living or not arises from a commonality it shares with other systems, and which is of a relational character. As noted earlier, physical measurements do not deal with such commonalities, but rather quite the reverse. Indeed, if we wanted to interrogate our Laplacian Spirit about the characteristics of a particular organism, he would find it impossible to give us an answer, or even to understand us; if all he knew were the positions and momenta of all the particles in the universe, he could never locate an organism in the particulate chaos, let alone tell us about it. From this it is clear that biology forces us to demand other kinds of information, and make other kinds of measurements, besides those with which physics is concerned. Moreover, these measurements must pertain to properties which classes of systems hold in common, and not to properties which simply allow us to distinguish one system from another.

The challenge of biology which we are discussing has been, if anything, intensified by several decades of explosive research in molecular biology, which represents the most recent attempt to enforce the reductionist-empiricist paradigm on biological systems. We have indeed learned a great deal about molecular constituents of particular organisms, but this knowledge has only served to widen the gap between what we know and what needs to be explained. The molecular biologist has asserted that all problems in biology have been "reduced" to physics and thereby solved; what has in fact happened is quite the reverse of this. Namely, by demonstrating the inability of conventional physical descriptions to deal effectively with the phenomena of that subclass of physical systems which happens to be organic, we thereby illuminate fundamental and heretofore unrecognized gaps in physics itself. In other words, far from biology losing its autonomy by disappearing into present-day physics, biology will necessarily force the physicist to fundamentally and radically revise the character and scope of his discipline.

All of this was clear to Ludwig von Bertalanffy more than forty years ago. He was perhaps the first to clearly articulate the various points we have made above; the fact that the study of living systems is a study of commonalities shared by systems of differing physical structures; that physics cannot encompass organic phenomena without fundamental modification and extension; and that the commonalities characteristic of biological systems are exemplary of other kinds of commonalities which could form the basis for a general theory of systems *per se*.

All this was manifested in von Bertalanffy's analysis of the concept of *equifinality*, particularly as it appears in developmental biology. By equifinality is meant the stubborn tendency of a developing system to reach the same final state despite such experimental interventions as amputations, randomizations or hybridizations. Phenomena of this character are quite remarkable, and utterly different from the behavior of familiar physical systems; study of them led the embryologist Driesch to despair of ever finding a physical explanation for them, and thus to a thoroughgoing vitalism. Indeed, the phenomena of development in general have always been puzzling to physicists, who on the basis of the Second Law of Thermodynamics argue that material systems should proceed from ordered to disordered states; not the other way around. Von Bertalanffy was among the first to realize, and argue forcefully, that biological systems are open systems, able to exchange matter and energy with their environments, and consequently the Second Law is not even directly applicable to them; he was among the first to perceive the distinctions between the equilibria of closed systems and the steady states of open systems; he realised that the behavior of open systems in the vicinity of stable steady states exhibited all the properties traditionally associated with equifinality; and finally, on this basis, he could account for the apparently purposive, telic behavior of developing systems in an entirely plausible and satisfying way.

Now let us consider the revolutionary implications of this analysis.

First, von Bertalanffy explicitly showed that a basic characteristic of biological systems could be approached entirely through a functional or relational criterion (i.e., that the system be *open*) rather than through reductionistic analysis of the specific particles of which any particular system happens to be composed.

Second, von Bertalanffy was perhaps the first to stress the relation between regulatory phenomena in biology, particularly those which appear to involve purposiveness, and the formal concept of stability.

Third, von Bertalanffy concretely exhibited the incapacity of the physics of the time to cope with developmental phenomena such as equifinality. For instance, thermodynamics at that time was (and largely still is) restricted to closed, isolated systems very near to thermodynamic equilibrium. Clearly, such a thermodynamics is helpless in dealing with organisms. Within recent years we have indeed seen numerous attempts to develop a thermodynamics of open, far-from-equilibrium systems (the best known is probably the work of Prigogine and his school). It might be remarked parenthetically that these attempts, which seek to extrapolate from conventional closed-system thermodynamics, are in many ways quite unsatisfactory; it is likely that entirely new ideas, perhaps drawn from abstract approaches to stability (for example, from catastrophe theory) will be required. In any case, we see even here how biology forces physics into new directions.

Fourth, and perhaps most important, was von Bertalanffy's recognition of the fact that any open system, in the vicinity of an asymptotically stable steady state, must behave similarly to a developing organism exhibiting equifinality. To employ a graphic (if not completely satisfactory) term, such behavior is a *metaphor* for equifinality. Therefore, we see that there is a sense in which we can learn something about a particular system S, such as a developing organism, by studying some other system S', perhaps completely different from S in any purely reductionistic or physical sense, but nonetheless manifesting the same *behavior*.

To me, this last proposition articulates the essential characteristic of system theory, which distinguishes it from everything that has gone before. Hitherto, the reductionistic-empirical approach to systems mandated that any given system of interest be analyzed into ultimate particulate units; understanding any behavior of the system meant resolving that behavior into the properties of these particulate units and the forces acting on them. On the other hand, von Bertalanffy stressed that in the organic realm, *many important classes of behaviors are shared by systems of the utmost physical diversity*. This fact has, in turn, two utterly revolutionary implications: first, that if such behaviors are to be analyzed, the appropriate analytic units cannot be the particles of the physicist, and hence, that the reductionist analytic paradigm is not the only tool available for the study of natural systems. Indeed, as we shall see, system theory shows that the reductionistic analytic paradigm is only one of the large family of analytic paradigms, each of which captures a distinct aspect of physical reality. The second implication, which arises from the first, is this: if two systems S and S' which are physically different (i.e., are constructed of different kinds of particles) nevertheless behave similarly, then there is a sense in which we can learn about S by studying S'. That is, *physically disparate systems can nevertheless be models, or analogs, of each other*.

To me, general system theory is precisely the study of the different analytic paradigms which can be applied to understand particular system behaviors, and the characterization of the circumstances in which a system S' may be a model of a system S. All of this was clear to von Bertalanffy long ago, and these insights animated every aspect of his work.

There is one further aspect of these considerations which must be pointed out before proceeding further. I have already mentioned the fact that the Newtonian approach to the world already inextricably tied the study of natural phenomena to the properties of certain mathematical or formal representations of those phenomena. Precisely the same thing is true of the other kinds of analytic but non-reductionist paradigms which comprise general system theory. Indeed, in the last analysis, *two distinct systems S, S' can behave similarly only to the extent that they comprise alternate realizations of a common mathematical or formal structure*. The study of this mathematical structure, through the techniques of mathematics itself, is simultaneously a study of all of the realizations of that structure; i.e., of an entire class of analogous but physically distinct systems. Consequently, general system theory also comprises, in addition to the various analytic paradigms applicable to natural system behaviors, the study of formal or mathematical systems as an essential part of the natural systems which realize them.

On the basis of what we have said so far, we can now begin to understand why the very existence of a general system theory is so profoundly alienating to many of those reared in the empiricist-reductionist tradition. It is also disturbing to those who feel that any form of analysis violates the integrity of a system. It is worth while taking some time to discuss these objections, which may with some justice be regarded as an autoimmune disease of the intellectual community, and which has been characterized by some bizarre and even sinister syndromes in the recent past, generally expressed in a variety of false dualisms: theory-experiment; holist-reductionist, etc., in which these terms are related as heresy is to faith, sin to virtue.

We have seen that the study of mathematical or formal systems, and the manner in which they can be realized in nature, is an essential part of system theory. There are many who simply reject out of hand the notion that there can be any relation between mathematics and the world of life, and hence who reject any study in which mathematical analysis plays a role. As the population biologist J. G. Skellam once wrote:[3]

> Because of their extreme simplicity and abstract nature, mathematical models contrast so sharply with the intricate and colorful situations which they are intended to represent, that many biologists quite naturally have felt them to be fundamentally different in kind, and have therefore found them either inacceptable or worthless as aids to their work.

Indeed, there are certain intellectual spheres, notably those in which some kind of ineffable "inspiration" plays a primary role, in which the very term "mathematical" is the worst kind of opprobrium. This fact often takes amusing forms. Let us for example ponder some critical diatribes directed against a few well-known musical compositions:[4]

> The First Symphony of Brahms seemed to us . . . mathematical music evolved with difficulty from an unimaginative brain. . . . This noisy, ungraceful, confusing and unattractive example of dry pedantry . . . abounds in headwork without a glimmer of soul . . . we might pore over a difficult problem in mathematics until the same result is reached. . .
>
> (In) the Franck Symphony . . . one hears the creaking of machinery; one watches the development with no greater emotion than would be evoked by a mathematical demonstrator at a blackboard.

> They (Reger's compositions) are like mathematical problems and solutions, sheer brain-spun and unlyrical works.
>
> In the Schoenberg *Variations* . . . one feels that Schoenberg has reached some such conclusions as "a straight line tangent to a circle is perpendicular to the radius drawn to the point of contact".
>
> The science of M. Berlioz is a sterile algebra.
>
> The music of Wagner imposes mental tortures that only algebra has a right to inflict.

The poet Coleridge perceived an incredible antithesis between mathematics and imagination:[1]

> I have often been surprised that Mathematics, the quintessence of truth, should have found admirers so few and so languid. Frequent considerations and minute scrutiny have at length unravelled the cause; viz. that though Reason is feasted, Imagination is starved, whilst Reason is luxuriating in its proper Paradise, Imagination is wearily travelling on a dreary desert.

It is clear from these few excerpts, and many others which might have been chosen, that mathematics is perceived by many as separated from the living world by an unbridgeable chasm. Mathematics is mechanical, abstract, remote, inapplicable. The false dualisms implicit in these views are clear.

The above responses to mathematical investigations in science may seem comic. At least, they might simply be put down to individual idiosyncracy and dismissed. But at least twice within the past tragic half-century, they have taken much more serious forms. There is nothing funny in the arguments that mathematical investigations of natural systems are a pernicious debasement of science; or that they are incompatible with some state-supported orthodoxy, and hence indicative of disloyalty or treason to the State. For instance, the growth of National Socialism in Germany was accompanied by what was called "Deutsche Physik" or Aryan Physics. This was primarily a reaction on the part of certain experimental physicists against the explosive development of mathematical physics during the first few decades of this century (primarily relativity, but also quantum theory). It was argued that "'those who studied equations instead of nature" were charlatans and worse; the true scientist could only be an experimentalist, who developed a kind of mystic rapport with nature through direct interaction with it. Likewise, in the Soviet Union, the growth of what came to be called Lysenkoism was closely linked with a phobia against mathematical investigation of biological systems, particularly in genetics. In both of these cases, we find a rebellion on the part of some empirical scientists who found themselves alienated by the new directions in which their science was moving, and who found it desirable to attack these developments by allying their viewpoint with a convenient ambient political and social philosophy. In both cases mathematical approaches were denounced as wicked; in Germany as materialistic, in Russia as idealistic; what was common here was the denunciation, and the state of mind which gave rise to it.

The other aspect of system theory which many empiricists find deeply disquieting revolves around the modelling relation between systems, which, as we have seen, implies that we can learn something important about a system of interest by studying a different system, which we may call a model system or analog. To many empiricists and experimentalists, who have devoted their lives to careful observation and experiment performed on a small number of specific systems, it seems somehow outrageous, or dishonest, to acquire information about these systems through general arguments; they regard it as akin to magic. Indeed, there is a sense in which the study of model systems does resemble the practice of sympathetic magic, and in which the creation of mathematical models resembles numerology. In the form of mysticism known as Kaballa, for example, specific numbers are associated with systems, through the fact that each system has a name, and the letters of the alphabet in which the name is written also have associated numerical values. Thus, a mathematical object (a number) is associated with a given system in a unique fashion. The basic hypothesis of Kaballa, the one which makes it mysticism rather than science, is that the name of a system is not arbitrary or accidental, but partakes of the reality of the system; thus particular properties of the system are manifested in corresponding numerical properties of the associated number, and can be studied thereby. The superficial resemblance of this procedure to mathematical modelling in science is clear; such modelling also proceeds by associating a mathematical object with a system, which we believe captures some aspect of the reality of the system; we seek then to learn about the system by studying the properties of the associated model.

The crucial difference between magic and science, as we have noted, resides in the *manner* in which models are generated, and through which specific properties of the modelled system are captured.

The animosities and misunderstandings aroused through system-theoretic modes of analysis, which as we have noted, revolve to one degree or another about the modelling relation between systems, might lead one to believe that such relations are rare or obscure. Nothing could be further from the truth; in fact the modelling relation permeates our lives and experience, even when it is not ordinarily perceived as such. This being the case, it may be helpful to turn to some concrete examples, drawn from a broad spectrum of disciplines, to see how widespread these relations really are.

Let us begin with mathematics itself. We are all more or less familiar with the construction of mathematical representations or models of natural systems. But it is perhaps not widely recognized that mathematicians often construct mathematical models of other mathematical systems. Probably the best example of this activity is found in that branch of mathematics called *algebraic topology*; it is worth while spending a moment reviewing how this comes about.

As we all know, topology is that branch of mathematics which studies the notion of continuity. The systems studied in topology are thus those mathematical structures on which continuous processes can be defined; these are *topological spaces*. One of the basic problems of topology is to determine whether two such spaces are indistinguishable from one another from the standpoint of the continuous processes which may be defined on them; if this is the case, the two spaces are said to be *homeomorphic*.

Now topology in general is a very difficult subject, and the problem of deciding whether two spaces are homeomorphic is a very difficult problem. By contrast, the study of algebraic systems, such as groups or rings, is often very much simpler. It was Poincaré who first showed how it was possible to canonically associate algebraic structures (groups) with topological spaces, in such a way that homeomorphic topological spaces had associated with them isomorphic groups. These associated groups can therefore be regarded as *topological invariants*, and they precisely constitute algebraic models of the topological spaces from which they are constructed. The isomorphism of these groups, which can be relatively easily decided, then provides a necessary condition for the homeomorphism of the corresponding spaces. Further, specific numbers can be effectively associated with such groups in an invariant way (e.g., the rank of the group). These numbers are themselves topological invariants, in fact they play the same role for topological spaces that physical observables play for natural systems, and the formal machinery for computing them are the analogs of measuring instruments.

The field of algebraic topology is that portion of topology concerned with the construction and interpretation of algebraic models of topological spaces. The goal of algebraic topology is to construct a sufficiently broad spectrum of such algebraic models so as to completely classify topological spaces; this would in particular allow us to always decide when two given spaces are homeomorphic. Stated otherwise: each algebraic model captures some feature of the topology from which it is constructed; we seek a sufficiently comprehensive set of models so that *every* feature of the topology can be recaptured from them. It is still very much an open question whether this goal can be reached.

In any case, the relation between a topological space and an associated group is a modelling relation, and we see how we learn about the space by studying an entirely different system, the group. Now this relation between spaces and their associated groups is something which can itself be studied in a mathematical context. Such a study was initiated by the mathematicians Eilenberg and MacLane in a landmark paper entitled "The General Theory of Natural Equivalences" which was published in 1945. From this paper has grown a new branch of mathematics, which has come to be called the Theory of Categories, which cuts across all others. We can see, from what has been said so far, that Category Theory is in fact a general theory of the modelling relations which exist between formal systems. It is not surprising that this theory is coming to play an increasingly important role in general systems theory (which, as I have argued above, is precisely concerned with the establishment of modelling relations).

It is interesting to note that Category Theory has elicited exactly the same reaction within the community of pure mathematicians that system-theoretic approaches have generated in the community of experimental or empirical scientists. Initially, category theory was dismissed out of hand by many mathematicians as at best a cumbersome language for stating well-known facts about algebraic topology. Later, when it became clear that category theory could answer important open questions in a variety of areas, it was conceded to be of some limited usefulness for specific applications, but still of no significance as an independent branch of mathematics. This attitude is still prevalent; indeed, just a few years ago, I served on the Executive Committee of the Department of Mathematics at SUNY Buffalo, when that department was in the process of hiring an eminent category theorist. A substantial group of senior faculty tried to block this appointment on the grounds that category theory was not real mathematics; they were dissuaded only with difficulty from writing to other departments to collect opinions on whether category theory was mathematics or not.

Before leaving this brief discussion of mathematics, we may note that the study of mathematical systems is one which cannot proceed according to a reductionist paradigm. We cannot learn much about a mathematical system, a group, say, by decomposing it into elements. Rather, we require higher-level analytic units, specifically related to the global properties which define the group. In general, I believe that we can learn much from the analogy between the mathematician's study of formal systems and the scientist's study of natural ones; the analogy itself may well be what is responsible for what Wigner called the *unreasonable* effectiveness of mathematics in the natural sciences.

Let us now turn to physics. As we saw earlier, the Newtonian world-view has both generated and validated the empirical, reductionistic approach which has dominated modern science for centuries. Thus it is perhaps ironic that one of the supreme achievements of physics should be of the nonreductionistic character typical of system-theoretic arguments. This is, of course, the mechano-optical analogy developed by Hamilton over one hundred years ago. Hamilton showed that two distinct and independent branches of physics, namely optics and mechanics, could be conceptually unified, not through any *reduction* of optics to mechanics or vice versa, but rather through the fact that both disciplines obeyed the same formal laws; i.e., realized the same formal system.

In more detail, it had been shown by Fermat that the facts of geometric optics could be understood by postulating that light will follow that path through an optical medium which minimizes (or more accurately, which extremises) the time of transit. Likewise, there was a long tradition, culminated by Hamilton himself, in which the Newtonian laws of motion of a material system were obtained by regarding the change of configuration of that system as proceeding along a path which minimized (or extremised) a mechanical quantity called *action*. Hamilton perceived that, on this basis, one could establish an exact analogy between geometric optics and particle mechanics; i.e., one could establish a dictionary by means of which any assertion in optics could be translated into an assertion about mechanics, and *vice versa*. Moreover, any behavior of a mechanical system could be *simulated* or *modelled* by constructing an appropriate optical system; conversely, any optical behavior could be simulated in some appropriate mechanical system. On this basis, Hamilton was led to what is now called the Hamilton-Jacobi partial differential equation for the spreading of wavefronts of constant action, which is not only one of the crowning developments of 19th century science, but also provides the prototype for modern theories of optimal control. But this is not all. Hamilton was, of course, aware that geometric optics, governed by Fermat's Principle, was a limiting case of a more microscopic theory called wave optics. Hamilton himself established a one-to-one correspondence between particle mechanics and geometric optics. If he had had the least reason for doing so, he might have been tempted to argue that sitting underneath particle mechanics there was also a more microscopic theory, related to particle mechanics in the same way that wave optics is related to geometric optics. Such a more microscopic mechanical theory would naturally be called *wave mechanics*. Indeed, by exploitation of this analogy, Hamilton was fully capable of deriving Schrödinger's equation, nearly a century before it was actually derived by Schrödinger, who exploited precisely this argument. Such has been the power of analogy in physics, the bastion of reductionism.

In the past century, especially since the advent of relativity and modern quantum theory, physics has become more and more concerned with properties which remain invariant to certain classes of modifications of the system manifesting these properties. More generally, it has become concerned with expressing the manner in which physical quantities transform under such circumstances. These developments belong to a venerable tradition, whose earliest modern antecedents go back to Galileo. Among other things, Galileo investigated the manner in which the mechanical properties of structures change when the structures are subjected to a change in scale. He found, of course, that mechanical properties do not scale linearly with geometric magnitudes in general, and that therefore structures which are *geometrically* similar will not in general be *mechanically* similar.

Thus, if we wish to construct a full-sized building which is mechanically similar to a small model, we cannot simply scale up that model. These observations of Galileo provide the point of departure for the dynamical modelling which plays such a crucial role in our technology.

This kind of dynamical modelling provides a prime example of how we may study a structure of interest by investigating another system (which in this context is called a scale model). Typically, such modelling is our only recourse in system design, for we wish to determine the characteristics of a system which does not, as yet, even exist. We might hope to study these characteristics from first principles, employing physical laws, but as a practical matter this is often unfeasible. Therefore the employment of a scale model, as a kind of analog computer, becomes mandatory. Thus, our own technology provides abundant evidence of the significance of modelling relations as an alternative to the reductionistic paradigm.

All these considerations lead us naturally back to biology. The kinds of scaling laws just mentioned apply as much to organisms as to other kinds of physical systems, and there is a long tradition of investigation of biological similarities, particularly in connection with growth and with evolution. It might be mentioned parenthetically that von Bertalanffy was, of course, much interested in these developments, and himself proposed a law of growth for a class of biological systems which now bears his name. More generally, the study of quantities which remain invariant while the underlying system is subjected to transformations (e.g., is growing or evolving), which as we have seen is characteristic of much of modern physics, is also the province of what in biology is called homeostasis. In this way, we come back full circle to the phenomena with which we started, for homeostasis is simply another language for describing equifinality, stability, and control.

Biology is itself replete with examples of analogs and model systems. Many of these, oddly enough, are widely employed in empirical studies. For instance, many enzymologists, interested in studying the catalytic mechanisms whereby enzymes function, often study completely inorganic systems (such as colloidal metal suspensions), which they term *enzyme models*. Membranologists, interested in the permeability mechanisms of biological membranes, construct and study a bewildering variety of films and ion exchangers which are collectively called *model membranes*. There is an extensive literature on so-called *neuromimes*; synthetic systems which can generate and propagate an action potential, and hence exhibit excitability. As we abundantly pointed out previously, the relation between such model systems and their biological prototypes cannot reside in any identity at the particulate level, but must reside in common functional or relational properties which are preserved invariant when we replace the prototype by the model. Thus, the wide employment of model systems already reveals the tacit recognition of analytic alternatives to reductionism.

A more startling fact about biological systems is that profound analogies are found between different levels of biological activities themselves. These are of the same character as the mechano-optical analogy between different classes of physical systems, which we have already discussed above; some of the biological analogies may ultimately turn out to be as significant. Let us mention just one: the analogy between the neural networks of higher organisms and the genetic networks of individual cells.

Neural networks have, of course, played a fundamental and many-sided role in the history of system theory. They were initially introduced by McCulloch and Pitts in 1943, as discrete, Boolean versions of the continuous-time, dynamical networks introduced earlier by Rashevsky and his collaborators. Technically, the passage to discrete time served to sidestep the then insuperable analytical problems arising from the systems of non-linear differential equations which characterize the continuous-time case; the study of neural nets requires only algebra and combinatorics. It quickly became clear that the neural networks not only furnished key insights into the operation of the brain, but also indicated how one might fabricate new systems which could behave in a brainlike fashion. Consequently, neural networks became one of the primary motivations of the theory of automata in all of its ramifications, especially in computation.

The theory of genetic networks, on the other hand, arose from an attempt to come to grips with the fact that cells with the same genes could nevertheless be very different from one another. This is obvious in the differentiated cells of a multicellular organism, which in fact constitute a clone; but it is also true in bacteria. Indeed, many bacteria can adaptively modify their chemical machinery very quickly in the face of varying environmental conditions. It was clear from such facts that merely specifying a cell's genome is not sufficient; superimposed on the genome must be some kind of control mechanism which will determine whether, and when, any specific gene will be expressed (i.e., turned on).

On the basis of their experiments with bacterial systems, which were then blithely extrapolated to the rest of the biosphere, the French microbiologists Jacob and Monod proposed such a control mechanism. In brief, these workers proposed the existence of a new formal unit of control, which they termed an *operon*. A genome was conceived as a set of these formal units. More than this: the operons in a genome were in constant communication; the vehicles for this communication were postulated to be chemical specificities, residing in binding sites in the control region of the operon. Such specific interactions serve to turn the operons on and off, depending on initial conditions and environmental circumstances, with concomitant expression of the corresponding genes.

The analogy between a network of operons and a network of neurons is clear. Indeed, some of the simple operon networks which appear in the original papers of Jacob and Monod, illustrating how various kinds of differentiation can take place in cells, are formally identical with neural networks proposed a decade earlier in relation to such mental phenomena as learning and conditioning. Some rich rewards have already been gained from exploitation of the analogy, but certainly many more still await discovery.

Several features of the operon system are worth pointing out, in connection with our earlier discussion of alternatives to the reductionist paradigm in biology. For instance: it must be stressed that the operon is a purely formal unit. It is not a physical particle, nor is it composed of such particles. It cannot be isolated physically from the cell to which it belongs. It is thus not like a neuron in a neural net, which we may at least hope to identify with a discrete anatomical structure; namely, a biological nerve cell. In short: the operon is a purely abstract unit of analysis, which cuts obliquely across the particulate structures provided by the reductionistic paradigm. It is, in fact, precisely the sort of analytic unit which we argued is required in dealing with functional activities manifested by large classes of similar organic systems.

The communication channels which turn a set of operons into a network are likewise not physical structures, as are the axons in the brain. They are again purely formal units of analysis, whose properties ultimately rest on specific active sites at the molecular level.

These sites themselves also constitute examples of non-particulate functional units; they cannot be physically extracted from the molecules which bear them, nor can they be characterized in any simple fashion in terms of quantum-theoretic descriptions of those molecules. They in fact constitute a new and separate class of microphysical systems; a set of specific functional variates which are linked together, but which are not physically separable into a specific particle. Such considerations indicate yet another way in which biological activities force reconsideration of physical descriptions, this time at the most fundamental microscopic levels.

The point of these various examples has been to exhibit the ubiquity of modelling relations between systems of different physical natures; to show how they have been (and can be) used to allow us to understand the properties of physically distinct systems in a unified context, and to point out how the existence of a modelling relation between systems indicates the further existence of an underlying mathematical structure which the given systems realize. We have also seen how the modelling relation forces us to recognize other sets of analytic units besides the particles provided in the reductionistic paradigm. Such units are no less physically real than the particles, and enter into explanation of natural phenomena on the same footing. It is thus meaningless to ask which mode of analysis is "primary" in some abstract sense. Indeed, it generally happens that several different analytic modes may be brought to bear on the same system. Indeed, I would define a complex system to be one which may be analyzed in more than one way. When this happens, it becomes important to establish relations between the analytic units involved; but in general this relation is not a reductionistic one.

Let me conclude this survey of system theory with a brief discussion of one more way in which the modelling relation is intertwined with the study of life; one which I think will be most important in the years to come. Specifically, I would like to discuss the concept of an *anticipatory system*.

As was noted previously, von Bertalanffy was the pioneer in showing how apparently purposive behavior in biological systems could be understood in terms of stability. We also saw how such ideas are related to the concept of homeostasis, and thence to the crucial idea of feedback. Now the essence of feedback control is that it is *error-actuated*; in other words, the stimulus to corrective action is the discrepancy between the system's actual present state and the state the system should be in. Stated otherwise, a feedback control system must already be departing from its nominal behavior before control begins to be exercised.

On the other hand, we know from introspection that many, if not most, of our own conscious activities are generated in quite a different way. We typically decide what to do *now* in terms of what we perceive will be the effects of our action at some *later* time. To take a simple example: if we are walking in the woods, and see a bear appear from behind a tree, we need not wait until he is upon us before vacating the area. Rather we leave immediately, because we can *anticipate* the unpleasant consequences of continuing our walk. The vehicle by which we anticipate is in fact a *model*, which enables us to pull the future into the present. The stimulus for our action is in fact not simply the sight of the bear; it is the prediction or output of our model under these conditions.

This simple example contains the essence of anticipatory behavior. It involves the concept of *feedforward*, rather than feedback. The distinction between feedforward and feedback is important, and is as follows. In a feedback system, as we have seen, control is error-actuated, and serves to correct a system performance which is already deteriorating. In a feedforward system, on the other hand, system behavior is *preset*, according to some model relating present inputs to their predicted outcomes.

The essence of a feedforward system, then, is that present change of state is determined by an anticipated future state, computed in accordance with some internal model of the world.

Despite the ubiquity of feedforward modes of control, at all biological levels, anticipatory behavior of this kind has hardly begun to be seriously studied. There are several reasons for this. First, feedforward behavior seems frankly telic, or goal directed. The goal is in fact built in as part of the model which transduces between predicted future states and present changes of state. But the very suggestion that a behavior is goal-directed is repellent to many scientists, who regard it as a violation of the Newtonian paradigm.

Secondly, anticipatory systems have always been rejected out of hand in formal approaches to system theory, because they appear to violate causality. We have always been taught that we must not allow present changes of state to depend on future states; the future cannot affect the present. But in fact the entire purpose of a dynamic model is precisely to pull the future into the present; if we have a good model, and act by virtue of its predictions, we thereby approximate to a true anticipatory system.

If indeed feedforward or anticipatory control is as ubiquitous as it seems to be, then a number of fundamental new questions are posed to us. Among them are the following. Can we truly say we understand the behavior of such a system if we do not know the model employed by the system? How is it possible to determine the character of that model, in terms of measurements or observations performed on the system? More generally, under what circumstances is it possible for a system to contain an internal model of its world? What relations must exist between a set of indicators (environmental signals) and system effectors which will allow an effective feedforward control model to be constructed? How can the behaviors of different systems, perceiving the same set of circumstances but equipped with different models, be integrated? (This last is essentially the problem of conflict and conflict resolution.)

The questions just raised bear directly on the present search for forecasting and planning technologies to guide our behavior in the political, social and economic realms. Tacit in this search is the perception that our society and its institutions can no longer function effectively in a cybernetic or reactive mode; it must somehow be transformed into a predictive or anticipatory mode. That is, it must become more like an organism, and less like a machine. Such problems fall directly within the purview of system theory, which as I have repeatedly emphasized, is concerned with modelling relations between systems in all their aspects. It is with such questions that I believe system theory will become increasingly involved in the coming decades.

In dealing with these problems, I reiterate my conviction that the properties of biological systems will provide crucial insights. Indeed, considered in an evolutionary context, biology represents a vast encyclopedia of how to effectively solve complex problems; and also of how not to solve them. Biology provides us with existence proofs, and specific examples, of co-operative rather than competitive activities on the part of large and diverse populations. These insights represent natural resources to be harvested; resources perhaps even more important to our ultimate survival than the more tangible biological resources of food and energy. But to reap such a harvest, we need to fabricate the proper tools. I believe we are indeed learning how to forge these tools, and that in these treat endeavors, the concepts of system theory pioneered by von Bertalanffy and others like him will play a pre-eminent role. We are faced with great challenges in the years to come; if we meet them with wisdom, these challenges will enable us to learn and grow. Ludwig von Bertalanffy was a wise man, and a noble man; if we can assimilate his insights, and continue making progress along the paths he illuminated, we will not go wrong.

References

1. F. Cajori, *History of Mathematical Notation*. Open Court, La Salle, Ill.
2. P. Frank, *Einstein: His Life and Times*. Alfred A. Knopf, New York.
3. *Proceedings of the 1962 Cullowhee Conference on Bio-Mathematics*. Cullowhee, North Carolina.
4. N. Slonimsky (editor), *Lexicon of Musical Invective: Critical Assaults on Composers Since Beethoven's Time*. University of Washington Press.

8

Cybernetic Explanation

Gregory Bateson

It may be useful to describe some of the peculiarities of cybernetic explanation.

Causal explanation is usually positive. We say that billiard ball B moved in such and such a direction because billiard ball A hit it at such and such an angle. In contrast to this, cybernetic explanation is always negative. We consider what alternative possibilities could conceivably have occurred and then ask why many of the alternatives were not followed, so that the particular event was one of those few which could, in fact, occur. The classical example of this type of explanation is the theory of evolution under natural selection. According to this theory, those organisms which were not both physiologically and environmentally viable could not possibly have lived to reproduce. Therefore, evolution always followed the pathways of viability. As Lewis Carroll has pointed out, the theory explains quite satisfactorily why there are no bread-and-butterflies today.

In cybernetic language, the course of events is said to be subject to *restraints* and it is assumed that, apart from such restraints, the pathways of change would be governed only by equality of probability. In fact, the "restraints" upon which cybernetic explanation depends can in all cases be regarded as factors which determine inequality of probability. If we find a monkey striking a typewriter apparently at random but in fact writing meaningful prose, we shall look for restraints, either inside the monkey or inside the typewriter. Perhaps the monkey could not strike inappropriate letters; perhaps the type bars could not move if improperly struck; perhaps incorrect letters could not survive on the paper. Somewhere there must have been a circuit which could identify error and eliminate it.

Ideally—and commonly—the actual event in any sequence or aggregate is uniquely determined within the terms of the cybernetic explanation. Restraints of many different kinds may combine to generate this unique determination. For example, the selection of a piece for a given position in a jigsaw puzzle is "restrained" by many factors. Its shape must conform to that of its several neighbors and possibly that of the boundary of the puzzle; its color must conform to the color pattern of its region; the orientation of its edges must obey the topological regularities set by the cutting machine in which the puzzle was made; and so on. From the point of view of the man who is trying to solve the puzzle, these are all clues,

---

From *The American Behavioral Scientist* **10**, 29. Copyright © 1967 by Sage Publications, Inc.

i.e., sources of information which will guide him in his selection. From the point of view of the cybernetic observer, they are *restraints*.

Similarly, from the cybernetic point of view, a word in a sentence, or a letter within the word, or the anatomy of some part within an organism, or the role of a species in an ecosystem, or the behavior of a member within a family—these are all to be (negatively) explained by an analysis of restraints.

The negative form of these explanations is precisely comparable to the form of logical proof by *reductio ad absurdum*. In this species of proof, a sufficient set of mutually exclusive alternative propositions is enumerated, e.g., "P" and "not P," and the process of proof proceeds by demonstrating that all but one of this set are untenable or "absurd." It follows that the surviving member of the set must be tenable within the terms of the logical system. This is a form of proof which the nonmathematical sometimes find unconvincing and, no doubt, the theory of natural selection sometimes seems unconvincing to nonmathematical persons for similar reasons—whatever those reasons may be.

Another tactic of mathematical proof which has its counterpart in the construction of cybernetic explanations is the use of "mapping" or rigorous metaphor. An algebraic proposition may, for example, be mapped on to a system of geometric coordinates and there proven by geometric methods. In cybernetics, mapping appears as a technique of explanation whenever a conceptual "model" is invoked or, more concretely, when a computer is used to simulate a complex communicational process. But this is not the only appearance of mapping in this science. Formal processes of mapping, translation or transformation are, in principle, imputed to *every* step of any sequence of phenomena which the cyberneticist is attempting to explain. These *mappings* or transformations may be very complex, e.g., where the output of some machine is regarded as a transform of the input; or they may be very simple, e.g., where the rotation of a shaft at a given point along its length is regarded as a transform (albeit identical) of its rotation at some previous point.

The relations which remain constant under such transformation may be of any conceivable kind.

This parallel, between cybernetic explanation and the tactics of logical or mathematical proof, is of more than trivial interest. Outside of cybernetics, we look for explanation, but not for anything which would simulate logical proof. This simulation of proof is something new. We can say, however, with hindsight wisdom, that explanation by simulation of logical or mathematical proof was expectable. After all, the subject matter of cybernetics is not events and objects but the *information* "carried" by events and objects. We consider the objects or events only as proposing facts, propositions, messages, percepts and the like. The subject matter being propositional, it is expectable that explanation would simulate the logical.

Cyberneticians have specialized in those explanations which simulate *reductio ad absurdum* and "mapping." There are perhaps whole realms of explanation awaiting discovery by some mathematician who will recognize, in the informational aspects of nature, sequences which simulate other types of proof.

Because the subject matter of cybernetics is the propositional or informational aspect of the events and objects in the natural world, this science is forced to procedures rather different from those of the other sciences. The differentiation, for example, between map and territory, which the semanticists insist that scientists shall respect in their writings must, in cybernetics, be watched for in the very phenomena about which the scientist writes.

Cybernetic Explanation

Expectably communicating organisms and badly programmed computers will mistake map for territory; and the language of the scientist must be able to cope with such anomalies. In human behavioral systems, especially in religion and ritual and wherever primary process dominates the scene, the name often *is* the thing named. The bread *is* the Body, and the wine *is* the Blood.

Similarly, the whole matter of induction and deduction—and our doctrinaire preferences for one or the other—will take on a new significance when we recognize inductive and deductive steps not only in our own argument but in the relationships among data.

Of especial interest in this connection is the relationship between *context* and its content. A phoneme exists as such only in combination with other phonemes which make up a word. The word is the *context* of the phoneme. But the word only exists as such—only has "meaning"—in the larger context of the utterance, which again has meaning only in a relationship.

This hierarchy of contexts within contexts is universal for the communicational (or "emic") aspect of phenomena and drives the scientist always to seek for explanation in the ever larger units. It may (perhaps) be true in physics that the explanation of the macroscopic is to be sought in the microscopic. The opposite is usually true in cybernetics: Without context, there is no communication.

In accord with the negative character of cybernetic explanation, "information" is quantified in negative terms. An event or object such as the letter K in a given position in the text of a message *might* have been any other of the limited set of 26 letters in the English language. The actual letter excludes (i.e., eliminates by restraint) 25 alternatives. In comparison with an English letter, a Chinese ideograph would have excluded several thousand alternatives. We say, therefore, that the Chinese ideograph carries more information than the letter. The quantity of information is conventionally expressed as the log to base 2 of the improbability of the actual event or object of its context.

Probability, being a ratio between quantities which have similar dimensions, is itself of zero dimensions. That is, the central explanatory quantity, information, is of zero dimensions. Quantities of real dimensions (mass, length, time) and their derivatives (force, energy, etc.) have no place in cybernetic explanation.

The status of energy is of special interest. In general in communicational systems, we deal with sequences which resemble stimulus-and-response rather than cause-and-effect. When one billiard ball strikes another, there is an energy transfer such that the motion of the second ball is energized by the impact of the first. In communicational systems, on the other hand, the energy of the response is usually provided by the respondent. If I kick a dog, his immediately sequential behavior is energized by his metabolism, not by my kick. Similarly, when one neuron fires another, or an impulse from a microphone activates a circuit, the sequent event has its own energy sources.

Of course, everything that happens is still within the limits defined by the law of energy conservation. The dog's metabolism might in the end limit his response but, in general, in the systems with which we deal, the energy supplies are large compared with demands upon them; and, long before the supplies are exhausted, "economic" limitations are imposed by the finite number of available alternatives, i.e., there is an economics of probability. This economics differs from an economics of energy or money in that probability—being a ratio—is not subject to addition or subtraction but only to multiplicative processes, such as fractionation. A telephone exchange at a time of emergency may be "jammed" when a large fraction of its alternative pathways are busy. There is, then, a low probability of any given message getting through.

In addition to the restraints due to the limited economics of alternatives, two other categories of restraint must be discussed: restraints related to "feedback" and restraints related to "redundancy."

We consider first the concept of "feedback":

When the phenomena of the universe are seen as linked together by cause-and-effect and energy transfer, the resulting picture is of complexly branching and interconnecting chains of causation. In certain regions of this universe (notably organisms in environments, eco-systems, thermostats, steam engines with governors, societies, computers, and the like), these chains of causation form circuits which are *closed* in the sense that causal interconnection can be traced around the circuit and back through whatever position was (arbitrarily) chosen as the starting point of the description. In such a circuit, evidently, events at any position in the circuit may be expected to have effect at *all* positions on the circuit at later times.

Such systems are, however, always *open*: (a) in the sense that the circuit is energized from some external source and loses energy usually in the form of heat to the outside; and (b) in the sense that events within the circuit may be influenced from the outside or may influence outside events.

A very large and important part of cybernetic theory is concerned with the formal characteristics of such causal circuits, and the conditions of their stability. Here I shall consider such systems only as sources of *restraint*.

Consider a variable in the circuit at any position and suppose this variable subject to random change in value (the change perhaps being imposed by impact of some event external to the circuit). We now ask how this change will affect the value of this variable at that later time when the sequence of effects has come around the circuit. Clearly the answer to this last question will depend upon the characteristics of the circuit and will, therefore, be *not random*.

In principle, then, a causal circuit will generate a nonrandom response to a random event *at that position in the circuit at which the random event occurred*.

This is the general requisite for the creation of cybernetic restraint in any variable at any given position. The particular restraint created in any given instance will, of course, depend upon the characteristics of the particular circuit—whether its overall gain be positive or negative, its time characteristics, its thresholds of activity, etc. These will together determine the restraints which it will exert at any given position.

For purposes of cybernetic explanation, when a machine is observed to be (improbably) moving at a constant rate, even under varying load, we shall look for restraints—e.g., for a circuit which will be activated by changes in rate and which, when activated, will operate upon some variable (e.g., the fuel supply) in such a way as to diminish the change in rate.

When the monkey is observed to be (improbably) typing prose, we shall look for some circuit which is activated whenever he makes a "mistake" and which, when activated, will delete the evidence of that mistake at the position where it occurred.

The cybernetic method of negative explanation raises the question: is there a difference between "being right" and "not being wrong"? Should we say of the rat in a maze that

Cybernetic Explanation

he has "learned the right path" or should we say only that he has learned "to avoid the wrong paths"?

Subjectively, I feel that I know how to spell a number of English words, and I am certainly not aware of discarding as unrewarding the letter K when I have to spell the word "many." Yet, in the first level cybernetic explanation, I should be viewed as actively discarding the alternative K when I spell "many."

The question is not trivial and the answer is both subtle and fundamental: *Choices are not all at the same level.* I may have to avoid error in my choice of the word "many" in a given context, discarding the alternatives, "few," "several," "frequent," etc. But if I can achieve this higher level choice on a negative base, it follows that the word "many" and its alternatives somehow must be conceivable to me—must exist as distinguishable and possibly labelled or coded patterns in my neural processes. If they do, in some sense, exist, then it follows that, after making the higher level choice of what word to use, I shall not necessarily be faced with alternatives at the lower level. It may become unnecessary for me to exclude the letter K from the word "many." It will be correct to say that I know positively how to spell "many"; not merely that I know how to avoid making mistakes in spelling that word.

It follows that Lewis Carroll's joke about the theory of natural selection is not entirely cogent. If, in the communicational and organizational processes of biological evolution, there be something like *levels*—items, patterns and possibly patterns of patterns—then it is logically possible for the evolutionary system to make something like positive choices. Such levels and patterning might conceivably be in or among genes or elsewhere.

The circuitry of the above mentioned monkey would be required to recognize deviations from "prose," and prose is characterized by pattern or—as the engineers call it—by redundancy.

The occurrence of the letter K in a given location in an English prose message is not purely random event in the sense that there was ever an equal probability that any other of the 23 letters might have occurred in that location. Some letters are more common in English that others, and certain combinations of letters are more common than others. There is, thus, a species of patterning which partly determines which letters shall occur in which slots. As a result: if the receiver of the message had received the entire rest of the message but had not received the particular letter K which we are discussing, he might have been able, with better than random success, to guess that the missing letter was, in fact, K. To the extent that this was so, the letter K did not, for that receiver, exclude the other 25 letters because these were already partly excluded by information which the recipient received from the rest of the message. This patterning or predictability of particular events within a larger aggregate of events is technically called "redundancy."

The concept of "redundancy" is usually derived, as I have derived it, by considering first the maximum of information which might be carried by the given item and then considering how this total might be reduced by knowledge of the surrounding patterns of which the given items is a component part. There is, however, a case for looking at the whole matter the other way round. We might regard patterning or predictability as the very essence and *raison d'être* of communication, and see the single letter unaccompanied by collateral clues as a peculiar and special case.

The idea that communication *is* the creation of redundancy or patterning can be applied to the simplest engineering examples. Let us consider an *observer* who is watching A send a message to B. The purpose of the transaction (from the point of view of A and B) is to create in B's message pad a sequence of letters identical with the sequence which formerly occurred in A's pad. But from the point of view of the observer this is the creation of redundancy. If he has seen what A had on his pad, he will not get any new information about the message itself from inspecting B's pad.

Evidently, the nature of "meaning," pattern, redundancy, information and the like, depends upon where we sit. In the usual engineers' discussion of a message sent from A to B, it is customary to omit the observer and to say that B received information from A which was measurable in terms of the number of letters transmitted, reduced by such redundancy in the text as might have permitted B to do some guessing. But in a wider universe, i.e., that defined by the point of view of the observer, this no longer appears as a "transmission" of information but rather as a spreading of redundancy. The activities of A and B have combined to make the universe of the observer more predictable, more ordered and more redundant. We may say that the rules of the "game" played by A and B explain (as "restraints") what would otherwise be a puzzling and improbable coincidence in the observer's universe, namely the conformity between what is written on the two message pads.

To guess, in essence, is to face a cut or slash in the sequence of items and to predict across that slash what items might be on the other side. The slash may be spatial or temporal (or both) and the guessing may be either predictive or retrospective. A pattern, in fact, is definable as an aggregate of events or objects which will permit in some degree such guesses when the entire aggregate is not available for inspection.

But this sort of patterning is also a very general phenomenon, outside the realm of communication *between* organisms. The reception of message material by *one* organism is not fundamentally different from any other case of perception. If I see the top part of a tree standing up, I can predict—with better than random success—that the tree has roots in the ground. The percept of the tree top is redundant with (i.e., contains "information" about) parts of the system which I cannot perceive owing to the slash provided by the opacity of the ground.

If then we say that a message has "meaning" or is "about" some referent, what we mean is that there is a larger universe of relevance consisting of message-plus-referent, and that redundancy or pattern or predictability is introduced into this universe by the message.

If I say to you "it is raining," this message introduces redundancy into the universe, message-plus-raindrops, so that from the message alone you could have guessed—with better than random success—something of what you would see if you looked out of the window. The universe, message-plus-referent, is given pattern or form—in the Shakespearean sense, the universe is *informed* by the message; and the "form" of which we are speaking is not in the message nor is it in the referent. It is a correspondence between message and referent.

In loose talk, it seems simple to locate information. The letter K in a given slot proposes that the letter in that particular slot is a K. And, so long as all information is of this very direct kind, the information can be "located": the information about the letter K is seemingly in that slot.

The matter is not quite so simple if the text of the message is redundant but, if we are lucky and the redundancy is of low order, we may still be able to point to parts of the text

Cybernetic Explanation

which indicate (carry some of the information) that the letter K is expectable in that particular slot.

But if we are asked: where are such items of information as that: a. "This message is in English"; and b. "In English, a letter K often follows a letter C, except when the C begins a word," we can only say that such information is *not* localized in any part of the text but is rather a statistical induction from the text as a whole (or perhaps from an aggregate of "similar" texts). This, after all, is meta-information and is of a basically different order—of different logical type—from the information that "the letter in this slot is K."

This matter of the localization of information has bedeviled communication theory and especially neurophysiology for many years and it is, therefore, interesting to consider how the matter looks if we start from redundancy, pattern or form at the basic concept.

It is flatly obvious that no variable of zero dimensions can be truly located. "Information" and "form" resemble contrast, frequency, symmetry, correspondence, congruence, conformity and the like in being of zero dimensions and, therefore, are not to be located. The contrast between this white paper and that black coffee is not somewhere between the paper and the coffee and, even if we bring the paper and coffee into close juxtaposition, the contrast between them is not thereby located or pinched between them. Nor is that contrast located between the two objects and my eye. It is not even in my head; or, if it be, then it must also be in your head. But you, the reader, have not seen the paper and the coffee to which I was referring. I have in my head an image or transform or name of the contrast between them; and you have in your head a transform of what I have in mine. But the conformity between us is not localizable. In fact, information and form are not items which can be localized.

It is, however, possible to begin (but perhaps not complete) a sort of mapping of formal relations within a system containing redundancy. Consider a finite aggregate of objects or events (say a sequence of letters, or a tree) and an observer who is already informed about all the redundancy rules which are recognizable (i.e., which have statistical significance) within the aggregate. It is then possible to delimit regions of the aggregate within which the observer can achieve better than random guessing. A further step toward localization is accomplished by cutting across these regions with slash marks, such that it is across these that the educated observer can guess, from what is on one side of the slash, something of what is on the other side.

Such a mapping of the distribution of patterns is, however, in principle, incomplete because we have not considered the sources of the observer's prior knowledge of the redundancy rules. If, now, we consider an observer with *no* prior knowledge, it is clear that he might discover some of the relevant rules from his perception of *less* than the whole aggregate. He could then use his discovery in predicting *rules* for the remainder—rules which would be correct even though not exemplified. He might discover that "H often follows T" even though the remainder of the aggregate contained no example of this combination. For this order of phenomenon a different order of slash mark—meta-slashes—will be necessary.

It is interesting to note that meta-slashes which demarcate what is necessary for the naive observer to discover a rule are, in principle, displaced relative to the slashes which would have appeared on the map prepared by an observer totally informed as to the rules of redundancy for that aggregate. (This principle is of some importance in aesthetics. To the aesthetic eye, the form of a crab with one claw bigger than the other is not simply asymmetrical. It first proposes a rule of symmetry and then subtly denies the rule by proposing a more complex combination of rules.)

When we exclude all things and all real dimensions from our explanatory system, we are left regarding each step in a communicational sequence as a *transform* of the previous step. If we consider the passage of an impulse along an axon, we shall regard the events at each point along the pathway as a transform (albeit identical or similar) of events at any previous point. Or if we consider a series of neurons, each firing the next, then the firing of each neuron is a transform of the firing of its predecessor. We deal with event sequences which do not necessarily imply a passing on of the same energy.

Similarly, we can consider any network of neurons, and arbitrarily transect the whole network at a series of different positions, then we shall regard the events at each transection as a transform of events at some previous transection.

In considering perception, we shall not say, for example, "I see a tree," because the tree is not within our explanatory system. At best, it is only possible to see an image which is a complex but systematic transform of the tree. This image, of course, is energized by my metabolism and the nature of the transform is, in part, determined by factors within my neural circuits: "I" make the image, under various restraints, some of which are imposed by my neural circuits, while others are imposed by the external tree. An hallucination or dream would be more truly "mine" insofar as it is produced without immediate external restraints.

All that is not information, not redundancy, not form and not restraints—is noise, the only possible source of *new* patterns.

9

Systems and Distinctions; Duality and Complementarity

Joseph A. Goguen and Francisco J. Varela

1. Introduction

The world does not present itself to us neatly divided into systems, subsystems, environments, and so on. These are divisions which we make ourselves, for various purposes, often subsumed under the general purpose evoked by saying "for convenience." It is evident that different people find it convenient to divide the world in different ways, and even one person will be interested in different systems at different times, for example, now a cell, with the rest of the world its environment, and later the postal system, or the economic system, or the atmospheric system.

The established scientific disciplines have, of course, developed different preferred ways of dividing the world into environment and system, in line with their different purposes, and have also developed different methodologies and terminologies consistent with their motivations.

In this paper, we present a framework within which a number of these various preferred views on systems can be unified. Of particular interest to us are the differences stemming from the study of natural systems (particularly biological and social systems) and man-made systems (such as engineering and computer systems). Contemporary system theory has developed extensively through experience in the later fields, but the insights derived from natural systems have remained by and large much less formally developed. We hold that the notions of cooperative interaction, self-organization, and autonomy—in brief, holistic notions—are relevant and basic to the study of natural systems. In the framework of this paper and its successors, these notions are not only made more explicit and applicable, but are also presented as complements to the more traditional notions of system theory, such as control and input-output behavioral description.

2. Distinction and Indication

A *distinction* splits the world into two parts, "that" and "this," or "environment" and "system," or "us" and "them," etc. One of the most fundamental of all human activities is the making of distinctions. Certainly, it is the most fundamental act of system theory, the very act of defining the system presently of interest, of distinguishing it from its environment.

Distinctions coexist with purposes. A particularly basic case is a system defining its own boundaries and attempting to maintain them; this seems to correspond to what we think of as self-consciousness. It can be seen in individuals (ego or identity maintenance) and in social units (clubs, subcultures, nations). In such cases, not only is there a distinction, but an *indication*, that is, a marking of one of the two distinguished states as being primary ("this," "I," "us," etc.); indeed, it is the very purpose of the distinction to create this indication.[1,2]

A less basic kind of distinction is one made by a distinctor for some purpose of his own. This is what we generally see explicitly in science, for example, when a discipline "defines its field of interests," or a scientist defines a system which he will study.

In either case, the establishment of system boundaries is inescapably associated with what we will call a *cognitive point of view*, that is, a particular set of presuppositions and attitudes, a perspective, or a frame in the sense of Bateson[3] or Goffman[4]; in particular, it is associated with some notion of value, or interest. It is also linked up with the cognitive capacities (sensory capabilities, knowledge background) of the distinctor. Conversely, the distinctions made reveal the cognitive capabilities of the distinctor. It is in this way that biological and social structures exhibit their coherence, and make us aware that they in fact have cognitive capacities or that they are "conscious" in some degree.

The importance for system theory of cognitive coherence (or cognitive point of view, or cognitive capability) is a theme which runs throughout this paper and its successors. Because of the focus on system theory, we shall feel free to invoke the idea of an "observer," one or more persons who embodies the cognitive point of view which created the system in question, and from whose perspective it is subsequently described.

A simple but fundamental property of the situation involving a system and an observer, is that he may choose to focus his attention either on the internal constitution of the system, or else on its environment, taking the system's properties as given. That is, an observer can make a distinction into an indication through the imposition of his value. If the observer chooses to pay attention to the environment, he treats the system as a simple entity with given properties and seeks the regularities of its interaction with the environment, that is, the constraints on the behavior of the system imposed by its environment.[5] This leads naturally to the problem of controlling the behavior of the system, as considered in engineering control theory. On the other hand, the observer may choose to focus on the internal structure of the system viewing the environment as background, for example, as a source of perturbations upon the system's autonomous behavior. From this viewpoint, the properties of the system emerge from the interactions of its components.

Biology has iterated this process of indication, creating a hierarchy of levels of biological study. The cell biologist emphasizes the cell's autonomy, and views the organism of which it is part as little more than a source of perturbations for which the cell compensates. But the physiologist views the cell as an element in a network of interdependencies constituting the individual organism: this corresponds to a wider view of environment, namely the ecology in which the individual participates. A population biologist

Duality and Complementarity

makes his distinctions at a still higher level, and largely ignores the cell. A similar hierarchy of levels can be found in the social sciences. It seems to be a general reflection of the richness of natural systems that indication can be iterated to produce a hierarchy of levels.

At a given level of the hierarchy, a particular system can be seen as an outside to systems below it, and as an inside to systems above it; thus, the status (i.e., the mark of distinction) of a given system changes as one passes through its level, in either the upward or the downward direction. The choice of considering the level above or below corresponds to a choice of treating the given system as autonomous or controlled (constrained). Figure 1 illustrates a variety of configurations of systems, subsystems, and marks, and Figure 2 illustrates the hierarchy of levels.

3. Recursion and Behavior

In system theory, the autonomy/control distinction appears more specifically as a recursion/behavior distinction. The behavioral view reduces a system to its input-output performance or behavior, and reduces the environment to inputs to the system. The effect of outputs on environment is not taken into account in this model of the system. The recursive view of a system emphasizes the mutual interconnectedness of its components.[6,2,7] That is, the behavioral view arises when emphasis is placed on the environment, and the recursive view arises when emphasis is placed on the system's internal structure.

If we stress the autonomy of a system S_i (see Figure 1) then the environmental influences become perturbations (rather than inputs) which are compensated for through the underlying recursive interdependence of the system's components (the s_{i-1}'s in the figures). Each such component, however, is treated behaviorally, in terms of some input-output description.

The recursive viewpoint is more sophisticated than the behavioral, since it involves the simultaneous consideration of three different levels, whereas the behavioral strictly speaking involves only two. This is because the behavioral model, in taking the environment's view of the system, does not involve making any new distinctions. But expressing interest in how the system achieves its behavior through the interdependent action of its parts adds a new distinction, between the system and its parts.

The following may help to make this seem less abstract. The most traditional way to express the interdependence of variables in a system is by differential equations. A (time varying) autonomous system can be represented by equations of the form

$$\dot{x}_i = F_i(x, t) \quad \text{for } 1 \leq i \leq n \tag{1}$$

where $x = \langle x_1, \ldots, x_n \rangle$ is the state vector of the system. The autonomous behavior of the system is described by a solution vector $x(t)$, which satisfies (1). This involves treating everything as happening on the same level, and all variables as being observable; in effect, environment is treated as part of the system (or ignored).

However, the effect of the environment on the system can be represented by a vector $e = \langle e_1, \ldots, e_k \rangle$ of parameters, giving

$$\dot{x}_i = F_i(x, e, t) \quad \text{for } 1 \leq i \leq n \tag{2}$$

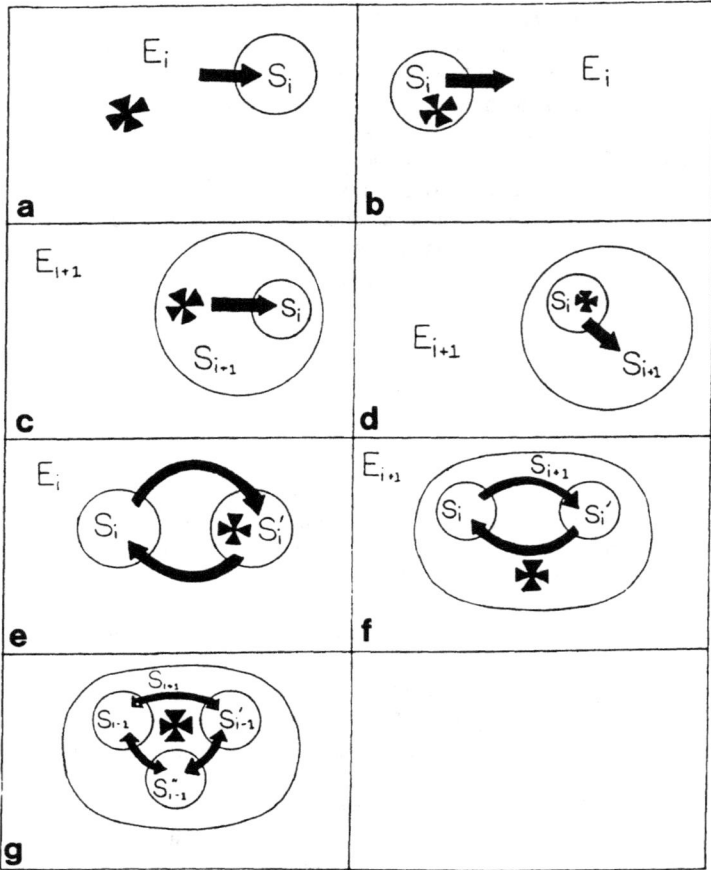

Figure 1. Various configurations of systems, subsystems and marks; each configuration represents a cognitive viewpoint, and the mark indicates its center. The arrows indicate the flow of signals and interactions. (a) Control of a system S_i by its environment E_i. (b) Autonomy of system S_i in its environment E_i. (c) Control of a subsystem S_i in a system S_{i+1}. (d) Autonomy of a subsystem S_i of a system S_{i+1}. (e) Feedback control of system S_i by system S'_i. (f) Communication between (coordination of) two subsystems S_i, S'_i of system S_{i+1}. (g) Coordination (ecology) of subsystems S_{i-1}, S'_{i-1}, S''_{i-1}.

which explicitly takes account of two levels. Solutions to the system (2) are now also parameterized by e, that is, they are of the form $x(e, t)$.

The situation of (2) can be elaborated in two directions. In control theory, it is usual to assume that the internal variables x of the system are either unobservable, or are of no direct interest, and that we have instead direct access to (or interest in) an output vector y of variables which are functionally dependent on x. The variables e are usually taken to be under the control of the observer, and the question is posed, how to use those variables to obtain certain desired values of y. The equations are thus of the form

Duality and Complementarity

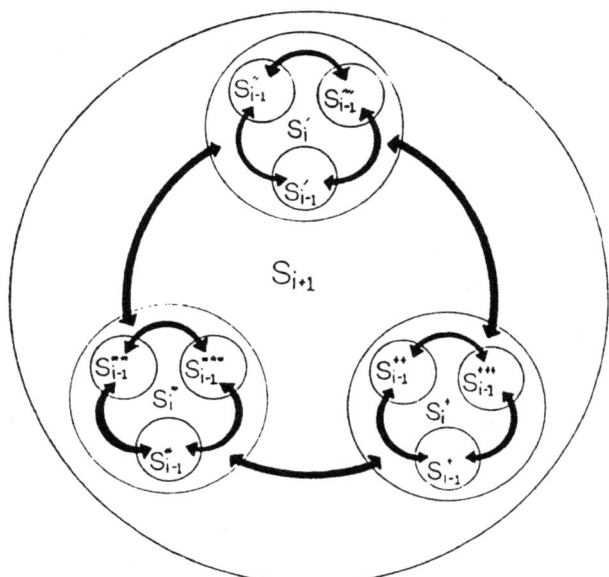

Figure 2. Diagrammatic evocation of a hierarchy of system levels: Systems S_{i-1}^*, S_{i-1}^{**}, S_{i-1}^{***}, ... of level $(i-1)$ constitute system S_i^* at level i; similarly, systems S_i^*, S_i^+, S_i', ... of level i constitute systems S_{i+1} at level $(i+1)$; and S_{i+1} together with other systems of level $(i+1)$ will constitute a system at level $(i+2)$; and so on, upward and downward.

$$\dot{x}_i = F_i(x, e, t) \quad (3)$$
$$y = H(x)$$

Strictly speaking, the equations span three levels, and can be used, for example, to infer information about the system's internal state, but the emphasis ("mark") is on the environment, which is identified with the observer. Behavior appears as an input-output function $y(e, t)$, the observable results of applying the inputs (also called "controls") e to the system.

An alternative elaboration of the situation of (2) views the vector e as not necessarily or particularly under the control of an observer, but rather as a source of perturbations upon (2). For example, the components e_j of e may be some coefficients which are regarded as constants in the original equation (1). A natural question to pose is the stability of the system under such perturbations, that is the relation of (1) to a perturbed system

$$\dot{x}_i = F_i(x, e, t) + \delta F_i(x, e, t) \quad (4)$$

in which δ (in a fairly intuitive notation) represents a "small change." It is known, for example, that changes in structural constants can cause the system to undergo a "catastrophic" change (in the sense of Thom[8]) into a new configuration.

The system (2) has in it nothing which intrinsically prefers the approach of either (3) or (4). This choice depends on the interest of the analyst.

Note that "recursion" (in the broad sense of feedback which affects the future) plays a role in all these formulations, but is more obscure in the control theory interpretation. On the other hand, the behavioral information, though still available, is more obscure in the stability interpretation (4). We are not, of course, claiming that either of these approaches is inherently better.

Historically speaking, some of the many possible approaches to systems have been much more developed than others. The most highly developed parts, in fact, center on the notions of control, input–output behavior, and state-transition. This is presumable because of the interest in applying these approaches in engineering.

It seems, however, that the notion of autonomy is particularly important for natural systems, e.g., biological and social systems, and the lack of a well-developed theory of autonomous systems is felt as a serious difficulty. An engineer designing an artifact will choose the inputs of interest to him for this application with some assurance that the choice will be adequate. But a biologist studying a cell is forced to acknowledge the autonomy of the cell; if the biologist's preferences for input and output variables do not match the cell's internal organization, the biologist's theory will not work. Furthermore, the hierarchy of levels seems to particularly assert its importance for natural systems, so that it is generally necessary to take account of at least three levels. Even when the lowest level is very well understood, the role which it plays, at the next higher level when interconnected with other systems, can be quite obscure. For example, an enzyme biochemist may be able to describe a particular metabolic loop very effectively by a transfer function, but be quite unable to specify how it fits into the overall metabolic process of the cell as a coherent whole.

This situation of being unable to understand how elements, even quite well understood elements, coordinate or somehow function effectively together at the next higher level, is quite common in the study of natural systems, and is a major source of our desire for a better developed theory of autonomous systems.

Some fragments of theory emphasizing the autonomy of systems do exist, but they are far less developed than is the behavioristic approach. First and foremost, the idea of stability derived from classical mechanics, has been extensively studied and used. As we said before, a set of interdependent differential equations can be used to represent the autonomous properties of a whole system. Rosen,[9] Iberall[10] and Lange[11] have applied this perspective to natural systems with various degrees of emphasis on autonomous behavior. More specific examples can be found in population biology,[12] cellular biology,[13,14] and more recently, in neurobiology.[15] Some thought has been given to cooperative interactions in this area of hierarchical multilevel systems. The idea of hierarchy is often presented from the point of view of the interdependency of different levels of systems descriptions.[16,17,18] Particular instances of hierarchical structure including multilevel cooperation can be found in Refs. 19–21. Goguen[22,23] presents a general theory of hierarchical systems of interdependent processes. Its basic ideas are interconnection, behavior, and level, and its theoretical framework is categorical algebra. A last area in which the idea of a whole system is somewhat explicit is that of self-organizing systems. Work in this area based on an information-theoretic approach includes von Foerster,[24] and Atlan.[25]

We do not intend to unite all these various threads of research together in a single framework. Rather, we emphasize the ways in which pairs of seemingly different points of view, such as autonomy/control, are complementary, in the sense of contributing to a better

4. Holism and Reductionism

If we think of the philosophy of science, the duality holism/reductionism comes to mind as analogous to the material previously discussed in this paper.

Most discussions place holism/reductionism in polar opposition.[26–28] This seems to stem from the historical split between empirical sciences, viewed as mainly reductionist or analytic, and the (European) schools of philosophy and social science that grope toward a dynamics of totalities.[29,30]

In the light of the previous discussion, both attitudes are possible for a given descriptive level, and in fact they are complementary. On the one hand, one can move down a level and study the properties of the components, disregarding their mutual interconnection as a system. On the other hand, one can disregard the detailed structure of the components, treating their behavior only as contributing to that of a larger unit. It seems that both these directions of analysis always coexist, either implicitly or explicitly, because these descriptive levels are mutually interdependent for the observer. We cannot conceive of components if there is no system from which they are abstracted; and there cannot be a whole unless there are constitutive elements.

It is interesting to consider whether one can have a measure for the degree of wholeness of a system. One can, of course, always draw a distinction, make a mark, and get a "system" but the result does not always seem to be equally a "whole system," a "natural entity," or a "coherent object" or "concept." What is it that makes some systems more coherent, more natural, more whole, than others? (See Refs. 31 and 32 for different ways to answer these questions.)

One thing to notice is, that in the hierarchy of levels, "emergent" or "immanent" properties appear at some levels. For example, let us consider music as a system or organization of notes (for the purpose of this example, we do not attempt to reduce notes to any lower level distinctions). Then harmony only arises when we consider the simultaneous or parallel sounding of notes, and melody only arises when we consider the sequential sounding of notes. That is, harmony and melody are emergent properties of a level of organization above that of the notes themselves. Similarly, form can only emerge at a still higher level of organization, relating different melodic units to one another. These properties, form, melody and harmony, are *systems properties*, arising from hierarchical organizations of notes into pieces of music; they are not properties of notes. (See Ref. 33 for further discussion of the hierarchical organization of music.) It also appears that "life" is an emergent property of the biological hierarchy of levels: it is nowhere to be found at the level of atoms and molecules; but it becomes clear at the level of cells through the autopoietic organization of molecules.[34,35] Language can be seen as an emergent property at a still higher level of this hierarchy.[36]

Thus, one point of view toward wholeness is that it co-occurs with interesting *emergent* properties at some level. A sequence of notes is whole if it is an interesting melody.

Similarly, the significant differences in behavior of cells and organism from the behaviors of atoms and molecules, marks the cells and organisms as whole systems.

Another point of view toward wholeness, is that it can be measured by the *difficulty of reduction*: Because it is very hard to reduce the behavior of organisms to the behaviors of molecules, we may say that organisms are whole systems. Similarly, it is very difficult (if not impossible) to reduce the effects of melodies to the effects of notes. One must consider properties of patterns of notes or molecules.

A third point of view is that a system is whole to the extent that its parts are *tightly interconnected*, that is, to the degree that it is difficult to find relatively independent subsystems. This is clearly related to the previous views. An interesting corollary of this view is that a system with a strongly hierarchical organization will be less whole than a system with a strongly heterarchical organization; that is, nets are more whole that trees. More precisely, given that the graph of connections of the parts of a system has no isolated subsystems, the more tree-like it is, the less whole it is, while still being (presumably) a system. The extreme is probably a pure linear structure, without any branching at all.

A fourth point of view, is that a system seems more whole if it is more *complex*, that is, more difficult to reduce to descriptions as interconnections of lower level components. It is necessary in this discussion to take account of the very modern point of view that the relatively more complex a system is to describe, the more *random* it is; in fact, Kolmogorov,[37] who gave the original foundation for probability theory in terms of measure theory,[38] has proposed to redefine probability theory and information theory in terms of complexity theory. Thus, for example, the wholeness of a living system is, in everyday encounters, construed as unpredictability. The more difficult it is to reduce a system to a simple input/output control the more likely it is we will deem it alive. In this sense complete autonomy is logically equivalent to complete randomness. Another example: a piece of music which is too complex (relative to our cultural expectations and inherent capacities) will sound random, chaotic, perhaps meaningless, but it will also sound whole. Here the extreme is white noise. (See Ref. 33 for a discussion of hierarchical complexity with applications to musical aesthetics.) This viewpoint toward wholeness involves measurement relative to some standard interpreting system, such as a human being. But given such a standard, this viewpoint can be deduced from the preceding viewpoints. For surely, if it is difficult to describe a system, it will also be difficult to reduce it to lower levels, and its parts will seem to be tightly interconnected. Quite possibly, its very complexity will appear as an emergent property.

On the other hand, a different cognitive viewpoint might well be better able to process what now seems like a very complex system, and thus see it as less whole. Once again, the relativity to cognitive capacity appears. This can be seen over and over in the history of science, as new techniques, tools, or ideas, suddenly make long standing problems soluble, or perhaps, uninteresting or even ill-posed.

These descriptive levels haven't been generally realized as complementary largely because there is a difference between publicly announced methodology and actual practice, in most fields of research in modern science. A reductionist attitude is strongly promoted, yet the analysis of a system cannot begin without acknowledging a degree of coherence in the system to be investigated; the analyst has to have an intuition that he is actually dealing with a coherent phenomenon. Although science has publicly taken a reductionist attitude, in *practice* both approaches have always been active. It is not that one has to have a holistic

Duality and Complementarity

view as opposed to a reductionist view, or *vice versa*, but rather that the two views of systems are complementary.

There is a strong current in contemporary culture advocating "holistic" views as some sort of cure-all. This is easily seen in discussions of environmental phenomena, health delivery, and value systems. The way in which holism is used in this paper is quite different. We simply see reductionism and holism as complementary, or "cognitively adjoint," for the descriptions of those systems in which we are interested. Reductionism implies attention to a lower level, while holism implies attention to a higher level. These are intertwined in any satisfactory description; and each entails some loss relative to our cognitive preferences, as well as some gain.

References and Notes

1. G. Spencer-Brown, *Laws of Form*. Bantam Books, New York, 1973.
2. F. Varela, "A Calculus for Self-Reference." *Int. J. Gen. Systems*, 2, 1975, pp. 5–24.
3. G. Bateson, "A Theory of Play and Fantasy." Reprinted in: G. Bateson, *Steps Towards an Ecology of Mind*. Ballantine, New York, 1972.
4. E. Goffman, *Frame Analysis*. Harper, Colophon Books, New York, 1974.
5. Calling S "the system" rather than "the environment" already indicates a preference for marking S; that is, the language incorporates the preference. But we may speak of "marking the environment" to suggest that there are in fact two distinct possibilities.
6. H. von Foerster, "Notes for an Epistemology of Living Things," *L'Unite de L'homme*, edited by E. Morin and M. Piatelli, Eds. du Seuil, Paris, 1974, pp. 401–417.
7. F. Varela and J. Goguen, "The Arithmetic of Closure." *J. Cybernetics*, 8, 1978.
8. R. Thom, *Stabilité Structurelle et Morphogénèse*. Benjamin, New York, 1972.
9. R. Rosen, *Dynamical Systems Theory in Biology*. John Willey, New York, 1972.
10. A. Iberall, *Towards a General Science of Viable Systems*. McGraw Hill, New York, 1972.
11. O. Lange, *Wholes and Parts*. Pergamon Press, New York, 1965.
12. R. May, *Model Ecosystems*. Princeton U. Press, Princeton, 1971.
13. M. Eigen, "Self-organization and the Evolution of Matter." *Naturwiss*, 58, 1971, pp. 465–523.
14. B. C. Goodwin, *Temporal Organization of Cells*. Academic Press, New York, 1968.
15. A. Katchalsky, V. Rowland and R. Blumenthal, "Dynamic Patterns of Brain Cell Assemblies." *Neurosciences Res. Prog. Bull.*, 12, No. 1, 1974.
16. H. Pattee, ed., *Hierarchy Theory*. George Brazillier, New York, 1972.
17. L. Whyte, A. Wilson and D. Wilson, *Hierarchical Structures*. Elsevier, New York, 1968.
18. M. Mesarovic, D. Macko, Y. Takahara, *Theory of Hierarchical Multilevel Systems*. Academic Press, New York, 1972.
19. S. Beer, *The Brain of the Firm*. Allen Lane, London, 1972.
20. L. Kohout and B. Gaines, "Protection as a General Systems Problem." *Int. J. Gen. Systems*, 3, 1976, pp. 3–24.
21. T. Baumgartner, T. Burns, L. Meeker and B. Wild, "Open Systems and Multi-Level Processes: Implications for Social Research." *Int. J. Gen. Systems*, 3, 1976, pp. 25–42.
22. J. A. Goguen, "Mathematical Representation of Hierarchically Organized Systems." *Global Systems Dynamics*, edited by E. Attinger, S. Karger, Basel, 1971, pp. 112–128.
23. J. A. Goguen, "Systems and Minimal Realization." *Proceedings of 1971 IEEE Conference on Decision and Control*, Miami Beach, 1972, pp. 42–46.

24. H. von Foerster, "On Self-Organizing Systems and Their Environments." *Self-Organizing Systems*, edited by M. Yovits and S. Cameron, Pergamon Press, London, 1966, pp. 31–50.
25. H. Atlan, *L'Organization Biologique et la Theorie de la Information*. Herman, Paris, 1972.
26. J. C. Smuts, *Holism and Evolution*. Macmillan, New York, 1925.
27. E. Jantsch, "Evolution: Self-Realization Through Transcendence;" *Evolution and Human Consciousness*, edited by E. Jantsch and C. H. Waddington, Addison-Wesley, New York, 1976, pp. 37–63.
28. E. Laszlo, *An Introduction to Systems Philosophy*. Harper Colophon, New York, 1972.
29. K. Kosik, *La Dialectique du Concrete*. F. Maspero, Paris, 1969.
30. G. Radnitzky, *Contemporary Schools of Metascience*. Gateway Editions, Chicago, 1973.
31. J. A. Goguen, "Objects." *International Journal of General Systems*, **1**, No. 4, 1975, pp. 237–243.
32. C. Linde, "The Organization of Discourse," *The English Language: English in its Social and Historical Context*, edited by T. Shopen, A. Zwicks, and P. Griffin, 1977.
33. J. A. Goguen, "Complexity of Hierarchically Organized Systems and the Structure of Musical Experiences." *UCLA Computer Science Dept., Quarterly*, Oct. 1975, pp. 51–88; *International Journal of General Systems*, **3**, No. 4, 1977, pp. 237–251.
34. H. Maturana and F. Varela, *De Maquinas y Seres Vivos*. Editorial Universitaria: Santiago de Chile (English version: Biological Computer Lab. Report 9.4, U. of Illinois, Urbana, Ill. 1975), 1973.
35. F. Varela, H. Maturana and R. Uribe, "Autopoiesis: The Organization of Living Systems, Its Characterization and a Model." *Biosystems*, **5**, 1974, pp. 187–196.
36. H. Maturana, "Biology of Language: the Epistemology of Reality." *Psychology and Biology of Language*, a symposium in the memory of E. Lenneberg, Ithaca, Cornell U., May 1977.
37. A. N. Kolmogorov, "Logical Bases for Information and Probability Theory." *IEEE Trans. Inform. Th.*, **14**, No. 5, 1968, pp. 662–664.
38. A. N. Kolmogorov, *Mathematical Foundations of Probability Theory*. Chelsea 2nd English Edition, New York, 1956.

10

The Challenges of System Theory

Robert Rosen

We all know, I think, that system theory is a revolutionary development in scientific thought. The blossoming of system theory over the past three or four decades betokens a massive paradigm shift, on a scale which has not been seen since the publication of Newton's *Principia*. However, whereas previous paradigm shifts have resulted in the relinquishing of one paradigm in favor of another, ultimately equally restrictive one, system theory offers us not one alternate paradigm, but many; it has shown us that there exists many primary modes of system analysis, each one independent of the others, and thus that there exists no single over-arching analytic procedure which encompasses all others. The mode we choose in dealing with a particular problem must be determined by the nature of the question we are asking, and must be guided by intelligence and insight. These last are qualities which have been excluded from science for too long by slavish adherence to a single mode of system analysis; system theory has re-introduced them, and indeed established them as the first prerequisite in the generation of scientific theory.

Thus, system theory is a revolutionary development. But the very fact that it exists indicates a wider and more profound revolution in the biological nature of man, which it may be worthwhile to briefly describe.

Man is a product of biological evolution; of natural selection acting and re-acting over the eons. Now evolution is one of the most conservative of natural processes; the number of strategic innovations which it has generated since the origin of life can be counted on the fingers of one hand. Evolution has concentrated on *tactical* variations, while its essential strategy has remained firmly fixed.

As a product of evolution, man reflects the innate conservatism of the force which shaped him. One of the primary ways in which this conservatism is manifested is the following: whenever confronted with a new problem, of whatever character, man follows a wired-in strategy which tells him, in effect, to do what worked last time. We are creatures governed by precedent. This is the strategy which evolution has built into us, and which evolution itself has heretofore followed.

Now we can perhaps begin to see the true revolutionary significance of system theory.

For system theory in effect says to us: what worked last time need not work this time. It says to us: there exist other options which we can explore, and among which we can choose the best. Clearly this could not happen unless evolution itself were loosening its grip on us. This could not happen unless evolution was modifying its own strategy; something which has been unthinkable for countless millennia. Thus the very existence of system theory, the science of alternatives, serves as a harbinger of an utterly momentous event occurring in the biosphere. I therefore perceive system theory as the cutting edge, not just of science, but of an entirely new direction of human evolution.

As I have tacitly indicated above, I equate the notion of paradigm shift with the notion of strategic innovation. I first learned about strategic innovation not so much from science where it is rare, but from literature where it is all-pervasive. In literary situations, a universal principle is to force an individual to confront a problem, equipped with strategic principles inadequate to deal with the problem. Conflict is thereby generated, which animates the entire work. The distinction between tragedy and comedy lies only in this: in tragedy the conflict kills you, while in comedy it only makes a fool of you.

Let us look at a few examples. I recall an old routine of Woody Allen (one of those comedians who always confronts the world with inadequate strategic resources). In this routine, he had visited London, and decided to buy a Rolls-Royce. He did not want to pay duty on such an expensive car, so he hit on the idea of disassembling it and packing the pieces into a large number of suitcases. By claiming that these parts were modern sculpture, he managed to get the suitcases through customs, and on arriving home, proceeded to reassemble his car. His first attempt yielded 200 sewing machines. His next attempt yielded a very large lawn mower. His third attempt produced 100 bicycles. Try as he would, he could never find his Rolls-Royce in his disassembled fragments. His inadequate strategy indeed made a fool of him, and that is why the routine is funny.

A most suggestive example of inadequate strategy is found in the Edgar Allan Poe short story "The Purloined Letter." Let me briefly review the story, for those not familiar with it. A compromising letter has been stolen from the queen of France. The thief is known; it is a bold, ambitious and clever minister, who is using the letter to blackmail the queen. In desperation, she confides the matter to the prefect of police, and promises an enormous reward for the return of the letter. The prefect assures her that this will be a simple matter, since the letter must necessarily be in the minister's home, available for instant use. However, the most extensive searches fail to disclose it. The prefect, baffled, consults Poe's detective Dupin (the prototype for Sherlock Holmes). Dupin asks what methods have been employed by the police; the prefect explains that the matter is really very simple in principle; it is only a matter of accounting for a given space. The minister's home had accordingly been divided into elementary units, and then each unit was thoroughly scrutinized. The walls were probed to detect cavities. Every cushion was probed by long needles. Every piece of furniture was examined microscopically. This was repeated three times, and yet the letter could not be found. Dupin looks into the matter, and in fifteen minutes he finds the letter.

How could he succeed so quickly, where the resources of the Parisian police failed utterly? He explains to his friend: the police are very able in their way; they are thorough, diligent and painstaking. Therefore, had the principle governing the concealment of the letter fallen within the scope of the methods employed to discover it, the letter would have been found. The fact that it was not found thus implies that the principle of the letter's concealment was outside the scope of the methods. As Dupin said:

> What, in the present case, has been done to vary the principles of action? What is all this sounding, and boring, and probing, and scrutinizing with the microscope? What is it all but the exaggeration of one particular set of principles, to which the prefect, in the course of long experience, has become accustomed?

In fact, of course, the letter had been turned inside out, re-addressed, and then set out in plain sight in a rack in the minister's sitting room. It was thus quickly found by Dupin.

This story impressed me when I first read it, because the virtues of the Parisian police, as described by Dupin, are the virtues of the good empirical scientist. The scientist seeks to discover a message hidden from him by nature. Could it then be that the failure to discover the message is not the result of inadequate diligence, but indicates the fundamental inapplicability of the basic strategy?

Similar situations exist in the generation of illusions by magicians. I remember reading an article by Houdini on mentalist illusions, such as the determination of the time indicated by the hands of a closed pocket watch. In this illusion, the closed watch is laid on a table by the mentalist's assistant. The heart of illusion resides in the fact that the position of the watch on the table, and its orientation, constitute a code previously established by the illusionist and his assistant. No amount of material scrutiny of the watch, or the table, or any other physical search, will reveal this code or even its existence. An empiricist confronted with such an illusion can guess that some information channel exists, but he will seek a material channel, and the more he seeks, the more baffled he will become. Indeed, this is why magicians and other illusionists like nothing better than to be studied by scientists.

As we can see from these few examples, and many others which could be provided, there is nothing more painful than to face the world with inadequate strategic resources, and even more, to have to admit this. That is why paradigm shifts in science are such wrenching affairs. However, there is one, and, as far as I know, only one, area of human thought which has never been afflicted by the need to make such an admission; never had to concede defeat; and never had to entirely retreat from a path once embarked on. That happy field is mathematics. To be sure, mathematics has had its share of upheavals (internally termed "foundation crises") but these are invariably concerned with tactical matters, and not with strategic ones. As examples of such tactical "foundation crises," we may cite the discovery of incommensurable magnitudes (i.e., irrational numbers) by the Pythagoreans; the proliferation of divergent series (ultimately resolved by Cauchy in the early 19th century) and the "paradoxes" in naive set theory (a tactical controversy which is still ongoing).

A tactical crisis in mathematics which is worth a second look is that arising from the notorious "parallel postulate" of Euclid. This postulate states that, given a line in a plane, and a point of the plane not on the line, there exists one and only one line through the point and parallel to the given line. By "parallel" we understand that the two lines will never meet, no matter how far they are prolonged.

The Greeks, with their keen esthetic sense, already had reservations about this postulate. It was not that they doubted its truth; rather, they felt it lacked the self-evident character demanded of a postulate. The problem was that the test for parallelism was open-ended; if two lines have not met after any finite prolongation, we still have no grounds for deciding whether they are parallel or not. Thus it was already felt in ancient times that the proposition in question should be a theorem, and not a postulate. Thus were born the ongoing attempts to "prove" the parallel postulate.

The problem was attacked by two main methods. The first sought to produce a direct proof of the postulate from the other Euclidean axioms and postulates. On close scrutiny, all such purported proofs turned out to be circular. The second approach sought to produce a logical contradiction by supposing the parallel postulate was false. On this basis, quite a number of results were derived which looked most peculiar when compared to their Euclidean counterparts, but no *logical* contradiction was forthcoming. Finally, in the early part of the 19th century, a number of mathematicians were driven to assert that no logical contradiction could be found, and hence that there were other geometries besides that characterized by Euclid.

This was a most revolutionary assertion. Euclid was then the supreme model of scientific exposition, and the geometry he described had become canonized as an essential feature of human thought, given *a priori*. To assert that there might be "other geometries" was thus a contradiction in terms; even Gauss hesitated to make such a claim because of "the outcry of the Boetians."

Many conservative mathematicians countered the claims regarding "non-Euclidean" geometries by pointing out that the failure to discover a logical contradiction in such a geometry did not imply that contradictions did not exist, any more than the failure of two finite line segments to meet implied that they were parallel. These objections were demolished by the discovery of "Euclidean models" of non-Euclidean geometry, by Beltrami, Klein and Poincare. The existence of such models constituted a relative consistency proof for non-Euclidean geometries; they showed that these geometries could be no less consistent than Euclidean geometry, and hence that any contradiction in these geometries must already be present in Euclid.

At this point, a number of basic questions arose. One of them was: what is the actual geometry of physical space? This question leads, more or less directly, to general relativity. An even deeper question for mathematicians was: what in fact is geometry, anyway? What is it that a geometer studies? This question was answered by Felix Klein, and became embodied in what is now called the Erlangen Program. Let us look briefly at what this entails.

Klein pointed out that every geometry is governed by positing some basic set of transformations of an underlying space. For instance, the central concept of Euclidean geometry is *congruence*, and the transformations basic to this geometry are the ones which bring congruent figures into coincidence. These are the familiar translations, rotations and reflections; the transformations which preserve lengths and angles. Thus, Euclidean geometry for Klein is the study of the group of rigid motions. In general, then, any geometry is simply the study of a particular group of transformations imposed on a set.

Now let us ask: what is the relation between two congruent figures in Euclidean geometry, or more generally, between two geometric objects which are inter-transformable by group operations characterizing a given geometry? The answer, in a nutshell, is this: two such objects are *models* of each other. More specifically, there is a mathematical transformation which allows us to compute (i.e., predict) the properties of one of them when the corresponding properties of the other are known.

Thus, geometries in the sense of Klein are mathematical structures which generate and study modelling relations. The converse proposition is also true; the establishment of modelling relations among classes of objects automatically generates a geometry in the sense of Klein. Hence, geometry and modelling are co-extensive.

Challenges of System Theory

Now let us observe that the establishment of modelling relations is the essential business of system theory. Hence we can assert that, in a certain definite sense, system theory and geometry are co-extensive.

We can pull all these threads together to draw a satisfying conclusion. We have already noted that mathematics, and hence geometry, is impervious to the kinds of paradigm shifts (i.e., strategic changes) which have affected every other aspect of human thought. Thus, system theory, although an essential ingredient of paradigm shifts in other spheres, shares with geometry a basic immunity to these shifts. Thus, *system theory is an unmoved mover*; a revolutionary advance which itself can never be challenged by revolution.

And with these happy thoughts, I bid you a good evening.

11

From Circuit Theory to System Theory

L. A. Zadeh

1. Introduction

The past two decades have witnessed an evolution of classical circuit theory into a field of science whose domain of application far transcends the analysis and synthesis of RLC networks. The invention of the transistor, followed by the development of a variety of other solid-state devices, the trend toward microminiaturization and integrated electronics, the problems arising out of the analysis and design of large-scale communication networks, the increasingly important role played by time-varying, nonlinear and probabilistic circuits, the development of theories of neuroelectric networks, automata and finite state machines, the progress in our understanding of the processes of learning and adaptation, the advent of information theory, game theory and dynamic programming, and the formulation of the maximum principle by Pontryagin, have all combined to relegate classical circuit theory to the status of a specialized branch of a much broader scientific discipline—system theory—which, as the name implies, is concerned with all types of systems and not just electrical networks.

What is system theory? What are its principal problems and areas? What is its relationship to such relatively well-established fields as circuit theory, control theory, information theory, and operations research and systems engineering? These are some of the questions which are discussed in this paper, with no claim that the answers presented are in any way definitive. Technological and scientific progress is so rapid these days that hardly any assertion concerning the boundaries, content and directions of such a new field as system theory can be expected to have long-term validity.

The obvious inference that system theory deals with systems does not shed much light on it, since all branches of science are concerned with systems of one kind or another. The distinguishing characteristic of system theory is its generality and abstractness, its concern with the mathematical properties of systems and not their physical form. Thus, whether a

From *IRE Proc.* **50**, 856. Copyright © 1962 by IRE (now IEEE).

system is electrical, mechanical or chemical in nature does not matter to a system theorist. What matters are the mathematical relations between the variables in terms of which the behavior of the system is described.

To understand this point clearly, we have to examine more closely the concept of a system. According to Webster's dictionary, a system is ". . . an aggregation or assemblage of objects united by some form of interaction or interdependence." In this sense, a set of particles exerting attraction on one another is a system; so is a group of human beings forming a society or an organization; so is a complex of interrelated industries; so is an electrical network; so is a large-scale digital computer, which represents the most advanced and sophisticated system devised by man; and so is practically any conceivable collection of interrelated entities of any nature. Indeed, there are few concepts in the domain of human knowledge that are as broad and all-pervasive as that of a system.

It has long been known that systems of widely different physical forms may in fact be governed by identical differential equations. For example, an electrical network may be characterized by the same equations as a mechanical system, in which case the two constitute analogs of one another. This, of course, is the principle behind analog computation.

While the analogies between certain types of systems have been exploited quite extensively in the past, the recognition that the same abstract "systems" notions are operating in various guises in many unrelated fields of science is a relatively recent development. It has been brought about, largely within the past two decades, by the great progress in our understanding of the behavior of both inanimate and animate systems—progress which resulted on the one hand from a vast expansion in the scientific and technological activities directed toward the development of highly complex systems for such purposes as automatic control, pattern recognition, data-processing, communication, and machine computation, and, on the other hand, by the attempts at quantitative analyses of the extremely complex animate and man-machine systems which are encountered in biology, neurophysiology, econometrics, operations research and other fields.

It is these converging developments that have led to the conception of system theory—a scientific discipline devoted to the study of general properties of systems, regardless of their physical nature. It is to its abstractness that system theory owes its wide applicability, for it is by the process of abstraction that we progress from the special to the general, from mere collections of data to general theories.

If one were asked to name a single individual who above anyone else is responsible for the conception of system theory, the answer would undoubtedly be "Norbert Wiener," even though Wiener has not concerned himself with system theory as such, nor has he been using the term "system theory" in the sense employed in this paper. For it was Wiener who, starting in the twenties and thirties, has introduced a number of concepts, ideas and theories which collectively constitute a core of present-day system theory. Among these, to mention a few, are his representation of nonlinear systems in terms of a series of Laguerre polynomials and Hermite functions, his theory of prediction and filtering, his generalized harmonic analysis, his cinema-integraph, the Paley–Wiener criterion, and the Wiener process. It was Wiener who, in the late forties, laid the foundation for cybernetics—the science of communication and control in the animal and the machine—of which system theory is a part dealing specifically with systems and their properties. It should be noted, however, that some of the more important recent developments in system theory no longer bear Wiener's imprint. This is particularly true of the theory of discrete-state systems, the state-space techniques for continuous systems, and the theory of optimal control, which is associated mainly with the names of Bellman and Pontryagin. We shall touch upon these developments later on in the paper.

Among the scientists dealing with animate systems, it was a biologist—Ludwig von Bertalanffy—who long ago perceived the essential unity of systems concepts and techniques in various fields of science and who in writings and lectures sought to attain recognition for "general systems theory" as a distinct scientific discipline. It is pertinent to note, however, that the work of Bertalanffy and his school, being motivated primarily by problems arising in the study of biological systems, is much more empirical and qualitative in spirit than the work of those system theorists who received their training in the exact sciences. In fact, there is a fairly wide gap between what might be regarded as "animate" system theorists and "inanimate" system theorists at the present time, and it is not at all certain that this gap will be narrowed, much less closed, in the near future. There are some who feel that this gap reflects the fundamental inadequacy of the conventional mathematics—the mathematics of precisely-defined points, functions, sets, probability measures, etc.—for coping with the analysis of biological systems, and that to deal effectively with such systems, which are generally orders of magnitude more complex than man-made systems, we need a radically different kind of mathematics, the mathematics of fuzzy or cloudy quantities which are not describable in terms of probability distributions. Indeed, the need for such mathematics is becoming increasingly apparent even in the realm of inanimate systems, for in most practical cases the *a priori* data as well as the criteria by which the performance of a man-made system is judged are far from being precisely specified or having accurately-known probability distributions.

System theory is not as yet a well-crystallized body of concepts and techniques which set it sharply apart from other better established fields of science. Indeed, there is considerable overlap and interrelation between system theory on the one hand and circuit theory, information theory, control theory, signal theory, operations research and systems engineering on the other. Yet, system theory has a distinct identity of its own which perhaps can be more clearly defined by listing its principal problems and areas. Such a list is presented below, without claims that it is definitive, complete and noncontroversial. (To avoid misunderstanding, brief explanations of the meaning of various terms are given in parentheses.)

Principal Problems of System Theory

1. System characterization (representation of input-output relationships in mathematical form; transition from one mode of representation to another).
2. System classification (determination, on the basis of observation of input and output, of one among a specified class of systems to which the system under test belongs).
3. Systems identification (determination, on the basis of observation of input and output, of a system within a specified class of systems to which the system under test is equivalent; determination of the initial or terminal state of the system under test).

4. Signal representation (mathematical representation of a signal as a combination of elementary signals; mathematical description of signals).
5. Signal classification (determination of one among a specified set of classes of signals or patterns to which an observed signal belongs).
6. System analysis (determination of input-output relationships of a system from the knowledge of input-output relationships of each of its components).
7. System synthesis (specification of a system which has prescribed input-output relationships).
8. System control and programming (determination of an input to a given system which results in a specified or optimal performance).
9. System optimization (determination of a system within a prescribed class of systems which is best in terms of a specified performance criterion).
10. Learning and adaptation (problem of designing systems which can adapt to changes in environment and learn from experience).
11. Reliability (problem of synthesizing reliable systems out of less reliable components).
12. Stability and controllability (determination of whether a given system is stable or unstable, controllable—subject to specified constraints—or not controllable).

Principal Types of Systems

1. Linear, nonlinear.
2. Time-varying, time-invariant.
3. Discrete-time (sampled-data), continuous-time.
4. Finite-state, discrete-state, continuous-state.
5. Deterministic (nonrandom), probabilistic (stochastic).
6. Differential (characterized by integro-differential equations), nondifferential.
7. Small-scale, large-scale (large number of components).

Some Well-Established Fields Which May Be Regarded as Branches of System Theory

1. Circuit theory (linear and nonlinear).
2. Control theory.
3. Signal theory.
4. Theory of finite-state machines and automata.

Comment 1: Note that *approximation* is not listed as a separate problem, as it is usually regarded in classical circuit theory. Rather, it is regarded as something that permeates all the other problems.

Comment 2: Note that *information theory* and *communication theory* are not regarded as branches of system theory. System theory makes extensive use of the concepts and results of information theory, but this does not imply that information theory is a branch of system theory, or vice versa. The same comment applies to such theories as *decision theory* (in statistics), *dynamic programming, reliability theory*, etc.

Comment 3: Note that there is no mention in the list of *systems engineering* and *operations research*. We regard these fields as being concerned specifically with the operation and management of large-scale man-machine systems, whereas system theory deals on an abstract level with general properties of systems, regardless of their physical form or the domain of application. In this sense, system theory contributes an important source of concepts and mathematical techniques to systems engineering and operations research, but is not a part of these fields, nor does it have them as its own branches.

It would be futile to attempt to say something (of necessity brief and superficial) about each item in the above list. Instead, we shall confine our attention to just a few concepts and problems which play particularly important roles in system theory and, in a way, account for its distinct identity. Chiefly because of limitations of space, we shall not even touch upon a number of important topics such as the design of learning and adaptive systems, the analysis of large-scale and probabilistic systems, the notion of feedback and signal flow graph techniques, etc. In effect, the remainder of this paper is devoted to a discussion of the concept of state and state-space techniques, along with a brief exposition of systems characterization, classification and identification. We have singled out the concept of state for special emphasis largely because one can hardly acquire any sort of understanding of system theory without having a clear understanding of the notion of state and some of its main implications.

2. State and State-Space Techniques

It is beyond the scope of this presentation to trace the long history of the evolution of the concept of state in the physical sciences. For our purposes, it will suffice to observe that the notion of state in essentially the same form it is used today was employed by Turing[1] in his classical paper, "On computable numbers, with an application to the Entscheidungs problem," in which he introduced what is known today as the Turing machine.

Roughly speaking, a Turing machine is a discrete-time ($t = 0, 1, 2, \ldots$) system with a finite number of states or internal configurations, which is subjected to an input having the form of a sequence of symbols (drawn from a finite alphabet) printed on a tape which can move in both directions along its length. The output of the machine at time t is an instruction to print a particular symbol in the square scanned by the machine at time t and to move in one or the other direction by one square. A key feature of the machine is that the output at time t and the state at time $t + 1$ are determined by the state and the input at time t. Thus, if the state, the input and the output at time t are denoted by s_t, u_t, and y_t, respectively, then the operation of the machine is characterized by

$$s_{t+1} = f(s_t, u_t), \qquad t = 0, 1, 2, \ldots, \tag{1}$$

$$y_t = g(s_t, u_t), \tag{2}$$

where f and g are functions defined on pairs of values y_t, s_t, and u_t. Note that (1) and (2) imply that the output symbols from any initial time t_0 on are determined by the state at time t_0 and the input symbols from time t on.

An important point about this representation, which was not pointed out by Turing, is that it is applicable not only to the Turing machine, but more generally to any discrete-time system. Furthermore, we shall presently see that it is a simple matter to extend (1) and (2) to systems having a continuum of states (i.e., continuous-state systems).

The characterization of a system by means of equations of the form (1) and (2) (to which we will refer as the *Turing representation* or, alternatively, as the *state equations* of a system) was subsequently employed by Shannon[2] in his epoch-making paper on the mathematical theory of communication. Specifically, Shannon used (1) and (2) to characterize finite-state noisy channels, which are probabilistic systems in the sense that s_t and u_t determine not s_{t+1} and y_t, but their joint probability distribution. This implies that the system is characterized by (1) and (2), with f and g being random functions, or, equivalently, by the conditional distribution function $p(s_{t+1}, y_t|s_t, u_t)$, where for simplicity of notation the same letter is used to denote both a random variable and a particular value of the random variable (e.g., instead of writing S_t for the random variable and s_t for its value, we use the same symbol, s_t, for both).

It was primarily Shannon's use of the Turing representation that triggered its wide application in the analysis and synthesis of discrete-state systems. Worthy of note in this connection is the important work of von Neumann[3] on probabilistic logics, which demonstrated that it is possible, at least in principle, to build systems of arbitrarily high degree of reliability from unreliable (probabilistic) components. Also worthy of note is the not-too-well-known work of Singleton[4] on the theory of nonlinear transducers, in which techniques for optimizing the performance of a system with quantized state space are developed. It should be remarked that the problem of approximating to a system having a continuum of states with a system having a finite or countable number of states is a basic and as yet unsolved problem in system theory. Among the few papers which touch upon this problem, those by Kaplan[5,6] contain significant results for the special case of differential system (a system which is characterized by one or more ordinary differential equations) subjected to zero input. A qualitative discussion of the related problem of ϵ-approximation may be found in a paper by Stebakov.[7]

There are two important notions that are missing or play minor roles in the papers cited above. These are the notions of equivalent states and equivalent machines which were introduced by Moore[8] and, independently and in a somewhat restricted form, by Huffman.[9] The theory developed by Moore constitutes a contribution of basic importance to the theory of discrete-state systems and, more particularly, the identification problem.

So far, our discussion of the notion of state has been conducted in the context of discrete-state systems. In the case of a differential system, the state equations (1) and (2) assume the form

$$\dot{\mathbf{s}}(t) = \mathbf{f}(\mathbf{s}(t), \mathbf{u}(t)), \qquad (3)$$

$$\mathbf{y}(t) = \mathbf{g}(\mathbf{s}(t), \mathbf{u}(t)), \qquad (4)$$

where $\dot{\mathbf{s}}(t) = d/dt\, \mathbf{s}(t)$, and $\mathbf{s}(t)$, $\mathbf{y}(t)$, and $\mathbf{u}(t)$ are vectors representing the *state*, the *input*, and *output* of the system at time t. Under various guises, such equations [particularly for the case where $\mathbf{u}(t) \equiv 0$] have long been used in the theory of ordinary differential equations, analytical dynamics, celestial mechanics, quantum mechanics, econometrics, and other fields. Their wide use in the field of automatic control was initiated largely in the late 1940s and early 1950s by Soviet control theorists, notably A. I. Lur'e, M. A. Aizerman, Ya. Z. Tsypkin, A. A. Fel'dbaum, A. Ya. Lerner, A. M. Letov, N. N. Krasovskii, I. G. Malkin,

L. S. Pontryagin, and others. In the Unites States, the introduction of the notion of state and related techniques into the theory of optimization of linear as well as nonlinear systems is due primarily to Bellman, whose invention of dynamic programming[10] has contributed by far the most powerful tool since the inception of the variational calculus to the solution of a whole gamut of maximization and minimization problems. Effective use of and or important contributions to the state-space techniques in the field of automatic control have also been made by Kalman,[11] Kalman and Bertram,[12] LaSalle,[13] Laning and Battin[14] (in connection with analog simulation), Friedland,[15] and others. It is of interest to note, however, that it was not until 1957 that a general method for setting up the state equations of an RLC network was described by Bashkow.[16] An extension of Bashkow's technique to time-varying networks was recently presented by Kinarawala.[17]

Despite the extensive use of the notion of state in the current literature, one would be hard put to find a satisfactory definition of it in textbooks or papers. A reason for this is that the notion of state is essentially a primitive concept, and as such is not susceptible to exact definition. However, it is possible, as sketched below, to define it indirectly by starting with the notion of *complete characterization* of a system. Specifically, consider a black box B and some initial time t_0. For simplicity, it is assumed that B is (1) deterministic (i.e., non-random), (2) time-invariant, (3) nonanticipative (not acting on the future values of the input), and (4) continuous-time (i.e., with t ranging over the real time axis). We assume that B is associated with three types of time functions:

1. A controllable variable u [i.e., a time function whose values can be chosen at will from a specified set (input space) for all $t \geq t_0$].
2. An initially controllable variable s [i.e., a time function whose value can be chosen at will at $t = t_0$ from a specified set (*state space*), but not thereafter].
3. An observable variable y (i.e., a time function whose values can be observed for $t \geq t_0$, but over which no direct control can be exercised for $t \geq t_0$).

Furthermore, we assume that this holds for all values of t_0.

If these assumptions are satisfied, then we shall say that B is *completely characterized* if for every $t \geq t_0$ the value of the output at time t, $\mathbf{y}(t)$, is uniquely determined by the value of s at time t_0 and the values of u over the closed interval $[t_0, t]$. Symbolically, this is expressed by

$$\mathbf{y}(t) = B(\mathbf{s}(t_0); \mathbf{u}_{[t_0,t]}), \tag{5}$$

where $u_{[t_0,t]}$ denotes the segment of the time function u extending between and including the end points t_0 and t; $\mathbf{s}(t_0)$ is the value assumed by $\mathbf{s}(t)$ at time t_0; and $B(\cdot\ ;\ \cdot)$ is a single-valued function of its arguments. [Note that B is a functional of $\mathbf{u}_{[t_0,t]}$ and an ordinary function of $\mathbf{s}(t_0)$; $\mathbf{s}(t)$ is usually a vector with a finite number of components.] It is understood that (5) must hold for all possible values of $\mathbf{s}(t_0)$ and $\mathbf{u}_{[t_0,t]}$ and that to every possible input–output pair $\mathbf{u}_{(t_0,t)}$, $\mathbf{y}_{(t_0,t)}$ there should correspond a state $\mathbf{s}(t_0)$ in the state space of B.

If B is completely characterized in the sense defined above, then $\mathbf{u}(t)$, $\mathbf{y}(t)$, and $\mathbf{s}(t)$ are, respectively, the values of the *input*, the *output*, and the *state* of B at time t. [The range of values of $\mathbf{s}(t)$ constitutes the *state-space* of B. A particular value of $\mathbf{s}(t)$, i.e., a particular state of B, will be denoted by $\mathbf{q}$.] In this way, the input, the output, and the state of B are defined simultaneously as by-products of the definition of complete characterization of B.

The intuitive significance of the concept of state is hardly made clear by the somewhat artificial definition sketched above. Essentially, s(t) constitutes a description of the internal condition in B at time t. Equation (5), then, signifies that, given the initial conditions at time t_0, and given the values of the input between and including t_0 and t, we should be able to find the output of B at time t if the system is completely characterized.

With (5) as the starting point, it is a simple matter to demonstrate that (5) can be replaced by an equivalent pair of equations:

$$\mathbf{s}(t) = \mathbf{f}(s(t_0); \mathbf{u}_{[t_0, t]}), \quad t \geq t_0, \quad (6)$$

$$\mathbf{y}(t) = \mathbf{g}(s(t); \mathbf{u}(t)), \quad (7)$$

the first of which expresses the state at time t in terms of the state at time t_0 and the values of the input between and including t_0 and t, while the second gives the output at time t in terms of the state at time t and the input at time t. Note that these relations are in effect continuous analogs of the Turing representation

$$\mathbf{s}_{t+k} = f(s_t, u_t, \ldots, u_{t+k-1}), \quad (8)$$

$$\mathbf{y}_t = g(s_t, u_t). \quad (9)$$

3. State and System Equivalence

One cannot proceed much further with the discussion of state-space techniques without introducing the twin notions of equivalent states and equivalent systems.

Suppose that we have two systems B_1 and B_2, with q_1 being a state of B_1 and q_2 being a state of B_2. As the term implies, q_1 and q_2 are *equivalent states* if, for all possible input time functions u, the response of B_1 to u starting in state q_1 is the same as the response of B_2 to u starting in state q_2. Following Moore, B_1 and B_2 are said to be *equivalent systems* if, and only if, to every state in B_2 there is an equivalent state in B_1, and vice versa. What is the significance of this definition? Roughly speaking, it means that if B_1 and B_2 are equivalent, then it is impossible to distinguish B_1 from B_2 by observing the responses of B_1 and B_2 to all possible inputs u, if the initial states of B_1 and B_2 are not known to the experimenter.

4. The Notion of Policy

Another important difference between circuit theory and system theory manifests itself in the way in which the input to a system (circuit) is represented. Thus, in circuit theory it is customary to specify the desired input to a network as a function of time. By contrast, in system theory it is a common practice—particularly in dealing with control problems—to express the input as a function of the state of the system rather than as a function of time. In many ways, this is a more effective representation, since it is natural to base the decision on what input to apply at time t on the knowledge of the state of the system at time t. Furthermore, in the latter representation (input in terms of state) we have feedback, whereas in the former (input in terms of time) we have not.

To say that the input depends on the state of the system means, more precisely, that the input at time t is a function of the state at time t, i.e.,

$$\mathbf{u}(t) = \pi(s(t)), \tag{10}$$

where π is a function defined on the state space with values in the input space. This function is referred to as a *policy function*, or simply a *policy*. In effect, a policy is a function which associates a particular input with each state of the system.

The notion of policy plays a key role in system theory and, especially, in control theory. Thus, a typical problem in control theory involves the determination of a policy for a given system B which is optimal in terms of a specified performance criterion for B. More specifically, the performance criterion associates with each policy π a number $Q(\pi)$, which is a measure of the "goodness" of π. The problem, then, is to find a policy π which maximizes $Q(\pi)$. Such a policy is said to be *optimal*. It is tacitly assumed that B is a deterministic system. Otherwise, $Q(\pi)$ would be a random variable, and the performance of B would generally be measured by the expected value of $Q(\pi)$.

As was stated previously, the most effective general technique for solving problems of this nature is provided by Bellman's dynamic programming. The basis for dynamic programming is the so-called *principle of optimality* which in Bellman's words reads: "An optimal policy has the property that whatever the initial state and the initial decision are, the remaining decisions must constitute an optimal policy with regard to the state resulting from the first decision."

Needless to say, one can always resort to brute force methods to find a policy π which maximizes $Q(\pi)$. The great advantage of dynamic programming over direct methods is that it reduces the determination of optimal π to the solution of a succession of relatively simple maximization or minimization problems. In mathematical terms, if the payoff resulting from the use of an optimal policy when the system is initially (say at $t = 0$) in state $s(0)$ is denoted by $R(s(0))$, then by employing the principle of optimality one can derive a functional equation satisfied by R. In general, such equations cannot be solved in closed form. However, if the dimension of the state vector is fairly small, say less than four or five, then it is usually feasible to obtain the solution through the use of a moderately-sized digital computer. The main limitation on the applicability of dynamic programming is imposed by the inability of even large-scale computers to handle problems in which the dimensionality of the state vector is fairly high, say of the order of 20. A number of special techniques for getting around the problem of dimensionality have recently been described by Bellman.[18]

A very basic problem in system theory which has been attacked both by the techniques of dynamic programming[19,20] and by extensions of classical methods of the calculus of variations[21] is the so-called *minimal-time* or *optimal-time* problem. This problem has attracted a great deal of attention since the formulation of the so-called maximum principle by Pontryagin[20] in 1956, and at present is the object of numerous investigations both in the United States and the Soviet Union. Stated in general terms (for a time-invariant, single-input continuous-time system) the problem reads as follows.

Given: (1) A system B characterized by the vector differential equation

$$\dot{\mathbf{x}}(t) = \mathbf{f}(x(t), u(t)), \tag{11}$$

where $x(t)$ and $u(t)$ represent, respectively, the state and the input of B at time t. (x is a vector; u is a scalar, and $\mathbf{f}$ is a function satisfying certain smoothness conditions.)

(2) A set of constraints on u, e.g., $|u(t)| \leq 1$ for all t or $|u(t)| \leq 1$ and $|\dot{u}(t)| \leq 1$ for all t.
(3) A specified initial state $x(0) = q_0$ and a specified terminal state q_1.

Find an input u (satisfying the prescribed constraints) which would take B from q_1 to q_2 in the shortest possible time. This, in essence, is the minimal-time problem.

In a slightly more general formulation of the problem, the quantity to be minimized is taken to be the cost of taking the system from q_0 to q_1, where the cost is expressed by an integral of the form

$$C(u; q_0, q_1) = \int_0^{t_1} f_0(x(t), u(t))dt. \tag{12}$$

In this expression, f_0 is a prescribed function, t_1 is the time at which B reaches the state q_1, and $C(u, q_0, q_1)$ denotes the cost of taking B from q_0 to q_1, when the input u is used.

It is not hard to see why the minimal-time (or, more generally, the minimal-cost) problem plays such an important role in system and, more particularly, control theory. Almost every control problem encountered in practice involves taking a given system from one specified state to another. The minimal-time problem merely poses the question of how this can be done in an optimal fashion.

Various special cases of the minimal time problem were considered by many investigators in the late 1940s and early 1950s. What was lacking was a general theory. Such a theory was developed in a series of papers by Pontryagin, Boltyanskii, and Gamkrelidze.[21,22]

The maximum principle of Pontryagin is essentially a set of necessary conditions satisfied by an optimal input u. Briefly stated let ψ be a solution of the variational system

$$\dot{\psi} = -\left(\frac{\partial f}{\partial x}\right)' \psi, \tag{13}$$

where $[\partial f/\partial x]'$ is the transpose of the matrix $[\partial f_i/\partial x_j]$, in which f_i and x_j are, respectively, the ith and jth components of f and x in the equation $\dot{x} = f(x, u)$. Construct a Hamiltonian function $H(x, \psi, u) = \psi \cdot \dot{x}$ (dot product of ψ and $\dot{x}$), with the initial values of ψ in (13) constrained by the inequality $H(x(0), \psi(0), u(0)) \geq 0$. The maximum principle asserts that if $\hat{u}(t)$ is an optimal input, then $\hat{u}(t)$ maximizes the Hamiltonian $H(x, \psi, u)$, with x and ψ held fixed for exact t.

An application of the maximum principle to a linear system characterized by the vector equation

$$\dot{x} = Ax + Bu, \tag{14}$$

where A is a constant matrix and B is a constant vector, leads to the result that an optimal input is "bang-bang," that is, at all times the input is as large amplitude-wise as the limits permit. More specifically, an optimal input is of the form

$$\hat{u}(t) = \text{sgn}[\psi(t) \cdot B], \tag{15}$$

where sgn stands for the function $\text{sgn}\,x = 1$ if $x > 0$, $\text{sgn}\,x = -1$ if $x \leq 0$, and $\text{sgn}\,x = 0$ if $x = 0$, and ψ is a solution of the adjoint equation

$$\dot{\psi} = -A'\psi \tag{16}$$

satisfying the inequality

$$\psi(0) \cdot [Ax(0) + Bu(0)] \geq 0. \tag{17}$$

This and other general results for the linear case were first obtained by Pontryagin and his co-workers. Somewhat more specialized results had been derived independently by Bellman, Glicksberg, and Gross.[25]

The main trouble with the maximum principle is that it yields only necessary conditions. The expression for $\hat{u}(t)$ given by (15) is deceptively simple; in fact, in order to determine $\hat{u}(t)$, one must first find a $\psi(t)$ which satisfies the differential equation (16), the inequality (17), and, furthermore, is such that $\mathbf{B}$ reaches $\mathbf{q}_1$ when subjected to the $\hat{u}$ given by (15). Even then, there is no guarantee that $\hat{u}$ is optimal, except when either the initial state or the terminal state coincides with the origin. Still another shortcoming of the method is that the optimal input is obtained as a function of time rather than the state of the system.

One could hardly expect the maximum or any other principle to yield complete and simple solutions to a problem as difficult as the minimal-time problem for nonlinear, continuous-state, continuous-time systems. Actually, complete solutions can be and have been obtained for simpler types of systems. Particularly worthy of note is the solution for the case of a linear discrete-time system which was recently obtained by Desoer and Wing.[26] Quite promising for linear continuous-time systems is the successive approximation technique of Bryson and Ho.[27] In the case of systems having a finite number of states, the minimal-time problem can be solved quite easily even when the system is probabilistic and the terminal state changes in a random (Markovian) fashion.[28]

Closely related to the minimal-time problem are the problems of reachability and controllability, which involve the existence and construction of inputs which take a given system for one specified state to another, not necessarily in minimum time. Important contributions to the formulation and solution of these problems for unconstrained linear systems were made by Kalman.[29] It appears difficult, however, to obtain explicit necessary and sufficient conditions for reachability in the case of constrained, linear, much less nonlinear, systems.

5. Characterization, Classification, and Identification

Our placement of system characterization, classification and identification at the top of the list of principal problems of system theory (see Section 1) reflects their intrinsic importance rather than the extent of the research effort that has been or is being devoted to their solution. Indeed, it is only within the past decade that significant contributions to the formulation as well as the solution of these problems, particularly in the context of finite-state systems, have been made. Nevertheless, it is certain that problems centering on the characterization, classification and, especially, identification of systems as well as signals and patterns, will play an increasingly important role in system theory in the years to come.

The problem of characterization is concerned primarily with the representation of input-output relationships. More specifically, it is concerned both with the alternative ways in which the input-output relationship of a particular system can be represented (e.g., in terms of differential equations, integral operators, frequency response functions, characteristic functions, state equations, etc.), and the forms which these representations assume for various types of systems (e.g., continuous-time, discrete-time, finite-state, probabilistic, finite-memory, nonanticipative, etc.). Generally, the input-output relationship is expressed in terms of a finite or at most countable set of linear operations (both with and without memory) and nonlinear operations (without memory). For example, Cameron and Martin[30] and Wiener[31] have shown that a broad class of nonlinear systems can be characterized by (ground-state) input-output relationships of the form

$$y(t) = \sum_{n=0}^{\infty} A_n X_n(t) \tag{18}$$

where the $X_n(t)$ represents products of Hermite functions of various orders in the variables $z_1, z_2, \ldots$, which in turn are linearly related to u (input) through Laguerre functions. Note that the operations involved in this representation are (1) linear with memory, viz., the relations between the z's and u; (2) memoryless nonlinear, viz., the relations between the X_n and $z_1, z_2, \ldots$; and (3) linear with no memory, viz., the summations. In this connection, it should be pointed out that the basic idea of representing a nonlinear input-output relationship as a composition of an infinite number of (1) memoryless nonlinear operations and (2) linear operations with memory, is by no means a new one. It had been employed quite extensively by Volterra[32] and Frechet[33] near the turn of the century.

A key feature of the Wiener representation is its orthogonality [meaning uncorrelatedness of distinct terms in (18)] for white noise inputs. This implies that a coefficient A_n in (18) can be equated to the average (expected) value of the product of $X_n(t)$ and the response of the system to white noise. In this way, a nonlinear system can be identified by subjecting it to a white noise input, generating the $X_n(t)$ functions, and measuring the average values of the products $y(t)X_n(t)$. However, for a variety of technical reasons this method of identification is not of much practical value.

The problem of system classification may be stated as follows. Given a black box B and a family (not necessarily discrete) of classes of systems $C_1, C_2 \ldots$, such that B belongs to one of these classes, say C_λ, the problem is to determine C_λ by observing the responses of B to various inputs. Generally, the inputs in question are assumed to be controllable by the experimenter. Needless to say, it is more difficult to classify a system when this it not the case.

A rather important special problem in classification is the following. Suppose it is known that B is characterized by a differential equation, and the question is: What is its order? Here, C can be taken to represent the class of systems characterized by a single differential equation of order n. An interesting solution to this problem was described by Bellman.[34]

Another practical problem arises in the experimental study of propagation media. Suppose that B is a randomly-varying stationary channel, and the problem is to determine if B is linear or nonlinear. Here, we have but two classes: C_1 = class of linear systems, and C_2 = class of nonlinear systems. No systematic procedures for the solution of problems of this type have been developed so far.

Finally, *the problem of system identification* is one of the most basic and, paradoxically, least-studied problems in system theory. Broadly speaking, the identification of a system B involves the determination of its characteristics through the observation of responses of B to test inputs. More precisely, given a class of systems C (with each member of C completely characterized), the problem is to determine a system in C which is equivalent to B. Clearly, the identification problem may be regarded as a special case of the classification problem in which each of the classes $C_1, C_2, \ldots$, has just one member. This, however, is not a very useful viewpoint.

It is obvious that such commonplace problems as the measurement of a transfer function, impulsive response, the A_n coefficients in the Wiener representation (18), etc., may be regarded as special instances of the identification problem. So is the problem of location of malfunctioning components in a given system B, in which case C is the class of all malfunctioning versions of B.

A complicating feature of many identification problems is the lack of knowledge of the initial state of the system under test. Another source of difficulties is the presence of noise in observations of the input and output. For obvious reasons, the identification of continuous-state continuous-time systems is a far less tractable problem than the identification of finite-state discrete-time systems. For the latter, the basic theory developed by Moore[8] provides very effective algorithms in the case of small-scale systems, that is, systems in which the number of states as well as the number of input and output levels is fairly small. The identification of large-scale systems calls for, among other things, the development of algorithms which minimize the duration of (or the number of steps in) the identifying input sequence. With the exception of an interesting method suggested by Bellman,[35] which combines dynamic programming with the minimax principle, little work along these lines has been done so far.

Another important area which is beginning to draw increasing attention is that of the identification of randomly-varying systems. Of particular interest in this connection is the work of Kailath[36] on randomly-varying linear systems, the work of Hofstetter[37] on finite-state channels, the work of Gilbert[38] on functions of a Markov process, and the generalization by Carlyle[39] of some aspects of Moore's theory to probabilistic machines. All in all, however, the sum total of what we know about the identification problem is far from constituting a body of effective techniques for the solution of realistic identification problems for deterministic, much less probabilistic, systems.

6. Concluding Remarks

It is difficult to do justice to a subject as complex as system theory in a compass of a few printed pages. It should be emphasized that our discussion was concerned with just a few of the many facets of this rapidly-developing scientific discipline. Will it grow and acquire a distinct identity, or will it fragment and become submerged in other better-established branches of science? This writer believes that system theory is here to stay, and that the coming years will witness its evolution into a respectable and active area of scientific endeavor.

References

1. A. M. Turing, "On computable numbers, with an application to the Entscheidungsproblem," *Proc. London Math. Soc.* ser. 2, vol. 42, pp. 230–265; 1936.

2. C. E. Shannon, "A mathematical theory of communication," *Bell Sys. Tech. J.*, vol. 27, pp. 379–423, 623–656; 1948.
3. J. von Neumann, "Probabilistic logics and the synthesis of reliable organisms from unreliable components," in *Automata Studies*, Princeton University Press, Princeton, N.J., pp. 43–98; 1956.
4. H. E. Singleton, "Theory of Nonlinear Transducers," Res. Lab. of Electronics, Mass. Inst. Tech., Cambridge, RLE Rept. No. 160; August, 1950.
5. W. Kaplan, "Dynamical systems with indeterminacy," *Am. J. Math.*, vol. 72, pp. 575–594; 1950.
6. W. Kaplan, "Discrete Models for Nonlinear Systems," presented at AIEE Winter General Mtg., New York, February, 1960; Conference Paper 60-109.
7. S. A. Stebakov, "Synthesis of systems with prescribed ϵ-behavior," *Proc. Conf. on Basic Problems in Automatic Control and Regulation*, Acad. Sci. USSR, Moscow, 1957; pp. 101–106.
8. E. F. Moore, "Gedanken-experiments on sequential machines," in *Automata Studies*, Princeton University Press, Princeton, N.J., pp. 129–153; 1956.
9. D. Huffman, "The synthesis of sequential switching circuits," *J. Franklin Inst.*, vol. 257, pp. 161–190; April, 1954.
10. R. E. Bellman, *Dynamic Programming*, Princeton University Press, Princeton, N.J.; 1957.
11. R. E. Kalman, "Analysis and synthesis of linear systems operating on randomly sampled data," Ph.D. dissertation, Columbia University, New York; August, 1957.
12. R. E. Kalman and J. E. Bertram, "General synthesis procedure for computer control of single and multi-loop nonlinear systems," *Trans. AIEE*, vol. 77 (*Application and Industry*), pp. 602–609; 1958.
13. J. P. LaSalle, "The time-optimal control problem," in *Contributions to the Theory of Nonlinear Oscillations*, Princeton University Press, Princeton, N. J., vol. 5, pp. 1–24, 1960.
14. J. H. Laning, Jr., and R. H. Battin, *Random Processes in Automatic Control*, McGraw-Hill, New York; 1956.
15. B. Friedland, "Linear modular sequential circuits," *IRE Trans. Circuit Theory*, vol. CT-6, pp. 61–68; March, 1959.
16. T. Bashkow, "The A matrix, new network description," *IRE Trans. Circuit Theory*, vol. CT-4, pp. 117–119; September, 1957.
17. B. Kinarawala, "Analysis of time-varying networks," 1961 *IRE International Convention Record*, pt. 4, pp. 268–276.
18. R. E. Bellman, "On the Reduction of Dimensionality for Classes of Dynamic Programming Processes," RAND Corp., Santa Monica, Calif., Paper P-2243; March, ,961.
19. C. A. Desoer, "Pontragin's maximum principle and the principle of optimality," *J. Franklin Inst.*, vol. 271, pp. 361–367; May, 1961.
20. L. I. Rozonoer, "The maximum principle of L. S. Pontryagin in the theory of optimal systems," *Automat. i Telemekh.*, vol. 20, nos. 10–12; October–December, 1959.
21. L. Berkovitz, "Variational Methods in Problems of Control and Programming," RAND Corp., Santa Monica, Calif., Paper P-2306; May, 1961.
22. L. S. Pontryagin, "Some mathematical problems arising in connection with the theory of optimal automatic control systems," *Proc. Conf. on Basic Problems in Automatic Control and Regulation*, Acad. Sci. USSR, Moscow, 1957.
23. V. G. Boltyanskii, R. V. Gamkrelidze, and L. S. Pontryagin, "On the theory of optimal processes," *Dokl. Akad. Nauk SSSR*, vol. 110, no. 5, pp. 7–10; 1956.
24. V. G. Boltyanskii, R. V. Gamkrelidze, and L. S. Pontryagin, "On the theory of optimal processes," *Invest. Akad. Nauk SSSR*, vol. 24, pp. 3–42; 1960.
25. R. E. Bellman, I. Glicksberg, and O. A. Gross, "Some Aspects of the Mathematical Theory of Control Processes," RAND Corp., Santa Monica, Calif., Rept. R-313; January, 1958.
26. C. A. Desoer and J. Wing, "A minimal time discrete system," *IRE Trans. Automatic Control*, vol. AC-6, pp. 111–125; May, 1961.

27. Y. C. Ho, "A Successive Approximation Technique for Optimal Control Systems Subject to Input Constraints," presented at JACC, Boulder, Colo., June, 1961; ASME Paper No. 61-JAC-10.
28. J. H. Eaton and L. A. Zadeh, "Optimal Pursuit Strategies in Discrete-State Probabilistic Systems," presented at JACC, Boulder, Colo., June, 1961; ASME Paper No. 61-JAC-11.
29. R. E. Kalman, "On the general theory of control systems," *Proc. 1st Internatl. Congress on Automatic Control*, Moscow, USSR, 1960.
30. R. H. Cameron and W. T. Martin, "The orthogonal development of nonlinear functionals in series of Fourier-Hermite functionals," *Ann. Math.*, vol. 48, pp. 385–392; April, 1947.
31. N. Wiener, *Nonlinear Problems in Random Theory*, John Wiley and Sons, Inc., New York; 1958.
32. V. Volterra, "Sopra le funzioni che dipendono da altre funzioni," *Rend. accad. sci. Lincei*, vol. 3, 1887.
33. M. Frechet, "Sur les fonctionnelles continues," *Ann. école norm. sup.*, vol. 27, pp. 193–216; 1910.
34. R. E. Bellman, "On the separation of exponentials," *Boll. d'unione math.*, vol. 15, pp. 38–39; 1960.
35. R. E. Bellman, "Sequential machines, ambiguity and dynamic programming," *J. Assoc. Comp. Mach.*, vol. 7, pp. 24–28; January, 1960.
36. T. Kailath, "Correlation detection of signals perturbed by a random channel," *IRE Trans. Information Theory*, vol. IT-6, pp. 361–366; June, 1960.
37. E. M. Hofstetter, "Large Sample Sequential Decision Theory," Res. Lab. of Electronics, Mass. Inst. Tech., Cambridge, RLE Rept. No. 359; December, 1959.
38. E. J. Gilbert, "On the identifiability problem for functions of finite Markov chains," *Ann. Math. Statistics*, vol. 30, pp. 688–697; 1959.
39. J. W. Carlyle, "Equivalent Stochastic Sequential Machines," Ph.D. dissertation, Dept. of Elec. Engrg., University of California, Berkeley; August, 1961.

12

Science in the Systems Age: Beyond IE, OR, and MS

Russell L. Ackoff

I believe we are leaving one cultural and technological age and entering another; that we are in the early stages of a change in our conception of the world, a change in our way of thinking about it, and a change in the technology with which we try to make it serve our purposes. These changes, I believe, are as fundamental and pervasive as were those associated with the Renaissance, the Age of the Machine that it introduced, and the Industrial Revolution that was its principal product. The socio-technical revolution we have entered may well come to be known as the *Resurrection*.

The Machine Age

The intellectual foundations of the Machine Age consist of two ideas about the nature of the world and a way of seeking understanding of it.

The first idea is called *reductionism*. It consists of the belief that everything in the world and every experience of it can be reduced, decomposed, or disassembled to ultimately simple elements, indivisible parts. These were taken to be atoms in physics; simple substances in chemistry; cells in biology; monads, directly observables, and basic instincts, drives, motives, and needs in psychology; and psychological individuals in sociology.

Reductionism gave rise to an *analytical* way of thinking about the world, a way of seeking explanations and, hence, of gaining understanding of it. For many, "analysis" was synonymous with "thought." Analysis consists, first, of taking what is to be explained apart—disassembling it, if possible, down to the independent and indivisible parts of which it is composed; secondly, of explaining the behavior of these parts; and, finally, aggregating these partial explanations into an explanation of the whole. For example, the analysis of a problem consists of breaking it down into a set of as simple problems as possible, solving

---

From *Operations Research* 21, 661. Copyright © 1973 by ORSA.

each, and assembling their solutions into a solution of the whole. If the analyst succeeds in decomposing the problem he faces into simpler problems that are independent of each other, aggregating the partial solutions is not required, because the solution to the whole is the sum of the solutions to its independent parts.

It should be noted—even if with unjustified brevity—that the concepts "division of labor" and "organizational structure" are manifestations of analytical thinking.

In the Machine Age, understanding the world was taken to be the sum, or resultant, of understandings of its parts that were conceptualized as independently of each other as was possible. This, in turn, made it possible to divide the labor of seeking to understand the world into a number of virtually independent disciplines.

The second basic idea was that of *mechanism*. All phenomena were believed to be explainable by using only one ultimately simple relation, *cause–effect*. One thing or event was taken to be the cause of another, its effect, if it was both necessary and sufficient for the other.

Because a cause was taken to be sufficient for its effect, nothing was required to explain the effect other than the cause. Consequently, the quest for causes was environment-free. It employed what we now call "closed-system" thinking. Laws—like that of freely falling bodies—were formulated so as to exclude environmental effects. Specially designed environments, called "laboratories," were used so as to exclude environmental effects on phenomena under study.

Environment-free causal laws permit no exceptions. Effects are completely determined by causes. Hence, the prevailing view of the world was *deterministic*. It was also *mechanistic*, because science found no need for teleological concepts—such as functions, goals, purposes, choice, and free will—in explaining any natural phenomenon; they were considered to be either unnecessary, illusory, or meaningless.

The commitment to causal thinking yielded a conception of the world as a machine; it was taken to be like a hermetically sealed clock—a self-contained mechanism whose behavior was completely determined by its own structure. The major question raised by this conception was: Is the world a self-winding clock, or does it require a winder? Most took the world to be a machine created by God to serve His purposes, a machine for doing God's work. Additionally, man was believed to have been created in God's image. Hence, it was quite natural for man to attempt to develop machines that would serve His purposes, that would do His work.

The conception of work that was used derived from the conception of the world as consisting ultimately of particles of matter with two intrinsic properties, mass and energy, and an extrinsic property, location in a space-time coordinate system. Work was taken to be the movement of mass through space or the application of energy to matter so as to transform either matter or energy, or matter into energy. Work that was to be mechanized was analyzed. Such analysis came to be called "work study." It was thus decomposed into work elements, indivisible tasks. To these, elementary machines—the wheel and axle, the inclined plane, and the lever—energized by other machines, were applied separately or in combination. Separate machines were developed to perform as many elementary tasks as possible. Men and machines, each performing one or a small number of elementary tasks repetitively, were organized into processing networks that became mass-production and assembly lines.

The Industrial Revolution brought about mechanization, the substitution of machines for man as a source of physical work. This process affected the nature of work left for men to do. They no longer did all the things necessary to make a product; they repeatedly performed a simple operation in the production process. Consequently, the more machines were used as a substitute for men at work, the more working men were made to behave like machines. The dehumanization of men's work was the irony of the Industrial Revolution.

The Systems Age

Although eras do not have precise beginnings and ends, the 1940s can be said to have contained the beginning of the end of the Machine Age and the beginning of the Systems Age. This new age is the product of a new intellectual framework in which the doctrines of reductionism and mechanism and the analytical mode of thought are being supplemented by the doctrines of *expansionism*, *teleology*, and a new *synthetic* (or systems) mode of thought.

Expansionism is a doctrine maintaining that all objects and events, and all experiences of them, are parts of larger wholes. It does not deny that they have parts, but it focuses on the wholes of which they are part. It provides another way of viewing things, a way that is different from, but compatible with, reductionism. It turns attention from ultimate elements to a whole with interrelated parts, to *systems*. Preoccupation with systems emerged during the 1940s. Only a few of the highlights of this process are noted here.

In 1941 the American philosopher(ess) Suzanne Langer[2] argued that, over the preceding two decades, philosophy had shifted its attention from elementary particles, events, and their properties to a different kind of element, the *symbol*. A *symbol* is an *element* whose physical properties have essentially no importance. Charles W. Morris,[3] another American philosopher, built on Langer's work a framework for the scientific study of symbols and the *wholes* of which they were a part, *languages*. By so doing he expanded the center of attention. In 1949 Claude Shannon,[6] a mathematician at Bell Laboratories, developed a mathematical theory that turned attention to a still larger phenomenon, *communication*. Another famous mathematician at the Massachusetts Institute of Technology, Norbert Wiener,[11] in his book *Cybernetics*, put communication into a still larger context, that of *control*. By the early 1950s, it became apparent that interest in control and communication were only aspects of an interest in a still larger phenomenon, *systems*, to which the biologist Ludwig von Bertalanffy[10] drew attention with his work. "Systems" has since been widely recognized as the new organizing concept of science. The concept is not new, but its organizing role in science is.

A system is a set of interrelated elements of any kind; for example, concepts (as in the number system), objects (as in a telephone system or human body), or people (as in a society). The set of elements has the following three properties.

1. The properties or behavior of each part of the set has an effect on the properties or behavior of the set as a whole. For example, every organ in an animal's body affects the performance of the body.
2. The properties and behavior of each part and the way they affect the whole depend on the properties and behavior of at least one other part in the set. Therefore, no part has an independent effect on the whole. For example, the effect that the heart has on the body depends on the behavior of the lungs.

3. Every possible subgroup of elements in the set has the first two properties. Each has an effect, and none can have an independent effect, on the whole. Therefore, the elements cannot be organized into independent subgroups. For example, all the subsystems in an animal's body—such as the nervous, respiratory, digestive, and motor subsystems—interact, and each affects the performance of the whole.

Because of these three properties, a set of elements that forms a system always has some characteristics, or can display some behavior, that none of its elements or subgroups can. Furthermore, membership in the set either increases or decreases the capabilities of each element, but it does not leave them unaffected. For example, parts of a living body cannot live apart from that body or a substitute. The power of a member of a group is always increased or decreased by such membership.

A system is more than the sum of its parts; it is an *indivisible whole*. It loses its essential properties when it is taken apart. The elements of a system may themselves be systems, and every system may be a part of a larger system.

Preoccupation with systems brings with it the *synthetic* mode of thought. In the analytic mode, it will be recalled, an explanation of the whole was derived from explanations of its parts. In synthetic thinking, something to be explained is viewed as part of a larger system and is explained in terms of its role in that larger system. For example, universities are explained by their role in the educational system, rather than by the behavior of their colleges and departments. The Systems Age is more interested in putting things together than in taking them apart.

Analytic thinking is, so to speak, outside-in thinking; synthetic thinking is inside-out. Neither negates the value of the other, but by synthetic thinking we can gain understanding that we cannot obtain through analysis, particularly of collective phenomena.

The synthetic mode of thought, when applied to systems problems, is called the *systems approach*. This way of thinking is based on the observation that, when each part of a system performs as well as possible, the system as a whole seldom performs as well as possible. This follows from the fact that the sum of the functioning of the parts is seldom equal to the functioning of the whole. This can be shown as follows.

Suppose we collect one each of every model of available automobile. Suppose further that we then ask some top-flight automotive engineers to determine which of these cars has the best carburetor. When they have done so, we note the result. Then we ask them to do the same for transmissions, fuel pumps, distributors, and so on through each part required to make an automobile. When this is completed, we ask them to remove the parts noted and assemble them into an automobile, each of whose parts is the best available. They will not be able to do so, because the parts will not fit together. Even if they could be assembled, in all likelihood they would not work well together. System performance depends critically on how the parts fit and work together, not merely on how well each performs independently.

Furthermore, a system's performance depends on how it relates to its environment, the larger system of which it is a part, and to other systems in that environment. For example, an automobile's performance depends on the weather, the road on which it is driven, and how well it and other cars are driven. Therefore, in systems thinking we try to evaluate the performance of a system by evaluating its functioning as a part of the larger system that contains it.

It will be recalled that in the Machine Age cause–effect was the central relation in terms of which all explanations were sought. At the turn of this century the distinguished American philosopher of science, E. A. Singer, Jr.,[7] noted that cause–effect was used in two different senses. First, it was used in the sense already discussed: a cause is a necessary and sufficient condition for its effect. Secondly, it was also used when one thing was taken as necessary but *not* sufficient for the other. He used as an example an acorn, which is necessary but insufficient for an oak; various soil and weather conditions are also necessary. Similarly, a mother—despite women's liberation—is only necessary, not sufficient, for her child. Singer chose to refer to this latter sense of cause–effect as *producer–product*. It can also be thought of as a probabilistic or nondeterministic cause–effect.

Singer[7] went on to show that studies of phenomena that use the producer–product relation were compatible with, but richer than, studies restricted to the use of deterministic causality. Furthermore, he showed that a theory of explanation based on producer–product permitted functional, goal-seeking, and purposeful behavior to be studied objectively and scientifically. These concepts no longer needed to be taken as meaningless or inappropriate for scientific study.

Later, biologist G. Sommerhoff[8] came independently to the same conclusions as Singer had. In the meantime, in a series of papers that laid the groundwork for cybernetics, Arturo Rosenblueth, Norbert Wiener, and J. H. Bigelow[4,5] showed the great value of conceptualizing machines and man/machine systems as functioning, goal-seeking, and purposeful entities. In effect, they showed that, whereas it had been fruitful in the past to study men as though they were machines, it was now at least equally fruitful to study machines, man/machine systems, and, of course, men as though they were goal-seeking or purposeful. Thus, in the 1950s *teleology*—the study of goal-seeking and purposeful behavior—was brought into science and began to dominate our conceptualization of the world.

For example, in mechanistic thinking behavior is explained by identifying what caused it, never its effect. In teleological thinking, behavior can be explained either by what produced it or by what it is intended to produce. For example, a boy's going to the store can be explained either by the fact his mother sent him, or by the fact that he intends to buy ice cream for supper. Study of the functions, goals, and purposes of individuals and groups has yielded a greater ability to evaluate and improve their performance than mechanistically oriented research did.

The Post-Industrial Revolution

The doctrines of expansionism and teleology, and the synthetic mode of thought are both the producers and the products of the Post-Industrial Revolution. But this revolution is also the product of three technological developments, two of which occurred during the (First) Industrial Revolution. One of these emerged with the telegraph in the first half of the nineteenth century, followed by the invention of the telephone by Alexander Graham Bell in 1876, and of the wireless by Marconi in 1895. Radio and television followed in this century. Such devices mechanized communication, the transmission of symbols. Since symbols are not made of matter, their movement through space does not constitute physical work. The significance of this fact was not appreciated at the time of the invention of communication machines.

The second technology emerged with the development of devices that can observe and record the properties of objects and events. Such machines generate symbols that we call *data*. The thermometer, odometer, speedometer, and voltmeter are familiar examples of such instruments. In 1937 there was a major advance in the technology of observation when it 'went electronic' with the invention of radar and sonar.

Instruments can observe what we cannot observe without mechanical aids, or magnitudes and differences too large or small for our senses. Note that such instruments, like communication machines, do not perform physical work.

The third and key technology emerged in the 1940s with the development of the electronic digital computer. This machine could manipulate symbols logically. For this reason, it is frequently referred to as a thinking machine.

These three technologies made it possible to observe, communicate, and manipulate symbols. By organizing them into a system, it became possible to *mechanize mental work*, to *automate*. This is what the Post-Industrial Revolution is all about.

Development and utilization of automation technology requires an understanding of the mental processes that are involved in observing, recording, and processing data, communicating them, and using them to make decisions and control our affairs. Since 1940 a number of interdisciplines have been developed to generate and apply knowledge and understanding of mental processes. These include communication and information sciences, cybernetics, systems engineering, operations research, and management and behavioral sciences. Such fields provide the software of the Post-Industrial Revolution.

The Organizing Problems of the Systems Age

Because the Systems Age is teleologically oriented, it is preoccupied with systems that are goal-seeking or purposeful; that is, systems that can display *choice* of either means or ends, or both. It is interested in purely mechanical systems only insofar as they can be used as instruments of purposeful systems. Furthermore, the Systems Age is most concerned with purposeful systems, some of whose parts are purposeful; these are called *social groups*. The most important class of social groups is the one containing systems whose parts perform different functions, that have a division of functional labor; these are called *organizations*. Systems-Age man is most interested in groups and organizations that are themselves parts of larger purposeful systems. All the groups and organizations, including institutions, that are part of society can be conceptualized as such three-level purposeful systems.

There are three ways in which such systems can be studied. We can try to increase the effectiveness with which they serve their own purposes, the *self-control* problem; the effectiveness with which they serve the purposes of their parts, the *humanization* problem; and the effectiveness with which they serve the purposes of the systems of which they are a part, the *environmentalization* problem. These are the three strongly interdependent organizing problems of the Systems Age.

Science in the Systems Age

Up to this point I have tried to deal with the question: What in the world is going on in the world? My response to this question provides a vantage point from which I would now like to look at science in general and at the management sciences in particular.

Since its inception, science has not only been taking the world apart, but it has also been taking itself apart, although not without reason or benefit. The decomposition of science could not have been avoided. The reason is revealed in the statement with which Colin Cherry[1] opened his book, *On Human Communication*:

> Leibnitz, it has sometimes been said, was the last man to know everything. Though this is most certainly a gross exaggeration, it is an epigram with considerable point. For it is true that up to the last years of the eighteenth century our greatest mentors were able not only to compass the whole science of their day, perhaps with mastery of several languages, but to absorb a broad culture as well.

The continuous accumulation of scientific knowledge that occurred during and after the eighteenth century made it necessary to divide and classify this knowledge. Scientific disciplines were the product of this effort. Science formally separated itself from philosophy only a little more than a century ago. It then divided itself into physics and chemistry. Biology emerged out of chemistry, psychology out of biology, and the social sciences out of psychology. This much was completed at the beginning of this century. But scientific fission continued. Disciplines proliferated. The National Research Council now lists more than one hundred and fifty of them.

Disciplines are categories that facilitate filing the content of science. They are nothing more than filing categories. Nature is not organized the way our knowledge of it is. Furthermore, the body of scientific knowledge can, and has been, organized in different ways. No one way has ontological priority. The order in which the disciplines developed was dictated to a large extent by what society permitted scientists to investigate, not by any logical ordering of subject matter. Scientists started to investigate the areas that least challenged deeply held social, cultural, religious, and moral beliefs of the time. The subject matter of science was chosen—and not always successfully—so as to maximize the probability of survival of scientists and science. As science gained prestige, it pressed against the social barriers that obstructed its development; one by one they were breached.

But scientists and philosophers wanted to invest the history of science with more logic than history itself provided. Therefore, they sought to rationalize the order of disciplinary development by invoking the concept of a hierarchy of the sciences. They argued that physics deals with objects, events, and properties of both that were ultimately simple, hence irreducible and directly observable. Each successive discipline, it was argued, dealt with increasingly complex functions and aggregations of these objects, events, and properties. Hence, each discipline except physics was taken to rise out of, and to be reducible to, the one that preceded it. Physics was taken to be basic and fundamental. Dependence between sciences was taken to flow in only one direction.

This hierarchical myth is still widely accepted in and out of science. It is the basis of a caste system in the community of science that is as severe and irrational as any that has existed in society.

It is still widely believed that the physical sciences alone deal with ultimate reality and that they have no need of the other disciplines in their effort to do so. This belief is maintained despite the fact that we can demonstrate that no concept used in any one discipline is ultimately fundamental and incapable of being illuminated by work in other disciplines.

Consider, for example, a concept used in physics that, perhaps more than any other, is thought to be its exclusive property: *time*. Physicists have dealt with time in one of two ways. They have either taken it to be a primitive concept that cannot be defined, and hence a concept whose meaning can only be grasped by direct experience of it; or they have dealt with it operationally, defining it by the operations used to measure it. Techniques of measuring time in physics all derive from use of the rotation of the earth around the sun as a basic unit. Clocks, sun dials, water clocks, sand clocks, and so on are instruments to divide this unit into equal parts. Thus, in physics time is dealt with as an ateleological astronomical concept. It is generally assumed that contributions to understanding it cannot be made by any other discipline.

This is not the case. Time can be considered teleologically, not as a property of the universe that is out there for us to take, but as a concept deliberately constructed by man to serve his purposes.

People develop alternative ways of individuating events. For example, a person may differentiate between breakfasts by their content, location, or by those with whom he had the meal. Some of these individuating properties may be adequate only in special circumstances. He may have the same breakfast with the same people at the same place on different days. Two events that occur to the same individual may be the same with respect to every property except one, time. Two events that occur to the same individual at the same time cannot be otherwise identical: they must differ in some respect; otherwise, they would not be two events.

Therefore, from a functional point of view, time is a property of events that is sufficient to enable a person to individuate any two changes in the same property of the same thing. Because we measure time using physical phenomena, we erroneously conclude it is a physical concept. The error becomes apparent in situations in which astronomical measures do not serve our purposes well. In measuring the rate of growth of plants, for example, C. W. Thornthwaite[9] found astronomical time inadequate. He sought a biological clock and found one in the pea plant; he used the time between appearances of successive nodes on the pea plant as units of time for his work. These units were of different duration when measured astronomically, but they made possible more accurate prediction and control of harvests than did hours and days. We measure time by using events that are identical in all respects save time; and, in principle, these can be of any type—which type we use is determined by our purposes.

As the application of science increased it became useful to organize its findings functionally around areas of application, into professions, as well as into disciplines. Old professions that preceded science borrowed from a number of scientific disciplines and new ones did so as well.

The disciplinary and professional classifications of scientific knowledge are orthogonal to one another, and hence can be represented by a matrix in which the disciplines form the rows and the professions the columns. New rows and columns can be expected to be added in the future.

Science in the Systems Age

As the problems to which science was addressed became more complex—and particularly as it began to address itself to problem complexes, systems of problems that I like to call *messes*—a new organization of scientific and technological effort was required. The first response to this need occurred between the two World Wars and took the form of *multidisciplinary* research. In such research the problem complex investigated was decomposed into unidisciplinary and uniprofessional problems that were taken to be solvable independently of each other. Hence, they were assigned to different disciplines or professions, separately solved, and the solutions were either aggregated or allowed to aggregate themselves. With the emergence of systems thinking, however, it was realized that the effect of multidisciplinary research on the treatment of the whole was frequently far from the best that could be obtained.

This realization gave rise to *interdisciplinary* research, in which the problem complex was not disassembled into disciplinary parts, but was treated as a whole by representatives of different disciplines working collaboratively. Operations research and the management sciences were among the interdisciplines born of this effort. So were cybernetics, the organizational sciences, the policy sciences, planning science, general systems research, and the communication sciences, among others.

Universities began to educate the young for such work. Those so educated were not of any one discipline but of the intersection of several. Hence, their loyalty was not directed to one discipline but to an interdisciplinary concept. But this did not last long. The interdisciplines sought recognition and status by emulating the disciplines and professions. Academic departments and professional societies were formed along conventional lines. The interdisciplines began to identify themselves with the instruments that they developed and used—that is structurally—rather than with what these instruments were used for—that is, functionally. They began to introvert, to look inward and contemplate their own methods and accomplishments, rather than the messes that had given rise to their activities. Jurisdictional disputes and efforts to individuate interdisciplines arose between activities created to eliminate just such disputes and individuation.

As the problem complexes with which we concern ourselves increase in complexity, the need for bringing the interdisciplines together increases. What we need may be called *metadisciplines*, and what they are needed for may be called *systemology*.

The formation of interdisciplines in the last three decades can now be understood as a transitional development, a beginning to an evolutionary synthesis of human knowledge, not only within science, but between science and technology, and, most importantly, between science and the humanities. Consider the distinction between science and the humanities. I believe that in the Systems Age science will come to be understood as the search for similarities among things that are apparently different; and the humanities will come to be understood as the search for differences among things that are apparently the same. The former seeks generality; the latter uniqueness. This makes science and the humanities like the head and tail of a coin; they can be looked at and discussed separately, but they cannot be separated. Consider why.

In the conduct of any inquiry, we must determine the ways in which the subject under study is similar to other subjects previously studied. Doing so enables us to bring to bear what we have already learned. But, in addition, it is also necessary to determine how the subject at hand differs from any we have previously studied: in what ways it is unique. Its uniqueness defines what we have yet to learn before we can solve the problems it presents.

Thus, the humanities define the aspects of messes that we still have to learn how to handle, and science provides ways both of handling or researching the aspects that have previously been dealt with, and of finding ways of approaching the aspects that have not been studied previously.

The effective study of large-scale social systems requires the synthesis of science and the humanities, of science and technology, and of the disciplines within science and the professions that use them.

Despite the need for integration, universities and professional and scientific societies preserve the autonomy of the parts of science and their application. What is needed is not a temporary association of autonomous interdisciplines such as we have here at Atlantic City, but a permanent integration of interdisciplines that yields a broader synthesis of methods and knowledge than any yet attained.

We need a fusion of interdisciplines that extends well beyond those represented here. Nevertheless, a fusion of these three would be a significant step in the right direction. But as far as I can tell—after considerable effort to merge two of them—they would rather die separately than live together. And they are dying despite their growth. Death is *not* a function of the number of cells in a body, but of their vitality: the membership of even a cemetery can expand continuously.

None of what I have said denies the usefulness of either disciplinary science or the professions. They will remain useful in dealing with problem areas that can be decomposed into problems that are independent of each other. But the major organizational and social messes of our time do not lend themselves to such decomposition. They must be attacked holistically, with a comprehensive systems approach.

Nor are my remarks intended to diminish the past accomplishments of IE, MS, and OR—they have been significant and I share with you a pride in them. But their accomplishments are becoming less significant because their development has not kept pace with the growing complexity of the situations with which managers and administrators are faced.

As currently conceived, taught, and organized, industrial engineering is not broad enough to engineer industry effectively by itself. This is obvious. Look at the wide variety of other types of engineers crawling all over industry. The management sciences are not broad enough to make management scientific. What percentage of the decisions of even the managers most dedicated to these sciences are based on science? Operations research is not broad enough to research effectively the operating characteristics of our social system that most urgently need research: discrimination, inequality within and between nations, the bankruptcy of education, the inefficiency of health services, increasing criminality, deterioration of the environment, war, and so on. This is not to say the IE, MS, and OR are no good, but it does say that they are *not good enough*. Each of them suffers not only from the lack of competencies that are required to deal with the messes that preoccupy those who manage most public and private systems, but also because they use Machine-Age concepts and methods in attempts to deal with Systems-Age problems.

Meetings such as this one should be dedicated to the marriage of movements, and to the conception and birth of ways of coping with complexity. But, instead, they are wakes at which interdisciplines are laid out and put on display in their best attire. Eulogies are delivered in which accounts are given about how messes were murdered by reducing them to problems, how problems were murdered by reducing them to models, and how models were murdered by excessive exposure to the elements of mathematics.

But those who attend a wake are not dead. They can still raise hell. And, if they do, even a corpse—like that of James Joyce's Finnegan—may respond and rise with a shout.

References

1. Colin Cherry, *On Human Communication*, Wiley, New York, 1957.
2. S. K. Langer, *Philosophy in a New Key*, Penguin Books, New York, 1948.
3. Charles Morris, *Signs, Language and Behavior*, George Braziller, New York, 1955.
4. A. Rosenblueth and N. Wiener, "Purposeful and Non-Purposeful Behavior," *Phil. of Sci.* **17**, 318–326 (1950).
5. A. Rosenblueth, N. Wiener, and J. H. Bigelow, "Behavior, Purpose, and Teleology," *Phil. of Sci.* **11**, 18–24 (1943).
6. C. E. Shannon and Warren Weaver, *The Mathematical Theory of Communication*, The University of Illinois Press, Urbana, 1949.
7. E. A. Singer, Jr., *Experience and Reflection*, C. West Churchman (ed.), University of Pennsylvania Press, Philadelphia, 1959.
8. G. Sommerhoff, *Analytical Biology*, Oxford University Press, London, 1950.
9. C. W. Thornthwaite, "Operations Research in Agriculture," *Opns. Res.* **1**, 33–38 (1953).
10. Ludwig von Bertalanffy, *General Systems Theory*, George Braziller, New York, 1968.
11. Norbert Wiener, *Cybernetics*, Wiley, New York, 1948.

13

Systems Profile: The Emergence of Systems Science

George J. Klir

Beginnings

In retrospect, it seems that the general shape of my systems profile was determined early in my professional career and under rather special circumstances. I refer here to the period 1952–1964: the 1950s were the years of my university studies (undergraduate studies in electrical engineering and graduate studies in computer science) and the early 1960s were the years during which I was primarily involved in industrial computer research. The place was Prague, the capital of Czechoslovakia.

My first exposure to complex systems occurred when I studied the various types of telephone exchanges during my junior year. Later, when my studies focused on computer architecture and logic design, I observed that telephone exchanges and computers belonged to the same category of systems—systems that process information, even though they do that for different purposes. This observation made me realize that computers were not only number crunching devices (which was a common misconception at that time), but general symbol manipulators.

The discovery that a new insight may be obtained when such diverse systems as telephone exchanges and computers are conceived from a broader perspective gave me a sense of the intellectual power of *generalization*. As a result, I was pleased to learn about Norbert Wiener's cybernetics when it eventually became 'legal' in Czechoslovakia. (Cybernetics was initially dismissed in the Soviet Union and its satellites as a 'bourgeois pseudoscience' and virtually forbidden as a subject of scientific discussion. It was later 'rehabilitated' in the post-Stalin period.) The generalization of concepts such as information, communication and control, upon which cybernetics is based, was in perfect harmony with my own thinking at that time. It was exciting to actually meet Wiener after one of his lectures in Prague and exchange ideas.

When I studied electrical engineering, I learned that sophisticated methods developed for analyzing electric circuits could also be applied, by established *isomorphisms*, to mechanical, magnetic, acoustical, and thermal systems, whose methodologies were far less developed. The recognition of these methodological possibilities led eventually to the creation of a new mathematical theory—the theory of generalized circuits, which captured similarities among several areas of physics. These similarities were also essential to the design, production, and use of various kinds of analog computers.

I was quite fascinated by the fact that problems regarding such diverse systems as electrical, mechanical, or thermal could be dealt with by the same general methodology. I began to realize that there were profound similarities between phenomena investigated by diverse disciplines of science or engineering, and that these similarities, when established, allowed us to *transfer knowledge from one discipline to another*. The boundaries between the disciplines started to lose some of their significance in my mind.

The period 1960–1964, when I was with the Institute for Computer Research in Prague, was perhaps the most important period regarding the shaping of my systems profile. At that time, computers were still at their infant age. There was a great and rapidly growing interest among scientists and professionals to learn about the capabilities of computers and their prospective role in dealing with problems in their disciplines. Since the number of people familiar with computers was rather small in Czechoslovakia at that time, I received numerous invitations to lecture on various topics regarding computers. In some cases, a research cooperation emerged from these lectures, enabling me to work with biologists, medical doctors, economists, linguists, psychologists, psychiatrists, and even parapsychologists and musicologists. In each case, I learned a lot about the systems and problems these scientists and professionals deal with. I observed that these systems and problems were not completely different from one discipline to another. I began to recognize some similarities that could be utilized for developing a highly general methodology for dealing with systems problems. Although this idea was rather primitive and vague at that time, it haunted me for many years; it resulted eventually in the formulation of my systems conceptual framework that is now referred to as the *General Systems Problem Solver* (GSPS).

One person had a profound influence upon my intellectual development in the period 1960–1964: my intellectual mentor, Antonin Svoboda. Officially, Svoboda was my dissertation advisor in the late 1950s and then, for a few years, my boss at the Institute. He remained my principal teacher and close friend until he passed away in 1980.

Svoboda was a child prodigy. He mastered calculus, differential equations, and other areas of higher mathematics by the age of fourteen, and was also an accomplished pianist. His formal education covered the fields of mathematics, physics, and electrical engineering, and he had extensive knowledge in several other fields, particularly biology, medicine, psychology, and philosophy of science.

During World War II, Svoboda was in exile, first in France and later in the United States, where he did research on analog computers at the M.I.T. Radiation Laboratory. After the war, he returned to Czechoslovakia, and founded the Institute for Computer Research. He educated several generations of computer scientists and was a chief architect of several highly original digital computers. Dissatisfied with the social and political environment in

Czechoslovakia, he and his family took a unique opportunity in 1964 and immigrated to the United States.

Although he is primarily known for his contributions to computer science, Svoboda was a systems thinker *par excellence*. This is well evidenced by the computers he designed. Their architecture reflects the underlying holistic approach and other sound systems principles: perfect compatibility of physical properties of the used technology with the superimposed logical structure, almost optimal balance of efficiency, economy, fault-tolerance and other desirable features, and excellent man-machine interface. In his methodological work, Svoboda had always strived for a symbiosis of the man and the computer, taking advantage of the complementarity of the two worlds. This effort made him interested not only in algorithms and their computational complexity, but also in the issue of representation, especially in the visual domain.

As his assistant for several years, I learned from Svoboda a great deal about systems design, the practice of genuine scientific research, and the underlying philosophical issues. Under his influence, I also developed high standards of professional ethics and the ability to be critical as well as tolerant. Above all, I learned from him some of the most fundamental systems principles.

Another influence exerted on my systems thinking in the early 1960s came from a small circle of six colleagues affiliated with the Institute for Computer Research; it was known under the acronym K. Vasspeg, which was formed from the names of the six participants: Klir, VAlach, Sehnal, Spiro, PElikan, Gecsei. This group of young computer researchers (all students of Svoboda) emerged quite naturally by common interests in the philosophical, mathematical, and methodological issues associated with cybernetics and systems research. The group met one evening each week for an informal discussion of some of these issues. The initial objective was to scrutinize fundamental concepts such as system, environment, information, control, organization, structure, invariance and change, and to arrive at a consensus on their meaning. Later, more sophisticated issues regarding adaptivity, self-organization, learning, pattern recognition, and the like were discussed.

The discussions at the meetings of K. Vasspeg were usually quite intensive. A diary was kept indicating, at each stage, points of agreement and disagreement. The group eventually developed a clear position on some fundamental issues of cybernetics and systems research, which was expressed in two papers published under the acronym K. Vasspeg. In one of the papers (published only in Czech), the group expressed some original views regarding the meaning of cybernetics[44]; in the other paper, it developed a sound conceptual framework for studying the phenomenon of self-organization.[45]

The first stage of my professional life culminated in Summer 1964, when I completed two books. One of them, *Cybernetic Modelling*,[29] was coauthored with Miroslav Valach (a fellow Vasspegian). It introduces a broad conceptual framework within which some basic types of systems and modelling principles regarding these systems are formulated. In addition, it contains numerous case studies of systems modelling. The second book, *Synthesis of Switching Circuits*,[30] I coauthored with Lev K. Seidl. Although focusing on special systems, switching circuits, the methodology presented in the book is consistently based upon the general conceptual framework adopted from *Cybernetic Modelling*. This framework became a nucleus of the GSPS. Some of its aspects have remained unchanged and are included in the current conception of the GSPS;[23] most were later refined by further research and only a few were totally abandoned.

On My Own

In Summer 1964, shortly after I completed the two books,[29,30] I received an offer from the Czechoslovak Government for a two-year academic position at a branch of the Baghdad University in Mosul, Iraq. Although the offer caught me by surprise, it was a development that I wholeheartedly welcomed. In spite of the stimulating activities of K. Vasspeg, life in Czechoslovakia was rather stifled and intellectually sterile. Censorship, dogmatism, distrust, and boredom were orders of the day. Travel to the West was virtually taboo, Western literature was either prohibited or almost impossible to obtain for economic reasons, the country was closed to information, and one was always exposed to the danger of committing unwittingly a 'political crime'. Almost everyone was fed up with this situation and I was no exception. This possibility of leaving the country was thus a blessing for me.

I quickly accepted the offer and went to Iraq in September 1964. My wife and infant daughter joined me four months later.

I left for Iraq with great enthusiasm. Psychologically, I badly needed to leave the depressive environment. I was also hungry for experiencing different cultures, geographies, climates, and the like. Above all, I was hungry for information, which was in short supply in Czechoslovakia.

Life was not easy in Mosul, due to rather primitive living conditions, but it was full of rich experiences. Almost everything was new. We were exposed to a culture totally different from our own; the remains of Babylonian and Assyrian architecture were truly fascinating; and we experienced a climate that was very unusual to us.

Perhaps the most important thing was the tremendous source of information that suddenly opened to me. Although Iraq was by no means free from censorship, it was by far more open to information than Czechoslovakia. Not only were Western newspapers and magazines readily available at newsstands, but one could order virtually any publication through some bookstores or directly by mail. I took this opportunity wholeheartedly. Gradually, as my income allowed, I ordered almost all the literature on systems research and cybernetics available at that time, as well as some key books on certain areas of mathematics, computer science, and philosophy.

I spent a great deal of time studying the newly acquired literature and, based on this new information, reflecting upon my own systems thinking. It was a lonely enterprise since there was no one in Mosul with whom I could discuss the emerging ideas. This situation was often frustrating, but it had one advantage. It allowed me to comprehend well the status of the systems movement at that time and to recognize that my own views on systems research and cybernetics were not fully compatible with any of the predominant views expressed in the literature. Contrary to some of these views, I felt that the study of general properties of systems would eventually become a legitimate science of its own.

The conception of the prospective science of systems was not very clear in my mind at that time. I felt, however, that an essential step in establishing such a science would be the development of an adequate *taxonomy of systems*. I also felt that the only indigenous sources for developing this taxonomy were the various disciplines of science, engineering, and other human activities. I set it as a goal for my own research to try to identify the notions of systems in many distinct disciplines, abstracting them from their specific contexts, to categorize the abstractions, and integrate the categories eventually into a coherent whole. It took me some 20 years to fully achieve this goal.[23]

When my two-year commitment at the Baghdad University was approaching its end, my wife and I made a decision not to return to Czechoslovakia. Instead, we went to Vienna, where we applied for immigration to the United States. After several months of uncertainty, during which I continued my lonely research work, we eventually obtained our immigration visas and left for the United States on 29 November 1966.

First Years in America

Our destination was Los Angeles, where I assumed an academic position in computer science at UCLA. My scientific loneliness finally ended. I made a happy reunion with Antonin Svoboda, also with UCLA, and established a close relationship with some of my new colleagues.

At first, I felt like Alice in Wonderland. I spent long hours in the library and various bookstores, fascinated by the tremendous range of available literature on every imaginable subject. I was enchanted by the advanced computer technology and I tried to participate in almost all of the seminars that took place at UCLA on various subjects. Soon I realized how little I could actually cover of what I desired. Although I quickly learned that one has to be highly selective in such a rich environment, time has remained my most precious commodity since our arrival in America; it is always in short supply.

At UCLA, I taught courses in computer science, primarily computer architecture and logic design. On the research side, however, I continued to work on the development of a broad conceptual framework for general systems methodology. I spent a great deal of time with Svoboda involved in fascinating discussions. Apart from that, unfortunately, none of my other colleagues were interested in the kind of research I pursued.

In Spring 1967, Frank Barnes, Chairman of the Electrical Engineering Department at the University of Colorado in Boulder, gave a seminar to our faculty. He became quite interested in my research and invited me to present my ideas in the form of a special graduate course at his Department in Summer 1967. I accepted with great enthusiasm the opportunity and challenge of teaching for the first time my own systems ideas.

During my stay in Boulder in Summer 1967, I worked very hard. The time was divided between teaching and preparing class notes for the course. Although most of my ideas were already well developed at that time, some still required further research. By the end of the summer session, the class notes formed almost a complete text. The course, which was taken by a small group of excellent graduate students, was very successful and I benefited a lot from my interactions with the students.

I returned to UCLA with confidence that I was on the right track. When I completed my text, I made a proposal to our faculty for developing a course based on my experience at Boulder. To my surprise, the faculty response was lukewarm at best.

In the meantime, publication of my new text *An Approach to General Systems Theory* was considered by Van Nostrand. I received a thorough and very constructive review of the manuscript by John Warfield, who waived his reviewer's anonymity. His review helped me to improve the manuscript considerably; the book was eventually published in 1969.[14] Since this first encounter, John and I have kept in contact and influenced each other on several occasions.

I made a few more attempts at UCLA to establish some activities in systems research, but all were in vain. It was thus natural that I accepted a generous offer from Harold Rothbart, Dean of the College of Science and Engineering of the Fairleigh Dickinson University in Teaneck, New Jersey, to join the University and have a leading role in developing a program in computer and systems sciences. It was not an easy decision since it implied a separation from my beloved teacher and friend Antonin Svoboda.

My stay at Fairleigh Dickinson University turned out to be rather short—just one academic year (1968-1969). Working with Dean Rothbart was delightful. He was a man of great vision. Our views and interests were quite compatible. However, his views were less compatible with the upper administration of the University as well as with the faculties of the various departments of the College itself. This resulted in considerable uncertainty regarding the prospective program in computer and systems sciences.

Suddenly, in the midst of this uncertainty, a new turn in my life opened. It started at a conference on systems education, where I met Donald Gause. We discussed our common research interests and he told me about the School of Advanced Technology, which was just established at SUNY-Binghamton. It was a professional school whose mission was to develop graduate programs in computer systems, applied mathematics, and general systems. Don felt that my interests and background were perfectly matched with this mission. Soon, Don and the Dean of the new School, Walter Lowen, visited me at Fairleigh Dickinson. The conclusion of the meeting was that, indeed, the match was perfect. I joined the School in Fall 1969 and have been there ever since.

Although my stay at Fairleigh Dickinson was so short, I established a strong and lasting professional relationship and close friendship with one of my colleagues there, Robert Orchard. This relationship has exerted a great influence on shaping my systems profile since 1968.

At SUNY-Binghamton

When I joined the School of Advanced Technology (SAT) at SUNY-Binghamton in Fall 1969, I finally felt at home. For the first time, my work in systems research was really appreciated and encouraged. In fact, I was hired to take a major role in developing a graduate program in general systems. There were also several colleagues with genuine interests in systems research at SAT, with whom I could discuss my ideas. My *Approach to General Systems Theory*[14] was just published and I used it as a text in one of the regular SAT courses.

I actually spent my first semester with SAT as a Fellow at the IBM Systems Research Institute (SRI) in New York City. The Institute offered a great variety of advanced courses on topics related to computer technology, some of which were not available anywhere else. It was thus a great opportunity for me to up-grade my background in computer science. It turned out that SRI became interested in my general systems framework. As a result, I commuted to SRI from Binghamton for the next five years to teach a course based on my

Approach to GST.[14] I also taught this course in the Systems Science Program at Portland State University in Oregon every summer during this period.

SAT emerged from the needs of the rapidly growing high technology industries in the Binghamton area. Its architect was Walter Lowen, the founding Dean of the School. His basic idea was to develop three graduate programs relevant to high technology industries (programs in computer systems, applied mathematics, and general systems) within a nondepartmentalized School. He employed some innovative ideas in founding the School, one of which made students with bachelor degrees in any field eligible for admission to the graduate programs. Above all, however, he possessed a great capability to identify people with desirable talents and provide them with an academic environment in which these talents could be developed to their full potential.

Several stages in the evolution of SAT can be recognized. The School remained nondepartmentalized only until 1976, when it was divided into the Department of Computer Science and the Department of Systems Science. At that time, Walter Lowen resigned as Dean and returned to research and teaching in the Systems Science Department. I was appointed Chairman of the Systems Science Department and have served in this capacity ever since. In 1983, under the leadership of Dean Lyle Feisel, the two departments of SAT were integrated into a larger unit, the currently existing Thomas J. Watson School of Engineering, Applied Science, and Technology.

Joining SAT was a liberation for me. Finally, I did not have to sacrifice my true research interests for some other 'legitimate' activities. Initially, I taught courses in the computer systems and general systems programs. I also did research in both areas, but my work on general systems methodology was dominant. This is best illustrated by my book on *Methodology of Switching Circuits*.[15] which I wrote as an application of general systems methodology to the analysis and synthesis of switching circuits. Since the mid 1970s, virtually all my research and teaching have been in systems science.

It has not been all that smooth sailing with SAT and, later, the Watson School: occasional clashes between some faculty members in the School, frequent budget inadequacies, and various other problems. By and large, however, these Schools have provided my colleagues and I with a reasonably supportive environment for our research in systems science. At SUNY-Binghamton, I was at last able to finalize my general systems conceptual framework, and investigate within it some fundamental issues of systems science. These investigations contributed to the development of systems science as a legitimate field of scientific inquiry. My work was recognized by my colleagues and the SUNY administration and I was appointed in 1984 a Distinguished Professor of Systems Science. For me, this was also a symbolic approval of the legitimacy of systems science.

My Search for a General Systems Framework

As already indicated, the search for a general systems framework has been the story of my professional life. It began as a conscious effort in the mid 1960s, during my Middle East episode, but I can recognize traces of this effort already in the early 1960s.

One issue of great interest in cybernetic circles around 1960 was the possibility of self-organization in systems. I remember vividly a lecture devoted to self-organizing systems, which I attended in Prague in 1960. The speaker (whose name I do not remember) began his lecture by saying: 'There is no doubt that all of you know the meaning of the term "system" and, therefore, I will not waste time and proceed directly to the explication of the term "self-organization".' How wrong he was! A long discussion after the lecture revealed that the term 'system' had different connotations for different people, and this multiplicity of meanings resulted in unsurmountable semantic difficulties.

The lecture and especially the discussion stimulated my curiosity about the possible meanings of the term "system." This curiosity was shared by all members of the group K. Vasspeg. The more we discussed the issue, the more difficult it became. Eventually, we arrived at a simple but very fundamental distinction between an *object*, loosely understood as a part of the world that is of interest to someone (say an *investigator*), and a *system* defined on that object, which reflects the *interaction* between the investigator and the object. In this sense, a system is always an abstraction that is based upon the distinctions made by the investigator. This view became a cornerstone of my inquiry into systems.

By being in contact with people in different disciplines, I realized that they dealt with several very different types of systems. As a consequence, I felt that it was not meaningful to try to develop a universal conception of a general system, which would subsume all these various types of systems as special cases; such a conception would be overly general and, consequently, pragmatically sterile. Instead, I felt, a taxonomy of systems should be developed that would capture, in an organized fashion, all recognized types of systems and, perhaps, would suggest some additional types. These feelings were later reinforced when I read the well-known paper "General Systems Theory—the Skeleton of Science" by Kenneth Boulding.[4]

In developing a useful taxonomy of systems, I started with the conceptions of systems employed in the various disciplines of science and engineering with which I was at least partly familiar. I abstracted these conceptions from their specific interpretations and categorized them. Later, I examined some additional disciplines for this purpose and began to compare my emerging types of systems with those of other systems researchers. I found some resemblance to the types of systems that were conceptualized by W. Ross Ashby.[1,2] As I described elsewhere,[19] his work has been for many years a great source of inspiration to me.

The initial form of my *taxonomy of systems* was published in 1969.[14] It consisted of five major epistemological categories of systems, each subsuming a set of types of systems distinguished from each other by various methodological distinctions. These five categories formed an *epistemological hierarchy*. Let me give a brief overview of this early version of the hierarchy.

The most primitive epistemological category consists solely of an *experimental frame*: a set of variables, their state sets with appropriate observation channels, time (as a parameter within which the variables are observed), and an appropriately defined time set. The term *source systems* is usually used for systems defined at this level to indicate that such systems are, at least potentially, sources of empirical data. Systems on different higher epistemological levels are distinguished from each other by the level of knowledge regarding the variables of the associated source system.

The next category of systems, epistemologically higher than the source system, is defined by a *given activity* of the variables within the experimental frame; the activity is basically a function from the time set into the overall state set. Still higher in the epistemological hierarchy are two categories of systems, which are called behavior systems and state-transition systems. A *behavior system* is defined by a time-invariant relation among selected present, past, or future states of the variables in the experimental frame. The states are selected by a mask through which we observe the given activity and take samples corresponding to different time instants. The name *mask* was coined by Svoboda, who introduced the concept in the context of switching circuits[41,42]; it is an important concept in my framework. A *state-transition system* is defined in terms of a set of overall states based upon a particular mask and a specification of conditions for transitions between these states. Epistemologically highest in this early version of the hierarchy is a *structure system*, which is defined as a set of interacting behavior or state-transition systems.

Although this early taxonomy of systems was still incomplete, it already proved useful in characterizing and dealing with many systems problems. For example, Robert Orchard conceptualized a 'problem solving roadmap', based upon the five epistemological categories of systems, which he used with great success for dealing with practical problems at Bell Laboratories.[33] Orchard also proposed an extension of the epistemological hierarchy by allowing a system in any of the categories to change according to some time-invariant procedure.[32] Systems of this type are now called *metasystems*. When a procedure of a metasystem is itself allowed to change according to a higher level time-invariant procedure, the system is called a meta-metasystem, or a metasystem of second order. In the same way, metasystems of third and higher orders are defined.

Several other extensions were incorporated into the original epistemological hierarchy before it assumed its current form. The main extensions were: (i) time was extended to the general concept of a *backdrop* to capture not only dynamic (time-dependent) variables, but also spatial variables, population-based variables, as well as variables that depend on more than one parameter; (ii) structure systems of second order, whose elements are themselves structure systems, were introduced, and recursively, higher order structure systems were also defined; (iii) structure systems and metasystems were defined not only in terms of behavior and state-transition systems, but in terms of any other system as well, and that even allowed elements of metasystems to be structure systems and elements of structure systems to be metasystems.

The current version of the epistemological hierarchy, as formulated in my book *Architecture of Systems Problem Solving*,[23] is a semilattice of epistemologically distinct types of systems. In general, a system at a higher level contains all information available in the corresponding systems at any lower level, but it contains some additional information. This epistemological hierarchy has recently been employed as a framework for knowledge formalization in expert systems by Orchard, Reese, and Tausner[33]; its role in expert systems was also recognized by Gaines.[10]

When I was developing the epistemological hierarchy, the skeleton of my systems framework, I simultaneously tried to formulate basic categories of systems problems. I conceptualized each category in terms of a transformation from one systems type to another that satisfies some requirement type (a set of objectives and constraints). Additional problem types were then represented by sequences of the basic types. This allowed me to capture the infinite variety of systems problems by a finite set of significant problem categories. A practical version of this categorization of systems problems is Orchard's problem solving roadmap.[33]

Once my conceptual framework began to take shape, I started research on methods for dealing with some of the problem types, especially those which I found methodologically undeveloped. Although this research could obviously cover only a small subset of problem types, it is important to realize that the framework is sufficiently broad to allow one to import methods developed within other frameworks. To facilitate this importation of well-developed methods into the GSPS framework is one of my current efforts. My ultimate objective is to develop a powerful expert system for dealing with systems problems.

Years of Activism in Systems Movement

My professional life has not been just research and teaching. I have also been an activist supporting, to the best of my capabilities, the systems movement. My activism was particularly intensive from 1974 to 1984.

The first outcome was the *International Journal of General Systems*—the first journal fully devoted to general systems methodology, applications and education, which I founded in 1974. It did not happen overnight. A five-year story, which is described in my first editorial,[16] preceded the actual foundation of the journal. Having served as editor of the journal since its beginning, I have tried to maintain high editorial standards and contribute thus to the respectability of the systems movement.

In 1974, I also received numerous invitations to lecture on my research. This gave me an opportunity to meet various groups involved in the systems movement. My travelling even intensified in 1975–1976, when I was a Fellow at the Netherlands Institute for Advanced Studies in Wassenaar (NIAS) during my first sabbatical. I visited many groups in virtually all countries in Western Europe as well as Japan, Israel, and Iran. Everywhere, I found a great interest in systems ideas and the emerging field of systems science. Through an exchange of information about the activities of each group, I felt that I acted as a catalyst, building a network of these groups and contributing to the general awareness of the rising systems movement.

Of all the groups I encountered during the mid 1970s, only three were established and active organizations supporting systems research and cybernetics: the Society for General Systems Research (SGSR), the Dutch Society for Systems Research, and the Austrian Society for Cybernetic Studies. I developed a strong relationship with all three and, consequently, became involved in preliminary discussions regarding the desirability of founding an international federation of the existing societies.

My main involvement in the systems movement began in 1977, when I organized a large, NATO-supported International Conference on Applied Systems Research at SUNY-Binghamton.[18] At that conference, I was asked to serve as Managing Director and Vice-President of the SGSR. I accepted and spent the next eight years in various administrative functions at the very center of the systems movement. Although this involvement was at the expense of my research, I felt that the systems movement was at a critical crossroad and I was in a position to influence it in the right direction.

The most important development during my tenure as Managing Director and, later, President of the SGSR was the foundation of the International Federation for Systems Research (IFSR), which originally consisted of the SGSR, and the Dutch and Austrian Societies. The IFSR was legally registered in Austria on 10 April 1980, under a special agreement with the Austrian Government. The Government provided the IFSR with offices and financial support. This was an important symbol: the international systems movement was recognized, supported, and finally had a home. It was also symbolic that the first IFSR offices were in Laxenburg, just across the street from the International Institute for Applied Systems Analysis (IIASA).

I was selected the First IFSR President. Colleagues who served with me on the initial IFSR Board were the key players who had made this achievement possible: Gerrit Broekstra, Brian Gaines, Franz Pichler, Robert Trappl, and Gerard de Zeeuw. The following are the main initiatives that the IFSR undertook during my tenure on the IFSR Board: publishing a Newsletter with a large circulation; sponsoring an International Conference on Systems Methodology (Washington, DC, 5–9 January 1982); establishing the prestigious W. Ross Ashby Memorial Lecture Series; preparing and publishing bibliographies in the field of basic and applied general systems research,[43] as a continuation of the first publication of this sort we prepared at SUNY-Binghamton[26]; and publishing its own journal—*Systems Research*, as well as its own *International Series on Systems Science and Engineering*.

Too many other things happened during my service in the SGSR and IFSR (1977–1984) that I cannot cover this period adequately in this short autobiographical article. To some extent, the main events are captured in my reports in the *SGSR Bulletin* (Vols 8–10) and, later, in the *IFSR Newsletter*.

Although I have considerably reduced my involvement in the systems movement since 1984, I am still active in some functions supporting the movement. I continue my editorship of the *International Journal of General Systems* and the *IFSR Book Series*, and I serve on the SGSR Board of Trustees. I have also taken a major role in the development of *Systems & Control Encyclopedia*,[39] a monumental eight volume information resource, the first of its kind in systems science.

Wholes and Parts

In 1975–1976, when I was a Fellow at NIAS during my first sabbatical, my initial goal was to advance and refine the GSPS framework, and to write an updated version of my *Approach to GST*.[4] This goal was not realized at that time since I was sidetracked by the notorious problem of the relationship between wholes and parts. I resumed work on the GSPS framework seven years later, again at NIAS, during my second sabbatical.

The whole-part relationship had been on my mind prior to 1975, since at least the days of K. Vasspeg. I was familiar with a paper by Ashby[3]—the only paper that discussed this relationship in an operational way. When I re-read the paper in Fall 1975 and also came across a related newer paper by Madden and Ashby,[31] I became so fascinated by the issues raised but not solved in these papers that I started to work on them at the expense of my intended work on the GSPS framework.

Using the existing GSPS framework, I first formulated the whole-part relationship in terms of a relationship between an overall behavior system and its various subsystems. Two problems emerged from this formulation. One is a simplification problem: the aim is to replace a given overall system with its subsystems, as small as possible, that are adequate to reconstruct the overall system, to an acceptable degree of approximation, solely from the information contained in the subsystems. This problem is now called a *reconstruction problem*. The second problem involves the identification of an unknown overall system from the knowledge of some of its subsystems that are given; it is called an *identification problem*.

These two problems involve a number of challenging philosophical, mathematical, and computational issues. They have been of great interest to me since I recognized them in 1975. My first paper on the subject[17] also stimulated the interest of many researchers, who later extended or complemented my own research, most notably Broekstra, Cavallo, Cellier, Conant, Hai, Higashi, Jones, Krippendorff, Mariano, Parviz, Pittarelli and Uyttenhove. When the issues involved became well characterized, Roger Cavallo and I coined the name *reconstructability analysis*, under which we intended to subsume all aspects of a methodology pertaining to the reconstruction and identification problems. This name is now generally accepted as a technical term.

My own contributions to reconstructability analysis are primarily foundational.[23,28] I developed a complete characterization of reconstruction hypotheses in terms of a lattice that is based on a natural refinement ordering among the hypotheses. A reconstruction hypothesis is basically a family of subsystems of a given overall system that does not contain any subsystem that itself is a subsystem of another one in the family. I also developed various procedures for searching the lattice as well as procedures for evaluating the reconstruction capabilities of the individual hypotheses. For the identification problem, I formulated the concept of a reconstruction family—the set of all overall systems implied by the given subsystems. I also introduced the notion of identification uncertainty and investigated the issue of selecting one overall system from the reconstruction family. Furthermore, I developed a conceptual basis for dealing with possible inconsistencies among the given subsystems.

Both the reconstruction and identification problems are computationally very hard. As a consequence, the applicability of reconstructability analysis has been limited to systems with a modest number of variables, say a dozen or less. Recent developments of some powerful heuristic procedures by Roger Conant[6] seem to indicate, however, that systems with hundreds and, possibly, even thousands of variables could be analyzed on existing large computers. Thus, 24 years after the publication of the seminal paper by Ashby[3] and 12 years after my first paper on reconstructability analysis,[17] some powerful tools for dealing with the whole-part relationship are finally emerging.

Inductive Modelling

In the process of developing the GSPS framework, various types of systems problems emerged. Some of them were well recognized and methodologically developed; other problems were new or, at least, methodologically underdeveloped. In my research, I have always been tempted to work on the latter problems.

Most of my work on systems methodology has been devoted to the various problems associated with data-driven systems modelling. These problems, which involve climbing up the epistemological hierarchy of systems, are usually subsumed under the name *inductive modelling*. I define inductive modelling as a blend of discovery and postulational approaches to modelling in which our background knowledge is employed for restricting the class of possible models while empirical evidence is utilized inductively for determining a particular model (or a set of admissible models) from the delimited class.

I have always considered systems modelling as the central problem area in systems science. In my opinion, it encompasses both systems inquiry and systems design. In both cases, we attempt to construct systems (at appropriate epistemological levels) that are adequate models of something of our interest: either some aspects of nature or some aspects of desirable man-made objects. In systems inquiry, we construct models for the purpose of understanding the phenomenon of inquiry, making adequate predictions or retrodictions, controlling the phenomenon in a desirable way, and making appropriate decisions; in systems design, we construct models for the purpose of prescribing operations by which a desirable artificial object can be made to satisfy objective criteria within given constraints.

I am aware of several reasons for my continuing interest in inductive modelling. I find that inductive modelling is largely neglected by other systems researchers. Consequently, it is a source of many challenging research problems, some intimately connected with advances in computer technology. Another reason is my fascination with the amazing capabilities of higher-level living organisms to learn and reason inductively, which enable them to function as anticipatory systems. My interest in inductive modelling is also a natural consequence of my instinctive distrust of postulated models. They are often deceivingly pleasing from mathematical, aesthetic, or other points of view while, at the same time, have little relevance to the real world. In the economic, social, and political spheres, a strict adherence to postulated models (utopias) combined with little respect for evidence has repeatedly led to tragic consequences and enormous human suffering. As a boy of 16, I had already experienced two such social tragedies in Czechoslovakia (the Nazi occupation in 1939 and the imposition of Stalinist policies in 1948).

One of my contributions to inductive modelling is the set of procedures I developed for the reconstruction problem. By solving this problem, a justifiable transition is made from a behavior system to a structure system. Since information in the derived structure system is not entailed in the given behavior system, the reconstruction problem is one component in the whole process of inductive modelling. Another contribution is the methodology I developed for the selection of optimal masks, which is essential for inferring appropriate behavior models from given data. For some peculiar reason, the important concept of a mask is largely neglected by other systems researchers.

In the early 1980s, I discovered a potentially powerful new principle for inductive modelling. This discovery was made accidentally when I inspected the results of simulation experiments, which were intended to characterize an heuristic procedure for solving the reconstruction problem. This principle, based on reconstructability analysis, reveals that a *justifiable novelty* (something not explicitly in the data) can be produced at the expense of pure conformation to data. I conceive this novelty production mechanism as follows: although the novel information is not available in the given data explicitly (by definition), it is encoded in their reconstruction properties; reconstructability analysis enables us to determine these reconstruction properties (by identifying the superior reconstruction hypotheses) and, consequently, it enables us to decode this implicit information.

I speculated about this *reconstruction principle of novelty production* during my SGSR Presidential Address in 1982.[20] Later, the principle was validated and its applicability delimited by extensive simulation experiments.[25] In general, this principle is a powerful tool for inductive modelling with small data sets. Further work is still needed, however, to develop its full potential.

Uncertainty, Information, Complexity

In my work on inductive modelling, I first considered only deterministic systems, capable of reproducing the given data exactly. Soon I realized that this restriction was not realistic in inductive modelling; I concluded that it would lead not only to overly complex models, but also to models with questionable credibility. This conclusion was later reinforced by related work of Brian Gaines[8,9] and Judea Pearl.[34]

After I rejected this restriction, I had to deal with uncertainty (in prediction or retrodiction) embedded in all nondeterministic models. First, I employed probability theory—the classical mathematical apparatus capable of characterizing situations under uncertainty. Later, I tried to explore new avenues opened by the concepts of fuzzy sets[47] and fuzzy measures.[40]

I observed that inductive modelling involved almost invariably an interplay between two types of problems: (i) *ampliative reasoning*—a reasoning in which conclusions are not entailed in the given premises; and (ii) *simplification*—a reduction of complexity of a given system to a desirable level. To deal with these problems, I conceived two complementary principles of uncertainty. For ampliative reasoning, it was a *principle of maximum uncertainty*: maximize, within the constraints of the given premises, the uncertainty associated with the conclusion. For the simplification problem, it was a *principle of minimum uncertainty*: minimize the increase in uncertainty when simplifying a system.

To make these principles operational, a measure of uncertainty was needed. A well-justified measure of uncertainty was readily available for probabilistic systems: The Shannon entropy.[38] I labored to develop my uncertainty principles in terms of the Shannon entropy, only to learn later that a great deal of work in this respect had been done by other researchers, especially by Ronald Christensen.[5]

In the early 1980s, I became interested in conceptualizing nondeterministic systems in terms of possibility theory.[48] Initially, I encountered great difficulty in determining a justifiable measure of uncertainty applicable to possibilistic systems. First, I investigated various *measures of fuzziness* of a possibility distribution.[12] Then, I searched for a possibilistic counterpart of the Shannon entropy. I discovered later that this effort was ill-conceived; I should have searched for a generalization of the Hartley measure of uncertainty instead.[11] Eventually, after a period of frustrating trials and errors, Masahiko Higashi and I derived a function that possessed all desirable properties of a possibilistic measure of uncertainty.[13] We called this function the *U-uncertainty*.

It was obvious to me when we derived the U-uncertainty that it could also be applied to an alternative interpretation of possibility theory, the theory of consonant (nested) belief and plausibility functions introduced by Glenn Shafer as a subset of his evidence theory.[37] Then, by a simple consideration, the U-uncertainty could be generalized to arbitrary belief and plausibility functions.[37] When investigating properties of the generalized U-uncertainty, I came to an unexpected conclusion: the U-uncertainty and the Shannon entropy measure totally different types of uncertainty. This conclusion was independently confirmed by

Ronald Yager.[46] It became evident that the U-uncertainty measured the degree of *nonspecificity* in characterizing uncertain situations, while the Shannon entropy measured the degree of *dissonance*.

Since 1985, I have concentrated almost exclusively on a thorough investigation of the various types of uncertainty. This period has been rich in important new results: Matthew Mariano and I conceptualized a branching property for a possibilistic measure of uncertainty and employed it to prove the uniqueness of the U-uncertainty[27]; Arthur Ramer proved the uniqueness of the generalized U-uncertainty (a measure of nonspecificity) in evidence theory[35]; the Shannon entropy was shown to bifurcate into two distinct measures when generalized from probability theory to evidence theory[7]; and some new properties of the various measures of uncertainty were discovered. Most of these results are included in a textbook I coauthored with Tina Folger.[24] It evolved from a set of classnotes for a course that I have taught at SUNY-Binghamton since the early 1980s. In this course, I intentionally combine fuzzy set theory with information theory.

The more I investigated the concept of uncertainty and the associated concept of information, the more obvious it became that these concepts were multidimensional. Consequently, the uncertainty principles must be formulated as multiple objective criteria optimization problems. A proper development of these principles and their applications to various types of systems problems is the kernel of my current research activities.

Systems Science Manifesto

When I started to work on the GSPS framework in the mid 1960s, I did it because I believed that such a framework was a necessary foundation for establishing and developing a prospective science of systems. The vision of such a science, now referred to as *systems science*, was the dominant force shaping my whole professional life. I must admit, however, that this vision was initially quite hazy. It became sufficiently clear only after some 15 years, when my epistemological hierarchy of systems types was virtually completed and I had several years of experience with our systems science program at SUNY-Binghamton.

Around 1980, I felt that my comprehension of systems science was sufficient to warrant publication of a statement expressing my views in this regard—sort of my personal systems science manifesto. A good opportunity opened in May 1981, when I was invited to visit the Soviet Academy of Sciences, and also to deliver a major lecture at a *Symposium on the Role of Science in the Post-Industrial Society* in France. I presented my views about systems science on both of these occasions. The actual manifesto was first published in Russian in 1983[21]; its somewhat modified form, focusing more on the connection of systems science with the information society, was published two years later in *Systems Research*.[22]

I argued in my manifesto, as I would argue now as well, that systems science is a legitimate science since it has its own domain of inquiry and knowledge pertaining to this domain, and its own methodology and metamethodology. The domain of systems science consists of those properties of systems and associated problems that emanate from the general notion of systemhood, as introduced recently by Robert Rosen.[36] To characterize the domain more specifically requires a comprehensive framework (such as the GSPS framework) by which the full scope of systems we recognize is conceptualized in terms of appropriate categories of general systems. The knowledge regarding each of these categories can be obtained in some cases mathematically, but more often experimentally, by designing, performing and analyzing experiments with systems of the given category on a computer or in some other experimental setting. In this respect, I view the computer as a systems science laboratory.

Systems science has also its own methodologies, which I view as coherent collections of methods for dealing with those types of systems problems that emanate from a particular conceptual framework. Furthermore, systems science has its own metamethodology. Its purpose is to determine characteristics of individual methods (such as computational complexity, performance, and range of applicability) and utilize these characteristics for selecting the right method for a given problem in a specific context.

In spite of all its science-like characteristics, I argue in my manifesto that systems science is not a science in the ordinary sense, but rather *a new dimension in science*. Each system developed as a model of some phenomenon in any traditional science represents knowledge pertaining to that science. Knowledge in systems science is not of this kind. Rather, it is knowledge regarding knowledge structures, i.e., certain specific categories of systems. Hence, experimental objects in systems science are not objects of the real world, but rather abstractions, which we are able to make real on the computer.

I do not claim that my systems science manifesto is endorsed by the entire systems science movement. I know, however, that the legitimacy of systems science is accepted by more people now than some 20 years ago, when the possibility of a science of systems first occurred to me. I am grateful that I have had the good fortune not only to observe the emergence of systems science during my lifetime, but to take an active part in this very process.

References

1. W. R. Ashby, *Design for a Brain*. John Wiley, New York (1952).
2. W. R. Ashby, *An Introduction to Cybernetics*. John Wiley, New York (1956).
3. W. R. Ashby, Constraint analysis of many-dimensional relations. *General Systems Yearbook* **9** (1964), 99–105.
4. K. L. Boulding, General systems theory—the skeleton of science. *Mgt Sci.* **2** (1956), 197–208. (Reprinted in *General Systems Yearbook* **1** (1956), 11–17.)
5. R. Christensen, *Entropy Minimax Sourcebook* (4 Vols). Entropy Ltd, Lincoln, MA (1980–81).
6. R. C. Conant, Extended dependency analysis of large systems. Part I: Dynamic analysis; Part II: Static analysis. *Int. J. Gen. Syst.* **14** (1988), 97–142.
7. D. Dubois and H. Prade, Properties of measures of information in evidence and possibility theories. *Fuzzy Sets Syst.* **24** (1987), 161–182.
8. B. R. Gaines, On the complexity of causal models. *IEEE Trans. Syst. Man. Cybernet.* **SMC-6** (1976), 56–59.
9. B. R. Gaines, System identification, approximation and complexity. *Int. J. Gen. Syst.* **3** (1977), 145–174.
10. B. R. Gaines, An overview of knowledge-acquisition and transfer. *Int. J. Man-Machine Studies* **26** (1987), 453–472.
11. R. V. L. Hartley, Transmission of information. *Bell Syst. Technical Jl* **7** (1928), 535–563.

12. M. Higashi and G. J. Klir, On measures of fuzziness and fuzzy complements. *Int. J. Gen. Syst.* **8** (1982), 169–180.
13. M. Higashi and G. J. Klir, Measures of uncertainty and information based on possibility distributions. *Int. J. Gen. Syst.* **9** (1983), 43–58.
14. G. J. Klir, *An Approach to General Systems Theory*. Van Nostrand Reinhold, New York (1969).
15. G. J. Klir, *Introduction to the Methodology of Switching Circuits*. Van Nostrand Reinhold, New York (1972).
16. G. J. Klir, Editorial. *Int. J. Gen. Syst.* **1** (1974), 1–2.
17. G. J. Klir, Identification of generative structures in empirical data. *Int. J. Gen. Syst.* **3** (1976), 89–104.
18. G. J. Klir (ed.), *Applied General Systems Research: Recent Developments and Trends*. Plenum Press, New York (1978).
19. G. J. Klir, Forward. In R. Conant (ed.), *Mechanisms of Intelligence: Ashby's Writings on Cybernetics*. Intersystems, Seaside, CA (1981).
20. G. J. Klir, On systems methodology and inductive reasoning: the issue of parts and wholes. *General Systems Yearbook* **26** (1981), 29–38.
21. G. J. Klir, Systems science: a new dimension in science (in Russian). In *Systems Research: Methodological Problems*, pp. 61–85. Nauka, Moscow (1983).
22. G. J. Klir, The emergence of two-dimensional science in the information society, *Syst. Res.* **2** (1985), 33–41.
23. G. J. Klir, *Architecture of Systems Problem Solving*. Plenum Press, New York (1985).
24. G. J. Klir and T. A. Folger, *Fuzzy Sets, Uncertainty and Information*, Prentice-Hall, Englewood Cliffs, NJ (1988).
25. G. J. Klir and B. Parviz, General reconstruction characteristics of probabilistic and possibilistic systems. *Int. J. Man-Machine Studies* **25** (1986), 367–397.
26. G. J. Klir, G. Rogers and R. G. Gesyps, *Basic and Applied General Systems Research: A Bibliography*. SUNY-Binghamton, Binghamton, NY (1977).
27. G. J. Klir and M. Mariano, On the uniqueness of possibilistic measure of uncertainty and information. *Fuzzy Sets Syst.* **24** (1987), 197–219.
28. G. J. Klir and E. C. Way, Reconstructability analysis: aims, results, and open problems. *Syst. Res.* **2** (1985), 141–163.
29. J. Klir and M. Valach, *Cybernetic Modelling* (in Czech). SNTL, Prague (1965). English translation: Iliffe Books, London, U.K. and D. van Nostrand, Princeton, NJ (1967).
30. J. Klir and L. K. Seidl, *Synthesis of Switching Circuits* (in Czech). SNTL, Prague (1966). English translation: Iliffe Books, London, U.K. and Gordon & Breach, New York (1968).
31. R. F. Madden and W. R. Ashby, On the identification of many-dimensional relations. *Int. J. Syst. Sci.* **3** (1972), 343–356.
32. R. A. Orchard, On an approach to general systems theory. In G. J. Klir (ed), *Trends in General Systems Theory*, pp. 205–250. Wiley-Interscience, New York (1972).
33. R. A. Orchard, E. J. Reese and M. R. Tausner, On the foundations of knowledge engineering. In M. S. Elzas, T. I. Oren and B. P. Zeigler (eds), *Modelling and Simulation Methodology: Knowledge Systems Paradigms*. North-Holland, Amsterdam (1988).
34. J. Pearl, On the connection between the complexity and credibility of inferred models. *Int. J. Gen. Syst.* **4** (1978), 255–264.
35. A. Ramer, Uniqueness of information measure in the theory of evidence. *Fuzzy Sets Syst.* **24** (1987), 183–196.
36. R. Rosen, Some comments on systems and systems theory. *Int. J. Gen. Syst.* **13** (1986), 1–3.
37. G. Shafer, *A Mathematical Theory of Evidence*. Princeton Univ. Press, Princeton, NJ (1976).
38. C. E. Shannon, The mathematical theory of communication. *Bell Syst. Tech. J.* **27** (1948), 379–423, 623–656.

39. M. G. Singh (ed.), *Systems & Control Encyclopedia: Theory, Technology, Applications* (8 Vols). Pergamon Press, Oxford (1987).
40. M. Sugeno, Fuzzy measures and fuzzy integrals: a survey. In M. M. Gupta, G. N. Saridis and B. R. Gaines (eds.), *Fuzzy Automata and Decision Processes*, pp. 89–102. North-Holland, Amsterdam (1977).
41. A. Svoboda, Synthesis of logical systems of given activity. *Trans. IEEE Electronic Computing* **EC-12** (1963), 904–910.
42. A. Svoboda, Behaviour classification in digital systems. *Inf. Proc. Mach.* **10** (1964), 25–42.
43. R. Trappl, W. Horn and G. J. Klir, *Basic and Applied General Systems Research: A Bibliography* (1977–1984). IFSR, Laxenburg, Austria (1984).
44. K. Vasspeg, What is cybernetics—and what it is not? (in Czech). *Vesmir* **43** (1964), 89.
45. K. Vasspeg, On organizing of systems. *Inf. Proc. Mach.* **11** (1965), 167–176.
46. R. R. Yager, Entropy and specificity in a mathematical theory of evidence. *Int. J. Gen. Syst.* **9** (1983), 249–260.
47. L. A. Zadeh, Fuzzy sets. *Inf. Control* **8** (1965), 338–353.
48. L. A. Zadeh, Fuzzy sets as a basis for a theory of possibility. *Fuzzy Sets Syst.* **1** (1978), 3–28.

14

Methodology in the Large: Modeling All There Is

Brian R. Gaines

Introduction

> . . . in the three decades between now and the twenty-first century, millions of ordinary, psychologically normal people will face an abrupt collision with the future. (Ref. 58, p. 18)

> Man not only exists but knows that he exists. In full awareness he studied his world and changes it to suit his purposes. He has learned how to interfere with 'natural causation,' insofar as this is merely the unconscious repetition of immutable similars. He is not merely cognizable as extant, but himself freely decides what shall exist. Man is mind, and the situation of man as man is a mental situation. (Ref. 31, p. 11)

> The quantification of nature, which led to its explication in terms of mathematical structures, separated reality from all inherent ends and, consequently, separated the true from the good, science from ethics. (Ref. 38, p. 122)

In April 1982 a conference concerned with Model Realism took place at Bad Honnef in Germany. It was remarkable because the organizer, Horst Wedde, attempted to force comparability between the different approaches to modeling proposed by asking all participants to illustrate their methodologies applied to one of three well-defined case histories. The papers and commentaries given at the conference are available in the book *Adequate Modeling of Systems*.[61] This paper is based on an evening address given by the author in a wine cellar as part of the lighter side of the conference. It attempts to put our endeavours to create increasingly *real* global models in the wider perspective of man's search for meaning, illustrating the general points made by quotations from the book

Groping in the Dark[40] which is based on comparisons of seven global models at the Sixth IIASA Symposium on Global Modelling.

Throughout recorded history the human mind has been inventive of a multitude of descriptive and prescriptive methodologies to explain observed phenomena, predict possible worlds, determine which ones should be made real, and bring this about. Until this decade all such methodologies suffered from the intrinsic tunnel vision of human computation which can see relatively few of the consequences of action. What we modeled and simulated were micro-worlds, often carved out from reality in the laboratory, and whilst this enabled great leaps forward to be made in the technologies of the artificial, it usually neglected the natural or subsumed it only by destroying it.

We can comprehend much of our technological universe only because it is designed to be comprehensible, a world of simplicity in a universe of chaos. The powerful models of linear systems theory work not because they reflect reality but rather because we have built worlds of mechanical and electronic systems which are linear, and hence can be modeled, designed and controlled. Outside technologically created reality linear systems theory has far less to offer in modeling the worlds of nature.

In recent years the computer has enhanced our capability to project the detail of possible worlds and widen our vision of the consequences of our actions. We bring together the physical and human variables, political and economic policies, and constraints of resources, and expect increasing model realism. With the ever-increasing pace of advances in computer technology we have begun to see the possibility of large-scale methodologies that eventually cope with *all there is*. One can envision a next generation of global models that encompass broad socioeconomic scenarios such as Kondratiev's[41] and Toffler's[59] *waves*, extending the simplified and aggregated *global models* of today[40] to a level of economic, psychological and physical detail as at great a depth as required.

However, the momentum of our new intellectual technologies should not blind us to the foundational weaknesses of our endeavours. A major lesson of all the global models of today is that we are the creators of reality, not its discoverers, and the presuppositions that we make increasingly determine the worlds that we find ourselves in. The utopias of yesterday[25] may seem foolish in their naive hopes but our computational scenarios of tomorrow, for all their quantitative detail, are only computing the consequences of the assumptions we have made according to the rules we have built in.

This presentation is concerned with placing our endeavours for model realism within the socioeconomic sphere within the much broader context of humankind's endeavours for model realism in general. One conclusion of the IIASA Symposium on Global Modelling was: "The methodological problems of global modellers are common to all social-system modellers."[40]

One conclusion to be drawn from this paper will be that *the methodological problems of global modellers are common to all modellers*, noting that the most outstanding characteristic of humankind is that we all are, and always will be, modellers.[57] We have also always been concerned with modeling *all there is* but it is only now, as we use machines with greater capacity than ourselves, that we can really appreciate that when, or if, we do this we may not have the capacity to understand what we have done let alone check its validity; whatever that means.

The next section expresses the main philosophical problems underlying our situation in a parable, and then goes on to sketch some of the solutions proposed for them in the past.

Pluralities of Realities

> ... though all our knowledge begins with experience, it by no means follows, that all arises out of experience. For, on the contrary, it is quite possible that our empirical knowledge is a compound of that which we receive through impressions, and that which the faculty of cognition supplies from itself. (Ref. 33, p. 1)
>
> Law is the Reflection of Appearance into identity with itself. (Ref. 28, Vol. II, p. 132)
>
> The expectation of my death as a definite departure (from the life world) also arises out of my existence in the intersubjective world. Others become older, die, and the world continues on (and I in it) . . . I become older; thus I know that I will die and I know that the world will continue. (Ref. 56, p. 47)

What is the method by which we can model reality, or, at least, realistically model? For an individual the world presents few philosophical problems. He, she or it, can follow his, her or its own process of knowledge acquisition. Each moment brings a little more information and a closer glimpse of reality. The summit of knowledge, to know all, is like a mountain peak; perhaps never to be climbed but the path through the foothills is inexorable and obvious.

Let us assume that the individual is myself happily meandering through the paths of experience, when I suddenly become aware of you. You, that is, not just as part of the scenery and the experience, but you as an analog of me. I make the conceptual leap that identifies you as having processes of knowledge comparable with mine. Now I begin to have problems. Your paths through the foothills are different. Are they better or worse than mine? If there are two of us, why not more? Suddenly I notice the others, me, you and them, all taking different directions.

After a while I realize that people are born and die, and that I also was born and will die. Will I reach the summit before I die? If I do what will I do with my knowledge of reality? I am now thankful for the others because I can pass the knowledge on. Reality will survive me. I also realize how much I am relying now on the experience and knowledge of others. Much of my own knowledge is now not from my own experience but from the experience of others. The search for reality has become a communal endeavour and reality itself, given the death of individuals, has become meaningful only in a communal context.

All these realizations have taken place within a framework of a common quest undertaken through a common approach. Then I notice that one of the others is not only taking a different direction but is exploring the world using entirely different methods. He floats by in a hot air balloon. Will he reach the mountain top and see the whole of reality before us?

At some point in my pondering I may discover one of the fundamental laws of reality, that *things may be enumerated in three different ways; there is zero, one or infinity of everything*. There was one of me, and then I found you, so there must be an infinite number of epistemological beings. There could also be none of us. That is an awesome possibility. What is left if we all go, and what happens to reality? There was one method of discovering reality, and then this fellow came by in a balloon, so now there are two, and hence an infinity. Suppose there are none. The mountain summit cannot be reached and our effort is meaningless. What then?

But what of reality itself? If there are many searchers and many methods of searching then may not there be many realities? And sure enough, as I watch, the sun comes over the mountain and I glimpse many more mountain peaks in the distance.

This parable of the wanderers in the foothills of the mountains captures most of the philosophical problems of reality. None of them have ever been solved. We can be confident that none of them ever will be. In recent years Niklas Luhmann has put the dilemmas that this entails in system-theoretic terms and used them as an underlying dialectic from which to derive the operation of society:

> The world is overwhelmingly complex for every kind of real system. . . . Its possibilities exceed those to which the system has the capacity to respond. A system locates itself in a selectively constituted 'environment' and will disintegrate in the case of disjunction between environment and 'world'. Human beings, however, and they alone, are conscious of the world's complexity and therefore of the possibility of selecting their environment—something which poses fundamental questions of self-preservation. Man has the capacity to comprehend the world, can see alternatives, possibilities, can realise his own ignorance, and can perceive himself as one who must make decisions. (Ref. 37, p. 6)

He also emphasizes the problems created for the modeller of there being other modellers:

> We invoke a whole new dimension of complexity: the subjective 'I-ness' of other human beings which we experience (perceive) and understand. Since other people have their own first-hand access to the world and can experience things differently they may consequently be a source of profound insecurity for me. (Ref. 37, p. 6)

This 'sense of insecurity' was profoundly expressed some 2300 years ago by a discussant of modeling problems who, in his simile of the man who went outside the cave and saw the *reality* behind the shadows on the wall, remarked about the others in the cave: 'if anyone tried to release them and lead them up, they would kill him'.[46] Some of the 'friendly remarks' by commentators at the modeling conferences may be taken as the modern, more civilized, version of such interaction.

It is not easy to accept the insecurity and apply it to one's own model of the world. We prefer to stick to our 'known' models and not regard them as assumptions:

> Many plans, programs and agreements, particularly complex international ones, are based on assumptions about the world that are either mutually inconsistent or inconsistent with physical reality. Much time and effort is spent designing and debating policies that are, in fact, simply impossible. (Ref. 40, p. 16)

The law of reality stated above will be used as the theme of the next three sections, showing how the three ways of enumerating methods of reality each have their major proponents, strengths and weaknesses.

ZERO: There is no Method of Reality

> . . . there is nothing in any object consider'd in itself, which can afford us a reason for drawing a conclusion beyond it. (Ref. 29, p. 139)

> Scepticism is *not* irrefutable, but obviously nonsensical when it tries to raise doubts where no questions can be asked. (Ref. 64, Sec. 6.51)

> I found Hume's refutation of inductive inference clear and conclusive. (Ref. 47, Chap 1, IV)

That there is no method of reality has been the sceptical position throughout the ages, leading in its most extreme forms to total nihilism. The first clear statement of nihilist scepticism is attributed to Gorgias, but probably predated him and has been many times rediscovered. Gorgias held:

1. Nothing exists.
2. Even if something did exist it could not be known.
3. Even if it were known this knowledge could not be communicated.

This is a particularly attractive argument sequence because it can so easily be extended:

4. Even if it were communicated this communication could not be understood.
5. Even if it were understood this understanding could not be utilized.
6. Even if it were utilized this utilization could not be beneficial.

And so on!

The clearest, and most convincing, statement of the sceptical position is that of the Roman philosopher Sextus Empiricus in his *Outlines of Pyrrhonism*,[13] who discusses a form of non-nihilist scepticism originating with Pyrrho of Elis but substantially developed by many subsequent philosophers into a methodology of thought and decision based on the *suspension of judgement*. Pyrrhonists based their suspension on some 10 rules which seem fresh and cogent today, e.g.: I suspend judgement; I determine nothing; to every argument an equal argument is opposed; regression *ad infinitum* is necessary in any form of explanation not based on dogmatism.

It is easy to dismiss the **ZERO** hypothesis as being absurd and offensive to commonsense. Wittgenstein countered Hume's doubts about information from the past relating to events in the future, noting that the pattern of reasoning involved was prior to that used by Hume and transcended it:

> If anyone said that information about the past could not convince him that something would happen in the future, I should not understand him. (Ref. 65, Sect. 481)

However, Hume, although often vilified, has never been answered in his own terms:

> To refute him has been, ever since he wrote, a favourite pastime among metaphysicians. For my part, I find none of their refutations convincing; nevertheless, I cannot but hope that something less sceptical than Hume's systems may be discoverable. (Ref. 53, Chap. XVII)

and new defences of the sceptical position are still being published.[60]

The sceptical position itself seems to offer only disillusionment, that we "sit down in forlorn Scepticism" because we have departed "from sense and instinct to follow the light of a superior Principle" and "a thousand scruples spring up in our minds concerning those things which before we seemed fully to comprehend" (Ref. 2, p. 45) and Russell terms Hume's scepticism: "the bankruptcy of eighteenth century reasonableness" (Ref. 53, p. 645).

It is, however, this same dissolution of illusion, the ripping of the veil of maya, the dynamic bankruptcy that leaves us with all false currency spent and only new beginnings before us, that is the vital force of scepticism as a genesis for knowledge. This dialectical significance of scepticism shows up in Popper's reply to Hume, based not on answer but acceptance. He re-establishes an empiricist epistemology on the possibility of laws being *falsified* but accepts the Humean position that they cannot be verified: "We must regard all laws or theories as hypothetical or conjectural; that is, as guesses" (Ref. 49, Chap. 1.6).

Descartes[10] rediscovered scepticism as the tool of ultimate doubt that removes all but the essence of reality. Sartre[54] continues in the Cartesian tradition with his emphasis on *néantisation* as the force behind the transcendent upsurge of consciousness that makes knowledge possible. As Catalano remarks in his commentary on *L'Etre et Néant*:

> when I ask, 'What is a tree?' I remove, or negate, the tree from the totality of nature in order to question it as a distinct entity. Also, when I question the nature of a tree, I must have a certain 'distance' within myself that allows the tree to reveal itself to me. It is this 'nothingness' within myself that both separates the tree as this thing within nature and allows me to be aware of the tree. It is this break with a causal series, which would tie being in with a being in a fullness of being, that is the nothingness within man and the source of nothingness within the world. (Ref. 7, p. 66)

This line of argument could be developed and exemplified further but enough has been said here to illustrate the role of what Margaret Wiley[63] has termed *Creative Scepticism* and illustrates with literary examples as well as those from Eastern and Western philosophy. It is not the nihilist scepticism of Gorgias that became the dogmatic scepticism of many later philosophers. This is self-defeating because the positive affirmation of non-existence is itself subject to scepticism. It is rather the Pyrrhonism propounded by Sextus Empiricus that suspends belief, searches out opposites, quests for truth through balance rather than dogma, and holds the manner of quest itself subject to doubt at the very moment that truth appears to have been found.

In practical modeling the **ZERO** hypothesis has a key role in allowing us to break out of self-consistent systems that somehow do not work or, more insidiously, that do work but not as well as they could. In general it is the *tried and trusted* rule which generates the biggest explosion of novelty under the fuse of doubt. It is the *strong point* of an argument that yields most under a sceptical attack. We should doubt that which we find most efficacious, and disbelieve that which seems most obvious.

In this day and age Kuhn's[35] *normal science* proceeds at such a rapid pace that the consequences of an argument, its verification through a wealth of exemplars, and its practical utilization through implementation in systems, are as good as over once begun. We consolidate innovation to form dogma at a pace that allows little scope for contemplative imagination. The circle is no longer open than it is complete again. With the advent of the computer this tendency becomes amplified since computers are generally programmed to be the ultimate dogmatist, propounding incessantly and without variation those dogmas that have been set into them through software. Even the fortuitous processes of evolution cannot be used to break out of such algorithmic dogma.[4] It is active scepticism that must, in Popper's words: 'replace routine more and more by a critical approach,'[50] and somehow we have to find ways to embed it in our reality-modeling systems.

Methodology in the Large

In the world of global modeling the key role of the active critical scepticism of the Science Policy Research Unit at Sussex is widely acknowledged and has been made a feature of the SARU model development (Ref. 40, p. 66). Donella Meadows notes in relation to the Forrester/Meadows model: 'We would recommend to model sponsors that every model be fully documented and then given to its strongest ideological enemies for testing' (Ref. 40, p. 131).

ONE: There Is One Correct Method of Reality

> When all the Conditions of a Fact are present, it enters into Existence. (Ref. 28, Vol. II, p. 105)
>
> Very good! What has the Absolute Idea and idealism to do with it? Amusing, this 'derivation' of . . . existence . . . (Ref. 36, p. 147)

The **ONE** hypothesis has its dynamics and its dangers fully equal to those of **ZERO**. The great significance of existence hypotheses and existence proofs and the key role they play in mathematics is always something of a surprise to those who meet it for the first time. To go from knowing nothing about A to knowing that A exists may seem a very small step on the path to those who wish to know that A actually *is*. However, an existence proof is often sufficient in its own right to lead to a derivation of the properties of A and even a construction of A itself. The line of argument involved is the of the form:

(i) A exists
(ii) Any A must P
(iii) B does P
(iv) No other entity does P
(v) Hence B is A.

It is interesting to note that the obvious temptation to put this into symbolic logic in the form of the classical predicate calculus must be resisted. This is because step (ii) is not adequately captured by the statement:

$$(\text{ii}') \quad \forall A P(A)$$

since we have the standard result, $\forall A P(A) \supset \exists A P(A)$, that is, (ii') presupposes (i), whereas (ii) itself is intended to be independent of the truth of (i). We can state that *all unicorns have horns* without having claimed that *a unicorn exists*. It is clearly desirable that this pattern of reasoning be adequately formalized, and Schock[55] has given an exposition of the problems involved and some of the solutions developed.

Returning to the argument sequence stated above, we can see that its significance lies in the fact that given only A exists, and that A has the property P, we may find out under some circumstances precisely what A actually is. Somehow the necessity of existence of A has generated a complete ontology of A. The danger is that a false hypothesis of existence can lead through a weak and obvious property to a strong ontological result. The strength of such fallacious reasoning is that the existence hypothesis itself appears to have little content; certainly too little to be responsible for that of the result derived from it.

The classic example[19] of the misapplication of the argument above is:

(I) There exists a largest positive integer.
(II) The square of any integer is greater than or equal to it. The square of the largest integer cannot be greater than it so that it must be equal to it.
(III) 1 squared equals 1.
(IV) No other positive integer squared equals itself.
(V) Hence the largest positive integer is 1.

Only the first step, (I) the existence hypothesis, is false in this line of argument. From the supposition that a largest positive integer exists we have managed to determine precisely what it must be.

Note also the key role of step (IV). In the example given step (IV) may be proved explicitly. However the **ONE** hypothesis gives us both existence and unicity without any further requirement for proof; steps (I) and (IV) in the argument are available for free. Essentially, the **ONE** hypothesis says that if we can find an agreed property that A must have to be termed A, and we can find an actual entity B that has that property, then there is no need to perform any further tests of B to verify that it is A, nor any need to look for alternatives to B to falsify that it is A. Without further activity we may say that B necessarily is A.

The **ONE** presupposition often turns up in modeling literature as an assumption of the existence of a *unique optimum solution* to a problem, e.g., we will determine the best model of this economy. There may be no such best entity because the decision criterion cannot be uniformly satisfied, and even if there is one it may not be unique. These various possibilities show up as an ambiguity in the use of the word *optimum*:

*Opt*1: an optimum solution is one such that no other is better.
*Opt*2: an optimum solution is one that is better than all others.
*Opt*3: an optimum solution is one that is better or equal to any other.

The three definitions coincide under conditions of unicity but not necessarily otherwise. To differentiate between them we have to enlarge our vocabulary and call *Opt*1 *admissible*[16] rather than optimum; the non-existence of better solutions may be due to incomparability. *Opt*2 would be called a *unique optimum*, leaving *Opt*3 as the correct precisiation of *optimum* (reading *correct* here as *agreed by convention* since any of these definitions may be taken as precisifying the colloquial term optimum).

In the control literature lack of appreciation of these distinctions has several times resulted in the publication of extensions to the Pontryagin maximum principle which purports to show that it is applicable to discontinuous decision spaces also. Such forms of *discrete maximum principle* are however incorrect; the proofs incorporate tacit assumptions of results from the continuous case that do not carry over to the discrete case. One of the most powerful features of continuity is the well-ordering it establishes in solution neighbourhoods, and this is what allows Pontryagin's formulation but no discrete equivalent.

As a further example of the role of unique existence hypotheses in modeling, categorial adjunctions may be seen as arising essentially through the unicity of a pair of reciprocal functors. The Goguen/Arbib/Ehrig behaviour/structure adjunction[17] encompassing a wide

Methodology in the Large

range of system *identification* schemes is dependent on the existence of a unique structure ascribable to an observed behaviour. Attempts to determine a similar adjunction for stochastic systems were doomed to failure because no comparable unique solution was definable. However, the metasystemic move to define the *solution* in terms of the *admissible space* of structures has allowed Ralescu[51] to express behaviour/structure transformations in the stochastic and fuzzy cases as adjunctions because the *admissible space* is itself unique.

It is the presupposition of **ONE** that most often leads to fruitless searches for solutions that result in the conclusion that *the problem is insufficiently well-defined*. What we mean by *well-defined* seems to be the existence of a unique solution. However, it should be clear that problems can be solved in some sense without necessitating unicity of solution, and thus that some problems may be solved even though they are *ill-defined*. Indeed, requiring them to be precisified to a state of well-definition in this sense may destroy the essence of the problem:

> complexity and precision bear an inverse relation to one another in the sense that, as the complexity of a problem increases, the possibility of analysing it in precise terms diminishes.[67]

However, although one may point to the problems that **ONE** causes, one should not be blind to its virtues. The defence of a false theory against a powerful attack on its strong points can generate precisely the environment in which new ideas are generated. Certainly many good ideas are not developed as early as they might be because their originators drop them prematurely, only to see others regenerate them later and show that superficial weaknesses overlay great strength. Defending weak positions is often infinitely more rewarding than buttressing up strong ones. As Kenneth Boulding[3] has noted, in system theory one must be 'willing to make a fool of oneself'.

In the search for models of reality, **ONE** has its virtues. Even if we are dissatisfied with all existent theories and prepared to defend none, it is the belief that there is **ONE** that keeps us looking. Unified field theory for gravitational and electromagnetic forces, an organic basis for schizophrenia, controlled energy from thermonuclear reactions, and so on, are all of them unsolved problems, but the belief that a solution exists has made them inspirations for major fields of endeavour and achievement.

In global modeling it is an act of faith that the socioeconomic infrastructure has a rationale that can be modeled and a coherence that can be projected into the future. The artefacts inherent in **ONE** existence phenomena are widely recognized, however, and the modeling presuppositions and different axiological bases for the different schools of global modeling have been extensively documented:

> Here are some other acts of faith that regularly appear in global models of both the mathematical and mental kinds:
>
> 1. The poor nations of the world are developing in the same pattern as the Western industrialized nations, but after a time lag.
> 2. Political leaders are above the global system, outside it, making the important decisions affecting it, and not affected by it.
> 3. The most important phenomena in the world are economic and can be described in monetary units. (Ref. 40, p. 285)

MANY: There Is an Indefinite Variety of Methods of Reality

> To believe in the one or in the many, that is the classification with the maximum number of consequences. (Ref. 30, p. 64)
>
> The only principle that does not inhibit progress is: *anything goes*. (Ref. 15, p. 23)

The pluralist hypothesis is that which best summarizes actual modeling practice. The decisions to be made are usually highly overdetermined and skill in practical modeling comes from the ability to balance and make most effective use of a variety of different bases. This is not necessarily a problem of multiple criteria, but most often one of multiple information sources each of which, in theory, provides sufficient information for modeling.

A good example of this is the long-standing controversy over distance perception[24]: *What are the cues that people use in determining the distance of an object?* Experimenter X claims that phenomenon A is the prime determinant and demonstrates this by removing all cues but A. Sure enough distance perception remains and is highly accurate. Experimenter Y claims that phenomenon B is the prime determinant and demonstrates this by removing all cues but B. Sure enough distance perception remains and is highly accurate. Sooner or later, after the most refined experimental designs to ensure that no cues of type B are slipping in to confound those of type A, or vice versa, it is realized that not only are A and B each individually completely adequate distance cues, but the people subconsciously switch from one to another depending on which is available. At this point the excitement of controversy dies down, perhaps even the scientific research (the **ONE** hypothesis is highly important as a social dynamic), and a few patient researchers are left determining *all* the different, interchangeable, bases for distance perception.

William James saw pluralism as a healthy *tough-minded* approach and a key variable in philosophical positions. In recent years Gellner's[23] promotion of egalitarianism amongst all processes for the *Legitimation of Belief*, and Feyerabend's[15] promotion of scientific anarchism in *Against Method* have been the most robust arguments for pluralism in both principle and practice. Some of the precepts are:

> Science is an essentially anarchistic enterprise: theoretical anarchism is more humanitarian and more likely to encourage progress than its law-and-order alternatives. (Ref. 15, p. 17)
>
> Proliferation of theories is beneficial for science, whilst uniformity impairs its critical power. (Ref. 15, p. 35)
>
> Facts are constituted by older ideologies, and a clash between facts and theories may be a proof of progress. (Ref. 15, p. 56)

The real problem, once a pluralist basis for some aspect of practical modeling is found, is to determine how the many different bases are brought together to determine a single model when more than one is available, i.e. how is overdetermination resolved? This is a difficult point which is often missed. **MANY** seems to lack the dialectical strength of both **ZERO** and **ONE** because it allows for all possibilities and hence does not bring them into essential conflict. In terms of explanation this may be so. Your explanation is consistent and adequate, so is mine. We are both good fellows who do not need to fight but can revel in mutual self-satisfaction. However, in terms of explanation even, a meta-problem immediately arises as to how two explanations can account for a single phenomenon. Are they ordered in that one can be derived from the other, but not vice versa? Are they unrelated, in which case is there a deeper underlying explanation from which both may be derived? One of the rules of the scientific game is that, like acausality, plurality is not allowed except as a matter of short-term expediency.

In terms of the natural world, there is no rule of the game that says that a plurality of bases is not allowed. Moreover, there is no rule either that says that these bases cannot conflict. Generally they do. Over-determination in a precise theory leads to multiple values for essentially single-valued variables and hence conflict, paradox, and, if the rules of the theory are precisely applied, a total breakdown of the basis for modeling. In distance perception the possibility of such conflicts between the multiple bases of perception leads to optical illusions.[26] The related phenomenon of *reasoning illusions* in the practical reasoning is neglected in work on formal logic because the classical predicate calculus has the theorem:

$$(\forall P, Q)(P \& \sim P \supset Q)$$

i.e. that a breakdown of the law of contradiction may be used to derive any conclusion, and hence there is nothing that can be usefully said about this (in the same way that nothing can be said about existence). However, in practical reasoning we seem able to avoid the Wittgensteinian trap of knowing all the consequences of our premises—by using a more appropriate logic rather than just by not working out all possibilities[52,20]—and the mechanisms for conflict resolution are a key component of our systems of practical reasoning.

Thus **MANY** does have its own means of generating dialectical conflict and it is the most subtle and important of all. We have to accept as a basis for practical reasoning that multiple accounts of equal standing will arise and can be in conflict. Decision-making under uncertainty is usually seen as leading to underdetermination, but in practice it most often leads to overdetermination. Conflict because we are overprovided with information is far more prevalent than other sources of uncertainty where we have too little.

The variety of different global models now available, with different results based on differing presuppositions and value systems may seem confusing to those who wish to use them. However, this variety is an important lesson in itself: 'Perhaps one of the greatest lessons to be learned as the number of global models increases is how human beings from various parts of the globe see the world differently and how they see it commonly' (Ref. 40, p. 103).

Choice in the Foundations of Modeling

> All testing, all confirmation and disconfirmation of a hypothesis takes place already within a system. And this system is not a more or less arbitrary and doubtful point of departure for all our arguments; no, it belongs to the essence of what we call an argument. The system is not so much the point of departure, as the element in which arguments have their life. (Ref. 66, Sec. 105)

> ... cybernetics is the science or the art of manipulating defensible metaphors; showing how they may be constructed and what can be inferred as a result of their existence. (Ref. 43, p. 13)

The foundations of modeling have been dramatized here because they are worthy of it. If we are unaware of the seething conflicts below any theories, methodologies and practical schemes that we erect then we are not only guilty of that false peace of mind that stems from ignorance, but we are also missing out on that major element of choice that comes through conflict. If there are different presuppositions possible even at a truly foundational level, all of which are of equal merit (in the sense that they can be defended one against the other), then we have freedom of action in moving between them. It is our choice to be sceptical, to defend a unifying theory, to give equal status to mutually contradictory schemes.

The realization of the extent of choice enables them to be taken lightly. Practical modeling is sometimes a game against nature but most often a game against other modellers, and real games are most often *won* by changing the rules. Even in the *hot-war* against nature itself, the rules under which we play are of our own contrivance. It was the decision to consider the *impossible* concept of *action-at-a-distance* that enabled Isaac Newton to forecast the motions of apples and planets. It was the decision to place Ernst Mach's eyes in the *impossible* vehicle of a photon that forced Albert Einstein to distort the *certain* constancies of space. If this seems more a prescription for rhetoric than for modeling-science then so be it. If rhetoric were not so neglected a science the powerful analogy by which the whole of science is seen as the *persuasion of nature* would be more often used. In modeling this becomes more than an analogy because it is the *persuasion of the world* through action that is the key to our success.

In summary, these sections have pointed to two key dialectical conflicts in the foundations of modeling. In terms of the form of argument outlined in (I) to (V) above, there is first a conflict over step (I), existence: the Gorgian sceptic denies it; the **ONE** and **MANY** hypotheses both affirm it; the Pyrrhonian sceptic transcends all of them by suspending judgement. Secondly, the **ONE** and **MANY** hypotheses themselves come into conflict over step (IV), unicity. This is the classic conflict between the tendency to unify and that to split apart.

In any system of modeling that we build, both of these dialectical possibilities will be present, and in good systems they will be explicitly stated. The greater the awareness that we have of them, the more control we have over the possible choices they give, the more versatile and powerful the modeling system will be. The less we are aware of them, the more likely we are to ascribe valuable differences in approach to incompatible differences in content, and to fail to take advantage of multiple perspectives. As Meadows remarks:

> the single worst problem in the field of modeling at present is the inexperience of modelers in examining their own assumptions, realizing their lack of objectivity, and understanding the relative strengths and weaknesses of their methods. (Ref. 40, p. 217)

The Computer in World 3

> By great good fortune, and just in time, we have to hand a device that can rescue us from the mass of complexity. That device is the computer. The computer will be to the

Methodology in the Large

> organisation revolution what steam power was to the industrial revolution. The computer can extend our organising power in the same way as steam extended muscle power. . . . Of course we have to ensure that the result is more human rather than less human. Similarly we have to use the computer to reduce complexity rather than to increase complexity, by making it possible to cope with increased complexity. (Ref. 9, p. 18)

If we accept that many of the problems of global modeling reduce to those ontological, epistemological and axiological problems that have concerned philosophers for some thousands of years, then we have to ask 'what is new?' The sources of novelty in our current situation seem to be two-fold and related: firstly that population growth has forced the socioeconomic world to become florid with complexity, a tightly interlinked network of dependencies; and secondly that we have the computer in this present era as a mind-tool with which to extend ourselves to investigate this complexity. If Plato had been able to program the dynamics of the Republic on a personal computer then we might all still be citizens of that state under an algorithmic philosopher-king!

However, is the computer really a significant new methodological factor, and, if so, how can we bring its role into the debate? We can view the problem of model realism as one of linking abstract theories with concrete phenomena through the mechanism of our minds. The scientific tools of the past have been vehicles for exploring the phenomena of the physical world in greater detail and with greater precision; of opening up a channel between the physical world and that of our minds. The computer as a system for data acquisition, processing and control enhances these tools and widens that channel. However, it also has a quite different role as a system for simulation and model exploration; of opening up a channel between the modeling world and that of our minds.

This view of the computer as providing a vehicle for the exploration of the world of our models[18] may be formalized through Popper's[48] notion of *3 worlds*. In his autobiography he introduces it (Ref. 50, p. 143) by quoting Bolzano's notion of 'truths in themselves' in contradistinction to 'those thought processes by which a man may . . . grasp truths', proposing that: 'thoughts in the sense of contents or statements in themselves and thoughts in the sense of thought processes belong to two entirely different "worlds" and making the three-fold distinction: 'If we call the world of "things"—of physical objects—the *first world* and the world of subjective experience the *second world* we may call the world of statements in themselves the *third world* (. . . world 3)' (Ref. 50, p. 144).

Popper notes: 'I regard books and journals and letters as typically third-world objects, especially if they develop and discuss a theory' (Ref. 50, p. 145) and stresses the key role of world 3 in the development of human civilization, giving two gedanken experiments on the destruction of civilization to illustrate the status of world 3—if:

> (1) all machines and tools are destroyed, also all our memories of science and technology, including our subjective knowledge of machines and tools, and how to use them. But *libraries and our capacity to learn* from them survive . . . our world civilization may be restored . . . from the World 3 that survives
> (2) in addition *all libraries are destroyed* . . . men would be reduced to the barbarism of primitive man in early prehistory, and civilization could be restored only by the same slow and painful process that has characterized the story of man through Paleolithic times. (Ref. 48, p. 334)

Popper emphasizes the distinct ontological status of world 3:

> I regard the third world as being essentially the product of the human mind. It is we who create third-world objects. That these objects have their own inherent or autonomous laws which create unintended and unforeseeable consequences is only an instance (although a very interesting one) of a more general rule, the rule that all our actions have such consequences. (Ref. 50, p. 148)

The computer provides a new dynamic for world 3 just as did the harnessing of energy in world 1.[18] It brings world 3 into the demesne of man just as did the steam, internal combustion and jet engines for world 1. That is, we can move about in, conquer, control and fabricate to our needs the lands and materials of world 3 using computers in a way that makes our previous efforts, all but a few, look feeble. Those few we shall look back upon in wonder as we do the construction in world 1 of the Egyptian pyramids equalled in world 3 by Greek philosophy. However, such *impossible* achievements prior to the harnessing of inhuman energy and inhuman intellect will be surpassed in achievement, if not in wonder, through our control mechanisms that give us control of the worlds in which they exist: the energetic engines of world 1 and the informatic engines of world 3.

The role of computers in world 3 can be seen most clearly by contrasting information within a library with that in a computer database. The library itself is passive, waiting for scholars and technicians to tap its stored information, but powerless to process that information in any way, to classify it, extend it and correlate it, except through human mediation. The database contains the same information as the library but may also itself be active through processes that interact with that information without necessary human mediation, sifting through the stored data structures, analyzing and comparing information and building new structures to enhance and extend those already present.

The library is like a museum of preserved flowers, a static record of unchanging knowledge, whereas the database can be a living garden subject to growth and evolution, changing even as we study it. Less poetically, an active database system incorporating a complete library is the closest structure to a global model of all there is. Research on inference-based knowledge structures in fifth generation computing systems[42] is making the first steps in this direction but we still have a long way to go before the potential of such a database begins to be realized.

In case our enthusiasm for computers and world 3 becomes too great it is worth quoting some alternative viewpoints:

> the reification of abstract phenomena can be interpreted in psychiatric terms as schizophrenia, that is, as a kind of logical disease in which man constructs an abstract world but treats it as if it were real and concrete. (Ref. 68, p. 51)

The independence of these remarks of modern technology is clear in the remarks of a rather earlier writer:

> Leave us to ourselves, without our books, and at once we get into a muddle and lose our way—we don't know whose side to be on or where to give our allegiance, what to love and what to hate, what to respect and what to despise. We even find it difficult to be human beings . . . and are always striving to be some unprecedented kind of generalized human being. . . . Soon we shall invent a method of being born from an idea. (Ref. 11, p. 123)

Methodology in the Large

And the father of modern computer and communication technology warned us of its potential for abuse:

> the motive which the gadget worshipper finds for his admiration of the machine . . . is the desire to avoid personal responsibility for a dangerous or disastrous decision by placing the responsibility elsewhere: on chance, on human superiors and their policies which one cannot question, or on a mechanical device which one cannot fully understand but which has a presumed objectivity. (Ref. 62, pp. 59–60)

Certainly in global modeling with large and complex systems of equations having their own existence within a computer we can easily be guilty of any of these. However, the literature shows a high awareness of just such problems and the warnings are already part of it:

> Computer modeling techniques are also based on deep, unexpressed, technical paradigmatic assumptions that are not congruent with mental models and often constrain what can be included. (Ref. 40, p. 12)

What comes out of the model of the computer as a vehicle to explore world 3 is a more subtle criticism of our modeling to date. Namely that this very interactive exploration of our model spaces is just what is missing. We cannot through computer printouts, even in graphical form, get that essential gut-feel for our simulations that comes only in traversing them interactively. We have to project ourselves into the computer and become aware of the interplays involved by varying conditions and riding the waves thus created. This will involve fast simulation, effective multimodal information presentation, and the ability to play within our models freely and understandably. We should also be able to take others along with us on these voyages so that there is mutual understanding of the phenomena involved and the development of new shared meanings. It is this free exploratory conversational interaction with computers that makes the relationship *emancipatory cognitive* rather than *technical cognitive*.[27]

Systemic Foundations for 3 Worlds

> I have been considering whether anything in the world exists, and have observed from the very fact that I examine the question it necessarily follows that I do exist. (Ref. 10, p. 145)

> a universe comes into being when a space is severed or taken apart. . . . The act is itself already remembered, even if unconsciously, as our first attempt to distinguish different things in a world where, in the first place, the boundaries can be drawn anywhere we please. At this stage the universe cannot be distinguished from how we act upon it, and the world may seem like shifting sand beneath our feet. . . . Although all forms, and thus all universes, are possible, and any particular form is mutable, it becomes evident that the laws relating such forms are the same in any universe. (Ref. 5, p. v)

The notion of 3 worlds is attractive in giving us a framework in which to explore model realism, the role of ourselves and the role of the computer. It may be given systemic foundations by adopting the **ONE** position defined previously and defining Popper's 3 worlds by minimal distinctions necessary to generate them.[22] This, as is usual with **ONE**, leads a number of steps of reasoning that are blatantly fallacious by certain criteria. However, from an overall position that is based on **MANY** and conventionalistic, each of these false/true steps is a bifurcation generating various flavors of philosophy. **ONE** always skates over a great deal of very thin ice; the choice of where to fall in is part of the argument.

The fundamental postulate is that:

FP *There exists distinction-making, some necessary.*

This traps us (through ontological fallacies) into:

W3 *There exist distinctions—world 3*
W2 *There exist distinction makers—world 2*
W1 *There exists a source for necessary distinctions—world 1.*

The worlds arise through a similar chain of reasoning that since something exists it must exist somewhere. To term them all *worlds* hides a category error that undoes much of the beautiful analyses of Kant, Hegel and Peirce. However, the choice is open: 'to refuse to allow a distinction is also a meaningful action. . . . It does not allow changes in oneself, the world and others'.[21] It is useful to flatten the categories for the moment, and regard them as all being at the same level.

What we have at the end of this chain of 'reasoning' is a systemic framework for the three worlds. It is capable of being reorganized to match any metaphysical foundations. For example, the Cartesian argument doubts all, intuits **FP** through desperation, derives **W2** first through egocentricity, then gets **W3** from what is thought and **W1** from what is 'am'. The Hegelian problem is to justify **FP**; that need not be done for the purposes of this paper, or for the purposes of global modeling.

This definition of the three worlds through generative distinctions has proved useful in analysing a range of philosophical, theological, psychological and physical arguments. For example, the ascription of *reality* to *necessity* in defining world 1 goes back to the notion of the real world as that of absolute essences. However, in psychological terms it models the reification of that which we take to be necessary, phenomena outside our minds and beyond our choice. Accepted dogma that controls our actions becomes as much part of the physical world as gravity. We bump into a wall and it hurts us. We bump into a superordinate moral imperative and it hurts us. We cannot see through the wall but neither can we see through our own missing constructs.

Thus our systemic model of 3 worlds encompasses the distinction which Popper makes and gives new insight into the different interpretations possible for *reality*. In particular it enables us to account for the variety of criteria of *necessity* that people lump together in thinking about reality. Empirical testing is by no means the only test of scientific reality, and is itself open to the criticism of being *theory-laden*.[34] We need the untestable presuppositions of our ideologies to get started in our modeling, and even what is meant to be testable may not be in practice. Much of reality is validated by not being tested or by being tested in such a way that it cannot fail to be validated. As global modellers:

> We are not eager to put our hypotheses to stringent tests against the real world. We are only rewarded for being right. We are laughed at for being wrong. So we design our tests and select our results to maximize the possibility of being right.[40]

Some Exploration of 3 Worlds

Knowledge is an attitude, a passion. Actually an illicit attitude. For the compulsion to know is a mania, just like dipsomania, erotomania, homicidal mania: it produces a character out of balance. It is not at all true that a scientist goes after truth. It goes after him. It is something he suffers from. (Ref. 32, p. 97, attributed to Kierkegaard by Wilkins)

THROUGH ME THE WAY INTO THE WOEFUL CITY.
THROUGH ME THE WAY TO THE ETERNAL PAIN.
THROUGH ME THE WAY AMONG THE LOST PEOPLE.
... ABANDON EVERY HOPE, YE THAT ENTER. (Ref. 8, Canto III)

It is now possible to attempt a grand synthesis with the material at our disposal and set up a superordinate model of the board on which we play our game of life. Figure 1 shows Popper's 3 worlds as a Venn diagram in which the key aspects of the modeling debate relate to the dynamics of motion between the seven areas of intersection in this diagram. For clarity these areas are labelled in a blatantly simple-minded way which takes world 1 to be reality, world 3 to contain models of it and world 2 to contain the mediating processes whereby we both experience reality and comprehend models of it.

The central region where all three worlds intersect is that of *models* in world 3 that are *known* because they are in world 2 and *correct* because they correspond to the *reality* of

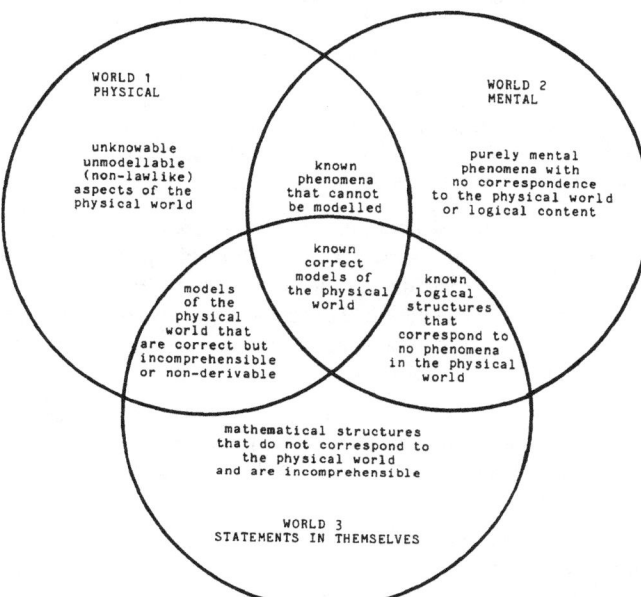

Figure 1. Venn diagram of Popper's three worlds.

world 1. This is the target region for our game and we aim to position our scientific persona in this region. This region is so sanctified by science that it deserves a truly glorious name and we may call it *paradise*.

The three intersections of two worlds outside this region each have interesting characteristics. Where world 1 intersects world 2 outside paradise is the domain of phenomena that can be experienced and known but not modeled. It encompasses the paranormal, luck, intuition, synchronicity and possibly some of the central dynamics of human existence such as agape and altruism. Where world 2 intersects world 3 outside paradise is the domain of formal structures that we can understand but correspond to no phenomena in world 1. It encompasses much pure mathematics but also art forms, myths, religions and cultural norms. Where world 3 intersects world 1 outside paradise is the domain of models of reality that are beyond our comprehension. It encompasses unified field theory and complete global models (unless both are myths properly belonging to the intersection of world 2 and world 3). In naming these regions we should take account of their transitory nature as way stations to paradise and see them as different forms of *purgatory*.

Most fascinating of all are those parts of the 3 worlds that intersect no others. World 1 outside purgatory and paradise is that of physical phenomena which cannot be experienced and cannot be modeled—some 99.9999 . . . % of the physical world. World 2 outside purgatory and paradise is that of mental events with no correspondence to reality and no logical content—some 99.9999 . . . % of our mental lives. World 3 outside purgatory and paradise is that of formal structures that are beyond the limits of our understanding and violate any correspondence principle—some 99.9999 . . . % of theorems that most axiom systems will generate. Such is the nature of *hell*.

Philosophies of science are concerned with the rules that define where a system of statements lies in this diagram. The positivists were concerned to restrict access to paradise to that which could meet very rigid entry conditions:

> We say that a sentence is factually significant to any given person, if, and only if, he knows how to verify the proposition which it purports to express—that is, if he knows what observations would lead him, under certain conditions, to accept the proposition as being true, or reject it as being false. (Ref. 1, p. 35)

The confirmationists allowed a creeping in or out of paradise as evidence built up for good or bad standing: 'A sentence is regarded as confirmable if observation sentences can contribute either positively or negatively to its confirmation' (Ref. 6, p. 59).

The falsifications on the other hand had an easy-going entry policy and concentrated on bouncing out those that proved unworthy, indeed they welcomed the most unlikely candidates as those most easy to throw out:

> The best hypothesis, in the sense of one most recommending itself to the enquirer, is the one which can be most readily refuted if it is false. This far outweighs the trifling merit of being likely. For after all, what is a likely hypothesis? It is one that falls in with our preconceived ideas.[45]

As scientists and global modellers we may all be seen as redemptionists searching through the three different forms of purgatory for candidates for paradise. One style is to attempt to drag the intersection of worlds 3 and 1 into world 2 by advancing our own comprehension. Another is to assume that there must exist rationales in world 3 for the phenomena in the intersection of worlds 1 and 2. Another is to search world 1 for phenomena corresponding to structures in the intersection of worlds 2 and 3. It is a significant question in evaluating any modeling school as to which approach, or mixture of approaches they are adopting. The existence presuppositions and corresponding artefacts are quite different: 'The global model you have decided you like best may tell you not so much about the world as about yourself' (Ref. 40, p. 106).

Conclusions

> Time present and time past
> Are both perhaps present in time future,
> And time future contained in time past.
> If all time is eternally present
> All time is unredeemable.
> What might have been is an abstraction
> Remaining a perpetual possibility
> Only in world of speculation. (Ref. 12, p. 13)

This paper has emphasized the many-faceted nature of reality and the choices that we have, that we are forced to make, before we can speak of model realism. One choice that has already been made by humankind is that we are not to be passive observers of reality but active participants in its development and change. Our technology has developed to a stage where our interference with natural reality is a major force for the future. Our improved models and powerful technologies give us increasing knowledge of reality and increasing power to change it. One message of global modeling has been that some highly possible futures held no place for people, largely through the actions of humankind.[44]

That is one of our choices, one of our distinctions, one of the value judgements we must make, whether to survive or not. Taking a **ONE** position and postulating our own existence in all possible futures is inherent in our natures but fallacious. Taking a similar position as a nation or as an individual and postulating our own supreme importance seems equally inherent in our natures yet equally fallacious. As we come to model *all there is* we are finding that the choices of today determine *all there may be* tomorrow, and that many 'obvious' choices that have served us well in the past have become inappropriate for the future.

Chris Evans advanced the argument that computer modeling of future scenarios in war games had been a major force in preventing world war 3, that human military experts when faced with the possibility of war are prone to assume that they can commence battle and win, but:

> when the same statistics are fed into the computer's unemotional, apolitical interior, what comes out is as true and objective an appraisal as can be made from the facts. Furthermore, whenever the data involves confrontation between nuclear powers, the unequivocal message that spills out—to both sides—is: You will lose! (Ref. 14, p. 212)

This argument involves great faith in both model realism and in human rationality but its logic is at the heart of most global modeling, that if we can project the implications of the present into the future then we can change our behavior to at least avoid the worst possible worlds.

Finally, at the end of a paper which is evidence of Goethe's remark that academics are just like dogs cocking their legs and dropping quotations on every possible occasion (Ref. 39, p. 186), let us end with a triply embedded quotation on the inevitability of our quest to model all there is. Evans quotes Korda's film of H. G. Wells *Things to Come*:

> there can be no rest, for once Man has taken the first step down the path of knowledge and understanding, he must take all those that follow. The alternative is to do nothing, to live with the insects in the dust. The choice is simple—it is the whole universe, or nothing. (Ref. 14, p. 245)

Acknowledgments

I am grateful to Susan Haack and Gordon Pask for their comments on an earlier draft of this paper: to Horst Wedde for giving me the opportunity to present it at Bad Honnef; to the referees for their stimulating comments: and to Mildred Shaw for her criticism of earlier drafts.

References

1. A. J. Ayer, *Language, Truth and Logic*. Victor Gollancz, London (1936).
2. G. Berkeley (1710), *The Principles of Human Knowledge*, G. J. Warnock, Ed. Collins, London (1962).
3. K. Boulding, General systems as a point of view. M. D. Mesarovic, Ed., *Views on General Systems Theory*, pp. 25–38. John Wiley, New York (1964).
4. H. J. Bremmermann, M. Rogson and S. Salaff, Search by evolution. M. Maxfield, A. Callahan and L. J. Fogel, Eds., *Biophysics and Cybernetic Systems*, pp. 157–167. Spartan Books, Washington D.C. (1965).
5. G. S. Brown, *Laws of Form*. George Allen & Unwin, London (1969).
6. R. Carnap, Intellectual autobiography. P. A. Schilpp, Ed., *The Philosophy of Rudolf Carnap*. Open Court, La Salle, Ill. (1963).
7. J. S. Catalano, *A Comment on Jean-Paul Sartre's "Being and Nothingness"*. Harper Torchbooks, New York (1974).
8. A. Dante, *The Divine Comedy—Inferno*, orig. *ca* 1300. trans. J. D. Sinclair. Bodley Head, London (1958).
9. E. De Bono, *Future Positive*. Maurine Temple Smith, London (1979).
10. R. Descartes, *Philosophical Writings*, orig. 1637, trans. E. Anscombe and P. T. Geach. Nelson, London (1954).
11. F. Dostoyevsky, *Notes from the Underground*, orig. 1864, trans. J. Coulson. Penguin, Middlesex (1972).
12. T. S. Eliot, *Four Quartets*. Faber & Faber, London (1959).
13. S. Empiricus, *Outlines of Pyrrhonism: I–III*, orig. *ca* 300 B.C. trans. R. G. Bury. W. Heinemann, London (1933).
14. C. Evans, *The Micro Millenium*. Viking Press, New York (1980).

15. P. Feyerabend, *Against Method*. NLB, London (1975).
16. B. R. Gaines, System identification, approximation and complexity. *Int. J. Gen. Syst.* **3** (1977), 145–174.
17. B. R. Gaines, General system identification—fundamentals and results. G. J. Klir, Ed., *Applied General Systems Research*, pp.91–104. Plenum Press, New York (1978).
18. B. R. Gaines, Computers and society: the basis for a systemic model. R. F. Ericson, Ed., *Improving the Human Condition: Quality and Stability in Social Systems*, pp. 523–530. Society for General Systems Research, Louisville, Kentucky (August 1979).
19. B. R. Gaines, General systems research: quo vadis? B. R. Gaines, Ed., *General Systems*. Vol. 24, pp. 1–9. Society for General Systems Research, Louisville, Kentucky (1980).
20. B. R. Gaines, Logical foundations for database systems. E. H. Mamdani and B. R. Gaines, Eds., *Fuzzy Reasoning and its Applications*, pp. 289–308. Academic Press, London (1981).
21. B. R. Gaines and M. L. G. Shaw, A programme for the development of a systems methodology of knowledge and action. W. J. Reckmeyer, Ed., *General Systems Research and Design: Precursors and Futures*, pp. 255–264. Society for General Systems Research, Louisville, Kentucky (January 1981).
22. B. R. Gaines and M. L. G. Shaw, Is there a knowledge environment? G. Lasker, Ed., *The Relation Between Major World Problems and Systems Learning*, pp.27–34. Society for General Systems Research, Louisville, Kentucky (May 1983).
23. E. Gellner, *Legitimation of Belief*. University Press, Cambridge (1974).
24. J. J. Gibson, *The Perception of the Visual World*. Houghton Mifflin, Boston (1950).
25. B. Goodwin, *Social Science and Utopia*. Humanities Press, New Jersey (1978).
26. R. L. Gregory, *The Intelligent Eye*. Weidenfeld & Nicolson, London (1970).
27. J. Habermas, *Knowledge and Human Interests*, trans. J. J. Shapiro. Heinemann, London (1968).
28. G. W. Fr. Hegel, *Science of Logic*, orig. 1812–1816, trans. W. H. Johnston and L. G. Struthers. George Allen & Unwin, London (1929).
29. D. Hume, *A Treatise of Human Nature*, orig. 1734, ed. L. A. Selby-Bigge. Clarendon Press, Oxford (1888).
30. W. James, *Pragmatism*, orig. 1907. Harvard University Press, Mass. (1975).
31. K. Jaspers, *Man in the Modern Age*, orig. 1931, trans. E. Paul and C. Paul. Routledge & Kegan Paul, London (1933).
32. H. F. Judson, *The Eighth Day of Creation*, Simon & Schuster, New York (1979).
33. I. Kant, *Critique of Pure Reason*, trans. J. M. D. Meiklejohn. George Bell, London (1897).
34. C. R. Kordig, *The Justification of Scientific Change*. D. Reidel, Holland (1971).
35. T. S. Kuhn, *The Structure of Scientific Revolutions*, 2nd edn. University of Chicago Press, Chicago (1970).
36. V. I. Lenin, *Conspectus of Hegel's Science of Logic*, Vol. 38, orig 1914–1916, trans. C. Dutt. Lawrence & Wishart, London (1961).
37. N. Luhmann, *Trust and Power*, orig. 1973, trans. H. Davis, J. Raffan and K. Rooney. John Wiley, Chichester (1979).
38. H. Marcuse, *One Dimensional Man: The Ideology of Industrial Society*. Sphere, London (1964).
39. D. Luke and R. Pick, Eds., *Goethe: Conversations and Encounters*. Oswald Wolff, London (1966).
40. D. Meadows, J. Richardson and G. Bruckmann, Eds., *Groping in the Dark*. John Wiley, London (1982).
41. G. Mensch, *Stalemate in Technology: Innovations Overcome the Depression*, orig. 1975, trans. C. Reade. Ballinger, Mass. (1979).
42. T. Moto-Oka, Ed., *Fifth Generation Computer Systems*. North-Holland, Amsterdam (1982).
43. G. Pask, *The Cybernetics of Human Performance and Learning*. Hutchinson, London (1975).
44. A. Peccei, *One Hundred Pages for the Future: Reflections of the President of the Club of Rome*. Pergamon Press, Oxford (1981).

45. C. S. Peirce, *Collected Papers I.*, C. Hartshorne and P. Weiss, Eds., orig. *ca* 1900. Harvard University Press, Mass (1958).
46. Plato, *The Republic*, orig. 380 B.C., trans. H. P. D. Lee. Penguin Books, Middlesex (1955).
47. K. R. Popper, *Conjectures and Refutations*. Routledge & Kegan Paul, London (1963).
48. K. R. Popper, Epistemology without a knowing subject. B. Van Rootselaar, Ed., *Logic Methodology and Philosophy of Science III*, pp. 333–373. North-Holland, Amsterdam (1968).
49. K. R. Popper, *Objective Knowledge*. Clarendon Press, London (1972).
50. K. R. Popper, Autobiography of Karl Popper. P. A. Schilpp, Ed., *The Philosophy of Karl Popper*, pp. 3–181. Open Court, La Salle, Ill. (1974).
51. D. Ralescu, A system theoretic view of social identification. R. F. Ericson, Ed., *Improving the Human Condition: Quality and Stability in Social Systems*, pp. 531–538. Society for General Systems Research, Louisville, Kentucky (1979).
52. N. Rescher and R. Brandom, *The Logic of Inconsistency*. Basil Blackwell, Oxford (1980).
53. B. Russell, *A History of Western Philosophy*. George Allen & Unwin, London (1946).
54. J. P. Sartre, *L'Etre et le Néant*. Gallimard, Paris (1943).
55. R. Schock, *Logics Without Existence Presuppositions*. Almquist & Wiksell, Stockholm (1968).
56. A. Schutz and T. Luckman, *The Structures of the Life World*, orig. *ca* 1950, trans. R. M. Zaner and H. T. Engelhardt. Heinemann, London (1973).
57. M. L. G. Shaw, *On Becoming a Personal Scientist*. Academic Press, London (1980).
58. A. Toffler, *Future Shock*. Pan, London (1970).
59. A. Toffler, *The Third Wave*. Bantam, New York (1980).
60. P. Unger, *Ignorance*. Clarendon Press, Oxford (1975).
61. H. Wedde, Ed., *Adequate Modeling of Systems*, pp. 100–111. Springer Verlag, Berlin (1983).
62. N. Wiener, *God & Golem Inc*. Chapman & Hall, London (1964).
63. M. L. Wiley, *Creative Scepticism*. George Allen & Unwin, London.
64. L. Wittgenstein, *Tractatus Logico-Philosophicus*, orig. 1921, trans. D. F. Pears and B. F. McGuiness. Routledge & Kegan Paul, London (1972).
65. L. Wittgenstein, *Philosophical Investigations*, orig. 1945, trans. G. E. M. Anscombe. Basil Blackwell, Oxford (1974).
66. L. Wittgenstein, *On Certainty*, orig. 1950, trans. D. Paul and G. E. M. Anscombe. Basil Blackwell, Oxford (1974).
67. L. A. Zadeh, Fuzzy languages and their relation to human intelligence. *Proceedings International Conference on Man and Computer*. Karger, Basel (1972).
68. A. C. Zijderveld, *The Abstract Society*. Allen Lane, London (1970).

15

Discrete and Continuous Models

Andrew G. Barto

Introduction

M. E. Van Valkenburg writes in the foreword to Steiglitz's *Introduction to Discrete Systems*,[1] that "Given the widespread availability of computers, there seems little doubt that the teaching of electrical engineering should undergo an evolution . . . In the emerging pedagogical approach, equations should be written in discrete form as difference equations, instead of in continuous form as differential equations. Indeed, equations should seldom be used, since principles should be stated directly in algorithmic form." Approaches having this character are increasingly being proposed not only in electrical engineering but in other fields where continuous mathematical methods have been traditional, and not only are pedagogical changes being suggested. Digital computing and discrete models are influencing our conception of real world systems and the role classical mathematical methods are to play in modelling them.

Discussions of this emerging approach occur frequently and continue to reveal strong biases which correspond to the technical backgrounds of the participants. Those whose experience is in the classical physical sciences or in traditional engineering disciplines where differential equations have been so remarkably successful, understandably have strongly developed intuitions in which continua and rates of change are powerful conceptual primitives. Those whose intuition has developed more directly under the influence of digital computing, on the other hand, find it very natural to think in terms of such concepts as algorithms, data structures, and automata. In many applications of these discrete concepts the availability of theoretical results is replaced by the computational power of digital computers.

Due to the relative isolation of these methodological traditions, both historically and in the educational process, there are aspects of continuous and discrete modelling techniques, and aspects of the relationship between these techniques, that are not generally recognized. Although the subject of numerical analysis focuses on the relationship between continuous and discrete methods, the perspective it provides lies thoroughly within the continuous

---

From *Int. J. Gen. Syst.* **4**, 163. Copyright © 1978 by Gordon and Breach Science Publishers Ltd.

tradition. While intimately concerned with digital computation, numerical analysis concentrates on going from a continuous model, e.g., a differential equation, to some discrete means of deducing that model's behavior. The issues that are important in numerical analysis are substantially different from those which arise from attempts to model *directly* in discrete form, bypassing continuous formulations entirely. In this direct approach the emphasis is on the relationship between some perceived real system and a discrete model, rather than between a continuous model and a discrete approximation of it. This approach has been convincingly put forward by Donald Greenspan and his associates in a series of articles[2,3,4,5] in which discrete models are proposed for the mechanical systems which physicists have classically modelled using differential equations. Not only are the discrete models easily simulated by computer, but they also preserve some of the theoretical attractiveness of the classical models. Although Greenspan's efforts have primarily been toward formulating (and simulating) discrete models of systems which are usually modelled by continuous methods, he points out, as do Zeigler and Barto,[6] that computational methods permit the exploration of entirely new classes of models.

The purpose of this article is to examine some of these issues from the point of view provided by system theory. One of the goals of the system sciences, as stated by Sutherland,[7] is to permit the properties of a particular problem rather than *a priori* methodological biases, to determine the analytical approach to be used, and going a step further, to prevent such biases from determining which problems are considered for analysis. These goals seem especially appropriate to the use of discrete and continuous models. Which approach is taken is usually determined by the background of a researcher or by the tradition prevalent within a discipline, rather than by which set of conceptual tools can provide the most expressive means of formalizing a model. Although the modelling formalism which most often comes to mind is the differential equation, system theory provides a framework for rigorously defining a much more general class of dynamical models. Since continuous methods have traditionally played the more visible role in mathematics, some of the material to follow, being well known in some quarters, is tutorial in nature. We hope our discussion dispels some of the apparent misapprehensions about the potential of discrete modelling techniques so that future discussions can focus on substantive criticisms of discrete modelling as it is now practiced.

The terms discrete and continuous apply to a wide range of structures and techniques, but for the purposes of this article they will refer to discrete and continuous representations of time as used in models of processes whose behavior unfolds over time. Models of temporal processes take the form of discrete-time or continuous-time dynamical systems. By the term discrete model, then, we shall mean a model formulated as a dynamical system having a countable time base (usually the integers). This kind of model might be formalized as a set of difference equations or as an automaton. The term continuous model will refer to a system whose time base is the uncountably infinite set of real numbers. Continuous models are usually, although not exclusively, formalized as differential equations. According to this terminology a discrete model can still involve uncountable sets, such as the real or complex numbers, as ranges of some or all of its descriptive variables. In other words, a discrete model (as we shall use the term) might be a discrete-time but continuous-state model. Distinctions between countable or finite sets and uncountable sets are clearly important when referring to other parts of a model besides the time base, especially since actual digital computers can manipulate only finite sets. However, the countable-uncountable distinction drawn between time bases classifies models in a way that most closely parallels in the classification arising in practice which distinguishes those who use differential equations from those who do not. We shall therefore focus on the use of different time bases, but much of what will be said is also applicable to the differences between discrete and continuous methods in general.

Modeling Traditions

On the surface, the distinction between countable and uncountable representations of time does not appear to present problems that are not already adequately treated by existing mathematical theory. Modern functional analysis subsumes both the discrete and continuous cases and makes explicit the algebraic and topological differences between spaces of countable and uncountable dimensionality. Whatever the usefulness of these results for modelling, the problems to which this article is addressed are not those of discrete and continuous *mathematics*, and we shall not touch on the logical status of the continuum nor, as we point out in the next section, on the direct empirical justification, or lack of it, for the use of continuous methods. Rather we address problems which have their roots in the isolation between discrete and continuous *traditions of model building*. The evolution of these traditions is attributable to the success of particular modelling formalisms for expressing hypotheses about observed regularities and to the successful development of techniques for deducing the consequences of these hypotheses. Discrete-time systems concepts evolved primarily as design tools for the digital technology emerging in the 1950's and as abstract models of computation. It became possible to develop methodologies for digital computing, both at hardware and software levels, that required no knowledge of the classical continuous models used so successfully in the physical sciences. The subsequent isolation is currently reflected in the very different characters of discrete-time and continuous-time theories, and in the fact that there are relatively few people who are comfortable with both.

The theory of discrete-time systems is very diverse, but it is generally characterized by an emphasis on the synthesis of a precisely specified process from a given set of primitives. For example, in switching theory one wants to construct a circuit having a specific behavior; in digital signal processing, one wants to design a filter with a certain frequency response; or, in computability theory, one wants a program using only certain instructions which computes a given function. At the risk of making too broad a generalization, the theoretical results thus involve the existence of systems with specific properties and synthesis procedures for completely describing these systems.

The emphasis in the continuous theory, on the other hand, has been on the behavioral analysis of general classes of systems having very high levels of mathematical structure. For example, systems defined by linear differential equations have much more formal structure than finite automata. Thus results exist which are both general and detailed and which relate the structure of these systems to their behavior and to the behaviors of structurally similar systems. With the exception the discrete counterparts of these linear results (about which we shall say more later), behavioral analysis methods for discrete-time systems apply to individual systems and not to general classes of discrete-time systems. One hopes, of course, that special classes of highly structured discrete-time systems will be identified, and that theories will be developed which do not as yet have continuous counterparts, but efforts in this direction are just beginning.

Even though discrete-time theory emphasizes synthesis and continuous-time theory emphasizes analysis, there are no mathematical or logical reasons for the separation of these theoretical orientations into two isolated traditions. In fact, there are numerous instances in which nearly identical mathematical structures are studied in virtual isolation from one another within different traditions. For example, automata (without the finiteness restriction) or sequential machines are, technically, essentially the same as what mathematicians call difference equations. The notations differ and usually difference equations involve more algebraic structure than the automata typically studied, but there is a one-one correspondence between these classes of structures. There is a similar correspondence between multidimensional difference equations of the kind used to approximate partial differential equations and the objects called cellular or tessellation automata. Another example is provided by the structures which are called recursive digital filters in the field of digital signal processing.[9] These objects are the same as linear sequential machines[10] which are the same as linear difference equations.

Despite their near formal identity, automata and difference equations are part of different traditions. Difference equations, although discrete, are more closely associated with continuous methods since they are usually studied as approximations to differential equations. Consequently the study of automata differs quite drastically from the study of difference equations. There are major differences in the relevant intuition, the theoretical results that are deemed important, and the subclasses that are deemed important, and the subclasses that are delineated in each area. However there are no logical obstacles to an integration of the intuition, theory, and applications of these areas. Similarly, there are no logical obstacles to the integration of discrete and continuous-time modelling formalisms and techniques. Although the system theory literature provides a framework for such integration as in Padulo and Arbib[11] and Zeigler,[12] it is not well-known among applications oriented model-builders and its consequences for modelling are just beginning to be explored.

The differences between discrete modelling and numerical analysis can be clearly seen in terms of the difference between modelling traditions. When a discrete model is derived to approximate the solution of a differential equation, the model is viewed within the tradition that surrounds the continuous modelling approach. The questions that are asked about discrete structures depend on their being viewed as approximations to continuous systems. For example, one is usually interested in error bounds and whether or not (and how) the behavior of the discrete model approaches the differential equation's solution as the step size, or mesh, converges to zero. These are questions about relationships between a continuous model and a discrete one, and the subject is usually subdivided on the basis of how discrete approximations are constructed given a differential equation rather than according to the behavioral properties of the discrete systems.

On the other hand, when a discrete model is formulated initially, it is not intended to be an approximation to a continuous model. In fact, it is often the intent of the model builder to represent aspects of a phenomenon that are considered as actually *being* discrete. The question of the desirability of a finer mesh or finer resolution level may not ever arise, and there may be no loss of information due to "undersampling" (a point we shall return to

Discrete and Continuous Models 381

later). Indeed, the model builder need not have any experience with continuous techniques and numerical analysis. Perhaps the most familiar instance of this is in automata theoretic models of digital computing devices. Since logic gates and digital memory devices stabilize at a time scale faster than that of the driving clock pulse, it is in most cases possible to completely ignore their behavior between clock pulses. Only the stabilized component states at the end of each time interval are relevant to the future behavior of the system. Going to a time "mesh" finer than the clock frequency introduces an entirely new order of complexity to the model that is simply not required for many design and analysis purposes. Indeed, the use of a continuous modelling formalism, even if solvable, would undoubtedly obscure the behavioral simplicities that are captured by the discrete-time model.

Thus, even though the direct formulation of a discrete-time model might result in a structure similar to one that could have been derived to approximate the solution of a differential equation, the fact that it was not so derived makes a major portion of numerical analysis irrelevant. Properties of discrete systems, e.g., stability, that are critical in numerical analysis may still be important, but the complex problem of determining whether or not they reflect properties of a differential equation's actual solution need not be faced. The crucial questions in the direct approach concern the validity of of the model in accounting for observed data.

Discrete-Time Models of Natural Systems

The example of discrete-time models of digital circuits raises several questions when it is suggested that similar methods may be generally applicable for modelling systems which aren't products of conscious human design. Some of these questions quickly lead to fundamental philosophical problems which, while being relevant in a wide sense to the entire enterprise of modelling, really need not be faced in discussing discrete and continuous modelling traditions. Some of the misapprehensions about discrete and continuous modelling can be attributed, we feel, to varying assessments of what a model's validity implies about "reality." When disagreement hinges on an issue of this kind, one becomes entangled in epistemological problems that have very long histories.[13] Model builders probably would not explicitly claim that time does or does not really flow continuously, but the degree of realism implied by such a claim does tacitly contribute to misunderstanding by being implicit in strong biases toward particular modelling traditions.

It can be argued, for example, that just because *nothing relevant* happens in a digital circuit between discrete time steps does not imply that *nothing at all* happens. In fact, talking about digital components stabilizing between time steps presupposes a temporal continuum, or, at least, a finer time scale at which the system's behavior can be described. This is indeed true for a digital circuit, and a similar argument might well be put forward whenever a system other than a purposefully synthesized digital system is modelled by a discrete-time system. If I am not mistaken, however, this argument is sometimes strengthened by modellers of the continuous tradition to the claim that approximations and/or omissions are *necessarily* present whenever the temporal dimension is represented by a structure other than the real numbers. Claiming that a continuous representation of time is in any sense necessary for the ultimate expression of natural regularities is tantamount to holding that nature itself is continuous, and we know much less about nature itself than whether it is discrete or continuous.

We're not saying that this realistic position is consciously held by modellers but only that it is reflected in remarks, often casually made, about discrete models. For example, after the description of a discrete-time model, it is not uncommon for an author to state that the discrete-time model represents an approximation to the differential equations which describe the actual dynamics of the system. It may well be true that such a differential equation model exists, but it would simply be another model useful for answering a possibly different set of questions. The term "dynamic" was originally associated only with continuous-time models since the differential equation was the only formalism available for modelling processes which unfold in time. We know now that the idea of dynamics is more general than the differential equation.

One possible reason for the belief that continuous representations are necessary for valid modelling was clearly expressed by A. N. Whitehead[14] in his description of a form of overstatement which he termed the "fallacy of misplaced concreteness." We have a natural tendency to overestimate the success of generalizations because we lose sight of the degree of abstraction involved. In other words, we tend to confuse models of experience with experience itself. Zeigler[12] makes a similar point when he cautions against confusing the "real system," by which he means the set of potentially acquirable data, and a "base model" of the real system. Any notion of structure or state of a system refers to some base model of the real system and not to the real system itself. Thus, what one means, for example, by the state of the real system is actually the state of a valid base model.

The disposition to reify abstractions is particularly strong when the abstractions have been very successful in accounting for observations, i.e., when models based on such abstractions have successfully undergone extensive validation testing and have displayed great predictive power. This is the case with differential equations. Models formulated as differential equations have been so successful that it is easy to overlook the considerable abstraction they entail. For example, it requires nontrivial assumptions to justify a differential equation model of a diffusion process which, modelled at a greater level of detail, is believed to arise from the interaction of discrete particles. Indeed, the classical diffusion equation would imply, if it were a valid model, that points in space which are arbitrarily far apart can instantaneously influence one another, an implication which contradicts other well established theory. The utility of the continuous formulation is obvious, but its range of validity should not be overestimated.

These remarks not only apply to dynamical models but also (and perhaps especially) to the set of real numbers. The very considerable abstraction involved in the construction of real numbers has unfortunately been obscured by treating real numbers as primitive concepts in introduction to calculus. Models formulated using real numbers and differential equations have had enormous success, but one unnecessarily limits conceptual range by concluding that models, to be valid, must be expressed within this tradition.

Another argument suggested by the digital circuit example is that since digital circuits manifest a simplicity when viewed at discrete instants of time by *design*, there is little reason to suppose that systems which are not so designed possess the properties necessary for valid discrete-time modelling. In other words, whether or not there is an underlying continuum, discrete-time models of naturally occurring systems are not likely to be valid even as far as providing *descriptions* of behavior at discrete times. This argument does not entail the same kind of overstatement as the first one. It is recognized that discrete methods *could* be applicable, but that by the nature of the phenomena to be modelled, discrete methods don't have the appropriate kind of expressive power. The evidence for this is not persuasive when one realizes that the kind of data with which continuous methods have found so much success is not the only kind of data we are capable of collecting. The inclination to reify models which are sufficiently valid has the additional consequence of limiting the kind of observations that are considered important, or worthy of investigation, or, in the extreme, worthy of perceiving. Not only is there a tendency to filter out data that is inconsistent with a popular model, but entire classes of observations which are neither inconsistent nor consistent with a model are excluded from consideration. Consequently, the apparent lack of properties which make discrete methods useful may be the result of observing only those aspects of a system which can be suitably modelled using differential equations. In Whitehead's words, "The concrete world has slipped through the meshes of the scientific net."[15]

In summary, our view is representative of the conception of modelling and model validity that is emerging from system theory. This view holds that a model's validity can only be discussed with respect to a particular set of attributes that are of interest to the modeller, that is, with respect to a specific "resolution level"[16] or a specific "experimental frame"[12] which characterizes the experimental access to a system. Thus, a model may be valid for some experimental frames and not for others, and there may be many valid models of whatever real system gave rise to a set of data. The validity of a discrete-time model does not, therefore, imply that time is discrete but only that the model faithfully summarizes a given set of data. Similarly, the validity of a continuous-time model does not imply that time is continuous. In fact, an analogy can be made between the development of this view and the change that has occurred in the way axioms and postulates are regarded. Wilder[17] discusses the gradual change, influenced by the invention of non-euclidean geometries, from the conception of axioms and postulates as *logical necessities* toward their being viewed as formal statements whose status as valid physical assertions is not a concern of mathematics. Similarly, it is possible to separate questions about the validity of dynamical models from questions about the ultimate nature of reality.

Discrete Functions

Introductions to discrete-time systems usually begin with a definition of a *discrete function*, or, using the terminology of signal processing, a *digital signal*. We shall briefly discuss discrete functions and their use in modelling in order to focus sharply on some of the central issues in discrete and continuous modelling. Let I be an finite set of contiguous integers such as, for example, $\{0,1,2,3\}$. A rule f that assigns to each $x \in I$ a value $y = f(x)$ (for our purposes y will be a real number) is a discrete function. We'll write $f: I \to R$, where R denotes the set of real numbers, to describe such a function. Thus, a discrete function is simply a real-valued function whose domain is a finite set of integers rather than the more familiar interval of real numbers. We've been somewhat arbitrary in choosing this definition since it would be useful to include discrete functions which take on values that are not real numbers or which have domains of higher dimension. It might also be useful to consider functions whose domains are abstract sets that are not connected in any way with the properties of the integers (e.g., order property). However our definition is sufficient for our purposes of comparing discrete functions with more familiar concepts.

The use of the notation $y = f(x)$ for representing values of a discrete function seems to be a very natural appropriation of the usual notation for functions of the reals. Actually, however, this notation represents an interesting and subtle alteration of the usual view taken of these structures. In more traditional approaches, the similarity between discrete functions and functions of the reals tends to be obscured by the practice of calling discrete functions *sequences* or sometimes simply *vectors*. For example, a vector $f = (f_0, f_1, f_2, f_3)$ is the same as the discrete function $f: I \to R$ where $I = \{0,1,2,3\}$ and $f(x) = f_x$. It's just more common in the discrete case to write arguments as subscripts and the function values as "coordinates." This very minor notational change is significant because it reflects a mingling of discrete and continuous mathematical traditions.

In modern functional analysis, real valued functions of the reals are viewed as vectors, an approach which represents the other side of this blending of traditions. A function $f: R \to R$ is viewed as a vector in an uncountably infinite dimensional linear space. Its values, $f(x)$, $x \in R$, are its coordinates in this space. The modern vector view of functions of the reals permits concepts which originated in a discrete or finite dimensional setting to be extended to uncountably infinite dimensional function spaces. A similar enrichment of discrete methods may occur by thinking of finite dimensional vectors as discrete functions.

In the study of finite-dimensional vectors, the index I is usually important only with respect to its size, which gives the dimension of the linear space, and as a set of labels for the coordinates (thus any other set of the same size would do). Any other structure that the set I may have, such as linear order or group structure, is important only in relatively advanced topics. Given the widespread knowledge of functions having intervals of the real numbers as domains, the functional notation, as opposed to the coordinate notation, makes it more natural to consider index sets (i.e., domains) that are more than just sets. In finite dimensional linear algebra, one does not discuss, for example, a "monotonically increasing vector" or a "linear vector" since these concepts depend on the index set I having various kinds of algebraic structure. A monotonic or a linear discrete function, on the other hand, can be directly understood by analogy with the corresponding concepts for functions of the reals.

This direct correspondence between the *concept* of a discrete function and the *concept* of a function of the reals should not be confused with any sort of correspondence between *particular* discrete functions and *particular* functions of the reals. In many applications it is very natural to regard a discrete function as having been derived by sampling from a function of the reals. For example, in digital signal processing, a digital signal (i.e., a discrete function) is often viewed as having been obtained from some underlying analog signal by a sampling process. In fact, the values of digital signals are usually called samples. But, as Steiglitz[1] points out, digital signals need not have been derived from any analog signal. He gives the example of a digital signal each value of which represents the total yardage of a football team in a particular game. There is one value per game and no analog signal is involved. Not only in practice are some discrete functions independent of any natural relationship with a function of the reals, but on the theoretical side, a discrete function can be viewed as an entirely independent, well-defined, and precisely manipulatable object. A discrete function is not, for example, a discontinuous and hence theoretically

Discrete and Continuous Models 385

troublesome function of the reals. The domain of a discrete function, as defined above, is a set of integers. The function is not defined for real numbers between the integers.

The above remarks can shed some light on the relevance to discrete and continuous modelling of what is known as the "sampling theorem." This result says, roughly, that if a real (or complex) valued function of the reals is smooth enough (i.e., lacks spectral components above a certain frequency), then sample values can be taken at small enough intervals so that no information is lost in the process, i.e., the exact original function can be recovered. Sampling at larger intervals loses information. The smoother the function is, the larger the sampling interval can be. This is an important result in many applications, but its significance as a universal principle should not be overestimated. It is tempting to conclude, by invoking the sampling theorem, that discrete systems are only capable of faithfully representing behavior that is sufficiently smooth. However, this conclusion cannot be justified. In its usual form, the sampling theorem refers to the process of sampling a function of the reals to produce a discrete function. For discrete functions that did not arise from such a sampling process, the issue of information preservation is not illuminated by the sampling theorem. Further, the relationship between smoothness and discreteness expressed by the sampling theorem is only one particular example of a range of other possible relationships between functions of the reals and discrete functions.

To understand this last remark it's necessary to characterize the kind of result the sampling theorem expresses. The process of sampling a function of the reals is clearly a many-to-one operation from the set of all functions of the reals to an appropriate set of discrete functions. In fact, an uncountably infinite number of functions of the reals are mapped to every discrete function by the sampling operation. The sampling theorem says that if the original function happens to be a member of the subset of sufficiently smooth functions, then it can be recovered from the resultant discrete function by a suitable operation. In other words, the sampling operation restricted to the set of "smooth functions" is inevitable and its inverse is known. Clearly, then, *any* function f of the reals can be recovered from its sampled version (whatever the sampling rate) provided that we have, in addition to the resultant discrete function, enough *other* information about f, namely that we know (1) that f belongs to a subset of functions on which the sampling operation is invertible and (2) we know how to compute the inverse operation for that subset. The set of "smooth" or "band-limited" functions is one such subset and is important in applications because membership in it is often a natural consequence of assumptions about the inertia of measuring instruments. Another such subset consists of functions that are non-zero only at the points to be sampled. Recovery in this case consists of simply converting the discrete function to the corresponding function of the reals whose value is zero between the sample points. No information is lost if we know that the original function f was a member of that subset. This is true even though such a function is never band-limited.

The assumption that a function has non-zero values only on a countable subset of the reals might also be a natural consequence of a modelling technique or measuring method. Discrete event systems, such as those discretely simulated using a simulation language like GPSS, can be thought of as continuous-time systems whose variables can change values only in discontinuous jumps. It's not generally a serious restriction to assume that the changes can occur at a known countable subset of the real numbers. Arrivals of customers at a bank might be represented by such a function.[12]

We emphasize, however, that it is important to distinguish between discrete-functions and functions of the reals which are non-zero at only discrete points. In fact, there is a discrete form of the sampling theorem based on the Discrete Fourier Transform[18] which sharply illustrates the magnitude of this distinction. A function of the reals which is non-zero only at discrete points is, as a function of the reals, extremely non-smooth. It has discontinuous jumps. However, it is possible to define what is meant by a sufficiently "smooth" discrete function in terms of its *discrete* Fourier components and apply the discrete sampling theorem. This result says that such a smooth discrete function can be sampled, say at every kth point, to produce a discrete function defined on a smaller domain from which the original function can be recovered by a suitable "smoothing" process. Thus, although a function of the reals having non-zero values at discrete points is very non-smooth, the corresponding discrete function obtained by sampling at these discrete points might be smooth in a discrete sense: its values at successive points not differing greatly. Of course our remarks about the possibility of other reconstruction techniques apply in the discrete case as well.

One of the most appealing aspects to the use of real valued functions of the reals is the ease with which their general properties can be grasped from graphical representations. Although graphs of discrete functions can be similarly displayed, they are not so well suited to the back-of-the-envelope figuring that plays an important part in using mathematical methods. The graph of a discrete function consists of all the ordered pairs $(x, f(x))$, $x \in I$, and might be displayed as in Figure 1. For small domains it's probably more natural to use the vector n-tuple notation. Thus, $(3, -4, 4.5)$ could be considered a kind of graph of the function $f: I \to R$ where $I = \{0,1,2\}$. One does not normally represent f's graph as in Figure 2. But this representation is clearly possible and is certainly more suggestive for vectors which are being viewed as discrete functions. Of course, digital computers can generate displays of discrete functions with no difficulty, and one can always pretend, in casual figuring, that a smooth curve represents the graph of a discrete function. It's noteworthy to relate a remark made by Greenspan[2] that if $I = \{k(10)^{-6} | k = 0, 1, \ldots, 10^6\}$, the graph of $f: I \to R$ given by $f(x) = x^2$, when drawn on a normal book page, is indistinguishable to the naked eye from the graph of $f(x) = x^2$ for x in the real interval $[0, 1]$.

This remark also illustrates the fact that it is perfectly feasible to use symbolic expressions to define discrete functions. For the formula $f(x) = x^2$, $x \in I$, to make sense it is only necessary that multiplication of elements of I is a meaningful operation and always results in an element of f's range. It is also possible to define operators on discrete functions in terms of symbolic manipulation of these formulae. Thus, by turning to discrete functions

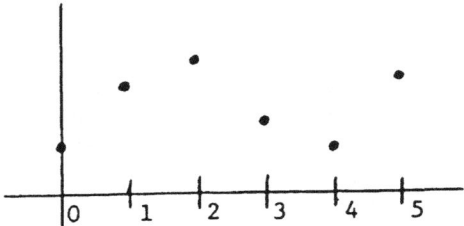

Figure 1. The graph of a discrete function.

Discrete and Continuous Models

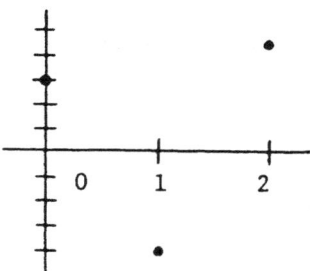

Figure 2. The vector $(3, -4, 4.5)$ could be graphed as a discrete function $f: I \to R$, where $I = \{0, 1, 2\}$.

one does not give up the possibility of concise symbolic expressions. One gains, however, the advantage that using symbolic expressions is not the only means of completely specifying functions as it is in the continuous case. Discrete functions can be completely defined by listing their values, e.g., storing the values in a computer so that "addresses" correspond to function arguments and "contents" correspond to function values, or by providing an algorithm whose input is a function argument and whose output is the corresponding value. The latter possibility is referred to, as in the opening quotation of this article, as replacing equations by algorithms. In a sense, of course, the description of an algorithm, or even the tabulation of a function's values, is a symbolic means of defining a function which, in principle, could be written as an equation. However, in its usual usage, the term equation refers to formulae consisting of constants, variables, the symbols for basic arithmetical operations, and a variety of other symbols to indicate differentiation, integration, etc. The primitive operations used in specifying algorithms (e.g., looping and conditional branching) permit the concise definition of functions which are impossible or very awkward to express by conventional algebraic means.

Symbolic and Computational Simulation

Differential equations and difference equations are both ways of specifying the constraints that are assumed to act locally in a system to produce its global behavior. For systems that describe temporal processes, these constraints act locally in time. A differential equation is used to formalize these local constraints as relations which must hold between the values of a system's attributes at any time and how these values are changing at that same time. For difference equations, the constraints are expressed as relations between present attribute values (and perhaps past values) and their values at the next discrete time step. In both the continuous and discrete cases, the objective is to determine what global behavior is implied by an initial condition and the uniform application of these local constraints at each point in time. The global behavior is the solution of the equation. The behavior is a function of the real numbers for a differential equation and a discrete function for a difference equation.

The term simulation usually refers to the process by which one determines a model's behavior from its structure, or, in the terminology used above, the method used to find a model's global behavior based on knowledge of an initial condition and the locally acting constraints. Since an equation's solution can be viewed as the behavior of a model specified by the equation, it is possible to think of solving an equation as a form of simulation. Although this term is usually applied only to discrete systems or to certain kinds of discrete approximations of continuous systems in which solutions are explicitly generated step-by-step, it is not misleading to think of symbolically solving an equation in closed form as a form of simulation. We can call this *symbolic simulation*. If, on the other hand, a computer is used to generate a model's behavior, we call the process *computational simulation*. It is more conventional to refer to computer simulation of models as computational *solution*, but the idea of simulation is more general than the concept of solution since the latter refers to equations, i.e., certain symbolic expressions, which are not the only means of specifying models. In many cases equations are not explicitly used since an algorithmic formulation may be more feasible or more intuitively appealing.

One of the major justifications for formulating models in terms of functions of the reals and derivatives is that there are many symbolic methods which can be useful for manipulating functions of the reals. Perhaps the most basic symbolic methods rest on the use of rules for differentiating or integrating functions by merely manipulating their formulae. The ability to symbolically determine and use derivatives and integrals helps make the notion of rate-of-change so powerful a conceptual primitive. In formulating a model as a set of differential equations, one hopes that symbolic methods will be useful in finding its behavior.

Although symbolic simulation techniques exist for certain kinds of discrete models, they are neither as well-known nor as well-developed as they are for continuous models. Simulation by computer is usually the method used to gain insight into a discrete model's behavior. This is especially true for automata theoretic models which are often expressed using set-theoretic language in which symbols for arithmetic operations have no meaning. The local constraints comprise a transition function which may be given by a table rather than by a formula. Simulation becomes, in effect, a series of table look-ups which can be performed exactly and quickly by digital computer. For discrete models having more algebraic structure, the transition function may be expressed by arithmetical formula which can be computed at each time step by means of arithmetic, or, as Greenspan[2] emphasizes, high-speed arithmetic. Other kinds of discrete models such as discrete-event models of queuing systems are also conveniently simulated computationally whether symbolic simulation techniques exist of not.

Computational simulation of discrete-time models is trivial compared to computationally approximating the behavior of a continuous model (although efficient computational simulation of large discrete models can involve many complexities). Hence, the fact that symbolic techniques are helpful for only the simplest, i.e., linear, differential equations is used to justify the direct formulation of discrete-time models and the complete by-passing of continuous models. Further, it is pointed out by Greenspan[2] that the experimental results which are to be modelled are originally discrete sets of data. "Theoreticians then analyze these data and, in the classical spirit, infer continuous models. Should the equations of these models be nonlinear, these would be solved today on computers by numerical methods, which results again in discrete data. Philosophically, the middle step of the activity

Discrete and Continuous Models 389

sequence . . . is consistent with the other two steps. Indeed, it would be simpler and more consistent to replace the continuous model inference by a discrete model inference. . . ."[2] Simulation is then performed arithmetically by a high speed computer. Many of the subtleties of numerical approximations of continuous models simply do not arise.

However persuasive this argument is, there is another aspect to the use of differential equations and symbolic methods which should be considered. A formula giving the solution of a differential equation does more than give a single behavioral trajectory of the model. By containing parameters used in the model description, a solution's formula can express the form of the behavior for a large class of initial conditions, forcing functions, and perhaps for a large class of related models. A computational simulation, on the other hand, produces a single trajectory for a single initial condition and forcing function. Symbolic methods give coherence to *classes* of models and help increase our understanding of systems by showing, concisely, what structural factors contribute to what behavioral characteristics. Indeed, in some cases it is possible to deduce certain properties of a system's general behavior without ever determining a single actual behavioral trajectory. A single computational simulation of a discrete model does not produce this kind of understanding unless it is one of a pattern of simulation runs designed to establish structural-behavioral correspondences through computational experimentation. This difference between computational and symbolic simulation is the same as the difference between arithmetic and algebra. Algebraic methods allow generalized arithmetic problems to be solved by separating the logical form of a problem from its specific arithmetical computations.

We do not mean to imply that symbolic techniques are applicable only to continuous-time models, and we shall discuss their use of certain types of discrete-time models later. Our point here is that the computational power so immediately available for discrete simulation makes it all too easy to obtain specific results without their contributing to an understanding of a system's dynamics. A specific discrete model's behavior is very easily generated, but it is a much more difficult task to develop a feeling of "why" the model behaved in a particular way. True, computational simulation can be applied where known symbolic methods are completely inadequate, but it is too pessimistic to argue that the only models which can be concisely understood are those specified by equations which have already been thoroughly treated by classical methods, i.e., by "simple" equations. It may in fact be tautological to say that symbolic solutions are possible only for simple (i.e., linear) equations since the simplicity of a process is directly related to concise expression of its regularities. To someone not knowing the linear theory, linear systems undoubtedly would appear to be quite complex.

High speed computational methods have the potential for helping us understand a much larger class of models by producing large numbers of specific results which can act as guides to theorizing. This is what von Neumann meant by the *heuristic use* of computers. He felt that by generating computer solutions to many specific equations, one might be able to discover general properties and develop a corresponding theory.[19] Although specific results rather than general theories are ultimately the aim of model building, the specific results which are most useful, judging from classical continuous modelling, are often those which follow from a general perspective about a class of related models.

Discrete-time models are very suitable for this approach to the use of computers, but it is also true that symbolic methods are applicable to those discrete models which are algebraically analogous to the continuous models which can be symbolically analyzed. In particular, linear difference equations can be understood as thoroughly as linear differential equations by the application of finite dimensional linear algebra (which is substantially simpler than the uncountably infinite dimensional linear algebra of functional analysis). It's true that the discrete linear theory lacks some of the subtlety of the continuous theory, but one wonders how much of this complexity is useful in modelling and how much has been generated by the theory itself. For example, the convergence of infinite sequences of functions of the reals is a central problem of functional analysis. Yet if experimental data is always a discrete set and only a finite number of experiments are ever performed, when does the convergence problem arise apart from in the analysis of an inferred continuous model? For many applications, the discrete linear theory suffices, but since the continuous theory is so readily available, it is automatically adopted whether its additional complexities are needed or not.

One example of this tendency to uncritically adopt continuous methods involves the use of methods based on Fourier analysis. These methods are often learned in a specific context so that an understanding of the principles used is tied to a specific formulation. Among the misconceptions arising in this way is the belief that functions need to be real or complex valued functions of one or several real variables in order for Fourier analysis techniques to be applicable. Less standard applications tend to be viewed as approximations to the continuous case or as necessarily derived from the continuous case. As a result, it is often felt that no real working knowledge can be achieved unless one feels comfortable with integrals, continuity, convergence properties, distributions and other concepts necessary for a thorough understanding of functions of real variables. However, as the discrete form of the Fourier Transform becomes more widely known it is becoming clear that a prior knowledge of the continuous theory is unnecessary.

The increased interest in discrete Fourier analysis is due to the widespread use of the Fast Fourier Transform (FFT) algorithm for computing the Discrete Fourier Transform (DFT). The DFT can be understood *completely* within an algebraic framework that involves none of the complexities (e.g., convergence and distributions) of the continuous Fourier Transform. *Any* real or complex valued discrete function of N points can be expressed *exactly* as a unique linear combination of the discrete functions $\phi_{(\cdot)}(x) = e^{2\pi i \omega x/N}$ where ω and x are integers between 0 and $N - 1$ (inclusive). If one is careful to remember that the convolution theorem for the discrete case refers to cyclic or circular convolution, then the DFT can be applied to discrete-time linear systems in the same way that the continuous Fourier Transform applies to continuous-time linear systems. Moreover, it is not necessary to know the continuous theory or even to know that the continuous theory exists.

Although the theory of discrete Fourier analysis is less complex than the continuous case, it developed much later than the continuous theory as a special case of the abstract theory of harmonic analysis of functions whose domains are topological groups.[20,21] The discrete case is the simplest special case of this theory since all the topological complexities disappear when attention is restricted to finite groups with the discrete topology (every function of such a group is technically a continuous function!). A relatively minor computational innovation (the FFT) has made what was formerly a trivial special case of a very abstract general theory into a subject that is now studied and applied without reference to the rest of theory.

When obvious advantages do not result from the formulation of a model within the continuous tradition, the continuous framework represents what might be called mathematical overkill. The conceptual subtlety of continuous models brings with it the possibility of applying powerful analysis techniques. But when these techniques are not used, the formulation entails unnecessary complexities—the choice of a continuous model being determined by the researcher's background, the tradition prevalent within a discipline, or because the modeller is unaware of or unfamiliar with other modelling techniques. On the other hand, there are instances in which discrete models and computational simulation are used when there exist applicable classical methods. For example, instead of exhaustively listing a function in tabular form, it may have been possible to express it concisely by a formula; or, instead of resorting to computational simulation, it may have been possible to formulate a model as a symbolically tractable system of difference or differential equations. This represents what might be called mathematical underkill since existing and pertinent methods are not used. Both mathematical overkill and underkill result from the relative isolation in which the discrete and continuous traditions have developed and can be minimized by an integration of methodologies which keeps the issues of model validity and predictive power in the forefront.

Behavioral Repertoire

We have emphasized that a discrete-time system need not be based, intuitively or formally, on an underlying continuous-time model. The answer to the question "what happens between time steps?" may be simply "nothing" or "nothing relevant." Yet if discrete modelling ever displaces classical continuous methods to a substantial degree, it will be pertinent to ask whether or not there are significant differences between the classes of system behavior that can be accounted for by discrete models, on the one hand, and differential equation models on the other. To fully answer this question would require, at least, a careful specification of what is meant by "accounted for" which we shall not attempt to do here. Rather, we shall point out some facts which bear on this question.

First, for the less problematic side of the issue, there are examples of behaviors which can be generated by very simple discrete-time models which, if similar behavior even could be produced by a differential equation, would probably require a much more complex specification. May[22] indicates that simple nonlinear differential equations describing population growth (e.g., the logistics equation) describe systems with very simple behavior, whereas the corresponding simple nonlinear difference equations have very complex behaviors, some of which are aptly described as "chaotic." Instead of regarding the behavioral regimes of these discrete-time models as artifacts of discrete approximations to continuous processes, one could view them, as May indicates, as possibly valid representations of actually observable phenomena. Thus, while not conclusively demonstrated, it is plausible that the repertoire of discrete-time models includes behaviors which cannot be produced by differential equations of similar complexity.

But what about the other side of the question? Are there classes of behavior exhibited by differential equation models which cannot be accounted for by discrete models? Clearly, since the behavior produced by a discrete-time model consists of discrete functions, a discrete-time model, by definition, cannot possess the temporal resolution level (assuming that high resolution is desirable which it often is not) of a continuous-time model. Thus, at the outset, we could conclude that *no* continuous behavior can be accounted for by a discrete model. However, in applications this lack of resolution is not the real problem since the time step can be made as small as necessary to represent the behavior with *enough* fine temporal detail as required. Since measuring instruments result in discrete data, the detail required is always less than that provided by a function defined on the entire continuum.

The major problem arises because of the notorious difficulty in providing a discretely specified model (e.g., a difference equation) whose behavior *exactly* agrees with the behavior of the continuous model at designated time samples. In the terms of numerical analysis, the simulation of such a discrete model would produce a numerical approximation that is "infinitely accurate at the mesh points," that is, the behavior of the discrete model would be equal to the sampled behavior of the continuous model. The discrete models constructed in numerical analysis behave only approximately like the continuous system and only do so in a restricted time period even if an ideal computer producing no round-off error is imagined. That infinitely accurate numerical approximations exist only in rare instances leads one to suspect that the behavioral repertoire of the class of continuous-time models may be richer than that of the class of discrete-time models, even putting aside the issue of temporal resolution. That is, given a continuous-time model there may not exist a discrete-time model whose structure generates, in a local manner, behavior which is equal to the sampled continuous behavior.

However, with a very general definition of well-specified discrete and continuous-time systems it can be shown that this is not true. For every continuous-time system there does exist a discrete-time system whose behavior agrees exactly with that of the continuous system at time samples which are integer multiples of an arbitrarily chosen positive number. This result, due to Zeigler,[12] says that such a discrete-time system exists, but it does not imply that it is easy to construct that system given only the continuous model's structure, e.g., given only a differential equation. The actual construction of the discrete system requires a knowledge of the continuous system's behavior, e.g., the solution of the differential equation, so that this result is not helpful for initially finding the solution. The rare cases where infinitely accurate numerical methods exist are those instances in which the discrete-time system can be constructed without this behavioral knowledge.

Nevertheless, the existence of such exact discrete models is important in the discrete modelling tradition where continuous models are not used at all. By restricting oneself to the use of discrete models one does not, *ab initio*, exclude the possibility of generating, from local rules and up to any discrete resolution, the full range of behaviors produced by continuous models. The problem of the construction of continuous-discrete model pairs does not arise since discrete models are formulated directly.

It is beyond the scope of this article to develop the theoretical framework in which this result can be rigorously proven, but we can indicate the character of the result by discussing a simple example. The initial value problem given by

$$\frac{dq}{dt} = aq, \qquad q(0) = q_0 \qquad (1)$$

has the solution

$$q(t) = q_0 e^{at}, \qquad t \geq 0. \qquad (2)$$

Discrete and Continuous Models

The corresponding discrete-time system is specified by the difference equation

$$q(t+h) = e^{ah}q(t), \quad q(0) = q_0. \tag{3}$$

The solution of (3) is the discrete function $q_0 e^{at}$ for $t \in \{kh \mid k = 0, 1, \ldots\}$. This can be seen by induction where the crucial step follows from the fact that by Eq. (3) $q(t + h) = e^{ah}(q_0 e^{at}) = q_0 e^{a(t+h)}$. Note that the coefficient e^{ah} in Eq. (3) is obtained from the solution of the differential equation and not from the equation itself.

The correspondence in this example depends on the specific property of exponential functions that $a^x \cdot a^y = a^{x+y}$. However this fact can be viewed as a specific form of a property possessed by other well-defined dynamical systems. This property is known as the *semigroup property* of state determined systems[23] or, in a more general form which includes input, as the *composition property*.[12] Very briefly, a system with this property can be characterized at any time t by a *state* q_t such that its state at any time $t + h$ is a function of q_t and h but not of t.

For the case of systems without input this implies *time-invariance*, but if input is considered it's possible for a system to have the composition property without being time invariant. See Zeigler.[12] Suppose a continuous-time system starts in the initial state q_0 at time $t = 0$ and at any time $t = h$ is in state q_h. If the system satisfies the composition property, then q_h is a function of q_0 and h. We can write

$$q_h = \delta(q_0, h).$$

If q_0 and q_h are indeed states of the system, then

$$q_{2h} = \delta(\delta(q_0, h), h)$$

and, in general

$$q_{(k+1)h} = \delta(q_{kh}, h).$$

Thus, the function δ can be used to iteratively generate a discrete function which agrees with the behavior of the continuous system at times which are integer multiples of h. For the system specified by the differential equation (1), $\delta(q_0, h) = q_0 e^{ah}$. Similar constructions are possible whether the system is linear or nonlinear. For the case with input, a continuous input segment defined from time $t = 0$ to $t = h$ is regarded as a single input symbol. This is quite natural for piecewise constant input functions (as in sampled-data systems) but can be extended to other kinds of input segments.[12]

Numerical analysts have not focused on discrete systems derived in this way from continuous systems since, as we have said, the behavior of the continuous system needs to be known for the derivation. If this behavior were known, a discretization would be unnecessary. Zeigler[12] remarks, however, that the behavior need only be known for the time interval $[0, h]$ which might be determined by a standard numerical method. Using this result for specifying a discrete-time system and then iteratively generating the behavior of the discrete system for longer periods may result in a decrease of error propagation. In addition, given a model consisting of interconnected components that in isolation from other components are described by differential equations whose solutions are known, it is possible to simulate the model using the discrete-time version of the components.[24]

A result having the character of that reported here is also useful in applications where synthesis rather than analysis is necessary, for example, if a discrete-time system is to be constructed whose behavior should approximate the known behavior of a continuous-time system. The theory of digital signal processing is partially concerned with this task, and the kind of construction described here is closely related to the technique called impulse invariant filtering.[9] A digital filter can be designed and implemented on a digital computer whose impulse response is equal (except for quantization error) to the sampled impulse response of an arbitrary continuous-time linear filter.

The reason for reporting this result here is that it shows that in a very strong sense the behavioral repertoire of discrete-time models is at least as rich as that of continuous-time models. The constructional difficulties occur only in going from a continuous model to the corresponding discrete model. If a discrete model were constructed to account for empirically observed behavior, rather than to match the behavior of a continuous model, these difficulties do not arise. Of course there remains the possibility that in a particular application a valid continuous model may be easier to construct than a valid discrete model, but restricting attention to discrete models does not further limit the kind of behavior that can be generated.

We can take this result further. Most standard numerical methods are based on approximations to the derivative, i.e., on discrete versions of the rate-of-change concept. The form of the "infinitely accurate" discrete systems described here indicates that simply *change* rather than rate-of-change is the appropriate conceptual primitive for discrete modelling. The difference equation (3) was formulated on the basis of asking "what change does the continuous system undergo from $t = 0$ to $t = h$?" A similar question might be asked about a system under observation in the natural world possibly resulting in the direct formulation of a valid discrete-time model. It might be argued here that examining mere change rather than rate-of-change is what postponed the understanding of motion provided by Newton. The quantities that remained invariant in simple mechanical systems were velocities or accelerations. However, since Newton's time we have learned a great deal about modelling dynamical processes and, in particular, about what is meant by the state of a system. The invariant property of a system is, more generally, a function that tells how states change to other states. The concept of instantaneous velocity is appropriate for modelling certain kinds of systems, but it is only one way of specifying such a function.

Conclusions

We have tried to articulate a point of view and to present some facts which would dispel common criticisms of discrete modelling as an alternative to modelling with differential equations. Most of these criticisms seem to be the result of unfamiliarity with discrete styles of mathematical thinking and a tendency to reify the abstractions used in models which have such long histories of success. There remains, however, a set of issues that cannot be so resolved. Rather than being criticisms of the *principle* or *potential* of discrete modelling, these issues pertain to current discrete modelling practices and to the fact that classical models of dynamical systems happen not to be expressed in discrete form.

Discrete and Continuous Models 395

Digital computers permit simulation of models whose complexity far transcends the current possibilities of symbolic techniques. In many cases, the purpose of such simulations is to experiment with alternative configurations of an existing or proposed physical construction without having to actually alter to construct it until numerous possibilities have been tried with the model. For example, models of industrial processes, traffic flow, and computer operating systems are often simulated for this purpose. Such models are not designed with an explicit goal of helping to "understand" the system. The system, in fact, may be one whose entire mechanism is regarded (perhaps mistakenly) as exposed and already understandable. This is why in much of the literature on this kind of modelling the term "simulation" rather than "modelling" is emphasized. The appropriate model is regarded as almost obvious while the emphasis is placed on the generation and analysis of its behavior. Theory, aside from statistical theory, plays very little role in this process.

The adoption of this style of modelling and simulation for purposes of unraveling observational patterns in order to "explain" or "understand" them immediately leads to difficulties. The terms explanation and understanding are admittedly problematic, but at the very least their meaning involves the ability to embed a particular model into a larger existing conceptual framework. In many scientific areas, the existing conceptual frameworks rely so heavily on continuous mathematics that discrete models, even if valid, tend to appear as merely descriptive models without adequate explanatory significance. Indeed, explanatory significance may in fact be lacking as long as contact with classical theory is not established or as long as sufficiently encompassing discrete theories are not constructed. The methodological biases produced by a prevailing modelling formalism are, in a sense, justified by their own prior existence. The often asked "What if digital computers were available to Newton?" is, after all, an academic question.

One can't, of course, conclude that continuous methods (including numerical analysis) must therefore remain the major tools in scientific modelling. On the contrary, the ease with which discrete modelling and simulation techniques can be applied in situations where classical methods are completely inadequate is precisely what is needed for further theoretical development. Computational experimentation can suggest the form of structural-behavioral correspondence in classes of systems that are not yet understood, but computational power alone is not a substitute for the careful simplification and theoretical generalization that have helped make classical methods so fruitful.

References

1. K. Steiglitz, *An Introduction to Discrete Systems*. John Wiley, New York, 1974.
2. D. Greenspan, *Discrete Models*. Addison-Wesley, Reading, Mass., 1973.
3. D. Greenspan, "An Algebraic, Energy Conserving Formulation of Classical Molecular and Newtonian n-Body Interaction." *Bulletin of the American Mathematical Society*, **79**, No. 2, March 1973.
4. D. Greenspan, "An Arithmetic Particle Theory of Fluid Mechanics." *Computer Methods in Applied Mechanics and Engineering*, **3**, North Holland, 1974.
5. D. Greenspan, "Computer Newtonian and Special Relativistic Mechanics." *Second U.S.A. Japan Computer Conference Proceedings*, 1975.

6. B. P. Zeigler and A. G. Barto, "Alternative Formalisms for Biosystem and Ecosystem Modelling." *Simulation Councils Proceedings Series*, **5**, No. 2, Dec. 1975.
7. J. W. Sutherland, *Systems: Analysis, Administration, and Architecture*. Van Nostrand Reinhold, New York, 1975.
8. A. G. Barto, "Cellular Automata as Models of Natural Systems." Ph.D. Thesis, Computer and Communication Sciences, The University of Michigan, 1975.
9. B. Gold and C. M. Rader, *Digital Processing of Signals*. McGraw-Hill, New York, 1969.
10. M. A. Harrison, *Lectures on Linear Sequential Machines*. Academic Press, New York, 1969.
11. L. Padulo and M. A. Arbib, *System Theory*. Saunders, Philadelphia, Pa., 1974.
12. B. P. Zeigler, *Theory of Modelling and Simulation*. John Wiley, New York, 1976.
13. B. F. Gaines, "System Identification, Approximation and Complexity." *International Journal of General Systems*, **3**, No. 3, 1977, pp. 145–174.
14. A. N. Whitehead, *Science and The Modern World*. Free Press, New York, 1967.
15. A. N. Whitehead, *Modes of Thought*. Free Press, New York, 1966, p. 18.
16. G. J. Klir, *An Approach to General Systems Theory*. Van Nostrand Reinhold, New York, 1969.
17. R. L. Wilder, *The Foundations of Mathematics*. John Wiley, New York, 1967.
18. J. W. Cooley, P. A. W. Lewis and P. D. Welch, "The Finite Fourier Transform." *IEEE Trans. on Audio and Electroacoustics*, **Au-17**, No. 2, June 1969.
19. J. von Neumann, *Theory of Self-Reproducing Automata*. Edited by A. W. Burks, University of Illinois Press, Urbana, 1966.
20. W. Rudin, *Fourier Analysis on Groups*. John Wiley, New York, 1962.
21. E. Hewitt and K. A. Ross, *Abstract Harmonic Analysis*. Springer-Verlag, Berlin, 1963.
22. R. M. May, "Biological Populations Obeying Difference Equations: Stable Points, Stable Cycles, and Chaos." *Journal of Theoretical Biology*, **51**, No. 2, June 1975, pp. 511, 524.
23. R. E. Kalman, P. L. Falb, and M. A. Arbib, *Topics in Mathematical Systems Theory*. McGraw-Hill, New York, 1969.
24. B. P. Zeigler, "Persistence and Patchiness of Preditor-Prey Systems Induced by Discrete Event Population Exchange Mechanisms." (Submitted to *Journal of Theoretical Biology*).

16

Reconstructability Analysis: An Offspring of Ashby's Constraint Analysis

George J. Klir

It is a great pleasure and a privilege for me to deliver this lecture and pay thus a tribute to W. Ross Ashby, a brilliant scholar who contributed so much to systems research and cybernetics.

The First W. Ross Ashby Memorial Lecture was presented by Heinz von Foerster in Washington, DC, on 6 January 1982. Heinz, who was closely associated with Ross for several years, gave his lecture a title 'Beginnings'. In his usual charming style, he described Ross Ashby as a great innovator in science, whose many contributions extended beyond the established paradigms of science and were beginnings of new directions in scientific thought.

A survey published in 1978[9] showed clearly that Ross Ashby was by far the most influential scholar in the area of systems research at that time. Seventeen participants in the survey declared his influence upon their own views as dominant. The second most influential person (listed by nine participants) was Ludwig von Bertalanffy, while Norbert Wiener and Anatol Rapoport tied for the third place (each listed by seven participants).

I was influenced by Ross Ashby so profoundly that I decided to dedicate one of my major recent books to him.[10] I also wrote a Foreword to a collection of Ashby's papers,[6] where I discuss the whole scope of his contributions and how they influenced me. In this lecture, I intend to focus on one area of my current research interests that emerged from Ashby's 'beginnings'. It is an area of systems science that has lately been referred to as *reconstructability analysis* (RA). My aim in this lecture is to discuss RA on the conceptual rather than technical level.

The purpose of RA is to deal with the various problems that emerge from the relationship between systems perceived as wholes and their various subsystems (i.e., parts of the wholes). RA is thus connected with the issues of wholeness, which have aroused philosophers since ancient times, and with the more recent controversy between reductionism and holism. In order to facilitate our discussion of RA, let me introduce the requisite concepts first.

A concept that is most fundamental not only in RA but in systems science at large is clearly the concept of a *system*: In general, a system is an abstraction distinguished on an object by an observer, which reflects the interaction between the observer and the object. In RA, this abstraction is conceived as a set of variables together with a characterization of the constraint (relationship, dependency, correlation) among the variables. Each variable is either an abstract image of some real-world attribute or some derivative of the image (e.g., a lagged variable); it is associated with a finite set of states (values), each representing a class of appearances of the corresponding attribute.

The constraint among variables of a system can be expressed in various ways. For some purposes, it is sufficient to express it simply in terms of a mathematical relation defined on the Cartesian product of the state sets. For other purposes, it is preferable to characterize the constraint by a probability measure on the Cartesian product or, alternatively, by a fuzzy measure of some other type.[1,12]

The constraint is assumed to be invariant with respect to the backdrop against which the variables are observed, most often time, space, or a population of individuals of the same kind. This does not mean, however, that the constraint must represent a description of the changeless; it may rather represent a changeless description of the changing.

Three additional concepts are important in RA: subsystem, structure system and overall system. We say that a system based on a subset of variables of another system is a *subsystem* of the latter. When several systems are considered as a unity, we call this collection of systems a *structure system*. Finally, we call the system that consists of all variables in a structure system an *overall system*.

Individual systems of a structure system, which are often called its elements, may share some variables. Elements of a structure system are coupled through these shared variables and, consequently, interact with each other. By definition, each element of a structure system is obviously a subsystem of the corresponding overall system.

Equipped with these essential and sufficiently broad concepts, we are now able to begin our general discussion of RA.

RA involves two problems, each of which may be broken down into several subproblems. One of these problems is concerned with a given structure system whose elements are viewed as subsystems of an unknown overall system. The aim is to make meaningful inferences about the overall system from information in the subsystems and, possibly, some additional background information. This problem is usually called the *identification problem*.

The second problem deals with a given overall system. The aim is to break the system into subsystems, as small as possible, that are adequate to reconstruct the overall system, to an acceptable degree of approximation, solely from the information contained in the subsystems. This problem is called the *reconstruction problem*.

The identification problem emerges when it is impossible or impractical to measure simultaneously all variables of the overall system of interest, or if the investigator of the overall system is dependent on specific subsystems of the overall system (developed usually by specialists in the various disciplines of classical science). The selection of the subsystems is under these circumstances beyond the control of the investigator: they are either given or

Reconstructability Analysis 399

their choice is restricted by various practical considerations. There is obviously no guarantee that the overall system is adequately represented by the subsystems. Hence, one of the aims of the identification problem is to determine the extent to which the given subsystems do adequately represent the overall system.

The identification problem belongs to the general class of problems of reasoning under incomplete information. This type of reasoning is fundamentally different from deductive reasoning in the sense that conclusions are not entailed in the given premises; it is often referred to as *ampliative reasoning*.

In general, ampliative inferences are not precise. In the identification problem, one can determine only a class of overall systems that are implied by the information in the given subsystems. The size of this class, which is usually called a *reconstruction family*, reflects the inadequacy of the subsystem representation of the overall system. The true overall system is always a member of the reconstruction family, but the subsystem information is not sufficient (except for special cases) to identify it uniquely. Hence, inferences one can make about the overall system from given subsystems are inherently imprecise.

The determination of the reconstruction family is basically a matter of formulating and solving appropriate algebraic equations. These equations characterize, within the mathematical formalism employed, the rules of projecting the unknown overall system into the given subsystems. That is, they relate the unknown entities of the overall system (e.g., overall states or state probabilities) to the corresponding entities of the given subsystems. In addition, the equations are always constrained by the requirements of the mathematical apparatus employed.

When the reconstruction family contains only one system, the identification of the overall system is perfect. In these cases, which are rather rare, the ampliative reasoning becomes deductive reasoning. It may also happen, of course, that the reconstruction family is empty. This means that the given structure system is inconsistent.

Two kinds of inconsistency in structure systems are distinguished in RA: local and global inconsistency. Local inconsistency pertains only to variables that are shared by two or more elements of the given structure system. Any two elements that share some variables are *locally inconsistent* if their characterizations of the shared variables are not exactly the same. A structure system is *globally inconsistent* if all pairs of its elements are locally consistent and, yet, the reconstruction family is empty.

Local inconsistencies are usually caused by the fact that the individual elements of a structure system are not perfect models of reality. They are developed from or validated by different data sets, each of which is incomplete, imprecise, or otherwise imperfect in some specific sense peculiar to it. Hence, local inconsistencies are a result of our ignorance regarding the subsystems involved and, consequently, it is meaningful to try to resolve them by a suitable method. At this time, however, research in this area is still in its infancy. Local as well as global inconsistencies can also be treated by rejecting the information that creates them. This amounts to the rejection of appropriate equations in the process of formulating the reconstruction family.

For some purposes, the uncertainty associated with inferences about the overall system is often acceptable, particularly when the reconstruction family is not prohibitively large. There are situations, however, in which we are forced to select one of the overall systems in the reconstruction family as a basis for making a decision and taking appropriate action.

This requires that we use some justifiable principle as an arbiter for choosing one member of the reconstruction family over the others.

One principle, well justifiable on epistemological grounds, requires that the chosen system be unbiased, i.e., implied solely by the information contained in the given subsystems. In general, the principle can be expressed as follows: in deciding which overall system to select, *use all but no more information than available*. Information is considered here solely in terms of the reduced uncertainty it produces in the overall system inferred. To make the principle operational, a unique and well justified measure of uncertainty must be determined for the mathematical formalism employed.

For systems conceptualized in terms of probability theory, for example, the measure of uncertainty is the Shannon entropy.[7] It has been proven in several alternative ways that the Shannon entropy is the only function that possesses all properties required for a probabilistic measure of uncertainty. The uniqueness of the Shannon entropy makes it qualified for the general information principle stated above. In this case, the general principle becomes the well known and broadly applied *principle of maximum entropy*.[4,5,14] According to this principle, we select that overall system for which the Shannon entropy reaches it maximum within the reconstruction family; the maximum is always unique in this case. Similar principles are now being investigated for other mathematical frameworks such as possibility theory or the Dempster-Shafer evidence theory.[12]

I would like to emphasize that the general information principle, which is well justified on epistemological grounds, may not be appropriate under some pragmatic considerations. In some situations, we may require, for example, to select such a member of the reconstruction family for which the maximum possible difference (defined in some specific way) from the actual (unknown) overall system is minimized. That is, we may employ a *principle of the least risk*.

Let me return now to the second problem pertaining to RA—the reconstruction problem. The principle motivation behind the reconstruction problem is to reduce the complexity of the system involved. That, in turn, is connected with our ability to comprehend and manage the system. For example, it is easier to monitor several small sets of variables than one large set of variables during a crisis situation, when decisions to take appropriate actions must be made quickly. Another aspect of systems manageability is expressed in terms of the size of computer memory required to store the system. When the system can be adequately represented by a set of subsystems, the reduction in the required memory size is often drastic.[10] Reconstructability of systems is also important in the context of regulation. In general, the variety of an overall system (expressed, e.g., by the number of its actual states) that is reconstructable from some of its subsystems is smaller than the variety of a comparable system that is not reconstructable. This reduction of variety, which is often very large, makes it easier to regulate the system in any desirable way. This is a direct consequence of the Ashby law of requisite variety, according to which the variety of the regulated system can be constrained only by at least the same variety of the regulating system.[1,2] That is, the variety of the regulating system must be at least as large as that of the regulated system. Hence, when we manage to reduce the variety of the regulated system by determining a set of subsystems that represent the system adequately, the required variety of the regulating system is reduced proportionally.

Practical reasons related to problems of measurement, manageability, efficiency, etc. are not the only reasons why it is desirable to represent overall systems by their subsystems.

A discovery that a system can be represented by a specific set of subsystems may provide the investigator with some knowledge that is not available, at least explicitly, in the corresponding overall system. For example, the subsystem configuration may give him information about causal relationships, the significance of the individual variables, the strength of dependencies among them, etc. In general, this additional knowledge may help the investigator to develop a better insight into the nature of the attributes studied.

The reconstruction problem belongs to the general class of problems of systems simplification. All these problems share the same *basic simplification principle*: a sound simplification of a system should minimize the loss of relevant information with respect to the required reduction of its complexity. As in the identification problem, the loss of information is measured here by the increase in uncertainty. An appropriate measure of uncertainty is thus used again as an arbiter. Among all comparable simplifications of the given system, we accept only those with minimum uncertainty. For probabilistic systems, for example, the simplification principle becomes the well established *principle of minimum entropy*.[4,5]

In the reconstruction problem, a system is simplified by being broken down into appropriate subsystems. A system with n variables has clearly 2^n subsystems and there are 2^{2^n} structure systems that can be formed from these subsystems. Each of these structure systems is a potential simplification of the overall system. Not all of them, however, must be considered in the reconstruction problem. We may disregard any structure system that contains at least one element that is a subsystem of another element. Such an element, which is a sub-subsystem of the overall system, does not contribute any information that is not included in the larger element and, hence, it is totally redundant in the context of the reconstruction problem.

When the redundant structure systems are excluded, the number of remaining structure systems (meaningful simplifications) is considerably smaller than 2^{2^n}. It still grows, however, extremely rapidly with the number of variables, reaching almost eight million for $n = 6$ and exceeding 2.4×10^{12} for $n = 7$. In RA, these meaningful simplifications are usually referred to as *reconstruction hypotheses*.

Given an overall system, its reconstruction hypotheses are ordered in a natural way by a relation of refinement. Given two reconstruction hypotheses, X and Y, X is viewed as a refinement of Y if and only if for each subsystem in X there is a larger or equal subsystem in Y; the term 'larger' is used here strictly in the sense of subset relationship between the sets of variables of the subsystems. The refinement ordering is only partial. It is known that the set of all reconstruction hypotheses for any overall system together with the refinement ordering form a lattice, which is usually called a *refinement lattice*.[10,13]

Given a set of reconstruction hypotheses of an overall system, we need to determine how much information about the overall system is contained in each of them. This can be done by reconstructing a hypothetical overall system for each reconstruction hypothesis, and then comparing it with the actual overall system. We must insure that the reconstruction method used utilizes all information available in each of the hypotheses. At the same time, however, we must be sure that no additional and unsupported (i.e., biasing) information is used in deriving the reconstructed system. Hence, we resort again to an appropriate principle of maximum uncertainty.

The comparison between the reconstructed and actual overall system is expressed in terms of a suitable distance function, preferably one based on the relevant measure of information. For probabilistic systems, for example, the *Shannon cross-entropy* is the appropriate information measure.

Our aim in the reconstruction problem is to maximize the refinement while simultaneously minimizing the distance. However, these two objective criteria conflict with each other. Any increase in the refinement implies that the distance increases or, at best, remains the same. The full solution set in the reconstruction problem is thus the set of all reconstruction hypotheses that are either equivalent or noncomparable in terms of the two objective criteria.

Research on RA has been viable since 1976, and one of my papers published that year is often considered the founding paper of RA.[8] The fact is, however, that the problems addressed by RA and some of the ideas upon which RA is based were recognized, at least, partially, by Ross Ashby much earlier.

In 1964, Ashby published a paper entitled 'Constraint analysis of many-dimensional relations',[3] in which he addressed the reconstruction problem. He introduced the concepts of a projection and cylindric extension for many-dimensional relations on finite sets. Using the concepts, he developed an algorithm to determine whether a given relation of dimension n can be reconstructed from its projections of some dimension $k < n$. The algorithm consists of forming cylindric extensions of the k-dimensional projections and, then, taking their set intersection. The n-dimensional relation reconstructed from some of its projections by this algorithm is the largest (i.e., the most nonspecific) of all relations in the reconstruction family implied by the projections. Hence, the algorithm conforms to the principle of maximum uncertainty.

Later, Ashby and Madden discussed the identification problem, again within the domain of many-dimensional relations.[15] While they discussed rather thoroughly conditions under which an unknown n-dimensional relation is uniquely identifiable from some of its projections, they failed to introduce and investigate the key concept of the reconstruction family.

It is peculiar that these important insights into the whole-part relationship developed by Ashby went virtually unnoticed for almost a decade. Also little noticed for many years was another aspect of Ashby's work that is closely connected with RA—his use of Hartley and Shannon information measures as general tools for analyzing systems.[6]

We realize now that almost all ingredients of RA were available in Ashby's writings well before RA became a subject of research interest. They were still 'beginnings' in his time and, as such, they were ill-understood and little appreciated.

Perhaps the most important contribution of RA is that it provides us with new insights regarding the nature of the relationship between wholes and parts. It goes far beyond the thinking emerging from both reductionism and holism. From the viewpoint of RA, these doctrines are only two extreme positions of the whole spectrum of methodological possibilities. RA recognizes that it is often essential, and sometimes even unavoidable, to reduce a complex system into appropriate subsystems in order to make it manageable. It makes us aware, however, that the choice of the subsystems is critical; some may represent the overall system quite well while others may be highly inadequate. We should not look for subsystems that look "natural," but rather for those which allow us to reconstruct the overall system with as high precision as possible.

Problems that are addressed by RA are computationally extremely difficult. As a consequence, systems with a modest number of variables (say 10 or fewer) can be handled by current computer technology. Due to the nature of the problems, however, it is reasonable to expect that this limit will be extended considerably with advances in computer technology making computation based on massive parallel processing a reality.

References

1. W. R. Ashby, *An Introduction to Cybernetics*. John Wiley, New York (1956).
2. W. R. Ashby, Requisite variety and its implications for the control of complex systems. *Cybernetica* **1** (1958), 1–17.
3. W. R. Ashby, Constraint analysis of many-dimensional relations. *General Systems Yearbook* **9** (1964), 99–105.
4. R. Christensen, Entropy minimax multivariate statistical modeling—1. Theory. *Int. J. Gen. Syst.* **11** (1985), 231–277.
5. R. Christensen, Entropy minimax multivariate statistical modeling—2. Applications. *Int. J. Gen. Syst.* **12** (1986), 227–305.
6. R. Conant (ed.), *Mechanisms of Intelligence: W. Ross Ashby's Writings on Cybernetics*. Intersystems, Seaside, CA (1981).
7. S. Guiasu, *Information Theory with Applications*. McGraw-Hill, New York (1977).
8. G. J. Klir, Identification of generative structures in empirical data. *Int. J. Gen. Syst.* **3** (1976), 89–104.
9. G. J. Klir (ed.), *Applied General Systems Research: Recent Developments and Trends* (Appendix B). Plenum Press, New York (1978).
10. G. J. Klir, *Architecture of Systems Problem Solving*. Plenum Press, New York (1985).
11. G. J. Klir, Where do we stand on measures of uncertainty, ambiguity, fuzziness, and the like? *Fuzzy Sets and Systems*, in press.
12. G. J. Klir and T. A. Folger, *Fuzzy Sets, Uncertainty, and Information*. Prentice-Hall, Englewood Cliffs, NJ (1987).
13. G. J. Klir and E. C. Way, Reconstructability analysis: aims, results, open problems. *Syst. Res.* **2** (1985), 141–163.
14. R. D. Levine and M. Tribus (eds). *The Maximum Entropy Formalism*. M.I.T. Press, Cambridge, MA (1979).
15. R. F. Madden and W. R. Ashby, On the identification of many-dimensional relations. *Int. J. Syst. Sci.* **3** (1972), 343–356.

17

Requisite Variety and Its Implications for the Control of Complex Systems

W. Ross Ashby

Recent work on the fundamental processes of regulation in biology (Ashby, 1956) has shown the importance of a certain quantitative relation called the law of requisite variety. After this relation had been found, we appreciated that it was related to a theorem in a world far removed from the biological—that of Shannon on the quantity of noise or error that could be removed through a correction-channel (Shannon and Weaver, 1949; theorem 10). In this paper I propose to show the relationship between the two theorems, and to indicate something of their implications for regulation, in the cybernetic sense, when the system to be regulated is extremely complex.

Since the law of requisite variety uses concepts more primitive than those used by entropy, I will start by giving an account of that law.

I

Variety

Given a set of elements, its *variety* is the number of elements that can be distinguished. Thus the set

$$\{gbcggc\}$$

has a variety of 3 letters. (If two observers differ in the distinctions they can make, then they will differ in their estimates of the variety. Thus if the set is

$$\{bcaaCaBa\}$$

its variety in shapes is 5, but its variety in letters is 3. We shall not, however, have to treat this complication.)

From *Cybernetica* **1**, 83. Copyright © 1958 by the Association for Cybernetics, Namur, Belgium.

For many purposes the variety may more conveniently be measured by the logarithm of this number. If the logarithm is taken to base 2, the unit is the *bit*. The context will make clear whether the number or its logarithm is being used as measure.

Regulation and the Pay-off Matrix

Regulation achieves a «goal» against a set of disturbances. The disturbances may be actively hostile, as are those coming from an enemy, or merely irregular, as are those coming from the weather. The relations may be shown in the most general way by the formalism that is already well known in the theory of games (Neumann and Morgenstern, 1947).

A set D of disturbances d_i can be met by a set R of responses r_j. The outcomes provide a table or matrix

		R			
		r_1	r_2	r_3	...
	d_1	z_{11}	z_{12}	z_{13}	...
	d_2	z_{21}	z_{22}	z_{23}	...
D	d_3	z_{31}	z_{32}	z_{33}	...
	d_4	z_{41}	z_{42}	z_{43}	...
	...	...	...	...	...

in which each cell shows an element z_{ij} from the set Z of possible outcomes.

It is not implied that the elements must be numbers (though the possibility is not excluded). The form is thus general enough to include the case in which the events d_i and r_j are themselves vectors, and have a complex internal structure. Thus the disturbances D might be all the attacks that can be made by a hostile army, and the responses R all the counter-measures that might be taken. What is required at this stage is that the sets are sufficiently well defined so that the facts determine a single-valued mapping of the product set $D \times R$ into the set Z of possible outcomes. (I use here the concepts as defined by Bourbaki, 1951).

The «outcomes» so far are simple events, without any implication desirability. In any real regulation, for the benefit of some defined person or organism or organisation, the facts usually determine a further mapping of the set Z of outcomes into a set E of values. E may be as simple as the 2-element set {good, bad}, and is commonly an ordered set, representing the preferences of the organism. Some subset of E is then defined as the «goal». The set of values, with perhaps a scale of preference, is often obvious in human affairs; but in the biological world, and in the logic of the subject, it must have explicit mention. Thus if the outcome is «gets into deep water», the valuation is uncertain until we know whether the organism is a cat or a fish.

In the living organisms, the scale of values is usually related to their «essential variables»—those fundamental variables that must be kept within certain «physiological» limits if the organism is to survive. Other organisations also often have their essential variables: in an economic system, a firm's profits is of this nature, for only if this variable keeps positive can the firm survive.

Requisite Variety

Given the goal—the «good» or «acceptable» elements in E—the inverse mapping of this subset will define, over Z, the subset of «acceptable outcomes». Their occurrence in the body of the table or matrix will thus mark a subset of the product set $D \times R$. Thus is defined a binary relation S between D and R in which «the elements d_i and r_j have the relation S» is equivalent to «r_j, as response to d_i, gives an acceptable outcome».

Control

In this formulation we have considered the case in which the regulator acts so as to limit the outcome to a particular subset, or to keep some variables within certain limits, or even to hold some variables constant. This reduction to constancy must be understood to include all those cases, much more numerous, that can be reduced to this form. Thus if a gun is to follow a moving target, the regulation implied by accuracy of aim may be represented by a keeping at zero of the difference between the gun's aim and the target's position. The same remark is clearly applicable to all cases where an unchanging (constant) relation is to be maintained between one variable that is independent and another variable that is controlled by the regulator.

Thus, as a special instance, if a variable y (which may be a vector) is to be controlled by a variable a, and if disturbance D has access to the system so that y is a function of both the control a and the value of disturbance D, then a suitable regulator that has access to the disturbance may be able to counter its effects, remove its effect from y, and thus leave y wholly under the control of a. In this case, successful regulation by R is the necessary and sufficient condition for successful control by a.

Requisite Variety

Consider now the case which, given the table of outcomes (the pay-off matrix), the regulator R has the best opportunities for success. (The other cases occur as degenerate forms of this case, and need not be considered now in detail.)

Given the table, R's opportunity is best if R can respond knowing D's value. Thus, suppose that D must first declare his (or its) selection d_i; a particular row in the table is thereby selected. When this has been done, and knowing D's selection, R selects a value r_j, and thus selects a particular column. The outcome is the value of Z at the intersection. Such a table might be:

		r_1	r_2	r_3
	d_1	c	a	d
	d_2	b	d	a
D	d_3	c	d	c
	d_4	a	a	b
	d_5	d	b	b

If outcomes a, b count as Good, and c, d as Bad, then if D selects d_1, R must select r_2; for only thus can R score Good. If D selects d_2, R may choose r_1 or r_3. If D selects d_3, then R cannot avoid a Bad outcome; and so on.

Nature, and other sources of such tables, provides them in many forms, ranging from the extreme at which every one of R's responses results in Good (these are distinctly rare!), to those hopeless situations in which every one of R's responses leads to Bad. Let us set aside these less interesting cases, and consider the case, of central importance, in which each column has all its elements different. (Nothing is assumed here about the relation between the contents of one column and those of another.) What this implies is that if the set D had a certain variety, the outcomes in any one column will have the same variety. In this case, if R is inactive in responding to D (i.e., if R adheres to one value r_j for all values of D), then the variety in the outcomes will be as large as that in D. Thus in this case, and if R stays constant, D can be said to be exerting full control over the outcomes.

R, however, aims at confining the actual outcomes to some subset of the possible outcomes Z. It is necessary, therefore, that R acts so as to lessen the variety in the outcomes. If R does so act, then there is a quantitative relation between the variety in D, the variety in R, and the smallest variety that can be achieved in the set of actual outcomes; namely, *the latter cannot be less than the quotient of the number of rows divided by the number of columns* (Ashby, 1956; S.II/5).

If the varieties are measured logarithmically, this means that if the varieties of D, R, and actual outcomes are respectively V_d, V_r, and V_0 then the minimal value of V_0 is $V_d - V_r$. If now V_d is given, V_0's minimum can be lessened *only by a corresponding increase in V_r*. This is the law of requisite variety. What it means is that restriction of the outcomes to the subset that is valued as Good demands a certain variety in R.

We can see the relation from another point of view. R, by depending on D for its value, can be regarded as a channel of communication between D and the outcomes (though R, by acting as a regulator, is using its variety subtractively from that of D). The law of requisite variety says that R's *capacity as a regulator cannot exceed its capacity as a channel for variety*.

The functional dependencies can be represented as in Fig. 1. (This diagram is necessary for comparison with Figs. 2 and 3.) The value at D threatens to transmit, via the table T to the outcomes Z, the full variety that occurs at D. For regulation, another channel goes through R, which takes a value so paired to that of D that T gives values at Z with reduced variety.

Nature of the Limitation

The statement that some limit cannot be exceeded may seem rash, for Nature is full of surprises. What, then, would we say if a case were demonstrated in which objective

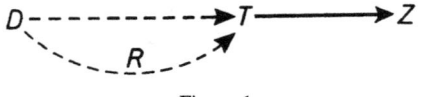

Figure 1

measurement show that the limit was being exceeded? Here we would be facing the case in which appropriate effects were occurring without the occurrence of the corresponding causes. We would face the case of the examination candidate who gives the appropriate answers before he has been given the corresponding questions! When such things have happened in the past we have always looked for, and found, a channel of communication which has accounted for the phenomenon, and which has shown that the normal laws of cause and effect do apply. We may leave the future to deal similarly with such cases if they arise. Meanwhile, few doubt that we may proceed on the assumption that genuine overstepping of the limitation does not occur.

Examples in Biology

In the biological world, examples that approximate to this form are innumerable, though few correspond with mathematical precision. This inexactness of correspondence does not matter in our present context, for we shall not be concerned with questions involving high accuracy, but only with the existence of this particular limitation.

An approximate example occurs when a organism is subject to attacks by bacteria (of species d_i) so that, if the organism is to survive, it must produce the appropriate anti-toxin r_j. If the bacterial species are all different, and if each species demands a different anti-toxin, then clearly the organism, for survival, must have at least as many anti-toxins in its repertoire of responses as there are bacterial species.

Again, if a fencer faces an opponent who has various modes of attack available, the fencer must be provided with at least an equal number of modes of defence if the outcome is to have the single value: attack parried.

Analysis of Sommerhoff

Sommerhoff (1950) has conducted an analysis in this matter that bears closely on the present topic. He did not develop the quantitative relation between the varieties, but he described the basic phenomenon of regulation in biological systems with a penetrating insight and with a wealth of examples.

He recognises that the concept of «regulation» demands variety in the disturbances D. His «coenetic variable» is whatever is responsible for the values of D. He also considers the environmental conditions that the organism must take into account (but as, in his words, these are «epistemically dependent» on the values of the coenetic variable, our symbol D can represent both, since his two do not vary independently). His work shows, irrefutably in my opinion, how the concepts of co-ordination, integration, and regulation are properly represented in abstract form by a relation between the coenetic variable and the response, such that the outcome of the two is the achievement of some «focal condition» (referred to as «goal» here). From our point of view, what is important is the recognition that without the regulatory response the values at the focal condition would be more widely scattered.

Sommerhoff's diagram (Fig. 2) is clearly similar. (I have modified it slightly, so as to make it uniform with Figs. 1 and 3.)

His analysis is valuable as he takes a great many biological examples and shows how,

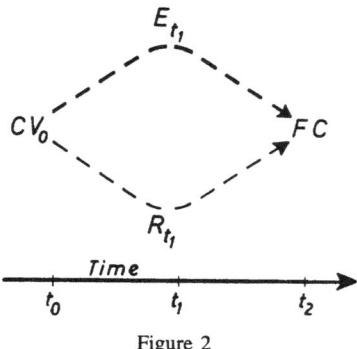

Figure 2

in each case, his abstract formulation exactly catches what is essential while omitting the irrelevant and merely special. Unfortunately, in stating the thesis, he did what I did in 1952—used the mathematical language of analysis and continuous functions. This language now seems unnecessarily clumsy and artificial; for it has been found (Ashby, 1956) that the concepts of set theory, especially as expounded by Bourbaki (1951), are incomparably clearer and simpler, while losing nothing in rigour. By the change to set theory, nothing in fact is lost, for nothing prevents the elements in a set from being numbers, or the functions from being continuous, and the gain in generality is tremendous. The gain is specially marked with biological material, in which non-numerical states and discontinuous functions are ubiquitous.

Let me summarise what has been said about «regulation». The concept of regulation is applicable when there is a set D of disturbances, to which the organism has a set R of responses, of which on any occasion it produces some one, r_j say. The physico-chemical orother nature of the whole system then determines the outcome. This will have some value for the organism, either Good or Bad say. If the organism is well adapted, or has the know-how, its response r_j, as a variable, will be such a function of the disturbance d_i that the outcome will always lie in the subset marked as Good. The law of requisite variety then says that such regulation cannot be achieved unless the regulator R, as a channel of communication, has more than a certain capacity. Thus, if D threatens to introduce a variety of 10 bits into the outcomes, and if survival demands that the outcomes be restricted to 2 bits, then at each action R must provide variety of at least 8 bits.

Ergodicity

Before these ideas can be related to those of the communication theory of Shannon, we must notice that the concepts used so far have not assumed ergodicity, and have not even used the concept of probability.

The fact that communication theory, during the past decade, has tended to specialise in the ergodic case is not surprising when we consider that its application has been chiefly to telephonic and other communications in which the processes go on incessantly and are

Requisite Variety

usually stationary statistically. This fact should not, however, blind us to the fact that many important communications are non-ergodic, their occurrence being specially frequent in the biological world. Thus we frequently study a complex biological system by isolating it, giving it a stimulus, and then observing the complex trajectory that results. Thus the entomologist takes an ant-colony, places a piece of meat nearby, and then observes what happens over the next twenty-four hours, without disturbing it further. Or the social psychologist observes how a gang of juvenile criminals forms, becomes active, and then breaks up. In such cases, even a single trajectory can provide abundant information by the comparison of part with part, but the only ergodic portion of the trajectory is that which occurs ultimately, when the whole has arrived at some equilibrium, in which nothing further of interest is happening. Thus the ergodic part is degenerate. It is to be hoped that the extension of the basic concepts of Shannon and Wiener to the non-ergodic case will be as fruitful in biology as the ergodic case has been in commercial communication. It seems likely that the more primitive concept of «variety» will have to be used, instead of probability; for in the biological cases, systems are seldom isolated long enough, or completely enough, for the relative frequencies to have a stationary limit.

Among the ergodic cases there is one, however, that is obviously related to the law of requisite variety. It is as follows.

Let D, R, and E be three variables, such that we may properly observe or calculate certain entropies over them. Our first assumption is that if R is constant, all the entropy at D will be transmitted to, and appear at, E. This is equivalent to

$$H_r(E) = H_r(D). \tag{1}$$

By writing $H(D,R)$ in two forms we have

$$H(D) + H_d(R) = H(R) + H_r(D).$$

Use of (1) gives

$$H(D) + H_d(R) = H(R) + H_r(E)$$
$$= H(R,E)$$
$$\leq H(R) + H(E);$$

i.e.,
$$H(E) \geq H(D) + H_d(R) - H(R). \tag{2}$$

The entropy of E thus has a certain minimum—the expression on the right of (2). If $H(E)$ is the entropy of the actual outcomes, then, for regulation, it may have to be reduced to a certain value. Equation (2) shows what can reduce it; it can be reduced:

(i) by making $H_d(R) = 0$, i.e., by making R a determinate function of D,
(ii) by making $H(R)$ larger.

If $H_d(R) = 0$, and $H(R)$ the only variable on the right of (2), then a decrease in $H(E)$ demands at least an equal increase in $H(R)$. This conclusion is clearly similar to that of the law of requisite variety.

A simple generalisation has been given (Ashby, 1956) in which, when R remains constant, only a certain fraction of D's variety or entropy shows in the outcomes or in $H(E)$. The result is still that each decrease in $H(E)$ demands at least an equal increase in $H(R)$.

With this purely algebraic result we can now see exactly how these ideas join on to Shannon's. His theorem 10 uses a diagram which can be modified to Figure 3 (to match

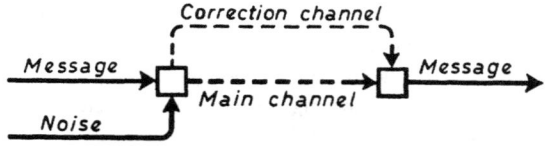

Figure 3

the two preceding Figures). Our "disturbance D," which threatens to get through to the outcome, clearly corresponds to the noise; and his theorem says that the amount of noise that can be prevented from appearing in the outcomes is limited to the entropy that can be transmitted through the correction channel.

The Message of Zero Entropy

What of the «message»? In regulation, the «message» to be transmitted is a constant, i.e., has zero entropy. Since this matter is fundamental, let us consider some examples. The ordinary thermostat is set at, say, 70 °F. «Noise», in the form of various disturbances, providing heat or cold, threatens to drive the output from the value. If the thermostat is completely efficient, this variation will be completely removed, and an observer who watches the temperature will see continuously only the value that the initial controller has set. The «message» is here the constant value 70.

Similarly, the homeostatic mechanism that keeps our bodies, in health, about 98 °F is set at birth to maintain this value. The control comes from the gene-pattern and has zero entropy, for the selected value is unchanging.

The same argument applies similarly to all the regulations that occur in other systems, such as the sociological and economic. Thus an attempt to stabilise the selling price of wheat is an attempt to transmit, to the farmers, a «message» of zero entropy; for this is what the farmer would receive if he were to ask daily «what is the price of wheat today»? The stabilisation, so far as it is successful, frees the message from the effects of those factors that might drive the price from the selected value.

Thus, all acts of regulation can be related to the concepts of communication theory by our noticing that the «goal» is a message of zero entropy, and that the «disturbances» correspond to noise.

The Error-Controlled Regulator

A case in which this limitation acts with peculiar force is the very common one in which the regulator is «error controlled». In this case the regulator's channel for information about the disturbances has to pass through a variable (the «error») which is kept as constant as possible (at zero) by the regulator R itself. Because of this route for the information, the more successful the regulator, the less will be the range of the error, and therefore the less will be the capacity of the channel from D to R. To go to the extreme: if the regulator is totally successful, the error will be zero unvaryingly, and the regulator will thus be cut off totally from the information (about D's value) that alone can make it successful—which is absurd. The error-controlled regulator is thus fundamentally incapable of being 100 percent efficient.

Living organisms encountered this fact long ago, and natural selection and evolution have since forced the development of channels of information, through eyes and ears for instance, that supply them with information about D before the chain of cause and effect goes so far as to cause actual error. At the present time, control by error is widely used in industry, in servomechanisms and elsewhere, as a means to regulation. Some of these regulations by error-control are quite difficult to achieve. Immersed in the intricacies of Nyquist's theorem, transfer functions, and other technical details, the design engineer may sometimes forget that there is another way to regulation. May I suggest that he would do well to bear in mind what has been found so advantageous in the biological world, and to consider whether a regulation which is excessively difficult to design when it is controlled by error may not be easier to design if it is controlled not by the error but by what gives rise to the error.

This is a first application to cybernetics of the law of requisite variety and Shannon's theorem 10.

It is not my purpose in this paper, however, to explore in detail how the limitation affects simple regulators. Rather I want to consider its effect in matters that have so far, I think, received insufficient attention. I want to indicate, at least in outline, how this limitation also implies a fundamental limitation on the human intellect, especially as that intellect is used in scientific work. And I want to indicate, in the briefest way, how we scientists will sometimes have to readjust our ways because of it.

II

The Limitations of the Scientist

In saying that the human intellect is limited, I am not referring to those of its activities for which there is no generally agreed valuation—I am not referring for instance, to the production of pictures that please some and displease others—for without an agreed valuation the concept of regulation does not exist. I refer rather to those activities in which the valuation is generally agreed on, and in which the person shows his capacity by whether he succeeds or fails in getting an acceptable outcome. Such is the surgeon, whose patient lives or dies; such is the mathematician, given a problem, which he does or does not solve; such is the manager whose business prospers or fails; such is the economist who can or cannot control an inflationary spiral.

Not only are these practical activities covered by the theorem and so subject to limitation, but also subject to it are those activities by which Man shows his «intelligence». «Intelligence» today is defined by the method used for its measurement; if the tests used are examined they will be found to be all of the type: from a set of possibilities, indicate one of the appropriate few. Thus all measure intelligence by the *power of appropriate selection* (of the right answers from the wrong). The tests thus use the same operation as is used in the theorem on requisite variety, and must therefore be subject to the same limitation. (D, of course, is here the set of possible questions, and R is the set of all possible answers.) Thus what we understand as a man's «intelligence» is subject to the fundamental limitation: it cannot exceed his capacity as a transducer. (To be exact, «capacity» must here be defined on a per-second or a per-question basis, according to the type of test.)

The Team as Regulator

It should be noticed that the limitation on «the capacity of Man» is grossly ambiguous, according to whether we refer to a single person, to a team, or to the whole of organised society. Obviously, that one man has a limited capacity does not impose a limitation on a team of n men, if n may be increased without limit. Thus the limitation that holds over a team of n men may be much higher, possibly n times as high, as that holding over the individual man.

To make use of the higher limitation, however, the team must be efficiently organised; and until recently our understanding of organisation has been pitifully small. Consider, for instance, the repeated attempts that used to be made (especially in the last century) in which some large Chess Club played the World Champion. Usually the Club had no better way of using its combined intellectual resources than either to take a simple majority vote on what move to make next (which gave a game both planless and mediocre), or to follow the recommendation of the Club's best player (which left all members but one practically useless). Both these methods are grossly inefficient. Today we know a good deal more about organisation, and the higher degrees of efficiency should soon become readily accessible. But I do not want to consider this question now. I want to emphasise the limitation. Let us therefore consider the would-be regulator, of some capacity that cannot be increased, facing a system of great complexity. Such is the psychologist, facing a mentally sick person who is a complexly interacting mass of hopes, fears, memories, loves, hates, endocrines, and so on. Such is the sociologist, facing a society of mixed races, religions, trades, traditions, and so on. I want to ask: given his limitation, and the complexity of the system to be regulated, what scientific strategies should he use?

In such a case, the scientist should beware of accepting the classical methods without scrutiny. The classical methods have come to us chiefly from physics and chemistry, and these branches of science, far from being all-embracing, are actually much specialised and by no means typical. They have two peculiarities. The first is that their systems are composed of parts that show an extreme degree of homogeneity: contrast the similarity between atoms of carbon with the dissimilarity between persons. The second is that the systems studied by the physicist and chemist have nothing like the richness of internal interaction that have the systems studied by the sociologist and psychologist.

Or take the case of the scientist who would study the brain. Here again is a system of high complexity, with much heterogeneity in the parts, and great richness of connexion and internal interaction. Here too the quantities of information involved may well go beyond the capacity of the scientist as a transducer.

Both of these qualities of the complex system—heterogeneity in the parts, and richness of interaction between them—have the same implication: the quantities of information that flow, either from system to observer or from part to part, are much larger than those that flow when the scientist is physicist or chemist. And it is because the quantities are large that the limitation is likely to become dominant in the selection of the appropriate scientific strategy.

As I have said, we must beware of taking our strategies slavishly from physics and chemistry. They gained their triumphs chiefly against systems whose parts are homogeneous and interacting only slightly. Because their systems were so specialised, they have developed specialised strategies. We who face the complex system must beware of accepting their strategies as universally valid. It is instructive to notice that their strategies have already broken down in one case, which is worth a moment's attention. Until about 1925, the rule «vary only one factor at a time» was regarded as the very touchstone of the scientific method. Then R. A. Fisher, experimenting with the yields of crops from agricultural soils, realised that the system he faced was so dynamic, so alive, that any alteration of one variable would lead to changes in an uncountable number of other variables long before the crop was harvested and the experiment finished. So he proposed formally to vary whole sets of variables simultaneously—not without peril to his scientific reputation. At first his method was ridiculed, but he insisted that his method was the truly scientific and appropriate one. Today we realise that the rule «vary only one factor at a time» is appropriate only to certain special types of system, not valid universally. Thus we have already taken one step in breaking away from the classical methods.

Another strategy that deserves scrutiny is that of collecting facts «in case they should come in useful some time»—the collecting of truth «for truth's sake». This method may be efficient in the systems of physics and chemistry, in which the truth is often invariant with time; but it may be quite inappropriate in the systems of sociology and economics, whose surrounding conditions are usually undergoing secular changes, so that the parameters to the system are undergoing changes—which is equivalent to saying that the systems are undergoing secular changes. Thus, it may be worthwhile finding the density of pure hafnium, for if the value is wanted years later it will not be changed. But of what use today, to a sociologist studying juvenile delinquency, would a survey be that was conducted, however carefully, a century ago? It *might* be relevant and helpful; but we could know whether it was relevant or not only *after* a comparison of it with the facts of today; and when we know these, there would be no need for the old knowledge. Thus the rule «collect truth for truth's sake» may be justified when the truth is unchanging; but when the system is not completely isolated from its surroundings, and is undergoing secular changes, the collection of truth is futile, for it will not keep.

There is little doubt, then, that when the system is complex, the scientist should beware of taking, without question, the time-honored strategies that have come to him from physics and chemistry, for the systems commonly treated there are specialised, not typical of those that face him when they are complex.

Another common aim that will have to be given up is that of attempting to «understand» the complex system; for if «understanding» a system means having available a model that is isomorphic with it, perhaps in one's head, then when the complexity of the system exceeds the finite capacity of the scientist, the scientist can no longer understand the system—not in the sense in which he understands, say, the plumbing of his house, or some of the simple models that used to be described in elementary economics.

Operational Research

It will now be obvious that the strategies appropriate to the complex system are those already getting well known under the title of «operational research». Scientists, guided doubtless by an intuitive sense of what is reasonable, are already breaking away from the classical methods, and are developing methods specially suitable for the complex system. Let me review briefly the chief characteristics of «operational» research.

Its first characteristic is that its ultimate aim is not understanding but the purely practical one of control. If a system is too complex to be understood, it may nevertheless still be controllable. For to achieve this, all that the controller wants to find is some action that gives an acceptable result; he is concerned only with what happens, not with why it happens. Often, no matter how complex the system, what the controller wants is comparatively simple: has the patient recovered?—have the profits gone up or down?—has the number of strikes gone up or down?

A second characteristic of operational research is that it does not collect more information than is necessary for the job. It does not attempt to trace the whole chain of causes and effects in all its richness, but attempts only to relate controllable causes with ultimate effects.

A third characteristic is that it does not assume the system to be absolutely unchanging. The research solves the problems of today, and does not assume that its solutions are valid for all time. It accepts frankly that its solutions are valid merely until such time as they become obsolete.

The philosopher of science is apt to look somewhat askance at such methods, but the practical scientist knows that they often achieve success when the classical methods bog down in complexities. How to make edible bread, for instance, was not found by the methods of classical science—had we waited for that we still would not have an edible loaf—but by methods analogous to those of operational research: if a variation works, exploit it further; ask not *why* it works, only *if* it works. We must be careful, in fact, not to exaggerate the part played by classical science in present-day civilisation and technology. Consider, for instance, how much empirical and purely practical knowledge plays a part in our knowledge of metallurgy, of lubricants, of house-building, of pottery, and so on.

What I suggest is that measurement of the quantity of information, even if it can be done only approximately, will tell the investigator where a complex system falls in relation to his limitation. If it is well below the limit, the classic methods may be appropriate; but should it be above the limit, then if his work is to be realistic and successful, he must alter his strategy to one more like that of operational research.

My emphasis on the investigator's limitation may seem merely depressing. That is not at all my intention. The law of requisite variety, and Shannon's theorem 10, in setting a limit to what can be done, may mark this era as the law of conservation of energy marked its era a century ago. When the law of conservation of energy was first pronounced, it seemed at first to be merely negative, merely an obstruction; it seemed to say only that certain things, such as getting perpetual motion, could not be done. Nevertheless, the recognition of that limitation was of the greatest value to engineers and physicists, and it has not yet exhausted its usefulness. I suggest that recognition of the limitation implied by the law of requisite variety may, in time, also prove useful, by ensuring that our scientific strategies for the complex system shall be, not slavish and inappropriate copies of the strategies used in physics and chemistry, but new strategies, genuinely adapted to the special peculiarities of the complex system.

References

Ashby, W. Ross, *Design for a brain*. 2nd. imp. Chapman & Hall, London, 1954.
Ashby, W. Ross, *An introduction to cybernetics*. Chapman & Hall, London, 1956.
Bourbaki, N., *Théorie des ensembles. Fascicule de résultats*. A.S.E.I. No. 1141. Hermann et Cie, Paris, 1951.
Neumann, J. (von) and Morgenstern, O., *Theory of games and economic behaviour*. Princeton, 1947.
Shannon, C. E. and Weaver, W., *The mathematical theory of communication*. University of Illinois Press, Urbana, 1949.
Sommerhoff, G., *Analytical biology*. Oxford, University Press, London, 1950.

18

Laws of Information Which Govern Systems

Roger C. Conant

1. Introduction

Information theory was created for the purpose of studying the communication of messages from one point to another, and since its appearance,[14] its focus has remained on the question, "how can the constraint between the two variables X (message sent) and Y (message received) be measured and maximized"? Although the theory was generalized to N dimensions,[10,2] and its relation to the analysis of variance noted,[9] not much use seems to have been made of the result, perhaps in part because the descriptors "N-dimensional Information Theory" or "Uncertainty Analysis" did not adequately represent what can actually be seen as the analysis of constraints in multivariable systems. In any statistically-analyzable system of several variables interacting in a lively way, some variables (or sets of them) exert effects on others. These effects are reflected statistically as *non*-independence of the variables involved, and it is this deviation from independence which we indicate by the term "constraint." We prefer this term to the term "dependence" because the latter suggests dependence of X on Y while the former is neutral as to direction. To the extent that the variables are *not* independent, they are "in communication" with one another, and information theory can be used to analyze the non-independence. In addition, the fluctuation of values taken by any variable can be viewed as a message it sends, a flow of information about itself to all other parts of the system which are "listening." The view of systems as networks of information transfer leads to quantitative conclusions about system behavior and structure which are somewhat novel and of wide applicability.

Just as ordinary information theory deals only with the statistics, and not the meaning of the messages sent and received, so its generalization to N dimensions employed here can be viewed simply as a tool for the statistical analysis of N-variable systems and particularly of the constraints implicit in the relations between the variables. As a statistical tool it is a good one, being applicable to systems whether linear or nonlinear, metric or nonmetric,

---

From *IEEE Trans. Syst. Man Cybernetics* **SMC-6**, 240. Copyright © 1976 by IEEE.

discrete or continuous. It is mathematically simple, and its measure of non-independence is zero in and only in the case of probabilistic independence, being positive otherwise. Information theory has two primary advantages over other techniques of statistical analysis: first, it allows measurement of *rates* of constraint (i.e., constraint per step or per second between dynamic variables with past history taken into account), and second, the measures possess additivity properties which allow *decomposition* of constraints (also possible in the analysis of variance which, however, is restricted to systems of variables having at least an interval scale). Some of the possible decompositions accord perfectly with a hierarchical view of systems; for example, the total constraint holding over all N variables equals the constraint holding *between* the subsystems (which constitute a partition of the N variables) plus the sum of the constraints holding *within* each subsystem, and the latter quantities can themselves be hierarchically decomposed. Various decomposition possibilities have been discussed by Ashby.[2,3] The fact that information theory fits neatly the hierarchical architecture which is so prevalent in systems of many sorts[13] seems very suggestive and indicates that the relation between information and system dynamics is a deep one. In fact, hierarchical architecture is very efficient for a system whose task is information processing,[7] and therefore, one could conclude on information-theoretic grounds alone that hierarchy would be a common form in dynamic systems.

To calculate the measures (entropies and transmissions) which are based on probabilities, one must either deduce the probabilities from known system properties or else estimate them from observation. In the latter case the bias and variance introduced by sampling can be taken into account by correction techniques.[11,8,12] In either case the system must be assumed ergodic over the period of interest.

After introducing the nomenclature and the basic definitions of N-dimensional information theory, the paper will discuss a Partition Law of Information Rates, some relations between information and hierarchy, including hierarchical decomposition of the terms in the Partition Law, a Law of Constraint Loss relevant to systems of parallel processors, and the importance of information blockage within elements of a system. A preview of the main points may be obtained by scanning Section VIII, the Summary.

A Caveat

Many interesting systems are not stationary, much less ergodic, and many cannot be observed sufficiently to allow accurate estimates of high-order multivariate probabilities. For these systems the formulas and identities of information theory may seem to be of questionable value. However, these formulas and identities often have common-sense, nonmathematical interpretations which are useful even for systems to which the rigorous versions cannot be applied. In this paper these informal interpretations will be stressed; readers who do not care to follow the mathematical details should therefore be able to get the essential points.

There are obvious dangers in applying information theory, designed for use under the severe mathematical constraints of stationarity and ergodicity, to real-world systems not thus constrained. The fact is, however, that there is currently no satisfactory quantitative theory of information applicable to such systems, and the choice faced by systems scientists is harsh: be content to say nothing about information, or try to use results from the formal theory by judicious interpretation and generalization. For this writer the latter course seems far preferable. For example, although human beings do not qualify as "channels" in the rigorous formal sense, useful measurements of human channel capacity can and have been made; as another example, there seems to be valuable insight to be gained by informally applying to humans the formal result that whatever output a channel produces beyond the limit of its own channel capacity must be pure noise, unrelated to input.

The ensuing discussion of implications of information-theoretic results will be suggestive rather than exhaustive and informal rather than technical. In encouraging application of the underlying principles to real-world systems which are nonperpetual and time-varying it is open to criticism, since such applications usually cannot be formally validated. Yet the writer believes that the system properties revealed by the results, and particularly by the Partition Laws, are not confined solely to ergodic systems and that when interpreted with care they can yield valuable insights into the behavior of general systems.

II. Notation

For our purposes a system S will be thought of simply as an ordered set of variables: $S = \{X_j | 1 \leq j \leq n\}$. No confusion will result here from thinking of S sometimes as a set, sometimes as a vector $\langle X_1, X_2, \cdots, X_n \rangle$, nor from failing to distinguish singleton sets from their sole elements. Accordingly, the symbols S and $X_1, X_2, \cdots, X_n$ will be used interchangeably for either the set or the vector, and the analogous convention will be followed for subsets (or subvectors). Thus using $=$ for equivalence,

$$S = \{X_1, X_2, \cdots, X_n\} = \langle X_1, X_2, \cdots, X_n \rangle = X_1, X_2, \cdots, X_n \tag{1}$$

and as a special degenerate case,

$$S = \{\} = \langle\rangle = \varnothing, \quad \text{the empty system.}$$

The system S receives a vector input E for its environment, which is considered to be all relevant variables which are not in S. Those variables X_j in S which can be directly observed from its environment constitute output variables; the set of these variables is denoted $S_0 = X_1, X_2, \cdots, X_k$. The remaining X_j in S are internal variables; the set of these is denoted $S_{\text{int}} = X_l, \cdots, X_n$. Thus, using $=$ for equivalence and commas both for set union and to distinguish members of sets or components of vectors,

$$S = S_0, S_{\text{int}} = X_1, X_2, \cdots, X_k, X_l, \cdots, X_n$$
$$= X_1, X_2, \cdots, X_n. \tag{2}$$

Figure 1 suggests the interpretation of this notation.

S can be partitioned into N disjoint subsystems S^i of variables X_j^i. Each S^i receives input E^i from *its* environment (those variables not in S^i) and has a set $S_0^i = X_1^i, X_2^i, \cdots, X_{k_i}^i$ of directly observable or output variables and a set $S_{\text{int}}^i = X_{l_i}^i, \cdots, X_{n_i}^i$ of internal variables. Note that a system variable may be in S_0^i but not in S_0, that is, it may be directly observable from another subsystem in S although not from the environment of S. Figure 1 and (2), if all symbols are superscripted with i, illustrate the notation for subsystems (except $k, l, n \rightarrow k_i, l_i, n_i$).

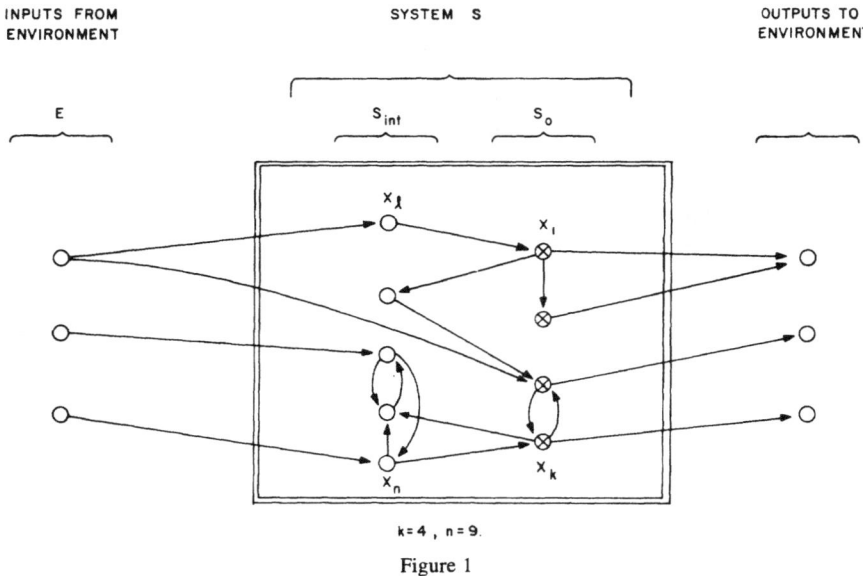

Figure 1

The subsystems can be partitioned similarly, but there is no need to do so here. By a hierarchical decomposition of S we mean this partitioning of S into subsystems, of the subsystems into sub-subsystems, and so on.

The S^i fall into two classes, those which contain one or more variables in S_o and those which do not. Those in the first class will be called output subsystems, with the class denoted $S_{OUT} = S^1, S^2, \cdots, S^k$; those in the second, internal subsystems, with the class denoted $S_{INT} = S^L, \cdots, S^N$. Thus as a consequence of our notation,

$$S^i \in S_{OUT} \Leftrightarrow \exists X_j^i \in S^i: X_j^i \in S_o \Leftrightarrow S^i \not\subset S_{int}$$

$$S^i \in S_{INT} \Leftrightarrow \forall X_j^i \in S^i: X_j^i \in S_{int} \Leftrightarrow S^i \subset S_{int}$$

$$S = S_{OUT}, \quad S_{INT} = S^1, S^2, \cdots, S^K, S^L, \cdots, S^N$$
$$= S^1, S^2, \cdots, S^N.$$

Of course, any of these sets may be empty, in general. Figure 2 illustrates the notation.

III. N-Dimensional Information Theory

A. Basic Definitions: Entropy, Transmission

N-dimensional information theory, sometimes called multivariate uncertainty analysis, has been adequately discussed in the literature,[10] as has its use as a statistical tool for

Laws of Information

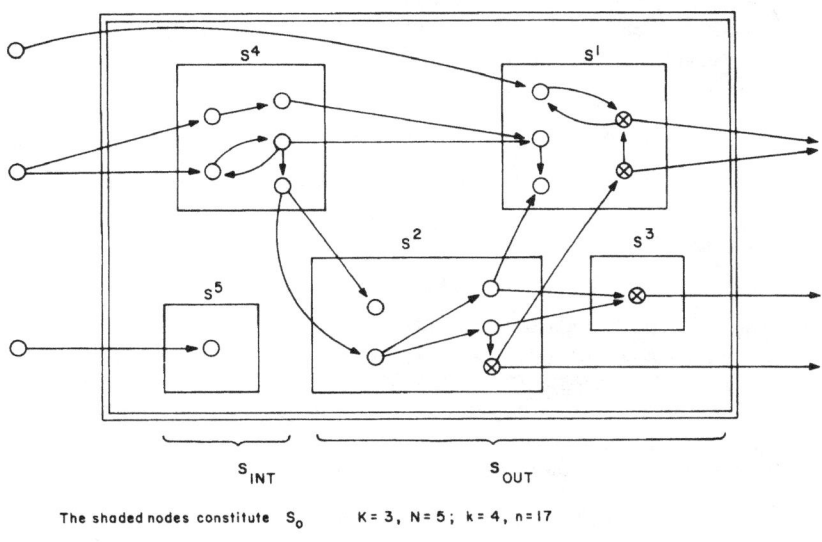

Figure 2

systems science.[9,2–4,6] The discussion here will therefore be brief, and the reader can consult the references for further clarification and discussion.

The basic quantity associated with a discrete variable X (which may be a quantized version of a continuous variable) is the *entropy* $H(x)$, which is a function of the probability distribution on X as follows:

$$H(X) = -\sum_{x} p(x) \log_2 p(x), \qquad (3)$$

where the summation is over all the values x taken by X. (A similar integral definition serves for continuous X.) $H(X)$ is an average of the unexpectedness of particular values x when unexpectedness is measured by $-\log p(x)$, and it is the average amount of information one obtains when informed of the value taken by X on an arbitrary occasion if the distribution is known *a priori*. It is related to the variance of X for variables on which variance is defined, but is more general since it does not require X to be continuous or metric or even to have an interval scale.[9] If X has a finite number n of values, $H(X)$ falls in the interval $[0, \log_2 n]$ with low values indicating that the probability distribution is concentrated and high values that it is diffuse; $H(X) = 0$ if and only if some x has probability 1, and $H(X) = \log_2 n$ if and only if all x are equally probable. The average uncertainty about what value X takes, on an arbitrary occasion, is the same as that for a distribution with $2^{H(X)}$ equally likely alternatives.

Similarly, $H(S)$, denoting the entropy of a system of variables, is defined by

$$H(S) = -\sum_s p(s) \log_2 p(s), \tag{4}$$

where the summation is over all possible values s of the vector $S = X_1, X_2, \cdots, X_n$ and p is the distribution on the n-tuples. Equation (3) is a special case of (4).

Conditional entropies such as $H_{X_1,X_2}(X_3)$ can be defined either through the conditional distribution

$$H_{X_1,X_2}(X_3) = -\sum_{x_1,x_2} p(x_1 x_2) \sum_{x_3} p(x_3 \mid x_1 x_2) \log_2 p(x_3 \mid x_1 x_2)$$
$$= -\sum_{x_1,x_2,x_3} p(x_1 x_2 x_3) \log_2 p(x_3 \mid x_1 x_2)$$

or equivalently through entropies

$$H_{X_1,X_2}(X_3) = H(X_1,X_2,X_3) - H(X_1,X_2)$$

with the quantities on the right defined by (4). $H_{X_1,X_2}(X_3)$ measures the average uncertainty of X_3 which remains for one who knows X_1 and X_2, i.e., the variability in X_1,X_2,X_3 not accounted for in X_1,X_2. $H_{X_1,X_2}(X_3)$ falls in the interval $[0,H(X_3)]$, being zero if X_3 is determined by X_1,X_2 in the sense that $p(x_3|x_1x_2) = 0$ or 1 only, and $H(X_3)$ if X_3 is independent of X_1,X_2 in the sense that $p(x_3|x_1x_2) = p(x_3)$ for all x_1,x_2.

Conditional entropies such as $H_{S^3}(S^4,X_2,S^6)$ are defined similarly, e.g.,

$$H_{S^3}(S^4,X_2,S^6) = H(S^3,S^4,X_2,S^6) - H(S^3).$$

If forms like this are rearranged, they illustrate the very important *additivity property* of the entropy functional, e.g.,

$$H(X_1,X_2,X_3) = H(X_1,X_2) + H_{X_1,X_2}(X_3)$$
$$= H(X_1) + H_{X_1}(X_2) + H_{X_1,X_2}(X_3),$$

which can be interpreted as partitioning the uncertainty of X_1,X_2,X_3 into the uncertainty of X_1, the uncertainty of X_2 which remains after knowledge of X_1, and the uncertainty of X_3 which remains after knowledge of X_1 and X_2. All entropies and conditional entropies are nonnegative.

The amount by which two variables are related, i.e., are not statistically independent, is measured by the *transmission* between them, denoted $T(X_1:X_2)$ and defined through probabilities or by

$$T(X_1:X_2) = H(X_1) + H(X_2) - H(X_1,X_2)$$
$$= H(X_1) - H_{X_2}(X_1)$$
$$= H(X_2) - H_{X_1}(X_2). \tag{5}$$

It is symmetric and measures the amount by which knowledge of one variable reduces uncertainty about the other, or the amount by which the joint uncertainty $H(X_1,X_2)$ is smaller than it would be with X_1 and X_2 independent $[= H(X_1) + H(X_2)]$. The transmission is therefore a measure of relatedness between variables, which accounts for its usefulness in systems science. $T(X_1:X_2)$ falls in the interval $[0, \min\{H(X_1),H(X_2)\}]$, being 0 if and only if X_1 and X_2 are statistically independent and maximum if and only if one variable determines the other. Generalizing (5) to n dimensions,

Laws of Information

$$T(X_1:X_2:\cdots:X_n) = \sum_{j=1}^{n} H(X_j) - H(X_1,X_2,\cdots,X_n) \tag{6}$$

defines the transmission between n variables. The first entity on the right is the value which the joint entropy would have if all variables were mutually independent, and the last is the actual joint entropy, so this transmission measures the global relatedness or constraint in a system of n variables. It is an upper bound on all transmissions between disjoint subsets of the n variables, such as $T(X_1:X_2:X_{n-1})$.

Transmissions between S^i, viewed as vector-valued variables, are defined by the obvious generalization of (6):

$$T(S^1:S^2:\cdots:S^N) = \sum_{i=1}^{N} H(S^i) - H(S^1,S^2,\cdots,S^N) \tag{7}$$

which measures the relatedness between the vectors (but not, of course, between the component variables within the S^i). *Conditional transmissions* such as $T_{S^1}(S^2:S^3)$, $T_{X_1}(X_2:X_3:X_4)$, etc. are defined by conditioning all terms in (7) on the same variable or vector, e.g.,

$$T_{S^1,X_2}(X_3:S^4) = H_{S^1,X_2}(X_3) + H_{S^1,X_2}(S^4) - H_{S^1,X_2}(X_3,S^4).$$

All transmissions and conditional transmissions are nonnegative, which with their additivity properties allows the total relatedness or constraint between groups of variables to be partitioned in a manner analogous to the partition of entropy or uncertainty discussed above. For example,

$$T(X_1:X_2:X_3:X_4) = T(X_1:X_2) + T(X_3:X_4) + T(X_1,X_2:X_3,X_4)$$

partitions the total constraint within a system of four variables into the constraint within $S^1 = X_1,X_2$, within $S^2 = X_3,X_4$, and between S^1 and S^2.

Ashby has developed many identities involving entropies and transmissions,[3] and the most useful for our purposes are given in the Appendix. The extremely valuable Rule of Uniform Subscripting, attributed by Ashby to Garner and McGill, states that if the same subscript is added to every quantity in an identity, the result is an identity. Thus, for example, from (5),

$$T(X_1:X_2) = H(X_1) - HX_2(X_1)$$

by subscripting with X_2 throughout and noting that duplicate subscripting (double conditioning) is redundant, we obtain the new identity

$$T_{X_2}(X_1:X_2) = H_{X_2}(X_1) - H_{X_2}(X_1).$$
$$= 0$$

The following are necessary conventions regarding the empty set $\emptyset$: for any variable or set of variables S,

$$H(\emptyset) = 0$$
$$H(S,\emptyset) = H(S)$$
$$H\emptyset(S) = H(S)$$

$$H_S(\varnothing) = 0$$
$$T(S:\varnothing) = 0. \tag{8}$$

B. Rate Definitions

Although the entropy defined above is a useful measure of the variability or spread of values taken by variables of a dynamic system, and although the transmission is a useful measure of relatedness, both have deficiencies when applied to analysis of dynamic systems in which the system's past history constrains its present values to an important extent. The basis of the deficiency is that this historical constraint is not reflected in the probabilities and joint probabilities which enter the calculations, and therefore, the usefulness of the history in reducing uncertainty is wasted. For example, suppose we are interested in the entropy of variable X_1, "time of tomorrow's sunrise." In Chicago this variable takes values over a three-hour range and therefore is associated with considerable uncertainty, yet nobody would seriously claim that the sunrise gives us information, since its values are so *predictable*; from the two preceding values, today's value can be precisely calculated before the sun rises. As another example, variable X_2, "the next letter to appear under my pencil as I write," has an entropy of about 4.1 bits (using English letter frequencies) which is reduced to about 1 bit if context is used (the uncertainty of X_2 for an observer who has read the preceding text).[15] Third, a delay device which accepts a random binary sequence and repeats it faithfully at the output three steps later: observation of simultaneous input-output pairs would discover a uniform joint distribution, giving zero transmission between input and output variables despite their very strong relation.

The deficiency of information theory in this regard is shared by all statistical techniques which use as data a collection of disassociated "snapshots" of the system. Variance analysis (information theory's stiffest competitor) does so, and so do auto- and cross-correlation techniques as can be seen from the invariance of these measures to cutting up the historical record and pasting it back together with time segments arbitrarily scrambled.

To appreciate the difference in uncertainty of a system with and without historical constraint, contrast an ordinary movie with a scrambled version produced by rearranging the frames into a random time sequence. Or compare the search for a particular listing in an ordinary telephone book with the search in an unalphabetized one (where the "historical" constraint is actually spacial).

The deficiency of information theory in dealing with dynamic systems can be overcome by defining new variables representing earlier versions of the current variables, but it is easier to define the entropy *rate*, i.e., the entropy of X conditional on all its prior values, which is the information carried per observation in a long sequence. The total uncertainty of a long sequence $\langle X(t), X(t+1), \cdots, X(t+m) \rangle$ is then (approximately) this entropy rate times the sequence length, and the *entropy rate* $\overline{H}(X)$ is defined accordingly:

$$\overline{H}(X) = \lim_{m \to \infty} \frac{1}{m} H(X(t), X(t+1), \cdots, X(t+m-1))$$

with units of bits per step.[14] The *conditional entropy rate* $\overline{H}_{X_1}(X_2)$ is defined by

Laws of Information

$$\overline{H}_{X_1}(X_2) = \overline{H}(X_1, X_2) - \overline{H}(X_1)$$

and is the uncertainty of X_2 per step, conditioned on *complete* knowledge of X_1—past, present, and future. Note that in the three-step delay device example above, $\overline{H}_{in}(out) = 0$ as is the case whenever one dynamic variable completely determines another; in that example the determination is mutual and $\overline{H}_{out}(in) = 0$ also. The *transmission rate* $\overline{T}(X_1:X_2)$ is defined by

$$\begin{aligned}\overline{T}(X_1:X_2) &= \overline{H}(X_1) + \overline{H}(X_2) - \overline{H}(X_1, X_2)\\ &= \overline{H}(X_1) - \overline{H}_{X_2}(X_1)\\ &= \overline{H}(X_2) - \overline{H}_{X_1}(X_2)\end{aligned}$$

and measures the constraint per step holding between the dynamic variables. It is an excellent measure of the true amount by which dynamic variables are related and is the quantity communication engineers try to maximize (with $X_1 \rightarrow$ message stream sent, $X_2 \rightarrow$ message stream received). Similarly, $\overline{T}(X^1:S^2)$, etc.

Rates corresponding to all the quantities discussed in Section IIIA will be indicated by overlining. All entropy and transmission rates are nonnegative. A useful rule analogous to the Rule of Uniform Subscripting is as follows:[7] if in an information theory identity every entropy and transmission is replaced by the corresponding rate (by overlining) the result is another identity.

The entropy and transmission rates are excellent tools for understanding dynamic systems, the entropy rate for measuring nondeterministic or unpredictable (from the observer's point of view) behavior of a variable or subsystem, and the transmission rate for measuring relatedness of variables or subsystems. For survival, any nontrivial organism (e.g., a biological organism, company, university, government, etc.) must accept information from its environment and execute appropriate behavior which is coordinated with that environment, and therefore $\overline{T}$(environment:behavior) which measures that relation is an important survival parameter of the organism and a limit on its ability to cope.[5,7] Since $\overline{T}$(input:output) of any system is bounded above by the system's channel capacity C,[14] we may presume that evolution acts to guarantee sufficient channel capacity in organisms which survive, or more vividly to guarantee extinction of those with insufficient capacity.

It is the purpose of the following sections to develop more laws concerning information rates and to suggest their implications for system architecture.

IV. Partition Law of Information Rates

A. Development and Explanation

Proofs of assertions in this and subsequent sections are all simple consequences of rules and identities given in the Appendix, and reference will be made to these by number, e.g., [A3].

To prove the law we equate two expressions [A4], [A7] for $T(E:S)$, recalling that $S = S_0, S_{int}$:

$$H(S) - H_E(S) = T(E:S_0) + T_{S_0}(E:S_{int})$$

then utilize the definition of $T(X_1:X_2:\cdots:X_n)$ [A5] to eliminate $H(S)$,

$$\sum_{j=1}^{n} H(X_j) - T(X_1:X_2:\cdots:X_n) - H_E(S) = T(E:S_0) + T_{S_0}(E:S_{int})$$

and lastly by rearranging and using rates [A2], obtain

$$\sum_{j=1}^{n} \overline{H}(X_j) = \overline{T}(E:S_0) + \overline{T}_{S_0}(E:S_{int}) + \overline{T}(X_1:X_2:\cdots:X_n) + \overline{H}_E(S). \tag{9a}$$

Equation (9) is called the (general form of the) Partition Law of Information Rates (PLIR) since it partitions the sum on the left into the nonnegative quantities on the right. As a notational convenience each term in (9a) will be given a name and symbol:

$F = \sum_{j=1}^{n} \overline{H}(X_j)$ total rate (of "information flow")

$F_t = \overline{T}(E:S_0)$ thruput rate

$F_b = \overline{T}_{S_0}(E:S_{int})$ blockage rate

$F_c = \overline{T}(X_1:X_2:\cdots:X_n)$ coordination rate

$F_n = \overline{H}_E(S)$ noise rate

and with these, the PLIR can be expressed as

$$F = F_t + F_b + F_c + F_n. \tag{9b}$$

The *total rate* F is the sum of the rates for the individual variables. It represents the total (nondeterministic) activity in S if intervariable relationships are ignored. Since each X_j is assumed to be a (possibly noisy) function of variables in E and S and can be viewed as an entity which computes that function, we can view F (somewhat loosely) as the total amount of "computing" going on in S. By (9), F bounds F_t, F_b, F_c, and F_n individually as well as their sum and consequently represents a global upper bound for all information rates.

The *thruput rate* F_t measures the input-output flow rate of S, or the number of bits per step passing through S as a communication channel, or the strength of the relation between the input and output of S. This rate is important for the survival of S as an organism[7]; survival, as a goal, generally dictates some minimum F_t. By definition, the channel capacity C_S, of S is the least upper bound for F_t. It is important to distinguish $\overline{H}(S_0)$, the output rate, from F_t, the thruput rate, since $\overline{H}(S_0)$ may exceed F_t if S is a generator of information (let us say by containing a random number generator); the thruput is only that portion of the output which is related to the input E.

The *blockage rate* F_b is the rate at which information about the input E is blocked within S and not allowed to affect the output. To see this expand F_b by [A4], [A1]:

$$F_b = \overline{H}_{S_0}(E) - \overline{H}_{S_0,S_{int}}(E).$$

Note that $S_0, S_{int} = S$. Next add and subtract $\overline{H}(E)$:

$$F_b = [\overline{H}(E) - \overline{H}_S(E)] - [\overline{H}(E) - \overline{H}_{S_0}(E)]$$
$$= \overline{T}(E:S) - \overline{T}(E:S_0).$$

$\overline{T}(E:S)$ measures the relation between E and the whole system S, or the effect of E on S, or the information which S carries about E. $\overline{T}(E:S_0)$ is similar but only for the output

Laws of Information

subsystem of S. Thus, F_b measures "information which S carries but S_O does not" about E, i.e., information carried by S_{int} about E which is not reflected in the output. The idea is that generally E affects S_{int} in ways not reflected in S_O and F_b is a measure of that blockage of information within S.

The blockage of information is an important function of information-processing organisms. Suppose, for example, you want to check pronunciation of a word by looking it up in a dictionary. To find the word you have to (marginally) read many irrelevant words before you arrive at the target word, and these bothersome words do not affect your output (pronunciation of target). They represent irrelevant information from the dictionary which is necessarily taken in but immediately forgotten. It is common for an organism to filter the information received from its environment, allowing only the "relevant" aspects to affect its output, and F_b is a measure of the rate at which irrelevant information from E is blocked or6discarded. Blockage of information will be discussed in more detail in Section VII.

The *coordination rate* F_c is a measure of the total coordination between all the variables in S. Just as $\overline{T}(X_1:X_2)$ measures the amount of constraint, relatedness, or coordination between two variables, so does F_c for all n variables. In general, it is not possible to break this global constraint into a sum of constraints each over fewer than n X_j's,[7] and when it is in fact possible to do so, the system may be simply a nominal conjunction of separate and independent subsystems. However, it is often possible to express F_c, through decomposition or partition rules (e.g., [A6][3]), in a way exposing the near-independence of subsystems. This is important since it is common for systems to be composed of several subsystems, within which the coupling is strong but between which the coupling is weak.[13] The terms in [A6] measure these couplings and provide, for example, for a calculation of the fraction of the total constraint F_c attributable to constraint between subsystems. Similar calculations can be used to detect subsystems of a complex system.[6] Hierarchical decompositions of this type will be discussed further in Section V.

The global constraint-rate F_c can be viewed as the rate of internal communication in S which allows S to act as a whole rather than a sum of independent parts. This internal signalling (nervous and chemical signals in biological organisms, memo-passing and telephone-calling in corporations, etc.) is required whenever S is faced with a "problem" of nontrivial complexity, and (a bit loosely) the minimum F_c needed to successfully cope with the problem can be interpreted as corresponding to the complexity of the problem posed, that is, the amount of global cooperation required between the parts of S for its successful solution.

The *noise rate* F_n is the amount of uncertainty per step about S, given complete knowledge of its input. Clearly it corresponds to internally-generated information, or some might view it as the rate of "free will," since it corresponds to behavior which has no apparent cause (at least not in the known environmental influences E). If the output rate $\overline{H}(S_O)$ exceeds the thruput rate $\overline{T}(E:S_O)$, the difference, which constitutes part of F_n, can be viewed as noise analogous to that on a communication channel—a message at the output not caused by a message at the input.

The thrust of this paper will be well served by making at this point a *deterministic assumption*: we will assume henceforth that all rates conditional on E are zero. This means that if the complete knowledge of the system's input is available (past, present, and future) there is no residual uncertainty about any of the system variables, and consequently these variables carry no information, either individually [$H_E(-) = 0$] or about each other [$\overline{T}_E(-)$

= 0]. The system S will be assumed deterministic, conveying and processing information from E but producing none by itself. With this deterministic assumption, $F_n = 0$, and F_t and F_b can be simplified [A4], [A1]:

$$F_t = \overline{H}(S_0)$$
$$F_b = \overline{H}_{S_0}(S_{int}).$$

With these simplifications, the PLIR becomes

$$\sum_{j=1}^{n} \overline{H}(X_j) = \overline{H}(S_0) + \overline{H}_{S_0}(S_{int}) + \overline{T}(X_1:X_2:\cdots:X_n)$$
$$F \quad = \quad F_t \quad + \quad F_b \quad + \quad F_c. \tag{10a}$$

Equation (10) will be called the Deterministic Partition Law of Information Rates (DPLIR).

The interpretations given above for the terms hold as well for the DPLIR, with the simplification that since output equals thruput under the deterministic assumption, F_t can also be considered the *output rate*. The DPLIR can be expressed in words as

(Total rate) = (thruput or output rate) + (blockage rate) + (coordination rate) (10b)

or more briefly (with "rate" understood),

$$\text{Total} = \text{thruput} + \text{blockage} + \text{coordination}. \tag{10c}$$

Each X_j may be viewed as the output of a communication channel from E to X_j, and under the deterministic assumption $\overline{H}(X_j)$ may be attributed entirely to E. For each X_j there is a maximal value of $\overline{H}(X_j)$ attainable by varying the statistical properties of E, called its channel capacity and denoted C_j. We will denote the sum of all the individual C_j by C_{sum}, and then clearly since C_j bounds $\overline{H}(X_j)$, C_{sum} bounds F.

B. Discussion

An easy conclusion from (10) is that the total rate is an upper bound on all of its three constituents. If minima for any or all of the three are specified, (10) specifies a lower bound for the total. Moreover, C_{sum} is an upper bound for C_s (which bounds F_t):

$$C_{sum} = \sum_{j=1}^{n} C_j = \sum_{j=1}^{n} \max\{\overline{H}(X_j)\}$$
$$\geq \max \sum_{j=1}^{n} \overline{H}(X_j) \geq \max\{F_t\} = C_s, \tag{11}$$

where the maxima are over all the ways the input statistics of E can be varied. Although it is well known that C_{sum}, the sum of capacities of the parts of a system, is an upper bound on the overall system's capacity to *transmit* information (F_t), it does not seem to be generally appreciated that this sum also limits the system's capacity to *block* information (F_b) and to *coordinate* internally (F_c). That it does so can be easily seen by substituting F_b and F_c for F_t in (11).

The DPLIR (and for that matter the PLIR) makes clear the sort of competition between thruput, blockage, and coordination which results if the total rate is fixed or bounded, let

Laws of Information   431

us say in a system having a fixed number of components each of fixed capacity (such as neurons in a human brain, individuals in an organization, primitive operations per second in a computer, etc.). For if the total is fixed, thruput can rise only at the expense of blockage and coordination, and similarly for the others.

With the total fixed, demands for high thruput or output can only be satisfied by reducing blockage and coordination, in other words by presenting S with information most of which is relevant for the output ($\Rightarrow$ low F_b) and posing problems for S which do not require much internal coordination ($\Rightarrow$ low F_c). The following is an example of $F_b = 0$, $F_c \approx 0$, $F_t \approx F$: the problem is to keypunch the listings in an ordinary telephone book, one card per listing, given n workers. The solution is to tear apart the book, distribute pages to the keypunch operators who then operate independently except for the final collation of outputs. Clearly demands for blockage ("Skip names containing an e") or coordination ("Integrate all the cards into one list, in phone-numerical order") reduce the output rate (proportional to the number of cards in the final result/overall time required).

Similarly, demands for high blockage when the total is fixed must induce low output and coordination rates. For an example of $F_c \approx 0$, $F_t \approx 0$, $F_b \approx F$, imagine instructing the above n workers to keypunch only entries with names containing both e and ti or whose telephone number is prime. Here practically all of the effort goes into discarding entries not to be punched, i.e., blocking.

Lastly, if the problem requires high coordination when the total is fixed, then blockage must be low and output will necessarily be small also. Here for an example imagine n workers assembling a jigsaw puzzle. Since every piece gets used, blockage is zero, and the larger the coordination required (roughly, the larger the puzzle) the smaller the output rate (which is inversely proportional to solution time).

The noise rate F_n (included in the PLIR but not in the DPLIR) competes with all of the above three when F is fixed, and clearly it should be minimized to maximize any of them. This yields the rather obvious conclusion that efficient processors should be deterministic.

It does not seem very surprising that there is a mutual antagonism in systems between the three goals of high output rate (F_t), quick discrimination and discard of irrelevant or misleading input (F_b), and ability to deal with complex and holistic problems (F_c), but it may be surprising that these demands are essentially independent and additive in the requirements they place on the information-processing or computing requirements of the system (F). Neither does it seem surprising that in a system of fixed capacity these requirements can be traded off against one another, if a suitable "recoding" can be found; the situation is reminiscent of the tradeoff between signal-to-noise ratio and inverse bandwidth in communication systems. However, it may be surprising that the implied tradeoff is such a simple one—$\Delta F_t + \Delta F_b + \Delta F_c = 0$, if F is fixed (otherwise $= \Delta F$).

The DPLIR makes it clear why the channel capacity C_s of S is such an inadequate parameter by which to gauge its information-processing ability, except in the case of systems devoted predominantly to the mere transport of information (such as conventional communication channels): C_s is a bound on F_t but is irrelevant to F_b and F_c. Since the information-processing tasks required of S may consist mostly of discerning the relevant inputs (F_b) and integrating these in "calculating" the appropriate response (F_c) with very little actual output (F_t), the C_s may be almost completely irrelevant. This situation is probably a very common one since important outputs often represent very small amounts of information. For a cornered raccoon, whether to flee or fight is a one-bit output/decision, similarly for a company, whether to market a new product or not, and for Congress, whether or not to pass a bill.

Corresponding to the definition of C_s as the highest possible value of F_t attainable by varying the statistical nature of the input E, we could define analogous bounds on F_b, F_c, and F as the blockage capacity B_s, the coordination capacity K_s, and the total capacity T_s of S, respectively. The capacity of S as an information-processor would then be much more adequately represented by the quadruple $\{C_s, B_s, K_s, T_s\}$ than by C_s alone. Of course, each of these is bounded by C_{sum} [by appropriate modifications of (11)]. Indeed this quadruple of capacities, and especially of the ratios between them, might characterize the "type" of problem which S is best equipped to handle. The conditions on E which realize each of these four capacities are, in general, different.

Under some circumstances it may be useful to consider the ratio of F to T_s, i.e., the ratio of actual total flow rate to the maximum of that quantity, as an indication of the informational efficiency of S, or alternatively of the optimality of the actual E with respect to S (since for *some* E, $F = T_s$ by definition of T_s). Although such a calculation is seldom feasible, estimates may be possible. An example is given in Section V.

If estimates can be made of the F (or of F_t, F_b, and F_c) needed by S to cope with a "problem," this estimate dictates a minimum for C_{sum}, and if channel capacities of available system components are known, this prescribes a minimum number of components needed in S. If the actual components in S constitute a C_{sum} in excess of the minimum, the excess represents underused capacity and reflects a mismatch between the actual E and the optimal E, or (given E) a mismatch between the actual S and the optimal S. Although such a mismatch should be minimized on grounds of economy, it is inevitable in any S which deals with different *types* of input at different times. The best that can be done is to minimize the average diseconomy; the wider the variety of inputs, the larger the average inefficiency. This affirms the common observation that general-purpose systems, which are expected to deal with a wide variety of problems, are slower and less efficient than special-purpose systems operating on a narrower variety of problems, if the comparison is made between systems of equal size.

It should not be inferred from the DPLIR that there is a partition of *variables* in S into the corresponding functions—that the activity of some variables is devoted to thruput, of others to blockage, and of still others to coordination. It is seldom possible to partition the variables that way, since F_t, F_b, and F_c represent global measurements on the whole system and cannot, in general, be localized. On the contrary, it may be that the whole system is devoted at one time to one of these functions and at another to another. For example, in the process of writing a technical paper these functions are executed more or less sequentially by the writer: collect the relevant points and weed out the irrelevant (blockage), then organize them into a consistent and coherent whole (coordination), then write (output).

From the PLIR and DPLIR we can draw several inferences about the system architecture which will make S perspective. We will suppose that the environment is given and poses a "problem" or sequence of problems for S. Our objective is to determine, within the class of systems which can successfully solve the problem, the form of those systems which will minimize F and by implication C_{sum} and at least loosely the number of variables or components in S. Informally this can be seen as the problem biological evolution "solves"—given the environment, find the most economical structure which can cope with it.

Although in some instances nondeterministic behavior can be useful (e.g., the random movements of a butterfly which confuse predators), that is exceptional, and we will assume that having $F_n = 0$ is more generally advantageous since it minimizes F.

The problem can be assumed to dictate a certain minimum F_t, that is, the system has to make *some* response. The blockage F_b required is really more dependent upon E than upon S, and little can be inferred about S other than that it should provide whatever minimum blockage is dictated jointly by E and the problem. To minimize F, then, requires reducing F_c to the minimum consistent with whatever is demanded by the problem. Since F_c represents coupling between the parts of S, this means that the parts of S should operate as nearly independently as possible, coordinating to only the minimum extent required by the problem. The best structure, in other words, is the loosest structure compatible with constraints imposed by the problem. In an informal way these guidelines for systems can be paraphrased as follows. From the class of system designs which solve the problem successfully select those which:

1. produce the minimum allowable output ("do not do anything unnecessary"),
2. perform as little blockage as possible ("try to take in a minimum of irrelevant input"),
3. reduce internal coordination to the minimum consistent with other requirements ("maximize freedom of the components").

These guidelines minimize F and therefore the *minimum* C_{sum} required. To make the actual $C_{sum} = \sum C$ differ as little as possible from this minimum, i.e., to make the system as economical as possible, we could add:

4. as far as possible, match components to tasks so that each component is operated at capacity ("let each component do what it does best, and work it as hard as you can").

Elsewhere arguments are presented to show that hierarchical structure is optimal for the latter guideline,[7] since that architecture allows (but of course does not necessarily guarantee) operation of every component in S at its full capacity.

V. Hierarchy

A. Decomposition of $\overline{H}$ and $\overline{T}$

There are a number of ways in which hierarchical structure is ideally suited for the information-processing functions which organisms must perform. ["Organism" is understood to mean any system—biological, mechanical, sociological, etc.—which has a dynamic, physical existence and to which the concept of survival (in the face of threats from the environment) is applicable.] Some of these have been discussed earlier.[7] In brief, such a structure allows efficient use of the information-processing capacity distributed throughout the system. If the system and environment are well matched, hierarchical structure allows all parts of the system to operate at their individual capacities, minimizes mere point-to-point transfer of information, and maximizes the prompt blockage of irrelevant input information.

It seems particularly interesting and suggestive, therefore, that the rules of N-dimensional information theory are so well adapted to the decomposition of S into subsystems, of these into sub-subsystems, and so on, the whole representing a *hierarchical* decomposition of S.

Consider $\overline{H}(S) = \overline{H}(X_1, X_2, \cdots, X_n)$, the *entropy* rate associated with the entire system of n variables. If the variables are grouped into N sub-subsystems S^i, there is an entropy rate $\overline{H}(S^i)$ associated with each. The entropy rate of S is smaller than the sum of the $\overline{H}(S^i)$, in general, since if the S^i are interrelated, knowledge of some reduces uncertainty about others; this relatedness is measured by the transmission rate between subsystems. The decomposition rule for $\overline{H}$ is [A5]:

$$\overline{H}(S) = \sum_{i=1}^{N} \overline{H}(S^i) - \overline{T}(S^1:S^2:\cdots:S^N) \tag{12a}$$

$$\begin{pmatrix} \text{Entropy rate} \\ \text{of system } S \end{pmatrix} = \begin{pmatrix} \sum_{i=1}^{N} \text{entropy rate of} \\ \text{subsystem } S^i \end{pmatrix} - \begin{pmatrix} \text{transmission rate} \\ \text{between subsystems} \end{pmatrix} \tag{12b}$$

This rule is in a form which appears throughout this section, specifically:

$$\text{quantity }(S) = \sum_{i=1}^{N} \text{quantity }(S^i) + \text{correction }(S^1, S^2, \cdots, S^N).$$

If S is viewed as the whole and the S^i as its parts, this form corresponds to the fact that in general "the whole is different from the sum of its parts." The partition which slices S into the S^i assigns each variable in S to a subsystem, and it also creates a "system of subsystems" which we will denote by $\mathscr{S}$. The variables in $\mathscr{S}$ are the vectors S^1 through S^N, each vector being viewed as a single undifferentiated variable. Thus in the general form, quantity (S^i) is associated with variables in S^i only and correction $(S^1, S^2, \cdots, S^N)$ is associated with variables in $\mathscr{S}$, i.e., with interactions between the subsystems seen as modular units.

The general form can of course be applied iteratively to the subsystems, and so on, and therefore is compatible with hierarchical decomposition of S. In the current instance, to decompose $\overline{H}(S^i)$ the same identity (12) could be used again, although as a consequence of our notation (which assumed that the subsystems of the S^i were the atomic elements X_j^i) we have [A5]:

$$\overline{H}(S^i) = \sum_{j=1}^{n_i} \overline{H}(X_j^i) - \overline{T}(X_1^i:X_2^i:\cdots:X_{n_i}^i).$$

The transmission rate between the n variables of S, which measures their global relatedness, can be expressed as the sum of transmission rates within and between subsystems as follows [A6]:

$$\overline{T}(X_1:X_2:\cdots:X_n) = \sum_{i=1}^{N} \overline{T}(X_1^i:X_2^i:\cdots:X_{n_i}^i) + \overline{T}(S^1:S^2:\cdots:S^N) \tag{13a}$$

$$\begin{pmatrix} \text{Coordination in} \\ \text{the system } S \end{pmatrix} = \begin{pmatrix} \sum_{i=1}^{N} \text{coordination in} \\ \text{subsystem } S^i \end{pmatrix} + \begin{pmatrix} \text{coordination in } \mathscr{S}, \\ \text{the system of subsystems} \end{pmatrix} \tag{13b}$$

Laws of Information

Each term in (13) can be further decomposed by reiteration of the same rule, for example,

$$\overline{T}(S^1:S^2:\cdots:S^N) = \overline{T}(S^1:S^2) + \overline{T}(S^3:S^4:S^5) + \overline{T}(S^6:\cdots:S^N)$$
$$+ \overline{T}(S^1,S^2:S^3,S^4,S^5:S^6,\cdots,S^N).$$

If the last term in (13) is small with respect to the first term, it is an indication that the coupling between subsystems is small compared to the coupling within them, i.e., that S may be "nearly" decomposable into *independent* subsystems. When it occurs, this is very important for the systems scientist since it often means that over the short run, the subsystems can be understood rather well in isolation and that only in the long run do the subsystem interactions play a dominant role. In his classic article Simon[13] argues that evolution favors organisms with this nearly-separable structure. His argument, based on stability considerations, leads to the same conclusion as the informational argument of the preceding section, viz., for maximum economy (which has an evolutionary advantage), minimize the internal coordination rate F_c, which considering the coordination rate within subsystems to be fixed means make the subsystems as nearly independent as possible. Since (as Simon shows) this near-independence of parts at all levels is associated with hierarchical structure, we have yet another reason, based on informational considerations, to expect hierarchy in complex organisms.

To this point there has been no suggestion of how to partition the variables of S into N groups (subsystems). Clearly the left side of (13) is not dependent on the partition, but of the various possible partitions (or ways to look at S as composed of subsystems) some will result in the last term of (13) being large, some in it being small, and the latter usually are considered "better" partitions since the implied subsystems are more nearly independent and presumably can be better understood in isolation. A complete hierarchical decomposition of S consists of a "good" partition into subsystems, plus a "good" partition of each subsystem into sub-subsystems, and so on down to atomic levels. It can be represented by a tree with S at the top (trunk) and the n atomic variables at the bottom (branches), with successive levels representing successively finer partitions of the system. For example, the identity (from [A6] applied twice)

$$T(X_1:X_2:X_3:X_4:X_5:X_6:X_7:X_8) = [T(X_1:X_2) + T(X_3:X_4) + T(X_1,X_2:X_3,X_4)]$$
$$+ [T(X_5:X_6) + T(X_7:X_8) + T(X_5,X_6:X_7,X_8)] + T(X_1,X_2,X_3,X_4:X_5,X_6,X_7,X_8)$$

represents a decomposition of S into $S^1 = X_1,X_2,X_3,X_4$ and $S^2 = X_5,X_6,X_7,X_8$, then decomposition of S^1 into $S^{11} = X_1,X_2$ and $S^{12} = X_3,X_4$, and finally decomposition of S^{11} into X_1 and X_2, etc. Another example is the standard hierarchical representation of the U.S. government: Government = Legislative, Judicial, Executive; Legislative = House, Senate; etc.

B. Decomposition of Terms in the Partition Law

Since decomposition rules exist for entropy and transmission rates, it could be expected that analogous rules would hold for all rates in the DPLIR, making it possible to discuss the thruput, blockage, coordination, and total rates of S in terms of corresponding rates within the S^i and within $\mathscr{S}$. These rules will be developed below for deterministic systems and subject to the following simplifying assumption: we will assume henceforth that $S_0 = S_0^1, S_0^2, \cdots, S_0^K$, i.e., that all the output variables of the K output subsystems are outputs of S. The rules for F and F_c do not depend upon that assumption but those for F_t and F_b are thereby considerably simplified.

For the *total* flow rate the rule is trivial:

$$F = \sum_{i=1}^{N} F^i \tag{14a}$$

$$\begin{pmatrix} \text{Total rate} \\ \text{for system} \end{pmatrix} = \begin{pmatrix} \sum_{i=1}^{N} \text{total rate for} \\ \text{subsystem } i \end{pmatrix} \tag{14b}$$

Proof.

$$\sum_{j=1}^{n} \overline{H}(X_j) = \sum_{i=1}^{N} \sum_{j=1}^{n_i} \overline{H}(X_j^i)$$

since the X on the left and right are identical variables merely labeled differently. Both sums contain exactly n terms.

For the *thruput* F_t the rule is

$$F_t = \sum_{i=1}^{K} F_t^i - \overline{T}(S_0^1 : S_0^2 : \cdots : S_0^K) \tag{15a}$$

$$\begin{pmatrix} \text{Thruput of} \\ \text{system } S \end{pmatrix} = \begin{pmatrix} \sum_{i=1}^{K} \text{thruput of output} \\ \text{subsystem } S^i \end{pmatrix} - \begin{pmatrix} \text{coordination between the out-} \\ \text{puts of the output subsystems} \end{pmatrix} \tag{15b}$$

Proof.

$$F_t = \overline{H}(S_0) = \overline{H}(S_0^1, S_0^2, \cdots, S_0^K) \qquad [\text{Def. of } S_0]$$

$$= \sum_{i=1}^{K} \overline{H}(S_0^i) - \overline{T}(S_0^1 : S_0^2 : \cdots : S_0^K) \qquad [\text{A5}]$$

The thruput of the system is less than the sum of output subsystem thruputs by the amount the latter are related.

For the *blockage* F_b the rule is

$$F_b = \sum_{i=1}^{N} F_b^i + \overline{H}_{S_0}(S_0^L, \cdots, S_0^N) - [\overline{T}(S^1 : S^2 : \cdots : S^N) - \overline{T}(S_0^1 : S_0^2 : \cdots : S_0^N)] \tag{16a}$$

$$\begin{pmatrix} \text{Blockage within} \\ \text{the system } S \end{pmatrix} = \begin{pmatrix} \sum_{i=1}^{N} \text{blockage within} \\ \text{subsystem } S^i \end{pmatrix} + \begin{pmatrix} \text{subsystem output information} \\ \text{blocked within } \mathcal{S} \end{pmatrix}$$

$$- \begin{pmatrix} \text{coordination between the subsystems which is} \\ \text{not reflected in coordination of their outputs,} \\ \text{i.e., coordination blocked within } \mathcal{S} \end{pmatrix} \tag{16b}$$

Proof.

$$\overline{H}_{S_0}(S_{\text{int}}) = \overline{H}(S) - \overline{H}(S_0) \tag{A3}$$

$$= \sum_{i=1}^{N} \overline{H}(S^i) - \overline{T}(S^1:S^2: \cdots : S^N) \qquad [A5]$$

$$- [\overline{H}(S_0^1, \cdots, S_0^N) - \overline{H}_{S_0}(S_0^L, \cdots, S_0^N)] \qquad [A3]$$

$$= \sum_{i=1}^{N} \overline{H}(S^i) - \overline{T}(S^1:S^2: \cdots : S^N) - \sum_{i=1}^{N} H(S_0^i)$$

$$+ \overline{T}(S_0^1:S_0^2: \cdots : S_0^N) + \overline{H}_{S_0}(S_0^L, \cdots, S_0^N) \qquad [A5]$$

$$= \sum_{i=1}^{N} \overline{H}_{S_0}(S^i_{\text{int}}) + \overline{H}_{S_0}(S_0^L, \cdots, S_0^N) \qquad [A3]$$

$$- [\overline{T}(S^1:S^2: \cdots : S^N) - \overline{T}(S_0^1:S_0^2: \cdots : S_0^N)]$$

Note that $\overline{H}_{S_0}(S_0^L, \cdots, S_0^N)$ is subsystem output information not carried in S_0 and therefore blocked within $\mathcal{S}$. Of the two transmission terms in (16), the first is always at least as large as the second so the quantity subtracted is nonnegative, but the proof of that fact will be deferred until Section VI.

For the *coordination* F_c the rule is

$$F_c = \sum_{i=1}^{N} F_c^i + \overline{T}(S^1:S^2: \cdots : S^N)$$

$$\begin{pmatrix} \text{Coordination within} \\ \text{the system } S \end{pmatrix} = \begin{pmatrix} \sum_{i=1}^{N} \text{coordination within} \\ \text{subsystem } S^i \end{pmatrix} + \begin{pmatrix} \text{coordination within } \mathcal{S}, \\ \text{the system of subsystems} \end{pmatrix}$$

which is (13), repeated here for convenience.

Some of the terms in these rules which arise from $\mathcal{S}$ involve the complete vectors S^i, and some only involve the output subvectors S_0^i. To create a decomposition rule involving only output subvectors (which therefore represent observable variables), we can define a new rate $F_{\text{int}} = F_b + F_c$, the *internal* rate, and obtain

$$F_{\text{int}} = \sum_{i=1}^{N} F^i_{\text{int}} + \overline{H}_{S_0}(S_0^L, \cdots, S_0^N) + \overline{T}(S_0^1:S_0^2: \cdots : S_0^N). \qquad (17)$$

As would be expected, when the coordination within $\mathcal{S}$, the system of subsystems, is zero, all of the rules given above are simplified since then each subsystem is independent of all others. It is easy to show in that case, i.e., when $\overline{T}(S^1:S^2: \cdots : S^N) = 0$, that

$$\overline{H}(S) = \sum_{i=1}^{N} \overline{H}(S^i)$$

$$F = \sum_{i=1}^{N} F^i$$

$$F_t = \sum_{i=1}^{K} F_t^i$$

$$F_b = \sum_{i=1}^{K} F_b^i + \sum_{i=L}^{N} \overline{H}(S^i)$$

$$F_c = \sum_{i=1}^{N} F_c^i$$

C. Example

To illustrate hierarchical decomposition and the calculation of quantities discussed above we will employ a simple binary device which accepts eight binary inputs $E_1, \cdots, E_8$ which are mutually independent and equiprobably 0 or 1, and emit as output their sum, modulo 2. See Fig. 3. The system is constructed from binary adders (modulo 2) whose output is 0 if the two inputs are equal and 1 if they are different. A one-second delay is assumed in each adder and therefore a three-second delay is implied for the device. A fresh set of eight inputs, independent of all previous inputs, is presented to the system each second.

Under these conditions, $\overline{H}(E_i) = 1$ bit per second (bit/s) for each E_i and $\overline{H}(E_1, \cdots, E_8)$ = 8 bit/s. Each of the X_j takes values 0 and 1 equiprobably and independently of its earlier values, and therefore $\overline{H}(X_j) = 1$ bit/s for each X_j (from (3)). We will calculate the quantities in the DPLIR for subsystems S^1, S^2, and S^3, then for S two ways.

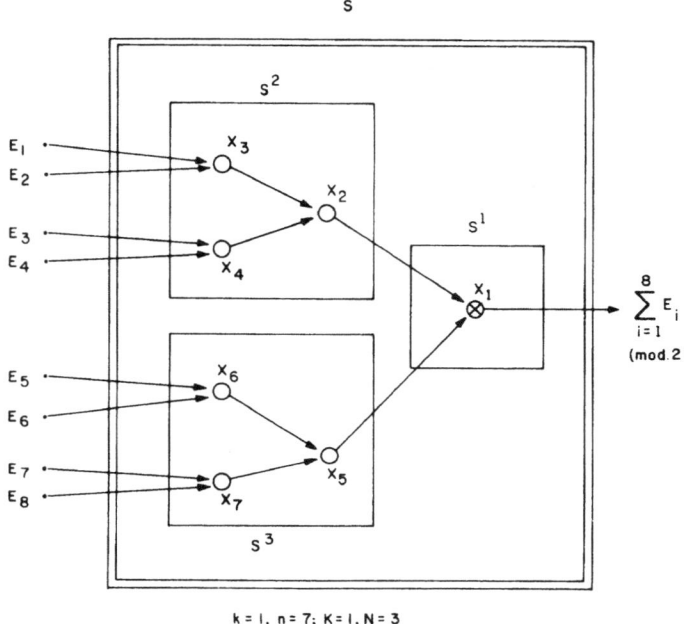

$k = 1, n = 7; K = 1, N = 3$

Figure 3

Laws of Information

Subsystem $S^1 = X_1$ is trivial: $F^1 = F_t^1 = 1$ and $F_b^1 = F_c^1 = 0$ [since there are no internal variables and $\bar{H}_{X_1}(\emptyset) = 0$, $T(X_1:\emptyset) = 0$ by (8)]. Thus the DPLIR says

$$F^1 = F_t^1 + F_b^1 + F_c^1$$
$$1 = 1 + 0 + 0.$$

For $S = X_3, X_4, X_2$, the total F^2 is $\sum \bar{H}(X_j^2) = 3$, and $F_t^2 = \bar{H}(X_2) = 1$. To calculate $F_b^2 = \bar{H}(X_3, X_4, X_2) - \bar{H}(X_2)$ we need $\bar{H}(X_3, X_4, X_2)$ which can be expressed as $\bar{H}(X_3) + \bar{H}(X_4) - \bar{T}(X_3:X_4) + \bar{H}_{X_3,X_4}(X_2)$. The $\bar{T}$ term is zero since X_3 and X_4 are independent, and the last term is zero since X_2 is a deterministic function of X_3, X_4. Thus $\bar{H}(S^2) = 2$ and $F_b^2 = 1$. Next, $F_c^2 = \sum \bar{H}(X_j^2) - \bar{H}(S^2) = 3 - 2 = 1$, and the DPLIR says

$$3 = 1 + 1 + 1.$$

The calculations for S^3 are identical.

For $S = X_1, X_2, \cdots, X_7$ we have $F = 7$ and $F_t = 1$. For the others we need $\bar{H}(S)$ which can be expressed as

$$\bar{H}(S) = \bar{H}(X_3) + \bar{H}(X_4) + \bar{H}(X_6) + \bar{H}(X_7) - \bar{T}(X_3:X_4:X_6:X_7) + \bar{H}_{X_3,X_4,X_6,X_7}(X_2,X_5,X_1)$$
$$= \bar{H}(X_3) + \bar{H}(X_4) + \bar{H}(X_6) + \bar{H}(X_7)$$
$$= 4.$$

Thus $F_b = 4 - 1 = 3$ and $F_c = 7 - 4 = 3$, and the DPLIR says

$$7 = 1 + 3 + 3.$$

To represent the four rates for S in terms of the rates for the S^i, we use (13)–(16). Thus

$$F = 7 = (1 + 3 + 3)$$

from (14),

$$F_t = 1 = (1) - 0$$

from (15),

$$F_b = 3 = (0 + 1 + 1) + 1 - [1 - 1]$$

from (16), and

$$F_c = 3 = (0 + 1 + 1) + 1$$

from (13). The calculation for F_b requires

$$\bar{H}_{S_0}(S_0^2, S_0^3) = \bar{H}(S_0, S_0^2, S_0^3) - \bar{H}(S_0) = 2 - 1 = 1$$

and

$$\bar{T}(S^1:S^2:S^3) = \bar{H}(S^1) + \bar{H}(S^2) + \bar{H}(S^3) - \bar{H}(S)$$
$$= 1 + 2 + 2 - 4 = 1$$

and

$$\bar{T}(S_0^1:S_0^2:S_0^3) = \bar{H}(X_1) + \bar{H}(X_2) + \bar{H}(X_3) - \bar{H}(X_1,X_2,X_3)$$
$$= 1 + 1 + 1 - 2 = 1.$$

Since the entropy of a binary variable cannot exceed 1 bit per observation, F_t cannot exceed 1 bit/s, and since F_t attains that limit in this case, the full channel capacity is being used: $F_t = C_s = 1$ bit/s. The blockage capacity B_s is the maximum attainable value of

$$\overline{H}(S) - \overline{H}(S_0) = \overline{H}(X_3, X_4, X_6, X_7) - \overline{H}(X_1)$$

and that maximum of 3 bit/s is realized in this case: $F_b = B_s = 3$ bit/s. The coordination capacity K_s is the maximum of $\overline{T}(X_1:X_2:X_3:X_4:X_5:X_6:X_7) = \sum \overline{H}(X_j) - \overline{H}(S)$, and that maximum of 3 bit/s is also realized: $F_c = K_s = 3$ bit/s. With seven binary variables, $F = \sum \overline{H}(X_j)$ cannot exceed 7 bit/s, and that is also realized. Thus all of the four system capacities defined above are realized for the input given. The input is therefore optimal for the system, or from another point of view the system is optimal (most economical) for the input. This case realizes the ideal of having every part of a system operating at full capacity. In Section VII it will be shown that for the assigned task no system of fewer binary adders can succeed.

By way of contrast, if the probabilities for the E_i are unbalanced to $p(0) = 0.89$, $P(1) = 0.11$, the terms in the DPLIR are

S^1: $0.987 = 0.987 + 0 + 0$
S^2 and S^3: $2.327 = 0.899 + 0.529 + 0.899$
S: $5.641 = 0.987 + 1.869 + 2.785$

The rates decompose thus:

$$F = 5.641 = (0.987 + 2.327 + 2.327)$$
$$F_t = 0.987 + (0.987) - 0$$
$$F_b = 1.869 = (0 + 0.529 + 0.529) + 0.811 - [0.987 - 0.987]$$
$$F_c = 2.785 = (0 + 0.899 + 0.899) + 0.987$$

In this case $\overline{H}(X_3) = \overline{H}(X_4) = \overline{H}(X_6) = \overline{H}(X_7) = 0.714$, $\overline{H}(X_2) = \overline{H}(X_5) = 0.899$, and $\overline{H}(X_1) = 0.987$. Nodes on the left are operated less efficiently than on the right, since their outputs deviate more from the optimum condition of equiprobability which realizes their capacity of 1 bit/s.

VI. Loss of Constraint

It is well known that all deterministic systems are information-losing in the sense that their outputs carry less information per second than their inputs, or at most the same amount $[\overline{H}(E) = \overline{H}(E,S) = \overline{H}(S) + \overline{H}_s(E) \geq \overline{H}(S) \geq \overline{H}(S_0)]$: $\overline{H}(E) \geq \overline{H}(S)$. The central point of this section is that deterministic systems are also constraint-losing in the sense that

$$\overline{T}(E^1:E^2: \cdots : E^N) \geq \overline{T}(S_0^1:S_0^2: \cdots : S_0^N).$$

Although it is not always true that the coordination rate $\overline{T}$ between output variables of a system is smaller than the corresponding rate between the input variables, it is true under important special conditions which we will discuss. That discussion will be facilitated by

Laws of Information

prior consideration of a related and important statistical question about the extent to which constraint between variables becomes obscured when the variables are observed through information-losing mappings.

The statistical question is this: suppose observations of a system of several discrete variables result in a multivariable frequency (or observed-probability) table, and transmissions between variables (two or more at a time) are calculated from it. Next suppose that in some or all of the variables, some of the values or categories of the variables are lumped together and the frequency table is partially collapsed or summed accordingly. For example, in a two-dimensional table, several rows would be deleted and a new row inserted which is their matrix sum; similarly in a three-dimensional table several planes would be replaced by a single new plane, their matrix sum. This operation of partially collapsing the frequency table along its X_j dimension corresponds to losing the distinction between some values of X_j, perhaps due to coarse observation. Now suppose transmissions between variables are calculated from the new, smaller frequency table. How much smaller are these T than the original T?

As posed the question is: how much does coarse observation of the variables reduce the apparent constraint between them? Since the answer obviously depends on the way the values are lumped or merged, a different version of the question is: if we want to merge values, for example in the interest of economical description, how shall that be done to minimize the reduction of the T and therefore preserve relational information as much as possible?

We will address this question in two dimensions and use induction to generalize the result. The operation of merging values of a variable X is performed by a many-one mapping $f: X \rightarrow Y$, where X is the input set and $Y = f(X)$ is the output set (possibly itself a subset of X). For the two-dimensional question we will suppose there are two variables X_1 and X_2 and a function $f_1: X_1 \rightarrow Y_1$ which maps X_1 convergently into Y_1. Using the fact that X_1 determines Y_1 [$H_{X_1}(Y_1) = 0$] and [A7], it is easy to obtain

$$T(X_1:X_2) - T(Y_1:X_2) = \text{loss of transmission due to } f_1$$
$$= T_{Y_1}(X_1:X_2) \tag{18a}$$
$$\geq 0. \tag{18b}$$

The fundamental result (18a) expresses the loss of apparent constraint which results from merging values of *one* variable. It can be used iteratively to express the loss which results from merging values of several variables. For example, if $f_2: X_2 \rightarrow Y_2$ merges values of X_2, it is easy to show the following:

$$T(X_1:X_2) - T(Y_1:Y_2) = \text{loss due to } [f_1 \text{ and } f_2]$$
$$= T_{Y_1}(X_1:X_2) + T_{Y_2}(Y_1:X_2) \tag{18c}$$
$$\geq 0. \tag{18d}$$

The term $T_{Y_1}(X_1:X_2)$ in (18a), representing the loss of constraint due to f_1, involves averaging over the values of Y_1. This averaging can be made explicit:

$$T_{Y_1}(X_1:X_2) = \sum_{y_1 \in Y_1} P(y_1)[H_{Y_1}(X_1) - H_{Y_1,X_2}(X_1)].$$

For any y_1 in Y_1 having a unique image in X_1, both entropies are zero. Consequently the summands are nonzero only for values y_1 with multiple images in X_1. Thus

$$T_{Y_1}(X_1:X_2) = \sum_{y_1 \in Y_1} P(y_1) T_{Y_1}(X_1:X_2) \tag{19}$$

such that $f_1^{-1}(y_1)$ is not unique. Thus to calculate the transmission lost, one can find the probability of each y_1 representing grouped values of X_1, multiply it by the transmission in the associated $X_1 \times X_2$ subtable collapsed by that grouping, then add all such products. For example, the frequency table for $X_1 \times X_2$ showing subtables is

		X_1				
		1	2	3	4	5
X_2	a	1	2	3	0	3
	b	2	4	0	3	0

and the frequency table for $Y_1 \times X_2$, after grouping by f_1 is

		Y_1		
		1–2	3–4	5
X_2	a	3	3	3
	b	6	3	0

Calculations:

$$p(1\text{–}2) = 9/18; \quad T_{1\text{–}2}(X_1:X_2) = 0 \quad \text{(first subtable)}$$

$$p(3\text{–}4) = 6/18; \quad T_{3\text{–}4}(X_1:X_2) = 1 \quad \text{(second subtable)}$$

$$T_{Y_1}(X_1:X_2) = (9/18)(0) + (6/18)(1) = 1/3$$

$$T(X_1:X_2) - T(Y_1:X_2) = 0.542 - 0.209 = 0.333 \text{ bits lost.}$$

Although more elegant representations are possible, (19) indicates how values can be grouped to minimize the loss of transmission: either by grouping values for which the probability is small, or by collapsing subtables (such as the first one in the example) having small transmission and therefore not contributing substantially to the total transmission $T(X_1:X_2)$.

The fact that the apparent constraint between functions of variables is always smaller than, or at most equal to, the constraint between the variables themselves will be referred to as the *Law of Constraint Loss* (LCL). It has been demonstrated in (18d) for two dimensions. The proof by induction for n dimensions is straightforward. Suppose the LCL holds for $n - 1$ dimensions:

$$T(X_1:X_2: \cdots : X_{n-1}) \geq T(Y_1:Y_2: \cdots : Y_{n-1})$$

then by (18d),

$$T(X_1,X_2, \cdots, X_{n-1}:X_n) \geq T(Y_1,Y_2, \cdots, Y_{n-1}:Y_n).$$

Adding these inequalities and using [A6], LCL is established for N dimensions:

$$T(X_1:X_2: \cdots : X_n) \geq T(Y_1:Y_2: \cdots : Y_n). \tag{20}$$

Laws of Information

The difference between the two transmissions can be expressed in forms similar to (18), omitted here because of their bulkiness.

The key element in establishing the LCL was the assumption that Y_1 was a function of X_1, i.e., $H_{X_1}(Y_1) = 0$. We have made the weaker assumption above that all subsystems S^i of S are determined by their environments $E^i(\overline{H}_{E^i}(S^i) = 0)$. A proof analogous to that for the LCL leads to the *Law of Coordination Rate Loss* (LCRL):

$$\overline{T}(E^1:E^2:\cdots:E^N) \geq \overline{T}(S^1:S^2:\cdots:S^N). \tag{21a}$$

This law was invoked in a proof in Section VB, although in a different form,

$$\overline{T}(S^1:S^2:\cdots:S^N) \geq \overline{T}(S_0^1:S_0^2:\cdots:S_0^N), \tag{21b}$$

which clearly is a corollary of the LCRL, since S^i determines S_0^i. Obviously,

$$\overline{T}(E^1:E^2:\cdots:E^N) \geq \overline{T}(S_0^1:S_0^2:\cdots:S_0^N). \tag{21c}$$

The coordination rate over the outputs of a set of processors cannot exceed the coordination rate over their inputs, if the processors obey the deterministic condition. It is in this sense that systems obeying the deterministic condition of the LCRL are constraint-losing; the mutual constraint binding the output variables (all of them or any subset) cannot exceed the mutual constraint binding corresponding inputs.

The assumption that S^i is determined by E^i for all i makes the law most useful for a system consisting of independent parallel processors, each determined by its own input in E only. To the extent that the subsystems of a system affect one another, the E^i in the law include outputs of other subsystems as well as variables in E, the environment of S. While the law still holds, the net effect is to dilute its impact (which is greatest when the E^i are all variables in E). For example, suppose that $E = E^1$ determines S_0^1, S_0^1 determines S_0^2, S_0^2 determines S_0^3, and S_0^3 determines $S_0^4 = S_0$. Then the LCRL says that

$$T(E:S_0^1:S_0^2:S_0^3) \geq T(S_0^1:S_0^2:S_0^3:S_0)$$

or

$$T(E:S_0^1,S_0^2,S_0^3) \geq T(S_0^1,S_0^2,S_0^3:S_0)$$

by [A6], which states that the internal vector S_0^1, S_0^2, S_0^3 is coordinated more strongly with the system input than with the system output, which is not very surprising.

In contrast with this highly serial example, consider a highly parallel one (perhaps from a pattern-recognition or image-enhancement scheme) in which a two-dimensional picture is the input to a processor which produces another such picture by scanning each cell in the input picture grid and displaying, on the corresponding cell in the output grid, a response depending *only* on that input cell and its history. The LCRL states that the output picture in such a scheme could not be more constrained or "orderly" than the input picture. The system could be changed to a "constraint increasing" one by allowing each output cell to depend upon a field of several input cells, as occurs in the human visual system.

Every S_0^i has an image in E, namely, a smallest set of variables in E which determines S_0^i. Denoting this image by I^i, the LCRL says that

$$\overline{T}(I^1:I^2:\cdots:I^N) \geq \overline{T}(S_0^1:S_0^2:\cdots:S_0^N).$$

The output coordination cannot exceed the coordination between the corresponding input images.

Suppose $\overline{T}(E^1:E^2:\cdots:E^N) = 0$ so that the inputs are totally independent. This implies that the S^i operate completely independently, and the LCRL states that their outputs must also be independent. If the outputs are not independent, the subsystems must in fact be interacting and nonindependent. For example, if students given different examinations turn in similar answers, there are grounds to suspect collusion.

As in the case of the LCL, the difference between input and output coordination-rates can be described analytically in forms similar to (18).

VII. Loss and Blockage of Information

Under the deterministic assumption, the blockage rate F_b was defined as $\overline{H}_{S_0}(S_{int})$. It represents the rate at which information about the input E is blocked *within* S and not allowed to affect the output. The identity (from [A3] and $S = S_0, S_{int}$)

$$\overline{H}(S) = F_b + F_t$$

indicates that of the information which crosses the interface between E and $S(\overline{H}(S))$, part is internally blocked (F_b) and the rest is output (F_t). To complete the picture we should consider also the rate at which information present in E is rejected at the interface, i.e., does not affect S and therefore does not contribute to $\overline{H}(S)$. This rejection rate is described by $\overline{H}_S(E)$, the average per-step conditional uncertainty of the environment E given complete knowledge of the behavior of S (which is determined by E). The overall loss of information consists of rejection at the interface and blockage within S. Thus (under deterministic assumption),

$$F_b = \overline{H}_{S_0}(S_{int}) \quad \text{blockage rate}$$
$$F_r = \overline{H}_S(E) \quad \text{rejection rate}$$
$$F_L = F_r + F_b \quad \text{loss rate}$$

and (using [A3] twice, and $S = S_0, S_{int}$)

$$\overline{H}(E) = F_r + F_b + F_t,$$

which says that part of the input information is rejected, part is blocked internally, and the rest is passed to the output of S. The loss rate F_L is the difference between the input and output rates:

$$F_L = \overline{H}(E) - F_t.$$

The distinction between F_r and F_b is important since blockage requires activity on the part of S while rejection does not. Rejection corresponds to simple insensitivity on the part of S to certain attributes of E, while blockage corresponds to actively discarding irrelevant information taken in. Under the normal condition in which the environment E presents more information than needed for output of the system S, it is clearly advantageous for the loss to be concentrated as much as possible in rejection rather than blockage, since active

Laws of Information

"computing" is required for blockage while for rejection the requirement is only that the "receptors" forming the interface between E and S be responsive to appropriate aspects of E and nonresponsive to the rest, i.e., act as a sort of passive filter. In human terms, rejection corresponds to simple deafness, blockage to selection of appropriate information and discard of the rest. Simple organisms can afford to reject almost all environmental information, but complex organisms cannot and have to retain sensitivity to much information which therefore must be quickly blocked. An example is the way the driver of a car hears the normal automotive noises but pays no attention to them (blockage), while he immediately notes *unusual* noises, evidence of his sensitivity to this environmental input.

The objective of many control and regulatory mechanisms is exactly the blockage of information which cannot be rejected; for example, thermostatic temperature regulation prevents information about environmental temperature fluctuations from reaching occupants of the room being controlled. This point has been elaborated elsewhere.[1,3] The common phenomenon termed "habituation" (diminished or vanishing response to repetitive stimuli) appears to be mainly blockage.

For F_L there is an hierarchical decomposition similar to that for F_b. Assuming as we did in the development for F_b that $S_0 = S_0^1, S_0^2, \cdots, S_0^K$, the decomposition is

$$F_L = \sum_{i=1}^{N} F_L^i + H_{S_0}(S_0^L, \cdots, S_0^N) - [T(E^1:E^2:\cdots:E^N) - T(S_0^1:S_0^2:\cdots:S_0^N)]$$

$$\begin{pmatrix} \text{Loss} \\ \text{by } S \end{pmatrix} = \begin{pmatrix} \sum_{i=1}^{N} \text{loss} \\ \text{by } S^i \end{pmatrix} + \begin{pmatrix} \text{subsystem output information} \\ \text{blocked within } \mathcal{S} \end{pmatrix}$$

$$- \begin{pmatrix} \text{coordination between the subsystem inputs} \\ \text{which is not reflected in coordination} \\ \text{of their outputs, i.e., lost coordination} \end{pmatrix}$$

[Without this assumption, the entropy term becomes $H_{S_0}(S_0^1, S_0^2, \cdots, S_0^N)$.] The proof is similar to that for F_b. The bracketed term can also be written as

$$[\{T(E^1:E^2:\cdots:E^N) - T(S^1:S^2:\cdots:S^N)\} + \{T(S^1:S^2:\cdots:S^N) - T(S_0^1:S_0^2:\cdots:S_0^N)\}]$$

since lost coordination = rejected coordination + blocked coordination. All of these quantities are nonnegative.

In any system S for which every input in E potentially has an effect on S_0 (e.g., subsystems S^3 and S^4 of Fig. 2, but not S^1, S^2, or S^5) the total loss rate F_L cannot exceed the sum of the subsystem losses, i.e.,

$$F_L \leq \sum_{i=1}^{N} F_L^i$$

since loss can occur only at subsystems and the information lost is additive only if independent, otherwise subadditive. Sometimes this fact can be used to determine the minimum number of elements required in S. For example the 8-bit binary adder of Section V C and Fig. 3 with $\overline{H}(E) = 8$ bit/s and $\overline{H}(S_0) = 1$ bit/s must have $F_L = 7$ bit/s. Since each 2-bit binary adder is capable of losing at most 1 bit/s, a minimum of seven of these are required.

The view of systems as *losers* of information, and of the usefulness of and requirement for that loss, is not widespread, perhaps because most users of information theory are primarily concerned with preserving information and maximizing its flow, for example, over communication channels. Outside the realm of formal communication systems, however, it is often desirable to *minimize* the output of information. It is usually considered desirable, for example, that scientific papers be as brief as feasible, that mathematical proofs omit obvious steps, that systems should produce the minimum acceptable output, take as little action as possible, etc. If the system objective is to minimize F, while $\overline{H}(E)$ is fixed, the objective becomes to maximize F_L. Since the portion of the loss attributed to blockage F_b requires computational resources, it is clear on grounds of system economy that as much as possible of the information loss should be accomplished through rejection F_r. This observation can be stated as another informal system guideline, actually an implementation of the minimum blockage guideline of Section IV B: Arrange for the sensors (input devices) of the system to respond only to those aspects of the environment which are potentially relevant. Although this guideline seems absurdly obvious, it is painfully common to find it breached, for example, by the gathering of data which one in retrospect sees could have had no effect on results of the study for which it was intended.

VIII. Summary

This paper has indicated a number of ways in which the laws of information theory govern the behavior of systems. In this section we will summarize some of the main points.

If each variable in a system is viewed as a message source which sends information about its values to the other variables which are "listening," then what is conventionally seen as a network of causes and effects can be viewed as a network of transmitters, channels, and receivers. The strength of the causal relations can then be quantified by information rates $\overline{T}$ between variables or over sets of variables. These rates are bounded by channel capacities. Information theory can therefore be used in the analysis of constraints or dependencies in multivariable systems, where it has particular advantages for dynamic systems and systems with hierarchical structure.

The Partition Law of Information Rates states that the total rate of information flow or processing in a system can be expressed as the sum of a rate devoted to blocking irrelevant information, a rate devoted to internal coordination, and the output rate (plus a noise rate, in nondeterministic systems). The simple additivity expressed in the law implies a competition for computational resources between blockage, coordination, and output, and suggests the tradeoffs possible between them in a system of fixed total capacity. If estimates for blockage, coordination, and output for a task are available and if the capacities of system components are known, the law provides a lower bound on the number of components needed to accomplish the task. The comparison of F with C_{sum} provides an indication of utilization of the system, since F attains its upper bound of C_{sum} only when all components are operating at full capacity.

Channel capacity was shown to be only one of several information-processing capacities of systems. For a deterministic system the channel capacity bounds the output rate, but for many tasks blockage or coordination are much more relevant than output and place more severe demands on the system.

Laws of Information

A set of informal guidelines was developed for the design of deterministic systems with information-processing functions. These assumed the desirability of a system minimally adequate to succeed at the tasks put to it, but maximally efficient. From the class of system designs which successfully meet design criteria, those should be selected which: (1) produce the minimum allowable output, (2) perform as little blockage as possible by avoiding irrelevant input, for example, by using input devices which respond only to relevant aspects of the environment, (3) reduce internal coordination as much as is possible without degrading performance, and (4) match components to tasks and vice versa so that each component is operated at or near its capacity.

The suitability of information theory for the study of hierarchical systems, indicated in earlier works,[6,7] was demonstrated by hierarchical decomposition of the terms in the Partition Law.

A method was proposed for calculation of the apparent reduction of constraint between discrete variables which results from reducing the number of values through grouping. This method also suggested how values can be grouped to minimize the loss of apparent constraint. A generalization of that result indicated a sense in which deterministic systems are "constraint-losers" and showed that the set of output variables can be no more constrained than the set of associated input images.

The paper closed with a brief discussion of information loss by systems. Since systems typically are confronted with much more information in their environment than is necessary or relevant for calculation of the appropriate responses they must make, they must typically reject or block most of it. It was argued on grounds of system economy that as much as possible of the loss should be in the form of rejection, i.e., nonresponsiveness by the system's sensors to irrelevant inputs.

Appendix

[A1] The Law of Uniform Subscripts states that from a valid identity of information theory, another valid identity is produced by uniformly subscripting all terms, i.e., conditioning all terms on the same variable.

[A2] The Law of Uniform Overlining states that from a valid identity of information theory involving H's and T's, but no rates, another valid identity is produced by replacing each H by $\bar{H}$ and each T by $\bar{T}$, i.e., substituting *rates* throughout.

[A3] $H(S^1, S^2) = H(S^1) + H_{S^1}(S^2)$.

[A4] $T(S^1 : S^2) = H(S^1) - H_{S^2}(S^1) = H(S^2) - H_{S^1}(S^2)$.

[A5] $T(S^1 : S^2 : \cdots : S^N) = \sum_{i=1}^{N} H(S^i) - H(S^1, S^2, \cdots, S^N)$

$$= \sum_{i=1}^{N} H(S^i) - H(S).$$

$$T(X_1^i : X_2^i : \cdots : X_{n_i}^i) = \sum_{j=1}^{n_i} H(X_j^i) - H(S^i),$$

a special case.

[A6] $T(X_1:X_2: \cdots : X_n) = \sum_{i=1}^{N} T(X_1^i:X_2^i: \cdots : X_{n_i}^i) + T(S^1:S^2: \cdots : S^N).$

[A7] $T(S^1:S^2,S^3) = T(S^1:S^2) + T_{S_2}(S^1:S^3).$

References

1. W. R. Ashby, *Introduction to Cybernetics*. London: Chapman and Hall, 1956.
2. W. R. Ashby, "Measuring the internal information exchange in a system," *Cybernetica*, vol 8, pp. 5–22, 1965.
3. W. R. Ashby, "Two tables of identities governing information flows within large systems," *Commun. Amer. Soc. Cybern.*, vol. 1, no. 2, pp. 2–7, 1969.
4. W. R. Ashby, "Information flows within co-ordinated systems," in *Progress of Cybernetics: Proc. 1st Int. Cong. Cybernetics*, J. Rose, Ed. London: Gordon and Breach, 1970, pp. 57–64.
5. R. C. Conant, "The information transfer required in regulatory processes," *IEEE Trans. Syst. Sci. Cybern.*, vol. SSC-5, no. 4, pp. 334–338.
6. R. C. Conant, "Detecting subsystems of a complex system," *IEEE Trans. Syst., Man, Cybern.*, vol. SMC-2, no. 4, pp. 550–553.
7. R. C. Conant, "Information flows in hierarchical systems," *Int. J. Gen. Syst.*, vol. 1, no. 1, pp. 9–18, 1974.
8. J. N. Cronholm, "A general method of obtaining exact sampling probabilities of the Shannon-Wiener measure of information $\hat{H}^*$," *Psychometrika*, vol. 28, no. 4, pp. 405–413, 1963.
9. W. R. Garner and W. J. McGill, "The relation between information and variance analyses," *Psychometrika*, vol. 21, no. 3, pp. 219–228, 1956.
10. W. J. McGill, "Multivariate information transmission," *Psychometrika*, vol 19, no. 2, pp. 97–116, 1954.
11. G. A. Miller, "On the bias of information estimates," in *Information Theory in Psychology*, H. Quastler, Ed. Glencoe, IL: The Free Press, 1955, pp. 95–100.
12. M. S. Rogers and B. F. Green, "Moments of sample information when the alternatives are equally likely," in *Information Theory in Psychology*, H. Quastler, Ed. Glencoe, IL: The Free Press, 1955, pp. 101–107.
13. H. Simon, "The architecture of complexity," in *Proc. Amer. Phil. Soc.*, vol 106, pp. 467–482, 1962. Reprinted in *Gen. Syst.*, vol. 10, pp. 63–76, 1965.
14. C. E. Shannon and W. Weaver, *The Mathematical Theory of Communication*. Urbana, IL: Univ. Illinois, 1949.
15. C. E. Shannon, "Prediction and entropy of printed English," *Bell Syst. Tech. J.*, vol. 30, pp. 50–64, 1951.

19

Science and Complexity

Warren Weaver

Science has led to a multitude of results that affect men's lives. Some of these results are embodied in mere conveniences of a relatively trivial sort. Many of them, based on science and developed through technology, are essential to the machinery of modern life. Many other results, especially those associated with the biological and medical sciences, are of unquestioned benefit and comfort. Certain aspects of science have profoundly influenced men's ideas and even their ideals. Still other aspects of science are thoroughly awesome.

How can we get a view of the function that science should have in the developing future of man? How can we appreciate what science really is and, equally important, what science is not? It is, of course, possible to discuss the nature of science in general philosophical terms. For some purposes such a discussion is important and necessary, but for the present a more direct approach is desirable. Let us, as a very realistic politician used to say, let us look at the record. Neglecting the older history of science, we shall go back only three and a half centuries and take a broad view that tries to see the main features, and omits minor details. Let us begin with the physical sciences, rather than the biological, for the place of the life sciences in the descriptive scheme will gradually become evident.

Problems of Simplicity

Speaking roughly, it may be said that the seventeenth, eighteenth, and nineteenth centuries formed the period in which physical science learned variables, which brought us the telephone and the radio, the automobile and the airplane, the phonograph and the moving pictures, the turbine and the Diesel engine, and the modern hydroelectric power plant.

The concurrent progress in biology and medicine was also impressive, but that was of a different character. The significant problems of living organisms are seldom those in which one can rigidly maintain constant all but two variables. Living things are more likely to present situations in which a half-dozen, or even several dozen quantities are all varying

---

From *American Scientist* **36**, 536. Copyright © 1948 by Sigma Xi.

simultaneously, and in subtly interconnected ways. Often they present situations in which the essentially important quantities are either non-quantitative, or have at any rate eluded identification or measurement up to the moment. Thus biological and medical problems often involve the consideration of a most complexly organized whole. It is not surprising that up to 1900 the life sciences were largely concerned with the necessary preliminary stages in the application of the scientific method—preliminary stages which chiefly involve collection, description, classification, and the observation of concurrent and apparently correlated effects. They had only made the brave beginnings of quantitative theories, and hardly even begun detailed explanations of the physical and chemical mechanisms underlying or making up biological events.

To sum up, physical science before 1900 was largely concerned with two-variable *problems of simplicity*; whereas the life sciences, in which these problems of simplicity are not so often significant, had not yet become highly quantitative or analytical in character.

Problems of Disorganized Complexity

Subsequent to 1900 and actually earlier, if one includes heroic pioneers such as Josiah Willard Gibbs, the physical sciences developed an attack on nature of an essentially and dramatically new kind. Rather than study problems which involved two variables or at most three or four, some imaginative minds went to the other extreme, and said: "Let us develop analytical methods which can deal with two billion variables." That is to say, the physical scientists, with the mathematicians often in the vanguard, developed powerful techniques of probability theory and of statistical mechanics to deal with what may be called problems of *disorganized complexity*.

This last phrase calls for explanation. Consider first a simple illustration in order to get the flavor of the idea. The classical dynamics of the nineteenth century was well suited for analyzing and predicting the motion of a single ivory ball as it moves about on a billiard table. In fact, the relationship between positions of the ball and the times at which it reaches these positions forms a typical nineteenth-century problem of simplicity. One can, but with a surprising increase in difficulty, analyze the motion of two or even of three balls on a billiard table. There has been, in fact, considerable study of the mechanics of the standard game of billiards. But, as soon as one tries to analyze the motion of ten or fifteen balls on the table at once, as in pool, the problem becomes unmanageable, not because there is any theoretical difficulty, but just because the actual labor of dealing in specific detail with so many variables turns out to be impracticable.

Imagine, however, a large billiard table with millions of balls rolling over its surface, colliding with one another and with the side rails. The great surprise is that the problem now becomes easier, for the methods of statistical mechanics are applicable. To be sure the detailed history of one special ball can not be traced, but certain important questions can be answered with useful precision, such as: On the average how many balls per second hit a given stretch of rail? On the average how far does a ball move before it is hit by some other ball? On the average how many impacts per second does a ball experience?

Earlier it was stated that the new statistical methods were applicable to problems of disorganized complexity. How does the word "disorganized" apply to the large billiard table with the many balls? It applies because the methods of statistical mechanics are valid only when the balls are distributed, in their positions and motions, in a helter-skelter, that is to say a disorganized, way. For example, the statistical methods would not apply if someone were to arrange the balls in a row parallel to one side rail of the table, and then start them all moving in precisely parallel paths perpendicular to the row in which they stand. Then the balls would never collide with each other nor with two of the rails, and one would not have a situation of disorganized complexity.

From this illustration it is clear what is meant by a problem of disorganized complexity. It is a problem in which the number of variables is very large, and one in which each of the many variables has a behavior which is individually erratic, or perhaps totally unknown. However, in spite of this helter-skelter, or unknown, behavior of all the individual variables, the system as a whole possesses certain orderly and analyzable average properties.

A wide range of experience comes under the label of disorganized complexity. The method applies with increasing precision when the number of variables increases. It applies with entirely useful precision to the experience of a large telephone exchange, in predicting the average frequency of calls, the probability of overlapping calls of the same number, etc. It makes possible the financial stability of a life insurance company. Although the company can have no knowledge whatsoever concerning the approaching death of any one individual, it has dependable knowledge of the average frequency with which deaths will occur.

This last point is interesting and important. Statistical techniques are not restricted to situations where the scientific theory of the individual events is very well known, as in the billiard example where there is a beautifully precise theory for the impact of one ball on another. This technique can also be applied to situations, like the insurance example, where the individual event is as shrouded in mystery as is the chain of complicated and unpredictable events associated with the accidental death of a healthy man.

The examples of the telephone and insurance companies suggests a whole array of practical applications of statistical techniques based on disorganized complexity. In a sense they are unfortunate examples, for they tend to draw attention away from the more fundamental use which science makes of these new techniques. The motions of the atoms which form all matter, as well as the motions of the stars which form the universe, come under the range of these new techniques. The fundamental laws of heredity are analyzed by them. The laws of thermodynamics, which describe basic and inevitable tendencies of all physical systems, are derived from statistical considerations. The entire structure of modern physics, our present concept of the nature of the physical universe, and of the accessible experimental facts concerning it rest on these statistical concepts. Indeed, the whole question of evidence and the way in which knowledge can be inferred from evidence are now recognized to depend on these same statistical ideas, so that probability notions are essential to any theory of knowledge itself.

Problems of Organized Complexity

This new method of dealing with disorganized complexity, so powerful an advance over the earlier two-variable methods, leaves a great field untouched. One is tempted to oversimplify, and say that scientific methodology went from one extreme to the other—from two variables to an astronomical number—and left untouched a great middle region. The importance of this middle region, moreover, does not depend primarily on the fact that the number of variables involved is moderate—large compared to two, but small compared to the number of atoms in a pinch of salt. The problems in this middle region, in fact, will often involve a considerable number of variables. The really important characteristic of the problems of this middle region, which science has as yet little explored or conquered, lies in the fact that these problems, as contrasted with the disorganized situations with which statistics can cope, show the essential feature of *organization*. In fact, one can refer to this group of problems as those of *organized complexity*.

What makes an evening primrose open when it does? Why does salt water fail to satisfy thirst? Why can one particular genetic strain of microorganism synthesize within its minute body certain organic compounds that another strain of the same organism cannot manufacture? Why is one chemical substance a poison when another, whose molecules have just the same atoms but assembled into a mirror-image pattern, is completely harmless? Why does the amount of manganese in the diet affect the maternal instinct of an animal? What is the description of aging in biochemical terms? What meaning is to be assigned to the question: Is a virus a living organism? What is a gene, and how does the original genetic constitution of a living organism express itself in the developed characteristics of the adult? Do complex protein molecules "know how" to reduplicate their pattern, and is this an essential clue to the problem of reproduction of living creatures? All these are certainly complex problems, but they are not problems of disorganized complexity, to which statistical methods hold the key. They are all problems which involve dealing simultaneously with a *sizable number of factors which are interrelated into an organic whole*. They are all, in the language here proposed, problems of *organized complexity*.

On what does the price of wheat depend? This too is a problem of organized complexity. A very substantial number of relevant variables is involved here, and they are all interrelated in a complicated, but nevertheless not in helter-skelter, fashion.

How can currency be wisely and effectively stabilized? To what extent is it safe to depend on the free interplay of such economic forces as supply and demand? To what extent must systems of economic control be employed to prevent the wide swings from prosperity to depression? These are also obviously complex problems, and they too involve analyzing systems which are organic wholes, with their parts in close interrelation.

How can one explain the behavior pattern of an organized group of persons such as a labor union, or a group of manufacturers, or a racial minority? There are clearly many factors involved here, but it is equally obvious that here also something more is needed than the mathematics of averages. With a given total of national resources that can be brought to bear, what tactics and strategy will most promptly win a war, or better: what sacrifices of present selfish interest will most effectively contribute to a stable, decent, and peaceful world?

These problems—and a wide range of similar problems in the biological, medical, psychological, economic, and political sciences—are just too complicated to yield to the old nineteenth-century techniques which were so dramatically successful on two-, three-, or four-variable problems of simplicity. These new problems, moreover, cannot be handled with the statistical techniques so effective in describing average behavior in problems of disorganized complexity.

These new problems, and the future of the world depends on many of them, requires

Science and Complexity

science to make a third great advance, an advance that must be even greater than the nineteenth-century conquest of problems of simplicity or the twentieth-century victory over problems of disorganized complexity. Science must, over the next 50 years, learn to deal with these problems of organized complexity.

Is there any promise on the horizon that this new advance can really be accomplished? There is much general evidence, and there are two recent instances of especially promising evidence. The general evidence consists in the fact that, in the minds of hundreds of scholars all over the world, important, though necessarily minor, progress is already being made on such problems. As never before, the quantitative experimental methods and the mathematical analytical methods of the physical sciences are being applied to the biological, the medical, and even the social sciences. The results are as yet scattered, but they are highly promising. A good illustration from the life sciences can be seen by a comparison of the present situation in cancer research with what it was twenty-five years ago. It is doubtless true that we are only scratching the surface of the cancer problem, but at least there are now some tools to dig with and there have been located some spots beneath which almost surely there is pay-dirt. We know that certain types of cancer can be induced by certain pure chemicals. Something is known of the inheritance of susceptibility to certain types of cancer. Million-volt rays are available, and the even more intense radiations made possible by atomic physics. There are radioactive isotopes, both for basic studies and for treatment. Scientists are tackling the almost incredibly complicated story of the biochemistry of the aging organism. A base of knowledge concerning the normal cell is being established that makes it possible to recognize and analyze the pathological cell. However distant the goal, we are now at last on the road to a successful solution of this great problem.

In addition to the general growing evidence that problems of organized complexity can be successfully treated, there are at least two promising bits of special evidence. Out of the wickedness of war have come two new developments that may well be of major importance in helping science to solve these complex twentieth-century problems.

The first piece of evidence is the wartime development of new types of electronic computing devices. These devices are, in flexibility and capacity, more like a human brain than like the traditional mechanical computing device of the past. They have "memories" in which vast amounts of information can be stored. They can be "told" to carry out computations of very intricate complexity, and can be left unattended while they go forward automatically with their task. The astounding speed with which they proceed is illustrated by the fact that one small part of such a machine, if set to multiplying two ten-digit numbers, can perform such multiplications some 40,000 times faster than a human operator can say "Jack Robinson." This combination of flexibility, capacity, and speed makes it seem likely that such devices will have a tremendous impact on science. They will make it possible to deal with problems which previously were too complicated, and, more importantly, they will justify and inspire the development of new methods of analysis applicable to these new problems of organized complexity.

The second of the wartime advances is the "mixed-team" approach of operations analysis. These terms require explanation, although they are very familiar to those who were concerned with the application of mathematical methods to military affairs.

As an illustration, consider the over-all problem of convoying troops and supplies across the Atlantic. Take into account the number and effectiveness of naval vessels available, the character of submarine attacks, and a multitude of other factors, including such an imponderable as the dependability of visual watch when men are tired, sick, or bored. Considering a whole mass of factors, some measurable and some elusive, what procedure would lead to the best over-all plan, that is, best from the combined point of view of speed, safety, cost, and so on? Should the convoys be large or small, fast or slow? Should they zigzag and expose themselves longer to possible attack, or dash in a speedy straight line? How are they to be organized, what defenses are best, and what organization and instruments should be used for watch and attack?

The attempt to answer such broad problems of tactics, or even broader problems of strategy, was the job during the war of certain groups known as the operations analysis groups. Inaugurated with brilliance by the British, the procedure was taken over by this country, and applied with special success in the Navy's anti-submarine campaign and in the Army Air Forces. These operations analysis groups were, moreover, what may be called mixed teams. Although mathematicians, physicists, and engineers were essential, the best of the groups also contained physiologists, biochemists, psychologists, and a variety of representatives of other fields of the biochemical and social sciences. Among the outstanding members of English mixed teams, for example, were an endocrinologist and an X-ray crystallographer. Under the pressure of war, these mixed teams pooled their resources and focused all their different insights on the common problems. It was found, in spite of the modern tendencies toward intense scientific specialization, that members of such diverse groups could work together and could form a unit which was much greater than the mere sum of its parts. It was shown that these groups could tackle certain problems of organized complexity, and get useful answers.

It is tempting to forecast that the great advances that science can and must achieve in the next fifty years will be largely contributed to by voluntary mixed teams, somewhat similar to the operations analysis groups of war days, their activities made effective by the use of large, flexible, and highspeed computing machines. However, it cannot be assumed that this will be the exclusive pattern for future scientific work, for the atmosphere of complete intellectual freedom is essential to science. There will always, and properly, remain those scientists for whom intellectual freedom is necessarily a private affair. Such men must, and should, work alone. Certain deep and imaginative achievements are probably won only in such a way. Variety is, moreover, a proud characteristic of the American way of doing things. Competition between all sorts of methods is good. So there is no intention here to picture a future in which all scientists are organized into set patterns of activity. Not at all. It is merely suggested that some scientists will seek and develop for themselves new kinds of collaborative arrangements; that these groups will have members drawn from essentially all fields of science; and that these new ways of working, effectively instrumented by huge computers, will contribute greatly to the advance which the next half century will surely achieve in handling the complex, but essentially organic, problems of the biological and social sciences.

The Boundaries of Science

Let us return now to our original questions. What is science? What is not science? What may be expected from science?

Science clearly is a way of solving problems—not all problems, but a large class of important and practical ones. The problems with which science can deal are those in which the predominant factors are subject to the basic laws of logic, and are for the most part measurable. Science is a way of organizing reproducible knowledge about such problems; of focusing and disciplining imagination; of weighing evidence; of deciding what is relevant and what is not; of impartially testing hypotheses; of ruthlessly discarding data that prove to be inaccurate or inadequate; of finding, interpreting, and facing facts, and of making the facts of nature the servants of man.

The essence of science is not to be found in its outward appearance, in its physical manifestations; it is to be found in its inner spirit. That austere but exciting technique of inquiry known as the scientific method is what is important about science. This scientific method requires of its practitioners high standards of personal honesty, open-mindedness, focused vision, and love of the truth. These are solid virtues, but science has no exclusive lien on them. The poet has these virtues also, and often turns them to higher uses.

Science has made notable progress in its great task of solving logical and quantitative problems. Indeed, the successes have been so numerous and striking, and the failures have been so seldom publicized, that the average man has inevitably come to believe that science is just about the most spectacularly successful enterprise man ever launched. The fact is, of course, that this conclusion is largely justified.

Impressive as the progress has been, science has by no means worked itself out of a job. It is soberly true that science has, to date, succeeded in solving a bewildering number of relatively easy problems, whereas the hard problems, and the ones which perhaps promise most for man's future, lie ahead.

We must, therefore, stop thinking of science in terms of its spectacular successes in solving problems of simplicity. This means, among other things, that we must stop thinking of science in terms of gadgetry. Above all, science must not be thought of as a modern improved black magic capable of accomplishing anything and everything.

Every informed scientist, I think, is confident that science is capable of tremendous further contributions to human welfare. It can continue to go forward in its triumphant march against physical nature, learning new laws, acquiring new power of forecast and control, making new material things for man to use and enjoy. Science can also make further brilliant contributions to our understanding of animate nature, giving men new health and vigor, longer and more effective lives, and a wiser understanding of human behavior. Indeed, I think most informed scientists go even further and expect that the precise, objective, and analytical techniques of science will find useful application in limited areas of the social and political disciplines.

There are even broader claims which can be made for science and the scientific method. As an essential part of his characteristic procedure, the scientist insists on precise definition of terms and clear characterization of his problem. It is easier, of course, to define terms accurately in scientific fields than in many other areas. It remains true, however, that science is an almost overwhelming illustration of the effectiveness of a well-defined and accepted language, a common set of ideas, a common tradition. The way in which this universality has succeeded in cutting across barriers of time and space, across political and cultural boundaries, is highly significant. Perhaps better than in any other intellectual enterprise of man, science has solved the problem of communicating ideas, and has demonstrated the world-wide cooperation and community of interest which then inevitably results.

Yes, science is a powerful tool, and it has an impressive record. But the humble and wise scientist does not expect or hope that science can do everything. He remembers that science teaches respect for special competence, and he does not believe that every social, economic, or political emergency would be automatically dissolved if "the scientists" were only put into control. He does not—with a few aberrant exceptions—expect science to furnish a code of morals, or a basis for esthetics. He does not expect science to furnish the yardstick for measuring, nor the motor for controlling, man's love of beauty and truth, his sense of value, or his convictions of faith. There are rich and essential parts of human life which are alogical, which are immaterial and non-quantitative in character, and which cannot be seen under the microscope, weighed with the balance, nor caught by the most sensitive microphone.

If science deals with quantitative problems of a purely logical character, if science has no recognition of or concern for value or purpose, how can modern scientific man achieve a balanced good life, in which logic is the companion of beauty, and efficiency is the partner of virtue?

In one sense the answer is very simple: our morals must catch up with our machinery. To state the necessity, however, is not to achieve it. The great gap, which lies so forebodingly between our power and our capacity to use power wisely, can only be bridged by a vast combination of efforts. Knowledge of individual and group behavior must be improved. Communication must be improved between people of different languages and cultures, as well as between all the varied interests which use the same language, but often with such dangerously differing connotations. A revolutionary advance must be made in our understanding of economic and political factors. Willingness to sacrifice selfish short-term interests, either personal or national, in order to bring about long-term improvement for all must be developed.

None of these advances can be won unless men understand what science really is; all progress must be accomplished in a world in which modern science is an inescapable, ever-expanding influence.

The Architecture of Complexity

Herbert A. Simon

A number of proposals have been advanced in recent years for the development of "general systems theory" which, abstracting from properties peculiar to physical, biological, or social systems, would be applicable to all of them. We might well feel that, while the goal is laudable, systems of such diverse kinds could hardly be expected to have any nontrivial properties in common. Metaphor and analogy can be helpful, or they can be misleading. All depends on whether the similarities the metaphor captures are significant or superficial.

It may not be entirely vain, however, to search for common properties among diverse kinds of complex systems. The ideas that go by the name of cybernetics constitute, if not a theory, at least a point of view that has been proving fruitful over a wide range of applications. It has been useful to look at the behavior of adaptive systems in terms of the concepts of feedback and homeostasis, and to analyze adaptiveness in terms of the theory of selective information. The ideas of feedback and information provide a frame of reference for viewing a wide range of situations, just as do the ideas of evolution, of relativism, of axiomatic method, and of operationalism.

In this paper I should like to report on some things we have been learning about particular kinds of complex systems encountered in the behavioral sciences. The developments I shall discuss arose in the context of specific phenomena, but the theoretical formulations themselves make little reference to details of structure. Instead they refer primarily to the complexity of the systems under view without specifying the exact content of that complexity. Because of their abstractness, the theories may have relevance—application would be too strong a term—to other kinds of complex systems that are observed in the social, biological, and physical sciences.

In recounting these developments, I shall avoid technical detail, which can generally be found elsewhere. I shall describe each theory in the particular context in which it arose. Then, I shall cite some examples of complex systems, from areas of science other than the initial application, to which the theoretical framework appears relevant. In doing so, I shall make reference to areas of knowledge where I am not expert—perhaps not even literate. I feel quite comfortable in doing so before the members of this society, representing as it does the whole span of the scientific and scholarly endeavor. Collectively you will have little difficulty, I am sure, in distinguishing instances based on idle fancy or sheer ignorance from instances that cast some light on the ways in which complexity exhibits itself wherever it is found in nature. I shall leave to you the final judgment of relevance in your respective fields.

I shall not undertake a formal definition of "complex systems." Roughly, by a complex system I mean one made up of a large number of parts that interact in a nonsimple way. In such systems, the whole is more than the sum of the parts, not in an ultimate, metaphysical sense, but in the important pragmatic sense that, given the properties of the parts and the laws of their interaction, it is not a trivial matter to infer the properties of the whole. In the face of complexity, an in-principle reductionist may be at the same time a pragmatic holist.

The four sections that follow discuss four aspects of complexity. The first offers some comments on the frequency with which complexity takes the form of hierarchy—the complex system being composed of subsystems that, in turn, have their own subsystems, and so on. The second section theorizes about the relation between the structure of a complex system and the time required for it to emerge through evolutionary processes: specifically, it argues that hierarchic systems will evolve far more quickly than non-hierarchic systems of comparable size. The third section explores the dynamic properties of hierarchically-organized systems, and shows how they can be decomposed into subsystems in order to analyze their behavior. The fourth section examines the relation between complex systems and their descriptions.

Thus, the central theme that runs through my remarks is that complexity frequently takes the form of hierarchy, and that hierarchic systems have some common properties that are independent of their specific content. Hierarchy, I shall argue, is one of the central structural schemes that the architect of complexity uses.

Hierarchic Systems

By a *hierarchic system*, or hierarchy, I mean a system that is composed of interrelated subsystems, each of the latter being, in turn, hierarchic in structure until we reach some lowest level of elementary subsystem. In most systems in nature, it is somewhat arbitrary as to where we leave off the partitioning, and what subsystems we take as elementary. Physics makes much use of the concept of "elementary particle" although particles have a disconcerting tendency not to remain elementary very long. Only a couple of generations ago, the atoms themselves were elementary particles; today, to the nuclear physicist they are complex systems. For certain purposes of astronomy, whole stars, or even galaxies, can be regarded as elementary subsystems. In one kind of biological research, a cell may be treated as an elementary subsystem; in another, a protein molecule; in still another, an amino acid residue.

Just why a scientist has a right to treat as elementary a subsystem that is in fact exceedingly complex is one of the questions we shall take up. For the moment, we shall accept the fact that scientists do this all the time, and that if they are careful scientists they usually get away with it.

Etymologically, the word "hierarchy" has had a narrower meaning than I am giving it here. The term has generally been used to refer to a complex system in which each of the subsystems is subordinated by an authority relation to the system it belongs to. More exactly, in a hierarchic formal organization, each system consists of a "boss" and a set of subordinate subsystems. Each of the subsystems has a "boss" who is the immediate subordinate of the boss of the system. We shall want to consider systems in which the relations among subsystems are more complex than in the formal organizational hierarchy just described. We shall want to include systems in which there is no relation of subordination among subsystems. (In fact, even in human organizations, the formal hierarchy exists only on paper; the real flesh-and-blood organization has many inter-part relations other than the lines of formal authority.) For lack of a better term, I shall use hierarchy in the broader sense introduced in the previous paragraphs, to refer to all complex systems analyzable into successive sets of subsystems, and speak of "formal hierarchy" when I want to refer to the more specialized concept.

Social Systems

I have already given an example of one kind of hierarchy that is frequently encountered in the social sciences: a formal organization. Business firms, governments, universities all have a clearly visible parts-within-parts structure. But formal organizations are not the only, or even the most common, kind of social hierarchy. Almost all societies have elementary units called families, which may be grouped into villages or tribes, and these into larger groupings, and so on. If we make a chart of social interactions, of who talks to whom, the clusters of dense interaction in the chart will identify a rather well-defined hierarchic structure. The groupings in this structure may be defined operationally by some measure of frequency of interaction in this sociometric matrix.

Biological and Physical Systems

The hierarchical structure of biological systems is a familiar fact. Taking the cell as the building block, we find cells organized into tissues, tissues into organs, organs into systems. Moving downward from the cell, well-defined subsystems—for example, nucleus, cell membrane, microsomes, mitochondria, and so on—have been identified in animal cells.

The hierarchic structure of many physical systems is equally clear-cut. I have already mentioned the two main series. At the microscopic level we have elementary particles, atoms, molecules, macromolecules. At the macroscopic level we have satellite systems, planetary systems, galaxies. Matter is distributed throughout space in a strikingly nonuniform fashion. The most nearly random distributions we find, gases, are not random distributions of elementary particles but random distributions of complex systems, i.e. molecules.

A considerable range of structural types is subsumed under the term hierarchy as I have defined it. By this definition, a diamond is hierarchic, for it is a crystal structure of carbon atoms that can be further decomposed into protons, neutrons, and electrons. However, it is a very "flat" hierarchy, in which the number of first-order subsystems belonging to the crystal can be indefinitely large. A volume of molecular gas is a flat hierarchy in the same sense. In ordinary usage, we tend to reserve the word hierarchy for a system that is divided into a *small or moderate number* of subsystems, each of which may be further subdivided. Hence, we do not ordinarily think of or refer to a diamond or a gas as a hierarchic structure. Similarly, a linear polymer is simply a chain, which may be very long, of identical subparts, the monomers. At the molecular level it is a very flat hierarchy.

In discussing formal organizations, the number of subordinates who report directly to a single boss is called his *span of control*. I will speak analogously of the *span* of a system, by which I shall mean the number of subsystems into which it is partitioned. Thus, a hierarchic system is flat at a given level if it has a wide span at that level. A diamond has a wide span at the crystal level, but not at the next level down, the molecular level.

In most of our theory construction in the following sections we shall focus our attention on hierarchies of moderate span, but from time to time I shall comment on the extent to which the theories might or might not be expected to apply to very flat hierarchies.

There is one important difference between the physical and biological hierarchies, on the one hand, and social hierarchies, on the other. Most physical and biological hierarchies are described in spatial terms. We detect the organelles in a cell in the way we detect raisins in a cake—they are "visibly" differentiated substructures localized spatially in the larger structure. On the other hand, we propose to identify social hierarchies not by observing who lives close to whom but by observing who interacts with whom. These two points of view can be reconciled by defining hierarchy in terms of intensity of interaction, but observing that in most biological and physical systems relatively intense interaction implies relative spatial propinquity. One of the interesting characteristics of nerve cells and telephone wires is that they permit very specific strong interactions at great distances. To the extent that interactions are channeled through specialized communications and transportation systems, spatial propinquity becomes less determinative of structure.

Symbolic Systems

One very important class of systems has been omitted from my examples thus far: systems of human symbolic production. A book is a hierarchy in the sense in which I am using that term. It is generally divided into chapters, the chapters into sections, the sections into paragraphs, the paragraphs into sentences, the sentences into clauses and phrases, the clauses and phrases into words. We may take the words as our elementary units, or further subdivide them, as the linguist often does, into smaller units. If the book is narrative in character, it may divide into "episodes" instead of sections, but divisions there will be.

The hierarchic structure of music, based on such units as movements, parts, themes, phrases, is well known. The hierarchic structure of products of the pictorial arts is more difficult to characterize, but I shall have something to say about it later.

The Evolution of Complex Systems

Let me introduce the topic of evolution with a parable. There once were two watchmakers, named Hora and Tempus, who manufactured very fine watches. Both of them were highly regarded, and the phones in their workshops rang frequently—new customers were constantly calling them. However, Hora prospered, while Tempus became poorer and poorer and finally lost his shop. What was the reason?

The watches the men made consisted of about 1,000 parts each. Tempus had so constructed his that if he had one partly assembled and had to put it down—to answer the phone say—it immediately fell to pieces and had to be reassembled from the elements. The better the customers liked his watches, the more they phoned him, the more difficult it became for him to find enough uninterrupted time to finish a watch.

The watches that Hora made were no less complex than those of Tempus. But he had designed them so that he could put together subassemblies of about ten elements each. Ten of these subassemblies, again, could be put together into a larger subassembly; and a system of ten of the latter subassemblies constituted the whole watch. Hence, when Hora had to put down a partly assembled watch in order to answer the phone, he lost only a small part of his work, and he assembled his watches in only a fraction of the man-hours it took Tempus.

It is rather easy to make a quantitative analysis of the relative difficulty of the tasks of Tempus and Hora: Suppose the probability that an interruption will occur while a part is being added to an incomplete assembly is p. Then the probability that Tempus can complete a watch he has started without interruption is $(1 - p)^{1000}$—a very small number unless p is .001 or less. Each interruption will cost, on the average, the time to assemble $1/p$ parts (the expected number assembled before interruption). On the other hand, Hora has to complete one hundred eleven subassemblies of ten parts each. The probability that he will not be interrupted while completing any one of these is $(1 - p)^{10}$, and each interruption will cost only about the time required to assemble five parts.

Now if p is about .01—that is, there is one chance in a hundred that either watchmaker will be interrupted while adding any one part to an assembly—then a straightforward calculation shows that it will take Tempus, on the average, about four thousand times as long to assemble a watch as Hora.

We arrive at the estimate as follows:

1. Hora must make 111 times as many complete assemblies per watch as Tempus; but,
2. Tempus will lose on the average 20 times as much work for each interrupted assembly as Hora [100 parts, on the average, as against 5]; and,
3. Tempus will complete an assembly only 44 times per million attempts ($.99^{1000} = 44 \times 10^{-6}$), while Hora will complete nine out of ten ($.99^{10} = 9 \times 10^{-1}$). Hence Tempus will have to make 20,000 as many attempts per completed assembly as Hora. $(9 \times 10^{-1}) / (44 \times 10^{-6}) = 2 \times 10^4$. Multiplying these three ratios, we get:

$$1/111 \times 100/5 \times .99^{10}/.99^{1000} = 1/111 \times 20 \times 20{,}000 \sim 4{,}000.$$

Biological Evolution

What lessons can we draw from our parable for biological evolution? Let us interpret a partially completed subassembly of k elementary parts as the coexistence of k parts in a small volume—ignoring their relative orientations. The model assumes that parts are entering the volume at a constant rate, but that there is a constant probability, p, that the part will be dispersed before another is added, unless the assembly reaches a stable state. These assumptions are not particularly realistic. They undoubtedly underestimate the decrease in probability of achieving the assembly with increase in the size of the assembly. Hence the assumptions understate—probably by a large factor—the relative advantage of a hierarchic structure.

Although we cannot, therefore, take the numerical estimate seriously the lesson for biological evolution is quite clear and direct. The time required for the evolution of a complex form from simple elements depends critically on the numbers and distribution of potential intermediate stable forms. In particular, if there exists a hierarchy of potential stable "subassemblies," with about the same span, s, at each level of the hierarchy, then the time required for a subassembly can be expected to be about the same at each level—that is proportional to $1/(1 - p)^s$. The time required for the assembly of a system of n elements will be proportional to $\log_s n$, that is, to the number of levels in the system. One would say—with more illustrative than literal intent—that the time required for the evolution of multi-celled organisms from single-celled organisms might be of the same order of magnitude as the time required for the evolution of single-celled organisms from macromolecules. The same argument could be applied to the evolution of proteins from amino acids, of molecules from atoms, of atoms from elementary particles.

A whole host of objections to this oversimplified scheme will occur, I am sure, to every working biologist, chemist, and physicist. Before turning to matters I know more about, I shall mention three of these problems, leaving the rest to the attention of the specialists.

First, in spite of the overtones of the watchmaker parable, the theory assumes no teleological mechanism. The complex forms can arise from the simple ones by purely random processes. (I shall propose another model in a moment that shows this clearly.) Direction is provided to the scheme by the stability of the complex forms, once these come into existence. But this is nothing more than survival of the fittest—i.e., of the stable.

Second, not all large systems appear hierarchical. For example, most polymers—e.g., nylon—are simply linear chains of large numbers of identical components, the monomers. However, for present purposes we can simply regard such a structure as a hierarchy with a span of one—the limiting case. For a chain of any length represents a state of relative equilibrium.

Third, the evolution of complex systems from simple elements implies nothing, one way or the other, about the change in entropy of the entire system. If the process absorbs free energy, the complex system will have a smaller entropy than the elements; if it releases free energy, the opposite will be true. The former alternative is the one that holds for most biological systems, and the net inflow of free energy has to be supplied from the sun or some other source if the second law of thermodynamics is not to be violated. For the evolutionary process we are describing, the equilibria of the intermediate states need have only local and not global stability, and they may be stable only in the steady state—that is, as long as there is an external source of free energy that may be drawn upon.

Because organisms are not energetically closed systems, there is no way to deduce the direction, much less the rate, of evolution from classical thermodynamic considerations. All estimates indicate that the amount of entropy, measured in physical units, involved in the formation of a one-celled biological organism is trivially small—about -10^{-11} cal/degree. The "improbability" of evolution has nothing to do with this quantity of entropy, which is produced by every bacterial cell every generation. The irrelevance of quantity of information, in this sense, to speed of evolution can also be seen from the fact that exactly as much information is required to "copy" a cell through the reproductive process as to produce the first cell through evolution.

The effect of the existence of stable intermediate forms exercises a powerful effect on the evolution of complex forms that may be likened to the dramatic effect of catalysts upon reaction rates and steady state distribution of reaction products in open systems. In neither case does the entropy change provide us with a guide to system behavior.

Problem Solving as Natural Selection

Let us turn now to some phenomena that have no obvious connection with biological evolution: human problem-solving processes. Consider, for example, the task of discovering the proof for a difficult theorem. The process can be—and often has been—described as a search through a maze. Starting with the axioms and previously proved theorems, various transformations allowed by the rules of the mathematical systems are attempted, to obtain new expressions. These are modified in turn until, with persistence and good fortune, a sequence or path of transformations is discovered that leads to the goal.

The process usually involves a great deal of trial and error. Various paths are tried; some are abandoned, others are pushed further. Before a solution is found, a great many paths of the maze may be explored. The more difficult and novel the problem, the greater is likely to be the amount of trial and error required to find a solution. At the same time, the trial and error is not completely random or blind; it is, in fact, rather highly selective. The new expressions that are obtained by transforming given ones are examined to see whether they represent progress toward the goal. Indications of progress spur further search in the same direction; lack of progress signals the abandonment of a line of search. Problem solving requires *selective* trial and error.

A little reflection reveals that cues signaling progress play the same role in the problem-solving process that stable intermediate forms play in the biological evolutionary process. In fact, we can take over the watchmaker parable and apply it also to problem solving. In problem solving, a partial result that represents recognizable progress toward the goal plays the role of a stable subassembly.

Suppose that the task is to open a safe whose lock has ten dials, each with one hundred possible settings, numbered from 0 to 99. How long will it take to open the safe by a blind trial-and-error search for the correct setting? Since there are 100^{10} possible settings, we may expect to examine about one-half of these, on the average, before finding the correct one—that is, fifty billion billion settings. Suppose, however, that the safe is defective, so that a click can be heard when any one dial is turned to the correct setting. Now each dial can be adjusted independently, and does not need to be touched again while the others are being set. The total number of settings that has to be tried is only 10×50, or five hundred. The task of opening the safe has been altered, by the cues the clicks provide, from a practically impossible one to a trivial one.

A considerable amount has been learned in the past five years about the nature of the mazes that represent common human problem-solving tasks—proving theorems, solving puzzles, playing chess, making investments, balancing assembly lines, to mention a few. All that we have learned about these mazes points to the same conclusion: that human problem solving, from the most blundering to the most insightful, involves nothing more than varying mixtures of trial and error and selectivity. The selectivity derives from various rules of thumb, or heuristics, that suggest which paths should be tried first and which leads are promising. We do not need to postulate processes more sophisticated than those involved in organic evolution to explain how enormous problem mazes are cut down to quite reasonable size.

The Sources of Selectivity

When we examine the sources from which the problem-solving system, or the evolving system, as the case may be, derives its selectivity, we discover that selectivity can always be equated with some kind of feedback of information from the environment.

Let us consider the case of problem solving first. There are two basic kinds of selectivity. One we have already noted: various paths are tried out, the consequences of following them are noted, and this information is used to guide further search. In the same way, in organic evolution, various complexes come into being, at least evanescently, and those that are stable provide new building blocks for further construction. It is this information about stable configurations, and not free energy or negentropy from the sun, that guides the process of evolution and provides the selectivity that is essential to account for its rapidity.

The second source of selectivity in problem solving is previous experience. We see this particularly clearly when the problem to be solved is similar to one that has been solved before. Then, by simply trying again the paths that led to the earlier solution, or their analogues, trial-and-error search is greatly reduced or altogether eliminated.

What corresponds to this latter kind of information in organic evolution? The closest analogue is reproduction. Once we reach the level of self-reproducing systems, a complex system, when it has once been achieved, can be multiplied indefinitely. Reproduction in fact allows the inheritance of acquired characteristics, but at the level of genetic material, of course; i.e., only characteristics acquired by the genes can be inherited. We shall return to the topic of reproduction in the final section of this paper.

On Empires and Empire-Building

We have not exhausted the categories of complex systems to which the watchmaker argument can reasonably be applied. Philip assembled his Macedonian empire and gave it to his son, to be later combined with the Persian subassembly and others into Alexander's greater system. On Alexander's death, his empire did not crumble to dust, but fragmented into some of the major subsystems that had composed it.

The watchmaker argument implies that if one would be Alexander, one should be born into a world where large stable political systems already exist. Where this condition was not fulfilled, as on the Scythian and Indian frontiers, Alexander found empire building a slippery business. So too, T. E. Lawrence's organizing of the Arabian revolt against the Turks was limited by the character of his largest stable building blocks, the separate, suspicious desert tribes.

The profession of history places a greater value upon the validated particular fact than upon tendentious generalization. I shall not elaborate upon my fancy, therefore, but will leave it to historians to decide whether anything can be learned for the interpretation of history from an abstract theory of hierarchic complex systems.

Conclusion: The Evolutionary Explanation of Hierarchy

We have shown thus far that complex systems will evolve from simple systems much more rapidly if there are stable intermediate forms than if there are not. The resulting complex forms in the former case will be hierarchic. We have only to turn the argument around to explain the observed predominance of hierarchies among the complex systems nature presents to us. Among possible complex forms, hierarchies are the ones that have the time to evolve. The hypothesis that complexity will be hierarchic makes no distinction among very flat hierarchies, like crystals, and tissues, and polymers, and the intermediate forms. Indeed, in the complex systems we encounter in nature, examples of both forms are prominent. A more complete theory than the one we have developed here would presumably have something to say about the determinants of width of span in these systems.

Nearly Decomposable Systems

In hierarchic systems, we can distinguish between the interactions *among* subsystems, on the one hand, and the interactions *within* subsystems—i.e., among the parts of those subsystems—on the other. The interactions at the different levels may be, and often will be, of different orders of magnitude. In a formal organization there will generally be more interaction, on the average, between two employees who are members of the same department than between two employees from different departments. In organic substances, intermolecular forces will generally be weaker than molecular forces, and molecular forces than nuclear forces.

In a rare gas, the intermolecular forces will be negligible compared to those binding the molecules—we can treat the individual particles, for many purposes, as if they were independent of each other. We can describe such a system as *decomposable* into the subsystems comprised of the individual particles. As the gas becomes denser, molecular interactions become more significant. But over some range, we can treat the decomposable case as a limit, and as a first approximation. We can use a theory of perfect gases, for example, to describe approximately the behavior of actual gases if they are not too dense. As a second approximation, we may move to a theory of *nearly decomposable* systems, in which the interactions among the subsystems are weak, but not negligible.

At least some kinds of hierarchic systems can be approximated successfully as nearly decomposable systems. The main theoretical findings from the approach can be summed up in two propositions: (*a*) in a nearly decomposable system, the short-run behavior of each of the component subsystems is approximately independent of the short-run behavior of the other components; (*b*) in the long run, the behavior of any one of the components depends in only an aggregate way on the behavior of the other components.

Let me provide a very concrete simple example of a nearly decomposable system.

	A1	A2	A3	B1	B2	C1	C2	C3
A1	—	100	—	2	—	—	—	—
A2	100	—	100	1	1	—	—	—
A3	—	100	—	—	2	—	—	—
B1	2	1	—	—	100	2	1	—
B2	—	1	2	100	—	—	1	2
C1	—	—	—	2	—	—	100	—
C2	—	—	—	1	—	100	—	100
C3	—	—	—	—	2	—	100	—

Figure 1. A hypothetical nearly decomposable system. In terms of the heat-exchange example of the test, A1, A2, and A3 may be interpreted as cubicles in one room, B1 and B2 as cubicles in a second room, and C1, C2, and C3 as cubicles in a third. The matrix entries then are the heat diffusion coefficients between cubicles.

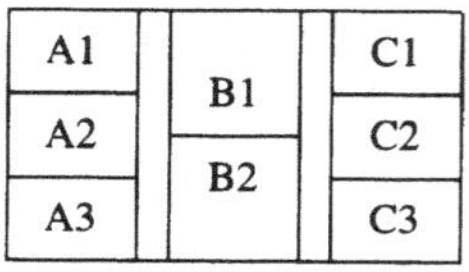

A1		C1
A2	B1	C2
A3	B2	C3

Consider a building whose outside walls provide perfect thermal insulation from the environment. We shall take these walls as the boundary of our system. The building is divided into a large number of rooms, the walls between them being good, but not perfect, insulators. The walls between rooms are the boundaries of our major subsystems. Each room is divided by partitions into a number of cubicles, but the partitions are poor insulators. A thermometer hangs in each cubicle. Suppose that at the time of our first observation of the system there is a wide variation in temperature from cubicle to cubicle and from room to room—the various cubicles within the building are in a state of thermal disequilibrium. When we take new temperature readings several hours later, what shall we find? There will be very little variation in temperature among the cubicles within each single room, but there may still be large temperature variations *among* rooms. When we take readings again several days later, we find an almost uniform temperature throughout the building; the temperature differences among rooms have virtually disappeared.

We can describe the process of equilibration formally by setting up the usual equations of heat flow. The equations can be represented by the matrix of their coefficients, r_{ij}, where r_{ij} is the rate at which heat flows from the ith cubicle to the jth cubicle per degree difference in their temperatures. If cubicles i and j do not have a common wall, r_{ij} will be zero. If cubicles i and j have a common wall, and are in the same room, r_{ij} will be large. If cubicles i and j are separated by the wall of a room, r_{ij} will be nonzero but small. Hence, by grouping all the cubicles together that are in the same room, we can arrange the matrix of coefficients so that all its large elements lie inside a string of square submatrices along the main diagonal. All the elements outside these diagonal squares will be either zero or small (see figure 1). We may take some small number, ϵ, as the upper bound of the extradiagonal elements. We shall call a matrix having these properties a *nearly decomposable matrix*.

Now it has been proved that a dynamic system that can be described by a nearly decomposable matrix has the properties, stated above, of a nearly decomposable system. In our simple example of heat flow this means that in the short run each room will reach an equilibrium temperature (an average of the initial temperatures of its offices) nearly independently of the others; and that each room will remain approximately in a state of equilibrium over the longer period during which an over-all temperature equilibrium is being established throughout the building. After the intra-room short-run equilibria have been reached, a single thermometer in each room will be adequate to describe the dynamic behavior of the entire system—separate thermometers in each cubicle will be superfluous.

Near Decomposability of Social Systems

As a glance at figure 1 shows, near decomposability is a rather strong property for a matrix to possess, and the matrices that have this property will describe very special dynamic systems—vanishingly few systems out of all those that are thinkable. How few they will be depends, of course, on how good an approximation we insist upon. If we demand that epsilon be very small, correspondingly few dynamic systems will fit the definition. But we have already seen that in the natural world nearly decomposable systems are far from rare. On the contrary, systems in which each variable is linked with almost equal strength with almost all other parts of the system are far rarer and less typical.

In economic dynamics, the main variables are the prices and quantities of commodities. It is empirically true that the price of any given commodity and the rate at which it is exchanged depend to a significant extent only on the prices and quantities of a few other commodities, together with a few other aggregate magnitudes, like the average price level or some over-all measure of economic activity. The large linkage coefficients are associated, in general, with the main flows of raw materials and semi-finished products within and between industries. An input-output matrix of the economy, giving the magnitudes of these flows, reveals the nearly decomposable structure of the system—with one qualification. There is a consumption subsystem of the economy that is linked strongly to variables in most of the other subsystems. Hence, we have to modify our notions of decomposability slightly to accommodate the special role of the consumption subsystem in our analysis of the dynamic behavior of the economy.

In the dynamics of social systems, where members of a system communicate with and influence other members, near decomposability is generally very prominent. This is most obvious in formal organizations, where the formal authority relation connects each member of the organization with one immediate superior and with a small number of subordinates. Of course many communications in organizations follow other channels than the lines of formal authority. But most of these channels lead from any particular individual to a very limited number of his superiors, subordinates, and associates. Hence, departmental boundaries play very much the same role as the walls in our heat example.

Physico-Chemical Systems

In the complex systems familiar in biological chemistry, a similar structure is clearly visible. Take the atomic nuclei in such a system as the elementary parts of the system, and construct a matrix of bond strengths between elements. There will be matrix elements of quite different orders of magnitude. The largest will generally correspond to the covalent bonds, the next to the ionic bonds, the third group to hydrogen bonds, still smaller linkages to van der Waals forces. If we select an epsilon just a little smaller than the magnitude of a covalent bond, the system will decompose into subsystems—the constituent molecules. The smaller linkages will correspond to the intermolecular bonds.

It is well known that high-energy, high-frequency vibrations are associated with the smaller physical subsystems, low-frequency vibrations with the larger systems into which the subsystems are assembled. For example, the radiation frequencies associated with molecular vibrations are much lower than those associated with the vibrations of the planetary electrons of the atoms; the latter, in turn, are lower than those associated with nuclear processes. Molecular systems are nearly decomposable systems, the short-run dynamics relating to the internal structures of the subsystems; the long-run dynamics to the interactions of these subsystems.

A number of the important approximations employed in physics depend for their validity on the near-decomposability of the systems studied. The theory of the thermodynamics of irreversible processes, for example, requires the assumption of macroscopic disequilibrium but microscopic equilibrium, exactly the situation described in our heat-exchange example. Similarly, computations in quantum mechanics are often handled by treating weak interactions as producing perturbations on a system of strong interactions.

Some Observations on Hierarchic Span

To understand why the span of hierarchies is sometimes very broad—as in crystals—sometimes narrow, we need to examine more detail of the interactions. In general, the critical consideration is the extent to which interaction between two (or a few) subsystems excludes interaction of these subsystems with the others. Let us examine first some physical examples.

Consider a gas of identical molecules, each of which can form covalent bonds, in certain ways, with others. Let us suppose that we can associate with each atom a specific number of bonds that it is capable of maintaining simultaneously. (This number is obviously related to the number we usually call its valence.) Now suppose that two atoms join, and that we can also associate with the combination a specific number of external bonds it is capable of maintaining. If this number is the same as the number associated with the individual atoms, the bonding process can go on indefinitely—the atoms can form crystals or polymers of indefinite extent. If the number of bonds of which the composite is capable is less than the number associated with each of the parts, then the process of agglomeration must come to a halt.

We need only mention some elementary examples. Ordinary gases show no tendency to agglomerate because the multiple bonding of atoms "uses up" their capacity to interact. While each oxygen atom has a valence of two, the O_2 molecules have a zero valence. Contrariwise, indefinite chains of single-bonded carbon atoms can be built up because a chain of any number of such atoms, each with two side groups, has a valence of exactly two.

Now what happens if we have a system of elements that possess both strong and weak interaction capacities, and whose strong bonds are exhaustible through combination? Subsystems will form, until all the capacity for strong interaction is utilized in their construction. Then these subsystems will be linked by the weaker second-order bonds into larger systems. For example, a water molecule has essentially a valence of zero—all the potential covalent bonds are fully occupied by the interaction of hydrogen and oxygen molecules. But the geometry of the molecule creates an electric dipole that permits weak interaction between the water and salts dissolved in it—whence such phenomena as its electrolytic conductivity.

Similarly, it has been observed that, although electrical forces are much stronger than gravitational forces, the latter are far more important than the former for systems on an astronomical scale. The explanation, of course, is that the electrical forces, being bipolar, are all "used up" in the linkages of the smaller subsystems, and that significant net balances of positive or negative charges are not generally found in regions of macroscopic size.

In social as in physical systems there are generally limits on the simultaneous interaction of large numbers of subsystems. In the social case, these limits are related to the fact that a human being is more nearly a serial than a parallel information-processing system. He can carry on only one conversation at a time, and although this does not limit the size of the audience to which a mass communication can be addressed, it does limit the number of people simultaneously involved in most other forms of social interaction. Apart from requirements of direct interaction, most roles impose tasks and responsibilities that are time consuming. One cannot, for example, enact the role of "friend" with large numbers of other people.

It is probably true that in social as in physical systems, the higher frequency dynamics are associated with the subsystems, the lower frequency dynamics with the larger systems. It is generally believed, for example, that the relevant planning horizon of executives is longer the higher their location in the organizational hierarchy. It is probably also true that both the average duration of an interaction between executives and the average interval between interactions is greater at higher than at lower levels.

Summary: Near Decomposability

We have seen that hierarchies have the property of near-decomposability. Intracomponent linkages are generally stronger than intercomponent linkages. This fact has the effect of separating the high-frequency dynamics of a hierarchy—involving the internal structure of the components—from the low frequency dynamics—involving interaction among components. We shall turn next to some important consequences of this separation for the description and comprehension of complex systems.

The Description of Complexity

If you ask a person to draw a complex object—e.g., a human face—he will almost always proceed in a hierarchic fashion. First he will outline the face. Then he will add or insert features: eyes, nose, mouth, ears, hair. If asked to elaborate, he will begin to develop details for each of the features—pupils, eyelids, lashes for the eyes, and so on—until he reaches the limits of his anatomical knowledge. His information about the object is arranged hierarchically in memory, like a topical outline.

When information is put in outline form, it is easy to include information about the relations among the major parts and information about the internal relations of parts in each of the suboutlines. Detailed information about the relations of subparts belonging to different parts has no place in the outline and is likely to be lost. The loss of such information and the preservation mainly of information about hierarchic order is a salient characteristic that distinguishes the drawings of a child or someone untrained in representation from the drawing of a trained artist. (I am speaking of an artist who is striving for representation.)

Near Decomposability and Comprehensibility

From our discussion of the dynamic properties of nearly decomposable systems, we have seen that comparatively little information is lost by representing them as hierarchies. Subparts belonging to different parts only interact in an aggregative fashion—the detail of their interaction can be ignored. In studying the interaction of two large molecules, generally we do not need to consider in detail the interactions of the nuclei of the atoms belonging to the one molecule with the nuclei of the atoms belonging to the other. In studying the interaction of two nations, we do not need to study in detail the interactions of each citizen of the first with each citizen of the second.

The fact, then, that many complex systems have a nearly decomposable, hierarchic structure is a major facilitating factor enabling us to understand, to describe, and even to "see" such systems and their parts. Or perhaps the proposition should be put the other way round. If there are important systems in the world that are complex without being hierarchic, they may to a considerable extent escape our observation and our understanding. Analysis of their behavior would involve such detailed knowledge and calculation of the interactions of their elementary parts that it would be beyond our capacities of memory or computation.

I shall not try to settle which is chicken and which is egg: whether we are able to understand the world because it is hierarchic, or whether it appears hierarchic because those aspects of it which are not elude our understanding and observation. I have already given some reasons for supposing that the former is at least half the truth—that evolving complexity would tend to be hierarchic—but it may not be the whole truth.

Simple Descriptions of Complex Systems

One might suppose that the description of a complex system would itself be a complex structure of symbols—and indeed, it may be just that. But there is no conservation law that requires that the description be as cumbersome as the object described. A trivial example will show how a system can be described economically. Suppose the system is a two-dimensional array like this:

Architecture of Complexity

$$\begin{array}{cccccccc}
A & B & M & N & R & S & H & I \\
C & D & O & P & T & U & J & K \\
M & N & A & B & H & I & R & S \\
O & P & C & D & J & K & T & U \\
R & S & H & I & A & B & M & N \\
T & U & J & K & C & D & O & P \\
H & I & R & S & M & N & A & B \\
J & K & T & U & O & P & C & D \\
\end{array}$$

Let us call the array $\left|\begin{smallmatrix} A & B \\ C & D \end{smallmatrix}\right|$ a, the array $\left|\begin{smallmatrix} M & N \\ O & P \end{smallmatrix}\right|$ m, the array $\left|\begin{smallmatrix} R & S \\ T & U \end{smallmatrix}\right|$ r, and the array $\left|\begin{smallmatrix} H & I \\ J & K \end{smallmatrix}\right|$ h. Let us call the array $\left|\begin{smallmatrix} a & m \\ m & a \end{smallmatrix}\right|$ w, and the array $\left|\begin{smallmatrix} r & h \\ h & r \end{smallmatrix}\right|$ x. Then the entire array is simply $\left|\begin{smallmatrix} w & x \\ x & w \end{smallmatrix}\right|$. While the original structure consisted of 64 symbols, it requires only 35 to write down its description:

$$S = \frac{wx}{xw}$$

$$w = \frac{am}{ma} \qquad x = \frac{rh}{hr}$$

$$a = \frac{AB}{CD} \qquad m = \frac{MN}{OP} \qquad r = \frac{RS}{TU} \qquad h = \frac{HI}{JK}$$

We achieve the abbreviation by making use of the redundancy in the original structure. Since the pattern $\begin{smallmatrix} A & B \\ C & D \end{smallmatrix}$, for example, occurs four times in the total pattern, it is economical to represent it by the single symbol, a.

If a complex structure is completely unredundant—if no aspect of its structure can be inferred from any other—then it is its own simplest description. We can exhibit it, but we cannot describe it by a simpler structure. The hierarchic structures we have been discussing have a high degree of redundancy, hence can often be described in economical terms. The redundancy takes a number of forms, of which I shall mention three:

1. Hierarchic systems are usually composed of only a few different kinds of subsystems, in various combinations and arrangements. A familiar example is the proteins, their multitudinous variety arising from arrangements of only twenty different amino acids. Similarly, the ninety-odd elements provide all the kinds of building blocks needed for an infinite variety of molecules. Hence, we can construct our description from a restricted alphabet of elementary terms corresponding to the basic set of elementary subsystems from which the complex system is generated.

2. Hierarchic systems are, as we have seen, often nearly decomposable. Hence only aggregative properties of their parts enter into the description of the interactions of those parts. A generalization of the notion of near-decomposability might be called the "empty world hypothesis"—most things are only weakly connected with most other things; for a tolerable description of reality only a tiny fraction of all possible interactions needs to be taken into account. By adopting a descriptive language that allows the absence of something to go unmentioned, a nearly empty world can be described quite concisely. Mother Hubbard did not have to check off the list of possible contents to say that her cupboard was bare.

3. By appropriate "recoding," the redundancy that is present but unobvious in the structure of a complex system can often be made patent. The most common recoding of descriptions of dynamic systems consists in replacing a description of the time path with a description of a differential law that generates that path. The simplicity, that is, resides in a

constant relation between the state of the system at any given time and the state of the system a short time later. Thus, the structure of the sequence, 1 3 5 7 9 11 . . ., is most simply expressed by observing that each member is obtained by adding 2 to the previous one. But this is the sequence that Galileo found to describe the velocity at the end of successive time intervals of a ball rolling down an inclined plane.

It is a familiar proposition that the task of science is to make use of the world's redundancy to describe that world simply. I shall not pursue the general methodological point here, but shall instead take a closer look at two main types of description that seem to be available to us in seeking an understanding of complex systems. I shall call these *state description* and *process description*, respectively.

State Descriptions and Process Descriptions

"A circle is the locus of all points equidistant from a given point." "To construct a circle, rotate a compass with one arm fixed until the other arm has returned to its starting point." It is implicit in Euclid that if you carry out the process specified in the second sentence, you will produce an object that satisfies the definition of the first. The first sentence is a state description of a circle, the second a process description.

These two modes of apprehending structure are the warp and weft of our experience. Pictures, blueprints, most diagrams, chemical structural formulae are state descriptions. Recipes, differential equations, equations for chemical reactions are process descriptions. The former characterize the world as sensed; they provide the criteria for identifying objects, often by modeling the objects themselves. The latter characterize the world as acted upon; they provide the means for producing or generating objects having the desired characteristics.

The distinction between the world as sensed and the world as acted upon defines the basic condition for the survival of adaptive organisms. The organism must develop correlations between goals in the sensed world and actions in the world of process. When they are made conscious and verbalized, these correlations correspond to what we usually call means-end analysis. Given a desired state of affairs and an existing state of affairs, the task of an adaptive organism is to find the difference between these two states, and then to find the correlating process that will erase the difference.

Thus, problem solving requires continual translation between the state and process descriptions of the same complex reality. Plato, in the *Meno*, argued that all learning is remembering. He could not otherwise explain how we can discover or recognize the answer to a problem unless we already know the answer. Our dual relation to the world is the source and solution of the paradox. We pose a problem by giving the state description of the solution. The task is to discover a sequence of processes that will produce the goal state from an initial state. Translation from the process description to the state description enables us to recognize when we have succeeded. The solution is genuinely new to us—and we do not need Plato's theory of remembering to explain how we recognize it.

There is now a growing body of evidence that the activity called human problem solving is basically a form of means-end analysis that aims at discovering a process description of the path that leads to a desired goal. The general paradigm is: given a blueprint, to find the corresponding recipe. Much of the activity of science is an application of that paradigm: given the description of some natural phenomena, to find the differential equations for processes that will produce the phenomena.

The Description of Complexity in Self-Reproducing Systems

The problem of finding relatively simple descriptions for complex systems is of interest not only for an understanding of human knowledge of the world but also for an explanation of how a complex system can reproduce itself. In my discussion of the evolution of complex systems, I touched only briefly on the role of self-reproduction.

Atoms of high atomic weight and complex inorganic molecules are witnesses to the fact that the evolution of complexity does not imply self-reproduction. If evolution of complexity from simplicity is sufficiently probable, it will occur repeatedly; the statistical equilibrium of the system will find a large fraction of the elementary particles participating in complex systems.

If, however, the existence of a particular complex form increased the probability of the creation of another form just like it, the equilibrium between complexes and components could be greatly altered in favor of the former. If we have a description of an object that is sufficiently clear and complete, we can reproduce the object from the description. Whatever the exact mechanism of reproduction, the description provides us with the necessary information.

Now we have seen that the descriptions of complex systems can take many forms. In particular, we can have state descriptions or we can have process descriptions; blueprints or recipes. Reproductive processes could be built around either of these sources of information. Perhaps the simplest possibility is for the complex system to serve as a description of itself—a template on which a copy can be formed. One of the most plausible current theories, for example, of the reproduction of deoxyribonucleic acid (DNA) proposes that a DNA molecule, in the form of a double helix of matching parts (each essentially a "negative" of the other), unwinds to allow each half of the helix to serve as a template on which a new matching half can form.

On the other hand, our current knowledge of how DNA controls the metabolism of the organism suggests that reproduction by template is only one of the processes involved. According to the prevailing theory, DNA serves as a template both for itself and for the related substance ribonucleic acid (RNA). RNA, in turn, serves as a template for protein. But proteins—according to current knowledge—guide the organism's metabolism not by the template method but by serving as catalysts to govern reaction rates in the cell. While RNA is a blueprint for protein, protein is a recipe for metabolism.

Ontogeny Recapitulates Phylogeny

The DNA in the chromosomes of an organism contains some, and perhaps most, of the information that is needed to determine its development and activity. We have seen that, if current theories are even approximately correct, the information is recorded not as a state description of the organism but as a series of "instructions" for the construction and maintenance of the organism from nutrient materials. I have already used the metaphor of a recipe; I could equally well compare it with a computer program, which is also a sequence of instructions, governing the construction of symbolic structures. Let me spin out some of the consequences of the latter comparison.

If genetic material is a program—viewed in its relation to the organism—it is a program with special and peculiar properties. First, it is a self-reproducing program; we have already considered its possible copying mechanism. Second, it is a program that has developed by Darwinian evolution. On the basis of our watchmaker's argument, we may assert that many of its ancestors were also viable programs—programs for the sub-assemblies.

Are there any other conjectures we can make about the structure of this program? There is a well-known generalization in biology that is verbally so neat that we would be reluctant to give it up even if the facts did not support it: ontogeny recapitulates phylogeny. The individual organism, in its development, goes through stages that resemble some of its ancestral forms. The fact that the human embryo develops gill bars and then modifies them for other purposes is a familiar particular belonging to the generalization. Biologists today like to emphasize the qualifications of the principle—that ontogeny recapitulates only the grossest aspects of phylogeny, and these only crudely. These qualifications should not make us lose sight of the fact that the generalization does hold in rough approximation—it does summarize a very significant set of facts about the organism's development. How can we interpret these facts?

One way to solve a complex problem is to reduce it to a problem previously solved—to show what steps lead from the earlier solution to a solution of the new problem. If, around the turn of the century, we wanted to instruct a workman to make an automobile, perhaps the simplest way would have been to tell him how to modify a wagon by removing the singletree and adding a motor and transmission. Similarly, a genetic program could be altered in the course of evolution by adding new processes that would modify a simpler form into a more complex one—to construct a gastrula, take a blastula and alter it!

The genetic description of a single cell may, therefore, take a quite different form from the genetic description that assembles cells into a multi-celled organism. Multiplication by cell division would require, as a minimum, a state description (the DNA, say), and a simple "interpretive process"—to use the term from computer language—that copies this description as a part of the larger copying process of cell division. But such a mechanism clearly would not suffice for the differentiation of cells in development. It appears more natural to conceptualize that mechanism as based on a process description, and a somewhat more complex interpretive process that produces the adult organism in a sequence of stages, each new stage in development representing the effect of an operator upon the previous one.

It is harder to conceptualize the interrelation of these two descriptions. Interrelated they must be, for enough has been learned of gene-enzyme mechanisms to show that these play a major role in development as in cell metabolism. The single clue we obtain from our earlier discussion is that the description may itself be hierarchical, or nearly decomposable, in structure, the lower levels governing the fast, "high-frequency" dynamics of the individual cell, the higher level interactions governing the slow, "low-frequency" dynamics of the developing multi-cellular organism.

There are only bits of evidence, apart from the facts of recapitulation, that the genetic program is organized in this way, but such evidence as exists is compatible with this notion.

To the extent that we can differentiate the genetic information that governs cell metabolism from the genetic information that governs the development of differentiated cells in the multi-cellular organization, we simplify enormously—as we have already seen—our task of theoretical description. But I have perhaps pressed this speculation far enough.

The generalization that in evolving systems whose descriptions are stored in a process language, we might expect ontogeny partially to recapitulate phylogeny has applications outside the realm of biology. It can be applied as readily, for example, to the transmission of knowledge in the educational process. In most subjects, particularly in the rapidly advancing sciences, the progress from elementary to advanced courses is to a considerable extent a progress through the conceptual history of the science itself. Fortunately, the recapitulation is seldom literal—any more than it is in the biological case. We do not teach the phlogiston theory in chemistry in order later to correct it. (I am not sure I could not cite examples in other subjects where we do exactly that.) But curriculum revisions that rid us of the accumulations of the past are infrequent and painful. Nor are they always desirable—partial recapitulation may, in many instances, provide the most expeditious route to advanced knowledge.

Summary: The Description of Complexity

How complex or simple a structure is depends critically upon the way in which we describe it. Most of the complex structures found in the world are enormously redundant, and we can use this redundancy to simplify their description. But to use it, to achieve the simplification, we must find the right representation.

The notion of substituting a process description for a state description of nature has played a central role in the development of modern science. Dynamic laws, expressed in the form of systems of differential or difference equations, have in a large number of cases provided the clue for the simple description of the complex. In the preceding paragraphs I have tried to show that this characteristic of scientific inquiry is not accidental or superficial. The correlation between state description and process description is basic to the functioning of any adaptive organism, to its capacity for acting purposefully upon its environment. Our present-day understanding of genetic mechanisms suggests that even in describing itself the multi-cellular organism finds a process description—a genetically encoded program—to be the parsimonious and useful representation.

Conclusion

Our speculations have carried us over a rather alarming array of topics, but that is the price we must pay if we wish to seek properties common to many sorts of complex systems. My thesis has been that one path to the construction of a non-trivial theory of complex systems is by way of a theory of hierarchy. Empirically, a large proportion of the complex systems we observe in nature exhibit hierarchic structure. On theoretical grounds we could expect complex systems to be hierarchies in a world in which complexity had to evolve from simplicity. In their dynamics, hierarchies have a property, near-decomposability, that greatly simplifies their behavior. Near-decomposability also simplifies the description of a complex system, and makes it easier to understand how the information needed for the development or reproduction of the system can be stored in reasonable compass.

In both science and engineering, the study of "systems" is an increasingly popular activity. Its popularity is more a response to a pressing need for synthesizing and analyzing complexity than it is to any large development of a body of knowledge and technique for dealing with complexity. If this popularity is to be more than a fad, necessity will have to mother invention and provide substance to go with the name. The explorations reviewed here represent one particular direction of search for such substance.

21

Complexity and System Descriptions

Robert Rosen

There is an enormous literature on the complexity of systems, and with attempts at specifying intrinsic measures of complexity. In this note, we will take an opposite view; that complexity is not an intrinsic property of a system, but rather manifests our capabilities to interact with the system. Since our capabilities to interact with systems around us are continually changing, so too do their apparent complexities. Some consequences of this viewpoint, bearing on system descriptions, and on the capacity of systems to make errors, will be dealt with below; a fuller discussion of these matters will be presented in another place.

Intuitively, we regard a system to be complex if we can interact with it significantly in many distinct ways, and if each of these different ways requires a different mode of description of the system to encompass it. Thus, organisms are complex; we can interact with them at many levels, and describe these interactions in many different ways. For ordinary purposes, a stone is a simple system; we typically interact with a stone in only a few ways, and basically a single mode of description is sufficient to describe these ways. If we adopt this viewpoint, that system complexity refers to our modes of interaction with the system, and to the corresponding number of different descriptions to which these modes of interaction give rise, then a number of questions immediately come to mind: (1) Is there one class of interactions, and consequently one mode of system description, from which all the others can be derived (the problem of reductionism), and (2) Is there any kind of measure of interactive capacity which will allow us to define complexity in other than subjective terms? In the subsequent discussion, we shall be concerned with some of the aspects of these questions.

As we have repeatedly emphasized elsewhere (Rosen, 1969, 1972), all system descriptions, however much they may vary in detail, ultimately consist of two kinds of notions: those dealing with the specification of an *instantaneous state* of a system, and those dealing with the manner in which this instantaneous state changes in time as a result of *forces* imposed on the system. As a typical example, we may consider the dynamical theory of particle mechanics, from which all other modes of system description have been derived. If

---

From *Systems, Approaches, Theories, Applications*, W. E. Hartnett, ed. Copyright © 1977 by D. Reidel Publishing Company.

we are given a system of N gravitating particles in ordinary 3-dimensional space, then the Newtonian formalism tells us that we can describe this system, at any instant of time, by specifying the displacements of the constituent particles from some convenient origin of coordinates, and by specifying their corresponding velocity or momenta. Thus, a set of $6N$ numbers suffices for a specification of the instantaneous state of the system. By this is meant that *any other* quantity pertaining to the system can be expressed as a function of the ones used to describe the instantaneous state, and hence is in a sense redundant. The displacements and momenta satisfy all the properties of a set of *state variables*; they represent a minimal set of observable quantities pertaining to the system, from which all other information pertaining to measurable quantities defined on the instantaneous states can be derived.

Newton's Laws also provide us with a dynamic description of the way in which the instantaneous states change with time, by identifying the rates of change of the momenta of the particles in our system with the forces imposed on the system. Knowing the forces, and knowing an initial state of the system, is thus sufficient to allow us to predict the state of the system at any time (at least in principle), and hence, to predict the value of any observable quantity associated with the system.

In this kind of formalism, we thus identify the abstract states of our system with a given mode of state description; i.e., with a subspace of Euclidean N-dimensional space. However, even in this formalism, there is a great deal of room for alternate descriptions. For instance, it is one of the fundamental assertions of physics that every real-valued function defined on the state space represents a potential dynamical variable of the system, and one which could be directly measured by some suitable measuring instrument. Conversely, of course, every measurable quantity pertaining to the system can be regarded as giving rise to such a function. Each such function inherits a dynamical equation, from the dynamical equations which govern the state variables. As we have shown elsewhere, however (Rosen, 1968), it can easily happen that a set of such observables will form a dynamical system in its own right; if we were to interact with the system in such a way as to observe those observables rather than the state variables, we would in effect see a completely different system. There even exist universal dynamical systems, which allow us to find observables inheriting any prescribed set of equations of motion posited in advance. Thus, even in this relatively straightforward situation, there are profound epistemological problems arising when we attempt to give an "intrinsic" character to the results of measurements or other kinds of interactions with the system.

Another aspect of the crucial concept of interaction between systems is discernible from the fact that the state description, which presumably contains within itself the answer to every question which can be asked about that system, pertains only to a perfectly isolated system. Given two systems perfectly specified in isolation, we cannot, *from those specifications alone*, decide how those systems are going to interact. To deal with the problem of interactions, we must invoke further information, usually in the form of some universal principle (e.g., mass action) which never forms part of the state description of individual systems. This is the primary reason why physics is successful in dealing with isolated systems, but cannot cope with interactions, except in the case in which they are infinitely weak (perturbations).

We have thus seen (a) that even a "simple" physical system can be interacted with in many ways (i.e., through many different sets of observables of the system); each of these

Complexity and System Descriptions

modes of interaction conveys some aspect of reality pertaining to the system, but the system will "look different" from mode to mode; and (b) the capacity of a system to manifest interaction is never considered a part of physical system descriptions, although it is crucial for such areas as biology. There is no reason why we should expect that two interacting systems will "see" each other through the same observables through which we find it convenient to interact with them separately, and consequently, there is no reason to expect that our conventional modes of state description will allow us to understand (let alone predict) the results of the interaction.

The above remarks bear heavily on the problem of reductionism, and on the more general problem of approaching complex systems through analytic means. By *analysis* we mean here the resolution of a system into a family of subsystems somehow "simpler" than the original system from which they were extracted, and attempting to infer the properties of the original system from the properties of the subsystems. The extraction of a subsystem corresponds formally to a process of *abstraction*, in which a number of degrees of freedom of the original system (i.e., potential interactive capabilities) are excluded, and only a limited number are retained. This process of abstraction can be physically implemented (as when a molecular biologist extracts a fraction of molecular species from a cell, thereby creating an abstract cell) or they can be purely formal (as when an ecologist represents a population of real organisms in terms of predation relations). The basic requirements of such abstractions are the following:

(1) The subsystems so obtained must be "simpler" than the original system from which they were abstracted;

(2) The subsystems must be obtained by "natural" means (i.e., utilizing familiar and justifiable procedures); and

(3) The properties of the subsystems so obtained must permit the determination of the properties of the original system.

The property (1) is obviously crucial; nothing is gained if we extract systems as intractable as the original system. This has long been recognized implicitly in scientific modes of analysis. Of equal importance is the property (3); any property of isolated subsystems not bearing on the properties of the original system is an *artifact*. The property (2), however, is a purely subjective matter, and refers only to the manner in which we find it convenient to interact with the original system. It thus stands on a different footing from (1) and (3).

Nevertheless, in many empirical modes of system analysis, the greatest weight is placed upon condition (2). It seems to be intuitively hoped that, by relying on procedures which satisfy (2), the conditions (1) and (3) will automatically be satisfied. At the very least, it is hoped that (1) + (2) will imply (3). However, from what we have already said, this is plainly absurd, in general. Indeed, what we learn from the above is that the crucial properties (1) and (3) which must be satisfied by any useful means of analysis of systems, must be allowed to determine what we are to regard as "natural." Indeed, "naturality" must not be allowed to be posited in advance, but only in terms of its bearing on the problems under discussion in a particular context.

A simple example may make this clear. In physics, the three-body problem is complex in a well-defined sense; the dynamical equations governing a system of three gravitating masses in an arbitrary configuration cannot be integrated directly. We could hope to approach this kind of problem by analysis into a family of "simpler" subsystems, which will allow us to solve the problem. Intuitively, the subsystems available to us are two-body systems and one-body systems. These are indeed "simpler" than the original system, and are abstracted from that system in "natural" ways. However, it is clear that we cannot solve a three-body problem in this fashion, for the act of decomposing the original system into isolated simpler subsystems destroys irreversibly the dynamics in which we were originally interested (here again we see the inability of physics to deal with arbitrary interactions). Thus, from the standpoint of solving the three-body problem, our apparently "natural" decompositions are useless; if analysis is to be successful in this kind of problem at all, the appropriate subsystems (i.e., those which satisfy (3)) must necessarily be of a kind which would appear most "unnatural" in terms of what we find it convenient to do physically to a system of particles.

An abstraction, in the sense in which we are using the term, owes its efficacy to the removal of degrees of freedom (i.e., interactive capabilities) present in the original system. It represents a simplification in that it faithfully represents what the original system would be like if it could only interact with the world through those degrees of freedom which are retained in the abstraction. Thus, if the abstraction exhibits similar interactive capabilities to those possessed by the original system, we are justified in assigning them to the properties retained in the abstraction. (However, we must note that an abstracted subsystem may exhibit new interactive capacities of its own, which do not pertain to the fact that it is in fact a subsystem of some larger system; always carefully distinguish between those properties which inhere in the original system, and those which arise in the model without reference to the fact that it is a model.)

Let us suppose that we have two different abstractions of the same complex system. How may these be compared or combined to give a fuller picture of the original system than either of them can given alone? We shall touch briefly on this question, and show how it relates to the notion of *structural stability*, which is presently attracting great interest in theoretical biology and elsewhere, largely through the influence of Thom (1972). Indeed, structural stability is a general framework for comparing one description of a class of objects against another description, usually in terms of measures of "closeness." Typically, if two objects "close" in one description are also "close" in the other, then the objects are called *generic*; generic objects are thus precisely those for which the two descriptions agree. Those objects which are not generic are called *bifurcation* objects; objects "close" to a bifurcation object in one description will not be "close" to that object in the other description. Generic objects thus do not require both descriptions; the descriptions are redundant on these objects. On the bifurcation objects, however, the two descriptions are conveying different information about the objects.

It is usual, in considerations of structural stability, to regard one of the two descriptions being compared as *intrinsic*; the second description is then compared with the intrinsic one. However, as we repeatedly emphasized (cf. Rosen, 1973), the topologies (i.e., the measures of "closeness") which we typically assign to systems and their states are not intrinsic, but are all contingent on how we choose to interact with a system. Thus, any description is at best conveying partial information about a limited fraction of the interactive capabilities of the system, and all are equally extrinsic.

Nevertheless, if we have an idea that a particular description is intrinsic, and that all other descriptions should be referred to it, we will typically find some bifurcation set where the two descriptions do not agree. Since we tend to prefer a description we regard as intrinsic, it would be reasonable for us to interpret the disagreement between the descriptions as a *mistake* or *error* on the part of the second description. Thus it is at this point in the analysis of complex systems that the concept of error arises in a natural way.

Error has always been troublesome for system theorists. Simple systems do not make errors; it is meaningless to regard a system of mechanical particles as behaving erroneously. Therefore there has always been a relation between complexity and the capability for error; just as with complexity, there has been an enormous literature generated in an attempt to obtain an intrinsic definition of error. We will argue, analogous to our argument regarding complexity itself, that error is not definable intrinsically, but only in terms of the spectrum of interactions available to a complex system. We shall also find that the conventional view, that error depends on complexity, is well-founded.

To put the case simply, we shall assert that error is measured by the deviation observed between the actual behavior of a complex system (with its many distinct interactive capabilities) and the behavior of a simple system or model, which exhibits only a restricted subfraction of the interactive capacities of the original system. The deviation between these behaviors arises precisely because the complex system can be, and in general is, doing many different things at once, and these different interactions interfere with each other in a way unpredictable in principle from the properties of the corresponding simplified system (since, of course, the vehicles for these interactions have been abstracted away, and hence permanently lost, in the very process of abstraction which led to the simplified system). Thus, a bridge can collapse because the particles of which it is composed may interact with each other in ways not consistent with the maintenance of cohesive properties; the genes of an organism can mutate because they can interact with elements of their environment in ways not compatible with their coding function, and so on. I would argue that, in fact, all forms of what we interpret as error arise through simultaneous manifestation of modes of interaction in complex systems, which happen to interfere with each other.

These observations have many implications for important areas of control theory and error correction at the technological and social levels, as well as for an understanding of biological and physical systems. For example, we may compare a model of our own regarding a particular biological and social process (i.e., a simple representation of a complex system) with that manifested by natural selection acting on that complex system. A comparison of the two descriptions will reveal bifurcation points, at which the selection behavior manifestly disagrees with our projections for system behavior. These are the points at which great care must be taken in exercising control. For instance, if we decide we want a system to behave optimally, or without error, in a single simplified mode, we can only achieve this at the cost of interfering in unpredictable ways with other modes of system interaction. We have discussed these problems, in a variety of contexts, in other work, to which we refer the reader for fuller discussions (Rosen, 1974). See also the papers of Pattee (1971). Further, it is important to note that it is in general not sufficient to attempt to replace the missing interactive capabilities which were abstracted away in the initial simplification of the system, by some generalized probability distribution imposed *ad hoc* on the degrees of freedom remaining in the simplified system. This is, of course, the usual procedure adopted in this regard, but from a more general perspective it is fraught with pitfalls.

We hope to have indicated, in the above brief survey, the epistemic character of the notions of system complexity, and the related problems of alternate description and of error in complex system behavior. These notions, we feel, are of great importance in attempting to deal with the spectrum of technological, social and biological problems which presently confront us.

References

Pattee, H.: 1971, in R. Buvet and C. Ponnamperuma (eds.), *Chemical Evolution and the Origin of Life*, North Holland: Amsterdam.
Pattee, H.: 1976, *Biogenesis: Evolution and Homeostasis*, Springer-Verlag: Heidelberg and New York.
Rosen, R.: 1968, *Bull. Math. Biophys.* **30**, 481–492.
Rosen, R.: 1969, in Saaty and Wegl (eds.), *Mathematics in the Sciences*, McGraw-Hill, New York.
Rosen, R.: 1970, *Dynamical System Theory in Biology*, Wiley-Interscience, New York.
Rosen, R.: 1973, *International Journal Systems Science*, Vol. 4, No. 1, 65–75.
Rosen, R.: 1974, *International Journal General Systems*, Vol. 1, 245–252.
Rosen, R.: 1974, *International Journal General Systems*, Vol. 1, No. 2, 93–103.
Thom, R.: 1972, *Stabilité Structurelle et Morphogénèse*, W. A. Benjamin, Inc. Reading, Mass.

22

New Perspectives on Complexity

Ilya Prigogine

I would like in this presentation to explain why the study of complexity has become so interesting today. The idea of studying complex systems is, of course, by no means new. Some 2,000 years ago, Aristotle had already studied domains as varied as marine biology and the political organization of towns. But, it is our present situation, rather, which is, in a sense, unique. When one thinks of complex phenomena, one immediately thinks of biology, society, economics, or areas of this kind, and when one thinks about simple phenomena, the repeatable experiments of physics and chemistry and the domain of planetary motion are what spring naturally to mind. The remarkable feature of our time is that the gap between these two sets of phenomena has narrowed dramatically.

We now begin to see new possibilities for understanding natural phenomena which we could not perceive before. The first 50 years of the century were dominated by the discovery of quantum theory and relativity, relating to the extremes of either very small or very large phenomena. In contrast, the last decade has been dominated by an extraordinary growth of physics on our own scale: that of the macroscopic physics of dynamic systems. This research is leading to a much better understanding of our "place" within nature, and we are now discovering extremely interesting new phenomena on our own spatial and temporal scale, without needing to go to the classical frontiers of science, namely microparticles or cosmology.

The shift from simplicity to complexity is not an ideological shift due simply to some a priori reasoning. Instead, it results from the fact that for many years, through such advances as quantum mechanics, for example, it seemed that we were going to be able to solve nearly everything. However, it gradually became clear that there remained many unsolved problems on our own scale, and this is why we became interested in a group of problems that, previously, was only studied by a very small number of specialists. Two branches of science have contributed enormously to this change of perception: non-equilibrium physics and the theory of classical dynamic systems.

Why have so many surprises arisen in the fields mentioned? Let me begin by giving a very simple example that is found in nearly every physics textbook. What happens when you heat a fluid from below? When the temperatures above and below are the same, you

---

From *The Science and Praxis of Complexity*. Copyright © 1985 by the United Nations University.

have, of course, uniformity. That means that each point in the system plays the same role. However, when you begin to heat the fluid from below, you establish a temperature gradient from the high temperature at the bottom to a lower temperature at the top. For very weak gradients, the heat is transmitted purely by conduction and the fluid remains very much as it was before. However, at a critical value of temperature gradient, convection processes suddenly take over and the famous Bénard cells appear (see Fig. 1). These convection cells display a remarkable spatial organization, with the fluid moving up the centre of each cell and down the sides, and the form of these cells, as can be seen from the figure, is hexagonal. This is a very remarkable phenomenon, involving the coherent behaviour of some molecules, and a privileged direction of rotation emerges. Euclidean space is destroyed in the fluid, since the role played by each point is no longer the same. A complex structural and functional organization of space emerges.

Of course, despite its complexity, we are still clearly a long way from the incredibly organized behaviour found in biological systems, but the important point is that we realize from this example that non-equilibrium is a source of organization, and that the flow of energy through the system can give rise to a new kind of structure—dissipative structure.[1]

Another very striking phenomenon of a similar nature which has been a great focus of study in recent years is that of chemical oscillation. That is to say that when some chemical reactions are forced to occur in a domain which is far from their chemical equilibrium, they begin to exhibit a temporal behaviour very much like a clock—we have a chemical clock! Essentially, what happens is this: near to chemical equilibrium the system is successfully described by the solution of the chemical kinetic equations which applies at equilibrium, a solution corresponding to the incoherent behaviour of the molecules and to a random mixing of the components. As the reaction rate is increased, however, at some point this solution becomes unstable and new branches of solution appear. They correspond to various states of spatio-temporal organization, from chemical clocks of various frequency to different patterns of spatial structure, either stationary or moving. It is here, at the point where the equilibrium branch of the solution becomes unstable, that the system begins to acquire some autonomy, to be different from its environment, which is of course one of the key properties of complex systems. What is the interest of this? For me it is one of the most important experiments of the twentieth century because it contradicts the simple ideas we held about intermolecular forces. For example, we always thought that molecules could only "sense" each other's presence if they were "touching" (i.e., if their individual force fields were in interaction); thus, each molecule would only "know" its neighbours. But, when the system undergoes an instability and a spatio-temporal organization appears, then the coherent behaviour implies that each molecule "senses" what billions of others are doing! In other words, we see a very long correlation—a macroscopic correlation over distances typical of our scale of existence, such as centimetres. Once again, non-equilibrium is the source of a spontaneous self-organization of the system. Speaking anthropomorphically, matter at equilibrium is "blind," it only "sees" at very short distances, while matter out of equilibrium develops a sensitivity to the outside world that is a sensitivity to distant events. This is one of the great surprises of the last decade.

There are many examples of this kind of bifurcation, as a result of which the system acquires a historical dimension. The earth's climate, for example, and I shall come back to this, is a historical object, as are the geological conditions in which we are living. In

New Perspectives on Complexity

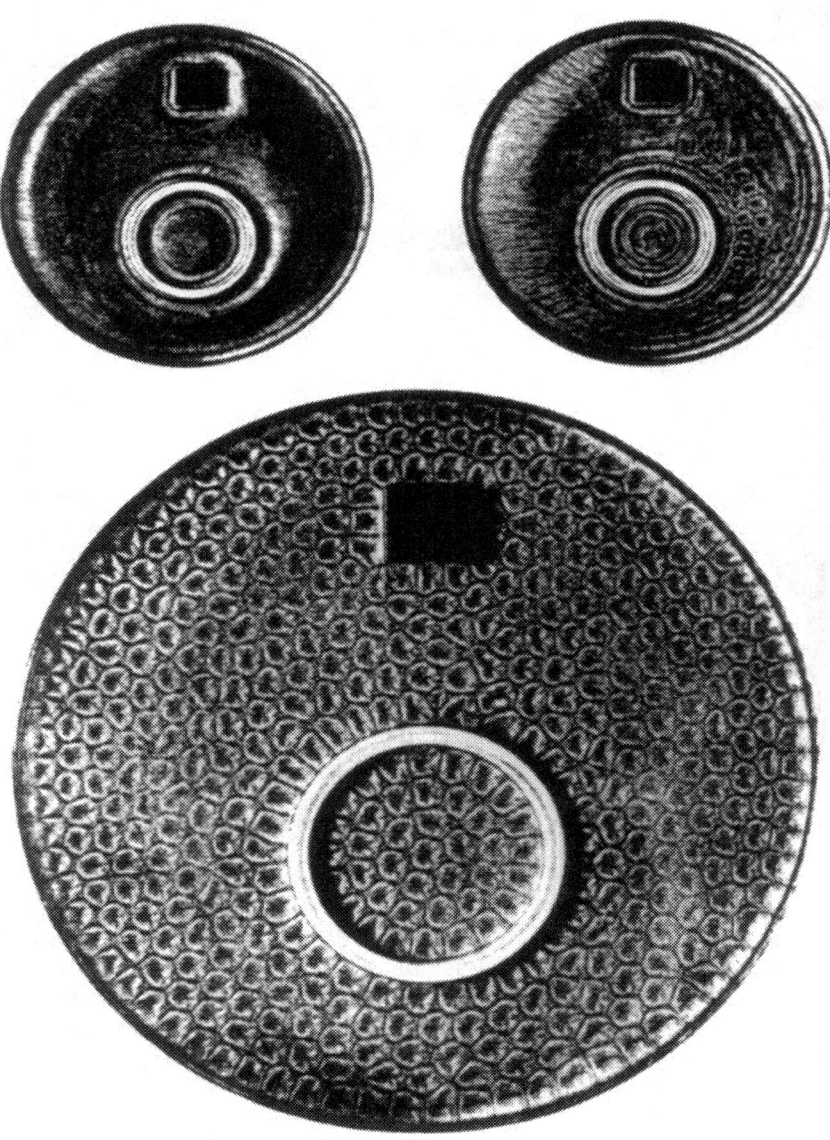

Figure 1. Spatial organization of convection cells

Figure 2. The evolution of spiral waves in the Belousov–Zhabotinsky reaction.

general, we begin to see that matter itself is to some extent a historical object, and a new and important role appears for stochasticity. In quantum theory, stochastic processes exist on the microscopic level, but the remarkable feature of the non-equilibrium systems which I have discussed here is that we now find that stochastic processes at the macroscopic level are important. At our own level, stochasticity and fluctuation begin to play a vital role in choosing which branch of solution the system will adopt, and this choice cannot be obtained from any deterministic procedure.

Let us now contrast and summarize to some extent the properties of these non-equilibrium, dissipative systems with those of non-dissipative dynamic systems. What is of primary importance in dissipative systems is that they have attractors. Of course, for 200 years or more we have known about attractors; if I take a pendulum and I wait until it finally stops, the point at which it stops is an attractor of the motion. If I make some small perturbation, then the pendulum will return to this same point. The system is stable, or as mathematicians say, "asymptotically stable." However, what is new in dissipative systems generally, and is very unexpected, is the existence of attractors that are much more complex than simply a point. The attractor may be a limit cycle (a stable orbit in variable space) or a chaotic attractor of very complex motion in a multi-dimensional space.

What is the difference between this and the behaviour of non-dissipative dynamic systems? The difference is that the world of pure dynamics does not know of any asymptotic stability, does not know of attractors! If I take a frictionless pendulum, then there is no way that it can "forget" some impressed change on its frequency or amplitude, but if, on the contrary, I run, then my heart will beat faster, but after a rest it will return to its regular pace. In other words, the worlds of dynamics and of thermodynamics are very different from each other. Not only is dissipation an additional term in the dynamics but it leads to a privileged direction in time, it orients the evolution of the system in time. In addition, it is a world of stability, and no life would be possible in the world of dynamics. Therefore, dissipation becomes an element of great structural importance that I consider as the great discovery of the last decade at the macroscopic level. However, we have now discovered that most dynamic systems are unstable in a much stronger sense than it was previously thought. The first person who seems to have understood this was Poincaré, but Poincaré lived before his time and his writings were not sufficiently understood. Only now are we beginning to rediscover what Poincaré had already conceived. We are beginning to understand that, in general, dynamic systems are so unstable that if we consider all the dynamic trajectories starting from a particular locality, however small, after some time the trajectories would have separated out to quite different regions. The trajectories of a dynamic system have a basic "divergence" property which undermines the conceptual construct that reposes on the idea of a single trajectory.

A very simple example of this is the so-called Baker transformation, which has been discussed in detail elsewhere.

What I would like to emphasize is that the concept of "trajectory," the very foundation on which determinism—our classical vision—was based, breaks down: if I start the system from any two points, however close, then it will move to quite different regions of the possible state space. So the idea of predicting exactly what a system will do disappears. Thus, there is now statistical uncertainty in the heart of a dynamic system. Such a system is called a Bernoulli system and is as uncertain as a Bernoulli game—a lottery, for example. This is an extraordinary and unexpected result because it shows us that the ideas of simplicity and determinism—which were the prototype of the idea of planetary motion, the truly "classical" dynamic system—were not correct in the sense that they do not represent archetypes, general situations. They represent extremely simplified systems, and it is only now that we begin to be able to deal with the much more general and usual types of situation.

The curious fact is that now we find that the same type of phenomena of instability, which characterize conservative dynamic systems, can occur in dissipative systems where the number of variables is greater than three. The presence of dissipative factors is not sufficient to ensure that the system will evolve to a steady state. This seems paradoxical and is why Ruelle referred to such phenomena as "strange attractors." It concerns the spontaneous onset of chaotic motion, as for the Baker transformation, but in dissipative systems.

Let me now, using these new concepts on very simple examples, briefly try to explain how these new ideas lead to a better understanding of nature around us. Let us consider three examples. The first is the question of why flowers blossom in spring. Flowers are subjected to a great deal of diurnal temperature changes, as day follows night and as the daily weather pattern produces considerable variations. Nevertheless, the flowers still succeed in blooming at a remarkably stable time, in a relatively concerted fashion, despite all the irregularities to which they are subjected. This phenomenon may be related to the extreme sensitivity of the mechanisms of bifurcation, for which even the smallest differences involving perhaps incredibly small amounts of energy can orient a selection of a bifurcation in one direction or another. In other words, the decision to "bloom" can be made dependent on the detection of some systematic drift, and the bifurcation would be able to respond to a very minute change in its value despite all the very large perturbations that it may undergo. The bifurcation offers a selection mechanism. In other words, we begin to understand why we sometimes find in nature more uniformity than we might have expected (Fig. 3).[2]

The second example refers to the opposite situation. Why are there climatic changes? Why are there glaciations? Why are there such dramatic changes in the quantity of polar ice? After all, the sun has always sent roughly the same amount of energy towards the earth, certainly over the last few millions of years. How is it that the earth has witnessed such dramatic changes in the ice cover and in its climatic pattern? This problem has, of course, interested many people, but it was only some two years ago that a reasonable explanation was put forward. It seems that very small perturbations deriving from small deviations in the earth's orbit can be amplified enormously through their interaction with the biosphere. This cannot be understood as an outcome of a deterministic, black box theory; it is rather the outcome of the autonomy of the earth's internal climatic fluctuations which can come into resonance with the small external perturbations. One can show that because of the balance of energy between the earth and the sun, there may be a principle that stipulates the

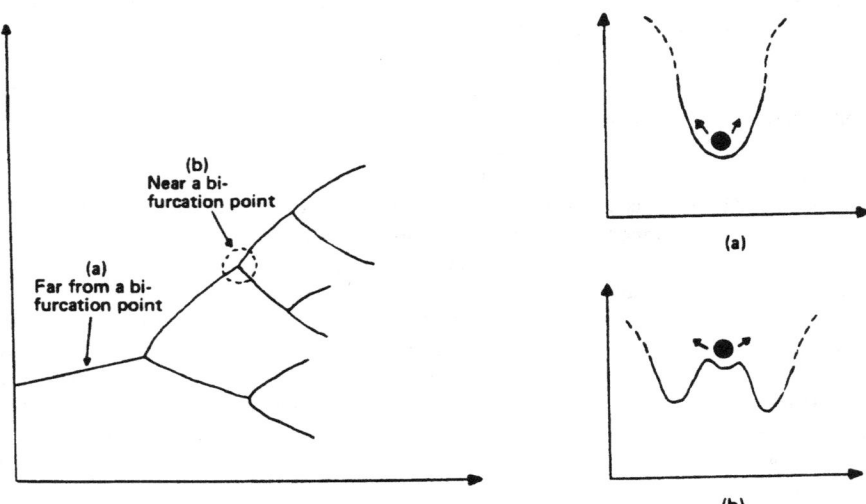

Figure 3. Near a bifurcation point the system becomes "sensitive" even to very weak fields.

following: given a cold climate and a warm climate, and a characteristic transition time to go from one to the other, if there is a small periodic exogenous effect and this period is related to that of transition, an enormous resonance can occur which will drive the climate from one type to another by an amplification of the initially small perturbations. This example underlines rather clearly how little we really know about the basic features of our environment.

The third example is the random theory of the causes of an explosion, which before have always been studied in a deterministic context. An explosion can occur when we have a reaction from x to y, which is stimulated by heat, and which also gives out heat as it occurs. As a critical temperature is approached the system remains inactive until, suddenly, very sharply, the explosion occurs. This is because, initially, the temperature is too low and there is no reaction. However, if a few molecules react, they heat the medium, which in turn leads to a catalytic self-acceleration of the phenomenon, which then leads to a "catastrophe." Two time scales are involved here. There is that over which the "external" temperature is increased, and there is that during which the explosion occurs. Initially, the temperature is relatively homogeneous, and after the explosion this is once again the case with every point being "hot." However, during the explosion the system becomes very inhomogeneous, with "hot" and "cold" regions and very large fluctuations of temperature. No deterministic theory, based as it must be on "averages," can explain the precise evolution of the system.[3]

These are a few examples which have been chosen in order to show some of the new possibilities.

I shall next turn to some very general questions. The first is: What is chance? What is randomness? In the classical view, randomness and determinism were very strong "opponents." There were two poles. At one, we had determinism, and planetary motion was perhaps the most important example. At the other was randomness, which was associated with irregularity, and the Brownian movement of small particles in a fluid was an important example. Today, we see a whole spectrum of possibilities, allowing us to speak of various degrees of randomness. Indeed, some random phenomena may be the outcome of a game involving differential equations, and we find the apparent paradox of randomness as the outcome of a deterministic game.

The interesting point here is that today we begin to be able to detect whether a process is random in the classical sense (completely random) or is random in this new more limited sense. An interesting study has been made recently which reviews the history of climate over a period of 700,000 years.[4] The profile of the temperature changes that have occurred during this period is extremely complicated, and the question arises whether it is possible to use the data available from the past to "reconstruct" the dynamics which gave rise to its particular history. In this case, the past is simply the succession of numbers of past temperatures, and our task would be to understand what type of laws could have given rise to this particular series of values.

The next question is: What is the dimension of the attractor around which these values turn? As I explained above, a dissipative system has attractors, and an attractor may be just a simple point, a periodic trajectory, or a "strange attractor." Can we identify the dimension of the attractor? If we could, then it would suggest a minimum model necessary to give rise to this type of behaviour. In other words, we may begin to understand the past, not just descriptively but from a dynamic point of view, one of mechanism and process.

New methods are now becoming available to estimate the dimensionality of the attractor, and this offers, therefore, an exciting perspective for the study of many complex systems.[5] If we were faced with a system having a point attractor or a simple limit cycle attractor, its dimensionality would be low, but if we have a completely stochastic phenomenon, such as Brownian motion, where there was strictly no correlation between what happened in the past and what will happen in the future, then the dimension of the attractor is very large.

When this technique is applied to the data concerning the temporal series of the temperature at a particular place on the earth's surface, over a period of some 700,000 years, the irregularities observed can be shown to be those of a strange attractor of dimension 3.3 and not the result of a purely stochastic game! It is a dynamic randomness that underlies the climatic variation. What does this mean? It means that this enormous complex system, which has given rise to the particular succession of temperatures observed over this period, can be understood as the outcome of a non-linear deterministic system of four variables. This is an extraordinary result, for it is not a random system. In a really random system the dimension of the attractor would have been infinite, but here we find 3.3, implying equations linking perhaps only four variables.

This suggests a path towards the understanding of human complexity. What are the types of dynamic systems which could generate the intricacies of human behaviour? Is it intrinsically based on a random system or is there a dynamic system of finite dimension which could generate it? How do the dynamics of the electrical fluctuations in the human brain differ from those of simpler beings, such as hydra, lobsters, etc.? Does the movement of the Dow-Jones index resemble that attributable to random noise or could it be generated by a simple, non-linear dynamic system? We may now begin to study the biological history of complexity with these new tools. Before, our ideas were too simple: they consisted of opposing randomness and determinism. But now we see that reality lies somewhere in between.

Let me, at this point, present a further example of how we can begin to explore complexity in a quantitative way. I spoke about explosions. The key point of interest here is that they proceed on two time scales: a long and a short one. Because of the superposition of the two time scales, we find abnormal fluctuations, that is, the law of large numbers does not hold. The time necessary for the explosion to take place is no longer given by a simple Gaussian curve. Another, more simple example of the same idea concerns the growth curve of children. At four years of age, the height distribution of a sample of children obeys a Gaussian form. However, when they reach 13 or 14 years old, their heights no longer fit this Gaussian curve because some have reached the age of puberty and others have not. The onset of puberty resembles, in this way, the "explosion" problem, mainly of course because puberty is the result of a period of "preparation" within the body.[6]

There are many other examples of such phenomena in social life, where the existence of two time scales is a very common feature. It is interesting to note that from the evidence of archaeology, there is almost never a long period of preparation for a new "civilization." The passage to some high level of organization occurs abruptly: pre-dynastic Egypt has very little in common with dynastic Egypt, and the transition occurred in a very short space of time.

Let me address, finally, problems of a more fundamental character. The first question that I wish to turn to concerns the nature of "time." Aristotle had already recognized that time was not simply motion; it was motion seen from the perspective of the temporal

New Perspectives on Complexity

horizon. What he meant by this was that there had to be something outside of this time, which he referred to as "soul," which could measure time using motion. In other words, a watch does not measure time. In point of fact, the watch simply has hands which "move," and indeed every 12 or 24 hours the hands come back to their original position! It is the person who uses the watch who really experiences time and simply employs the watch to measure it. That is what Aristotle meant by a horizon of temporality. The remarkable feature is that, owing to recent advances in the study of dynamic systems, we can now define the quantity which leads to a measure of time. This new time is no longer a motion but is related to instability and to the unfolding of topology. Think, as an example, of a drop of ink which has been put into a glass of water. As time goes on, it spreads in the liquid, taking on a whole variety of forms as it goes. It is the succession of these forms which generates an internal time, related to the irreversible processes occurring in the system. Once we have an internal time, then of course it can be "measured" using the motion of a watch. From these new ideas a new unification can be arrived at between non-dissipative and dissipative dynamic systems.

What, then, is the difference between a deterministic world and a stochastic one? What is the difference between the simplified world of classical science and the real world in which we seem to exist? In classical science, be it quantum or classical dynamics which are used, the object has its full symmetry with respect to time. The past and the future with respect to this time are identical. Indeed, being deterministic, the present determines both the past and the future. Therefore, there can be no "becoming." That is true for the view in which the internal time has its full symmetry. However, in dissipative systems, this symmetry of the internal time is broken. In such systems the past and the present are there, but the future is not. Let us draw an analogy. Take a biological system such as the human species. It evolved through various stages: fish, reptile, mammal . . . The past is, in a sense, conserved in our present, but the future is quite open. We do not know whether we will progress to some "super" human being, or degenerate back to the state of monkeys, perhaps of a new type. In other words, the objects of our world have a broken time symmetry. In the real world of "complexity," the future of the objects is no longer determined. There is no longer the full time symmetry which was supposedly present in the description at the classical or quantum microscopic level, and it is not present in the world to which we have access.

The fact that we come to a world which is open, in which the past is present and cumulative, in which the present is there but the future is not, is one of the answers to the question of the meaning of learning and of the ethical value of science. "Learning" is no longer simply learning about the past, it is discovery. And what is the future? The future does not exist yet, the future is in construction, a construction which is going on in all existing activities. Space-time itself becomes a result of this construction. Irreversibility changes the structure of space-time and introduces new relations between history and the world of the present in which we are involved. This is a view which makes the problem of "being" and of "becoming" quite different. This is because the world in which we live, the world of "becoming" has a broken time symmetry and is propagated by laws which have themselves a broken time symmetry. In the classical view, objects had a full time symmetry and were equally oriented towards the past or the future as well as being propagated by laws which also possessed this full time symmetry. I think that we are now in a world which is much more satisfactory because, in a sense, the classical world was in opposition to our internal experience. Today, these new ideas lead to a concordance between the scientific view and our internal and real experience, and therefore the resulting world view is perhaps more open and more tolerant of different cultural origins, recognizing more fully, as it does, a new coherence between subjective experience and the scientific viewpoint.

References

1. G. Nicolis and I. Prigogine, *Self-Organization in Non-Equilibrium Systems: From Dissipative Structures to Order through Fluctuations* (Wiley Interscience, New York, 1977): I. Prigogine and I. Stengers, *Order Out of Chaos* (Bantam Books, Toronto, New York, London, Sydney, 1984).
2. D. K. Kondepudi and G. W. Nelson, "Chiral Symmetry Breaking in Non-Equilibrium Systems," *Physical Review Letters*, vol. 50, no. 14 (1983): 1023–1026.
3. F. Baras et al., "Stochastic Theory of Adiabatic Explosion," *J. Stat. Phys.*, vol. 32, no. 1 (1983).
4. C. Nicolis and G. Nicolis, *Nature* (in press).
5. P. Grassberger and I. Procaccia, "Characterization of Strange Attractors," *Phys. Rev. Let.*, vol. 50, no. 5 (1983): 346–349.
6. S. Pahaut (paper in progress).

23

The Physics of Complexity

Robert Rosen

I was privileged to have known W. Ross Ashby personally, albeit briefly. We had the opportunity to interact rather intensively over a six-week period in the mid-1960s, when we both participated in a Summer Colloquium on Theoretical Biology, sponsored by NASA, which then had an interest in such things. I have very vivid memories of those days, and of Ashby himself, and accordingly I am most honored to be invited to present this Ashby Memorial Lecture.

What I propose to do is to critically review Ashby's ideas about the brain, about biology and about complexity in general, in the light of some three decades of subsequent experience acquired since the first publication of Ashby's two great books, *Design for a Brain* (1952) and *An Introduction to Cybernetics* (1956). It is relatively easy to do this, since Ashby, like Woodger, Waddington and many other English theorists, had an enormous gift for writing lucidly and explicitly about even the most complicated matters. Thanks to this crystalline style, often lacking in other writers (including myself, I am afraid), one always knew where Ashby stood, and exactly what was being assumed at each stage of any discussion.

Ashby's general approach to biological problems was not reductionistic, but it was Cartesian and Newtonian. That is to say, Ashby was a mechanist. Indeed, in many ways he represented a kind of culmination of the mechanistic approach to organisms and their behaviors; it is precisely for this reason that one can learn so much from him.

His general approach to problems of biological organization was set down many times, but never more clearly than in his book, *Design for a Brain*.[1] The problem which Ashby set himself was set forth at the outset with his customary clarity (Ref. 1, p. 1):

> I hope to show that a system can be both mechanistic in nature and yet produce behavior that is adaptive. I hope to show that the essential difference between the brain and any machine yet made is that the brain makes extensive use of a method hitherto little used in machines. I hope to show that by the use of this method a machine's behavior may be made as adaptive as we please, and that the method may be capable of explaining even the adaptiveness of Man.

From *Syst. Res.* **2**, 171. Copyright © 1985 by International Federation for Systems Research.

To pursue this problem, we must first characterize what we mean by "mechanistic," and how a mechanistic object or machine is to be studied. For Ashby, the concept of "machine" is co-extensive with that of "material system," or with what I myself have called a "natural system." It could be an atom or an organism or an ecosystem or an automobile; the only requirement is that it populate the external world of events, rather than the internal, subjective world of ideas, impressions and symbols.

Such a "machine" is to be studied objectively through real or idealized processes of measurement. Such measurement processes lead naturally to the idea of what Ashby calls a "variable" (i.e., what are now generally called "observables"). He defines this as follows (Ref. 1, p. 14): "A variable is a measurable quantity which at every instant has a definite numerical value." But, as Ashby recognizes, every real machine presents us at the outset with an infinity of such variables. We cannot directly study such an infinity; thus to study any machine through its variables, we must make a choice. We must select a finite number of such variables and forget about all the rest, i.e., forget about almost all of the machine. The result of our choice, then is to create an abstract object, consisting of a finite number of variables associated with a machine; it is such an abstract object which Ashby calls a *system* (Ref. 2, pp. 39–40).

A *state* of such an abstract system is a set of numbers; namely, the values which all of the system's variables assume at a particular instant. The behavior of the original machine (i.e., its temporal sequence of events) will thus reflect itself as a sequence of state transitions in any abstract system we create by concentrating on any finite set of its variables. But, as Ashby points out, we can now identify a special subclass of systems within the infinitude of ways of selecting a finite set of variables out of the original infinity which a machine presents to us. This is the subclass of what Ashby calls *state-determined systems*; finite sets of variables for which, at any instant, the state transition is completely determined by the present state. In fact, he makes the following explicit postulate (Ref. 1, p. 28): ". . . given a (finite) set of variables (i.e., a system), he (the scientist) can always find a larger (finite) set that (1) includes the given variables, and (2) is state-determined." Ashby remarks that "the assumption that such a larger set exists is implicit in almost all science, but, being fundamental, it is seldom mentioned explicitly" (Ref. 1, p. 28).

Thus, of all the infinity of abstract images of machines (i.e., systems), we are most interested in the state-determined ones, which Ashby points out, share with the machine itself the property that "if its internal state is known, and its surrounding conditions, then its behavior follows necessarily." And of the state-determined systems, we are most interested in those which are *simplest* in some sense.

Thus Ashby posits quite a string of abstractions; from machine to system, to state-determined system, to simplest state-determined system. He asks, rhetorically, why the study of such abstract things should be of value for biology, with its enormous complexity and variability. His answer is central to his entire scientific enterprise, and has two interrelated facets: (a) such abstract systems can be studied *precisely and exactly*, so that in principle they can be completely understood; (b) less abstract systems, and ultimately the machine itself, while not corresponding exactly to the systems we have studied, are nevertheless *close* to one or another of them. In Ashby's own words (Ref. 1, p. 29):

> (We) must try to be exact in certain selected cases, these cases being selected because we can be exact. With these exact cases known, we can then face the multitudinous

cases that do not quite correspond, using the rule that if we are satisfied there is some *continuity* in the systems' properties, then in so far as each is near *some* exact case, so will its properties be *near* to those shown by the exact case. [Emphasis added.]

This idea is really the crux of Ashby's mechanistic epistemology, and we shall return to it a number of times as we proceed.

In concluding this brief review of Ashby's ideas, we must mention explicitly that Ashby viewed his mechanistic approach as the only valid scientific alternative to teleology. Indeed, he regarded teleology as fundamentally antithetical to true science. Thus, he says (Ref. 1, p. 9):

It will be assumed throughout that a machine or an animal behaved in a certain way at a certain moment because its physical and chemical nature at that moment allowed it no other action . . . our purpose is to explain the origin of behavior which *appears* to be teleologically directed.

In other words, by showing explicitly how a mechanism can manifest apparently telic behavior, Ashby wished to show that concepts like "goals" or "ends" were at best superfluous and at worst mystical and unscientific.

What I propose to do now is to briefly indicate how the mechanism assumed by Ashby represents a direct embodiment of 17th-century Newtonian mechanics. These Newtonian ideas, and the epistemology underlying them, have permeated all of our ideas about systems and their behaviors ever since; in fact, they are tacitly assumed to be the *only* way that systems can be studied. However, by precisely isolating these epistemological presuppositions, it is possible to see explicitly that alternatives indeed exist; to make a case that the Newtonian picture is in fact unduly restrictive, and must be extensively modified if we are to progress.

It must be clearly recognized at the outset that the influence of Newtonian mechanics has radiated in two main directions; a reductionistic direction and a paradigmatic direction. The former argues that, in so far as every material system can be regarded as a system of mass points, the mechanics of Newton (or some extension, like quantum theory) in principle contains the solution of every scientific problem. All we need to do to understand *any* material system is to characterize its particles and the forces acting on them, formulate the necessary equations of motion, and integrate them. Ashby himself does not embrace this reductionistic aspect, but he does accept its paradigmatic aspect; that the *language* in which Newton described his theory of systems of mass points is the universal language for talking about systems in general, even if they have not been, or cannot be, reduced to systems of mass points. Indeed, Ashby's state-determined systems are nothing but a paraphrase of the Newtonian *language*, adapted to inherently non-mechanical situations. The essence of this language, as we shall see, is that systems have states, and that their behaviors are represented by dynamical laws superimposed on these states. These states are the cognates of mechanical phases; the dynamical laws are the cognates of mechanical forces (more precisely, of impressed forces). In one form or other, every mode of system description known to me is a technical adaptation or modification of this basic presupposition.

The Newtonian ideas were regarded in their time as the supreme embodiment of the concept of Natural Law. Thus, we must digress for a moment to discuss this concept.

The idea of Natural Law has two quite separate facets. On the one hand, there is implicit in it a belief that the sequence of events manifested in the external world is not utterly capricious or arbitrary or chaotic, but rather that there exists some *relation* between them. The relation between events in the external world can be summed up in a single word: *causality*. Thus, the first facet of a belief in natural law consists of a belief in a *causal order* relating events we perceive in the external world. We could not do science, and in fact we probably could not stay sane, without a belief in causal order.

But that is only one facet of our belief in Natural Law. The other one is that this causal order relating events can be (at least in part) grasped and articulated by the human mind. This means ultimately that the causal order relating events can be translated or mirrored by corresponding relations between *propositions* describing these events. But propositions are mental constructs of a linguistic, symbolic character; relations between propositions cannot be causal. Nevertheless, there does exist a relation between propositions, playing the same role as causality does in the external world; that relation is the logical one of *implication*. Thus, the other half of our belief in Natural Law is this: that the causal order relating events in the external world can be imaged by implications between propositions describing these events. Indeed, I would argue that the whole task of theoretical science is to bring causal order into congruence with implicative order within an appropriately constructed formal image.

The formal images of causal order belong, in the broadest sense, to mathematics. I would argue that mathematics is nothing but the study of implication in formal systems; it is the art of extracting conclusions (theorems) from premises (hypotheses). When we have properly brought such a mathematical system into congruence with some causal structure in the external world, the theorems of that system thereby become *predictions* about the causal order.

A great deal of theoretical science is concerned with characterizing the class of mathematical systems which can be images of causal structures in the external world. One of the achievements of Newtonian mechanics was to posit a class (or in mathematical language, a category) of such formal images; a category of dynamical systems. In Ashby's language, this is essentially the category of "state-determined systems," and as has already been noted, this has been the arena for all of system theory ever since. Many of the deep problems of theoretical science deal precisely with this category of mathematical images, and the mathematical relations which exist between them; the problem of reductionism, for instance, involves nothing else.

Now let us return to the concept of the causal order between events, which as we have argued is one essential part of our belief in Natural Law. We may first note that, as a result of the pervasive belief in the universality of the Newtonian ideas, many perceptive scientists and philosophers (including Bertrand Russell) have argued that the very notion of causality is obsolete and pre-scientific, and should be expunged from science. These writers have noted that the word "cause" does not appear any more as a technical term in mechanics (or in physics in general), and that *therefore* it has no meaning. This position only reflects the complete tacit acceptance of the view that the order between events has already been completely imaged in a formal category of "state-determined systems," and henceforth we need not concern ourselves further with the imaging process itself, nor even, for that matter, with the events themselves. And since causality pertains to relations between events, and not between propositions, it does indeed disappear from explicit view when we forget about events and only consider their formal images. But this "forgetting the events" is itself a process of abstraction, and as we shall see, we do it at our peril.

The first, and still the most influential, treatment of the concept of cause as a relation between events was provided two millenia ago by Aristotle. Aristotle was the only one of the great philosophers who was primarily a biologist, and this fact colored his thinking in a unique way. Let us briefly review what he argued. In his view, the entire business of science was to grasp "the why of things"; since the answer to "why" is "because," he was thus naturally led to consider the notion of cause in terms of the ways of answering the question "why?" In a nutshell, he argued that there were four distinct and inequivalent ways of saying "because," and these led him to posit four corresponding *categories of causation*. In modern parlance, these are: (a) material causation, which roughly has to do with the physical basis of an event; (b) formal cause, which concerns what we would now call program; (c) efficient cause, which we should now call a program-determined operator on material cause; (d) final cause, which concerns *telos* or end. For over a millenium, Aristotle's views dominated what there was of science; science was the study of causes. This situation persisted until Newton replaced it with a return to even older views of the pre-Socratic Greek atomists.

These categories of causation, or at least most of them, do have mathematical images in the Newtonian picture. For instance, if we regard "the state of a system at time t" as an *effect*, and its mathematical image as obtained from integrating the equations of motion, then *material cause* translates into *initial conditions*; *formal cause* translates into *structural or constitutive parameters*; *efficient cause* translates into the *integral operator* which generalizes what the engineers call transfer function. But there is no *final cause* in this picture. Indeed, as Ashby noted, the concept of final cause would have to involve a notion of the future acting on the present; of future state or input affecting present change of state; of *anticipation*. This is resolutely excluded, once we have decided that the category of dynamical (i.e., "state-determined") systems constitutes the only acceptable class of mathematical images of the external world. Since Newton's time, this assumption has been made automatically, and it is essentially for this reason that telos, and its associated notion of anticipation, have been routinely excluded from science. The Newtonian picture we have adopted simply cannot accommodate them and survive.

Let us look again at the relation between the Newtonian picture and the Aristotelian categories of causation. The essential point is that, *in the Newtonian picture, these categories are isolated into independent mathematical elements of the total dynamics*. For instance, if "initial state" is identified with "material cause," then the concept of state space segregates the category of material causation from the other categories, and enables us to manipulate material causation while leaving all the other categories of causation unaffected. Likewise the formal cause, which is segregated into a parameter space, and with efficient cause, segregated into either an input-dependent family of integral operators, or in differential form, into the dynamical laws themselves.

Using these ideas, it can be shown rigorously that the Newtonian language, which we have accepted tacitly and uncritically from the outset as the universal vehicle for system description, is equivalent to asserting that the categories of causation are entirely isolated from each other; that we can modify any one of them separately, leaving everything else fixed. When looked at in this light, perhaps that language does not look quite so universal, after all.

Even if this independence of causal categories is accepted, however, the categories of causation are still inequivalent. This means, in more precise terms, that for example the same effect cannot be produced by a variation in initial conditions alone (material cause), and by a variation in constitutive parameters alone (formal cause), and by a variation in environmental controls (material cause) alone. Or, what is the same thing, that a variation in one category of causation cannot be offset by corresponding variations in the others. The problem of determining under what circumstances the categories of causation are equivalent in this sense translates mathematically into a problem of stability of mappings in parameterized families, and as is well known, not all such families can be stable; there will in general occur *bifurcations*. Indeed, bifurcations *must* occur whenever we compare an equivalence relation (similarity of behavior) with a topology (nearness of parameters). But this is precisely the kind of situation which arises with Ashby's assertion about "continuity of system properties," based on the argument that any system will be "near" a simple state-determined one. Thus, even if we accept the Newtonian language, with its segregation of the categories of causation into independent mathematical structures, the inequivalence of these categories raises crucial theoretical questions which have never really been addressed. And indeed, in so far as to be "simple" is non-generic, we may expect as a general rule that bifurcations will occur precisely around these "simple" systems, which we are attempting to use as models for *all* systems. In the cases which Ashby cites to justify this whole approach (ideal gases, frictionless oscillators, etc.) this is exactly what happens; closed, isolated, conservative systems and the like are inherently so degenerate and nongeneric that literally *anything* can happen when we open them up.

The infatuation of contemporary physics for this kind of degeneracy and nongenericity goes quite a long way in explaining the scandalous absence of any important relation between even the most powerful theories of physics and the most marginal biological phenomena. Indeed, viewed in this light, it is most ironic that theoretical physics should fancy itself as concerned with universal laws, and in quest for these should have disdained biology as dealing merely with an insignificant class of inordinately specialized systems from which no universal principles could possibly be expected. And doubly ironic is the abject acquiescence of many molecular biologists in this view, seeking to bury themselves and their field in a specious reflected association with remote and inapplicable universal laws. In fact, the situation is quite the reverse; contemporary physics is not the general nor biology the particular. Indeed, if physics is ever to become in fact what it presently claims to be, namely the science of material nature in all of its manifestations, then it must come to terms with the realities of biology, and in doing so will be forced to transform itself out of all present recognition. Some slight inkling of what will be involved in this can already be seen through contemplation of what the concept of the "open system" has done to thermodynamics; where after nearly half a century there is still no physics capable of dealing with even the most rudimentary biological (or even physical) situations. But that is another story.

Now let us return to the main line of the argument, and look briefly at what happens when we abandon the requirement that the categories of causation must be represented in independent mathematical structures. This means, in Ashby's terminology, that we give up the idea that each "variable" of a system can be classified as belonging exclusively to the category of material cause, or exclusively to the category of formal cause, or exclusively to the category of efficient cause. In other words, we allow that some, and perhaps all, variables simultaneously participate in two or more of these causal categories. Then what happens?

What happens, of course, is that we must allow a wider class of mathematical images of physical reality than the dynamical systems, or "state-determined" systems, to which system theory has hitherto restricted itself. In this new mathematical world, it generally happens that the value of a system observable, the value of its rate of change, etc., are independently determined, instead of all being derivable from a single rate law as in the Newtonian picture. These mathematical images become more like webs of informational interactions, no level of which can be derived from any of the others. In particular, there is no "state space" which can be fixed once and for all. The class of all of these new mathematical images form a category, in which the category of "state-determined" systems sits as a very small subcategory, just as the rational numbers sits as a set of measure zero in the set of all real numbers. However, just as in that case, it turns out that the members of the big new category can be regarded as limits of sequences of dynamical systems. That is, there is a sense in which the behavior of one of these webs can be *approximated*, albeit only locally and temporarily, by an appropriate "state-determined" system. So there is still a notion of approximability, but it is very different from the one Ashby visualized so long ago. The fact that the new approximability is only local and temporary explains a great deal about why we have been able to go as far as we have with the non-generic Newtonian picture, and why we have never been able to go further with it. The situation is similar to that faced by the early cartographers, who were attempting to map a sphere with pieces of planes; here, the Newtonian language should be thought of as the planes, and the new images, of layers of independent informational structures, as the spheres. Locally, the difference between sphere and plane disappears, but as we attempt to map out larger and larger regions on the sphere, we have to keep changing our planes. The sphere is in some sense a limit of the local planar pieces, but these pieces are related by a global condition (i.e., the topology of the sphere) which cannot be found locally. And the requirement that we must continually pass to other planes as we attempt to map more and more remote regions can, depending on how we look at it, be regarded as *error* (the discrepancy between planar and spherical surface) or as *emergence* (of the curvature of the sphere).

I have elsewhere proposed that this new category of presumptive mathematical images of physical reality be called a category of *complex systems*, while the subcategory of "state-determined" systems be called the category of *simple systems or mechanisms*. There are many reasons for choosing this terminology; among them, it is a corollary of their structure that a complex system, in the above sense, possesses a multitude of simple system descriptions, which cannot be combined into a single "master description" of this type. I had earlier taken this to be the very definition of complexity.

Viewed in this light, then, physics and all of its manifold system-theoretic variants comprise a *science of simple systems*. And organisms are not simple systems. Thus, I can visualize a *science of complex systems*, from which both contemporary physics and biology, in two distinct ways, emanate.

In closing, let me indicate one corollary of passing to the more general framework of what I have called complex systems. Namely, by loosening the Newtonian shackles, we can introduce a category of final causation in a perfectly respectable, non-mystical way. In other words, the concept of *anticipation* is meaningful in the category of complex systems. This fact alone, perhaps, is sufficient justification for looking seriously at this new world.

We have thus come a long way from the world of mechanisms which Ashby studied so long and so thoroughly. I am sure that he would be aghast at much of what I have said, but I am equally sure that he would take it seriously. And indeed, much of my motivation for probing beyond the limits of the Newtonian paradigm arises from my knowledge of Ashby and of his work; had the problems with which he dealt been solvable within the paradigm he was using, he would have surely solved them.

References

1. W. R. Ashby, *Design for a Brain*, 2nd edn. Chapman & Hall. London (1960).
2. W. R. Ashby, *An Introduction to Cybernetics*. John Wiley, New York (1956).

24

The Simplification of Science and the Science of Simplification

Gerald M. Weinberg

In order to understand the successes of science, we can do no better than to examine physics—and particularly mechanics—for these sciences are often taken to be ideal models. The beauty of the mechanical model of the world was well expressed by Deutsch,[1] who said that mechanism

> ... implied the notion of a whole which was completely equal to the sum of its parts; which could be run in reverse; and which would behave in exactly identical fashion no matter how often these parts were disassembled and put together again, and irrespective of the sequence in which the disassembling or reassembling would take place. It implied consequently that the parts were never significantly modified by each other, nor by their own past, and that each part once placed in its appropriate position with its appropriate momentum, would stay exactly there and continue to fulfill its completely and uniquely determined function.

Yet the beauty of this system is a bit dulled when we realized that mechanical systems ordinarily have only a handful of identifiable parts—most often two, but sometimes ten or perhaps even thirty or forty if they are highly constrained, as are the parts of a bridge. For if there are too many parts, we may write down equations which relate the behaviors of the different parts of the system, but we cannot solve the equations, even by approximate methods. High-speed computers have extended the range of mechanical systems whose equations can be solved (approximately), but only by a relatively small amount.

If the formal methods of mechanics are so limited, why is mechanics considered to be a model for the sciences? We must—if we are to have the answer—consider not the formal methods but the informal ones by which complex mechanical systems are reduced to simpler ones, for only then can they be subjected to the working of the formal methods. Consider, for example, Newton's achievement in explaining the motions of the bodies in the solar system. Rapoport,[2] in speaking about this problem, pointed out, "Fortunately for the success of the mechanistic method, the solar system . . . constituted a special tractable case of several bodies in motion."

From *Trends in General Systems Theory*, G. J. Klir, ed. Copyright © 1972 by John Wiley & Sons, Inc.

Although Rapoport's analysis is correct and pertinent as far as it goes, it does not penetrate deeply enough into the heart of Newton's success. The solar system, in the first place, does not consist of "several bodies in motion." We now know that there are thousands upon thousands of celestial bodies in our solar system plus other matter not in "bodies." Any analysis of planetary motions, however, begins by ignoring most of these bodies because they are "too small" to have a significant effect on the calculations. Although this seems a natural step—so natural that textbooks on mechanics do not ordinarily mention it—it is a step which happens to work only in very special circumstances. Yet any other circumstances are not considered proper systems for mechanists to think about.

Consider, for instance, the pineal body, a tiny piece of tissue in the brain. Can physiologists ignore this in their attempts to understand the behavior of the human body? Perhaps they can—the question is quite alive—and perhaps they cannot; but in any case no physiologist would think of arguing that, because the mass of the pineal body is small with respect to the mass of the brain, it can be ignored on that account. The DNA in a living cell is a miniscule amount of the cell material, if measured according to mass; but understanding of cellular biology would be impossible without considering its role. The queen bee in a hive is only one of thousands of bees, and constitutes only a small fraction of the total mass of the hive, but no ethologist dare ignore her role in hive behavior.

Mechanics, then, is the study of those systems for which the approximations of mechanics work successfully. It is strictly a matter of empirical evidence, not of theory, that the human body cannot be understood by considering only the gravitational attractions between its parts.

But how important is the step of ignoring "small" bodies—the asteroids, comets, satellites, and other pieces of space flotsam—to the calculation of planetary orbits? Consider the equations describing a system with only two objects. We must describe how each object behaves by itself, which gives us two equations, or relationships, one for each object. We must consider how the behavior of each body is related to that of the other, which gives us another relationship. Finally, we must consider how things will behave if neither of the bodies is present (the "field" equation), making four relationships in all. In general, if we have n bodies, the number of relationships is 2^n. For 10 bodies, this means about 1000 relationships ($2^{10} = 1024$), and for 100,000 bodies, about $10^{30,000}$. Thus, by "ignoring small masses," we reduce the number of relationships from perhaps $10^{30,000}$ to approximately 1000, for 10 bodies. One thousand equations we might conceivably write down, even if we could not solve them.

Now, we have just promised to speak to general audiences, and here we find ourselves talking about solving a system of 1000 equations. But our readers do not have to know how to solve such a system of equations; they must understand only how much effort is involved.

Why are we interested in the amount of effort required to solve equations? In Newton's day, the impact of mechanics on thought was strong. Many philosophers believed, with Laplace, that, given precise observations on the position and velocity of every particle in the universe, one could calculate the entire future of the universe. They realized, of course, that they would need a large computing machine. But they lacked even the simplest computers. How could they possibly put a measure on the required computation?

Only in our lifetime have real computers come into existence, and with them philosophical thought has undergone a revolution. Anyone who doubts that there has been a

Simplification and Science

revolution should read the debate on teleology between Rosenbluth, Wiener, and Bigelow, on the one hand, and Richard Taylor, on the other. (The four articles of this debate may be found in Part V.A. of Buckley.[3]) Although this debate took place after World War II, Taylor, the philosopher, was still able to advance arguments which today could be refuted by any undergraduate student in computer science. One aspect of this revolution, of course, was the concern with quantifying computational complexity and power.

Even now the question of quantifying complexity is not a closed one, but for our arguments we do not need exact measures. Instead, we merely want to estimate how the amount of computation increases as the size of the problem increases. Experience has shown that, unless some simplifications can be made, the amount of computation involved increases at least as fast as the square of the number of equations. Thus, if we double the number of equations, we will have to find a computer four times as powerful to solve them in the same amount of time. Naturally, the time often goes up faster than this, particularly if some technical difficulty arises, such as a decrease in the precision of results. For our present arguments, however, we may conservatively use the square law of computation to estimate how much more computing is required for one set of equations than for another.

In practice, then, there is an upper limit to the size of the system of equations which can be solved. Clearly, $10^{30,000}$ equations are far beyond that limit. And in Newton's day, without computers at all, the practical limit of computations was well below 1000 second-order differential equations, especially since Newton had just invented differential equations. Newton needed all the simplifying assumptions—explicit or implicit—he could get away with, just as do physiologists and psychologists today. We may note, in this regard, that old-time physicists now say that the "youngsters" no longer do "real physics," because they use the computer to solve large sets of equations, rather than applying physical "intuition" to reduce the equations to a form that can be solved with a pencil on the back of the proverbial envelope.

Thinking about the practical problem of computation, then, can give us a new point of view about what mechanics is, or what any science is. Practical computation demands that implicit assumptions be brought out into the open; hence it is no coincidence that computer programmers are attracted to a systems theory which devotes itself to studying how people make assumptions. An excellent example of the kind of experience that computing people have relates to another assumption already made in our reduction of the solar system problem to 1000 equations.

We have assumed, as one always assumes in mechanics, that only certain kinds of interactions are important. In this case, the only important interaction was gravitational, which meant that each relationship gave only one equation. How do we know that only gravitational attraction is important in this system? How do we know that we can ignore magnetic effects, electrostatic forces, light pressure, force of personality, and so forth? One answer is that this would not be a problem in mechanics if those other forces were important, but that is merely begging the question. How do we know that it is a problem in mechanics?

As before, we know that it is a problem in mechanics because when we try these approximations they give us satisfactory answers—that is, answers which match observational data. If we had a problem for which they did not work, it would never make its way into the mechanics textbooks. Our practical computing example of this quandary is the calculations that were made of the orbit of the Echo satellite, which was a large, inflated

Mylar sphere. After a few months, it was found that the classical solution of the gravitational equations was not doing a satisfactory job of predicting Echo's orbit. After much perplexing labor, the programmers realized that Echo, because of its small density, was much larger than any "normal" solar body of the same mass. Consequently, the pressure of the sun's light radiating on its surface could not be implicitly ignored, as it is in all "ordinary" orbital calculations. No, mechanics does not tell us which systems are "mechanical."

And yet, even having reduced—by applying deeply buried assumptions—the number of equations to 1000, we still may not be able to say that we have solved a particular mechanical system. The equations may still prove intractable, even for a large computer. We need further simplifications. Newton supplied an important one in his law of universal gravitation, which stated that the force of attraction (F) between two (point) masses is given by the equation

$$F = \frac{Gm_1m_2}{r_{12}^2},$$

where m_1 is the mass of the first body, m_2 is the mass of the second, r_{12} is the distance between them, and G is a universal constant. From the viewpoint of simplification, this equation says more implicitly than explicitly, for it states that no other equation is needed. It says, for instance, that the force of attraction between two bodies is in no way dependent on the presence of a third body, so that only pairs of bodies need be considered in turn and then all of their effects may be added up.

A psychologist, for one, would be tickled pink if he could consider only summed pair interactions. This simplification would mean that, to understand the behavior of a family of three, he would study the behavior of the father and mother together, the father and son together, and the mother and son together. When all three got together, their behavior could be predicted by summing their pairwise behaviors. Unfortunately, it is only in mechanics and a few other sciences that superposition of pairwise interactions can be successful.

In the case of the solar system, pairwise superposition reduces 1000 equations to about 45—that being the number of ways in which 10 things can be taken in pairs. From a computational point of view, we have reduced the size of our task by the square of 1000/45—or about 400 times, at least. We might be willing to stop at this point, although Newton (perhaps because he lacked the computers that we have) went still further.

As it happens, the solar system has one body (the sun) whose mass is much larger than any of the other masses—larger, in fact, than the mass of all of the other bodies together. Because of this dominant mass, the pair equations not involving the sun's mass yield forces small enough to be ignored, at least considering the accuracy of the data Newton was trying to explain. (Discrepancies in this assumption led to the discovery of at least one planet that Newton did not know.) This simplification, which is made possible by the solar system, rather than by mechanics, reduces the number of equations to about 10, instead of 45—giving an estimated 20 times reduction in computation.

But Newton went even further, for he observed that the dominant mass of the sun enabled him to consider each planet together with the sun as a separate system from each of the others. Such a separation of a system into noninteracting subsystems is an extremely important technique known to all developed sciences—and to systems theorists as well. To understand the power of such a separation, we need only recall the square law of computation. If solving a system of n equations takes n^2 units of computation, n separate single equations taken one at a time will require only n of the same units.

At this point, Newton stopped simplifying and solved the equations analytically. He had actually made quite a few other simplifications, such as his consideration of each of the solar bodies as a point mass. In each of these cases, he and his contemporaries were generally more aware of—and more concerned about—the simplifying assumptions than are many present-day physics professors who lecture about Newton's calculations. Consequently, students find it hard to understand why Newton's calculations of planetary orbits is ranked as one of the highest achievements of the human mind.

But the general systems theorists understands. He understands because it is his chosen task to understand the simplifying assumptions of a science—those assumptions which delimit its field of application and magnify its power of prediction. He wants to go right to the beginning of the process by which a scientist forms his models of the world, and to follow that process just as far as it will help him in suggesting useful models for other sciences.

And why is the general systems theorist interested in the simplifications of science—in the science of simplifications? For exactly the same reason as Newton. The systems theorist knows that the square law of computation puts a limit on the power of any computing device, and he believes that the human brain is in some sense a computing device. Thus he knows that, if we are to survive in this complex world, we shall need all the help we can get. Newton was a genius, but not because of the superior computational power of his brain. Newton's genius was, on the contrary, his ability to simplify, idealize, and streamline the world so that it became, in some measure, tractable to the brains of perfectly ordinary men. By studying the methods of simplification which have succeeded and failed in the past, the general systems theorist hopes to make the progress of human knowledge a little less dependent on genius.

References

1. Deutsch, K., "Mechanism, Organism, and Society." *Philosophy of Science*, Vol. 18, pp. 230–252, 1951.
2. Rapoport, A., "Mathematical Aspects of General Systems Analysis." *General Systems Yearbook*, Vol. XI, pp. 3–11, 1966.
3. Buckley, W. F., *Modern Systems Research for the Behavioral Scientist*. Chicago: Aldine Publishing Co., 1967.

25

Introductory Remarks at Panel Discussion

W. Ross Ashby

What is systems theory? Today the answer is becoming clear, and I would like to describe how the subject seems to be taking form and developing logical coherence.

It dates from about 1940. In that year Howard Aiken completed Mark I and proved to the world that chains of cause and effect, with each effect becoming the next cause, could be extended to an unlimited length. In its length, the trajectory could contain an unlimited quantity of complex processing. At the end of the 1930s, too, the radio engineers succeeded in taming "feedback," so that they could now understand and control unlimited regenerative or circulating cause-and-effect actions. A little later, information theory provided a technique by which large numbers of causes and effects could be counted, even though they were coded into forms that previously had not been recognized as countable. In the 1940s then, the study of systems in which causes and effects acted with great multiplicity rose to an entirely new height.

The study of interacting parts goes back, of course, as far as Newton; and the solution of a set of simultaneous ordinary differential equations,

$$\frac{dx_i}{dt} = f_i(x_1, \ldots, x_n), \qquad i = 1, \ldots, n,$$

studies, in some sense exhaustively, the interactions between the n variables x_i. In practice, such equations were manageable only when n was very small, with 5 as a practical maximum, and the theory of big systems with rich interactions tended to be evaded. Science has, in fact, triumphed for 200 years largely because it exploited the many interesting systems in which interaction is small: molecules in a gas at low pressure, so that collisions are rare; crystal structure when the atoms are so little perturbed that the vibrations are almost independent; through the range to neurophysiology, studying reflexes that have only the slightest effect on one another.

From *Views in General Systems Theory*, M. D. Mesarovic, ed. Copyright © 1964 by John Wiley & Sons, Inc.

Since 1940, however, a serious attempt has been made, aided by the new techniques, to grapple with the problems of the dynamic system that is both large and richly connected internally, so that the effects of interaction are no longer to be ignored, but are, in fact, often the focus of interest. The neurophysiologist no longer deals only with a bundle of unconnected reflexes. The economist wants to consider models which have something like the richness of interaction shown in the real world. The traffic engineer is no longer content to study the case of the crossroads to which cars come only at long intervals!

So has arisen *systems theory—the attempt to develop scientific principles to aid us in our struggles with dynamic systems with highly interacting parts.*

In "developing scientific principles," however, we must go cautiously. In making use of our heritage of scientific experience, we must take care that we do not unwittingly follow some old rule which is actually obsolete in the next context. I think there is such a danger, which I would like to discuss briefly.

If we study "interactions" generally, and make a preliminary quantitative estimate of what to expect, we often find that the number comes out very large. Then we are apt to characterize it as "astronomical" and to say, "We can't tackle this size just now, but the coming diode is twice as fast, so then we shall be able to do it." I want to suggest that this attitude of mind is seriously in error.

First, what is an "astronomical" number? For example, the time since the earth solidified, in microseconds, is about 10^{23}; the number of atoms in the whole visible universe is about 10^{73}; in fact, all the actual, physically existent astronomical numbers are less than 10^{100}. If we go to the limit, and assume that an atomic event occupies 10^{-10} second, and then ask what is the total number of atomic events that have occurred anywhere in the universe, ever since the earth solidified, we find the number to be about 10^{100}.

In the same spirit, Bremermann[1] has shown that, even if we take single atomic states as markers (i.e., "digits") for computation, the known physical laws make it impossible for any computer made of matter to process more than about 10^{47} bits per gram per second. Let such a computer be as big as the earth and go on for all geological time, it is physically incapable of processing more than about 10^{73} bits. Let me epitomize with:

Everything material stops at 10^{100}

This number is commonly considered large, but we must develop a better sense of proportion in these matters. This number has been obtained by processes that are essentially multiplicative, whereas in systems theory many of our most important quantities grow combinatorially, and this rate is commonly far faster. Here is a simple example that will make the point.

Suppose we have a square block of lamps, for displaying visual patterns, measuring 20 by 20 lamps, and suppose each lamp is either off or on. Obviously there is nothing extravagant about this set of objects. Since each lamp can be off or on, the block can show 2^{400} pictures—about 10^{120}. Suppose now that we are thinking of dividing these pictures into two sets according to some criterion—so that we can say, "This set has the property P, the remainder do not"—from how many properties is the property P picked out? Since each picture may have the property P or not, the number is

$$2^{(10^{120})}$$

Introductory Remarks

As this is $10^{(10^{119.5})}$, we can write it approximately in more convenient form as
$$10^{(10^{120})}$$

How big is this number? To call it "astronomical" is seriously misleading, for the word suggests that it is among the physically achievable numbers, and this is not so. We can get some intuitive grasp of it in the following way:

First notice that $10^{10} - 10^7$ is practically 10^{10}, for it is actually 9,990,000,000. Thus a number 10^K is practically immune to subtraction unless the number subtracted has an exponent within 2 or 3 units of K. Now consider $10^{(10^{80})}$. This number, written out, would be a 1 followed by 10^{80} zeros. As there are only 10^{73} atoms in the universe, there are not enough atoms to carry its zeros; thus $10^{(10^{80})}$ is so large that it cannot be written, in ordinary notation, in our universe.

Finally, what is $10^{(10^{120})} \div 10^{(10^{80})}$? Its exponent is $10^{120} - 10^{80}$, and this (as we saw) is practically 10^{120}. Thus, the number $10^{(10^{120})}$—the number of properties definable on our 20 × 20 block of lamps—is so large that it is not appreciably affected when divided by a number itself so large that it cannot be written in our universe. This is large indeed! By comparison with it, the (properly) "astronomical" is hardly distinguishable from the infinitesimal. In particular, the number 10^{100}, given above as the absolute limit to the physically achievable, is now seen to be a restriction of great severity.

What does this mean? It seems to me to have a clear moral. These numbers of *combinatorial* size tend to occur as soon as we start to consider such topics as:

Combinations	Relations
Ordering	Patterns
Subsets	Constraints
Properties	Partitions
Types	Connection-patterns

Every step from the primary set of elements to one of these topics jumps the size to an exponential or factorial function of the original number, and these functions grow far faster than the merely linear or quadratic. To talk, for instance, of *ordering* the *relations* between the *relations* possible on a set jumps the function to something like $(e^{(e^n)})$! If the set is as trivially small as 5—the five types of cloud, for instance—this number is already far beyond the limiting 10^{100}.

Even the elementary question, "I wonder what this machine will do?" carries its sting. If it is of n parts, each of which has only two possible states, its number of states is 2^n. To speculate on its trajectory is to ask (given the initial state) as to which of the $2^n - 1$ will come next; then which of the remaining $2^n - 2$; and so on. The variety we face is of the order of $(e^n)!$—a function that increases far faster than the exponential, itself often considered explosive!

The systems theorist may thus be defined as a man, with resources not possibly exceeding 10^{100}, who faces problems and processes that go vastly beyond this size. What is he to do?

At this point, it seems to me, he must make up his mind whether to accept this limit or not. If he does not, let him attack it and attempt to find a way of defeating it. If he does accept it, let him accept it wholeheartedly and consistently. My own opinion is that this limit is much less likely to yield than, say, the law of conservation of energy. The energy law is essentially empirical, and may vanish overnight, as the law of conservation of mass did, but the restriction that prevents a man with resources of 10^{100} from carrying out a process that genuinely calls for more than this quantity rests on our basic ways of thinking about cause and effect, and is entirely independent of the particular material on which it shows itself.

If this view is right, systems theory must become based on methods of simplification, and will be founded, essentially, on the *science of simplification*. Many sciences, of course, have used simplifications, and R. A. Fisher[2] said, without qualification, ". . . the object of statistical methods is the reduction of data." But simplifications today are often used apologetically because of the fear that it will be called an "over"-simplification, though when a simplification becomes an "over"-simplification is often clear to no one.

The science of simplification has, I think, been well started by the mathematicians in their studies of *homomorphisms*, but much remains to be done, and many questions essential for applications have to be answered. For example: Is every act of simplification, as it occurs in the daily work of the scientist, *always* an application of an equivalence relation over the primary set, or are there other methods? Does Bourbaki's[3] formulation of a property being "compatible" (with an equivalence relation) correspond always to "this is not an over-simplification"? Can the methods of simplification be classified and studied systematically? (here we remember, of course, that the "ways of grouping" are themselves apt to increase with combinatorial speed). The science of simplification clearly has its own techniques and its own sophistication. *The systems theorist of the future, I suggest, must be an expert in how to simplify.*

References

1. Bremermann, H. J., "Optimization Through Evolution and Re-combination," in *Self-Organizing Systems* (Yovits, M. C., Jacobi, G. T., and Goldstein, G. D., eds.), Spartan Books, Washington, D.C., 1962.
2. Fisher, R. A., "On the Mathematical Foundations of Theoretical Statistics," *Phil. Trans. Roy. Soc. Lond.*, A, **222** (1922), pp. 309–368.
3. Bourbaki, N., *Théorie des ensembles; fascicule de résultats*, ASEI 1141, Hermann & Cie., Paris, 1958 (3rd edition).

26

Every Good Regulator of a System Must Be a Model of That System

Roger C. Conant and W. Ross Ashby

1. Introduction

Today, as a step towards the control of complex dynamic systems, models are being used ubiquitously. Being modelled, for instance, are the air traffic flows around New York, the endocrine balances of the pregnant sheep, and the flows of money among the banking centres.

So far, these models have been made mostly with the idea that the model might help, but the possibility remained that the cybernetician (or the Sponsor) might think that some other way was better, and that making a model (whether digital, analogue, mathematical, or other) was a waste of time. Recent work (Conant, 1969), however, has suggested that the relation between regulation and modelling might be much closer, that modelling might in fact be a *necessary* part of regulation. In this article we address ourselves to this question.

The answer is likely to be of interest in several ways. First, there is the would-be designer of a regulator (of traffic round an airport say) who is building, as a first stage, a model of the flows and other events around the airport. If making a model is *necessary*, he may proceed relieved of the nagging fear that at any moment his work will be judged useless. Similarly, before any design is started, the question: How shall we start? may be answered by: A model *will* be needed; let's build one.

Quite another way in which the answer would be of interest is in the brain and its relation to behaviour. The suggestion has been made many times that *perhaps* the brain operates by building a model (or models) of its environment; but the suggestion has (so far as we know) been offered only as a possibility. A proof that model-making is necessary would give neurophysiology a theoretical basis, and would predict modes of brain operation that the experimenter could seek. The proof would tell us what the brain, as a complex regulator for its owner's survival, *must* do. We could have the basis for a theoretical neurology.

From *Int. J. Syst. Sci.* 1, 89. Copyright © 1970 by Taylor & Francis, Ltd.

The title will already have told this paper's conclusion, but to it some qualifications are essential. To make these clear, and to avoid vaguenesses and ambiguities (only too ready to occur in a paper with our range of subject) we propose to consider exactly what is required for the proof, and just how the general ideas of regulation, model, and system are to be made both rigorous and objective.

2. Regulation

Several approaches are possible. Perhaps the most general is that given by Sommerhoff (1950) who specifies five variables (each a vector or n-tuple perhaps) that must be identified by the part they play in the whole process.

(1) There is the total set Z of events that may occur, the regulated and the unregulated; e.g., all the possible events at an airport, good and bad. (Set Z in Ashby's (1967) reformulation in terms of set theory.)

(2) The set G, a subset of Z, consisting of the "good" events, those ensured by effective regulation.

(3) The set R of events in the regulator R (e.g., in the control tower). [We have found clarity helped by distinguishing the regulator as an object from the set of events, the values of the variables that compose the regulator. Here we use italic and Roman capitals respectively.]

(4) The set S of events in the rest of the system S (e.g., positions of aircraft, amounts of fuel left in their tanks) [with italic and Roman capitals similarly].

(5) The set D of primary disturbers (Sommerhoff's "coenetic" variable); those that, by causing the events in the system S, tend to drive the outcomes out of G (e.g., snow, varying demands, mechanical emergencies).

(Figure 1 may help to clarify the relations, but the arrows are to be understood for the moment as merely suggestive.) A typical act of regulation would be given by a hunter firing at a pheasant that flies past. D would consist of all those factors that introduce disturbance by the bird's coming sometimes at one angle, sometimes another; by the hunter being, at the moment, in various postures; by the local wind blowing in various directions; by the lighting being from various directions. S consists of all those variables concerned in the dynamics of bird and gun other than those in the hunter's brain. R would be those variables in his brain. G would be the set of events in which shot does hit bird. R is now a "good regulator" (is achieving "regulation") if and only if, for all values of D, R is so related to S that their *interaction* gives an event in G.

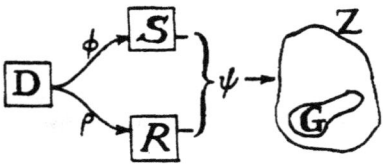

Figure 1

Every Good Regulator of a System Must Be a Model of That System

This formulation has withstood 20 years' scrutiny and undoubtedly covers the great majority of cases of accepted regulation. That it is also rigorous may be shown (Ashby, 1967) by the fact that if we represent the three mappings by which each value (Fig. 1) evokes the next:

$$\phi: \quad D \to S$$
$$\rho: \quad D \to R$$
$$\psi: \quad S \times R \to Z$$

then "R is a good regulator (for goal G, given D, etc., ϕ, and ψ)" is equivalent to

$$\rho \subset [\psi^{-1}(G)] \cdot \phi \qquad (1)$$

to which we must add the obvious condition that

$$\rho\rho^{-1} \subset 1 \subset \rho^{-1}\rho$$

to ensure that ρ is an actual mapping, and not, say, the empty set! (We represent composition by adjacency, by a dot, or by parentheses according to which best gives the meaning.)

It should be noticed that in this formulation there is no restriction to linearity, to continuity, or even to the existence of a metric for the sets, though these are in no way excluded. The variables, too, may be partly functions of earlier real time; so the formulation is equally valid for regulations that involve "memory," provided the sets D, etc., are defined suitably.

Any concept of "regulation" must include such entities as the regulator R, the regulated system S, and the set of possible outcomes Z. Sometimes, however, the criterion of success is not whether the outcome, after each interaction of S and R, is within a goal-set G, but is whether the outcomes, on some numerical scale, have a root-mean-square sufficiently small.

A third criterion for success is to consider whether the entropy $H(Z)$ is sufficiently small. When Z can be measured on an additive scale they tend to be similar: complete the constancy of outcome $\Leftrightarrow H(Z) = 0 \Leftrightarrow$ r.m.s. $= 0$ (though the mathematician can devise examples to show that they are essentially independent). But the entropy measure of scatter has the advantage that it can be applied when the outcome can only be classified, not measured (e.g., species of fish caught in trawling, amino-acid chain produced by a ribosome). In this paper we shall use the last measure, $H(Z)$, and we define "successful regulation" as equivalent to "$H(Z)$ is minimal."

3. Error- and Cause-Controlled Regulation

The reader may be wondering why error-controlled regulation has been omitted, but there has been no omission. Everything said so far is equally true of this case: for if the cause-effect linkages are as in Fig. 2, R is still receiving information about D's values, as in Fig. 1, but is receiving it after a coding through S. The matter has been discussed fully by Conant (1969). There he showed that the general formulation of Fig. 1 (which represents only that R must receive information from D by *some* route) falls into two essentially distinct classes according to whether the flow of information from D to Z is conserved or lost. Regulation by error-control is essentially information-conserving, and the entropy of Z

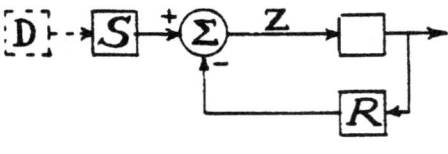

Figure 2

cannot fall to zero (there must be some residual variation). When, however, the regulator R draws its information directly from D (the cause of the disturbance) there need be no residual variation: the regulation may, in principle, be made perfect.

The distinction may be illustrated by a simple example. The cow is homeostatic for blood-temperature, and in its brain is an error-controlled centre that, if the blood-temperature falls, increases the generation of heat in the muscles and liver—but the blood-temperature must fall first. If, however, a sensitive temperature-recorder is inserted in the brain and then a stream of ice-cold air driven past the animal, the temperature rises without any preliminary fall. The error-controlled reflex acts, in fact, only as a reserve: ordinarily, the nervous system senses, at the skin, that the cause of a fall has occurred, and reacts to regulate before the "error" actually occurs. Error-controlled regulation is in fact a primitive and demonstrably inferior method of regulation. It is inferior because with it the entropy of the outcomes Z cannot be reduced to zero; its success can only be partial. The regulations used by the higher organisms evolve progressively to types more effective in using information about the causes (at D) as the source and determiner of their regulatory actions. *From here on, in this paper, we shall consider "regulation" of this more advanced, cause-controlled type* (though much of what we say will still be true of the error-controlled).

4. Models

Defining "regulation," as we have seen, is easy in that one is led rapidly to one of a few forms, closely related and easily distinguished in practical use. The attempt to define a "model," however, leads to no such focus. We shall obtain a definition suitable for this paper, but first let us notice what happens when one attempts precision. We can start with such an unexceptional "model" as a table-top replica of Chartres cathedral. The transformation is of the type, in three dimensions:

$$y_1 = kx_1$$
$$y_2 = kx_2$$
$$y_3 = kx_3$$

with k about 10^{-2}. But this example, so clear and simple, can be modified a little at a time for forms that are very different. A model of Switzerland, for instance, might well have the vertical heights exaggerated (so that the three k's are no longer equal). In two dimensions, a (proportional) photograph from the air may be followed by a Mercator's projection with distortion, that no longer leaves the variables separable. So we can go through a map of a subway system, with only the points of connection valid, to "maps" of a type describable only mathematically.

In dynamic systems, if the transformation converts the real time t to a model time t' also in real time we have a "working" model. An unquestionable "model" here would be a flow of electrons through a net of conducting sheets that accurately models, in real time, the flow of underground water in Arizona. But the model sailing-boat no longer behaves proportionately, so that a complex relation is necessary to relate the model and the full-sized boat. Thus, in the working models, as in the static, we can readily obtain examples that deviate more and more from the obvious model to the most extreme types of transform, without the appearance of any natural boundary dividing model from non-model.

Can we follow the mathematician and use the concept of "isomorphism"? It seems that we cannot. The reason is that though the concept of isomorphism is unique in the branch where it started (in the finite groups) its extension to other branches leads to so many new meanings that the unicity is lost.

As example, suppose we attempt to apply it to the universe of binary relations. R, a subset of $E \times E$, and S, a subset of $F \times F$, are naturally regarded as "isomorphic" if there exists a one–one mapping σ of E onto F such that $S = \sigma R \sigma^{-1}$ (Riguet, 1948, 1951; Bourbaki, 1958). But S and R are still closely related, and able to claim some "model" relationship if the definition is weakened to

$$\exists \sigma, \tau: S = \sigma R \tau^{-1}$$

(with τ also one–one). Then it can be weakened further by allowing ϕ (and τ) to be a mapping generally or even a binary relation. The sign of equality similarly can be weakened to "is contained in." We have now arrived at the relation given earlier (1) under "regulation":

$$\rho \subset A \cdot \phi$$

which evidently implies some "-morphic" relation between ρ and ϕ (with A assumed given).

In this paper we shall be concerned chiefly with isomorphism between two dynamic systems (S and R in Fig. 1). We can therefore try using the modern abstract definition of "machine with input" as a rigorous basis.

To discuss iso-, and homo-, morphism of machines, it is convenient first to obtain a standard representation of these ideas in the theory of groups, where they originated. The relation can be stated thus:

Let the two groups be, one of the set E of elements e_i, with group operation (multiplication) δ, so that $\delta(e_i, e_j) = e_k$, and other similarly of δ' on elements F. Then the second is a homomorph of the first if and only if there exists a mapping h, from E to F, so that, for all $e_i, e_j \in E$:

$$\delta'[h(e_i), h(e_j)] = h[\delta(e_i, e_j)] \qquad (2)$$

If h is one–one onto F, they are isomorphic. This basic equation form will enable us to relate the other possible definitions.

Hartmanis and Stearns' (1966) definition of machine M' being a homomorphism of M follows naturally. Let machine M have a set S of internal states, a set I of input-values (symbols), a set O of output-values (symbols), and let it operate according to δ, a mapping of $S \times I$ to S, and λ, a mapping of $S \times I$ to O. Let machine M' be represented similarly by S', I', O', δ', λ'. Then M' is a homomorphism of M if and only if there exist three mappings:

h_1, of S to S'
h_2, of I to I'
h_3, of O to O'

such that, for all $s \in S$ and $i \in I$:

$$h_1[\delta(s,i)] = \delta'[h_1(s), h_2(i)]$$
$$h_3[\lambda(s,i)] = \lambda'[h_1(s), h_2(i)]. \qquad (3)$$

This definition corresponds to the natural case in which corresponding inputs (to the two machines) will lead, through corresponding internal states, to corresponding outputs. But, unfortunately for our present purpose, there are many variations, some trivial and some gross, that also represent some sort of "similarity." Thus, a more general form, representing a more complex form of relation, would be given if the mappings

h_1, of S to S', and h_2, of I to I'

were replaced by one mapping

h_4 of $I \times S$ to $I' \times S'$.

(More general because h_4 may or may not be separable into h_1 and h_2.) Then the criterion would be,

$$\forall i, s: \delta'[h_4(s,i)] = h_4[\delta(s,i)], \qquad (4)$$

a form not identical with that at (3).

There are yet more. The "Black Box" case ignores the internal states S, and treats two Black Boxes as identical if equal inputs give equal outputs. Formally if μ and μ' are the mappings from input to output, then the second Box is a homomorphism of the first if and only if there exists a mapping h, of I to I', such that:

$$\forall i \in I: \mu'[h(i)] = h[\mu(i)]. \qquad (5)$$

Here it should be remembered that equality of outputs is only a special case of correspondence. Also closely related are two Black Boxes such that the second is "decoder" to the first: the second, given the first's output, will take this as input and emit the original input:

$$\forall i \in I: \mu'\mu(i) = i. \qquad (6)$$

This is an isomorphism. In the homomorphic relation, the input i and the final output $\mu'\mu(i)$ would both be mapped by h to the same class:

$$\forall i \in I: h\mu'\mu(i) = h(i). \qquad (7)$$

These examples may be sufficient to show the wide range of abstract "similarities" that might claim to be "isomorphisms." There seems, in short, to be as many definitions possible to isomorphism as to model. It might seem that one could make practically any assertion one likes (such as that in our title) and then ensure its truth simply by adjusting the

Every Good Regulator of a System Must Be a Model of That System

definitions. We believe, however, that we can mark out one case that is sufficiently a whole to be worth special statement.

We consider the regulatory situation described earlier, in which the set of regulatory events R and the set of events S in the rest of the system (i.e., in the "reguland" S, which we view as R's opponent) jointly determine, through a mapping ψ, the outcome events Z. By an optimal regulator we will mean a regulator which produces regulatory events in such a way that $H(Z)$ is minimal. Then under very broad conditions stated in the proof below, the following theorem holds:

Theorem. The simplest optimal regulator R of a reguland S produces events R which are related to the events S by a mapping $h: S \rightarrow R$.

Restated somewhat less rigorously, the theorem says that the best regulator of a system is one which is a model of that system in the sense that the regulator's actions are merely the system's actions as seen through a mapping h. The type of isomorphism here is that expressed (in the form used above) by

$$\exists h: \forall i: \rho(i) = h[\sigma(i)], \tag{8}$$

where ρ and σ are the mappings that R and S impose on their common input I. This form is essentially that of (5) above.

Proof. The sets R, S, and Z and the mapping $\psi: R \times S \rightarrow Z$ are presumed given. We will assume that over the set S there exists a probability distribution $p(S)$ which gives the relative frequencies of the events in S. We will further assume that the behaviour of any particular regulator R is specified by a conditional distribution $p(R|S)$ giving, for each event in S, a distribution on the regulatory events in R. Now $p(S)$ and $p(R|S)$ jointly determine $p(R,S)$ and hence $p(Z)$ and $H(Z)$, the entropy in the set of outcomes. $[H(Z) \equiv -\sum_{z_k \in Z} p(z_k) \log p(z_k).]$ With $p(S)$ fixed, the class of optimal regulators therefore corresponds to the class of optimal distributions $p(R|S)$ for which $H(Z)$ is minimal. We will call this class of optimal distributions π.

It is possible for there to be very different distributions $p(Z)$ all having the same minimal entropy $H(Z)$. To consider that possibility would merely complicate this proof without affecting it in any essential way, so we will suppose that every $p(R|S)$ in π determines, with $p(S)$ and ψ, the same (unique) $p(Z)$. We now select for examination an arbitrary $p(R|S)$ from π.

The heart of the proof is the following lemma:

Lemma. $\forall s_j \in S$, the set $\{\psi(r_i, s_j) : p(r_i, s_j) > 0\}$ has only one element. That is, for every s_j in S, $p(R|s_j)$ is such that all r_i with positive probability map, with s_j under ψ, to the same z_k in Z.

Proof of Lemma. Suppose, to the contrary, that $p(r_1|s_j) > 0$, $p(r_2|s_j) > 0$, $\psi(r_1, s_j) = z_1$, and $\psi(r_2, s_j) = z_2 \neq z_1$. Now $p(r_1, s_j)$ and $p(r_2, s_j)$ contribute to $p(z_1)$ and $p(z_2)$, respectively, and by varying these probabilities (by subtracting Δ from $p(r_1, s_j)$ and adding Δ to $p(r_2, s_j)$) we could vary $p(z_1)$ and $p(z_2)$ and thereby vary $H(Z)$. We could make Δ either positive or negative, whichever would make $p(z_1)$ and $p(z_2)$ more unequal. One of the useful and fundamental properties of the entropy function is that any such increase in imbalance in $p(Z)$ necessarily decreases $H(Z)$. Consequently, we could start with a $p(R|S)$ from the class π, which minimizes $H(Z)$, and produce a new $p(R|S)$ resulting in a lower $H(Z)$; this contradiction proves the lemma.

Returning to the proof of the theorem, we see that for any member of π and any s_j in S, the values of R for which $p(R|s_j)$ is positive all give the same z_k. Without affecting $H(Z)$, we can arbitrarily select one of those values of R and set its conditional probability to unity and the others to zero. When this process is repeated for all s_j in S, the result must be a member of π with $p(R|S)$ consisting entirely of ones and zeros. In an obvious sense this is the *simplest* optimal $p(R|S)$ since it is in fact a mapping h from S into R. Given the correspondence between optimal distributions $p(R|S)$ and optimal regulators R, this proves the theorem.

The Theorem calls for several comments. First, it leaves open the possibility that there are regulators which are just as successful (just as "optimal") as the simplest optimal regulator(s) but which are unnecessarily complex. In this regard, the theorem can be interpreted as saying that although not all optimal regulators are models of their regulands, the ones which are not are all unnecessarily complex.

Second, it shows clearly that the search for the best regulator is essentially a search among the mappings from S into R; only regulators for which there is such a mapping need be considered.

Third, the proof of the theorem, by avoiding all mention of the inputs to the regulator R and its opponent S, leaves open the question of how R, S, and Z are interrelated. The theorem applies equally well to the configurations of Fig. 1 and Fig. 2, the chief difference being that in Fig. 2 R is a model of S in the sense that the events R are mapped versions of the events S, whereas in Fig. 1 the modelling is stronger; R must be a homo- or isomorph of S (since it has the same input as S and a mapping-related output).

Last, the assumption that $p(S)$ must exist (and be constant) can be weakened; if the statistics of S change slowly with time, the theorem holds over any period throughout which $p(S)$ is essentially constant. As $p(S)$ changes, the mapping h will change appropriately, so that the best regulator in such a situation will still be a model of the reguland, but a time-varying model will be needed to regulate the time-varying reguland.

5. Discussion

The first effect of this theorem is to change the status of model-making from optional to compulsory. As we said earlier, model-making has hitherto largely been suggested (for regulating complex dynamic systems) as a possibility: the theorem shows that, in a very wide class (specified in the proof of the theorem), success in regulation implies that a sufficiently similar model must have been built, whether it was done explicitly, or simply developed as the regulator was improved. Thus the would-be model-maker now has a rigorous theorem to justify his work.

To those who study the brain, the theorem founds a "theoretical neurology." For centuries, the study of the brain has been guided by the idea that as the brain is the organ of thinking, whatever it does is right. But this was the view held two centuries ago about the human heart as a pump; today's hydraulic engineers know too much about pumping to follow the heart's method slavishly: they know what the heart ought to do, and they measure its efficiency. The developing knowledge of regulation, information-processing, and control is building similar criteria for the brain. Now that we know that any regulator (if it conforms to the qualifications given) must model what it regulates, we can proceed to measure how efficiently the brain carries out this process. There can no longer be question about *whether* the brain models its environment: it must.

References

Ashby, W. Ross, 1967, *Automaton Theory and Learning Systems*, edited by D. J. Stewart (London: Academic Press), p. 23–51.
Bourbaki, N., 1958, *Théorie des Ensembles; Fascicule de Résultats*, 3rd edition (Paris: Hermann).
Conant, Roger C., 1969, *I.E.E.E. Trans. Systems Sci.*, **5**, 334.
Hartmanis, J., and Stearns, R. E., 1966, *Algebraic Structure Theory of Sequential Machines* (New York: Prentice-Hall).
Riguet, J., 1948, *Bull. Soc. Math. Fr.*, **76**, 114; 1951, Thèse de Paris.
Sommerhoff, G., 1950, *Analytical Biology* (Oxford University Press).

27

Principles of the Self-Organizing System

W. Ross Ashby

Questions of principle are sometimes regarded as too unpractical to be important, but I suggest that that is certainly not the case in *our* subject. The range of phenomena that we have to deal with is so broad that, were it to be dealt with wholly at the technological or practical level, we would be defeated by the sheer quantity and complexity of it. The total range can be handled only piecemeal; among the pieces are those homomorphisms of the complex whole that we call "abstract theory" or "general principles." They alone give the bird's-eye view that enables us to move about in this vast field without losing our bearings. I propose, then, to attempt such a bird's-eye survey.

What Is "Organization"?

At the heart of our work lies the fundamental concept of "organization." What do we mean by it? As it is used in biology it is a somewhat complex concept, built up from several more primitive concepts. Because of this richness it is not readily defined, and it is interesting to notice that while March and Simon (1958) use the word "Organizations" as title for their book, they do not give a formal definition. Here I think they are right, for the word covers a multiplicity of meanings. I think that in future we shall hear the *word* less frequently, though the *operations* to which it corresponds, in the world of computers and brain-like mechanisms, will become of increasing daily importance.

The hard core of the concept is, in my opinion, that of "conditionality." As soon as the relation between two entities A and B becomes conditional on C's value or state then a necessary component of "organization" is present. Thus *the theory of organization is partly co-extensive with the theory of functions of more than one variable*.

We can get another angle on the question by asking "what is its converse?" The

---

From *Principles of Self-Organization*, H. von Foerster and G. W. Zopf, eds. Copyright © 1962 by Pergamon Press.

converse of "conditional on" is "not conditional on," so the converse of "organization" must therefore be, as the mathematical theory shows as clearly, the concept of "reducibility." (It is also called "separability.") This occurs, in mathematical forms, when what looks like a function of several variables (perhaps very many) proves on closer examination to have parts whose actions are *not* conditional on the values of the other parts. It occurs in mechanical forms, in hardware, when what looks like one machine proves to be composed of two (or more) sub-machines, each of which is acting independently of the others.

Questions of "conditionality," and of its converse "reducibility," can, of course, be treated by a number of mathematical and logical methods. I shall say something of such methods later. Here, however, I would like to express the opinion that the method of Uncertainty Analysis, introduced by Garner and McGill (1956), gives us a method for the treatment of conditionality that is not only completely rigorous but is also of extreme generality. Its great generality and suitability for application to complex behavior, lies in the fact that it is applicable to any arbitrarily defined set of states. Its application requires neither linearity, nor continuity, nor a metric, nor even an ordering relation. By this calculus, the *degree* of conditionality can be measured, and analyzed, and apportioned to factors and interactions in a manner exactly parallel to Fisher's method of the analysis of variance; yet it requires no metric in the variables, only the frequencies with which the various combinations of states occur. It seems to me that, just as Fisher's conception of the analysis of variance threw a flood of light on to the complex relations that may exist between variations on a metric, so McGill and Garner's conception of uncertainty analysis may give us an altogether better understanding of how to treat complexities of relation when the variables are non-metric. In psychology and biology such variables occur with great commonness; doubtless they will also occur commonly in the brain-like processes developing in computers. I look forward to the time when the methods of McGill and Garner will become the accepted language in which such matters are to be thought about and treated quantitatively.

The treatment of "conditionality" (whether by functions of many variables, by correlation analysis, by uncertainty analysis, or by other ways) makes us realize that the essential idea is that there is first a product space—that of the *possibilities*—within which some sub-set of points indicates the actualities. This way of looking at "conditionality" makes us realize that it is related to that of "communication"; and it is, of course, quite plausible that we should define parts as being "organized" when "communication" (in some generalized sense) occurs between them. (Again the natural converse is that of independence, which represents non-communication.)

Now "communication" from A to B necessarily implies some constraint, some correlation between what happens at A and what at B. If, for given event at A, all possible events may occur at B, then there is no communication from A to B and no constraint over the possible (A, B)-couples that can occur. Thus the presence of "organization" between variables is equivalent to the existence of a *constraint* in the product-space of the possibilities. I stress this point because while, in the past, biologists have tended to think of organization as something extra, something *added* to the elementary variables, the modern theory, based on the logic of communication, regards organization as a restriction or constraint. The two points of view are thus diametrically opposed; there is no question of either being exclusively right, for each can be appropriate in its context. But with this opposition in existence we must clearly go carefully, especially when we discuss with others, lest we should fall into complete confusion.

This excursion may seem somewhat complex but it is, I am sure, advisable, for we have to recognize that the discussion of organization theory has a peculiarity not found in the more objective sciences of physics and chemistry. The peculiarity comes in with the product space that I have just referred to. Whence comes this product space? Its chief peculiarity is that *it contains more than actually exists in the real physical world*, for it is the latter that gives us the actual, constrained *subset*.

The real world gives the subset of what *is*; the product space represents the uncertainty of the *observer*. The product space may therefore change if the observer changes; and two observers may legitimately use different product spaces within which to record the same subset of actual events in some actual thing. The "constraint" is thus a *relation* between observer and thing; the properties of any particular constraint will depend on both the real thing and on *the observer*. It follows that *a substantial part of the theory of organization will be concerned with properties that are not intrinsic to the thing but are relational between observer and thing*. We shall see some striking examples of this fact later.

Whole and Parts

"If conditionality" is an essential component in the concept of organization, so also is the assumption that we are speaking of a whole composed of parts. This assumption is worth a moment's scrutiny, for research is developing a theory of dynamics that does *not* observe parts and their interactions, but treats the system as an unanalyzed whole (Ashby, 1958a). In physics, of course, we usually start the description of a system by saying "Let the variables be $x_1, x_2, \ldots, x_n$," and thus start by treating the whole as made of n functional parts. The other method, however, deals with unanalyzed states, $S_1, S_2, \ldots$ of the whole, without explicit mention of any parts that may be contributing to these states. The dynamics of such a system can then be defined and handled mathematically; I have shown elsewhere (Ashby, 1960a) how such an approach can be useful. What I wish to point out here is that we can have a sophisticated *dynamics*, of a whole as complex and cross-connected as you please, that makes no reference to any parts and that therefore does *not* use the concept of organization. Thus the concepts of dynamics and of organization are essentially independent, in that all four combinations, of their presence and absence, are possible.

This fact exemplifies what I said, that "organization" is partly in the eye of the beholder. Two observers studying the same real material system, a hive of bees say, may find that one of them, thinking of the hive as an interaction of fifty thousand bee-parts, finds the bees "organized," while the other, observing whole states such as activity, dormancy, swarming, etc., may see *no* organization, only trajectories of these (unanalyzed) states.

Another example of the independence of "organization" and "dynamics" is given by the fact that whether or not a real system is organized or reducible depends partly on the point of view taken by the observer. It is well known, for instance, that an organized (i.e., interacting) linear system of n parts, such as a network of pendulums and springs, can be seen from another point of view (that of the so-called "normal" coordinates) in which all the (newly identified) parts are completely separate, so that the whole is reducible. There is therefore nothing perverse about my insistence on the relativity of organization, for advantage of the fact is routinely taken in the study of quite ordinary dynamic systems.

Finally, in order to emphasize how dependent is the organization seen in a system on the observer who sees it, I will state the proposition that: given a whole with arbitrarily given behavior, a great variety of arbitrary "parts" can be seen in it; for all that is necessary, when the arbitrary part is proposed, is that we assume the given part to be coupled to another suitably related part, so that the two together form a whole isomorphic with the whole that was given. For instance, suppose the given whole, W of 10 states, behaves in accordance with the transformation:

$$W \downarrow \quad \begin{array}{c} p\,q\,r\,s\,t\,u\,v\,w\,x\,y \\ q\,r\,s\,q\,s\,t\,t\,x\,y\,y \end{array}$$

Its kinematic graph is

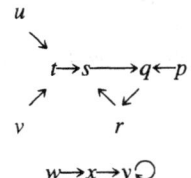

and suppose we wish to "see" it as containing the part P, with internal states E and input states A:

$$\left. \begin{array}{c|cc} \downarrow & \multicolumn{2}{c}{E} \\ & 1 & 2 \\ \hline A \quad 1 & 2 & 1 \\ 2 & 1 & 1 \end{array} \right\} P$$

With a little ingenuity we find that if part P is coupled to part Q [with states (F, G) and input B] with transformation Q:

$$\left. \begin{array}{c|cccccc} & \multicolumn{6}{c}{(F, G)} \\ \downarrow & 1,1 & 1,2 & 1,3 & 2,1 & 2,2 & 2,3 \\ \hline B \quad 1 & 2,1 & 1,2 & 1,2 & 2,1 & 1,2 & 1,2 \\ 2 & \cdot & 2,3 & \cdot & 2,1 & 2,2 & 2,2 \end{array} \right\} Q$$

by putting $A = F$ and $B = E$, then the new whole W' has transformation

$$W': \quad \downarrow \quad \begin{array}{ccccc} 1,1,1 & 1,1,2 & 1,1,3 & 1,2,1 & \text{etc.} \\ 2,2,1 & 2,1,2 & 2,1,2 & 1,2,1 & \text{etc.} \end{array}$$

which is *isomorphic with* W under the one–one correspondence

$$\downarrow \quad \begin{array}{ccccc} 1,1,1 & 1,1,2 & 1,1,3 & 1,2,1 & \text{etc.} \\ w & s & p & y & \text{etc.} \end{array}$$

Thus, subject only to certain requirements (e.g., that equilibria map into equilibria) *any dynamic system can be made to display a variety of arbitrarily assigned "parts,"* simply by a change in the *observer's* view point.

Machines in General

I have just used a way of representing two "parts," "coupled" to form a "whole," that anticipates the question: what do we mean by a "machine" in general?

Here we are obviously encroaching on what has been called "general system theory," but this last discipline always seemed to me to be uncertain whether it was dealing with *physical* systems, and therefore tied to whatever the real world provides, or with mathematical systems, in which the sole demand is that the work shall be free from internal contradictions. It is, I think, one of the substantial advances of the last decade that we have at last identified the *essentials* of the "machine in general."

Before the essentials could be seen, we had to realize that two factors must be *excluded as irrelevant*. The first is "materiality"—the idea that a machine must be made of actual matter, of the hundred or so existent elements. This is wrong, for examples can readily be given (e.g., Ashby, 1958a) showing that what is essential is whether the system, of angels and ectoplasm if you please, *behaves* in a law-abiding and machine-like way. Also to be excluded as irrelevant is any reference to energy, for any calculating machine shows that what matters is the *regularity* of the behavior—whether energy is gained or lost, or even created, is simply irrelevant.

The fundamental concept of "machine" proves to have a form that was formulated at least a century ago, but this concept has not, so far as I am aware, ever been used and exploited vigorously. A "machine" is that which behaves in a machine-like way, namely, that its internal state, and the state of its surroundings, defines uniquely the next state it will go to.

This definition, formally proposed fifteen years ago (Ashby, 1945) has withstood the passage of time and is now becoming generally accepted (e.g., Jeffrey, 1959). It appears in many forms. When the variables are continuous it corresponds to the description of a dynamic system by giving a set of ordinary differential equations with time as the independent variable. The *fundamental* nature of such a representation (as contrasted with a merely convenient one) has been recognized by many earlier workers such as Poincaré, Lotka (1925), and von Bertalanffy (1950 and earlier).

Such a representation by differential equations is, however, too restricted for the needs of a science that includes biological systems and calculating machines, in which discontinuity is ubiquitous. So arises the modern definition, able to include both the continuous and the discontinuous and even the discrete, without the slightest loss of rigor. The "machine with input" (Ashby, 1958a) or the "finite automaton" (Jeffrey, 1959) is today defined by a set S of internal states, a set I of input or surrounding states, and a mapping, f say, of the product set $I \times S$ into S. Here, in my opinion, we have the very essence of the "machine"; all known types of machine are to be found here; and all interesting deviations from the concept are to be found by the corresponding deviation from the definition.

We are now in a position to say without ambiguity or evasion what we mean by a machine's "organization." First we specify which system we are talking about by specifying its states S and its conditions I. If S is a product set, so that $S = \Pi_i T_i$ say, then the parts i are each specified by its set of states T_i. *The "organization" between these parts is then specified by the mapping f.* Change f and the organization changes. In other words, the possible organizations between the parts can be set into one–one correspondence with the set of possible mappings of $I \times S$ into S. Thus "organization" and "mapping" are two ways of looking at the same thing—the organization being noticed by the observer of the actual system, and the mapping being recorded by the person who represents the behavior in mathematical or other symbolism.

"Good" Organization

At this point some of you, especially the biologists, may be feeling uneasy; for this definition of organization makes no reference to any *usefulness* of the organization. It demands only that there be conditionality between the parts and regularity in behavior. In this I believe the definition to be right, for the question whether a given organization is "good" or "bad" is quite independent of the prior test of whether it is or is not an organization.

I feel inclined to stress this point, for here the engineers and the biologists are likely to think along widely differing lines. The engineer, having put together some electronic hardware and having found the assembled network to be roaring with parasitic oscillations, is quite accustomed to the idea of a "bad" organization; and he knows that the "good" organization has to be searched for. The biologist, however, studies mostly animal species that have survived the long process of natural selection; so almost all the organizations he sees have already been selected to be good ones, and he is apt to think of "organizations" as *necessarily* good. This point of view may often be true in the biological world but it is most emphatically not true in the world in which we people here are working. We *must* accept that

1. most organizations are bad ones;
2. the good ones have to be sought for; and
3. what is meant by "good" must be clearly defined, explicitly if necessary, *in every case*.

What then is meant by "good," in our context of brain-like mechanisms and computers? We must proceed cautiously, for the word suggests some evaluation whose origin has not yet been considered.

In some cases the distinction between the "good" organization and the "bad" is obvious, in the sense that as everyone in these cases would tend to use the same criterion, it would not need explicit mention. The brain of a living organism, for instance, is usually judged as having a "good" organization if the organization (whether inborn or learned) acts so as to further the organism's survival. This consideration readily generalizes to all those cases in which the organization (whether of a cat or an automatic pilot or an oil refinery) is judged "good" if and only if it acts so as to keep an assigned set of variables, the "essential" variables, within assigned limits. Here are all the mechanisms for homeostasis, both in the original sense of Cannon and in the generalized sense. From this criterion comes the related one that an organization is "good" if it makes the system stable around an assigned equilibrium. Sommerhoff (1950) in particular has given a wealth of examples, drawn from a great range of biological and mechanical phenomena, showing how in all cases the idea of a "good organization" has as its essence the idea of a number of parts so interacting as to achieve some given "focal condition." I would like to say here that I do not consider that Sommerhoff's contribution to our subject has yet been adequately recognized. His identification of *exactly* what is meant by coordination and integration is, in my opinion, on a par with Cauchy's identification of exactly what was meant by convergence. Cauchy's discovery was a real discovery, and was an enormous help to later workers by providing them with a concept, rigorously defined, that could be used again and again, in a vast range of contexts, and always with exactly the same meaning. Sommerhoff's discovery of how to represent *exactly* what is meant by coordination and integration and good organization will, I am sure, eventually play a similarly fundamental part in our work.

His work illustrates, and emphasizes, what I want to say here—*there is no such thing as "good organization" in any absolute sense*. Always it is relative; and an organization that is good in one context or under one criterion may be bad under another.

Sometimes this statement is so obvious as to arouse no opposition. If we have half a dozen lenses, for instance, that can be assembled this way to make a telescope or that way to make a microscope, the goodness of an assembly obviously depends on whether one wants to look at the moon or a cheese mite.

But the subject is more contentious than that! The thesis implies that there is no such thing as a brain (natural or artificial) that is good in any absolute sense—it all depends on the circumstances and on what is wanted. Every faculty that a brain can show is "good" only conditionally, for there exists at least one environment against which the brain is handicapped by the possession of this faculty. Sommerhoff's formulation enables us to show this at once: whatever the faculty or organization achieves, let that be *not* in the "focal conditions."

We know, of course, lots of examples where the thesis is true in a somewhat trivial way. Curiosity tends to be good, but many an antelope has lost its life by stopping to see what the hunter's hat is. Whether the organization of the antelope's brain should be of the type that does, or does not, lead to temporary immobility clearly depends on whether hunters with rifles are or are not plentiful in its world.

From a different angle we can notice Pribram's results (1957), who found that brain-operated monkeys scored higher in a certain test than the normals. (The operated were plodding and patient while the normals were restless and distractable.) Be that as it may, one cannot say which brain (normal or operated) had the "good" organization until one has decided which sort of temperament is wanted.

Do you still find this non-contentious? Then I am prepared to assert that there is not a single mental faculty ascribed to Man that is good in the absolute sense. If any particular faculty is *usually* good, this is solely because our terrestrial environment is so lacking in variety that its usual form makes that faculty usually good. But change the environment, go to really different conditions, and possession of that faculty may be harmful. And "bad," by implication, is the brain organization that produces it.

I believe that there is not a single faculty or property of the brain, usually regarded as desirable, that does not become *un*desirable in some type of environment. Here are some examples in illustration.

The first is Memory. Is it not good that a brain should have memory? Not at all, I reply—only when the environment is of a type in which the future often *copies* the past; should the future often be the *inverse* of the past, memory is actually disadvantageous. A well known example is given when the sewer rat faces the environmental system known as "pre-baiting." The naïve rat is very suspicious, and takes strange food only in small quantities. If, however, wholesome food appears at some place for three days in succession, the sewer rat will learn, and on the fourth day will eat to repletion, and die. The rat without memory, however, is as suspicious on the fourth day as on the first, and lives. Thus, in *this* environment, memory is positively disadvantageous. Prolonged contact with this environment will lead, other things being equal, to evolution in the direction of diminished memory-capacity.

As a second example, consider organization itself in the sense of connectedness. Is it not good that a brain should have its parts in rich functional connection? I say, No—not *in general*; only when the environment is itself richly connected. When the environment's parts are *not* richly connected (when it is highly reducible, in other words), adaptation will go on faster if the brain is also highly reducible, i.e., if its connectivity is small (Ashby, 1960d). Thus the *degree* of organization can be too high as well as too low; the degree we humans possess is probably adjusted to be somewhere near the optimum for the usual terrestrial environment. It does not in any way follow that this degree will be optimal or good if the brain is a mechanical one, working against some grossly nonterrestrial environment—one existing only inside a big computer, say.

As another example, what of the "organization" that the biologist always points to with pride—the development in evolution of specialized organs such as brain, intestines, heart and blood vessels. Is not this good? Good or not, it is certainly a specialization made possible only because the earth has an atmosphere; without it, we would be incessantly bombarded by tiny meteorites, any one of which, passing through our chest, might strike a large blood vessel and kill us. Under such conditions a better form for survival would be the slime mould, which specializes in being able to flow through a tangle of twigs without loss of function. Thus the development of organs is not good unconditionally, but is a specialization to a world free from flying particles.

After these actual instances, we can return to theory. It is here that Sommerhoff's formulation gives such helpful clarification. He shows that in all cases there must be given, and specified, first a *set of disturbances* (values of his "coenetic variable") and secondly a goal (his "focal condition"); the disturbances threaten to drive the outcome outside the focal condition. The "good" organization is then of the nature of a *relation* between the set of disturbances and the goal. Change the set of disturbances, and the organization, without itself changing, is evaluated "bad" instead of "good." As I said, there is no property of an organization that is good in any absolute sense; all are relative to some given environment, or to some given set of threats and disturbances, or to some given set of problems.

Self-Organizing Systems

I hope I have not wearied you by belaboring this relativity too much, but it is fundamental, and is only too readily forgotten when one comes to deal with organizations that are either biological in origin or are in imitation of such systems. With this in mind, we can now start to consider the so-called "self-organizing" system. We must proceed with some caution here if we are not to land in confusion, for the adjective is, if used loosely, ambiguous, and, if used precisely, self-contradictory.

To say a system is "self-organizing" leaves open two quite different meanings.

Principles of the Self-Organizing System

There is a *first meaning* that is simple and unobjectionable. This refers to the system that starts with its parts separate (so that the behavior of each is independent of the others' states) and whose parts then act so that they change towards forming connections of some type. *Such a system is "self-organizing"* in the sense that it changes from *"parts separated"* to *"parts joined."* An example is the embryo nervous system, which starts with cells having little or no effect on one another, and changes, by the growth of dendrites and formation of synapses, to one in which each part's behavior is very much affected by the other parts. Another example is Pask's system of electrolytic centers, in which the growth of a filament from one electrode is at first little affected by growths at the other electrodes; then the growths become more and more affected by one another as filaments approach the other electrodes. In general such systems can be more simply characterized as *"self-connecting,"* for the change from independence between the parts to conditionality can always be seen as some form of "connection," even if it is as purely functional as that from a radio transmitter to a receiver.

Here, then, is a perfectly straightforward form of self-organizing system; but I must emphasize that there can be no assumption at this point that the organization developed will be a good one. If we wish it to be a "good" one, we must first provide a criterion for distinguishing between the bad and the good, and then we must ensure that the appropriate selection is made.

We are here approaching the *second meaning* of "self-organizing" (Ashby, 1947). "Organizing" may have the first meaning, just discussed, of *"changing from unorganized to organized."* But it may also mean *"changing from a bad organization to a good one,"* and this is the case I wish to discuss now, and more fully. This is the case of peculiar interest to *us*, for this is the case of the system that changes itself from a bad way of behaving to a good. A well known example is the child that starts with a brain organization that makes it fire-seeking; then a change occurs, and a new brain organization appears that makes the child fire-avoiding. Another example would occur if an automatic pilot and a plane were so coupled, by mistake, that positive feedback made the whole error-aggravating rather than error-correcting. Here the organization is bad. The system would be "self-organizing" if a change were *automatically* made to the feedback, changing it from positive to negative; then the whole would have changed from a bad organization to a good. Clearly, *this* type of "self-organization" is of peculiar interest to us. What is implied by it?

Before the question is answered we must notice, if we are not to be in perpetual danger of confusion, that *no machine can be self-organizing in this sense*. The reasoning is simple. Define the set S of states so as to specify which machine we are talking about. The "organization" must then, as I said above, be identified with f, the mapping of S into S that the basic drive of the machine (whatever force it may be) imposes. Now the logical relation here is that f determines the changes of S:—f is *defined* as the set of couples (s_i, s_j) such that the internal drive of the system will force state s_i to change to s_j. To allow f to be a function of the state is to make nonsense of the whole concept.

Since the argument is fundamental in the theory of self-organizing systems, I may help explanation by a parallel example. Newton's law of gravitation says that $F = M_1 M_2 / d^2$, in particular, that the force varies inversely as the distance to power 2. To power 3 would be a different law. But suppose it were suggested that, not the force F but the *law* changed with the distance, so that the power was not 2 but some function of the distance, $\phi(d)$. This suggestion is illogical; for we now have that $F = M_1 M_2 / d^{\phi(d)}$, and this represents not a law that varies with the distance but *one* law covering all distances; that is, were this the case we would *redefine* the law. Analogously, were f in the machine to be some function of the state S, we would have to redefine our machine. Let me be quite explicit with an example. Suppose S had three states: a, b, c. If f depended on S there would be three f's: f_a, f_b, f_c, say. Then if they are

↓	a	b	c
f_a	**b**	a	b
f_b	c	**a**	a
f_c	b	b	**a**

then the transform of a must be under f_a, and is therefore b, so the whole set of f's would amount to the *single* transformation:

±	a	b	c
	b	a	a

It is clearly illogical to talk of f as being a function of S, for such talk would refer to operations, such as $f_a(b)$, which cannot in fact occur.

If, then, no machine can properly be said to be self-organizing, how do we regard, say, the Homeostat, that rearranges its own wiring; or the computer that writes out its own program?

The new logic of mechanism enables us to treat the question rigorously. We start with the set S of states, and assume that f changes, to g say. So we really have a *variable*, $\alpha(t)$ say, a function of the time that had at first the value f and later the value g. This change, as we have just seen, cannot be ascribed to any cause in the set S; so it must have come from some outside agent, acting on the system S as input. If the system is to be in some sense "*self-organizing*," the "self" must be enlarged to include this variable α, and, to keep the whole bounded, the cause of α's change must be in S (or α).

Thus the appearance of being "self-organizing" can be given only by the machine S being coupled to another machine (of one part):

$$\boxed{S} \rightleftarrows \boxed{\alpha}$$

Then the part S can be "self-organizing" within the whole $S + \alpha$.

Only in this partial and strictly qualified sense can we understand that a system is "*self-organizing*" without being self-contradictory.

Since no system can correctly be said to be self-organizing, and since the use of the phrase "self-organizing" tends to perpetuate a fundamentally confused and inconsistent way of looking at the subject, the phrase is probably better allowed to die out.

The Spontaneous Generation of Organization

When I say that no system can properly be said to be self-organizing, the listener may not be satisfied. What, he may ask, of those changes that occurred a billion years ago, that

Principles of the Self-Organizing System

led lots of carbon atoms, scattered in little molecules of carbon dioxide, methane, carbonate, etc., to get together until they formed proteins, and then went on to form those large active lumps that today we call "animals"? Was not this process, on an isolated planet, one of "self-organization"? And if it occurred on a planetary surface can it not be made to occur in a computer? I am, of course, now discussing the origin of life. Has modern system theory anything to say on this topic?

It has a great deal to say, and some of it flatly contradictory to what has been said ever since the idea of evolution was first considered. In the past, when a writer discussed the topic, he usually assumed that the generation of life was rare and peculiar, and he then tried to display some way that would enable this rare and peculiar event to occur. So he tried to display that there is *some* route from, say, carbon dioxide to the amino acid, and thence to the protein, and so, through natural selection and evolution, to intelligent beings. I say that this looking for special conditions is quite wrong. The truth is the opposite—*every* dynamic system generates its own form of intelligent life, is self-organizing in this sense. (I will demonstrate the fact in a moment.) Why we have failed to recognize this fact is that until recently we have had no experience of systems of medium complexity; either they have been like the watch and the pendulum, and we have found their properties few and trivial, or they have been like the dog and the human being, and we have found their properties so rich and remarkable that we have thought them supernatural. Only in the last few years has the general-purpose computer given us a system rich enough to be interesting yet still simple enough to be understandable. With this machine as tutor we can now begin to think about systems that are simple enough to be comprehensible in detail yet also rich enough to be suggestive. With their aid we can see the truth of the statement that *every isolated determinate dynamic system obeying unchanging laws will develop "organisms" that are adapted to their "environments."*

The argument is simple enough in principle. We start with the fact that systems in general go to equilibrium. Now most of a system's states are nonequilibrial (if we exclude the extreme case of the system in neutral equilibrium). So in going from *any* state to one of the equilibria, the system is going from a larger number of states to a smaller. In this way it is performing a selection, in the purely objective sense that it rejects some states, by leaving them, and retains some other state, by sticking to it. Thus, as every determinate system goes to equilibrium, so does it select. We have heard *ad nauseam* the dictum that a machine cannot select; the truth is just the opposite: every machine, as it goes to equilibrium, performs the corresponding act of selection.

Now, equilibrium in simple systems is usually trivial and uninteresting; it is the pendulum hanging vertically; it is the watch with its main-spring run down; the cube resting flat on one face. Today, however, we know that when the system is more complex and dynamic, equilibrium, and the stability around it, can be much more interesting. Here we have the automatic pilot successfully combating an eddy; the person redistributing his blood flow after a severe haemorrhage; the business firm restocking after a sudden increase in consumption; the economic system restoring a distribution of supplies after a sudden destruction of a food crop; and it is a man successfully getting at least one meal a day during a lifetime of hardship and unemployment.

What makes the change, from trivial to interesting, is simply the *scale* of the events. "Going to equilibrium" *is* trivial in the simple pendulum, for the equilibrium is no more than a single point. But when the system is more complex; when, say, a country's economy goes back from wartime to normal methods then the stable region is vast, and much interesting activity can occur within it. The computer is heaven-sent in this context, for it enables us to bridge the enormous conceptual gap from the simple and understandable to the complex and interesting. Thus we can gain a considerable insight into the so-called spontaneous generation of life by just seeing how a somewhat simpler version will appear in a computer.

Competition

Here is an example of a simpler version. The competition between species is often treated as if it were essentially biological; it is in fact an expression of a process of far greater generality. Suppose we have a computer, for instance, whose stores are filled at random with the digits 0 to 9. Suppose its dynamic law is that the digits are continuously being multiplied in pairs, and the right-hand digit of the product going to replace the first digit taken. Start the machine, and let it "evolve"; what will happen? Now under the laws of this particular world, even times even gives even, and odd times odd gives odd. But even times odd gives even; so after a mixed encounter *the even has the better chance of survival*. So as this system evolves, we shall see the evens favored in the struggle, steadily replacing the odds in the stores and eventually exterminating them.

But the evens are not homogeneous, and among them the zeros are best suited to survive in this particular world; and, as we watch, we shall see the zeros exterminating their fellow-evens, until eventually they inherit this particular earth.

What we have here is an example of a thesis of extreme generality. From one point of view we have simply a well defined operator (the multiplication and replacement law) which drives on towards equilibrium. In doing so it *automatically* selects those operands that are *specially resistant* to its change-making tendency (for the zeros are uniquely resistant to change by multiplication). This process, of progression towards the specially resistant form, is of extreme generality, demanding only that the operator (or the physical laws of any physical system) be determinate and unchanging. This is the general or abstract point of view. The biologist sees a special case of it when he observes the march of evolution, survival of the fittest, and the inevitable emergence of the highest biological functions and intelligence. Thus, when we ask: What was necessary that life and intelligence should appear? the answer is not carbon, or amino acids or any other special feature but only that the dynamic laws of the process should be *unchanging*, i.e., that the system should be *isolated. In any isolated system, life and intelligence inevitably develop* (they may, in degenerate cases, develop to only zero degree).

So the answer to the question: How can we generate intelligence synthetically? is as follows. Take a dynamic system whose laws are unchanging and single-valued, and whose size is so large that after it has gone to an equilibrium that involves only a small fraction of its total states, this small fraction is still large enough to allow room for a good deal of change and behavior. Let it go on for a long enough time to get to such an equilibrium. Then examine the equilibrium in detail. You will find that the states or forms now in being are peculiarly able to survive against the changes induced by the laws. Split the equilibrium in two, call one part "organism" and the other part "environment": you will find that this "organism" is peculiarly able to survive against the disturbances from this "environment."

Principles of the Self-Organizing System

The *degree* of adaptation and complexity that this organism can develop is bounded only by the size of the whole dynamic system and by the time over which it is allowed to progress towards equilibrium. Thus, as I said, every isolated determinate dynamic system will develop organisms that are adapted to their environments. There is thus no difficulty in principle, in developing synthetic organisms as complex or as intelligent as we please.

In *this* sense, then, *every* machine can be thought of as "self-organizing," for it will develop, to such degree as its size and complexity allow, some functional structure homologous with an "adapted organism." But does this give us what we at this Conference are looking for? Only partly; for nothing said so far has any implication about the organization being good or bad; the criterion that would make the distinction has not yet been introduced. It is true, of course, that the developed organism, being stable, will have its own essential variables, and it will show its stability by vigorous reactions that tend to preserve its own existence. To *itself*, its own organization will *always*, by definition, be good. The wasp finds the stinging reflex a good thing, and the leech finds the blood-sucking reflex a good thing. But these criteria come *after* the organization for survival; having seen *what* survives we then see what is "good" for that form. What emerges depends simply on what are the system's laws and from what state it started; there is no implication that the organization developed will be "good" in any absolute sense, or according to the criterion of any outside body such as ourselves.

To summarize briefly: there is no difficulty, in principle, in developing *synthetic organisms as complex, and as intelligent as we please*. But we must notice two fundamental qualifications; first, their intelligence will be an adaptation to, and a specialization towards, their particular environment, with no implication of validity for any other environment such as ours; and secondly, their intelligence will be directed towards keeping their own essential variables within limits. They will be fundamentally selfish. So we now have to ask: In view of these qualifications, can we yet turn these processes to our advantage?

Requisite Variety

In this matter I do not think enough attention has yet been paid to Shannon's Tenth Theorem (1949) or to the simpler "law of requisite variety" in which I have expressed the same basic idea (Ashby, 1958a). Shannon's theorem says that if a correction-channel has capacity H, then equivocation of amount H can be removed, *but no more*. Shannon stated his theorem in the context of telephone or similar communication, but the formulation is just as true of a biological regulatory channel trying to exert some sort of corrective control. He thought of the case with a lot of message and a little error; the biologist faces the case where the "message" is small but the disturbing errors are many and large. The theorem can then be applied to the brain (or any other regulatory and selective device), when it says that the amount of regulatory or selective action that the brain can achieve is absolutely bounded by its capacity as a channel (Ashby, 1958b). Another way of expressing the same idea is to say that any quantity K of appropriate selection demands the transmission or processing of quantity K of information (Ashby, 1960b). *There is no getting of selection for nothing*.

I think that here we have a principle that we shall hear much of in the future, for it dominates all work with complex systems. It enters the subject somewhat as the law of conservation of energy enters power engineering. When that law first came in, about a hundred years ago, many engineers thought of it as a disappointment, for it stopped all hopes of perpetual motion. Nevertheless, it did in fact lead to the great practical engineering triumphs of the nineteenth century, because it made power engineering more realistic.

I suggest that when the full implications of Shannon's Tenth Theorem are grasped we shall be, first sobered, and then helped, for we shall then be able to focus our activities on the problems that are properly realistic, and actually solvable.

The Future

Here I have completed this bird's-eye survey of the principles that govern the self-organizing system. I hope I have given justification for my belief that these principles, based on the logic of mechanism and on information theory, are now essentially *complete*, in the sense that there is now no area that is grossly mysterious.

Before I end, however, I would like to indicate, very briefly, the directions in which future research seems to me to be most likely to be profitable.

One direction in which I believe a great deal to be readily discoverable is in the discovery of new types of dynamic process. Most of the machine-processes that we know today are very specialized, depending on exactly what parts are used and how they are joined together. But there are systems of more net-like construction in which what happens can only be treated statistically. There are processes here like, for instance, the spread of epidemics, the fluctuations of animal populations over a territory, the spread of wave-like phenomena over a nerve-net. These processes are, in themselves, neither good nor bad, but they exist, with all their curious properties, and doubtless the brain will use them should they be of advantage. What I want to emphasize here is that they often show very surprising and peculiar properties; such as the tendency, in epidemics, for the outbreaks to occur in waves. Such peculiar new properties may be just what some machine designer wants, and that he might otherwise not know how to achieve.

The study of such systems must be essentially statistical, but this does not mean that each system must be individually stochastic. On the contrary, it has recently been shown (Ashby, 1960c) that no system can have greater efficiency than the determinate when acting as a regulator; so, as regulation is the one function that counts biologically, we can expect that natural selection will have made the brain as determinate as possible. It follows that we can confine our interest to the lesser range in which the sample space is over a set of mechanisms each of which is individually determinate.

As a particular case, a type of system that deserves much more thorough investigation is the large system that is built of parts that have many states of equilibrium. Such systems are extremely common in the terrestrial world; they exist all around us, and in fact, intelligence as we know it would be almost impossible otherwise (Ashby, 1960d). This is another way of referring to the system whose variables behave largely as part-functions. I have shown elsewhere (Ashby, 1960a) that such systems tend to show habituation (extinction) and to be able to adapt progressively (Ashby, 1960d). There is reason to believe that some of the well-known but obscure biological phenomena such as conditioning, association, and Jennings' (1906) law of the resolution of physiological states may be more or less simple and direct expressions of the multiplicity of equilibrial states. At the moment I am investigating the possibility that the transfer of "structure," such as that of three-dimensional space, into a dynamic system—the sort of learning that Piaget has specially considered—may be an *automatic* process when the input comes to a system with many equilibria. Be that as it may, there can be little doubt that the study of such systems is likely to reveal a variety of new dynamic processes, giving us dynamic resources not at present available.

A particular type of system with many equilibria is the system whose parts have a high "threshold"—those that tend to stay at some "basic" state unless some function of the input exceeds some value. The general properties of such systems are still largely unknown, although Beurle (1956) has made a most interesting start. They deserve extensive investigation; for, with their basic tendency to develop avalanche-like waves of activity, their dynamic properties are likely to prove exciting and even dramatic. The fact that the mammalian brain uses the property extensively suggests that it may have some peculiar, and useful, property not readily obtainable in any other way.

Reference to the system with many equilibria brings me to the second line of investigation that seems to me to be in the highest degree promising—I refer to the discovery of *the living organism's memory store*: the identification of its physical nature.

At the moment, our knowledge of the living brain is grossly out of balance. With regard to what happens from one millisecond to the next we know a great deal, and many laboratories are working to add yet more detail. But when we ask what happens in the brain from one hour to the next, or from one year to the next, practically nothing is known. Yet it is these longer-term changes that are the really significant ones in human behavior.

It seems to me, therefore, that if there is one thing that is crying out to be investigated it is the physical basis of the brain's memory-stores. There was a time when "memory" was a very vague and metaphysical subject; but those days are gone. "Memory," as a *constraint* holding over events of the past and the present, and a *relation* between them, is today firmly grasped by the logic of mechanism. We know exactly what we mean by it behavioristically and operationally. What we need now is the provision of adequate resources for its investigation. Surely the time has come for the world to be able to find resources for *one* team to go into the matter?

References

1. W. Ross Ashby, The physical origin of adaptation by trial and error, *J. Gen. Psychol.* **32**, pp. 13–25 (1945).
2. W. Ross Ashby, Principles of the self-organizing dynamic system, *J. Gen. Psychol.* **37**, pp. 125–8 (1947).
3. W. Ross Ashby, *An Introduction to Cybernetics*, Wiley, New York, 3rd imp. (1958a).
4. W. Ross Ashby, Requisite variety and its implications for the control of complex systems, *Cybernetica*, **1**, pp. 83–99 (1958b).
5. W. Ross Ashby, The mechanism of habituation. In: *The Mechanization of Thought Processes*. (Natl. Phys. Lab. Symposium No. 10) H.M.S.O., London (1960).
6. W. Ross Ashby, Computers and decision-making, *New Scientist*, **7**, p. 746 (1960b).
7. W. Ross Ashby, The brain as regulator, *Nature, Lond.* **186**, p. 413 (1960c).

8. W. Ross Ashby, *Design for a Brain: the Origin of Adaptive Behavior*, Wiley, New York, 2nd ed. (1960d).
9. L. von Bertalanffy, An outline of general system theory, *Brit. J. Phil. Sci.* **1**, pp. 134–65 (1950).
10. R. L. Beurle, Properties of a mass of cells capable of regenerating pulses, *Proc. Roy. Soc.* **B240**, pp. 55–94 (1956).
11. W. R. Garner and W. J. McGill, The relation between information and variance analyses, *Psychometrika* **21**, pp. 219–28 (1956).
12. R. C. Jeffrey, Some recent simplifications of the theory of finite automata. Technical Report 219, Research Laboratory of Electronics, Massachusetts Institute of Technology (27 May 1959).
13. H. S. Jennings, *Behavior of the Lower Organisms*, New York (1906).
14. A. J. Lotka, *Elements of Physical Biology*, Williams & Wilkins, Baltimore (1925).
15. J. G. March and J. A. Simon, *Organizations*, Wiley, New York (1958).
16. K. H. Pribram, Fifteenth International Congress of Psychology, Brussels (1957).
17. C. E. Shannon and W. Weaver, *The Mathematical Theory of Communication*, University of Illinois Press, Urbana (1949).
18. G. Sommerhoff, *Analytical Biology*, Oxford University Press, London (1950).

28

Anticipatory Systems in Retrospect and Prospect

Robert Rosen

Introduction

I have come to believe that an understanding of anticipatory systems is crucial, not only for biology, but also for any sphere in which decision making based on planning is involved. These are systems which contain predictive models of themselves and their environment, and employ these models to control their present activities.

As a general rule, a scientific exposition expunges any trace of the genesis and gestation of the ideas being expounded. I have always thought that this omission is a false economy; it is often helpful to understand the context in which ideas are generated and the manner by which they develop. Accordingly, I shall discuss not only the basic problems involved in the systematic study of anticipatory behavior, but also the circumstances in which I originally tried to formulate these problems, and how these problems relate to other important parts of theoretical science. It is my hope that such an approach will suggest the motivations behind them, and thereby make the ideas themselves more transparent.

Center for the Study of Democratic Institutions

The study of anticipatory systems was developed in 1972, when I was in residence as a visiting fellow at the Center for the Study of Democratic Institutions. The Center was a unique institution in many ways, as was its founder and dominating spirit, Robert M. Hutchins. Like Hutchins, it resisted pigeonholing and easy classification. Indeed, as with the Tao, anything one might say about either was certain to be wrong. Despite this I shall try to characterize some of the spirit of the place and of the remarkable man who created it.

From *Gen. Syst. Yearb.* 24, 11. Copyright © 1979 by Society for General Systems Research (now International Society for the Systems Sciences).

The Center's spirit and *modus operandi* revolved around the concept of the Dialog. The Dialog was indispensable in Hutchins' thought because he believed it to be the instrument through which an intellectual community is created. He felt that the reason why an intellectual community is necessary is that it offers the only hope of grasping the whole. The whole, for him, meant nothing less than discovering the means and ends of human society: The real questions to which we seek answers are: What should I do? What should we do? Why should we do these things? What are the purposes of human life and of organized society? The operative word here is *ought*; without a conception of *ought* there could be no guide to politics, which, as he often said, quoting Aristotle, "is architectonic." That is to say, he felt that politics, in the broadest sense, is ultimately the most important thing in the world. Thus for Hutchins the Dialog and politics were inseparable from one another.

For Hutchins, the intellectual community was both means and end. He said:

> The common good of every community belongs to every member of it. The community makes him better because he belongs to it. In political terms the common good is usually defined as peace, order, freedom and justice. These are indispensable to any person, and no person could obtain any one of them in the absence of the community. An intellectual community is one in which everybody does better intellectual work because he belongs to a community of intellectual workers. As I have already intimated, an intellectual community cannot be formed of people who cannot or will not think, who will not think about anything in which the other members of the community are interested. Work that does not require intellectual effort and workers that will not engage in a common intellectual effort have no place in [the intellectual community]. (Hutchins, 1977)

He viewed the Dialog as a continuation of what he called "the great conversation." In his view,

> The great conversation began with the Greeks, the Hebrews, the Hindus and the Chinese, and has continued to the present day. It is a conversation that deals—perhaps more extensively than it deals with anything else—with morals and religion. The questions of the nature and existence of God, the nature and destiny of man, and the organization and purpose of human society are the recurring themes of the great conversation. . . . (Douglas, 1977)

More specifically, regarding the Dialog at the Center, he said,

> Its members talk about what ought to be done. They come to the conference table as citizens, and their talk is about the common good . . . It does not take positions about what ought to be done. It asserts only that the issues it is discussing deserve the attention of citizens. The Center tries to think about the things it believes its fellow citizens ought to be thinking about. (Douglas, 1977)

The Dialog was institutionalized at the Center. Almost every working day at 11:00 A.M. the resident staff would assemble around the large green table to discuss a precirculated paper prepared by one of us, or by an invited visitor. At least once a month, and usually more often, a large-scale conference on a specific topic, organized by one or another of the resident senior fellows, and attended by the best in that field, would be held. Every word of these sessions was recorded, and often found its way into the Center's extensive publication program, through which the Dialog was disseminated to a wider public.

It might be wondered why a natural scientist such as myself was invited to spend a year at an institution of this kind, and even more, why the invitation was accepted. On the face of it, the Center's preoccupations were far removed from natural science. There were no natural scientists among the Center's staff of senior fellows, although several were numbered among the Center's associates and consultants. The resident population, as well as most of the invited visitors, consisted primarily of political scientists, journalists, philosophers, economists, historians, and a full spectrum of other intellectuals. Indeed, Hutchins himself, originally trained in law and preoccupied primarily with the role of education in society, was widely regarded as contemptuous of science and of scientists. Immediately upon assuming the presidency of the University of Chicago, for instance, he became embroiled in a fulminating controversy on curriculum reform, in which many of the faculty regarded his position as anti-scientific, mystical, and authoritarian. At an important conference on science and ethics, he said, "Long experience as a university president has taught me that professors are generally a little worse than other people, and scientists are a little worse than other professors." (*Science and Ethics*, 1953)

However, this kind of sentiment was merely an expression of the well-known Hutchins irony. His basic position had been clearly stated as early as 1931:

> Science is not the collection of facts or the accumulation of data. A discipline does not become scientific merely because its professors have acquired a great deal of information. Facts do not arrange themselves. Facts do not solve problems. I do not wish to be misunderstood. We must get the facts. We must get them all. . . . But at the same time we must raise the question whether facts alone will settle our difficulties for us. And we must raise the question whether . . . the accumulation and distribution of facts is likely to lead us through the mazes of a world whose complications have been produced by the facts we have discovered. (Hutchins, 1931)

Elsewhere, he said,

> The gadgeteers and data collectors, masquerading as scientists, have threatened to become the supreme chieftains of the scholarly world.
>   As the Renaissance could accuse the Middle Ages of being rich in principles and poor in facts, we are now entitled to inquire whether we are not rich in facts and poor in principles.
>   Rational thought is the only basis of education and research. Whether we know it or not, it has been responsible for our scientific success; its absence has been responsible for our bewilderment . . . Facts are the core of an anti-intellectual curriculum.
>   The scholars in a university which is trying to grapple with fundamentals will, I suggest, devote themselves first of all to a rational analysis of the principles of each subject matter. They will seek to establish general propositions under which the facts they gather may be subsumed. I repeat, they would not cease to gather facts, but they would know what facts to look for, what they wanted them for, and what to do with them after they got them. (Hutchins, 1933)

To such sentiments, one could only say amen. In my view, Hutchins was here articulating the essence of science, as I understand it.

Research at the Center

However, I had more specific intellectual reasons for accepting an invitation to spend a year at the Center, as, I think, the Center had for inviting me to do so. It may be helpful to describe them here. My professional activities have been concerned with the theory of biological systems, roughly motivated by trying to discover what it is about certain natural systems that makes us recognize them as organisms and characterize them as being alive. It is precisely on this recognition that biology as an autonomous science depends, and it is a significant fact that it has never been formalized. I am persuaded that our recognition of the living state rests on the perception of homologies between the behaviors exhibited by organisms, homologies which are absent in nonliving systems. The physical structures of organisms play only a minor and secondary role in this. The only requirement which physical structure must fulfill is that it allow the characteristic behaviors themselves to be manifested. Indeed, if this were not so, it would be impossible to understand how a class of systems as utterly diverse in physical structure as that which comprises biological organisms could be recognized as a unity at all. The study of biological organization from this point of view was pioneered by my major professor in my days as a graduate student at the University of Chicago, Nicolas Rashevsky (who, through no coincidence, idolized Robert Hutchins). Rashevsky called this study "relational biology."

The relational approach to organisms is in many ways antithetical to the more familiar analytic experimental approach which has culminated in biochemistry and molecular biology. Particularly in these latter areas, the very first step in any analytic investigation is to destroy the characteristic biological organization possessed by the system under study, leaving a purely physical system to be investigated by standard physical means of fractionation, purification, etc. The essential premise underlying this procedure is that a sufficiently elaborate characterization of structural detail will automatically lead to a functional understanding of behaviors in the intact organism. That this has not yet come to pass is, according to this view, only an indication that more of the same is still needed, and does not indicate a fault in principle. The relational approach, on the other hand, treats as primary that which is discarded first by physico-chemical analysis, i.e., the organization and function of the original system. In relational biology, it is the structural, physical detail of specific systems which is discarded, to be recaptured later in terms of *realizations* of the relational properties held in common by large classes of organisms, if not universally throughout the biosphere. Thus it is perhaps not surprising that the relational approach seems grotesque to analytic biologists, all of whose tools are geared precisely to the recognition of structural details.

In any case, one of the novel consequences of the relational picture is the following: That many, if not all, of the relational properties of organisms can be realized in contexts which are not normally regarded as biological. For instance, they might be realized in chemical contexts which, from a biological standpoint, would be regarded as exotic. This is why relational biology has a bearing on the possibility of extraterrestrial life which is inaccessible to purely empirical approaches. Or they might be realized in technological contexts, which are the province of a presently ill-defined area between engineering and biology often called bionics. Or, what is more germane to the present discussion, they may be realized in the context of human activities, in the form of social, political, and economic systems which determine the character of our social life.

The exploration of this last possibility was, I think, the motivation behind extending an invitation to me to visit the Center, as it was my primary motivation for accepting that invitation. The invitation itself was extended through John Wilkinson, one of the Center's senior fellows, whose interest in the then novel ideas embodied in the structur-

Anticipatory Systems

alism of Levi-Strauss clearly paralleled my own concern with relational approaches in biology.

It is plain, on the face of it, that many tantalizing parallels exist between the processes characteristic of biological organisms and those manifested by social structures or societies. These parallels remain, despite a number of ill-fated attempts to directly extrapolate particular biological principles into the human realm, as embodied, for example, in social Darwinism. Probably their most direct expression is found in the old concept of society as a *superorganism*: that the individuals who make up a society are related to one another as are the constituent cells of a multicellular organism. This idea was explored in greatest detail by the zoologists Alfred Emerson and Thomas Park, who studied insect societies, and who carefully established the striking degree of homology or convergence between social and biological organizations. (Coincidentally, both Emerson and Park were professors of zoology at the University of Chicago.)

What would it mean if common modes of organization could be demonstrated between social and biological structures? It seemed to me that, in addition to the obvious conceptual advantages in being able to effectively relate apparently distinct disciplines, there were a number of most important practical consequences. For instance, our investigation of biological organisms places us almost always in the position of an external observer, attempting to characterize the infinitely rich properties of life entirely from watching their effects without any direct perception of underlying causal structures. Thus, we may watch a cell in a developing organism differentiate, migrate, and ultimately die. We can perceive the roles played by these activities in the generation and maintenance of the total organism. But we cannot directly perceive the causal chains responsible for these various activities, and for the cell's transition or switching from one to another. Without such a knowledge of causal chains, we likewise cannot understand the mechanisms by which the individual behaviors of billions of such cells are integrated into the coherent, adaptive behavior of the single organism which these cells compose.

On the other hand, we are ourselves all *members* of social structures and organizations. We are thus direct participants in the generation and maintenance of these structures, and not external observers. Indeed it is hard for us to conceive what an external observer of our society as a whole would be like. As participants, we know the forces responsible for such improbable aggregations as football games, parades on the Fourth of July, and rush hours in large cities. But how would an external observer account for them?

It is plain that a participant or constituent of such an organization must perceive and respond to signals of which an external observer cannot possibly be aware. Conversely, the external observer can perceive global patterns of behavior which a participant cannot even imagine. Certainly, if we wish to understand the infinitely subtle and intricate processes by which biological organisms maintain and adapt themselves, we need information of both types. Within the purely biological realm, we seem eternally locked into the position of an external observer. But if there were some way to effectively relate biological processes to social ones, and if, more specifically, both biological and social behaviors constituted alternate realizations of a common relational scheme, it might become possible to utilize our social experience as a participant to obtain otherwise inaccessible biological insights. Indeed, this capacity for transferring data and information from a system in which it is easy to obtain to a similar system in which it is hard to obtain is a unique characteristic of the relational approach. This was my basic hope: that I as a theoretical biologist could learn something new about the nature of organisms by judiciously exploiting the cognate properties of social systems.

The other side of that coin was equally obvious: that by exploiting biological experience, obtained from the standpoint of an external observer, we could likewise develop entirely new insights into the properties of our social systems. At that time, however, my detailed knowledge of the human sciences was essentially nil. To explore the possibilities raised above would require what appeared to me to be a major educational effort, and one which at first sight seemed far removed from my own major interests and capabilities.

It was at this point that I perceived the benefits of the community of scholars which Robert Hutchins had created. At the Center I could explore such ideas, while at the same time it was possible for me to learn in the most painless possible fashion how the political scientist, the anthropologist, the historian, and the economist each viewed his own field and its relation to others. In short, the Center seemed to provide me with both the opportunity and the means to explore this virgin territory between biology and society, and to determine whether it was barren or fertile. I thus almost in spite of myself found that I was fulfilling an exhortation of Rashevsky, who had told me years earlier that I would not be a true mathematical biologist until I had concerned myself (as he had) with problems of social organization. At the time, I had dismissed these remarks of Rashevsky with a shrug; but I later discovered (as did many others who tried to shrug Rashevsky off) that he had been right all along.

Thus, I expected to reap a great deal of benefit from my association with the Center. But, as I stressed above, the Center was an intellectual community, and to participate in it, I was expected to contribute to the problems with which the other members of that community were concerned. Initially, it appeared that I would have no tangible contribution to make to such problems as the constitutionalization of the oceans, the role of the presidency, or the press as an institution. Gradually, however, I perceived the common thread running through these issues and the others under intense discussion at the Center, and it was a thread which I might have guessed earlier. As I have noted, Hutchins' great question was: What should we do now? To one degree or another, that was also what the economists, the political scientists, the urban planners, and all the others wanted to know. However different the contexts in which these questions were posed, *they were all alike in their fundamental concern with the making of policy, and the associated notions of forecasting the future and planning for it*. What was sought, in each of these diverse areas, was in effect a technology of decision making. But underlying any technology there must be an underlying substratum of basic principles: a science, a theory. What was the theory underlying a technology of policy generation?

Predictive Models

This was the basic question I posed for myself. It was a question with which I could feel comfortable, and through which I felt I could make an effective, if indirect, contribution to the basic concerns of the Center. Moreover, it was a question with which I myself had had extensive experience, though not in these contexts. For the forecasting of the future is

Anticipatory Systems

perhaps the basic business of theoretical science; in science it is called *prediction*. The vehicle for prediction must, to one degree or another, comprise a *model* for the system under consideration. And the making of models of complex phenomena, as well as the assessment of their meaning and significance, had been my major professional activity for the preceding 15 years. In some very real sense, then, the Center was entirely concerned with the construction and deployment of predictive models, and with the use of these predictive models to regulate and control the behaviors of the systems being modelled. Therefore, the basic theory which must underlie the technologies of policy making in all these diverse disciplines is the theory of modelling: the theory of the relation between a system and a model of that system.

This in itself was a pleasing insight. And it led to some immediate consequences which were also pleasing. For instance: Why did one need to make policies in the first place? It was clear that the major purpose of policy making and of planning was to eliminate or control *conflict*. Indeed, in one form or another, much attention at the Center was devoted to instances of conflict, whether it be between individuals or institutions. That was what law, for example, was all about. In each specific case, it appeared that the roots of conflict lay not so much in any particular objective situation, but rather in the fact that differing *models* of that situation had been adopted by the different parties to the conflict. Consequently, different predictions about that situation were made by these parties, and incompatible courses of action adopted thereby. Therefore, a general theory of policy making (or, as I would argue, a general theory of modelling) would have as a corollary a theory of conflict and, I hoped of conflict resolution.

I proceeded by attempting to integrate these thoughts with my overall program, which as I noted above was to establish homologies between modes of social and biological organization. Accordingly, I cast about for possible biological instances of control of behavior through the utilization of predictive models. To my astonishment, I found them everywhere, at all levels of biological organization. Before going further, it may be helpful to consider a few of these examples.

At the highest level, it is of course clear that a prominent if not overwhelming part of our own everyday behavior is based on the tacit employment of predictive models. To take a transparent example: If I am walking in the woods, and I see a bear appear on the path ahead of me, I will immediately tend to vacate the premises. Why? I would argue: Because I can *foresee* a variety of unpleasant consequences arising from failing to do so. The stimulus for my action is not *just* the sight of the bear, but rather the output of the model through which I predict the consequences of direct interaction with the bear. I thus change my *present* course of action, in accordance with my model's prediction. Or, to put it another way, my present behavior is not simply *reactive*, but rather is *anticipatory*.

Similar examples of anticipatory behavior at the human level can be multiplied without end, and may seem fairly trivial. Perhaps more surprising is the manifestation of similar anticipatory behavior at lower levels, where there is no question of learning or of consciousness. For instance, many primitive organisms are negatively phototropic: they move towards darkness. Now darkness in itself has no physiological significance; in itself it is biologically neutral. However, darkness can be *correlated* with characteristics which are not physiologically neutral, e.g., with moisture or with the absence of sighted predators. The relation between darkness and such positive features composes a *model* through which the organism predicts that by moving toward darkness, it *will* gain an advantage. Of course this is not a conscious decision on the organism's part: the organism has no real option, because the model is, in effect, "wired in." But the fact remains that a negatively phototropic organism changes state in the present in accord with a prediction about the future, made on the basis of a model which associates darkness (a neutral characteristic in itself) with some quality which favors survival.

Another example of such a "wired-in" model may be found in the wintering behavior of deciduous trees. The shedding of leaves and other physiological changes which occur in the autumn are clearly an adaptation to winter conditions. What is the cue for such behavior? It so happens that the cue is not the ambient temperature, but rather is day length. In other words, the tree possesses a model, which *anticipates* low temperature on the basis of a shortening day, regardless of what the present ambient temperature may be. Once again, the adaptive behavior arises because of a wired-in predictive model which associates a shortening day (which in itself is physiologically neutral) with a *future* drop in temperature (which is not physiologically neutral). In retrospect, given the vagaries of weather, we can see that the employment of such a model, rather than a direct temperature response, is the clever thing to do.

A final example, this one at the molecular level, illustrates the same theme. Let us consider a biosynthetic pathway, which we may represent abstractly in the form

$$A_0 \xrightarrow{E_1} A_1 \xrightarrow{E_2} A_2 \xrightarrow{E_3} \ldots \xrightarrow{E_n} A_n$$

in which each metabolite A_1 is the substrate for the enzyme E_1. A common characteristic of such pathways is a *forward activation* step, as for example where the initial substrate A_0 activates the enzyme E_n (i.e., increases its reaction rate). Thus, a sudden increase in the amount of A_0 in the environment will result in a corresponding increase in the activity of E_n. It is clear that the ambient concentration A_0 serves as a *predictor*, which in effect "tells" the enzyme E_n that there *will be* a subsequent increase in the concentration of A_{n-1} of its substrate, and thereby preadapts the pathway so that it will be competent to deal with it. The forward activation step thus embodies a model, which relates the present concentration of A_0 to a subsequent concentration of A_{n-1}, and thereby generates an obviously adaptive response of the entire pathway. It does not take too much imagination to see in this single example the prototype for the elaborate labyrinth of neural and endocrine controls which weld an organism together.

I remarked above that I was astonished to find this profusion of anticipatory behavior at all levels of biological organization. It is important here to understand why I found the situation astonishing, for it bears on the developments to be reported subsequently and raises some crucial epistemological issues.

We have already seen, in the few examples presented above, that an anticipatory behavior is one in which a change of state in the present occurs as a function of some predicted future state, and that the agency through which the prediction is made must be, in the broadest sense, a model. I have also indicated that obvious examples of anticipatory behavior abound in the biosphere at all levels of organization, and that much, if not most, conscious human behavior is also of this character. Further, it is true that organic behaviors at all of these levels have been the subject of incessant scrutiny and theoretical attention for a long time. It might then be expected that such behavior would be well understood, and that there would indeed be an extensive body of theory and of practical experience which could be immediately applied to the problems of forecasting and policy making which dominated the Center's interests. But in fact, nothing could be further from the truth. The surprise was not primarily that there was no such body of theory and experience, but rather that almost no systematic efforts had been made in these directions; and, moreover, almost no one recognized that such an effort was urgently required. In retrospect, the most surprising thing to me was that I myself had not previously recognized such a need, despite my overt concerns with modeling as a fundamental scientific activity, and despite my explicit preoccupation with biological behavior extending over many years. Indeed, I might never have recognized this need, had it not been for the fortuitous chain of circumstances I have described above, which led me to think seriously about apparently remote problems of policy making in a democratic society. Such are the powers of compartmentalization in the human mind.

In fact, the actual situation is somewhat worse than this. At its deepest level, the failure to recognize and understand the nature of anticipatory behavior has not simply been an oversight, but is the necessary consequence of the entire thrust of theoretical science since earliest times; for the basic cornerstone on which our entire scientific enterprise rests is the belief that events are not arbitrary, but obey definite laws which can be discovered. The search for such laws is an expression of our faith in causality. Above all, the development of theoretical physics, from Newton and Maxwell through the present, represents simultaneously the deepest expression and the most persuasive vindication of this faith. Even in quantum mechanics, where the discovery of the uncertainty principle of Heisenberg precipitated a deep reappraisal of causality, there is no abandonment of the notion that microphysical events obey definite laws. The only real novelty is that the quantum laws describe the statistics of classes of events rather than individual elements of such classes.

The temporal laws of physics all take the form of differential equations, in which the rate of change of a physical quantity at any instant is expressed as a definite function of the values of other physical quantities at that instant. Thus, from a knowledge of the values of all the relevant quantities at some initial instant t_0, the values of these quantities at the succeeding instant $t_0 + dt$ are determined. By iterating this process through an integration operation, the values of these quantities, and hence the entire behavior of the system under consideration, may be determined for all time. Carrying this picture to its logical conclusion, Laplace could say:

> An intelligence knowing, at a given instant of time, all forces acting in nature, as well as the momentary position of all things of which the universe consists, would be able to comprehend the motions of the largest bodies of the world as well as the lightest atoms in one single formula . . . To him nothing would be uncertain; both past and future would be present in his eyes. (Weyl, 1949)

This picture of causality and law, arising initially in physics, has been repeatedly generalized, modified, and extended over the years, but the basic pattern remains identifiable throughout. And one fundamental feature of this picture has remained entirely intact, indeed, itself elevated to the status of a natural law. That feature is the following: *In any law governing a natural system, it is forbidden to allow present change of state to depend upon future states*. It is perfectly clear from the above discussion why such a commandment is natural, and why its violation would appear tantamount to a denial of causality in the natural world.

A denial of causality thus appears as an attack on the ultimate basis of which science itself rests. This is also the reason why arguments from final causes have been excluded from science. In the Aristotelian parlance, a final cause is one which involves a purpose or goal. The explanation of systems behavior in terms of final causes is the providence of *teleology*. As we shall see, the concept of an anticipatory system has nothing much to do with teleology. Nevertheless, the imperative to avoid even the remotest appearance of telic explanation in science is so strong that all modes of system analysis conventionally exclude the possibility of anticipatory behavior from the very outset.

And yet, let us consider the behavior of a system which contains a predictive model, and which can utilize the predictions of its model to modify its present behavior. Let us suppose further that the model is a "good" model; that its predictions approximate future events with a sufficiently high degree of accuracy. It is clear that such a system will behave *as if it were* a true anticipatory system, i.e., a system in which present change of state does depend on future states. In the deepest sense, it is evident that this kind of system will not in fact violate our notions of causality in any way, nor need it involve any kind of teleology. But since we explicitly forbid present change of state to depend on future states, we will be driven to understand the behavior of such a system in a purely *reactive* mode, i.e., one in which present change of state depends only on present and past states.

This is indeed what has happened in attempting to come to grips theoretically and practically with biological behavior. Without exception (in my experience), all models and theories of biological systems are reactive in the above sense. As such, we have seen that they *necessarily* exclude all possibility of dealing directly with the properties of anticipatory behavior of the type I have been discussing.

How is it, then, that the ubiquity of anticipatory behaviors in biology could have been overlooked for so long? Should it not have been evident that the "reactive paradigm," as we may call it, was grossly deficient in dealing with systems of this kind? To this question there are two answers. The first is that many scientists and philosophers have indeed repeatedly suggested that something fundamental may be missing if we adopt a purely reactive paradigm for consideration of biological phenomena. Unfortunately, these authors have generally been able only imperfectly to articulate their perception, couching it in such terms as "will," "Geist," "élan," "entelechy," and others. This has made it easy to dismiss them as mystical, vitalistic, anthropomorphic, idealistic, or with similar unsavory epithets, and to confound them with teleology.

The other answer lies in the fact that the reactive paradigm is *universal*, in the following important sense. Given any mode of system behavior which can be described sufficiently accurately, *regardless of the manner in which it is generated*, there is a purely reactive system which exhibits precisely this behavior. In other words, any system behavior can be *simulated* by a purely reactive system. It thus might appear that this universality makes the reactive paradigm completely adequate for all scientific explanations, but this does not follow, and in fact is not the case. For instance, the Ptolemaic epicycles are also universal, in the sense that any planetary trajectory can be represented in terms of a sufficiently extensive family of them. The reason that the Copernican scheme was considered superior to the Ptolemaic lies not in the existence of trajectories which cannot be represented by the epicycles, but arises entirely from considerations of *parsimony*, as embodied for instance in Occam's Razor. The universality of the epicycles is nowadays regarded as an extraneous mathematical artifact irrelevant to the underlying physical situation, and it is for this reason that a representation of trajectories in terms of them can only be regarded as simulation, and not as an explanation.

In fact, the universality of the reactive paradigm is not very different in character from the universality of the epicycles. Both modes of universality ultimately arise from the mathematical fact that any function can be approximated arbitrarily closely by functions canonically constructed out of a suitably chosen "basis set" whose members have a special form. Such a basis set may for instance comprise trigonometric functions, as in the familiar Fourier expansion; the polynomials $1, x, x^2, \ldots$ form another familiar basis set. From this it follows that if any kind of system behavior can be described in functional terms, it can also be *generated* by a suitably constructed combination of systems which generate the elements of a basis set, and this entirely within a reactive mode. But it is clear that there is nothing unique about a system so constructed: We can do the same with *any* basis set. All these systems are different from one another, and may be likewise different from the initial system whose behavior we wanted to describe. It is in this sense that we can only speak of simulation, and not of explanation, of our system's behavior in these terms.

Nevertheless, I believe that it is precisely the universality of the reactive paradigm which has played the crucial role in concealing the inadequacy of the paradigm for dealing with anticipatory systems. Indeed, it is clear that if we are confronted with a system which contains a predictive model, and which uses the predictions of that model to generate its behavior, we cannot claim to understand the behavior unless the model itself is taken into account. Moreover, if we wish to construct such a system, we cannot do so entirely within the framework appropriate to the synthesis of purely reactive systems.

On these grounds, I was thus led to the conclusion that an entirely new approach was needed, in which the capability for anticipatory behavior was present from the outset. Such an approach would necessarily include, as its most important component, a comprehensive theory of models and of modeling. I therefore found myself led to an attempt to develop the principles of such an approach, and to describe its relation to other realms of mathematical and scientific investigation.

Anticipatory Behavior

With these and similar considerations in mind, I proceeded to prepare a number of working papers on anticipatory behavior, and the relation of this kind of behavior to the formulation and implementation of policy. Some of these papers were later published in the *International Journal of General Systems*. The first one I prepared was entitled, "Planning, Management, Policies and Strategies: Four Fuzzy Concepts," and it already contained the seeds of the entire approach I developed to deal with these matters. For this reason, and to indicate the context in which I was working at the Center, it may be helpful to cite some of the original material directly. The introductory section began as follows:

> It is fair to say that the mood of those concerned with the problems of contemporary society is apocalyptic. It is widely felt that our social structure is in the midst of crises, certainly serious, and perhaps ultimate. It is further widely felt that the social crises we perceive have arisen primarily because of the anarchic, laissez-faire attitude taken in the past towards science, technology, economics and politics. The viewpoint of most of those who have written on these subjects revolves around the theme that if we allow these anarchies to continue we are lost; indeed, one way to make a name nowadays is to prove, preferably with computer models, that an extrapolation of present practices will

lead to imminent cataclysm. The alternative to anarchy is management; and management implies in turn the systematic implementation of specific plans, programs, policies, and strategies. Thus it is no wonder that the circle of ideas centering around the concept of *planning* plays a dominant role in current thought.

However, it seems that the net effect of the current emphasis on planning has been simply to shift the anarchy we perceive in our social processes into our ideas about the management of these processes. If we consider, for example, the area of "economic development" of the underdeveloped countries (a topic which has been extensively considered by many august bodies), we find (a) that there is no clear idea of what constitutes "development"; (b) that the various definitions employed by those concerned with development are incompatible and contradictory; (c) that even among those who happen to share the same views as to the ends of development, there are similarly incompatible and contradictory views as to the means whereby the end can be attained. Yet in the name of developmental planning, an enormous amount of time, ink, money and even blood is in the process of being spilled. Surely no remedy can be expected if the cure and the disease are indistinguishable.

If it is the case that planning is as anarchic as the social developments it is intended to control, then we must ask whether there is, in some sense, a "plan for planning" or whether we face an infinite and futile anarchic regress. It may seem at first sight that by putting a question in this form we gain nothing. However, what we shall attempt to argue in the present paper is that, in fact, this kind of question is "well-posed" in a scientific sense; that it can be investigated in a rigorous fashion and its consequences explored. Moreover, we would like to argue that, in the process of investigating this question, some useful and potentially applicable insights into planning itself are obtainable. (Rosen, 1974c)

After a brief review of the main technical concepts to be invoked (which was essential for the audience at the Center), I then proposed a specific context in which anticipatory behavior could be concretely discussed:

We are now ready to construct our model world, which will consist of a class of systems of definite structure, involving anticipation in an essential way, and in which the fuzzy terms associated with "planning" can be given a concrete meaning.

Let us suppose that we are given a system S, which shall be the system of interest, and which we shall call the *object system*. S may be an individual organism, or an ecosystem, or a social or economic system. For simplicity we shall suppose that S is an ordinary (i.e., nonanticipatory) dynamical system.

With S we shall associate another dynamical system M, which is in some sense a *model* of S. We require, however, that if the trajectories of S are parameterized by real time, then the corresponding trajectories of M are parameterized by a time variable which goes faster than real time. That is, if S and M are started out at time $t = 0$ in equivalent states, and if (real) time is allowed to run for a fixed interval T, then M will have proceeded further along its trajectory than S. In this way, the behavior of M *predicts* the behavior of S; by looking at the state of M at time T, we get information about the state that S will be in at some time later than T.

We shall now allow M and S to be coupled; i.e., allow them to interact in specific ways. For the present, we shall restrict ourselves to ways in which M may affect S; later we shall introduce another mode of coupling which will allow S to affect M (and which will amount to updating or improving the model system M on the basis of the activity of S). We shall for the present suppose simply that the system M is equipped with a set E of *effectors*, which allow it to operate either on S itself, or on the environmental inputs to S,

Anticipatory Systems

in such a way as to change the dynamical properties of S. We thus have a situation of the type diagrammed in Figure 1. If we put this entire system into a single box, that box will appear to us to be an adaptive system in which prospective future behaviors determine present changes of state. It would be an anticipatory system in the strict sense if M were a perfect model of S (and if the environment were constant or periodic). Since in general M is not a *perfect* model, for reasons to be discussed in Section V below, we shall call the behavior of such systems *quasi-anticipatory*.

We have said that "M sees" into the future of S, because the trajectories of M are parameterized faster than those of S. How is this information to be used to modify the properties of S through the effector system E? There are many ways in which this can be formalized, but the simplest seems to be the following. Let us imagine the state space of S (and hence of M) to be partitioned into regions corresponding to "desirable" and "undesirable" states. As long as the trajectory in M remains in a "desirable" region, no action is taken by M through the effectors E. As soon as the M-trajectory moves into an "undesirable" region (and hence, by inference, we may expect the S-trajectory to move into the corresponding region at some later time, calculable from a knowledge of how the M- and S-trajectories are parameterized) the effector system is activated to change the dynamics of S in such a way as to keep the S-trajectory out of the "undesirable" region.

From this simple picture, a variety of insights into the nature of "planning," "management," "policies," etc., can already be extracted. (Rosen, 1974c)

The structure depicted in Figure 1 possesses properties which relate directly to the generation and implementation of plans; I then proceeded to sketch these properties:

A. *Choice of M.*

The first essential ingredient in the planning process in these systems involves the choice of the model system M. There are many technical matters involved in choosing M, which will be discussed in more detail in Section V below. We wish to point out here that the choice of M involves paradigmatic aspects as well, which color all future aspects of the "planning" process. One simple example of this may suffice. Let us suppose that S is a simple early capitalist economic system. If we adopt a model system which postulates a large set of small independent entrepreneurs, approximately equivalent in productive capability and governed by "market forces," we find that the system S is essentially stable; coalitions are unfavored and any technical innovations will rapidly spread to all competitors. On the other hand, if we adopt a model system M in

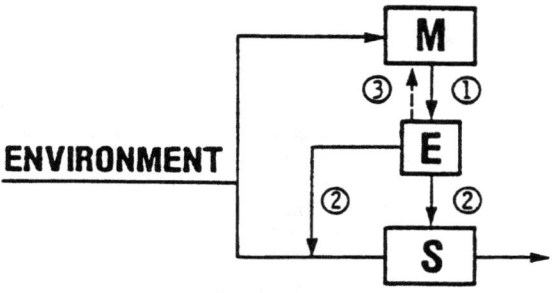

Figure 1

which there are positive feedback loops, then we will see the same situation as *unstable*, much as an emulsion of oil and water is unstable. That is, initially small local accretions of capital will tend to be amplified, and the initially homogeneous population of many small entrepreneurs will ultimately be replaced by a few enormous cartels. This, in a highly oversimplified way, seems to represent the difference between laissez-faire capitalism and Marxian socialism, proceeding from two different model systems of the same initial economic system S, and hence predicting two entirely different futures for S.

B. *Selection of the Effector System E.*
Once the model M has been chosen, the next step of the planning or management process for S is to determine how we are to modify the dynamics of S according to the information we obtain from M. The problem involves several stages. The first stage involves a selection of "steering" variables in S, or in the environment of S, through which the dynamical properties of S can be modified. In general, several different kinds of choices can be made, on a variety of different grounds. In empirical terms, this choice will most often be made in terms of the properties of the model system M; we will consider how M can be most effectively steered, and use the corresponding state variables of S (if possible) for the control of S. Thus again the initial choice of the model system M will again tend to play a major role in determining the specifics of the planning process.

C. *Design of the Effector System E.*
Once having chosen the control variables of S, we must now design a corresponding effector system. This is a technological kind of problem governed by the nature of the control variables of S and their response characteristics. We may wish, for example, to employ only controls which are easily reversible.

D. *Programming of the Effector System E.*
The final aspect of the planning process involves the actual programming of the effector system; i.e., the specification of a dynamics on E which will convert the input information from M (i.e., information about the future state of S) into a specific modification of the dynamics of S. This transduction can be accomplished in many ways, and involves a mixture of "strategic" and "tactical" considerations.

E. *Identification of "Desirable" and "Undesirable" Regions.*
Ultimately the programming of the effectors E will depend heavily on the character of the regions we consider "desirable" and those we consider "undesirable." This choice too is arbitrary, and is in fact independent of the model system M which we have chosen. It represents a kind of constraint added from the outside, and it enters into the planning process in an equally weighty fashion as does the model M and the effector system E.

F. *Updating the States of M.*
In Figure 1 we have included a dotted arrow (labeled (3)) from the effector system back to the model. This is for the purpose of resetting the states of the model, according to the controls which have been exerted on the system S by the effector system. Unless we do this, the model system M becomes useless for predictions about S subsequent to the implementation of controls through E. Thus, the effector system E must be wired into M in a fashion equivalent to its wiring into S.

The enumeration (A)–(F) above seems to be a useful atomization of the planning process for the class of systems we have constructed. Within this class, then, we can proceed further and examine some of the consequences of planning, and in particular the ways in which planning can go wrong. We shall sketch these analyses in the subsequent sections. (Rosen, 1974c)

The notion of "how planning could go wrong" was of course of primary interest to the Center; indeed, for months I had heard a succession of discouraging papers dealing with little else. It seemed to me that by elaborating on this theme I could establish a direct contact between my ruminations and the Center's preoccupations. My preliminary discussion of these matters ended as follows:

> We would like to conjecture further that, for any specific planning situation (involving an object system S, a model M, and suitably programmed effectors E), each of the ways in which planning can go wrong will lead to a particular kind of syndrome in the total system (just as the defect of any part of a sensory mechanism in an organism leads to a particular array of symptoms). It should therefore be possible, in principle, to develop a definite diagnostic procedure to "trouble-shoot" a system of this kind, by mimicking the procedures used in neurology and psychology. Indeed, it is amusing to think that such planning systems are capable of exhibiting syndromes (e.g., of "neurosis") very much like (and indeed analogous to) those manifested by individual organisms. (Rosen, 1974c)

Side-Effects

Such considerations as these led naturally to the general problems connected with system error, malfunction, or breakdown, which have always been hard to formulate, and are still poorly understood. Closest to the surface in this direction, especially in the human realm, were breakdowns arising from the incorporation of incorrect elements into the diagram shown in Figure 1 above: faulty models, inappropriate choice of effectors, etc. I soon realized, however, that there was a more profound aspect of system breakdown, arising from the basic nature of the modelling process itself, and from the character of the system interactions required in the very act of imposing controls. These were initially considered under the heading of *side-effects*, borrowing a medical terminology describing unavoidable and usually unfortunate consequences of employing therapeutic agents (an area which of course represents yet another branch of control theory). As I used the term, I meant it to connote unplanned and unforeseeable consequences on system behavior arising from the implementation of controls designed to accomplish other purposes; or, in a related context, the appearance of unpredicted behavior in a system built in accordance with a particular plan or blueprint. Thus the question was posed: Are such side-effects a *necessary* consequence of control? Or is there room for hope that, with sufficient cleverness, the ideal of the *magic bullet*, the miraculous cure which specifically restores health with no other effect, can actually be attained?

Since this notion of side-effects is so important, let us consider some examples. Of the medical realm we need not speak extensively, except to note that almost every therapeutic agent, as well as most diagnostic agents, create them, sometimes spectacularly so, as in the thalidomide scandal of some years past. We are also familiar with ecological examples, in which man has unwittingly upset "the balance of nature" through injudicious introduction or elimination of species in a particular habitat. Well-known instances of this are the introduction of rabbits into Australia, to give the gentlemen farmers something to hunt on the weekend; or the importation of the mongoose into Santo Domingo, in the belief that because the mongoose kills cobras it would also eliminate local poisonous snakes such as the fer-de-lance. Examples from technology also abound. For instance, we may cite the presence of unsuspected oscillatory modes in the Tacoma Bay Bridge, which ultimately caused it to collapse in a high wind; or the Ohio Turnpike, which was built without curves on the theory that curves are where accidents occur (this led to the discovery of road hypnosis). Norbert Wiener warned darkly of the possibility of similar disastrous side-effects in connection with the perils of relying on computers to implement policy. He analogized this situation to an invocation of magical aids as related in innumerable legends and folk-tales; specifically, such stories as "The Sorcerer's Apprentice," "The Mill Which Ground Nothing but Salt," "The Monkey's Paw," and "The Midas Touch." And of course, many of the social and economic panaceas introduced in the past decades have not only generated such unfortunate side-effects, but have in the long run served to exacerbate the very problems they were intended to control. The ubiquity of these examples, and the dearth of counter-examples, suggests that there is indeed something universal about such behavior, and that it might be important to discover what it is.

My first clumsy attempts to come to grips with this underlying principle, at this early stage, were as follows (the several references to an "earlier paper" in these excerpts refer to a paper written later than, but published before, the one being cited):

> There is, however a class of planning difficulties which do not arise from such obvious considerations, and which merit a fuller discussion. This class of difficulties has to do with the problem of *side effects*; as we shall see, these will generally arise, even if the model system is perfect and the effectors perfectly designed and programmed, because of inherent system-theoretic properties. Let us see how this comes about.
>
> In a previous paper we enunciated a conjecture which I believe to have general validity: namely; that in carrying out any particular functional activity, a system S typically only uses a few of its degrees of freedom. This proposition has several crucial corollaries, of which we noted two in the preceding paper:
>
> 1. The same structure can be involved simultaneously in many different functional activities, and conversely.
>
> 2. The same functional activity can be carried out (or "realized") by many different kinds of structures.
>
> We stressed in that paper how the fact that *all* of the state variables defining any particular system S are more or less strongly *linked* to one another via the equations of motion of the system, taken together with the fact that the many state variables not involved in a particular functional activity were free to interact with other systems in a non-functional or dysfunctional way, implied that any particular functional activity tends to be modified or lost over time. This, we feel, is a most important result, which bears directly on the "planning" process under discussion. The easiest way to see this is to draw another corollary from the fundamental proposition that only a few degrees of freedom of a system S are involved in any particular functional activity of S.
>
> 3. Any functional activity of a system S can be *modelled* by a system whose structure is simple compared to that of S (simply by neglecting the non-functional degrees of freedom of S). Indeed, it is largely because of this property that science is possible at all. Conversely,
>
> 4. No one model is capable of capturing the full potentialities of a system S for interactions with arbitrary systems.

The corollary (4) is true even of the best models, and it is this corollary which bears most directly on the problem of *side-effects*. Let us recall that S is by hypothesis a real system, whereas M is only a model of a particular functional activity of S. There are thus many degrees of freedom of S which are not modelled in M. Even if M is a good model, then, the capability for dealing with the non-functional degrees of freedom in S have necessarily been abstracted away. And these degrees of freedom, which continue to exist in S, are generally *linked* to the degrees of freedom of S which are modelled in M, through the overall equations of motion which govern S.

Now the planning process requires us to construct a real system E, which is to interact with S through a particular subset of the degrees of freedom of S (indeed, through a subset of those degrees of freedom of S which are modelled in M). But from our general proposition, only a few of the degrees of freedom of E can be involved in this interaction. Thus both E and S have in general many "non-functional" degrees of freedom, through which other, non-modelled interactions can take place. Because of the linkage of all observables, the actual interaction between E and S specified in the planning process will in general be affected. Therefore, we find that the two following propositions are generally true:

a. An effector system E will in general have other effects on an object system S than those which are planned;

b. The planned modes of interaction between E and S will be modified by these effects.

Both of these propositions describe the kind of thing we usually refer to as *side effects*. As we see, such side effects are unavoidable consequences of the general properties of systems and their interaction. They are by nature unpredictable, and are inherent in the planning process no matter how well that process is technically carried out. As we pointed out in our previous paper, there are a number of ways around this kind of difficulty, which we have partially characterized, but they are only applicable in special circumstances. (Rosen, 1974b)

The basic principle struggling to emerge here is the following: The ultimate seat of the side-effects arising in anticipatory control, and indeed of the entire concept of error or malfunction in system theory as a whole, rests on the discrepancy between the behavior actually exhibited by a natural system, and the corresponding behavior predicted on the basis of a model of that system. For a model is necessarily an *abstraction*, in that, degrees of freedom which are present in the system are absent in the model. In physical terms, the system is *open* to interactions through these degrees of freedom, while the model is necessarily *closed* to such interactions. The discrepancy between system behavior and model behavior is thus a manifestation of the difference between a closed system and an open one. This observation, in fact, provides the key to a rigorous theory of system failure, and of error in general, which is of basic importance for many purposes.

My initial paper on anticipatory systems concluded with several observations which I hoped would be suggestive to my audience. The first was the following: That it was unlikely that side-effects could be removed by simply augmenting the underlying model, or by attempting to control each side-effect separately as it appeared. The reason for this is that both of these strategies face an incipient infinite regress, similar to that pointed out by Gödel in his demonstration of the existence of unprovable propositions within any consistent and sufficiently rich system of axioms. Oddly enough, the possibility of avoiding this infinite regress was not entirely foreclosed. This followed in a surprising way from some of my earliest work in relational biology, which was mentioned earlier:

There are many ramifications of the class of systems developed above, for the purpose of studying the planning process, which deserve somewhat fuller consideration than we have allowed. In this section we shall consider two of them: (a) how we can update and improve the model system M, and the effector system E, on the basis of information about the behavior of S itself and (b) how can we avoid a number of apparent infinite regresses which seem to be inherent in the planning process?

These two apparently separate questions are actually forms of the same question. We can see this as follows. If we are going to improve, say, the model system M, then we must do so by means of a set of effectors E'. These effectors E' must be controlled by information pertaining to the effect of M and S; i.e., by a model system M' of the system (S + M + E). In other words, we must construct for the purpose of updating and improving M a system which looks exactly like Figure 1, except that we replace M by M', E by E', and S by S + M + E. But then we may ask how we can update M'; in this way we see an incipient infinite regress.

There is another infinite regress inherent in the discussion given of side effects in the preceding section. We have seen that the interaction of the effectors E with the object system S typically give rise to effects in S unpredictable in principle from the model system M. However, these effects too, by the basic principle that only a few degrees of freedom of S and E are utilized in such interactions, are capable of being modelled. That is, we can in principle construct a new model system M_1 of the interaction between S and E, which describes interactions not describable in M. If these interactions are unfavorable, we can construct a new set of effectors, say E_1, which will steer the system S away from these side effects. But just as with E, the system S will typically interact with E_1 in ways which are in principle not comprehensible within the models M or M_1; these will require another model M_2 and corresponding new effectors E_2. In this way we see another incipient infinite regress forming. Indeed, this last infinite regress is highly reminiscent of the "technological imperative" which we were warned against by Ellul and many others. Thus the question arises: can such infinite regresses be avoided?

These kinds of questions are well-posed, and can be investigated in system-theoretic terms. We have considered questions like these in a very different connection; namely, under what circumstances is it possible to add a new functional activity to a biological organization like a cell? It turns out that one cannot simply add an arbitrary function and still preserve the organization; we must typically keep adding functions without limit. But under certain circumstances, the process does indeed terminate; the new function is included (though not *just* the new function in general) and the overall organization is manifested in the enlarged system. On the basis of these considerations, I would conjecture that (a) it is possible in principle to avoid the infinite regresses just alluded to in the planning process, and in particular to find ways of updating the model M and the effectors E; (b) *not every way of initiating and implementing a planning process allows us to avoid the infinite regress*. The first conjecture is optimistic; there are ways of avoiding this form of the "technological imperative." The second can be quite pessimistic in reference to our actual society. For if we have in fact embarked on a path for which the infinite regresses cannot be avoided, then we are in serious trouble. Avoiding the infinite regresses means that developmental processes will stop, and that a stable steady-state condition can be reached. Once embarked on a path for which the infinite regresses cannot be avoided, no stable steady-state condition is possible. I do not know which is the case in our own present circumstances, but it should at least be possible to find out.

I hope that the above few remarks on the planning process will provide food for thought for those more competent to investigate such problems than I am. (Rosen, 1974c)

The theoretical principle underlying this analysis of failure in anticipatory control systems is not wholly negative. In fact, it also underlies the phenomena of *emergence* which characterize evolutionary and developmental processes in biology. It may be helpful to cite one more excerpt of a paper originally prepared for the *Center Dialog*, which dealt with this aspect:

> It may perhaps be worth noting at this point that the above phenomenon is responsible for many of the evolutionary properties exhibited by organisms, and many of the developmental characteristics of social organizations. Consider, for example, the problems involved in understanding, e.g., the evolution of a sensory mechanism such as an eye. The eye is a complicated physiological mechanism which conveys no advantage until it actually sees, and it cannot see until it is complicated. It is hard to imagine how one could even get started towards evolving such a structure, however valuable the end-result may be, and this was one of the major kinds of objections raised to Darwinian evolution. The response to this objection is essentially as follows: the proto-eye in its early stages was in fact not involved in the function of seeing, but rather was primarily involved in carrying out some other functional activity, and it was on this other activity that selection could act. If we now suppose that this other activity involved photosensitivity in an initially accidental way (simply because the physical structure of the proto-eye happened to also be photosensitive), it is easy to imagine how selection pressure could successively improve the proto-eye, with its accidental sensory capacity, until actual seeing could begin, and so that selection could begin to act on the eye directly as an eye. When that happened, the original function of the eye was lost or absorbed into other structures, leaving the eye free to evolve exclusively as a sensory organ.
>
> This "Principle of Function Change" is manifested even more clearly by the evolution of the lung as an organ of respiration. Many fish possess swim bladders, a bag of tissue filled with air, as an organ of equilibration. Being a bag of tissue, the swim bladder is vascularized (possesses blood vessels). When air and small blood vessels are in contact, there will necessarily be gas exchange between the blood and the air, and so a respiratory function is incipient in this structure, designed initially for equilibration. It is easy to imagine how successive increases in vascularization of this organ, especially in arid times, could be an advantage, and thus how selection could come to act on this structure as a lung. This Principle of Function Change is thus one of the cornerstones of evolution (and indeed of any kind of adaptive behavior), and it depends essentially on the fact that the same structure is capable of simultaneously manifesting a variety of functions. (Rosen, 1974a)

Thus the basic problem of avoiding infinite regresses in anticipatory control systems could be reformulated as follows: Can we design systems which are proof against a principle of function change? This was the circle of ideas which I was led to place on the table at the Center, and to pursue in subsequent work since then.

Anticipatory Systems: New Questions

For my own part, I continue to believe that the properties of anticipatory systems raise new questions for the scientific enterprise of the most basic and fundamental kind. These questions have led me to reformulate and refocus all of my previous work in the foundations of theoretical biology, and in the relation of biology to the physical and human sciences.

Indeed, there is no aspect of science which can be untouched by so fundamental an activity as a reconsideration of the reactive paradigm itself.

What are the problems which must be solved before the rough circle of ideas described above may be converted into a rigorous theory of anticipatory behavior? The first and most basic of them is simple: What is a model? What is the nature of the relation between two systems which allows us to assert that one of them is a model for the others? The essence of this property is that we may learn something new about a system of interest by studying a different system, i.e., a model. Roughly, the essence of a modelling relation consists of specifying an *encoding* (and a corresponding decoding) of particular system characteristics into corresponding characteristics of another system, in such a way that implication in the model corresponds to causality in the system. Thus in a precise sense a theorem about the model becomes a prediction about the system. When these remarks are rigorously pursued, the result is a general theory of the modelling relation. By itself, it has many important implications, particularly to more general situations of metaphor, to the way in which distinct models of a given system are related to each other, and to the manner in which distinct systems with a common model may be compared.

A most important class of modelling relations arises from dynamical considerations. Dynamic models are the crucial ingredient of predictive control, since these involve time in an essential way. To study them properly requires a careful reinvestigation of the concept of time itself. It turns out that time is itself a complex entity, in the sense that it admits many distinct models or images. For instance, the familiar "time" of particle mechanics, which as Newton said, "flows equably and of itself" represents one particular kind of image of time, which is tacitly *defined* by its differential from the familiar Hamiltonian relations

$$dx_i = \frac{\partial H}{\partial p_i} dt, \qquad dp_i = -\frac{\partial H}{\partial x_i} dt$$

This image of time is vastly different from that involved in statistical processes. In fact, the statistical time required in establishing, say, the Onsager reciprocity relations in statistical thermodynamics is formally incompatible with Hamiltonian time. Still another kind of time is sequential or logical time, which involves a notion of *logical* precedence. The interrelationships between all these different kinds of time are extremely interesting: They enter in a fundamental way into the character of dynamic models and in the nature of the predictions arising from them.

We pointed out earlier that a natural system is always more open than any model of it. This means that, under appropriate circumstances, the behavior predicted by a model will diverge from that actually exhibited by the system. This provides the basis for a theory of error and system failure on the one hand, and for an understanding of emergence on the other. It is crucial to understand this aspect in any comprehensive theory of control based on predictive models.

We may further ask: How does a system generate such predictive models? What is their ontogenesis? Here we may invoke some general ontogenetic principles, based on selection mechanisms, to achieve some understanding of this point. And finally, given a system which employs a predictive model to determine its present behavior, how should we observe the system so as to determine the nature of the model it employs? This last question raises fundamental new questions of importance to the empirical sciences, for it turns out that most of the observational techniques we traditionally employ actually destroy our capability to make such underlying models visible.

Conclusion

Thus it appears that the study of anticipatory behavior in general, and the manner by which we may properly analyze and synthesize anticipatory systems, raises deep new questions for theoretical science, empirical science, and technology. We have only begun to scratch the surface of these questions, and grasp a few of their implications. I hope in the preceding exposition to have motivated an approach to them, and to have indicated why I believe them to be important. Insofar as our own survival as individuals, and as a species, depends on intelligent planning, it is perhaps not too much to say that answering them may play a vital role in determining whether we survive or not.

References

Douglas, J. H. Eulogy for R. M. Hutchins. *Center Magazine*, 1977, 10(5), 23–25.
Hutchins, R. M. *Commencement address*. University of Chicago, 1931.
Hutchins, R. M. *Commencement address*. University of Chicago, 1933.
Hutchins, R. M. The intellectual community. *Center Magazine*, 1977, 10(11), 1–8.
Rosen, R. On the design of stable and reliable institutions. *International Journal of General Systems*, 1974, 1, 61–66. (*a*)
Rosen, R. Temporal aspects of political change. *International Journal of General Systems*, 1974, 1, 93–103. (*b*)
Rosen, R. Planning, management, policies and strategies: Four fuzzy concepts. *International Journal of General Systems*, 1974, 1, 245–252. (*c*)
Science and ethics. Center occasional paper. Center for the Study of Democratic Institutions, 1953.
Weyl, H. *Philosophy of mathematics and natural science*. Princeton, New Jersey: Princeton University Press, 1949, pp. 209–210.

29

Autopoiesis: The Organization of Living Systems, Its Characterization and a Model

F. G. Varela, H. R. Maturana, and R. Uribe

1. Introduction

Notwithstanding their diversity, all living systems must share a common organization which we implicitly recognize calling them "living." At present there is no formulation of this organization, mainly because the great developments of molecular, genetic and evolutionary notions in contemporary biology have led to the overemphasis of isolated components, e.g., to consider reproduction as a necessary feature of the living organization and, hence, not to ask about the organization which makes a living system a whole, autonomous unity that is alive regardless of whether it reproduces or not. As a result, processes that are history dependent (evolution, ontogenesis) and history independent (individual organization) have been confused in the attempt to provide a single mechanistic explanation for phenomena which, although related, are fundamentally distinct.

We assert that reproduction and evolution are not constitutive features of the living organization and that the properties of a unity cannot be accounted for only through accounting for the properties of its components. In contrast, we claim that the living organization can only be characterized unambiguously by specifying the network of interactions of components which constitute a living system as a whole, that is, as a "unity." We also claim that all biological phenomenology, including reproduction and evolution, is secondary to the establishment of this unitary organization. Thus, instead of asking "What are the necessary properties of the components that make a living system possible?" we ask "What is the necessary and sufficient organization for a given system to be a living unity?" In other words, instead of asking what makes a living system reproduce, we ask what is the organization reproduced when a living system gives origin to another living unity? In what follows we shall specify this organization.

From *Bio. Syst.* **5**, 187. Copyright © 1974 by Elsevier Scientific Publishers Ireland, Ltd.

2. Organization

Every unity can be treated either as an unanalyzable whole endowed with constitutive properties which define it as a unity, or else as a complex system that is realized as a unity through its components and their mutual relations. If the latter is the case, a complex system is defined as a unity by the relations between its components which realize the system as a whole, and its properties as a unity are determined by the way this unity is defined, and not by particular properties of its components. It is these relations which define a complex system as a unity and constitute its organization. Accordingly, the same organization may be realized in different systems with different kinds of components as long as these components have the properties which realize the required relations. It is obvious that with respect to their organization such systems are members of the same class, even though with respect to the nature of their components they may be distinct.

3. Autopoietic Organization

It is apparent that we may define classes of systems (classes of unities) whose organization is specifiable in terms of spatial relations between components. This is the case of crystals, different kinds of which are defined only by different matrices of spatial relations. It is also apparent that one may define other classes of systems whose organization is specifiable only in terms of relations between processes generated by the interactions of components, and not by spatial relations between these components. Such is the case of mechanistic systems in general, different kinds of which are defined by different concatenations (relations) of processes. In particular this is the case of living systems whose organization as a subclass of mechanistic systems we wish to specify.

The autopoietic organization is defined as a unity by a network of productions of components which (i) participate recursively in the same network of productions of components which produced these components, and (ii) realize the network of productions as a unity in the space in which the components exist. Consider for example the case of a cell: it is a network of chemical reactions which produce molecules such that (i) through their interactions generate and participate recursively in the same network of reactions which produced them, and (ii) realize the cell as a material unity. Thus the cell as a physical unity, topographically and operationally separable from the background, remains as such only insofar as this organization is continuously realized under permanent turnover of matter, regardless of its changes in form and specificity of its constitutive chemical reactions.

4. Autopoiesis and Allopoiesis

The class of systems that exhibit the autopoietic organization, we shall call autopoietic systems.

Autonomy is the distinctive phenomenology resulting from an autopoietic organization: the realization of the autopoietic organization is the product of its operation. As long as an autopoietic system exists, its organization is invariant; if the network of productions of components which define the organization is disrupted, the unity disintegrates. Thus an autopoietic system has a domain in which it can compensate for perturbations through the realization of its autopoiesis, and in this domain it remains a unity.

In contradistinction, mechanistic systems whose organization is such that they do not produce the components and processes which realize them as unities and, hence, mechanistic systems in which the product of their operation is different from themselves, we call allopoietic. The actual realization of these systems, therefore, is determined by processes which do not enter in their organization. For example, although the ribosome itself is partially composed of components produced by ribosomes, as a unity it is produced by processes other than those which constitute its operation. Allopoietic systems are by constitution non-autonomous insofar as their realization and permanence as unities is not related to their operation.

5. Autopoiesis: The Living Organization

The biological evidence available today clearly shows that living systems belong to the class of autopoietic systems. To prove that the autopoietic organization is the living organization, it is then sufficient to show, on the other hand, that an autopoietic system is a living system. This has been done by showing that for a system to have the phenomenology of a living system it suffices that its organization be autopoietic (Maturana and Varela, 1973).

Presently, however, it should be noticed that in this characterization, reproduction does not enter as a requisite feature of the living organization. In fact, for reproduction to take place there must be a unity to be reproduced: the establishment of the unity is logically and operationally antecedent to its reproduction. In living systems the organization reproduced is the autopoietic organization, and reproduction takes place in the process of autopoiesis; that is, the new unity arises in the realization of the autopoiesis of the old one. Reproduction in a living system is a process of *division* which consists, in principle, of a process of fragmentation of an autopoietic unity with distributed autopoiesis such that the cleavage separates fragments that carry the same autopoietic network of production of components that defined the original unity. Yet, although self-reproduction is not a requisite feature of the living organization, its occurrence in living systems as we know them is a necessary condition for the generation of a historical network of successively generated, not necessarily identical, autopoietic unities, that is, for evolution.

6. A Minimal Case: The Model

We wish to present a simple embodiment of the autopoietic organization. This model is significant in two respects: on the one hand, it permits the observation of the autopoietic organization at work in a system simpler than any known living system, as well as its spontaneous generation from components; on the other hand, it may permit the development of formal tools for the analysis and synthesis of autopoietic systems.

The model consists of a two-dimensional universe where numerous ○ elements ("substrate"), and a few ★ ("catalysts") move randomly in the spaces of a quadratic grid. These elements are endowed with specific properties which determine interactions that may result in the production of other elements ◎ ("links") with properties of their own and also capable of interactions ("bonding"). Let the interactions and transformations be as in Schema I.

Interaction [1] between the catalyst ★ and two substrate elements 2 ○ is responsible for the composition of an unbonded link ◎. These links may be bonded through Interaction [2] which concatenates these bonded links to unbranched chains of ◎. A chain so produced may close upon itself, forming an enclosure which we assume to be penetrable by the ○'s, but not for ★. Disintegration (Interaction [3]) is assumed to be independent of the state of links ◎, i.e., whether they are free or bound, and can be viewed either as a spontaneous decay or as a result of a collision with a substrate element ○.

In order to visualize the dynamics of the system, we show two sequences (Figs. 1 and 2) of successive stages of transformation as they were obtained from the print-out of a computer simulation of this system. (Details of computation are given in the Appendix. To facilitate appreciation of the developments, Figs. 1 and 2 are drawn from the print-outs with change of symbols used in the computations.)

If an ◎-chain closes on itself enclosing an element ★ (Fig. 1), the ◎'s produced within the enclosure by Interaction [1] can replace in the chain, via [2], the elements ◎ that decay as a result of [3] (Fig. 2). In this manner, a unity is produced which constitutes a network of productions of components that generate and participate in the network of productions that produced these components by effectively realizing the network as a distinguishable entity in the universe where the elements exist. Within this universe these systems satisfy the autopoietic organization. In fact, element ★ and elements ○ produce element ◎ in an enclosure formed by a bidimensional chain of ◎'s; as a result the ◎'s produced in the enclosure replace the decaying ◎'s of the boundary, so that the enclosure remains closed for ★ under continuous turnover of elements, and under recursive generation of the network of productions which thus remains invariant (Figs. 1 and 2). This unity cannot be described in geometric terms because it is not defined by the spatial relations of its components. If one stops all the processes of the system at a moment in which ★ is enclosed by the ◎-chain, so

[1] Composition: ★ + 2 ○ → ★ + ◎

[2] Concatenation: ◎−◎−...−◎ + ◎ → ◎−◎⋯...−◎
(Bonding)
$$\underbrace{\hphantom{◎-◎-...-◎}}_{n} \qquad \underbrace{\hphantom{◎-◎-...-◎}}_{n+1}$$
$n = 1, 2, 3, ...$

[3] Disintegration: ◎ → 2 ○

Schema I

Autopoiesis

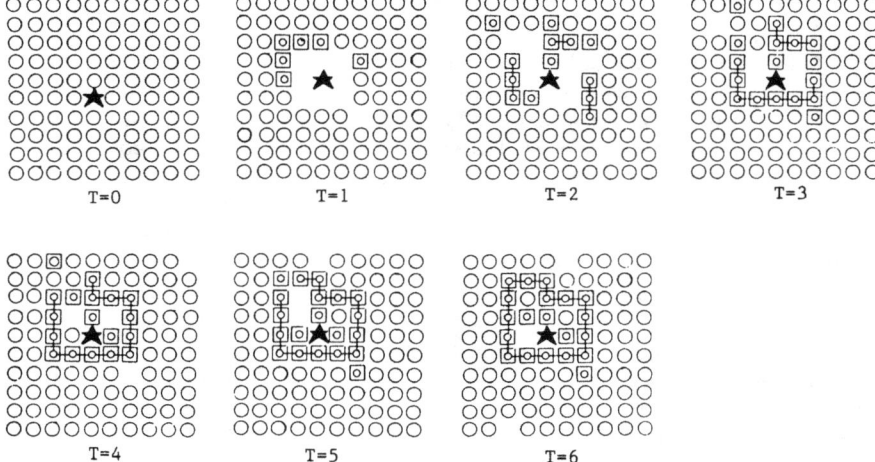

Figure 1. The first seven instants (0→6) of one computer run, showing the spontaneous generation of an autopoietic unity. Interactions between substrate ○ and catalyst ★ produce chains of bonded links ▣, which eventually enclose the catalyst, thus closing a network of interactions which constitutes an autopoietic unity within this universe.

that spatial relations between the components become fixed, one indeed has a system definable in terms of spatial relations, that is, a crystal, but not an autopoietic unity.

It should be apparent from this model that the processes generated by the properties of the components (Schema I) can be concatenated in a number of ways. The autopoietic organization is but one of them, yet it is the one that by definition implies the realization of a dynamic unity. The same components can generate other, allopoietic organizations; for example, a chain which is defined as a sequence of ▣'s, is clearly allopoietic since the production of the components that realize it as a unity do not enter into its definition as a unity. Thus, the autopoietic organization is neither represented nor embodied in Scheme I, as in general no organization is represented or embodied in the properties that realize it.

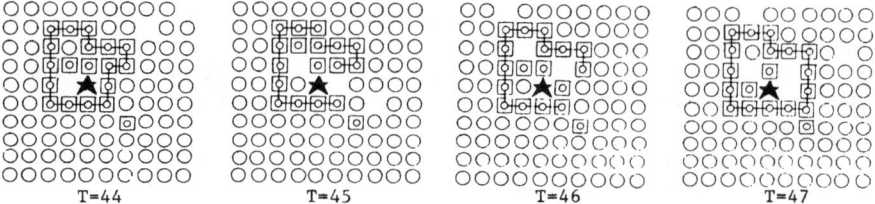

Figure 2. Four successive instants (44–47) along the same computer run (Fig. 1), showing compensation in the boundary broken by spontaneous decay of links. Ongoing production of links reestablishes the unity under changes of form and turnover of components.

7. Tessellation and Molecules

In the case described, as in a broad spectrum of other studies that can generically be called tessellation automata (von Neumann, 1966; Gardner, 1971), the starting point is a generalization of the physical situation. In fact, one defines a space where spatially distinguishable components interact, thus embodying the concatenation of processes which lead to events among the components. This is of course what happens to the molecular domain, where autopoiesis as we know it takes place. For the purpose of explaining and studying the notion of autopoiesis, however, one may take a more general view as we have done here, and revert to the tessellation domain where physical space is replaced by any space (a two-dimensional one in the model), and molecules by entities endowed with some properties. The phenomenology is unchanged in all cases: the autonomous self-maintenance of a unity while its organization remains invariant in time.

It is apparent that in order to have autopoietic systems, the components cannot be simple in their properties. In the present case we required that the components have specificity of interactions, forms of linkage, mobility and decay. None of these properties are dispensable for the formation of this autopoietic system. The necessary feature is the presence of a boundary which is produced by a dynamics such that the boundary creates the conditions required for this dynamics. These properties should provide clues to the kind of molecules we should look for in order to produce an autopoietic system in the molecular domain. We believe that the synthesis of molecular autopoiesis can be attempted at present, as suggested by studies like those on microspheres and liposomes (Fox, 1965; Bangham, 1968) when analyzed in the present framework. For example: a liposome whose membrane lipidic components are produced and/or modified by reactions that take place between its components, only under the conditions of concentration produced within the liposome membrane, would constitute an autopoietic system. No experiments along these lines have been carried out, although they are potential keys for the origin of living systems.

8. Summary

We shall summarize the basic notions that have been developed in this paper:

A. There are mechanistic systems that are defined as unities by a particular organization which we call autopoietic. These systems are different from any other mechanistic system in that the product of their operation as systems thus defined is necessarily always the system itself. If the network of processes that constitutes the autopoietic system is disrupted, the system disintegrates.

B. The phenomenology of an autopoietic system is the phenomenology of autonomy: all changes of state (internal relations) in the system that take place without disintegration are changes in autopoiesis which perpetuate autopoiesis.

C. An autopoietic system arises spontaneously from the interaction of otherwise independent elements when these interactions constitute a spatially contiguous network of productions which manifests itself as a unity in the space of its elements.

D. The properties of the components of an autopoietic system *do not* determine its properties as a unity. The properties of an autopoietic system (as is the case for every system) are determined by the constitution of this unity, and are, in fact, the properties of the *network* created by, and creating, its components. Therefore, to ascribe a determinant value to any component, or to any of its properties, because they seem to be "essential," is a semantic artifice. In other words, all the components, and the components' properties, as well as the circumstances which permit their productive interactions, are necessary when they participate in the realization of an autopoietic network, and none is determinant of the constitution of the network or of its properties as a unity.

9. Key

The following is a six-point key for determining whether or not a given unity is autopoietic:

1. Determine, through interactions, if the unity has identifiable boundaries. If the boundaries can be determined, proceed to 2. If not, the entity is indescribable and we can say nothing.

2. Determine if there are constitutive elements of the unity, that is, components of the unity. If these components can be described, proceed to 3. If not, the unity is an unanalyzable whole and therefore not an autopoietic system.

3. Determine if the unity is a mechanistic system, that is, the component properties are capable of satisfying certain relations that determine in the unity the interactions and transformations of these components. If this is the case, proceed to 4. If not, the unity is not an autopoietic system.

4. Determine if the components that constitute the boundaries of the unity constitute these boundaries through preferential neighborhood relations and interactions between themselves, as determined by their properties in the space of their interactions. If this is not the case, you do not have an autopoietic unity because you are determining its boundaries, not the unity itself. If 4 is the case, however, proceed to 5.

5. Determine if the components of the boundaries of the unity are produced by the interactions of the components of the unity, either by transformation of previously produced components, or by transformations and/or coupling of non-component elements that enter the unity through its boundaries. If not, you do not have an autopoietic unity; if yes, proceed to 6.

6. If all the other components of the unity are also produced by the interactions of its components as in 5, and if those which are not produced by the interactions of other components participate as necessary permanent constitutive components in the production of other components, *you have an autopoietic unity in the space in which its components exist*. If this is not the case and there are components in the unity not produced by components of the unity as in 5, or if there are components of the unity which do not participate in the production of other components, you do not have an autopoietic unity.

Acknowledgments

The authors wish to express their gratitude to the members of the Biological Computer Laboratory of the University of Illinois, Urbana, particularly to Richard Howe, Heinz Von

Foerster, Paul E. Weston and Kenneth L. Wilson, for their continuous encouragement, discussions, and help in clarifying and sharpening the presentation of our notions.

References

Bangham, D.D., 1968, Membrane models with phospholipids, *Progr. Biophys, Mol. Biol.*, **18**, 29.
Fox, S., 1965, A theory of macromolecular and cellular origins, *Nature*, **205**, 328.
Gardner, M., 1971, On cellular automata, self-reproduction, the Garden of Eden, and the game "life," *Sci. Amer.*, **224**(2), 112.
Maturana, H.R. and F. G. Varela, 1973, De maquinas y seres vivos, (Editorial Universitaria, Santiago).
von Neumann, J., 1966, *The theory of self-reproducing automata*, ed. A. Burks (University of Illinois Press, Urbana).

Appendix

Conventions

We shall use the following alphanumeric symbols to designate the elements referred to earlier:

Substrait: ○ → S
Catalyst: * → K
Link: ◻ → L
Bonded link: ─◻─ → BL

The algorithm has two principal phases concerned, respectively, with the motion of the components over the two dimensional array of positions, and with production and disintegration of the L components out of and back into the substrate S's. The rules by which L components bond to form a boundary complete the algorithm.

The "space" is a rectangular array of points, individually addressable by their row and column positions within the array. In its initial state this space contains one or more catalyst molecules K with *all* remaining positions containing substrate S.

In both the motion and production phases, it is necessary to make random selections among certain sets of positions neighboring the particular point in the space at which the algorithm is being applied. The numbering scheme of Figure 3 is then applied, with location 0 in the figure being identified with the point of application (of course, near the array boundaries, not all of the neighbor locations identified in the figure will actually be found).

Regarding motion, the components are ranked by increasing "mass" as S, L, K. The S's may not displace any other species, and thus are only able to move into "holes" or empty spaces in the grid, though they can pass through a *single* thickness of bonded link BL's to do so. On the other hand the L and K readily displace S's, pushing them into adjacent holes, if these exist, or else exchanging positions with them, thus passing freely through the substrate S. The most massive, K, can similarly displace free L links. However, neither of these can

Autopoiesis

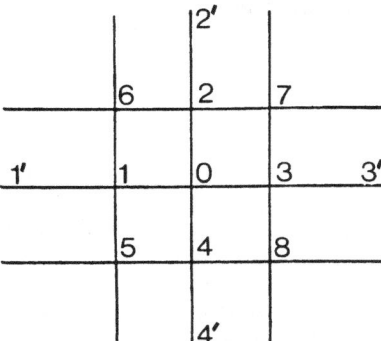

Figure 3. Designation of coordinates of neighboring spaces with reference to a space with designation "0."

pass through a bonded link segment, and are thus effectively contained by a closed membrane. Concatenated L's, forming bonded link segments, are subject to no motions at all.

Regarding production, the initial state contains no bonded links at all; these appear only as the result of formation from substrate S's in the presence of the catalyst. This occurs whenever two adjacent neighboring positions of a catalyst are occupied by S's (e.g., 2 and 7, or 5 and 4 in Fig. 3). Only one L is formed per time step, per catalyst, with multiple possibilities being resolved by random choice. Since two S's are combined to form one L, each such production leaves a new hole in the space, into which S's may diffuse.

The disintegration of L's is applied as a uniform probability of disintegration per time step for each L whether bonded or free, which results in a proportionality between failure rate and size of a chain structure. The sharply limited rate of "repair," which depends upon random motion of S's through the membrane, random production of new L's and random motion to the repair site, makes the disintegration a very powerful controller of the maximum size for a viable boundary structure. A disintegration probability of less than about .01 per time step is required in order to achieve any viable structure at all (these must contain roughly ten L units at least to form a closed structure with any space inside).

Algorithm

1. Motion, first step
   1.1. Form a list of the coordinates of all holes h_i.
   1.2. For each h_i, make a random selection, n_i, in the range 1 through 4, specifying a neighboring location.
   1.3. For each h_i in turn, where possible, move occupant of selected neighboring location in h_i.
   1.31. If the neighbor is a hole or lies outside the space, take no action.

1.32. If the neighbor n_i contains a bonded L, examine the location n'_i. if n'_i contains an S, move this S to h_i.

1.4. Bond any moved L, if possible (Rules, 6).

2. Motion, second step

2.1. Form a list of the coordinates of free L's, m_i.

2.2. For each m_i, make a random selection, n_i, in the range 1 through 4, specifying a neighboring location.

2.3. Where possible, move the L occupying the location m_i into the specified neighboring location.

2.31. If location specified by n_i contains another L, or a K, then take no action.

2.32. If location specified by n_i contains an S, the S will be displaced.

2.321. If there is a hole adjacent to the S, it will move into it. If more than one such hole, select randomly.

2.322. If the S can be moved into a hole by passing through bonded links, as in step 1, then it will do so.

2.323. If the S cannot be moved into a hole, it will exchange locations with the moving L.

2.33. If the location specified by n_i is a hole, then L simply moves into it.

2.4. Bond each moved L, if possible.

3. Motion, third step

3.1. Form a list of the coordinates of all K's, c_i.

3.2. For each c_i, make a random selection n_i, in the range 1 through 4, specifying a neighboring location.

3.3. Where possible, move the K into the selected neighboring location.

3.31. If the location specified by n_i contains a BL or another K, take no action.

3.32. If the location specified by n_i contains a free L, which may be displaced according to the rules of 2.3, then the L will be moved, and the K moved into its place. (Bond the moved L, if possible).

3.33. If the location specified by n_i contains an S, then move the S by the rules of 2.32.

3.34. If the location specified by n_i contains a free L, not movable by rules 2.3, exchange the positions of the K and the L. (Bond L if possible).

3.35. If the location specified by n_i is a hole, the K moves into it.

4. Production

4.1. For each catalyst c_i, form a list of the neighboring positions n_{ij}, which are occupied by S's.

4.11. Delete from the list of n_{ij} all positions for which neither adjacent neighbor position appears in the list (i.e., "1" must be deleted from the list of n_{ij}'s, if neither 5 nor 6 appears, and a "6" must be deleted if neither 1 nor 2 appears).

4.2. For each c_i with a non-null list of n_{ij}, choose randomly one of the n_{ij}, let its value be p_i, and at the corresponding location, replace the S by a free L.

4.21. If the list of n_{ij} contains only one which is adjacent to p_i, then remove the corresponding S.

4.22. If the list of n_{ij} includes both locations adjacent to p_i, randomly select the S to be removed.

4.3. Bond each produced L, if possible.

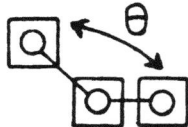

Figure 4. Definition of "Bond-Angle" θ.

5. Disintegration
   5.1. For each L, bonded or unbonded, select a random real number, d, in the range (0,1).
   5.11. If $d \leq P_d$ (P_d an adjustable parameter of the algorithm), then remove the corresponding L, attempt to re-bond (Rules, 7).
   5.12. Otherwise proceed to next L.

6. Bonding
   This step must be given the coordinates of a free L.
   6.1. Form a list of the neighboring positions n_i, which contain free L's, and the neighboring positions m_i, which contain singly bonded L's.
   6.2. Drop from the m_i any which would result in a bond angle less than 90°. (Bond angle is determined as in Figure 4).
   6.3. If there are two or more of the m_i, select two, form the corresponding bonds, and exit.
   6.4. If there is exactly one m_i, form the corresponding bond.
   6.41. Remove from the n_i any which would now result in a bond angle of less than 90°.
   6.42. If there are no n_i, exit.
   6.43. Select one of the n_i, form the bond, and exit.
   6.5. If there are no n_i, exit.
   6.6. Select one of the n_i, form the corresponding bond, and drop it from the list.
   6.61. If the n_i list is non-null, execute steps 6.41 through 6.43.
   6.62. Exit.

7. Rebond
   7.1. Form a list of all neighbor positions m_i occupied by singly bonded L's.
   7.2. Form a second list, p_{ij}, of pairs of the m_i which can be bonded.
   7.3. If there are any p_{ij}, choose a maximal subset and form the bonds. Remove the L's involved from the list m_i.
   7.4. Add to the bond m_i any neighbor locations occupied by free L's.
   7.5. Execute steps 7.1 through 7.3, then exit.

30

The Self-Reproducing System

W. Ross Ashby

High among the interesting phenomena of organization shown by life is that of reproduction. We are naturally led to ask. How can a system reproduce itself? And we go headlong into a semantic trap unless we proceed cautiously. In fact, the answer to the question, "How does the living organism reproduce itself?" is "It doesn't."

No organism reproduces *itself*. The only thing that ever has had such a claim made for it was the phoenix, of which we are told that there was only one, that it laid just one egg in its life, and that out of this egg came itself. What then *actually* happens when ordinary living organisms reproduce? We can describe the events with sufficient accuracy for our purpose here by saying:

1. There is a matrix (a womb, a decaying piece of meat, a bacteriological culture tube perhaps).
2. Into it is introduced a form (an ovum, a fly's egg, a bacterium perhaps).
3. A complex dynamic interaction occurs between the two (in which the form may be quite lost).
4. Eventually the process generates more forms, somewhat like the original one.

In this process we must notice the fundamental part played by the matrix. There is no question here of the ovum reproducing *itself*. What we see is the interaction between one small part of the whole and the remainder of the whole. Thus the outcome is a function of the *interaction* between two systems. The same is true of other forms. The bacterium needs a surrounding matrix which will supply oxygen and food and accept the excretion of CO_2, etc. An interaction between the two then occurs such that forms somewhat resembling the initial bacterium eventually appear.

So, before we start to consider the question of the self-reproducing system we must recognize that *no organism is self-reproducing*. Further, we would do well to appreciate that Rosen[2] has recently shown that the idea of a self-reproducing automaton is logically self-contradictory. He uses an argument formally identical with that used by me[1] to show that a self-organizing system is, strictly, impossible. In each case the idea of a self-acting machine

---

From *Aspects of the Theory of Artificial Intelligence*, C. A. Muses, ed. Copyright © 1962 by Plenum Press.

implies that a mapping must be able to alter itself—i.e., that it is within its own domain. Mathematics and logic can do nothing with such a concept. It is in the same class as the fantasy that can see a man getting behind himself and pushing himself along.

I make these remarks, not in order to confuse or to obstruct, but simply to make sure, by clearing away sources of confusion, that we do really find the right approach to our topic. Though the adjective "self-reproducing" is highly objectionable semantically and logically, it does of course refer to a highly interesting process that we know well, even if we sometimes use inappropriate words to describe it.

I propose, then, to consider the question re-formulated thus:

A given system is such that, if there occurs within it a certain form (or property or pattern or recognizable quality generally), then a dynamic process occurs, involving the whole system, of such a nature that eventually we can recognize, in the system, further forms (or properties or patterns or qualities) closely similar to the original.

I ask what we can say about such systems.

Can a Machine Do It?

Having got the question into its proper form, we can now turn to the question whether a machine can possibly be self-reproducing. In a sense the question is pointless, because we know today that all questions of the type "Can a machine do it?" are to be answered "Yes." Nevertheless, as we are considering self-reproduction, a good deal more remains to be said in regard to the more practical details of the process. Our question then is: Does there exist a mechanism such that it acts like the matrix mentioned, in that, given a "form," the two together lead eventually to the production of other forms resembling the first?

I propose to answer the question largely by a display of actual examples, leaving the examples to speak for themselves.

The first example I would like to give is a formal demonstration in computer-like terms showing the possibility. Let us suppose a computer has only ten stores, numbered 0 to 9, each containing a two-digit decimal number, such as 72, 50, 07, or perhaps 00. The "laws" of this little world are as follows: Suppose it has just acted on store $S - 1$. It moves to store S, takes the two digits in it, a and b say, multiplies them together, adds on 5 and the store-number S, takes the right-hand digit of the result, c say, and then writes the original two digits, a and b, into store c. It then moves on to the next store and repeats the process; and so on indefinitely.

At first sight, this "law" might seem to give just a muddle of numbers. At store No. 3 say, with 17 in the store, it multiplies together 1 and 7, adds 5 to the product, getting 12, adds the store number 3, getting 15, takes the right-hand digit, getting 5, and puts 17 into store 5. It then goes on to its next store, which is No. 4. There seems to be little remarkable in this process. On the other hand, a 28 in a store has a peculiar property. Suppose it is in store 7. $2 \times 8 = 16$, $16 + 5 = 21$, $21 + 7 = 28$, 28 gives 8, so 28 goes into store 8. When we work out the next step we find that 28 goes again into store 9, and so on into store after store. Thus, *once a 28 turns up in the store it spreads until it inhabits all the stores*. Thus the machine, with its program, is a dynamic matrix such that, if a "28" gets into it, the mutual interaction will lead to the production of more 28's. In this matrix, the 28 can be said to be self-reproducing.

Self-Reproducing System 573

The example just given is a formal demonstration of a process that meets the definition, but we can easily find examples that are more commonplace and more like what we find in the real world. Suppose, for instance, we have a number of nearly assembled screw drivers that lack only one screw for their completion. We also have many of the necessary screws. If now a single complete screw driver is provided, it can proceed to make more screw drivers. Thus we have again the basic situation of the matrix in which if one form is supplied a process is generated that results in the production of other examples of the same form.

On this example, the reader may object that a great deal of prefabrication has been postulated. This is true, of course, but it does not invalidate the argument, because the amount of prefabrication that occurs can vary over the widest limits without becoming atypical; and some prefabrication has to be allowed. After all, the living things that reproduce do not start as a gaseous mixture of raw elements.

(The same scale of "degrees of prefabrication" sometimes confuses the issue when a model maker claims that he has "made it all himself." This phrase cannot be taken in any absolute sense. If it were to be taken literally, the model maker would first have to make all the screws that he used, but before that he must have made the metal rods from which the screws were produced, then he must have found the ores out of which the metal was made, and so on. As there is practically no limit to this going backward, the rule that a model maker "must make it all himself" must be accompanied by some essentially arbitrary line stating how much prefabrication is allowed.)

The two examples given so far showed only reproduction at one step. Living organisms repeat reproduction: fathers breed sons, who breed grandsons, who breed great-grandsons, and so on. This possibility of extended reproduction simply depends on the scale of the matrix. It can be present or absent without appreciably affecting the fundamentals of the process.

Further Examples

The subject of self-reproduction is usually discussed on far too restricted a basis of facts. These tend to be on the one hand simply the living organisms, and on the other hand machines of the most rudimentary type, such as the watch and the motor car. In order to give our consideration more range, let us consider some further examples. Those I give below will be found to be sometimes unorthodox but every one of them, I claim, does accord with the basic definition—that the bringing together of the first form and matrix leads to the production of later forms similar to the first.

Example 3. A factory cannot start producing because the power is not switched on. The only thing that can switch the power on is a spanner (wrench) of a certain type. The factory's job is to produce spanners of that type.

Example 4. A machine that vibrates very heavily when it is switched on can be started by a switch that is very easily thrown on by vibration. Such a system, if at rest and then given a heavy vibration, is liable to go on producing further heavy vibrations. Thus, the form "vibration," in this matrix, is self-reproducing.

Example 5. Two countries, A and B, were at war. B discovered that country A was a dictatorship so intense that every document bearing the dictator's initials (X.Y.Z.)

had to be obeyed. Country B took advantage of this and ruined A's administration by bombing A with pieces of paper bearing the message: "Make ten copies of this sheet, with the initials, and send to your associates. X.Y.Z." In such a matrix, such a form is self-reproducing.

Example 6. A number of chameleons are watching one another, each affected by the colors it sees around it. Should one chameleon go dark it will increase the probability of "darkness" appearing around it. In this matrix, the property "darkness" tends to be self-reproducing.

Example 7. In a computer, if the order 0101010 should mean "type 0101010 into five other stores taken at random," then in this matrix the form 0101010 is self-reproducing.

Example 8. A computer has single digit decimal numbers in its various stores. It is programed so that it picks out a pair of numbers at random, multiplies them together, and puts the right-hand digit into the first store. In this condition, as any zero forces another zero to be stored, the zero is self-reproducing.

Example 9. Around any unstable equilibrium, any unit of deviation is apt to be self-reproducing as the trajectory moves further and further away from the point of unstable equilibrium. Thus, if a river in a flat valley happens to be straight, the occurrence of one meander tends to lead to the production of yet other meanders. Thus in this matrix the form "meander" is self-reproducing.

Example 10. A similar example occurs when a ripple occurs in a soft roadway. Under the repeated impact of wheels, the appearance of one tends to lead to the appearance of others. In this matrix, "ripple" is self-reproducing.

Example 11. (Due to Dr. Beurle) A cow prefers to tread down into a hole rather than up onto a ridge. So, if cows go along a path repeatedly, a hollow at one point tends to be followed by excessive wear at one cow's pace further on, and thus by a second hollow. And this tends to be followed by yet another at one pace further on. Thus, in this matrix, "hollow" is self-reproducing.

Example 12. Well known in chemistry is the phenomenon of "autocatalysis." In this class is the dissociation of ethyl acetate (in water) into acetic acid and alcohol. Here, of course, the dissociation is occurring steadily in any case, but the first dissociation that produces the acid increases the *rate* of the later dissociations. So, in this matrix, the appearance of one molecule of acetic acid tends to encourage the appearance of further molecules of the same type.

Example 13. In the previous example the form has been a material entity, but the form may equally well be a pattern. All that is necessary is that the entity, whatever it is, shall be unambiguously recognizable. In a supersaturated solution, for instance, the molecular arrangement that one calls "crystalline" is self-reproducing, in the sense that in this matrix, the introduction of one crystalline form leads to the production of further similar forms.

Example 14. With a community of sufficiently credulous type as matrix, the introduction of one "chain letter" is likely to lead to the production of further such forms.

Example 15. In another community of suitable type as matrix, one person taking up a particular hobby (as form) is likely to be followed by the hobby being taken up by other people.

Example 16. Finally, I can mention the fact that the occurrence of one yawn is likely to be followed by further occurrences of similar forms. In this matrix, the form "yawn" is self-reproducing.

Reproduction as a Specialized Adaptation

After these examples we can now approach the subject more realistically. To see more clearly how special this process of reproduction is, we should appreciate that reproduction is not something that belongs to living organisms by some miraculous linkage, but is simply a specialized means of adaptation of a specialized class of disturbances. The point is that the terrestrial environments that organisms have faced since the dawn of creation have certain specialized properties that are not easily noticed until one contrasts them with the completely nonspecialized processes that can exist inside a computer. Chief among these terrestrial properties is the extremely common rule that if two things are far apart they tend to have practically no effect on one another. Doubtless there are exceptions, but this rule holds over the majority of events. What this means is that when disturbances or dangers come to an organism, they tend to strike locally. Perhaps the clearest example would be seen if the earth had no atmosphere so that the organisms on it were subject to a continuous rain of small shotlike particles traveling at very high speeds. Under such a rain the threat by each particle is local, so that a living form much increases its chances of survival if replicates of the form are made and dispersed. The rule of course is of extremely wide applicability. Banks that may have a fire at one place make copies of their records and disperse them. If a computing machine were liable to sudden faults occurring at random places, there would be advantage in copying off important numbers at various stages in the calculation so as to have dispersed replicates. Thus, the process of reproduction should be seen in its proper relation to other complex dynamic processes as simply a specialized form of adaptation against a special class of disturbances. It is all that and nothing more. Should the disturbances not be localized there is no advantage in reproduction. Suppose, for instance, that the only threat to a species was the arrival of a new virus, that was either overwhelmingly lethal or merely slightly disturbing. Under such conditions the species would gain nothing by having many distinct individuals. The same phenomenon can be seen in industry. If an industry is affected by economic circumstances or by new laws, so that either all the companies in it survive, or all fail, then there is no advantage in the multiplicity of companies; a monopoly can be as well adapted as a multiplicity of small companies.

Fundamental Theory

After this survey we have at least reached a point where we can see "reproduction" in its proper nature in relation to the logic of mechanism. We see it simply as an adaptation to a particular class of disturbances. This means that it is at once subject to the theoretical formulations that Sommerhoff[3] has displayed do decisively. The fact that it is an adaptation means that we are dealing essentially with an invariant of some dynamic process. This means that we can get a new start, appropriate to the new logic of mechanism, that will on the one hand display its inner logic clearly, and on the other hand state the process in a form ready to be taken over by machine programming or in any related process. We start then with the fundamental concept that the dynamic process is properly defined by first naming the set S of states of the system and then the mapping f of that set into itself which corresponds to the dynamic drive of the system. *Reproduction is then one of the invariants that holds over the compound of this system and a set of disturbances that act locally.* If then f is such that some parts within the whole are affected individually, "reproduction" is simply a process by which these parts are invariant under the change-inducing actions of the dynamic drive f.

It must be emphasized that reproduction, though seeming a sharply defined process in living organisms, is really a concept of such generality that precise definition is necessary in all cases if it is to be clear what we are speaking of. Thus, in a sense every state of equilibrium reproduces itself; for if $f(x) = x$, then the processes f of the machine so act on x that at a moment later we have x again. This is exactly the case of the phoenix. It is also "self-reproduction" of a type so basic as to be uninteresting, but this is merely the beginning. It serves as a warning to remind us that processes of self-reproduction can occur, in generalized dynamic systems, in generalized forms that *far exceed in variety and conceptual content anything seen in the biological world.* Because they are nonbiological the biologist will hesitate to call them reproducing, but the logician, having given the definition and being forced to stick to it, can find no reason for denying the title to them. What we have in general is a set of parts, over some few of which a property P is indentifiable. This property P, if the concept is to be useful, must be meaningful at various places over the system. Then we show that "self-reproduction of P" holds in this system if along any trajectory the occurrence of P is followed, at the states later in the trajectory, by their having larger values for the variable "number of P's present."

It should be noted that because self-reproduction is an adaptation, which demands (as Sommerhoff has shown) a relation between organism and environment, and because the property P must be countable in its occurrences over the system, we must be dealing with a system that is seen as composed of parts. I mention this because an important new development in the study of dynamics consists of treating systems actually as a whole, the parts being nowhere considered. This new approach cannot be used in the study of reproduction because, as I have just said, the concept of reproduction demands that we consider the system as composed of parts.

The new point of view which sees reproduction simply as a property that may hold over a trajectory at once shows the proper position of an interesting extension of the concept. Reproduction, as I said, is a form of invariant. In general, invariants are either a state of equilibrium or a cycle. So far, we have considered only the equilibria, but an equally important consideration is the cycle. Here we reach the case that would have to be described by saying that A reproduces B, then B reproduces C, and then C reproduces A. Such a cycle is of course extremely common in the biological world. Not only are there the quite complicated cycles of forms through the egg, pupa, imago, and so on that the insects go through, there is of course also the simple fact that human reproduction itself goes regularly round the cycle: ovum, infant, child, adult, ovum, and so on.

A further clarification of the theory of the subject can be made. Let us define "reproduction" as occurring when the occurrence of a property increases the probability that that property will again occur elsewhere; this of course is positive reproduction. We can just as easily consider "negative" reproduction, when the occurrence of a property decreases the probability that the property will occur elsewhere. Examples of this do not appear to be common. We can of course at once invent such a system on a general-purpose computer; such "negative reproduction" would occur if, say, the instruction 00000 were to mean "replace all zeroes by ones." I have found so far only one example in real systems—namely, if, under electrodeposition, a whisker of metal grows toward the electrode, the chance of another whisker growing nearby is diminished. Thus "whiskers" have a negative net reproduction.

This observation gives us a clear lead on the question: Will self-reproducing forms be common or rare in large dynamic systems? The *negatively* self-reproducing forms clearly have little tendency to be obtrusive—they are automatically self-eliminating. Quite otherwise is it with the positively self-producing forms; for now, if the system contains a single form that is *positively* self-reproducing, that form will press forward toward full occupation of the system.

Suppose now we make the natural assumption that the larger the system, if assembled partly at random, the larger will be the number of forms possible within it. Add to this the fact that if any one is self-reproducing, then self-reproducing forms will fill the system, and we see that there is good reason to support the statement that *all sufficiently large systems will become filled with self-reproducing forms.*

This fact may well dominate the design of large self-organizing systems, forcing the designer to devote much attention to the question: "What self-reproducing forms are likely to develop in my system?" just as designers of dynamic systems today have to devote much attention to the prevention of simple instabilities.

References

1. W. Ross Ashby, *Principles of the Self-organizing System*, Symposium on Self-organizing Systems, University of Illinois, June 7–10, 1960, Pergamon Press, 1962.
2. R. Rosen, "On a Logical Paradox Implicit in the Notion of a Self-reproducing Automaton," *Bull. Math. Biophysics*, Vol. 21, pp. 387–394, 1959.
3. G. Sommerhoff, *Analytical Biology*, Oxford University Press, London, 1950.

31

Universal Principles of Measurement and Language Functions in Evolving Systems

H. H. Pattee

The ability to construct measuring devices and to predict the results of measurements using models expressed in formal mathematical language is now generally accepted as the minimum requirement for any form of scientific theory. The modern cultural development of these skills is usually credited to the Newtonian epoch, although traces go back at least 2000 years to the Milesian philosophers. In any case, from the enormously broader evolutionary perspective, covering well over three billion years, the inventions of measurement and language are commonly regarded as only the most recent and elaborate form of intelligent activity of the most recent and elaborate species.

In this discussion I argue that such a narrow interpretation of measurement and language does not do justice to their primitive epistemological character, and that only by viewing them in an evolutionary context can we appreciate how primitive and universal are the functional principles from which our highly specialized forms of measurement and formal languages arose. I present the view that the generalized functions of language and measurement form a semantically closed loop which is a necessary condition for evolution, and I point out the irreducible complementarity of construction and function for both measuring devices and linguistic strings. Finally, I discuss why current theories of measurement, perception, and language understanding do not satisfy the semantic closure requirement for evolution, and I suggest approaches to designing adaptive systems which may exhibit more evolutionary and learning potential than do existing artificial intelligence models.

My approach is to generalize measurement and linguistic functions by examining both the most highly evolved cognitive systems and the simplest living systems that are known to have the potential to evolve, and abstracting their essential and common measurement and linguistic properties. I want to emphasize that when I speak of molecular language strings and molecular measuring devices I am not constructing a metaphor. Quite the contrary, I mean to show that our most highly evolved languages and measuring devices are only very specialized and largely arbitrary realizations of much simpler and more universal functional principles by which we should define languages and measurements.

Generalized Measurement

The classical scientific concept of measurement requires a distinct physical measuring device that selectively interacts with the system being measured, resulting in output that has a symbolic interpretation, usually numbers. Most scientists regard the output of a number as an essential requirement and, indeed, numbers are required if the language of science is restricted to mathematics. If the laws are expressed by equations of motion, then the initial conditions must be numbers if we are to use the equations to predict other numbers. However, without questioning the enormous advantages of numbers and formal mathematical representations of laws, it is obvious that measurements are possible without numerical outputs (e.g., Nagel, 1932). For example, timing, navigating, surveying, weighing, and even counting were once accomplished by iconic, mimetic, or analog representations. Today the trend is away from the outputs of traditional laboratory measuring devices with visible numerical scales and toward transducers that feed computers and robots directly. In all cases the type of output from a measurement is chosen according to the particular functional requirements of the system as a whole.

The essential point is that while the selection of *input patterns*, the choice of *output actions*, and the *relation* of input to output in any measuring device is largely arbitrary, the only fundamental requirements for useful measurements are the *precision and reproducibility, or local invariances, of the input-output relation*, and the *functional value* of the entire operation to the system doing the measuring. The requirement of reproducibility means that the measuring device must be *isolatable* from the system being measured, and *resettable*, so that the measurement process can be repeated an arbitrary number of times to give the same output for the same input pattern. However, such an abstract description of measurement is incomplete, since it omits the crucial requirement of *system function*, or the *value* of the measurement.

From the abstract definition of measurement alone we would conclude that any relatively fixed or constrained set of particles in a physical system qualifies as a measuring device if we interpret pattern as simply the initial conditions of the free particles and action as the alteration of their free trajectories after collision with the constrained set. Thus, we might say that a rock in a stream maps the input flow pattern to the output action of turbulence, or say that in crystal growth the constraint of a dislocation on a crystal surface maps the patterns of molecular collisions to the specific action of binding more of its own constituents. However, we do not normally call these cases measurement processes. By contrast, the pattern recognition required for specific substrate binding and catalytic action of cellular enzymes I would call a measurement, even by the most rigorous definitions that apply to highly specialized, artificial devices. How do I justify this? Clearly, the enzyme's action is more complicated than crystal growth, but I do not see the level of complexity of the measuring device as the only criterion; for example, calipers are a simple, artificial constraint, that we may use to measure size. The only distinction I find convincing is that of *system function* or, more specifically, that of pattern-action mapping that supports the persistence or survival of the system, and subsidiarily of the measuring constraints that make up the system. In other words, there must be *functional closure*. It is necessary that the enzyme serves a function in the cell for its pattern recognition and catalytic action to be called a measurement. This is still too broad a definition, since it gives no clues as to the characteristics of function, other than survival, that are required of measurement. We must specify some further conditions on this mapping from patterns to actions that are necessary for efficient or effective measurements. Are there also conditions on the way that successful *systems* of measuring devices interact? Let us consider what is common to some extreme examples of successful measuring devices.

Measurement as a Classification

The most important, and yet the most deceptive, aspect of our highly evolved artificial measurements is the feeling we have as intelligent observers that we know what attribute we are measuring independently of the measuring constraints. This is a half-truth. We usually have an abstract concept of what attributes we wish to measure and design the constraints of the measuring device so that its output action expresses these attributes and minimizes all others. Since the output action is designed to be very simple, we often tacitly assume that the corresponding input patterns are very simple. For example, we think of temperature as a simple property of a gas, but our thinking does not change the complex molecular collisions of the gas. This is actually a useful deception in building classical models, although it leads to erroneous results in quantum mechanics. In fact, the measuring device necessarily interacts physically with all of the system's innumerable degrees of freedom, and it is precisely because of the innumerable internal constraints of that particular device that only a few degrees of freedom are available for the output actions.

It is primarily this property of mapping *complex input* patterns to *simple output* actions that distinguishes useful measurement functions from merely complex physical constraints. Without this complex-to-simple or many-to-one mapping process we would not be able to identify equivalence classes of events and, consequently, we would not be able to construct simple models of the world. I would go further and claim that *this classification property of measurement is an epistemological necessity*. Without classification, knowledge of events would not be distinguished from the events themselves, since they would be isomorphic images of each other. This also implies symmetry in time, and measurement must be an irreversible process.

From a broad biological perspective, the entire nervous system has evolved for the principle function of quickly and reliably mapping the ineffably complex configurations and motions of the environment to a very few vital actions; that is, run, fight, eat, sleep, mate, play, etc. Although these actions can be decomposed into complex subroutines, the decision is still which of only a small number of actions to employ. The entire organism can therefore qualify as an extreme case of a generalized measuring device. Let me return now to the other extreme of evolution and consider measuring devices at the molecular level.

At the cellular level we have the example of the single enzyme molecule. The action of an enzyme, like the action of an artificial measuring device, may be described very simply. Generally, it is the catalysis of one particular covalent bond and, consequently, we might think of the corresponding input pattern simply as one particular substrate molecule. But this would miss the essential property of an effective measuring device, which is to reduce the complexity of input interactions by means of its internal constraints. When we speak of an enzyme as highly specific it is another way of saying that it recognizes or distinguishes very complex input patterns.

This ability to recognize complex input patterns and, as a consequence, execute a simple action requires physical constraints of a special type. Since the many-to-one mapping is arbitrary, the constraints must arbitrarily couple the *configurations* available for fitting the input pattern to the *motions* of the device that produces the output actions. In physics these are called nonholonomic or nonintegrable constraints. A holonomic constraint is a restriction on the configurations of a set of particles, such as occurs in forming a crystal from a solution of molecules. This freezing-out of configurational degrees of freedom necessarily freezes-out the corresponding motions of the crystallized molecules, so that we see the constrained system as a rigid solid. A nonholonomic constraint may be defined as a restriction on the motions of the particles *without* a corresponding restriction in the particle configurations. In other words, a formal expression of a nonholonomic constraint appears as a peculiar equation of motion for selected velocity components, where certain configurational variables serve as initial conditions. However, we cannot generally eliminate any configurational variables of the system by using these relations because of the nonintegrability of the equations of constraint. This results in a flexible or allosteric configuration. What we call machines are made up of holonomic, rigid parts that are coupled by nonholonomic, moving linkages. In such machines more configurations of the parts are allowed kinematically than are allowed in the dynamic motions of the parts (e.g., Pattee, 1972b). In proteins it is these nonholonomic constraints that couple the complex configurations or patterns of the substrate to the allosteric motions causing catalytic actions.

The complexity of patterns that can be usefully distinguished clearly depends, in part, on the complexity of the internal constraints of the measuring device that fits the pattern. What is not so clear, but equally important for recognition, is that the output action must be simple and repeatable. In fact, we can imagine a complex fit that requires complex constraints without any corresponding simple action, as in a pile of gravel. We also speak of complex actions resulting from complex constraints, as in the weather. But it is only when complex interactions result in simple, repeatable actions that we speak of recognizing patterns. Enzymes require hundreds of amino acid residues to fold into a structure which we say fits the substrate; but any solid has molecules that physically fit each other just as well, yet we do not generally picture solids as pattern recognizers. It is only the simple catalytic action that establishes the fit interaction as a pattern candidate; but I would again argue that the only objectifiable existence of patterns is ultimately established by some form of system closure. That is, *the distinguishing property of measurement constraints is that their pattern-action mapping supports the system that is necessary to synthesize these constraints.* Since this is such a fundamental condition, let me discuss it in more detail. We shall see that for evolution to be possible, functional closure must be more complex than just autocatalytic cycles.

Function Requires Construction

Returning now to the human level we can say that the primary function of measurement is to map the ineffably complex interactions of the physical world into attributes which are necessary for our survival in this world. To realize this function, it is obviously necessary for us to pay attention to these attributes. This justifies the epistemological illusion of thinking about the world in terms of these measured attributes; that is, in terms of the simple *outputs* of the measuring devices rather than the complex inputs. In the everyday use of observations and measurements there is no survival value in analyzing the inner details of measuring devices. In other words *performance* of measurements does not benefit from *analysis* of the constraints of the measuring device.

In fact, if one analyzes the measurement constraints using a microphysical description, the measurement *function* unavoidably disappears into a measurement-free physical system with more degrees of freedom. On the other hand, *it is from this more detailed physical system that the complex measurement constraints must have been synthesized in the first place*. This means that we must have *control* over physical details of constructing measurement devices even though we do not want or need knowledge of these details while we actually perform measurements. The measurement activity therefore requires both *functional primitives*, in the sense that any analysis of the constraints of the measurement device necessarily obliterates the essential classification action, and *constructional primitives*, in the sense that knowledge of the function of the device can result in no necessary rules for synthesizing the device's constraint.

This apparently improbable interrelation between genes and enzymes is the simplest case of what I call *semantic closure* (Pattee, 1982). By general semantic closure I mean the relation between two primitive constraints, the generalized measurement-type constraints that map complex patterns to simple actions and the generalized linguistic-type constraints that control the sequential construction of the measurement constraints. The relation is semantically closed by the necessity for the linguistic instructions to be read by a set of measuring devices to produce the specified actions or meaning. The semantic closure principle is supportable from several levels:

1. As an empirically based generalization from the facts of molecular biology.
2. As a theoretical requirement based on the logic of heritable systems (e.g., von Neumann, 1966; Polanyi, 1968).
3. As an epistemological condition necessary for the distinction between matter and symbol (Pattee, 1982).

It may also be stated, as a complementarity principle, that the properties of measurement and language cannot be adequately defined individually, but form an irreducible, complementary pair of concepts.

Generalized Language

There is common agreement on many of the universals of language *structure* (e.g., Hockett, 1966). Natural and formal languages are discrete, one-dimensional (1-D) strings of elements from a small alphabet. The strings are further constrained by lexical and syntactic rules which may be very simple or very complex. These rules may be precise and explicit, as in formal languages, or ambiguous and difficult to formulate as in natural languages. Language strings are constructed and read sequentially, although all natural languages also have the essential metalinguistic ability to reference themselves out of sequence; that is, to construct strings that refer to other strings in the language. From what is known of the structure of the gene it appears to qualify fully as a natural language system (e.g., Pattee, 1972a).

When it comes to language *function* it is more difficult to find simple generalizations, let alone common agreement. Language unquestionably has many functions; for example, memory, instruction, communication, modeling, thought, problem-solving, prediction, planning, etc. What I am proposing is not inconsistent with any of these functions. However, my criteria for functions in both measurement and language are based on the most *primitive* conditions for evolvable systems. These include:

1. The ability to construct and coordinate measuring devices and other functional structures under the control of a heritable description (i.e., genetic control).
2. The ability to modify function by changing the description (i.e., mutability).
3. A heritable process for evaluating the description-construction system as a whole (i.e., natural selection).

The impressive techniques of molecular biology have shown us in some detail how present cells accomplish these processes, so in a phenomenological sense they are no longer considered problems by biologists. However, there remain the essential mysteries of how such cellular systems came to exist and how multicellular systems develop. That is, how does such a coordinated set of linguistic instructions and measuring constraints evolve from a nonliving physical world and how are such intricate multicellular morphologies constructed and maintained by these linguistic and measurement devices? I comment on approaches to these problems in the last section but, since I have no solutions, for the present I assume the existence of cells and simply generalize from the structure of linguistic constraints and how they function in the cell system.

We considered the basic function of measurement devices as mapping complex patterns to simple actions. In cells, these actions are typically the catalysis of a single bond; in effect, the smallest change that can be made in the constraints of a molecular system. However, this small change is only made if a complicated set of other constraints is satisfied, namely the recognition of the substrate molecule. In a linguistic device the functions are quite different. The function of the linear sequence is to control the sequence of actions necessary to construct the measuring device, but it does this through the simplest possible type of constraint, the chain of single bonds. At the other extreme of language constraints, we find that one principle function of our spoken and written languages is to give instructions; and it is an impressive fact, often taken for granted, that by forming discrete strings from about 30 types of simple marks we can effect the construction of almost any conceivable pattern, whether it is in the brain, in the actions of the body, or in the construction of artifacts. How these transformations take place from the simple, 1-D string of constraints to the physical structures and actions represented by these strings is almost a total mystery for natural language, even though we find we are able to know the meaning from the strings. But by contrast, at the molecular level we know in great detail how the genetic strings are transformed into the structures and actions represented by the strings, but we have no way of deriving the meaning of any string; that is, of how to tell from the genetic message alone what the function is of the protein it describes.

A generalized language might therefore be characterized as a simple chain of constraints that controls the construction of complex patterns. If we try to formalize this further, as we did with measurement, we might be tempted to say that linguistic devices map a domain of 1-D constraints to a range of n-D patterns. However, this would be a misleading abstraction. In the case of the measuring device, it is the actual constraints of the device itself that recognize the input pattern and are physically responsible for the output action. Therefore, by saying that the device *maps* the input pattern to output action we mean that it is responsible for dynamically *executing* the mapping. On the other hand, a string of constraints in a language is dynamically inactive. Language strings are pure configurations; that is, they have no significant motions or velocity components. Thus, symbol strings are rate independent in the sense that their meaning, or what they control, does not depend on how fast they are read.

Semantic Closure

We explained earlier how the *action* of measurement constraints is functionally primitive, since analysis of the details of the constraints interferes with the measurement function. In a similar way the *meaning* of a linguistic string is functionally primitive, since analysis of the mechanisms of production of the string interferes with the meaning. In practice, we look at the results of a measurement and do not confuse ourselves with the constructional details of the measuring device. Similarly, when we generate linguistic strings we focus on the meaning, not the mechanics of production.

This complementary primitiveness of measurement and linguistic meaning is not only an observable fact of biology but also, I believe, an epistemological necessity. As we said earlier, it is essential that we be able to directly picture the world from what we perceive or from the outputs of measuring devices without having to also know the physical details of the perceiving or measuring constraints as parts of the nonmeasuring interactions of the physical world. This requirement of semantic primitiveness of perception and measurement accounts for what I call the epistemic illusion of the reality of the world, which is not involved with the complex and largely arbitrary constraints that execute perception and measurement. The alternative possibility, that we must analyze these measurement constraints, only leads to irrelevant details at best or an infinite regress at worst. In a complementary sense it is essential that we be able to directly grasp the meaning of linguistic strings without becoming involved with the complex and largely arbitrary details of the constraints that generate and interpret strings. Just as in the case of measurement, this requirement of semantic primitiveness of language accounts for the epistemic illusion that strings have an intrinsic meaning independent of the dynamical constraints that generate them or that they ultimately control. Only through semantic closure do these two primitives complement each other and form an autonomous, evolvable system. The semantic closure principle allows us to treat the *action* of a measuring device as primitive because the details of its construction are accounted for by a linguistic string, while the *meaning* of the linguistic string can be treated as primitive because the details of interpretation are accounted for by a set of measuring devices. For me, it is this fundamental relation between the *relative primitives* of measurement and language constraints that distinguishes evolvable or epistemic systems from normal physical systems. In the final sections I elaborate on why this closure principle offers more promise for models of evolvable systems than other approaches. Before doing this, let me summarize the properties of generalized measurement and language.

Properties of Generalized Measurement

1. Measuring devices are localized, isolatable, resettable structures with repeatable actions.
2. Measuring devices have no intrinsic output actions, but may be triggered to simple actions by specific input patterns (nonholonomic constraints).
3. Measurement constraints obey all physical laws, but are not derivable from laws (generated by system function).
4. Measuring devices are constructed sequentially under the control of linguistic constraints, but a complete, finite set of measuring devices is necessary to read linguistic strings (semantic closure).
5. Measuring devices execute a many-to-one mapping from complex input patterns to simple output actions (classification).
6. Measuring devices do not occur in isolation, but form functional, coherent sets within a system.
7. The value and quality of any measurement is a system property determined by the survival of the system in which it functions.
8. Beyond these properties, the domain of input patterns, the range of output actions, the choice of mapping, and many other aspects of measuring devices are largely arbitrary.

Properties of Generalized Language

1. Language structures are discrete, 1-D strings made up of a small number of types of elements.
2. Language strings have no intrinsic actions, but may trigger action in measuring devices (nonholonomic constraints).
3. Linguistic strings obey all physical laws, but are not derivable from laws (generated by system function).
4. A complete, but finite, set of measuring devices is necessary to read and interpret linguistic strings.
5. Linguistic instructions are necessary to control the synthesis of this interpreting set (semantic closure), as well as the synthesis of other functional components of a system.
6. Language strings are transcribed sequentially, independently of rate, but they may reference themselves out of sequence (metalanguage).

7. The value and meaning of any linguistic string is a system property determined by the survival of the system that it controls.
8. Beyond these properties, the physical structure, the choice of alphabet, the units of meaning, and many other aspects of language strings are largely arbitrary.

Models of Evolution

How can these primitive closure requirements for measurement and language be incorporated into a model of an evolving or learning system? How would such a model differ from previous models? Let us begin with the second question. Many more or less literal simulations of genetically controlled, self-reproducing systems have been studied, beginning with von Neumann's self-reproducing automaton in which he first explicitly recognized the need for a genetic description as well as a universal constructor that must read and execute this description if evolution is to produce increasingly complex systems. However, von Neumann (1966) was more interested in the logical or linguistic aspects of the model than in the physical aspects of pattern recognition and measurement. He was well aware of this neglect of the physical aspects of the problem (". . . one has thrown half the problem out the window and it may be the more important half."), but at that time (ca. 1948) the Turing concept of computation was well-developed, while molecular biology was still a great mystery.

Many later simulations of evolution have been attempted for the purpose of improving the adaptation or optimization process in formal or artificial systems (e.g., Bremermann, 1962; Fogel *et al.*, 1966; Klopf and Gose, 1969; Holland, 1975; Barto, 1984). Only a few models of evolution have been constructed to help conceptualize and test the postulates of neo-Darwinian theory (e.g., Moorehead and Kaplan, 1967; Conrad and Pattee, 1970). In all but one of these models, the process of natural selection is accomplished by fitness criteria which are explicit and preestablished by the programmer. In the Conrad and Pattee model no explicit fitness criteria were introduced. Instead, a set of general conditions or rules of interaction between organism and environment were defined, such as conservation of metabolic resources. However, the nature of the environment with respect to the organisms was preestablished; that is, no genetically modifiable measurement constraints were introduced in this model. Thus, in the existing models of evolution the environment has been represented as a fixed, objective framework that produces the selection pressures on the populations of organisms. Our present complementary view of language and measurement requires the epistemic condition that the organism can only respond directly to the simple output of measurements of the environment. As we have seen, these simple outputs are a consequence of complex constraints resulting from genetically controlled syntheses. However, there is no explicit relation of the gene string to the input-output mapping of the measuring device. Gene strings that construct measuring devices cannot be thought of as programs that manipulate data structures in a computer. In the latter case, every program instruction must be completely explicit. Explicit actions require that all types of inputs, outputs, and hardware operations be preestablished. By contrast, in the organism it is the genetic instructions that construct the hardware that determines all the inputs, outputs, and actions. Genetic consequences are therefore entirely implicit. One cannot assign an element of the gene to an element of action, yet this is the central requirement of a program or effective procedure in computation. Furthermore, simply to say that the architecture of present computers is totally unlike the architecture of organisms is a misleading understatement, since even the concept of architecture plays an entirely different role in organisms to that in computers. For these reasons any form of computational metaphor for organisms must be treated with skepticism.

Up until quite recently the predominant view of genetic control has been very much like the view of computation as an explicit program control of data strings in memory. The alternative view that morphogenesis depends both on autonomous dynamics (archetypes) and internal constraints (chreods) for which genes provide only local switching forces is well known (Waddington, 1968), but for many years lacked empirical evidence and a conceptually clear, formal model. Currently, such topological and dynamical models of morphogenesis are more popular largely because of the application of elegant mathematical formalizations of the singularities, bifurcations, degeneracies, and instabilities of dynamical systems. These mathematical and physical theories of continuous systems arose from completely distinct concepts and methodologies to those of the computational models of morphogenesis, yet they have also led to models for the growth of many types of biological patterns as well as impressive claims for more general powers (e.g., Thom, 1975; Prigogine, 1980; Eigen and Schuster, 1979; Haken, 1981). However, in spite of these significant contributions to mathematical and physical theory, biologists usually perceive the excitement over these formal models as coming more from the physicists and mathematicians who are impressed with the complex patterns that can be generated from such simple equations and boundary conditions. The problem is that molecular genetics is itself so well-established at the foundations of biology that dynamical models are not likely to be useful until they can incorporate these linguistic constraints into their models of evolution and development. While some of these dynamical models have helped clarify measurement constraints (e.g., Prigogine, 1980), none of them has directly contributed to the genotype-phenotype closure relation that is necessary for evolution.

It is also instructive to review current theories of cognitive activities at the other end of the evolutionary scale where the subjects of interest are perception, action, learning, language, knowledge, and other forms of intelligent activity. It is significant that here also we find two opposing schools, one based on explicit linguistic strings and the other on implicit measurement dynamics. The first school arose from logic and computation theory, and is now dominated by the paradigm of the computer as the universal symbol system that can model cognitive tasks such as pattern recognition, classification, learning, and understanding natural language. It is the claim of the computationalists or information processors that these tasks can be understood as purely linguistic or string-processing activities without reference to measurement or any physical dynamics, except as pre-established input and output transducers for the strings. These computational modelers appear to have a principled commitment to the epistemic illusion characteristic of linguistic constraints that strings contain implicate meaningful information, and that by processing these strings with a sufficiently clever rewriting of the rules, this meaning can be explicated (e.g., Newell, 1980; Pylyshyn, 1980). In a somewhat less principled way, the information processors are committed to the complementary epistemic illusion of measurement that only the simple output action need be entered into their models, and that the origin of the complex dynamical constraints that generate these simple outputs need not be considered as a part of their cognitive process.

The opposing school, which arose from the ecological physics approach of J. J. Gibson (1979) takes the other extreme of basing their models on a principled avoidance of linguistic constraints, which they argue are neither essential for mapping perception to action nor for the construction of measurement constraints. Ecological physics models are based on extensions of the dynamical singularity theories of physics (e.g., Turvey and Carello, 1981; Turvey and Kugler, 1984), and understandably emphasize perception-action models rather than genetic control or language understanding.

Both the information processing and the ecological physics schools of cognitive modeling appear to have committed themselves to their exclusive methodological principles without serious consideration of the empirical facts of development and evolution. In effect, the information processors are committed to the principle that discrete strings possess intrinsic meanings independent of the physical dynamics that generate the strings, while the ecological physicists are committed to the principle that physical dynamics possess intrinsic meanings independent of the genetic strings that have constructed the dynamical constraints. One simple, but very frustrating, fact of evolution is that natural selection does not follow the physical or logical principles of most other scientific models, but operates only through opportunistic and even haphazard experiments. Survival depends on balancing many highly interrelated, qualitative system properties such as speed, reliability, efficacy, recovery from error, efficiency, and adaptability. Thus, although it may be technically efficient for us to recognize shape by computation on a string of data obtained by an arbitrary scanning of the shape, the enzyme is much quicker using direct 3-D template recognition with no computation whatsoever; and although it is technically possible to cast a machine from a 3-D template with no string processing, the enzyme is constructed more reliably by sequentially processing a gene string. At the cognitive level why should this opportunistic strategy be different? We can recognize the number of rocks in a pile directly if there are less than 6 or 7, but we must count them sequentially if there are more. In a fraction of a second we directly recognize our complex friends in a crowd, but may have to follow long strings of inductions to identify a simple mineral in a rock. The brain, like the cell, has clearly evolved the power both to directly perceive patterns (measure) and to process strings (compute).

To me, the effort to model the brain as exclusively one or the other type of constraint may be useful engineering—in principle it can be done—but that is not our problem. Our problem with the nervous system is to understand the functional interrelation of direct perceptions and language necessary for efficacious action and learning, just as the problem with the cell is to understand the functional interrelations of gene strings and cellular dynamical constraints necessary for development and evolution. These interrelations are certainly very complex and largely unknown, but what is perhaps the most fundamental evolutionary fact we already know, and that is the meaninglessness of strings or dynamics taken in isolation. From the evolutionary perspective it is only the semantic closure of genotypic language strings and phenotypic measurement dynamics that defines any biological organism in the first place. Whether any physical strings or dynamical constraints can be said to form a language or a measuring device, or whether either has function or meaning can only be decided in terms of its origin and function in the life of the organism.

Conditions for Artificial Evolving Systems

I now come to the question of how this semantic closure property of measurement and language can be incorporated into an artificial system. Although language and measurement are complementary primitives they do not relate symmetrically. We pointed out that measurement constraints are dynamically active without linguistic inputs, even though they may have been constructed under linguistic constraints. Measurement devices physically execute the mapping from input patterns to output actions. This means that a system of measuring constraints, once constructed, can perform complex dynamical tasks without further linguistic control. In other words, the specific actions of measurement systems do not require a program to run them. By contrast, a linguistic constraint has no intrinsic dynamics, it is rate independent, and it therefore can execute no rule or action by itself. Every action of a linguistic system must therefore have an external rule or program step to execute it. In effect this is how computation is defined. Only a string that is mapped into another string *by means of* an effective procedure can qualify as formal computation; but a measurement by itself is not an effective procedure since it has no explicit input. Conrad and Hastings (1985) have proposed naming such direct transformation a new computational primitive, but since there is no explicit input, they must use a nonstandard definition of computation. Gibbsonians often refer to measurement constraints as "smart machines" that accomplish their function without computation, to contrast them with string processing that requires smart programming if any useful output is to result.

I am proposing that any model of an evolutionary process must clearly represent and functionally distinguish language and measurement constraints (i.e., the genotype and phenotype) and must preserve the properties and relations of each. This includes the construction of the measuring devices under the constraints of the linguistic strings and the reading of these strings by measuring devices. It must also include the ability of the strings to gradually or suddenly modify the inputs, outputs, and mappings of the measuring devices and must allow the representation of measuring devices to function under an autonomous dynamics once they have been constructed. This latter condition is difficult to fulfill in an artificial model since the function of a measuring device depends on its interactions with an environment. If we try to simulate the natural environment, the model becomes very complex and yet is incomplete. On the other hand, if we invent too simple an artificial environment, the measurement mapping becomes trivial. The engineering approach is to have the model adapt to the real natural environment, but this requires the construction of real measuring devices under genetic control. This may be practical, but one could question its status as an explanatory model, or even a model at all, since it would appear to be a real evolving system. One more pedagogic-type model might utilize an artificial environment that could be gradually modified in the hope of inducing new measurements by the organism. What are the simplest conditions under which we can expect such emergent behavior?

It appears obvious that the simulation of language constraints on a computer is simpler than the simulation of measurements. However, there is an enormous difference between natural languages and artificial programming languages, which is easily recognized, but not understood. Typically, computer languages do not tolerate mutations or recombinations, whereas genes and natural languages depend on such changes for evolution and creative expression. One difference which may be significant is the lack of complementary measurement constraints in current computer architectures. Since linguistic constraints have no intrinsic dynamics, the computer does nothing unless given a program step. Furthermore, this step must be explicit; that is, the mapping from the domain of program steps to the range of output actions must be unconditionally defined in advance. This total dependency on linguistic inputs results in a total intolerance to the absence of inputs or to inputs with syntactical error. It also follows from the requirement of explicitness in the program steps that errors are also explicit; that is, changes in input-output mappings cannot be gradual. Natural systems, on the other hand, operate with measurement constraints under autonomous dynamics that do not require linguistic inputs for their function. Furthermore, this function depends only implicitly on the linguistic strings that controlled their construction; that is, the mapping from strings to measurement function cannot be specified as a sequence of unconditionally defined steps as in a program. Each linguistic step contributes to the final function only in conjunction with the contributions of other steps so that no single linguistic input step can be assigned an unconditional consequence in the output action. This input-output relation can be observed most directly in the folding transformation that converts the linguistic string constraints of the polypeptide's primary structure into the 3-D globular structure of a functioning enzyme. The significant result of this transformation is that a mutation or recombination of the linguistic string may result in all degrees of functional change, from virtually no change, to gradual or continuous change, to discontinuous change, to a new function. This same variability in meaning occurs in natural language where a single change in a letter or word may result in no change of meaning, a shift of meaning, or an entirely new meaning.

The nature of this relation between description and function or between language and meaning is certainly the most crucial and yet the most puzzling aspect of any epistemic or evolutionary system. It is a problem as old as philosophy and even now it is not clear that a complete explanatory model is possible. My only conclusion from this discussion is that unless an artificial system contains representations of the constraints of both generalized language and generalized measurement, as well as the complementary relations between them that I have described as semantic closure, the model is not likely to evolve similarly to living systems or to contribute significantly to the theory of evolution.

References

Barto, A. (1984). *Simulation Experiments with Goal-seeking Adaptive Elements. Final Report, June 1980–August 1983* (Avionics Lab., Air Force Wright Aeronautical Lab, Wright-Patterson Air Force Base, Ohio 45433).

Bremermann, H. J. (1962). Optimization through evolution and recombination, in M. C. Yovits, G. T. Jacobi, and G. D. Goldstein (Eds.) *Self-Organizing Systems* (Washington, DC: Spartan Books).

Conrad, M. and Hastings, H. M. (1985). Scale change and the emergence of information processing primitives. *J. Theoret. Biol.* (in press).

Conrad, M. and Pattee, H. H. (1970). Evolution experiments with an artificial ecosystem. *J. Theoret. Biol.* **28**:393–409.

Eigen, M. and Schuster, P. (1979). *The Hypercycle: A Principle of Natural Self-Organization* (Heidelberg, Berlin, New York: Springer).

Fogel, L. J., Owens, A. J., and Walsh, M. J. (1966). *Artificial Intelligence Through Simulated Evolution* (New York: Wiley).
Gibson, J. J. (1979). *The Ecological Approach to Visual Perception* (Boston: Houghton-Mifflin).
Haken, H. (1981). Synergetics: is self-organization governed by universal principles?, in E. Jantsch (Ed.) *The Evolutionary Vision*. AAAS Selected Symposium 61 (Boulder, CO: Westview Press).
Hockett, E. (1966). The problem of universals in language, in J. H. Greenberg (Ed.) *Universals of Language* (Cambridge: MIT Press) pp. 1–29.
Holland, J. (1975). *Adaptation in Natural and Artificial Systems* (Ann Arbor, MI: Michigan University Press).
Klopf, A. H. and Gose, E. (1969). An evolutionary pattern recognition network, *IEEE Trans. of Systems Science and Cybernetics* 5:247–50.
Moorehead, P. S. and Kaplan, M. M. (Eds.) (1967). *Mathematical Challenges to the Neo-Darwinian Interpretation of Evolution* (Philadelphia: The Wistar Institute Press).
Nagel, E. (1932). Measurement. Reprinted in Danto, A. and Morganbesser, S. (1960). *Philosophy of Science* (New York: World Publishing Co.).
von Neumann, J. (1966). *Theory of Self-Reproducing Automata*. Edited and completed by A. W. Burks (Urbana, IL: University of Illinois Press).
Newell, A. (1980). Physical symbol systems, *Cognitive Science* 4:135–83.
Pattee, H. (1972a). The nature of hierarchical controls in living matter, in R. Rosen (Ed.) *Foundations of Mathematical Biology*, Vol. 1 (New York: Academic Press) pp. 1–22.
Pattee, H. (1972b). Physical problems of decision-making constraints. *Int. J. Neuroscience* 3:99–106.
Pattee, H. (1982). Cell psychology: an evolutionary approach to the symbol-matter problem. *Cognition and Brain Theory* 5(4):325–341.
Polanyi, M. (1968). Life's irreducible structure. *Science* 160:1308–1312.
Prigogine, J. (1980). *From Being to Becoming: Time and Complexity in the Physical Sciences* (San Francisco: W. H. Freeman & Co).
Pylyshyn, Z. (1980). Computation and cognition: issues in the foundations of cognitive science. *The Behavioral and Brain Sciences* 3:111–169.
Thom, R. (1975). *Structural Stability and Morphogenesis* (Reading, MA: W. A. Benjamin).
Turvey, M. and Carello, C. (1981). Cognition: the view from ecological realism. *Cognition* 10:313–321.
Turvey, M. and Kugler, P. (1984). An ecological approach to perception and action, in H.T.A. Whiting (Ed.) *Human Motor Actions: Burnstein Reassessed* (Amsterdam: North-Holland Publishing Co).
Waddington, C. H. (1968). The basic ideas of biology, in C. H. Waddington (Ed.) *Towards a Theoretical Biology I. Prolegomena* (Edinburgh: Edinburgh University Press) pp. 1–32.

32

The GST Challenge to the Classical Philosophies of Science

Mario Bunge

The great majority of philosophers of science have ignored general systems theories (henceforth GSTs). And those few who have taken notice of GSTs have too often drawn on popularizations and on careless philosophical formulations, and as a result have come to the conclusion that GSTs constitute a new version of the old holistic metaphysics and the old antianalytic epistemology associated with that metaphysics.

This neglect, on the part of philosophers, of the technical literature in the various GSTs, is deplorable for a number of reasons. The main one, however, is that GSTs present a serious challenge to the two most popular philosophies of science, namely empiricism (or inductivism or confirmationism), as represented by the late Rudolf Carnap, and rationalism (or deductivism or refutationism), as championed by Sir Karl Popper.[1] Indeed none of these philosophers ever had GSTs in mind and as a consequence there is no room for GSTs in their philosophies. Worse, according to either of these philosophies, GSTs are nonscientific, for they yield no precise predictions that can be checked by observation or experiment. This sounds odd and even insulting to the practitioners of GSTs, none of whom seem to doubt that what they are doing is science.

This is a serious situation—not for GSTs but for philosophy. However, the situation is not new: actually not even the great scientific theories, praised but not analyzed by the devotees of the standard philosophies of science, conform to the latter. Indeed, those great theories are far too general to be able to yield predictions without further ado, so by themselves they are untestable. The birth of GSTs has simply highlighted a philosophical crisis that existed before—albeit that it was unnoticed except occasionally.

In the face of a serious crisis in a discipline there is but one possible course of action: to examine the presuppositions, to reexamine the object of study, and to make a fresh start no matter which idols may fall. The present paper is addressed to the crisis outlined above. As for the new philosophy of science which the author believes to be called for by current science, it is sketched elsewhere.[2-7]

From *Int. J. Gen. Syst.* **4**, 29. Copyright © 1977 by Gordon and Breach Science Publishers, Ltd.

1. The Classical Paradigm of a Scientific Hypothesis

Our problem is to find out what kind of animals GSTs are and, in particular, whether they are scientific theories. But this task requires that we first answer the question What constitutes a scientific theory? Or, equivalently, What are the necessary and sufficient conditions for a theory to qualify as a Scientific theory? Or again, What is the criterion of scientificity in the case of theories?

Now, theories are systems of statements (propositions, formulas) closed under deduction—i.e., sets such that every statement that can be deduced from any statements in the theory is already contained in the latter. And scientific theories are hypothetical-deductive systems, i.e., self-contained sets of hypotheses, or statements going beyond observation. (Think of classical electrodynamics, or of the selective theory of evolution, or of the theory of social mobility.) We may therefore start by asking the more restricted question What is a scientific hypothesis?

We have all been taught that a hypothesis is scientific to the extent to which it entails empirically testable consequences. But whether it passes the test of observation or of experiment is essential to determining not its scientific character but its truth value. A hypothesis may be false yet scientific, or true yet unscientific. Archimedes' law of the lever, a true scientific hypothesis, can be used to predict that a grandfather or his grandchild will lose balance if they ride a seesaw. On the other hand the hypothesis of telepathy does not allow one to predict anything, whence it is unscientific. And the hypothesis that the reader has at least one enemy is testable and moreover probably true, but hardly of interest to science.

In sum, we have been conditioned to accept:

Definition 1. A theory is *scientific* if and only if it entails empirically testable consequences. In obvious symbols,

$$St =_{df} (\exists e) \quad (e \text{ is empirically testable and } t \vdash e),$$

where $\vdash$ stands for the relation of deducibility or entailment.

Once formulated, all the schools seemed to agree on this definition. The only difficulty was determining what is to be meant by "testable." While the empiricists equated "testable" with "confirmable" (or having possible examples), rationalists equated it with "refutable" (or having possible counterexamples). Thus, whereas empiricists like Carnap and Feigl regarded psychoanalytic hypotheses as scientific because of the many cases that seem to confirm them, rationalists like Popper[8] held that they are not, precisely because it is so easy to confirm them and so hard to refute them. But aside from this divergence both empiricists and rationalists agreed that direct empirical testability is the trademark of scientific hypotheses.

2. A First Crack in the Classical Paradigm

On the basis of Definition 1 inductivists posed the ambitious inverse problem: finding out, given a bit e of empirical evidence, which among the hypotheses entailing e is the

GST Challenge to Classical Philosophies of Science 595

most probable [that is, finding the h which maximizes the conditional probability $\Pr(h|e)$]. Rationalists pointed out that this was the wrong problem to pose, for the likeliest hypotheses are the least bold ones, namely those framed precisely to account for e and nothing else— i.e., the *ad hoc* hypotheses, incapable of undergoing generalization. As a reaction they demanded instead that the least probable hypotheses be preferred. And, just as the inductivists built systems of inductive logic (Carnap[9]) based on the assumption that the more probable a hypothesis the better, so the deductivists constructed theories of corroboration based on the negation of that assumption.

Both empiricists and rationalists turned out to be wrong on two counts. Firstly, they assigned probabilities to statements. It is impossible to do so—except of course by fiat, i.e., conventionally. Facts can be assigned probabilities on condition that they be handled by some stochastic theory involving a random mechanism such as blind shuffling. But propositions are not facts: they cannot be shuffled, particularly if they belong to orderly wholes such as hypothetical-deductive systems. Hence both inductive logic (*pace* von Wright[10]) and the theories of corroboration are empty formalisms. None of them is in a position to make meaningful probability assignments to a formula, say, in quantum mechanics or in the genetics of populations. Both sets of theories are wholly artificial. Scientific research is after not maximally probable (or maximally improbable) hypotheses but maximally true, deep and systematizable ones.

That was not the only flaw in the classical paradigm. The second flaw is even more apparent and it consists in inferring empirical (observational or experimental) consequences from hypotheses alone, which is impossible. It is false that every scientific h implies some e. For example, the ideal gas law does not predict by itself what the exact value of the volume is unless the exact value of the pressure is given. Thus from $pV = k$ and $p = 1$ atmosphere we infer that $V = k$ liters. No empirical information, no prediction. Hence Definition 1 should be replaced by:

Definition 2. A theory is scientific if and only if, jointly with empirical data, it entails empirically testable consequences. I.e.,

$$St =_{df} (\exists d)(\exists e)$$

[d is a datum relevant to t and e is empirically testable and $(t \text{ and } d) \vdash e$].

Such was no fatal objection to the rationalists. On the contrary, it bolstered their view that scientific hypotheses are no mere inductive syntheses: they stated explicitly that scientific hypotheses are *noninstantial*, i.e., do not entail observational cases by themselves. As for the empiricists, they still have to learn that no t entails e's, whence the search for the best t that covers a given e is as illusory as the search for the promised land.

Finally, both refutationists and inductivists have still to face the fact that statements, unlike facts, cannot be assigned probabilities, or that, if they are assigned probabilities by some conventional rule, then such probabilities are subjective (because of the arbitrariness of the assignment rule), hence just as repugnant to the empiricist as to the rationalist.

The upshot so far in our discussion is this. Empiricism is not a viable philosophy of scientific experience, which it largely ignores. And rationalism can be upheld only on condition that it give up all attempts at building probabilistic theories of scientific inference. However, there is still worse to come: the next round will knock out rationalism as well.

3. The Challenge Presented by Generic Scientific Theories

The specific hypotheses of science, such as Snell's law of refraction, certainly qualify as scientific according to Definition 2. How about the extremely high level hypotheses, such as the general principles of classical mechanics, of electrodynamics, or of the selective theory of evolution? They are seldom studied by philosophers, so it should come as no surprise if we find that they do not fit Definition 2 either. A single counterexample to the latter will suffice to ruin it.

Consider the basic equation of motion of classical (i.e., continuum) mechanics, namely

$$\rho \frac{d^2 X}{dt^2} = K + \text{div}\, T,$$

where X and t are the position and time coordinates respectively, ρ the mass density, K the body force density, and T the stress tensor. For putting this equation to empirical tests no data will suffice, not even in the simplest of cases, i.e., that in which $\rho = \text{const}$ and $\text{div}\, T = \text{const}$. Indeed, in this simple case we would still need to know the initial position and the initial velocity of every one of the infinitely many point particles. And in the general case we would have to know, in addition, the exact values of ρ and T at each point in the region of the manifold occupied by the body.

Hence, as it stands, the equation of motion cannot be solved even with the adjunction of a million empirical data. What one does is not to adjoin it data but to make *further hypotheses*, namely, about initial streamlines, about the mass distribution, and about the stress distribution. In other words one builds a conceptual and hypothetical *model* of the body of interest. (Sometimes such a model may be suggested by data, as is the case with photoelasticity. But such suggestions are always imprecise and usually difficult to express mathematically.) Once the equation of motion has been supplemented with such subsidiary hypotheses, it can be integrated to yield the lines of flow. (Even so, chances are that no integration in a closed form is possible unless one makes additional assumptions of a simplifying nature.)

The solutions obtained at the end of the process we have just summarized can finally be contrasted with the empirical evidence—provided the latter has been couched in the language of the theory that is being tested. The same holds, *mutatis mutandis*, for every other scientific theory of the generic as opposed to the specific kind. In all such cases the testing process, far from conforming to Definition 2, comes closer to satisfying:

Definition 3. A theory is scientific if and only if, jointly with subsidiary assumptions and empirical data, it entails empirically testable consequences. I.e.,

$$St =_{df} (\exists s)(\exists d)(\exists e)$$

[s is a subsidiary assumption and d is a datum relevant to t and e is empirically testable and $(t \text{ and } s \text{ and } d) \vdash e$].

Will this refinement do? Not yet: there is still the matter of the compatibility of t with the body of background knowledge. Indeed nobody but a crackpot wastes his time conducting empirical tests on conjectures that collide head-on with the bulk of scientific knowledge—as is the case with, e.g., the hypotheses of teletransportation and of communication with the dead. That is, before anything else one subjects the hypothesis of interest to the test of compatibility with standard scientific theories and rejects it out of hand if it is inconsistent with the best knowledge at hand. Moreover, one requires compatibility with equally firm philosophical hypotheses, such as that no thing is isolated and utterly unknowable.

The upshot is that one more reform in our definition of a scientific conjecture is called for.

Definition 4. A theory is scientific if and only if (i) it is compatible with the bulk of scientific knowledge and (ii) jointly with subsidiary hypotheses and empirical data, it entails empirically testable consequences. I.e., if B stands for the bulk of scientific knowledge and "/" for the relation of incompatibility,

$$St =_{df} \overline{\neg}(B/t) \text{ and } (\exists s)(\exists d)(\exists e)$$

[s is a subsidiary assumption and d is a datum relevant to t and e is empirically testable and (t and s and d)$\vdash e$].

Actually even this refinement is insufficient for coping with standard scientific theories of the generic type, for it makes no provision for the theories and procedures underlying the production of the data d and the fresh empirical evidence e—bits of information which are anything but straightforward. However, Definition 4 will do as a good approximation for our ultimate goal, which is to see how GSTs fare. (For details on the prior theoretical and philosophical tests, as well as for the theoretical underpinnings of high precision empirical tests, see Bunge,[2] Vol. II, Chaps. 14 and 15, and Bunge,[5] Chap. 10.)

4. The Challenge Presented by GSTs

It would seem that all of the standard scientific theories, in particular the high grade ones such as quantum mechanics, conform to Definition 4—if not exactly, at least reasonably well. On the other hand the pseudoscientific conjectures, such as those of telekinesis and of one's fond memories from intrauterine times, do not conform to the definition because they are incompatible with physics, as in the first case, and with neurophysiology, as in the second. (Telekinesis violates energy conservation, and intrauterine memories are impossible because of the lack of maturity of the foetus' nervous system.) So, it seems that Definition 4 allows one to keep in science all that is worth keeping, and to shed all that is not normally regarded as scientific. (The partisans of absolute and permanent scientific revolutions will of course object to the condition (i) of compatibility with the bulk of scientific knowledge. Never mind, for only crackpot theories claim to effect such radical and total revolutions. The compatibility condition, which amounts indirectly to a battery of instant empirical tests, does not prevent genuine scientific revolutions. On the contrary, it makes them testable and credible.)

Have we reached a satisfactory solution to the problem of determining the necessary and sufficient conditions for qualifying a theory as scientific? If this question had been raised before the GST revolution, the answer would have been an unqualified "yes." But the situation has altered radically since World War II, for since then we have reaped a rich crop of theories even more general than the general frameworks of classical mechanics, electrodynamics, or quantum mechanics: theories not just general like the preceding ones but *hypergeneral*. I refer of course to GSTs, such as the statistical theory of information, game theory, control theory, automata theory, and even the general lagrangian framework, the general classical theory of fields, and the general quantum theory of fields (misnamed "axiomatic field theory").

All of these theories are phenomenological, i.e., mechanism-free: they are black box or grey box theories but not translucid box theories. Hence they can describe the behavior of certain systems but not explain how they work. Moreover they are stuff-free—i.e., they make no detailed assumptions concerning the nature of the components of the systems concerned. Whence their extreme generality. They are in fact so general as to be unable to yield any predictions, not even when enriched with empirical data. To convince yourself that this is so, try to predict the behavior of an information system, or of an automaton, without resorting to any extra knowledge concerning their composition, the materials they are made of, and so forth. Clearly, such hypergeneral theories defy the standard philosophies of science even more openly than the standard generic scientific theories such as quantum electrodynamics or the selective theory of evolution.

Take for example the so-called law of requisite variety, a cybernetic analog of Shannon's Theorem 10. A possible formulation of that law might be this: The information-theoretic entropy in the output is at least as great as the excess of the entropy of the external disturbance over the entropy of the control device. Ashby[11] (pp. 208–209) made the shrewd comment that, although this formula does exclude certain events, it has *nothing to fear from experience* for it is independent of the properties of matter. In this case the function of experiment is not to check the theorem but to feed it with data. Moreover, if a given set of empirical data seems to refute this formula of the theory—or any other formula of it—then the indicated course is to redraw the boundary between the system and its environment until a system is circumscribed that conforms to the given formula. For example, if the above law of requisite variety seems to fail because the system is too noisy, then we include noise among the external disturbances and so agreement between theory and fact is restored (pp. 216–217).

Likewise with the theories of information and communication: if a system fails to conform to them, the system is flunked, not the theory, for an information-processing device is, *by definition*, one that fits these theories. In particular, "any device, be it human or electrical or mechanical, must conform to the theory [of communication] if it is to perform the function of communication" (Miller[12]). Consequently the concepts of degree of confirmation, corroboration, and testability are pointless with regard to this theory: the whole of inductive logic is irrelevant to it and so is the refutationist methodology.

Similar considerations apply to automata theory. (See, e.g., Harrison[13] and Bunge,[4] Ch. 8, for an axiomatization of the theory.) This theory supplies a precise definition of a sequential machine and enables one to study machine homomorphisms, behavioral equivalence among machines, the composition of machines, and even the entire lattice of machines. It is not a theory in abstract mathematics, because it concerns a certain genus of concrete system interacting with its environment, although it is totally uncommitted as to the precise nature of either. Any real system that happens to conform to the theory, regardless of its physics and chemistry, will qualify as an automaton. And those concrete systems that do not fit the description just do not qualify.

Some concrete systems may be *forced* to behave as automata, thus providing a cheap confirmation of the theory. For example, a pigeon may be trained to behave like a two state automaton, and, if it fails to learn the trick, the theory is unscathed and the experimenter may have pigeon pie for supper. If no real system is found or built or even thought to be technically feasible, the automata theorist won't be deterred provided he can show that his theoretical automaton is a good model for possible machines. (This is the case with Turing machines, which, being equipped with infinitely long tapes, are strictly speaking unrealizable.)

In sum, while automata theory is *applicable* (by specification) and moreover guides much of advanced engineering design and even some psychological research (Suppes[14]), it is *irrefutable*. It is not even confirmable in the traditional manner of predicting and checking: the theory makes no specific prediction, it prohibits hardly any event, and it suggests no experiments other than *gedankenexperimente*. In short, automata theory, like information theory and every other member of the variegated set of GSTs, is *empirically untestable* in any of the traditional ways discussed in Section 1. The same holds even for some of the applications of GSTs, such as the cybernetic model of the reticular formation or RF: "there is no experiment that could invalidate our claims; our concept [actually a theoretical model] has not yet produced any risky predictions; it does not forbid any measurable RF event; and we have not yet proposed any real alternatives" (Kilmer *et al.*,[15] p. 321).

Now, if GSTs are unable to yield precise predictions even when enriched with subsidiary assumptions and empirical data, then they are not covered by Definition 4 in Section 3. Why then should GSTs be regarded as scientific? But this question deserves a special section.

5. The Methodological Status of GSTs

GSTs, as we have just seen, are neither confirmable nor refutable the way standard scientific theories are. However, they are not dogmas above criticism. In fact GSTs are *corrigible* if not exactly refutable in the light of empirical evidence. To begin with, they can be improved upon formally, i.e., logically or mathematically. For example they can be overhauled and made mathematically more powerful with the explicit use of new mathematical theories. (See, e.g., Klir[16] and Padulo and Arbib.[17]) Or they can be made more complex in an attempt to fit better their intended referents. Thus if the goal is to model a learning system then, since learning is largely a stochastic process, the theory of probabilistic automata may be found more useful than the theory of deterministic automata. In a sense, then, GSTs are *confirmable*.

Surely GSTs are not confirmed in the classical way, i.e., through prediction and empirical checking. Nevertheless they are confirmed, although in a special way, namely by being shown either to fit a whole family of specific theories (i.e., theories concerning specific systems) or to take part in the design of viable systems. The former may be called *conceptual confirmation*, the latter *practical confirmation*, and either kind differs from the usual *empirical confirmation*. Actually all GSTs are confirmed both conceptually and practically without ever being confirmed empirically, i.e., by contrasting their predictions with empirical evidence. Thus general network theory is confirmed conceptually by being shown to capture the traits common to all nets, whether physical or informational. And it is confirmed practically by being used in the design of nets of some kind.

In other words, GSTs are not just generalizations of specific theories, for they can be applied. And, to the extent that they are applied successfully, they are shown to be *suitable and fruitful* without being true, let alone false. (A false dogma, such as racism, can be used for certain purposes.) Moreover, GSTs are doubly confirmable (conceptually and practically) but they can never be falsified. At most they can be shown to be irrelevant to the problem at hand. They either "apply" or they don't.

This does not entail that GSTs can be applied to particular cases, hence tested for usefulness, without further ado. Even though they are nonspecific and therefore empirically untestable, their application to specific situations requires some substantive knowledge of the latter: only this can provide a suitable interpretation of the theory in question. In other words, GSTs are so many *general frameworks that must be adjoined specific models* of the system of interest before they can be of any use. For example, in the case of information theory we must be able to identify at least the sources of information and of noise, the channel, and the receiver; and we must have a code for the system of signals. (Incidentally, because none of these conditions is fulfilled in the case of molecules, the use of information theory in molecular biology is purely metaphorical.) Likewise in the case of cybernetics we must be able to identify the system, its regulator, the environment, and the disturbances originating in the latter; besides, we must know what the goal (or the set of final states) is, for otherwise we won't even suspect what is to be regulated—let alone how to achieve the regulation. In fact, a cybernetic problem looks like this: Given a system—which includes its inputs and outputs—together with its environment and the desired subset of output values (i.e., the goal), design a control system that, coupled to the main system, will keep its output within preassigned bounds. Without all these items of specific information—unobtainable without the help of observations or experiments supported by specific theories—one would be unable to pose the problem, hence to solve it.

In sum every application of a GST calls for the building of a specific model of the system of interest—the model being built, of course, with the concepts of the theory if it is to be coupled to the latter. In other words, a member of the GST class *becomes a specific theory* of the standard type, perhaps a scientific theory, when enriched with specific information concerning the system to which it is applied. And *this* specific theory is of course subject to the strict canons of empirical testability. That is, Definition 4 refers not to any GST, but to specifications or applications of GSTs. (And we know well enough that a large number of such applications do not live up to those standards of scientificity. But this is another story that will have to be told some day.)

The upshot of this section is: GSTs, though neither confirmable nor refutable in the usual way—i.e., through the empirical checking of predictions—are confirmable in a *sui generis* way. Indeed GSTs are confirmed (a) by fitting whole families of specific theories (*conceptual confirmation*) and (b) by helping in the building of specific theories that are tested the classical way (*indirect confirmation*). Yet this does not solve the problem of the scientific status of GSTs. Let us now turn to this problem.

6. The Scientific Status of GSTs

No matter what philosophers with their *a priori* criteria of scientificity may decide, GSTs are usually regarded as scientific. They are so classed because (a) they are precise (by virtue of being formulated mathematically), (b) they are not at variance with our antecedent scientific knowledge, and (c) when applied (by specification) they often yield either scientific knowledge—in particular scientific theories proper—or guides for efficient action—e.g., in the field of management. Surely GSTs do not provide detailed accounts of any real systems. In particular they do not explain and do not predict the behavior of any real systems. But they are no less useful for that reason. Indeed, GSTs are *generic frameworks* helping one to think of entire genera of entities in a variety of domains, from biology and psychology through hardware and human engineering to city planning and politics. True, GSTs solve no particular problems without further ado—but on the other hand they help in the discovery and formulation of new problems and they clarify basic ideas in all fields of inquiry (Rapoport,[18] p. 74). In short, GSTs are respectable members of the body of scientific knowledge—even though some scientists have misgivings because GSTs have not delivered all the goods promised by their most enthusiastic proponents.

But if GSTs are declared scientific then we must modify our previous canons of scientificity: *we must change our methodology of science*. To begin with we must replace Definition 4 by a more tolerant definition making room for conceptual and vicarious confirmability as an alternative to strict empirical testability. Let us try the following:

Definition 5. A theory is scientific if and only if (i) it is compatible with the bulk of scientific knowledge, and either (ii) jointly with subsidiary hypotheses and empirical data, it entails empirically testable consequences, or (iii) jointly with subsidiary hypotheses and empirical data, it entails theories that in turn entail empirically testable consequences as in (ii).

I.e., using the symbols occurring in Definition 4,

$$St =_{df} t \text{ fits Definition 4}$$

or $\neg(B/t)$ and $(\exists s)(\exists d)(\exists t')[s$ is a subsidiary assumption and d is a datum relevant to t and t' is a theory and $(t$ and s and $d) \vdash t'$ and t' conforms to Definition 4].

Shorter: t is scientific iff, when enriched with suitable subsidiary assumptions and empirical data, it becomes *empirically testable either directly or vicariously*, i.e., through some (specific) theory.

If we now look back on our previous definitions—except for thoroughly inadequate Definition 1—we realize that we have distinguished three levels of exigency. While we demand of all theories that they be compatible with the bulk of scientific knowledge, in some cases we require that they entail testable predictions when enriched with data, in others when they are adjoined data and subsidiary assumptions, and in still others we require that, when enriched with both data and subsidiary assumptions, they entail theories that are empirically testable.

In short, if we wish to keep GSTs within science then we must adopt an enlarged criterion of testability allowing for vicarious testability. But if we do so then we must give up the standard philosophies of science, none of which tolerates such an extension of the scientificity criterion. Would anything be gained by sticking to any of these philosophies? Let us see.

7. GSTs Sandwiched between Science and Metaphysics

When finding out that GSTs do not conform to the classical standards of scientificity, as embodied in Definition 4, we had two options. One was to give up the definition and adopt a broader one making room for GSTs. This is what we accomplished through Definition 5. The other option was to stick to conventional wisdom and declare GSTs to be *nonscientific*. In this case, since all of the genuine GSTs are mathematical in form, they must belong either in pure mathematics or in exact (i.e., mathematical) philosophy. (Exact philosophy is philosophy tamed by mathematics. It can still be wild with regard to having scientific evidence or it can be utterly irrelevant to science but it is formally correct because it makes explicit use of logical or mathematical tools. For a recent sample of exact philosophy see Bunge.[19]) Let us explore these latter two possibilities.

That GSTs are mathematical theories is sometimes held by mathematicians. However, every one of the GSTs is concerned with concrete though rather faceless entities. Moreover these theories are used in designing concrete systems such as communication networks or learning systems. And they are not on the same footing with mathematical theories, such as linear algebra or probability theory, which have a much higher degree of cross-disciplinarity. Instead GSTs, when applied, are employed as *broad schemata* or models of the things to be designed or controlled. In short, although GSTs are hypergeneral, they are not nearly as universal as mathematical theories. They are mathematical in form, not in content.

Hence if we insist on regarding GSTs are non-scientific we are left with the second possibility, namely that they belong in exact philosophy—in particular in mathematical ontology, or the formal theory of extremely wide genera of concrete things. Pause and ponder before smiling. GSTs share three basic traits with theories in exact ontology: (a) they are *mathematical* in form, (b) they concern *genera* (not just species) of concrete, material, real things, and (c) they are *empirically untestable* except in a devious way. Why then not accept GSTs in the fold of ontology? There is but one reason for refraining from doing so, namely the belief that there *must* be an unsurpassable frontier between science and metaphysics—a belief shared of course by empiricists and rationalists alike even though they differ about the demarcation criterion.

Here again we have the choice of either following tradition or facing the facts and making our own decision. The fact, as I see it, is that GSTs are *both* scientific *and* ontological precisely because they share the above mentioned defining traits: exactness, hypergenerality, and empirical irrefutability. It may be rejoined that GSTs are, in addition, compatible with the bulk of scientific knowledge, whereas a number of theories in exact ontology are totally alien to science. (Thus the most fashionable among them consist in speculation about possible worlds, without caring for investigating the basic traits of the real world.) Agreed. But this shows only that there are two sorts of theory in present-day exact ontology: those which are and those which are not contiguous with the bulk of scientific knowledge.

We may then speak of *scientific ontology* when referring to theories that concern the most pervasive traits of reality, that are systematic (i.e., theories proper rather than bags of opinions), that make explicit use of mathematical logic or abstract algebra or any other branch of mathematics, that are congenial (not just compatible) with the science of the day, and that elucidate some of the key concepts occurring in philosophy or in the foundations of scientific theories—such as those of system, process, interaction, life, mind, or society.

The theories in scientific ontology are of course those satisfying Definition 5. For example, an ontological theory of space that explains in exact (mathematical) terms spatial relations as certain relations among physical things, is entitled to be called *scientific* if it ends up with a space that is metricizable, so that the physicist can add to it any metric function he needs for his theories. Another example: an ontological theory of society that defines the latter as a concrete system endowed with a structure consisting of a family of equivalence classes (one for each social group) is entitled to be called *scientific* because the sociologist can use it as a framework or matrix for working out his more specialized theories, such as the theories of stratification, of mobility, etc.

In sum, our enlarged criterion of scientificity (Definition 5) houses not only GSTs but also theories in scientific ontology. There is a large overlap between the two areas, the extent of which depends upon the distinction between them, which is largely a matter of convention. If terms were to help we might say that, whereas GSTs are hypergeneral, theories in scientific ontology, or SO, are superhypergeneral. The two are so many rungs in a ladder of generality that goes like this:

Hyperspecific scientific theories—e.g., the theory of the simple pendulum.
Specific scientific theories—e.g., classical particle mechanics.
Generic scientific theories—e.g., continuum mechanics.
Hypergeneral scientific theories—e.g., general lagrangian dynamics in GST.
Superhypergeneral scientific theories—e.g., an SO theory of change.

Since all five categories are scientific, and the last two are philosophic as well, the border between science and philosophy has disappeared in our perspective. And, since there is no frontier left, there is no occasion for frontier skirmishes and no point in looking further for a demarcation criterion. So, one more old philosophical debate has been bypassed in the advancement of science.

8. Conclusion

GSTs have defied the scientificity criteria upheld by the most influential philosophies of science of our time, namely empiricism and rationalism. These hypergeneral theories found philosophy unprepared for their degree of generality. Actually within the classical sciences there already existed extremely general theories that challenged the accepted slogan *Deduce-and-check*. Indeed the generic scientific theories, such as quantum mechanics and the theory of evolution, are untestable without further ado: we must enrich them not only with data but also with extra assumptions before we can deduce testable consequences. GSTs have, then, just given the *coup de grâce* to the standard epistemologies and methodologies.

The collapse of the standard philosophies of science has forced us to revise the testability criteria. This revision has led to proposing a broader definition of the concept of scientific theory, namely Definition 5. This definition makes room for vicarious empirical testability—i.e., testability through the intermediary of specific theories. GSTs are testable this way. But so are theories in scientific ontology. So this places GSTs in philosophy as well as in science.

This is not to say that GSTs and scientific ontology coincide: they have an appreciable overlap but they may be distinguished if one wishes. For one thing, GSTs, as conceived here, take for granted a number of notions that ontology makes it its business to elucidate—such as those of thing and property, space and time, causality and chance, natural law and history. For another, some GSTs are a bit too specific for philosophical purposes. For example, ontology is not particularly interested in the properties of linear macrosystems as opposed to nonlinear ones. Moreover ontology cannot restrict its attention—as GSTs have done so far—to classical, i.e., nonquantal and nonrelativistic, systems. This is why we call SO *superhypergeneral*. But the differences between GSTs and theories in SO are perhaps less interesting than their similarities in subject matter and in method. Likewise, although there are certainly differences between a hypergeneral theory such as the general classical field theory, and a generic theory such as classical electrodynamics, both categories are species of the genus science.[20]

Should any GST practitioner feel uncomfortable with this reclassification of his field because of the bad name so many philosophical schools have earned for themselves, he should take into account that being called a philosopher is the price he has to pay for calling himself a scientist.

References

1. K. R. Popper, *The Logic of Scientific Discovery*. Hutchinson, London, 1959.
2. M. Bunge, *Scientific Research*, 2 volumes. Springer-Verlag, New York, 1967.
3. M. Bunge, *Foundations of Physics*. Springer-Verlag, New York, 1967.
4. M. Bunge, *Method, Model and Matter*. D. Reidel, Dordrecht, 1973.
5. M. Bunge, *Philosophy of Physics*. D. Reidel, Dordrecht, 1973.
6. M. Bunge, *Sense and Reference*. D. Reidel, Boston, 1974.
7. M. Bunge, *Interpretation and Truth*. D. Reidel, Boston, 1974.
8. K. R. Popper, *Conjectures and Refutations: the Growth of Scientific Knowledge*. Routledge & Kegan Paul, London, 1962.
9. R. Carnap, *Logical Foundations of Probability*. University of Chicago Press, Chicago, 1950.
10. G. H. von Wright, *The Logical Problem of Induction*. 2nd ed. Basil Blackwell, London, 1957.
11. W. R. Ashby, *An Introduction to Cybernetics*. John Wiley, New York, 1956.
12. G. A. Miller, *The Psychology of Communication*. Basic Books, New York, 1967, p. 46.
13. M. Harrison, *Introduction to Switching and Automata Theory*. McGraw-Hill, New York, 1965.
14. P. Suppes, "Stimulus-response theory of finite automata." *Journal of Mathematical Psychology* **6**, 1969, pp. 327–355.

15. W. L. Kilmer, W. S. McCulloch and J. Blum, "Some mechanisms for a theory of the reticular formation." In: *System Theory and Biology*, edited by M. D. Mesarović, Springer-Verlag, New York, 1968.
16. G. J. Klir, ed., *Trends in General Systems Theory*. Wiley-Interscience, New York, 1972.
17. L. Padulo and M. Arbib, *System Theory*. W. B. Saunders, Philadelphia, 1974.
18. A. Rapoport, "The uses of mathematical isomorphism in general systems theory." In: *Trends in General Systems Theory*, Ref. 16.
19. M. Bunge, ed., *Exact Philosophy: Problems, Tools, and Goals*. D. Reidel, Dordrecht, 1973.
20. D. G. B. Edelen, *The Structure of Field Space*. University of California Press, Berkeley and Los Angeles, 1962.

33
Some Systems Theoretical Problems in Biology

Robert Rosen

1. Introduction

Biology begins with the recognition of what we call living organisms as a separate class of entities, distinguished in structure and properties from the rest of the natural world. The intuitions on which this recognition is based are a mixture of introspections and experience, which despite great effort have never been completely formalized; that is, no one has ever been able to put forward a finite set of structural propositions that are satisfied by exactly those physical systems which our intuition tells us are organisms. Nevertheless, most of us take our intuitions on this matter seriously enough to believe that we can make a useful, scientifically significant distinction between living and nonliving, organic and inorganic. The absence of formalization means, however, that we cannot sharply specify the boundaries which separate the living from the nonliving. We encounter such boundaries when we ask, as some people do, whether viruses are alive, or whether it is possible to construct machines which can "live" in some sense, or whether there are other kinds of physicochemical systems (e.g., on the planet Jupiter) which we would want to classify as "living systems."

The absence of a formal characterization of a definite class of "living organisms" by means of a finite set of either-or propositions has long bothered biologists and philosophers of science. It seems to me that the difficulty arises simply from the fact that our biological intuitions are in fact not based primarily on the kinds of structural or metric considerations which, for example, dominate physics but are rather of a *relational* or *functional* character. The same difficulty of definition in fact arises whenever we try to specify in purely structural terms a class whose elements are defined functionally. To give one example; Wittgenstein asks:

> How should we explain to someone what a game is? I imagine that we should describe *games* to him, and we might add, "This *and similar things* are called *games*."

From *The Relevance of General Systems Theory*, E. Laszlo, ed. Copyright © 1972 by Ervin Laszlo.

Replace the word "game" by the word "organism" and we have exactly the biological situation.

We can see already that such relationally or functionally defined classes, and the intuitions to which they correspond, lean heavily on behavioral or dynamical analogies exhibited by the members of the class, or better, in the way in which we ourselves interact with the members of the class. And as we shall see, a study of such classes depends heavily on metaphors and metaphorical arguments. In the following remarks I shall sketch one manner in which we can attempt to arrive at some understanding of the behavior of systems which our intuition tells us are "living organisms"; how such understandings are to be related to our understanding of other areas of knowledge; and how the methods used to achieve this understanding can be applied to the study of other kinds of systems. This paper, then, constitutes an attempt to indicate at least some of the general system-theoretic concepts which are involved in attempting to construct a language suitable for the study of biological phenomena and biological organization.

Experimental biologists usually consider remarks of this kind to be of a very general character, and so in a certain sense they are. But in another sense they represent only a very limited and circumscribed area of application for these system-theoretic ideas. I have attempted in a few places to indicate some of the ways in which these ideas can be extended to other classes of organized systems, although the details of such extensions must be carried out by those with more detailed competence in these fields than I possess.

Any attempt to develop the general system-theoretic framework of even the limited area represented by biological systems must necessarily be permeated by the prophetic influence of Ludwig von Bertalanffy, and the present work is no exception. Indeed, the discerning reader will find on almost every page a clear indebtedness to Von Bertalanffy's pioneering work in general systems theory. The arguments I present against naive reductionism, the power of systems analogies in unifying apparently disparate branches of biology (and of systems in general), and the remarkable regulatory properties built into open systems, in particular, represent paraphrases of arguments Von Bertalanffy has expounded many times in the past, recast into the technical language which I believe most appropriate for drawing further specific biological inferences. The main claim to novelty in my own exposition, in fact, is in its promotion of the intensive technical development of system-theoretic tools appropriate to treat deep questions of biological structure and organization. This development should be viewed as a complement to the extensive approach most recently argued by Von Bertalanffy, which exhibits biological phenomena as merely one class of realizations of system-theoretic organizations, pervading the physical, engineering, and social sciences. Indeed, it is my belief that, now that the general conceptual framework of system theory has been laid (owing largely to the pioneering and heroic efforts of Von Bertalanffy), the next urgent task is to proceed to the parallel intensive study of many different classes of systems, aided and guided by the general homologies which the general theory of systems teaches us must hold between them.

2. Generalities on the Modeling of Biological Systems

The only consensus found among biologists about their subject is that biological systems are complicated, by any criterion of complexity that one may care to specify.

Therefore, if we are to achieve any kind of insight into the behavior of organisms, we must find some way of circumventing their inherent complexity; we must simplify them or abstract from them in some way; we must make models. Here the term "model" is to be taken in the widest sense; a molecular biologist, preparing a precisely definable fraction of the contents of the cell, is performing an abstraction; his resultant fraction is a simplified or abstract cell; it is a model. However, sharp controversies have arisen regarding the nature of biological modeling, the kinds of modeling which are acceptable, and the meaning and interpretation of biological models in this sense. At one pole we find the extreme reductionists, whose position will be considered in detail below; at the other we find holists who claim that any attempt to reduce the inherent complexity of organisms thereby automatically destroys their organic character, and that therefore any information pertaining to a model cannot pertain to the organism itself and must be erroneous.

A proper understanding of modeling in biology must, I feel, begin with an understanding of the interrelationships of physics and biology, which are profound and many-faceted. On the one hand, biological organisms are composed of atoms and molecules, and hence they simply *are* physical systems. The physicist is concerned with understanding the behavior of all assemblages of physical particles, including those that comprise organisms. And it is the fundamental principle of reductionism in biology that we have no real understanding of biological activities unless and until this understanding is expressed directly in terms of the interactions between the physical particles of which the organism is composed, i.e., in terms acceptable and recognizable to the physicist. This view, then, implicitly denies that there is any useful distinction between the organic and inorganic; between biology and physics.

A second and rather more subtle relation between physics and biology, which impinges even on holistic and systemic attempts to model biological systems, is that *the very machinery of system description*, the only tool we possess for this purpose, was developed for the analysis of simple physical systems (originating in Newtonian mechanics) and that despite extensive generalizations and refinements we still have no other conceptual tools available to describe systems and their behavior than those which proved convenient for physics.

A third relation, which plays a decisive though implicit role in motivating the reductionist viewpoint, is a counterpart of the preceding; namely that the only *experimental* tools available for the study of biological systems are also of a physical character. We have already mentioned that it is the manner in which we interact with systems which defines their character for us; in experimental biology we are constrained to interact with biological systems by means of techniques and tools invented by the physicist for studying inorganic nature. This bias on the manner in which we can observe biological systems automatically constrains us to a highly physical view of these systems, selectively emphasizing those aspects of biological systems which our observing procedures, drawn from physics, are geared to detect.

Thus both the experimental tools with which we observe biological systems, and the conceptual constructs by means of which we attempt to describe them, are drawn from a non-biological science, not concerned specifically with the complexity and highly interactive character typical of biological organisms. Therefore, in order to orient ourselves properly with regard to understanding how the modeling of biological systems is to be effectively accomplished, we must understand more specifically the nature of the biases which our physical tools, both experimental and theoretical, impose on us. We therefore turn now to a discussion of these matters.

3. Systems and Their Descriptions

In both physics and biology, and indeed in all other sciences of systems, there are essentially two ways in which we can attempt to obtain meaningful information regarding system behavior and system activities. We can either passively watch the system in its autonomous condition and catalogue appropriate aspects of system activity, or else we can actively interfere with the system by perturbing it from its autonomous activity in various ways, and observe the response of the system to this interference.

In systems for which the passive, autonomous aspect is paramount, a kind of system description is appropriate which we shall call an *internal description*. Typically such a description begins with a characterization of what the system is like at an instant of time; such a characterization is said to define a *state* of the system. The totality of all the possible states of the system, meaning the totality of different aspects the system can assume for us at an instant of time, forms a set called the *state space* of the system. In physics these states are typically defined through the measurement of certain numerical-valued observables of the system; these are called *state variables*, and typically have the property that if two states are at all different in any observable way, they differ in the values assigned to them by one or more of the state variables; if two states are identical in the values assumed on them by the state variables, they are identical in all other observables as well.

In Newtonian mechanics it is a consequence of Newton's laws that a system consisting of N particles may be described by a set of only 6N state variables; out of the infinity of system observables these are conventionally taken to consist of three variables of spatial displacement for each particle in each of the three spatial dimensions, and three variables of velocity or momentum for each particle in the direction of the corresponding displacements. Thus the state space (or phase space) for such a system can be identified with a subset of ordinary Euclidean 6N-dimensional space, and each state with a point of this space. But it must be carefully noted that there is nothing unique about a set of state variables.

The fundamental problem of system description is to determine how the internal states change with time under the influence of the *forces* acting on the system. In physics such dynamical problems are formulated in terms of differential equations, which specify the rate at which each of the state variables is changing with time. The solution of a dynamical problem thus involves the integration of a set of differential equations, with each solution specified uniquely when the initial state of the system and the particular set of forces acting upon it are known. The temporal evolution of the system thus takes the form of a curve, or *trajectory*, in the state space.

The other kind of system description is called an *external* description, sometimes graphically called a *black-box* description. In this situation we make no attempt to identify a set of state variables for the system. Rather we have at our disposal a family of perturbations which we can apply to the system, variously called system *forcings* or *inputs*, and one or more observables which we use to index the effect of applying a particular forcing or input to the system. Such system observables are generally called system *outputs*, or *responses*. In general, in this approach, it is desired to determine what the system response will be to an arbitrary input.

These two approaches are, of course, closely related conceptually. By the way in which internal state variables are defined, any system observable (and in particular the system outputs) must be already a function of the state variables themselves. Each forcing or input in our repertoire must correspond to a set of equations of motion of the system, and hence the system response to each particular forcing can be calculated by integrating the corresponding equations of motion. But in dealing with any particular problem, it is often most cumbersome to try to find an appropriate set of state variables, and we can proceed simply by an input-output analysis without talking about state variables at all. On the other hand, given a particular input-output analysis, it is theoretically possible to formally find a set of state variables for the box itself; what these formal state variables mean is usually not obvious.

It is one of the goals of science to be able to match up the two kinds of system description we have described. The external description is a functional one; it tells us what the system does, but not in general how it does it. The internal description, on the other hand, is a structural one; it tells us how the system does what it does, but in itself contains no functional content. We would like to be able to pass effectively back and forth between the two kinds of system description; i.e., we would like to be able to infer the system function (the external description) from a knowledge of system structure (the internal description), and conversely, knowing the system functions, we would like to be able to determine at least something about its structure.

In actual practice, theoretical physics is dominated by internal descriptions; the natural systems with which the physicist deals are generally of a simple type to which the concept of "function" is not appropriate. External descriptions begin to become important when we discuss artificial systems, especially the regulation and control of machines which we build for ourselves. Since in engineering we do things in the fashion simplest for us, our regulatory and control systems are related in a rather transparent way to corresponding internal descriptions.

In biology the situation is quite different, for a variety of reasons (some of which will be explored shortly). The crux of the matter is that a biological system is built on quite different (and largely unknown) principles from those systems which we build for ourselves, and our descriptions of organisms possess a curious mixture of internal and external characteristics. Many biological activities are in fact defined and observed only functionally, in terms of an input-output formalism. On the other hand, we can, as noted previously, employ many observational techniques (borrowed from physics) to obtain a wide variety of structural information. But there is no reason to expect that the structural information we find easy to measure should be related in a simple way to the external functional descriptions in terms of which so many biological phenomena are defined. Stated another way, the internal state variables which we find easily accessible bear no simple relation to the functional activities carried out by a biological system; and conversely, the external descriptions appropriate to the functional behavior of biological systems bear no simple relation to the structural observables which our physical techniques can measure. In the next few sections we shall explore some of the ramifications of this peculiar situation.

4. The Structural Characterization of Functional Properties

The remark closing the preceding section has an important bearing on the reductionist hypothesis, which asserts that the basic problems of biological systems can all be effectively understood in terms of the internal descriptions of physics (using as state variables the observables defined through the use of observing systems likewise drawn from physics). The question is then: how does a physicist approach a physicochemical system too complex to be studied as a whole? As indicated previously, he must abstract from or simplify the system in some way. The customary way is to physically fractionate the system; break it up by physical means into a spectrum of simpler subsystems, if necessary iterating the process by fractionating the individual fractions, until we are left with a family of subsystems each of which is simple enough to be studied as a whole. He then will attempt to assemble the information he has obtained regarding each of the fractions into information about the original system with which we began. Implicit in this are two crucial hypotheses, of a system-theoretic character: namely, that any physicochemical system, however complex, can be resolved into a spectrum of fractions such that (a) each of the fractions, in isolation, is capable of being completely understood, and, most important, that (b) *any* property of the original system can be reconstructed from the relevant properties of the fractional subsystems.

This last hypothesis is demonstrably false for many systems, including most of those of biological interest. A simple physical counterexample is a system of three gravitating masses in space (three-body problem). We can surely fractionate a three-body system into various two-body and one-body systems, each of which is simple enough to be completely understandable in isolation. But the crucial stability properties of a general three-body system can never be reconstructed from a knowledge of two-body or one-body systems, however comprehensive. The basic reason for this is that the fractionation techniques employed are not compatible with (or do not *commute* with) the dynamical properties of the original system; we irreversibly destroy this dynamics, the very object of interest, by the process of fractionation itself.

Thus when we apply a prespecified set of fractionation techniques to an unknown system, there is no reason why the fractions so obtained should be simply related to properties of the original system. Yet this is exactly what happens when a molecular biologist fractionates a cell and attempts to reconstruct its functional properties from the properties of his fractions.

This is not at all to say that fractionation per se will give no information about the properties of a complex system (although holists will go that far). What we must do to accomplish this is seek fractionations compatible with the system dynamics, in a definite, well-defined sense. We have argued that the fractionation techniques imported into biology from physics will not in general be compatible with the dynamics of biological systems. This does not imply that such fractionations do not exist; they may well exist, but they will generally be of a different character from those which have heretofore been important in analytical biology. They will be, in some sense, "function-preserving," as in the following simple example. We all know that a bird's wing is a combination propeller and airfoil, with both functions inextricably intertwined. This is different from the case of an artificial system like an airplane; we can physically fractionate an airplane into physically distinct parts which preserve such functions, but such procedures fail in the case of the bird's wing.

This example illustrates on one hand the different principles of construction on which biological structures and engineering structures are built, and at the same time illustrates that the fractionation procedures appropriate to biological organization must be of a different character than those appropriate to simpler physicochemical systems. Basically this is because in biological systems the same physical structure typically is simultaneously involved in a wide variety of functional activities.

The situation with regard to physicochemical fractionations of arbitrary systems is actually much worse than this, as will appear in the next section. But we have already shown enough to demonstrate that a simple reductionist hypothesis cannot be true for at least many of the functionally defined properties of the greatest biological interest.

5. System Analogies

As I said in the preceding section, a set of structurally meaningful state variables for a biological system is most difficult to identify, particularly if we restrict ourselves, as we usually do, to those quantities defined by purely physical observation techniques. We may always have recourse to an external description, i.e., to an input-output analysis; this is always appropriate to a system defined primarily in functional terms to begin with. But such black-box descriptions, though they are very useful (and allow us to make predictions about our system) carry with them only a limited understanding. Only an internal description, or something very much like it, can allow us to say that we fully understand the behavior of our system.

There is a sort of halfway house between internal and external descriptions which allows us to go a bit further than we can with external descriptions alone; this depends on the concept of system, and what I have called elsewhere the construction of dynamical metaphors for biological activity. Let us begin with the notion of analogue, which has long been employed by experimental biologists in the study of complex systems, under the generic term, *model systems*. Thus we find enzymologists attempting to learn about enzymes by studying systems ("enzyme models") which are not enzymes; we find physiologists attempting to learn about the properties of biological membranes by studying collodion films, thin glass, artificial lipid bilayers and other types of "model membranes"; we find neurophysiologists attempting to learn about the nervous system by studying a variety of artificial switching mechanisms or other forms of "neuromimes"; and all kinds of scientists attempting to understand the dynamics of their system of interest, whatever its character, by modeling on an analogue computer, i.e., an electrical system so constructed as to mimic the original dynamics.

The reason that the use of model systems is possible at all is that the same dynamical or functional properties can be exhibited by large classes of systems, of the utmost physical or chemical diversity. Two systems which are physically different but dynamically equivalent will be called *analogues* of one another (the terminology obviously drawn from analogue computation, which embodies this concept in a particularly transparent way). If our interest is in the system dynamics, then this dynamics can be studied equally well (and often better) in any convenient system analogous to our original system.

Modeling by system analogy has obviously a completely different basis than the kind of fractionation we discussed in the preceding section. Systems analogy shows us that dynamical or functional properties can be studied essentially independently of specifics of physicochemical structure, while fractionation, or other reductionist techniques, are bound up with these specifics in an essential way.

Analogies of this kind are common even in theoretical physics. The mechano-optical analogy of Hamilton and Jacobi or more generally the organization of whole branches of physics around analogous variational principles is well known. Indeed, the judicious exploitation of such variational principles is one of the most impressive unifying agencies which exists in physics, potentially binding all of physics together in terms of functional or dynamical analogies, instead of attempting a unification on the basis of the structural fact that every physical object is built out of the same set of elementary particles. We shall see that the concept of system analogy plays an equally striking unifying role in biology. Such a unification is the sole attractive aspect of biological reductionism; I shall suggest that one can hope to achieve unification on functional terms while avoiding reductionist pitfalls.

The concept of system analogy is a most interesting one mathematically, and even opens up new vistas in classical physics. System analogy is most conveniently defined in terms of internal descriptions; two systems are analogous if, roughly, there is a 1–1 mapping between their state spaces which commutes with the system dynamics. But in physical systems, we have not only the state variables, but the full set of system observables (i.e., real-valued functions on the state space) available to us. Once a set of state variables, and the equations of motion, of a system are specified, every observable of this system inherits a particular dynamics. It is generally possible to find sets of such observables which define dynamical systems in their own right—such systems are in effect subsystems of our original system. It turns out in fact that there exist physical systems which are *universal* in the sense that we can build a dynamical system out of appropriate observables of the universal system which is analogous to any arbitrary dynamical system.

This kind of result has many profound implications. For one thing, we have already mentioned that we apprehend a system in terms of those system observables which are in some sense easy for us to measure. The same system would present itself to us quite differently if we interacted with it differently; i.e., if other observables of the system were made easy for us to measure. Indeed, a universal system could be made to appear as an arbitrary dynamical system, simply by interacting with it in an appropriate way. This may open novel possibilities for simulation. And, returning to the notion of fractionation: we can fractionate such a universal system in such a way that the isolated fractions have *arbitrary* dynamical properties. This shows in a particularly graphic way the difficulties inherent in attempting to infer system dynamics from a study of fractional subsystems separated by conventional physicochemical means.

Let us return to the statement made previously that any functional or dynamical property of a given system can be studied equally well on any one of the system analogues, or even entirely in the abstract. Such an abstract functional property, exhibited by each of the system analogues which *realize* the abstract system, is what we call a "dynamical metaphor." For example, there are a number of important biological properties which follow simply from the fact that biological systems are open systems in the dynamical sense. To understand such properties we do not need to know which open system, in complete structural terms, is in fact before us, but merely that it is open. This is a situation familiar even in mathematics; if a particular property of a group, for example, follows simply from the group axioms, then it is redundant, and indeed incorrect, to prove the result by invoking the specific properties of the group elements comprising the specific group before us. In this way we can begin to carry out what we may call "functional fractionations," which allow us to see what follows already from the simplest dynamical properties of a metaphor, and what properties require the invoking of more specific dynamical or structural assumptions.

Such dynamical metaphors are playing an increasingly important role in our understanding of biological processes, different as they are from conventional structural modeling. The use of model systems has already been mentioned, as has the employment of open systems as metaphors for switching systems, threshold elements, equifinality in development and regeneration, etc. Another popular dynamical metaphor is the employment of a single metastable steady state as a metaphor for the establishment of polarities or gradients in differentiating systems.

There is one difficulty in the study of dynamical metaphors which must be mentioned. A dynamical metaphor, by its very nature, refers to a class of analogous systems, which may be of the utmost physical diversity. A typical biologist, on the other hand, is interested in the specific system before him, and asks for specific structural implications of any theoretical scheme that he may test on his system. Obviously the dynamical metaphors are not, by themselves, geared to provide us with specific structural information about individual systems in the class. Thus it is difficult to make explicit contact with the structural information available about individual biological systems, which after all comprises the vast bulk of our biological knowledge. We have suggested elsewhere that dynamical metaphors, appropriately supplemented with further conditions (in particular, with constraints arising from considerations of optimal design) allow us to pick individual systems out of a large class of analogous systems (namely those which satisfy the additional constraint of optimality), and about these individual systems we can make a great many more specific structural inferences.

6. Hierarchical Systems in Biology

I have stated several times previously that biological systems are constructed along different principles from the simple physical systems and engineering artifacts with which we are most familiar. One of the most obvious of these differences is the pronounced hierarchical character of biological systems; the separation of biological activities into distinct levels of organization. I shall call a system *hierarchically organized* if it satisfies the following two conditions: (a) the system is engaged simultaneously in a variety of separate distinguishable activities, and (b) different system descriptions are necessary to describe these several activities. It is this second condition which characterizes biological systems, with their stratification into many levels of organization.

We must say a word about what is meant by "different system descriptions." We pointed out above that the same system always admits at least an external description and an internal description, and that these are different. However, this is not the kind of difference which is meaningful for hierarchical organization. What is meant is that the system requires several essentially different internal descriptions (each of which carries with it a corresponding external description) to account for the various activities of the system. A simple physical example should make this clear. We can regard a gas in two quite different ways: on the one hand it can be regarded as a structureless fluid, describable in terms of the thermodynamic state variables (pressure, volume, temperature, etc.). On the other hand, a gas can be regarded as a very large number of small Newtonian particles, admitting an internal description in terms of the state variables appropriate to the dynamics of Newtonian systems; displacements and their corresponding velocities or momenta. These two state descriptions are essentially different; they refer to different structural levels of organization of the gas, and are made apprehensible to us in quite different ways.

Biological systems are very highly stratified in this sense, into levels ranging from the submolecular to the ecological. A great deal of theoretical biology in the past was devoted to an attempt to find an "anchor" level in the hierarchy; a level which was biologically meaningful, understandable in its own terms, and most important, would allow us to infer the properties of all the levels above it and below it in the hierarchy. For many years it was thought that the cellular level was such a level; this was the deeper significance of the cell theory of Schleiden and Schwann. With the advent of biochemistry and molecular biology many biologists regarded the biochemical level as the most appropriate "anchor" in the hierarchy. Indeed, the essential content of the reductionist hypothesis was that it asserted that it was possible to infer the properties of any level in the hierarchy from the biochemical level.

In the preceding sections we have seen some of the acute problems of the reductionist hypothesis; here we consider the question of whether it is, in fact, possible to pass *effectively* from the biochemical level to higher levels of biological organization. This is a question which has received ample attention in recent years, from such authors as Elsasser, Wigner, Polanyi, Pattee, and others. This is not the place to go into specifics of these arguments; it need only be stated that it is at best exceedingly problematic whether one can indeed effectively traverse organizational levels when one starts at the bottom. Even in physics, the tool used for passing between the dynamic and thermodynamic descriptions of a gas is statistical mechanics, a tool of the greatest difficulty and subtlety, which has hardly been fully mastered and is at best of limited applicability. And although it seems on the surface that statistical mechanical ideas can be readily imported into biology, the several attempts to do so have run into the gravest technical and conceptual difficulties.

There is, however, one aspect of the hierarchical organization of biological systems which bears mentioning at this point. Namely, it appears that the dynamical properties which emerge at successively higher levels of biological organization are *analogous*, in the strict sense employed in the preceding section, to those at the lower levels. One particularly striking instance of such analogies occurs between the biochemical and genetic control networks found by Jacob and Monod, and the neural networks in the central nervous system. The exploitation of this analogy may ultimately prove as fruitful for biology as the mechano-optical analogy has been for physics.

7. Implications for "Structural" Studies of Complex Systems

The main points which have been made in the above discussions are:

a. That the only way in which we know how to approach complex systems is to simplify or abstract from them in some way;

b. That such simplification amounts to splitting our system into subsystems, which are simple enough to be characterized in isolation, and such that our knowledge of the isolated subsystems can be effectively employed to give us information about the original system;

Systems Theoretical Problems in Biology

c. That in biology, the abstractions offered by physical reductionism do not in general satisfy proposition b, in that they are not generally compatible with the dynamics of the original system.

We believe that the first two of these propositions are universally applicable to a study of complex systems of whatever type; social, economic, political, linguistic, etc. What we seek in the study of such systems is a spectrum of "atomic" subsystems which can be understood in isolation, and whose essential properties are preserved when a set of such "atomic" subsystems are recombined. What point (c) above tells us is that we must avoid preconceptions as to the nature of such "atomic" subsystems; that those subsystems which seem a priori to be the most natural candidates for this purpose may in fact not be so; and that we must let the overall system dynamics decide this for us.

Actually, the identification of such "atomic" subsystems implies far more than this. For, by their very nature, such subsystems can be juxtaposed or recombined to produce new kinds of systems, different from those with which we originally started. That is, we may use these subsystems as elements from which new kinds of systemic organization can be synthesized by a set of canonical rules for the juxtaposition of our "atomic" subsystems. In such a situation, the systems with which we started are displayed as special cases of a generally much larger class of systems, all constructible from a family of "atomic" subsystems by means of a definite set of formal rules of combination or juxtaposition. This kind of analysis followed by resynthesis is typical of many fields within pure mathematics and the applied sciences; it is for example the basis of the numerous "canonical form" theorems for algebraic or topological structures.

It often happens, however, that we wish ultimately to restrict our attention to those synthetic reconstructions of our "atomic" subsystems which correspond to "natural" objects. We thus desire a set of rules which can characterize, out of the class of all such synthetically reconstructed systems, the subclass of those which are "natural." This is in general a much harder problem even than isolating our "atomic" subsystems in the first place. It amounts to exhibiting a set of rules (a "grammar," if you like) whereby the natural systems can be effectively exhibited, or at least effectively recognized. We pointed out the difficulties in carrying out such a program for biology at the outset of these remarks, when we noted that the class of "natural" biological organisms has never been successfully characterized within the class of all physical systems (implicitly taking for our "atomic systems" the set of real physical atoms). But this does not mean that we can never carry out this specification with any choice of atomic subsystems. We know already, indeed, that for biology the choice of such subsystems will be quite different from those the reductionist hypothesis gives us.

The entire process we have just sketched, beginning with the isolation of atomic subsystems, their recombination to generate a large class of systems, and the rules for selecting a subclass of systems of interest out of these, are implicit in the notion of "structuralism" or structural analysis for the study of complex systems. We have amply seen that the word "structural" has to be interpreted in a very wide sense; indeed in biology the relevant "structures" are always defined in *functional* terms.

It may be helpful to itemize the procedures involved in such a "structural analysis" of biological systems. This itemization is rather complicated, but once the essential aspects are systematically set down it will be recognized that exactly the same procedure is implicit in the structural study of all other kinds of organized complex systems.

We begin by supposing that we have already identified a class of "atomic subsystems" satisfying the hypothesis point (b). We may as well suppose that these are completely abstract systems, because by hypothesis any real biological system can be decomposed into real subsystems which realize such atomic subsystems; but in general different biological systems will give us different (but analogous) atomic subsystems realizing the same abstract systems. Let us designate this set of abstract atomic subsystems by the symbol A.

We now suppose that these abstract atomic subsystems can be combined or juxtaposed by a definite set of canonical operations of rules of composition, to form a large set of abstract systems, which we may suggestively designate as *abstract words*, and denote as $A^\#$. $A^\#$ is thus the set of abstract systems *generated* from A by the employment of the canonical composition rules.

Finally, we wish to identify or select out of $A^\#$ a subset, B, corresponding to the "abstract biological systems." The words of $A^\#$ not in B are the "abstract nonbiological systems." The set of rules we use to make such a selection or identification of the elements of B we may suggestively call a "grammar."

We thus have a sequence of operations going from the set A of abstract atomic systems to the set $A^\#$, the set of abstract words, to the set B (the set of abstract biological systems), which may be represented by the following diagram:

$$A \xrightarrow[\text{rules}]{\text{juxtaposition}} A^\# \xrightarrow[\substack{\text{rules} \\ (\text{"grammar"})}]{\text{selection}} B$$

Now the set A of abstract atomic systems can in principle be *realized* in physical terms in many different ways. Suppose that such sets of specific realizations are designated as

$$R_1(A), R_2(A), \ldots, R_n(A), \ldots$$

For each i, the real systems in $R_1(A)$ realize the abstract systems in A; hence there is a natural mapping of A into $R_1(A)$ associating to each abstract atomic subsystem its realization, and a natural mapping of each $R_i(A)$ into each $R_j(A)$, associating to each system in $R_i(A)$ its analogue in $R_j(A)$.

The rules of juxtaposition of abstract atomic subsystems, by which $A^\#$ is generated from A, may now be realized in terms of specific physical operations or processes in each $R_i(A)$, *perhaps in many different ways*; i.e., using different physical processes to combine the systems in $R_i(A)$. Thus in general each $R_i(A)$ can give rise to many sets of juxtaposed systems or words, which we may designate as

$$R_{i1}^\#(A), R_{i2}^\#(A), \ldots, R_{ik}^\#(A), \ldots$$

Each element of $R_{ik}^\#(A)$, for all i, k, is a realization of some word of $A^\#$; hence there is a again a natural mapping of each of the sets $R_{ik}^\#(A)$ into each of the others which associates analogous words (two words being analogous if they realize the same word in $A^\#$).

Further, we may identify in each $R_{ik}^\#(A)$ a number of sets $B_{ik1}, B_{ik2}, \ldots, B_{ikj}, \ldots$, these being selected according to different physical "grammars" on the set of words $R_{ik}^\#(A)$.

We thus have many different candidates for "real" biological systems, specified by the diagram

$$R_i(A) \longrightarrow R_{ik}{}^\#(A) \longrightarrow B_{ikl}$$

All such diagrams are connected by mappings into every other such diagram, which identify analogous but physically different systems. Presumably the study of "real" organisms is just one of these; whether the other diagrams are equally real (i.e., whether we can realize biological organization with novel physicochemical structures), or whether other such diagrams are excluded on some kind of physical grounds, is an open question.

This formalism applies equally well to other kinds of organization, even nondynamical ones like linguistics. Here we can assume that there is only one set $R_o(A)$ of realizations of the set A, comprising the linguistic "atoms" (morphemes or phonemes), and only one rule of juxtaposition leading to the set $R_{oo}{}^\#(A)$ of linear sequences of linguistic atoms. But there are in general many "grammars" leading to different but analogous sets $B_{ooj}(i = 1, 2, \ldots)$ of "natural languages."

8. Evolutionary Problems

Before concluding this brief paper, it is necessary to add a further word regarding the evolution of biological structures in time. I have, in the preceding analysis, been concerned entirely with "physiological processes," those which take place during the lifetimes of single organisms. We have neglected developmental problems, and most particularly, we have neglected evolutionary problems, which are concerned with the way in which the class of organisms changes over long periods. Since a "structural" analysis pertains only to the class of biological systems at single instants of time (i.e., is a static description of the biological world, considered in evolutionary terms) there is an essential dynamical element missing from our discussion; in the terms used above, we have specified the instantaneous states of the biological world, but not the forces acting on them to produce changes of state, nor the equations of motion to which these forces give rise.

The way in which such equations of motion, corresponding to evolutionary processes, can be constructed and investigated is a vast and difficult problem, somewhat simplified in biology by the curious analogies which exist between evolutionary and developmental processes. In purely descriptive terms, evolutionary processes can be regarded as a temporal dependence of the "grammatical" rules whereby a set B is selected from the set $A^\#$. But such temporal dependence requires its own kind of "structural analysis," and how to go about making such an analysis in any kind of evolutionary situation is, to my knowledge, a completely open question.

34
Economics and General Systems

Kenneth E. Boulding

In my own recollections the Society for General Systems Research, as it later came to be called, originated in a conversation around the lunch table at the Center for Advanced Study in the Behavioral Sciences in Palo Alto, California, in the fall of 1954. The four men sitting around the table who became the founding fathers of the Society were Ludwig von Bertalanffy, Anatol Rapoport, Ralph Gerard, and myself—a biologist, an applied mathematician and philosopher, a physiologist, and an economist. Economics, therefore, can certainly claim to have been in at the beginning of that enterprise, although this may have been largely an accident of my own personal interests. Certainly one cannot claim that the interaction between general systems and economics has been very extensive since that date, though the contributions of each to the other may be more than many people recognize. In the intervening years, however, the social sciences in general systems have been represented more by sociologists, such as Buckley,[1] and psychologists, such as the late Kenneth Berrien.[2] Almost the only other economist I can think of who has played much of a role in the development of general systems is Alfred Kuhn,[3] whose interest, like my own, has been primarily in going beyond economics to developing an integrated social science.

Just why the economics profession has viewed general systems with such a massive indifference I really do not know. Like the physicists, the economists are so bound up within the elegant framework of their own system that they find it hard to break out into a broader interest. Economics, indeed, may be a good example of a principle I have sometimes enunciated, that "nothing fails like success." The very success of economics, and especially of econometrics, in formulating systematic quantitative theories and methodologies may have prevented the profession from looking outside its own boundaries for further insights and models. I regard this as unfortunate, as in my view certainly the general systems approach to knowledge has important contributions to make to economics, as it does to virtually all other fields.

In the initial manifesto of what was then called the Society for the Advancement of General Systems Theory, which was published in the program of the Berkeley meeting of the American Association for the Advancement of Science in December 1954, a "general system" was defined as any theoretical system which was of interest to more than one

---

From *The Relevance of General Systems Theory*, E. Laszlo, ed. Copyright © 1972 by Ervin Laszlo.

discipline.[4] On this criterion many and perhaps all of the theoretical systems of economics would qualify as general systems, for they are certainly relevant to other disciplines. The theory of the general equilibrium of the prices and outputs of commodities, for instance, as originally developed by Walras,[5] and made in part operational by Leontief[6] in his input-output analysis in the 1930's, is clearly a special case of a general system of the utmost importance, for it is a special case of the general equations of ecological equilibrium. In the simplest formulation of this system, we suppose a number n of interacting populations, each composed of the individuals belonging to a single species. The species here may be biological species, such as humming-birds, or commodity species, such as automobiles, or even mineral species, such as available nitrogen in the soil or mineral nutrients in a pond. They may also be psychological species, such as the demand functions of different individuals for different commodities.

The simplest set of equations for a system of this kind simply states that each population has an equilibrium value which is a function of the existing values of all other populations. This gives us n equations and n unknowns immediately, and if this set of equations has a solution in positive values for each population, an equilibrium is at least possible. Whether it will actually be attained or not, of course, depends on the dynamics of the system, for the dynamic processes of the system may change these equations in the course of time. For instance, if any biological population in the course of the dynamics of the system falls to zero, it will never recover, and all the other equations will have to be changed, and the final equilibrium likewise will have to be changed. The system could also be formulated in dynamic terms by supposing that the rate of increase of any population, which would be negative, of course, in the case of a decrease, would also be a function of the size of all other populations. This gives us a set of simultaneous differential equations, which, again, may have a solution in terms of a path for all populations. This path may or may not move toward an equilibrium in which the rate of growth of all populations is zero. This is the model which clearly underlies the Walrasian equilibrium in economics and it is quite a small step to move from the special case of commodity equilibrium to the general case of ecological equilibrium. Economics may certainly, therefore, claim to have made a fundamental contribution in this regard.

Arising out of this interest of economists in general equilibrium, it is not surprising that it was an economist, Paul Samuelson, in his *Foundations of Economic Analysis*,[7] who gave the first clear exposition of the principle that the equilibrium of any system had to be derived from its dynamic path and that we could not find out about the stability of an equilibrium from an inspection of the equations of equilibrium alone. A stable equilibrium was itself a property of the dynamic path of a system, so that even the simplest properties of any equilibrium system depended in the last analysis on the dynamic process of which it was, in a sense, a special position. Equilibrium is simply a dynamic process in which the dynamic path of the system leads to a reproduction in successive states of some initial equilibrium state. "Staying the same" is simply a special case of "changing." Stability has to be seen as a subspecies of change.

Another area in which economics has made an important theoretical contribution to other disciplines is in the theory of behavior, where the great contribution of economics has been the theory of maximizing behavior, that is, the assumption that the behavior of an organization could be explained on the grounds that it was trying to maximize some internal variable. In the most general case, what is maximized is simply "utility" or an ordinal preference function. All we mean by this is that behavior consists of doing what the behaving organism or organization "thinks is best at the time." One difficulty with this theory is that it becomes too general and hence without much content, stating little more than the organizations and organisms do what they do simply because, if they didn't think that what they did was the best thing to do, they wouldn't do it.

The economist's concept of "revealed preference" is simply one way of finding a pattern in nonrandom behavior. In any behavior that is not random, we will be able to find some sort of revealed preference, that is, we can postulate a preference function from which we can deduce the behavior which is actually observed. If an amoeba, for instance, "chooses" a piece of food rather than an adjacent stone, we say it is because it has a preference function on which food ranks higher than nonfood. To say that when we have said this we have not said very much may be right. Nevertheless, even if we have not said very much, we have said something. This is at least one way of describing nonrandom behavior, and those who think it is merely empty should at least accept the challenge to describe it in some other way that seems more useful. The whole concept indeed of a cybernetic system with a detector-selector-effector mechanism implies that the selector has some principles according to which selection is made, which is precisely what we mean by a preference function.

I must still confess to some qualms about this formulation, mainly because there are clearly cases in which it seems anthropocentric, to say the least, to suppose that the selection process of the system involves preferences. In the general evolutionary theory of natural selection, for instance, it is only by stretching the language, perhaps beyond the point of legitimate strain, that we can say that the species which in fact actually survive are "preferred" by the selection process. If we had a satisfactory model of natural selection we might find that this would also throw a great deal of light on "artificial selection," that is, the phenomenon of choice and decision. The awful truth is, however, that we do not have any adequate models of natural selection. There is no general mathematical model of the evolutionary process, and it is by no means clear that this is even possible. We do have something like special patterns of selective processes, as, for instance, in the theories of Sewell Wright,[8] but we do not really have anything that can be called a general model of the whole evolutionary process. Until we do, the economic models of selection by preference and choice have a great deal to recommend them and, provided that we recognize that the language is dangerously anthropomorphic, we can apply these principles even to organisms as simple, or complex, as the amoeba.

The economic model of behavior perhaps has had a greater impact on the applied psychological sciences of management science and strategic science than it has on psychology directly. The concept of optimization is quite fundamental to management science and to operations research, even though it is not always clear what is optimized. Extensions of this into Herbert Simon's "satisficing"[9] are simply special cases of the optimizing principles, which state that any position of the organization in which the maximand, that is, the criteria of success, is below a certain level is regarded as unsatisfactory and any position above this is regarded as satisfactory. This is easily seen to be a special case of the general maximizing principle if we visualize the preference function as being more like a mesa with a flat top than it is a peak with a sharp maximum. In many cases a mesa view of the preference function seems to be realistic, that is, there is a considerable area of choice within which we are relatively indifferent, but beyond this area preference may fall off very fast in "cliffs." Game theory, likewise, can be seen as an extension of the simpler model of maximization, with cases in which the outcome of one person's choice depends on the choice of another person. It is certainly no accident that the classic work in game theory by Von Neumann and Morgenstern should have been entitled *The Theory of Games and Economic Behavior*.[10] Here we see economics moving toward more general systems, though again almost at the cost of ceasing to be economics.

Theories of economic behavior have had a substantial impact on political science in the last generation, which has been moving more and more in the direction of political economy. A number of authors—Lindblom and Dahl,[11] Anthony Downs,[12] and Riker[13]—have used what are essentially economic models in the interpretation of political behavior with considerable success.

A problem which emerges out of economics, but which has a highly general significance, is the problem of suboptimization, that is, under what circumstances does the attainment of some kind of optimum in part of a system preclude the attainment of an optimum for the whole? Welfare economics and the theory of perfect competition is one of the few areas of the social sciences which deal with this problem, yet it is a problem of great generality. Many of the failures of organizations, for instance, are a result of suboptimization, which could almost be defined as finding the best way of doing something which should not be done at all, or more generally, finding the best way of doing something particular without taking account of the costs which this solution imposes on other segments of the system.

Another field in which economics has made a substantial contribution toward a general system, at least in the social sciences, is in the field of international systems. I regard this indeed as one of my own major contributions. My book *Conflict and Defense*[14] is an attempt essentially to apply a body of general theory, much of which comes out of economics, and especially out of the theory of oligopoly and the interaction of firms, to the problem of the interaction of states in the international system. It is indeed a general theory of viability and survival, which comes out of economics and which has applications not only to other social systems but also perhaps in the biological field. It is closely related, for instance, to the theory of territoriality. It is relevant also to the theory of the niche and to the determinants of niches. Thus any organization in competition with others will find that its advantage in the interaction diminishes as it goes away from some kind of "home base," so that at some point the advantages of any further expansion fall to zero. This is what I have called the "boundary of equal advantage" between two organizations, but the concept could easily be generalized. It is these boundaries of equal advantage which really define the niches of an ecological system. Economics has made an important contribution here in location theory, especially in the work of Losch,[15] who demonstrated that, even if we start off with resources and population distributed uniformly in the geographical field, the sheer pressures of maximizing behavior will force the field into clusters and structures and will indeed create what are in effect niches in what previously had been a uniform field. This is a demonstration of great importance, which has received surprisingly little attention from the biologists.

When we look at the other side of the coin, the impact of general systems on economics, we find unfortunately that the record is meager. In the last generation certainly economics has pursued its own way, with very few influences from outside. The attempt of Parsons and Smelzer,[16] for instance, to produce a sociological contribution to economics, seems to have had virtually no influence on the economics profession itself. The impact of psychology has also been confined to a few pioneering individuals, like George Katona[17] at Michigan, and even here the main impact has been from the empirical rather than from the theoretical side. Even my own interest in these matters, I think, has been regarded as an amiable eccentricity by most of my fellow economists. This may be, as I suggested earlier, because general systems has made the greatest impact on those disciplines which felt the lack of a systematic theoretical core, which economics did not. Nevertheless it seems to me that this isolation of economics from one of the most interesting movements in thought in the last twenty or thirty years has been most unfortunate, and as a result economists have missed many opportunities for learning things which would have been useful to them even in their strictly professional capacities.

Thus one would expect that cybernetic theory, which has been so important in the development of general systems that some people have almost identified the two, I think quite falsely, would have had some impact on economics. Feedback mechanisms, for instance, are of crucial importance in a great many economic systems. If one is looking for an explanation of economic cycles, of fluctuations, either of particular speculative markets, such as the stock market, or in the economy in general, the feedback model is extremely useful. It is capable of explaining not only regular fluctuations, in the case of equilibrating (negative) feedback, but is also capable of explaining disequilibrium processes, as in the case of destabilizing (positive) feedback. Yet there has been astonishingly little use of this model. Perhaps as a result of this failure to take a significant intellectual tool simply because economists have not made it themselves, economists seem virtually to have lost interest in the theory of fluctuations, and what work has been done in this has been of an extremely mechanical nature, using, for instance, spectral analysis, which is an elegant way of detecting probably nonexistent cycles and throws no light on the real structures and processes which underlie fluctuations. There is very little appreciation among economists, for instance, that the Great Depression from 1929 to 1933 could be explained in very large measure by a destabilizing feedback process in which a decline in investment produced a decline in profits and that a decline in profits produced a further decline in investment, which again produced a further decline in profits, and so on, until by 1932 investment was almost zero and profits were negative. There has been some work of this kind on inventory cycles, which really is something like a feedback mechanism, but this is highly specialized and not widely used.

Another area in which general systems should have made a much larger impact on economics than it seems to have done is in the theory of the optimum size of the organization. One of the real triumphs of general systems theory, for which Von Bertalanffy must take a great deal of credit, is the demonstration that the processes which lead to the formation of organizations tend to exhibit "equifinality" in the rather special sense of self-limiting growth. This is a consequence of the principle of "allometry" which is based fundamentally on the thesis, which is really an identity, that as the growth of any particular structure proceeds, in the absence of change in the pattern of the structure itself, volumes will grow eight times as fast and areas will grow four times as fast as linear dimensions. This is why the whale has to live in the ocean, simply because, if it were a land animal, its legs would have to be bigger in cross section than the animal itself.

The same principles apply to social organizations, though the relationships here are not so simple. Here the main limiting factor is clearly the lines of communication, particularly up and down the hierarchy, which grow at a much slower rate than the total organization.

Hence organizations eventually limit their own growth simply by the sheer difficulty in getting communications from the "surface" of the organization, where it is in contact with its environment, into the decision-makers who are not in direct contact with the environment, but have to make decisions in the light of increasingly less realistic images of the world. Acephalous, nonhierarchical organizations, like a democratic family or a commune, or even a producers' cooperative, have even sharper limits on scale, simply because the number of people who have to talk to each other increases much faster than the number of people in the organization. Groups employing participatory democracy have the same tendency for fission as does the amoeba, for very much the same reason.

The economic significance of this principle is, of course, that the firm, like any other organization, has an optimum size and that this optimum depends in part, at any rate, on the internal diseconomies of scale setting in beyond a certain point, mainly due to the difficulty of maintaining communications in a large organization, even when it is hierarchical. Centralization produces failure to optimize because of the breakdown of the communication network. Decentralization on the other hand produces suboptimization, from the point of view of the organization as a whole, and likewise produces a failure to optimize. The cycles of centralization and decentralization that we see in nearly all large organizations such as the Soviet Union, the Catholic Church, and General Electric, almost certainly arise out of this principle. If there is something wrong with every alternative, one tends to try a succession of wrong things in the hope that maybe one of them will turn out, which it never does. This principle of internal returns to scale is very important in explaining why some industries, like agriculture, have rather small firms and hence can maintain something like perfect competition, while other industries, like automobiles, have very large firms and hence produce either monopoly or oligopoly.

Another general system of great potential importance in economics is that of population analysis. Any process involving a set containing elements in which the date of entry into the set, or "birth," and therefore the age at any particular moment, of any element can be identified, and in which also the elements leaving the set, that is, "death," can be identified, deserves the name of "population." If there are functional relationships between the age structure of the population at a moment of time and the total number of births and deaths, the population can be projected into the future by a fairly simple system of what are essentially differential equations. Populations do not have to be biological; we can perfectly well consider the total number of automobiles, for instance, as a population, and the future population of automobiles can be projected in very much the same way that we project populations of human beings or of deer. I did this in fact in 1954–55.[18] This is indeed the key to a good deal of economics that goes by the name of capital theory, capital being simply the total population of valuable objects that exist at a moment of time. The "period of production," which has been a prominent concept of capital theory, is virtually the same thing as the expectation of life at birth of the elements in a population. The principle that in an equilibrium population the total population is equal to the annual number of births or deaths multiplied by the average expectation of life at birth appears in economics rather crudely as the capital-income ratio, which has been important in certain development models, though often used quite illegitimately because of the failure to recognize that disequilibrium populations rather than equilibrium populations were involved.

Finally, one might look at the failures of economics in the present generation and see how far these might have been due to the failure of economics to use insights from other disciplines, and especially of course from general systems. The great success of economics in this period from, say, the end of the second world war, has been mainly due to the capacity of the Keynesian system to suggest policies which at least prevented large-scale unemployment of the kind that we had in the 1930's. Even though the level of employment in the United States has not been by any means wholly satisfactory, hovering as it has in the last twenty-five years between about 3 percent and 6 percent, it is certainly much better than the 25 percent unemployment we had in 1932. I have often drawn the contrast between the twenty years that followed the first world war, from say, 1919 to 1939, with the twenty years that followed the second world war, let us say, from 1945 to 1965. The first period was a disaster. The recovery of Europe from the first world war was very halting and even though the prosperity of the United States in the 1920's was quite real, the Great Depression was an unmitigated disaster, and the thirties slid almost inevitably into the second world war. By contrast, the twenty—now twenty-seven—years after the second world war were quite successful; there was no great depression, there was a substantial economic development, especially in the richer countries, and since 1961 at least we seem to have moved further from the "third world war" to the point even where there is quite a high probability that it will never happen at all.

Some of this success is due to the Keynesian economics, which operated very much within the traditional framework of economics itself and did not draw on either theoretical models or information from other disciplines. So far the economists might claim that economics was self-sufficient, and that its successes were striking enough so that it did not really need anything from other disciplines, least of all from the amorphous body of ideas called general systems. Nevertheless there are a couple of flies in this moderately sweet-smelling ointment. One is the very uneven success which economists have had with the advice that they have given to the poor countries seeking to develop. There are a few notable exceptions, but on the whole the tropical world has not done conspicuously well economically in the period since 1945, even though it has seen the virtual liquidation of the European empires, with the exception of the Portuguese. It seems quite reasonable to associate this relative failure of economics in the field of development with its inability to go beyond rather mechanical models, which abstract too much from the enormous complexity of real societies. It is simply not enough to use gross economic aggregates. The developmental process is essentially a process in human learning. It involves enormously subtle relationships of status, authority, threats, and persuasion, as well as exchange. The plain fact is that we have not yet produced an adequate and total model of the developmental process in any society, and especially in those tropical countries of ancient and complex cultures, and our knowledge of the total processes of these societies is skimpy indeed.

A somewhat related problem has been the failure of economics to deal with the problem of policy toward deteriorating cultures and deteriorating cities, especially in the rich countries. The failure of economics here, I think, is due again to its obsession with rather simple models of exchange and its refusal to recognize that many of these problems involve one-way transfers, that is, a grants economy, which depends in turn on an enormously complicated human learning process in matters of status, identity, community, benevolence and malevolence, and so on. The very fact that there is an Association for the Study of the Grants Economy[19] suggests that conventional economics have failed in this regard again perhaps because it was too shut up within its own particular system and frame of reference.

Perhaps the most visible and spectacular failure of economics in this period has been its failure to deal with the problem of how to get full employment without inflation. This is a failure of the Keynesian model itself. In the United States, for instance, we have had almost continuous inflation, at least as measured by price indices and national income deflators, since about 1939, and there are no signs of our being able to control this. President Nixon's wage-price policy of August 1971 indeed is a little reminiscent of King Canute ordering the tide to go back, and, while it may have a temporary psychological effect, it is extremely unlikely that it will provide a solution to the problem. The reason for the failure of economics, here also, perhaps is that it has lived too much within its own framework and has neglected to study the larger system of price determination with models that might have been derived, for instance, from epidemiology theory, or from the theory of the spread of fashion, or from the theory of communication networks, none of which has been at all familiar to economists. I suspect indeed that we are not going to solve this problem until we are able to identify the leading communication chains in the structure by which the inputs of information, which lead to changes in prices and wages, actually function.

It is easy to give economists good advice; it is very hard to make them take it. Nevertheless the defects in economics at the moment are so glaring that one hopes that a new generation will develop, with an appreciation of what can be learned in economic models from other sciences and from general systems, so that the general systems economist will not always remain as a voice crying in the wilderness.

References

1. Walter Buckley, *Modern Systems Research for the Behavioral Scientist*. Aldine Atherton, Chicago, 1968.
2. F. Kenneth Berrien, *General and Social Systems*. Rutgers University Press, New Brunswick, New Jersey, 1968.
3. Alfred Kuhn, *Study of Society: A Unified Approach*. Richard D. Irwin, Homewood, Illinois, 1963.
4. See the "original manifesto" of the Society for General Systems Research, American Association for the Advancement of Science meeting program, Berkeley, California, 1954.
5. Léon Walras, *Eléments d'economie politique pure*. 4th Edition, 1900, Lausanne (trans. by William Jaffe, *Elements of Pure Economics*. Richard D. Irwin, Homewood, Illinois, 1954).
6. W. W. Leontief, *The Structure of American Economy*, 1919–1939. Oxford University Press, New York, 1951.
7. Paul A. Samuelson, *Foundations of Economic Analysis*. Harvard University Press, Cambridge, Massachusetts, 1947.
8. Sewell Wright, *Evolution and the Genetics of Populations*, 2 vols. University of Chicago Press, Chicago, 1968 and 1969.
9. Herbert A. Simon, *Models of Man*. John Wiley, New York, 1957.
10. J. von Neumann and Oskar Morgenstern, *Theory of Games and Economic Behavior*. Princeton University Press, Revised Edition, Princeton, New Jersey, 1953.
11. R. A. Dahl and C. E. Lindblom, *Politics, Economics and Welfare*. Harper & Bros., New York, 1953.
12. Anthony Downs, *An Economic Theory of Democracy*. Harper, New York, 1957.
13. W. Riker, *Study of Local Politics*. Random House, New York, 1959.
14. K. E. Boulding, *Conflict and Defense*. Harper & Bros., New York, 1962.

15. August Lösch, *The Economics of Location* (tr. by W. H. Woglom). Yale University Press, New Haven, 1954.
16. T. Parsons and N. J Smelser, *Economy and Society. A Study in the Integration of Economic and Social Theory*. Routledge, London, 1956.
17. George Katona, *Psychological Analysis of Economic Behavior*. McGraw-Hill, New York, 1951.
18. K. E. Boulding, "An application of population analysis to the automobile population of the United States." *Kyklos*, **2** 1955, pp. 109–124.
19. K. E. Boulding, *The Economy of Love and Fear: A Preface to Grants Economics*. Wadsworth, Belmont, California, 1973.

35

Can Systems Theory Generate Testable Hypotheses? From Talcott Parsons to Living Systems Theory

James Grier Miller

Introduction

Talcott Parsons and I first met in 1939 when we attended clinical conferences presided over by Dr. Stanley Cobb at the Department of Psychiatry of the Massachusetts General Hospital. We remained friends throughout our lives, meeting fairly often both professionally and socially. We liked each other and always maintained a cordial relationship. As he reminded me from time to time, in our younger years we were both greatly influenced by the ideas of Walter B. Cannon, who was my Professor of Physiology in medical school, and Lawrence J. Henderson, who had a direct personal influence on both of us. Henderson's espousal of Pareto, both in the courses he taught and in his conversation, had a particularly vigorous impact.

I accepted Parsons' invitation, in his role as Chairman, to join the Department of Social Relations at Harvard at its very beginning and was, in 1946 and 1947, the only clinician on that faculty. My book *Living Systems*[22] was reviewed by Parsons in a lengthy and generous article.[27] This was the last paper Parsons wrote before his death.

Both Parsons and I were systems theorists, but our approaches to integrative theory differed significantly. In 1965[21] I distinguished between abstracted systems and concrete systems. The first sort were emphasized by Parsons and his followers and the second sort were emphasized by me and some others. It is an axiom of systems theory that all systems have certain common attributes. They all consist of units coupled in particular relationships. The units are similar in some of their properties. The units of abstracted systems are relationships abstracted or selected by an observer. These relationships are observed to inhere and interact in selected concrete, usually living, systems. The units of concrete systems, on the other hand, are other concrete systems (components, parts, or members). The relationships of concrete systems are spatial, temporal, causal, or results of information transmissions. In many ways the units and relationships of concrete systems are the reverse of the units and relationships of abstracted systems.

I have always maintained that it is possible to carry on scientific work in terms of abstracted systems. My contention has been that, in scientific propositions generally, it is practical and easy to refer to concrete objects with nouns and to their relationships with verbs or predicates. The reverse is less practical and more difficult. The straightforward language we learned as children states that the ball rolls or the child loves the mother. It is confusing, and achieves nothing important, to reverse this common mode of speech and say that fatherhood is assumed by another man when a woman with children remarries after a divorce, or that the presidency was occupied first by Lincoln and later by Kennedy.

Parsons accepted the abstracted-concrete distinction but always maintained that abstracted systems are scientifically more useful than concrete systems. Largely because of this great influence from the late 1940s to the 1960s in the macrotheory of sociology and other social sciences, the systems theory that became dominant in these fields was generally stated in abstracted systems terms. This was not true in biology, so there was a gulf between social and biological theory.

Parsons' systems theory and mine differ in another important way: our categorizations of subsystem processes. His general theory, as stated by Parsons and Smelser,[26] identifies four subsystems which must address four problems or "independent functional imperatives." The four subsystems are: the latent pattern-maintenance and tension-management subsystem; the goal-attainment or polity subsystem; the adaptive or economy subsystem; and the integrative subsystem. Each of these subsystems, in turn, differentiates into subsystem processes. Unfortunately, abstracted systems theory has produced no large body of quantitative research to support it, because the theoretical statements which characterize it are difficult to test by any form of data collection.

My theory identified not four but 19 subsystems, each of which is a process found in all living systems at each of seven biological and social levels in an evolutionary sequence: cells, organs, organisms, groups, organizations, societies, and supranational systems.

Living systems, in my view, are a subset of the class of concrete systems. They are accumulations of matter and energy in specific regions of physical space-time. They obey the natural laws that govern all concrete systems. In addition, all living systems contain complex organic macromolecules, including DNA, RNA, and proteins, which confer upon them special properties associated with life.

The subsystems of living systems process various sorts of matter, energy, and information in input, internal, and output processes (Table 1). These activities are required if a system is to maintain a steady state of negative entropy over a significant period, which is the fundamental definition of life. This is possible because they are able to adjust a set of essential subsystems to take from their environment substances of high negative entropy, use them for growth, reproduction, repair, and other processes, and return to the environment substances of lower negative entropy.

Experience has shown that it is not difficult to conduct quantitative research investigating propositions concerning variables in these subsystems and the total system.

Table 1. The 19 Critical Subsystems of a Living System

Subsystems which process both matter-energy and information

1. *Reproducer*, the subsystem which is capable of giving rise to other systems similar to the one it is in.

2. *Boundary*, the subsystem at the perimeter of a system that holds together the components which make up the system, protects them from environmental stresses, and excludes or permits entry to various sorts of matter-energy and information.

Subsystems which process matter-energy	Subsystems which process information
3. *Ingestor*, the subsystem which brings matter-energy across the system boundary from the environment.	11. *Input transducer*, the sensory subsystem which brings markers bearing information into the system, changing them to other matter-energy forms suitable for transmission within it.
	12. *Internal transducer*, the sensory subsystem which receives, from subsystems or components within the system, markers bearing information about significant alterations in those subsystems or components, changing them to other matter-energy forms of a sort which can be transmitted within it.
4. *Distributor*, the subsystem which carries inputs from outside the system or output from its subsystems around the system to each component.	13. *Channel and net*, the subsystem composed of a single route in physical space, or multiple interconnected routes, by which markers bearing information are transmitted to all parts of the system.
5. *Converter*, the subsystem which changes certain inputs to the system into forms more useful for the special processes of that particular system.	14. *Decoder*, the subsystem which alters the code of information input to it through the input transducer or internal transducer into a "private" code that can be used internally by the system.
6. *Producer*, the subsystem which forms stable associations that endure for significant periods among matter-energy inputs to the system or outputs from its converter, the materials synthesized being for growth, damage repair, or replacement of components of the system, or for providing energy for moving or constituting the system's outputs of products or information markers to its suprasystem.	15. *Associator*, the subsystem which carries out the first stage of the learning process, forming enduring associations among items of information in the system.

Continued

Subsystems which process matter-energy	Subsystems which process information
7. *Matter-energy storage*, the subsystem which retains in the system, for different periods of time, deposits of various sorts of matter energy.	16. *Memory*, the subsystem which carries out the second stage of the learning process, storing various sorts of information in the system for different periods of time.
	17. *Decider*, the executive subsystem which receives information inputs from all other subsystems and transmits to them information outputs that control the entire system.
	18. *Encoder*, the subsystem which alters the code of information input to it from other information processing subsystems, from a "private" code used internally by the system into a "public" code which can be interpreted by other systems in its environment.
8. *Extruder*, the subsystem which transmits matter-energy out of the system in the forms of products or wastes.	19. *Output transducer*, the subsystem which puts out markers bearing information from the system, changing markers within the system into other matter-energy forms which can be transmitted over channels in the system's environment.
9. *Motor*, the subsystem which moves the system or parts of it in relation to part or all of its environment or moves components of its environment in relation to each other.	
10. *Supporter*, the subsystem which maintains the proper spatial relationships among components of the system, so that they can interact without weighting each other down or crowding each other.	

Cross-Level Hypotheses

Numerous hypotheses can be stated concerning the structure and process of living systems. The term "hypothesis" is used here in the restricted sense of a proposition that can be confirmed or disconfirmed empirically rather than in the more general sense of any expression which is capable of being believed, doubted, or denied. As employed here, "hypothesis" and "proposition" are essentially synonymous.

A fundamental procedure in science is to make generalizations from one system to another on the basis of some similarity between them. One or more variables of a given system may be observed to change in the same direction as the same variable of another system, under comparable conditions. If these variations are so similar that they can be expressed by the same mathematical model or function, a *formal identity* exists between the two systems. If different models or functions are required to express the variations, there is a formal *disidentity*. The systems may be at the same or different levels.

A formal identity among concrete systems is demonstrated by a procedure composed of three logically independent steps: (a) recognizing an aspect of two or more systems which has comparable status in those systems, (b) hypothesizing a quantitative identity between them, and (c) empirically demonstrating that identity within a certain range of error by collecting data on a similar aspect of each of the two or more systems being compared.

By such a procedure, a set of observations at one level of living systems can be associated with findings at another, to support generalizations that are far from trivial. It may be possible to use the same conceptual system to represent two quite different sorts of concrete systems, or to make models of them with the same mathematical constructions. It may even be possible to formulate useful generalizations which apply to all living systems at all levels. A comparison of systems is complete only when statements of their formal identities are associated with specific statements of their interlevel, intertype, and interindividual disidentities. The confirmation of formal identities and disidentities is done by empirical study.

Empirical research on hypotheses of three degrees of generality is possible in the study of living systems. The first predicts that a formal identity and/or disidentity is present *across individual systems of the same species or type*. A research that tests the hypothesis that individual rats differ among themselves in the rates at which they characteristically learn to run mazes is of this sort.

Cross-type or cross-species generalization is the second degree of generality of hypotheses. Such generalizations are more powerful than the first sort. Also they involve more variance. Researches in comparative anatomy, comparative physiology, comparative psychology, and comparative sociology are of this sort. The relative size of skulls of rats, cats, dogs, dolphins, apes, and human beings, for instance, can be compared with the average measures of intelligence of each species to confirm the hypothesis that animals with larger skulls—and larger brains—have greater cognitive capacity. Of course there is also variance among individual members of a single species.

The third type of scientific generalization is from one level to another. These *cross-level generalizations* will, ordinarily, have greater variance than the other sorts since they include variance among types and among individuals. They can, however, be made. They can enable insights and discoveries made at one level to give rise to general principles which apply to multiple levels of living systems and, in some cases, to nonliving systems and technologies as well. As understanding of systems in general increases, fragments of knowledge may coalesce into patterns and science as a whole may become more coherent. Indeed, evaluating the validity of cross-level hypotheses may become a very significant innovation in the study of living systems.

Scientists undertake cross-level research less frequently than cross-individual or cross-type. There are a number of reasons for this. Scientists are trained to be specialists in one or another of the many disciplines and subdisciplines into which the subject-matter of science is traditionally divided. They limit the scope of their expertise on the grounds that one person cannot be informed in depth and expert on a broader range of phenomena. The usual patterns of rewards to scientists, including academic promotion and tenure, recognition in national scientific societies, grant awards, prizes, and so forth reinforce specialism. What is more, the traditional cautiousness of scientists makes them skeptical of any generalization from, say, a cell, to a system as different as an organism or a society. Scientists working at one level of living systems are rarely current with the literature at other levels. This is demonstrated by the paucity of references to researches at other levels in the texts or bibliographies of scientific papers. As a result, when a scientist discovers a principle at one level, he typically does not think it likely enough to be relevant to another level to take the trouble to find whether similar principles have been reported in the literature at other levels.

General living systems theory in no way denies the importance of specialists' knowledge or the advancement of understanding of phenomena in limited scientific areas. Specialized researches will properly continue to be the chief means by which science advances. At the same time, however, there should be studies of generalities or similarities, as well as differences, among different classes of phenomena.

The more general a scientific law is, the more powerful it can be because it applies to a larger class of phenomena. It is important, therefore, that there be more scientific effort devoted to testing a wide range of cross-level hypotheses. After all, the most recalcitrant problems now facing societies are not limited to single disciplines or single levels. Resolution of environmental problems, for example, often depends on knowledge derived from biology, the earth sciences, chemistry, biochemistry, engineering, economics, other social sciences, and the law. Increasingly, teams composed of experts in a number of fields work together.

One difficulty that arises in such interdisciplinary cooperation is the use of different terms, units, and dimensions, some of them applied only to a particular special field. The relationships among them are often unclear and it is difficult to translate them so that people who are not specialists in that field will understand them. One solution to this dilemma is to attempt to use the dimensions of the natural sciences as neutral reference points to which all other dimensions can be precisely related.

It is helpful, also, to use neutral terms to recognize similarities that exist in systems of different types and levels. These should be as acceptable as possible when applied at all levels and to all types of living systems. For example, "sense organ" is one term for the component that brings information into a system at the level of the organism, but "input transducer" is also satisfactory, and it is a more acceptable term to engineers and at the society level (e.g., a diplomat or foreign correspondent). Such usage may irritate some specialists used to the traditional terminology of their fields. A language that intentionally uses words that are acceptable in other fields is, of necessity, not the jargon of the specialty. Whoever uses it may be suspected of not being informed about the specialty. The specialist languages, however, limit the horizons of thought to the borders of the disciplines. They mask important intertype and interlevel generalities.

The Rise of Cross-Level Research

Although cross-level research has been uncommon, there is at least one continuing sequence of studies going back to early years of this century which deals with two levels of living systems.[33,36] These researches measure problem-solving and learning variables of individual persons and of groups. This sequence of studies probably was carried out only because persons and groups were both considered proper subjects for one particular discipline—psychology. Had they been objects of study in two fields, the sequence might never have begun. They were not conceptualized explicitly as cross-level studies.

During the 1950s and 1960s few cross-level propositions were given consideration. A notable exception is the work of Guetzkow,[10] who applied concepts developed in the study of groups to nation-states in an attempt to develop a general theory of intergroup relations applicable to social and community groups, political parties, cities, states, and both regional and international organizations.

Also in the 1960s, Berelson and Steiner published a number of cross-level propositions that they considered testable.[3]

In 1953[18] I suggested 10 cross-level formal identities worth examining. In 1955[19] I produced a list of more precisely stated cross-level hypotheses, some original and some suggested by other scientists. In 1965[21] I listed 165 such hypotheses, and in 1978 (Ref. 22, pp. 89–119) I listed 173. Some have been confirmed by research at one level and have been demonstrated to have parallels at another level. Others appear to be testable at, and possibly relevant to, multiple levels.

By 1984 several cross-level studies had been published by scientists in various fields and in several countries. Other such researches were being planned or carried out.

These studies are of several sorts. Some are experiments designed to test a particular hypothesis at more than one level. Some marshal data from a number of sources as support for a cross-level hypothesis. In some cases several scientists in different disciplines worked independently, with results that tend to confirm the same hypothesis. Finally, some apply a mathematical model to data from a variety of systems to demonstrate its generality. Some of the scientists stated that they were evaluating a cross-level hypothesis, but others made no explicit statement of such a purpose and, indeed, may have had no such intent.

Below is a list—almost certainly incomplete—of such cross-level studies:

Cross-Level Study 1: Miller et al.

Information Input Overload. The first cross-level hypothesis tested in the laboratory in experiments explicitly designed to measure the relationship between two comparable variables—one independent and one dependent—at multiple levels of biological and social living systems, concerned the effects of an overload of input information at the levels of the cell, organ, organism, group, and organization. As of 1984 no other integrated experimental research across more than two levels has been published. Societies and supranational systems (like the European Economic community) were not included in our research because no experimental procedure was devised to collect controlled empirical data upon such large systems.

These experiments were done at the University of Michigan in the late 1950s, published in 1960,[20] and republished in expanded form in 1978 (Ref. 22, pp. 121–202).

The "information input overload" hypothesis can be stated as follows:

As the information input to a single channel of a living system—measured in bits per second—increases, the information output—measured similarly—increases almost identically at first but gradually falls behind as it approaches a certain output rate, the channel capacity, which cannot be exceeded in the channel. The output then levels off at that rate, and finally, as the information input rate continues to go up, the output decreases gradually toward zero as breakdown or the confusional state occurs under overload.

Experiments at each level were conducted by specialists in relevant fields. Measurements of information input and output rates were made on single fibers from the sciatic nerves of frogs, optic tracts of several white rats (retina or optic nerve to optic cortex), human subjects working alone, human subjects in groups of three, and laboratory "organizations" made up nine subjects. In one form of organization (Oz-2) there were three groups of three each. Two of these groups simultaneously received information inputs. In each of these two parallel groups two members served as input transducers, each person receiving a different sequence of visual inputs. Each of these four persons serving as components of the input transducer subsystem pressed the appropriate button that corresponded to each signal he saw. This sent a signal over an electronic channel to a display before the third member of his group. This third member of each group output the signals he received by an electronic channel to a display before one member of a three-person group in the next room. Each of these forwarded the signals he received to the final person in the second room, the organization's decider and output transducer. He compared the signals he received from the two groups in his room, made a decision about them for the total organization, and output this decision signal by an electronic channel to an on-line device that recorded his decision and the time when it was received. This provided, in extremely simple form, the multiple-echelon structure that differentiates an organization from a face-to-face group. Data from the five biological and social levels yielded information input-output curves, the similar shapes of which—a cross-level formal identity—confirmed the hypothesis (Fig. 1). Measured in bits per second, the average rate of information transmission at each higher level was lower than at the previous level, an orderly disidentity across levels. In addition, a number of adjustment processes to the increasing overload were found to be common to several or all levels. The more complex systems, in general, used more adjustment processes, continuing to use those that were found at lower levels but also adding new ones in addition.

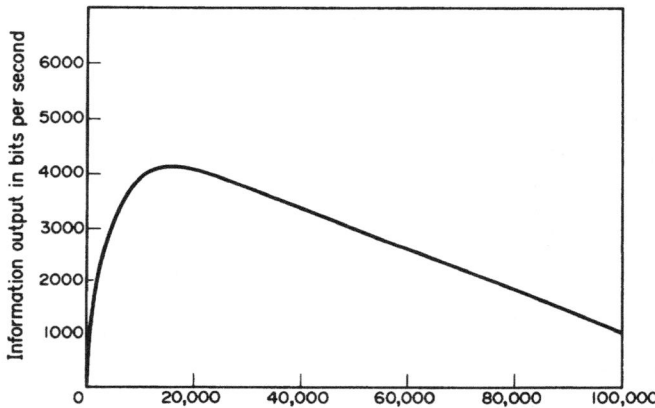

Figure 1. Information input–output curve (level of the cell).

Cross-Level Study 2: Rapoport et al.

Structural Models of Information-Processing Networks. About the same time that the above study was being done, my colleague Rapoport and a group working with him did experiments on the structure of communication networks. He had previously done several theoretical papers about mathematical models of nets in neural tissues.[31] These papers led his group to do research on the following cross-level hypothesis:

The structures of the communication networks of living systems at various levels are so comparable that they can be described by similar mathematical models of nonrandom nets.

Nonrandom nets are those in which flows are more probable over some channels than over others. They are "biased." Biases that affect the probability of one neuron communicating with another include distance; a reciprocity bias, which concerns the probability that if one component contacts another, the other will also contact the first; and a popularity or "field force" bias, the probability that some will be more "attractive" and be contacted more frequently than others.

A similar model was found to predict with reasonable accuracy some aspects of a sociogram representing the friendship network in organizations (elementary and junior high schools).[32] Studies by Dodd of diffusion of information in towns have shown that a similar biased net model would probably be applicable there as well.[5]

Cross-Level Study 3: Odum

The Maximum Power Principle. Odum, an engineer and ecologist, has carried out an extensive investigation of a proposition he calls the maximum power principle.[25] This principle was derived from Lotka's observation that systems prevail which develop designs that maximize the flow of useful energy.[16]

Odum has expanded this idea and for years studied the relationships between efficiency and power in open systems, living and nonliving. In an early article with Pinkerton, he stated the principle in the form of a hypothesis, as follows:

Natural systems tend to operate at that efficiency which produces a maximum power output. This efficiency is always less than the maximum efficiency.[24]

A major design principle of natural systems, he finds, is feedback from storage to energy inflow pathways. This stimulates energic inflow and functions as a reward. In this way, processes that are doing useful work are reinforced. Such feedback designs are *autocatalytic*. They maximize power, consequently generating more entropy. Autocatalytic processes depend upon sufficient concentrations of potential energy—energy available for doing work. If the energy source is weak, the system minimizes energy flow and entropy generation. The maximum power principle suggests that natural selection operates to select system designs that maximize power. Systems compete for available energy.

Systems of this sort develop hierarchies with successive transformations of energy in which energy increases in "quality" as units increase in size. The chains and webs through which matter is cycled in natural systems are of this sort, as are the food chains and food webs of ecological systems. Larger sources of energy support longer chains that lead to larger systems. The larger systems at the top of the chain have less energy available to them than do the microorganisms at the bottom but it is of higher quality. In any such chain one unit is the producer, which transforms energy and stores it for use by the next system unit, the consumer, which also has production processes.

Producer-consumer hierarchies often maximize power by oscillating or pulsing their feedback to the larger systems in which they are included. Frequency of pulses is related to the amount of energy stored. Ecosystems often have long periods of gradual buildup followed by sudden pulsed consumption and recycling. This may be catastrophic to the smaller systems included in them. Fire climax ecosystems that alternative growth and fire are examples. Larger systems oscillate over longer time-scales.

Odum models many sorts of systems in diagrams that use the symbols of his "energy circuit language." Among them are chemical and biochemical processes, populations of microorganisms, the life cycles of organisms, limited ecological systems such as marshes or swamps, large ecosystems, the whole biosphere, economic systems, and cultural systems. He also translates Forrester's world model (see below) into energy circuit language.

Cross-Level Study 4: Mountcastle, Stevens, Trappl

Weber Function. The Weber function is a statement of the relationship between the intensity of input and the intensity of output in an input transducer or sense organ of a living system. It has been tested in separate researches at the levels of cell and organism. The Weber function can be stated as a hypothesis as follows:

The intensity output signal of a system varies as a power function of the intensity of its input, the form of the power function being $\Psi = k(\Phi - \Phi_0)^n$, where Ψ is the intensity of the output signal, Φ is the physical magnitude of the input energies, Φ_0 is a constant, the physical magnitude of the minimum detectable or threshold input energies, k depends on the choice of measurement units, and the exponent n varies with different sensory modalities of the information processed.

At the organism level, Stevens found that this Weber function applied to the relationship between the intensity of sensory input and the subjects' reports of the intensity of their sensations.[35] For each sensory modality, the exponent n was different. It ranged from 0.33 for brightness of white light to 1.45 for heaviness of lifted weights, 1.6 for warmth, and 3.5 for electric shock to the finger.

Mountcastle investigated the "touch-spot" input transducer cells in the hairy skin of cats and monkeys.[23] These are minute domelike elevations of the skin from which exquisitely sensitive afferent fibers carry information toward the central nervous system. An apparatus controlled both the amount of indentation applied by mechanical force and the timing of stimulation. Mountcastle recorded the pulses along the nerve fibers that followed mechanical pressure on the sensory cells. Such fibers respond to stimulation by a maintained steady rate of discharge. He found that the input-output relationship of each of these cells was accurately described by a power function.

Events at these two levels are clearly not independent. Mountcastle suggests that the characteristics of input transducer components determine the nature of the organism's responses (Ref. 23, pp. 96, 102, 109). That is, the internal processing that occurs between inputs to a person's sensory cells and his report about them transmits the information in linear fashion.

A related study by Trappl et al. demonstrated that the relationship between the optic generator potential of the electroretinogram of enucleated rabbit eyes and light intensity was a power function.[38] Eight different logarithmically scaled luminances were used. The function was nearly identical to the psychophysical function that relates luminance to subjective brightness.

Cross-Level Study 5: Forrester et al.

Dynamic Models. Forrester and his colleagues have developed dynamic models which have been used to simulate the interactions and feedbacks among critical variables in complex systems. At the level of the organization he has studied industrial dynamics of various sorts of corporations[7] and urban dynamics.[8] At the level of the supranational system or the total planet composed of both living and nonliving components, he has developed world models.[9]

With such models it is possible to trace the consequences of such events as a rise in retail sales, a decrease in federal grants to a city, or a growth in world population. Both short-term and long-term impacts of such occurrences on all parts of the system can be examined. These models also have been used to extrapolate trends into the future and to compare the effects of alternative policies upon a system.

In each model, key variables of a particular real system are selected as state variables, "levels." Forrester uses this term in a different sense than we do. He defines the term as a variable representing a state or condition of a system at any time, an accumulation of the net quantity resulting from past flow rates that add to or subtract from the state it measures.

He does not have an overall conceptual system that determines what "levels" should be considered for inclusion in every model of any particular real system. Rather, he selects "levels" which he considers to be most relevant to the problem under consideration. For instance, he may choose the following six "levels" for a model of an industrial organization: materials, orders, money, personnel, capital equipment, and information, each of which passes through a network in the system (Ref. 7, pp. 60–72, 82). For another sort of organization, an urban community, he might select for a particular study the following nine "levels": new enterprise, mature business, declining industry; premium housing, worker housing, underemployed housing; managerial-professional personnel, laborers, and underemployed (Ref. 8, pp. 15–17). For a model of a higher level of system, the entire world, he might select as "levels" the following five: population, capital investment, natural resources, fraction of capital devoted to agriculture, and pollution (Ref. 9, pp. 18–24). Of course, the availability of data may determine what levels are chosen or the degree of aggregation of the data that is employed in any particular model.

Positive and negative feedbacks are major influences in controlling the varying values of "levels" in a model over time. In the total system, feedbacks among levels are intricate and often produce unexpected and unintended effects.

The interactions of variables in the simulations are modeled in a set of equations. A special-purpose compiler computer program, "Dynamo," tabulates data and generates graphical plots that depict the "levels," flows and feedbacks of the system being simulated.

Forrester and his associates have clearly demonstrated that his sort of dynamic models and simulations give important insights into both organizational and world processes. In this sense they undeniably have cross-level applicability. If he were to select his "levels" on the basis of some conceptual system which he thought applied to all systems at the levels that concern him, it would then be possible to test cross-level hypotheses with such dynamic models. So far, however, no such uniformity of "levels" among his models of organizations, communities, or the entire world has been attempted.

Cross-Level Study 6: Prigogine et al.

Thermodynamic Models of Self-Organizing Systems. Prigogine's work hypothesizes a fundamental formal identity among all self-organizing systems. They develop, and maintain over time, orderly structural organization far from thermodynamic equilibrium. This equilibrium is defined as entropic random distribution of components that, according to the second law of thermodynamics, is the ultimate state of thermodynamic systems.[29,30] This is an important definition of life, very like that which appears earlier in this paper.

Self-organizing, autocatalytic systems are complex, with a large number of heterogeneous, interacting components. They are open to input and output flows of matter and energy which, under proper feedback conditions, may cause fluctuations that destabilize the system so that "branching" or "bifurcation" of its states occurs. This constitutes a transition between the ordinary progression of thermodynamic states and new, stable, and more complex configurations. These are "dissipative" structures that can be maintained far from equilibrium only with sufficient flows of matter and energy. How such structures develop cannot be accurately predicted from the initial state of the system. Only probability statements can be made about future states.

Some chemical reactions have the characteristics of dissipative structures. Cells, which involve thousands of coordinated and regulated chemical reactions, also exhibit such order. Biological organization requires a hierarchy of structures and functions of increasing complexity. Prigogine and his colleagues are concerned with how transitions between levels occur, such as from molecular to supramolecular, or from cellular to supracellular systems. They have extended their studies to social as well as biological systems.

The actions of these different sorts of systems are represented in models based on the thermodynamics of irreversible processes. System processes are analyzed by linear and nonlinear differential equations and nonlinear mathematical techniques such as bifurcation theory, catastrophe theory, and stability theory, when appropriate. The object of such models is to interpret biological order in terms of the laws of macroscopic physics which are, essentially, the laws of thermodynamics.

Systems modeled by Prigogine and his colleagues include biochemical reactions at subcellular and cellular levels, for example, chemical reactions in which oscillations between states occur, like those in the glycolytic cycle of cells. Other cellular models deal with the synthesis of protein and regulation of the cellular cycle.

In evolution, branching and transition to new forms of order occur. Chemical processes, such as the formation of polymers like those that must have been formed in prebiotic evolution, can be modeled in the same way that Prigogine analyzes other chemical transitions. Laws that describe the growth, decay, and interaction of biological populations appear to Prigogine to be very closely analogous to certain laws of chemical kinetics, particularly to those of competing biopolymers.

Thermodynamic models of dissipative structures far from equilibrium have also been used by Prigogine and his collaborators to simulate large-scale social processes. The U.S. Department of Transportation has sponsored their program of research into how the development of public transportation routes such as freeways, other roads, canals, and railroads influence the spatial structure and the function of urban and regional systems. These were modeled as dynamic, nonlinear entities.[1]

A model of the Bastogne region of Belgium by a group established by Prigogine and led by Allen used basic economic and census data as well as information about employment and residence to study the region's spatial arrangement. The effects of transportation planning decisions upon community structural evolution and reorganization were investigated. Six types of variables were used to model or simulate the community: location of industry, which requires much space per employee; location of government, insurance, and finance activities, which require less space per employee; location of the service activities; location of the retail sector; location of low-income, blue-collar residents; and location of the medium or high income, white-collar residents. Using these variables and a selected set of parameters, simulation generated a variety of possible community spatial configurations. In some of these, industrial employment was centralized and in others it was decentralized.

A series of experiments were conducted to evaluate the quality of vehicular traffic in Austin, Texas. They made use of a kinetic, "two-fluid" model developed by Herman and Prigogine.[11] This assumes that the average running speed in a street network is proportional to the nth power of the fraction of the vehicles that are moving and that the fractional stop time of a test vehicle circulating in a network is equal to the average fraction of the vehicles stopped during the same period. The two "fluids" consist of moving cars and cars stopped as a result of traffic conditions.

One experiment involved eight automobiles, each with a driver and observer. Each car circulated so that starting places were random and each followed a car that happened to be in the test area until it either parked or left the area and then followed another one. Odometer readings and the times at the beginning and end of each two-mile trip, as well as the times for stops and starts, were recorded. Similar observations were also made in Dallas, where the timing plans of traffic signals were varied, and in Houston.

Cross-Level Study 7: Leontief et al.

Input-Output Models. An input-output economic model developed by Leontief has been applied by him to multiple levels of social systems—regional, national,[13] and global.[14]

He identifies various sectors in an economy and assumes that comparable sectors exist in the economies at the levels of regions of countries, nations, regions of the world, and the entire world. He apparently accepts it as a fact that economic systems are not basically rearranged structurally with fundamentally altered processes at each of these levels.

Consequently he is able to make comparable quantitative input-output economic analyses across multiple levels of social systems.

In all his studies he has identified a number of economic sectors. The number of these has differed widely from one study to another—from as few as 13 to as many as 192.[13] He has attempted to make his sector classifications compatible with those of the *Standard Industrial Classification Manual.*[6] The differences in number of sectors depend primarily upon the degree and type of aggregation of data selected for a given study. To carry out cross-level researches it is, of course, necessary to use similar classifications of sectors and degrees of aggregation of data. If this is done, however, it is possible to analyze cross-level identities and disidentities in the structures and dynamic changes that occur in the consistent patterns of interdependence of sectors which Leontief finds in all economic systems.

In addition to its use as a general-purpose model of economies, input-output analysis has been employed to study the possible futures of developing countries as well as world military spending.

Leontief's input-output models consist of sets of differential equations that represent the producing and consuming sectors of the economies being modeled. These sectors are connected with each other by flows of goods and services. Input and consumption coefficients describe the combination of goods and services needed by each sector to produce a unit of its output, or, in the case of households, a unit of their expenditures and income. Stocks of capital goods and natural resources are also included. Data for the models are drawn from regional or national statistics, or from sources like the United Nations, which collect data about national economies.

The complex systems of equations in the model are solved by computer. The resulting data are presented in tables that facilitate finding the inputs to all sectors from each of the others and the destinations of the outputs of each. By varying the estimates of total stocks and inputs, the model can be solved for different sets of initial conditions and the effects of policies can be calculated. Like the dynamic models of Forrester, these can be projected into the future to predict long-term outcomes.

In their global model Leontief and his colleagues divide the world into 15 regions, each with the same 45 sectors of economic activity (Ref. 14, p. 2). Each region is treated separately but a set of complex linkages brings them together. Among these are imports and exports, capital transfers, aid transfers, and foreign interest payments. Starting with data from 1970, this model was run with alternative projections for the years 1980, 1990, and 2000. Eight alternative sets of assumptions about, for example, growth rates of populations, growth rate of per capita income, and pollution abatement created eight "scenarios" for the future of the world economy. They indicated that no insurmountable physical barriers exist in this century to sustained economic growth in the world's developing regions. What hindrances there are appear to be political, social, or institutional.

Cross-Level Study 8: Lewis

Decision-Making with Conflicting Command Signals. A research by Lewis on individual human organisms and on groups was designed as a cross-level test of a hypothesis that relates to the decider subsystem of living systems[15]:

Can Systems Generate Testable Hypotheses? 645

When a system is receiving conflicting command signals from several suprasystems, it intermittently being a component of all of them, the more different the signals are, the slower is its decision making.

Subjects in these experiments played a game similar to the arcade game "Star Trek," a war game against an extragalactic enemy. A computer presented "commands" from five officers, all of the same rank and all superiors of the subjects. The subjects had to decide which command to follow if conflicting commands were received. Responses and other data on the experiments were stored by a computer.

Subjects could "fire," "warp," "scan," "dock," or "wait." Eight patterns of commands with differing amounts of conflict were used, ranging from total agreement upon which action to take to total disagreement among the five officers. Subjects were permitted to take actions only if they were ordered by at least one officer.

In the experiments with individual subjects, after each decision, the computer presented its result and a summary of all the results in the game. Each subject was told how many ships he had lost, how many enemy ships he had destroyed, how many sectors of space he had explored, how much energy and time he had left, and so forth.

After 40 turns, a final score was calculated, based on his success in defending his own ships and in destroying the enemy. The game was scored so that the greater the number of officers that agreed on a command, the more likely it was to be successful. Each subject was paid according to his success.

In the group experiment, three subjects worked together to arrive at a decision. Otherwise the procedure was like that for single subjects.

Both the individual and group experiments yielded highly significant results that supported the hypothesis. Decision time was longer for groups than for individual persons in all command situations. Lewis suggested that the work could be extended to the organization level, using at least three groups of three subjects, with one group receiving recommendations from the other two and acting as the final decider.

Cross-Level Study 9: Marchetti

Society as a Learning System. In 1981 Marchetti published an article on a cross-level formal identity comparing the growth of a child's vocabulary, a learning process at the organism level, with three examples of learning processes at the society or supranational level.[17] Each of these learning processes was demonstrated by sets of data which produced comparable growth curves. The first of these was a study by Whiston plotting the growth of a child's vocabulary from about 20 months to about 70 months along the time axis vs. the fraction of the set of 2500 words finally learned that were known at roughly equal time intervals between 20 and 70 months.[39] The data approximated a straight line growth curve.

Marchetti believes that societies or the entire world are systems which demonstrate similar learning growth curves. He collected data for three examples: (1) Discovery of a set of about 50 stable chemical elements over the time between about 1735 and 1840. He plotted the year of their discovery against the fraction of the total set discovered at roughly equal intervals over that period. These data produced a similar straight line learning curve. (2) The evolution of efficiency in steam engines between about 1700 and 1970. He plotted, over this period at roughly equal time intervals, the efficiency of what currently was the best commercial engine, against its inefficiency (that is, the maximum possible thermodynamic efficiency minus the efficiency of the current engine). Again the data fitted a straight line learning curve. (3) Somewhat similar curves were shown for the efficiency of iron making in the U.K. and electrical production in the world between approximately 1925 and 1970.

In this article Marchetti also reported certain conclusions about the cyclic nature of social learning processes, but he stated no cross-level hypothesis about these cycles.

Cross-Level Study 10: Staw, Sandelands, and Dutton

Effect of Threat on Flexibility of Performance. Staw, Sandelands, and Dutton surveyed the literature at levels of organism, group, and organization for evidence bearing upon a "threat-rigidity hypothesis" which they state as follows (Ref. 34, p. 502):

> ... a threat results in changes in both the information and control processes of a system, and, because of these changes, a system's behavior is predicted to become less varied or flexible.

A threat is defined in this study as an environmental event that has impending negative or harmful consequences for a system. An inadequately variable or flexible response is considered "rigid." The authors further propose that such behavior may be maladaptive when the threatening situation includes major environmental changes, since flexibility and diversity of response have survival value in such conditions and prior, well-learned responses are often inappropriate. If no major change is threatened, rigid but previously successful responses may be appropriate.

Organism Level Effects. At this level, threatening situations produce psychological stress, anxiety, and physiological arousal in human subjects. Threatening research situations have included the possibility of electric shock, feedback about failure on preceding tasks, excess pacing of tasks, and formal, unfriendly experimental settings. Standard tests of anxiety, such as the Taylor Manifest Anxiety Scale, can determine the degree of anxiety present.

Among findings that tend to support the threat-rigidity hypothesis are changes in perception. Subjects under stress in some such experiments had reduced ability to identify and discriminate among visual stimuli and were more likely than nonstressed subjects to use previously held "internal hypotheses" about the identity of unfamiliar stimuli. Similar effects were found with anxiety and arousal. Anxious subjects were less able to discriminate visual detail than those who did not have anxiety. Under arousal, sensitivity to peripheral cues was decreased, narrowing the range of cues processed.

Problem-solving and learning are also affected. Stressed subjects have been found to adhere to a previously learned solution even when it was not appropriate to the problem at hand. They are also less flexible than nonstressed subjects in their choices of solution methods.

Natural disasters have provided further insights into threat-rigidity effects. Clinical studies and observations have found narrowing of the perceptual field, limitations of the amount of information input, rigidity of response, and primitive forms of reaction in disaster victims. They may fail to heed warnings or follow directions, or may "freeze up" and be incapable of action.

Group Level Effects. Investigations of group responses to threat have been concerned with its effects upon group cohesiveness, that is, upon the tendency of group members to remain together; upon group leadership and control; and upon pressures toward uniformity within groups. In the research surveyed, threats to attainment of group goals included competition from other groups and radical change in rules for a group game that made it impossible for the group to predict what behavior would lead to success.

When a threat was attributed to an external source and the group thought it could be successful in meeting it, group cohesiveness was found to increase. Group members supported their leadership, and there was pressure upon members not to deviate from the majority. The search for consensus involved restricting information, ignoring divergent solutions, and centralizing control so that the opinions of dominant members prevailed. These results appear to support the threat-rigidity hypothesis.

When the threat was attributed to internal deficiencies of the group itself and success appeared unlikely, the rigid behavior did not appear. In this experimental condition, cohesiveness is reduced and leadership becomes unstable. New leaders and new forms of consensus may arise which promise improved group achievement.

Organization Level Responses. Threats at this level have included scarcity of resources, competition, and reduction in the size of markets. Three sorts of responses have been found: (1) Because of overload of communication channels, reliance on prior knowledge, and reduction in the complexity of communication, the information-processing capacity of an organization may be restricted. (2) Centralization of authority and increased formalization of procedures may lead to a constriction in control. (3) There may be increased efforts to conserve resources within the system through cost-cutting and efforts for greater efficiency. Depending upon the circumstances, these responses may be functional or dysfunctional.

Straw, Sandelands, and Dutton conclude that the evidence they have gathered supports the threat-rigidity hypothesis at the levels of the organism, the group, and the organization. They state that the relatively consistent findings at these three levels support a common explanation of reactions of systems to threat that can be stated best in the terms of general systems theory rather than the specialized languages of psychological or sociological theory.

Conclusion

The generalizing power of cross-level studies of living systems, particularly quantitative empirical experiments, is clear. Beginning in the 1950s, research explicitly designed to compare similar structures or processes in two or more biological and/or social systems has been conducted. In the past decade, interest in such work has accelerated but so far fewer than 20 individuals or groups have published cross-level studies. Applications of findings from cross-level studies have been made in the fields of organizational behavior,[4] psychopathology,[37] group psychology,[2] management science,[28] and sociology.[12]

If systems science is to become established as an experimental field, and if applications of systems science are to become more reliable, a fabric of fundamental cross-level research about many more types of biological and social structures and processes will be essential.

References

1. P. M. Allen, M. Sanglier, F. Boon, J. L. Deneubourg and A. DePalma, *Models of Urban Settlement and Structure as Dynamic Self-Organizing Systems.* U.S. Department of Transportation/RSPA/DPB-10/6, Washington, D.C. (1981).
2. A. P. Balutis, The role of the staff in the legislature: the case of New York. *Public Admin. Rev.* **35** (1975), 357.
3. B. Berelson and G. A. Steiner, *Human Behavior: An Inventory of Scientific Findings.* Harcourt, Brace, & World, New York (1964).
4. L. L. Cummings and T. A. DeCotiis, Organizational correlates of perceived stress in a professional organization. *Publ. Personnel Management* **2** (1973), 277.
5. S. C. Dodd, A test of message diffusion of chain tags. *Am. J. Sociol.* **61** (1956), 425–432.
6. Executive Office of the President, Bureau of the Budget, Division of Statistical Standards, Technical Committee on Industrial Classification, *Standard Industrial Classification Manual.* Government Printing Office, Washington (1945).
7. J. W. Forrester, *Industrial Dynamics.* M.I.T. Press–John Wiley, New York (1961).
8. J. W. Forrester, *Urban Dynamics.* M.I.T. Press, Cambridge (1969).
9. J. W. Forrester, *World Dynamics.* Wright-Allen Press, Cambridge, MA (1971).
10. H. Guetzkow, Isolation and collaboration: a partial theory of inter-nation relations. *J. Conflict Resolution* **1** (1957), 62.
11. R. Herman and I. Progogine, A two-fluid approach to town traffic. *Science, Wash.* **204** (1979), 148–151.
12. O. E. Klapp, Opening and closing in open systems. *Behav. Sci.* **20** (1975), 251–257.
13. W. Leontief et al., *Studies in the Structure of the American Economy: Theoretical and Empirical Explorations in Input-Output Analysis.* Oxford University Press, New York (1953).
14. W. Leontief et al., *The Future of the World Economy.* Oxford University Press, New York (1977).
15. L. F. Lewis, II, Conflicting commands versus decision time: cross-level experiment. *Behav. Sci.* **26** (1981), 79–84.
16. A. J. Lotka, *Elements of Physical Biology.* Williams and Wilkins, Baltimore (1926).
17. C. Marchetti, *Society as a Learning System: Discovery, Invention, and Innovation Cycles Revisited.* International Institute for Applied Systems Analysis, RR-81-29, Laxenburg, Austria (1981).
18. J. G. Miller, Introduction. In Members of the Committee on Behavioral Sciences, University of Chicago (eds), *Symposium: Profits and Problems of Homeostatic Models in the Behavioral Sciences*, pp. 1–11. Chicago Behavioral Science Publications (1), University of Chicago, Chicago (1953).
19. J. G. Miller, Toward a general theory for the behavioral sciences. *Am. Psychol.* **10** (1955), 513–531.
20. J. G. Miller, Information input overload and psychopathology. *Am. J. Psychiat.* **116** (1960), 695–704.
21. J. G. Miller, Living systems: cross-level hypotheses. *Behav. Sci.* **10** (1965), 380–411.
22. J. G. Miller, *Living Systems.* McGraw-Hill, New York (1978).
23. V. B. Mountcastle, The neural replication of sensory events in the somatic afferent system. In J. C. Eccles (ed.), *Brain and Conscious Experience*, pp. 85–115. Springer-Verlag, New York (1966).
24. H. T. Odum and R. C. Pinkerton, Time's speed regulator: the optimum efficiency for maximum power output in physical and biological systems. *Am. Sci.* **43** (1955), 331–343.

25. H. T. Odum, *Systems Ecology: An Introduction*. John Wiley, New York (1983).
26. T. Parsons and N. J. Smelser, *Economy and Society*, pp. 6–53. Free Press, Glencoe, IL (1946).
27. T. Parsons, Concrete systems and "abstracted" systems. *Contemp. Sociol.* **8** (1979), 696–705. See also T. Parsons, Concrete systems and "abstracted" systems. *Behav. Sci.* **25** (1980), 46–55.
28. G. J. B. Probst, *Kybertetische Gezetzes Hypothesen als Basis fur Gestaltungs- und Lenkungsregeln im Management*. Paul Haupt, Bern and Stuttgart (1981).
29. G. Nicolis and I. Prigogine, *Self-Organization in Nonequilibrium Systems: From Dissipative Structures to Order through Fluctuations*. John Wiley, New York (1977).
30. I. Prigogine, *From Being to Becoming: Time and Complexity in the Physical Sciences*. W. H. Freeman, San Francisco (1980).
31. A. Rapoport, Cycle distribution in random nets. *Bull. Math. Biophys.* **10** (1984), 145–157. See also A. Rapoport, Nets with distance bias. *Bull. Math. Biophys.* **13** (1951), 85–91; A. Rapoport, Contribution to the theory of random and biased nets. *Bull. Math. Biophys.* **19** (1957), 257–278; A. Rapoport, Nets with reciprocity bias. *Bull. Math. Biophys.* (1958) 191–201.
32. A. Rapoport and W. J. Horvath, A study of a large sociogram. *Behav. Sci.* **6** (1961), 279–291. See also C. C. Foster, A. Rapoport and C. J. Orwant, A study of a large sociogram II. Elimination of free parameters. *Behav. Sci.* **8** (1963), 56–65; C. C. Foster and W. J. Horvath, A study of a large sociogram. III. Reciprocal choice probabilities as a measure of social distance. *Behav. Sci.* **16** (1971), 429–435.
33. M. E. Shaw, A comparison of individuals and small groups in the rational solution of complex problems. *Am. J. Psychol.* **44** (1932), 491–504.
34. B. M. Staw, L. E. Sandelands and J. E. Dutton. Threat-rigidity effects in organizational behavior: a multilevel analysis. *Admin. Sci. Quart.* **26** (1981), 501–524.
35. S. S. Stevens, The psychophysics of sensory function. In W. A. Rosenblith (ed.), *Sensory Communication*, pp. 1–33. John Wiley, New York (1961).
36. D. W. Taylor and W. L. Faust, Twenty questions: efficiency in problem solving as a function of size of group. *J. Exp. Psychol.* **44** (1952), 360–368.
37. A. Toffler, *Future Shock*. Random House, New York (1970).
38. R. Trappl, A. v. Lutzow, L. Wundsch and H. Bornschein, A second-order model of the optic generator potential and its relation to Stevens' power function. *Pfluggers Archiv.* **372** (1977), 165–168.
39. T. G. Whiston, Life is logarithmic. In J. Rose (ed.), *Advances in Cybernetics and Systems*. Gordon and Breach, London (1974).

Author Index

Ackoff, R. L., 195, 211, 268, 325
Aiken, H., 507
Allen, T. F. H., 182, 195
Allen, W., 304
Arbib, M. A., 46, 187, 195, 202, 227, 362, 380, 396, 599, 604
Aristotle, 24, 490–491, 497
Arnold, V. I., 187, 195
Ashby, W. R., 36, 47, 102–105, 114, 125–126, 136, 155, 161, 167, 172, 195, 197–198, 211–212, 226–227, 249, 344, 347–348, 352, 397, 402–403, 405, 408, 410–411, 417, 420, 425, 448, 493–500, 507, 511–513, 519, 521, 523, 525, 528–529, 533–535, 571, 577, 598, 604
Atkin, R. H., 182, 195
Atlan, 298, 302
Auble, G. T., 182, 202
Auger, P., 142, 195
Aulin, A., 102, 182–184, 195, 226

Bangham, D. D., 564, 566
Barnsley, M. F., 187, 195–196
Barto, A. G., 196, 211, 377–378, 396, 587, 591
Bashkow, T., 315, 322
Barnes, F., 341
Bateson, G., 182, 196, 211, 285, 294, 301
Becker, K.-H., 106, 187, 196
Beckett, J. A., 182, 196
Beer, S., 182, 196, 268, 301
Bell, A. G., 329
Bell, D., 218, 226
Bellman, R., 135, 196, 311, 319–323
Bennett, S., 37, 196
Berelson, B., 637, 648

Berge, C., 187, 196
Berkeley, G., 230, 374
Berlinski, D., 165–168, 173, 196, 215
Bernard, C., 180
Berrien, F. K., 182, 196, 621, 628
Bertalanffy, L. von, 26, 33, 36, 167, 172, 196, 215, 250, 263, 268–269, 272–274, 279, 281, 283, 311, 327, 335, 525, 536, 608, 621, 625
Beurle, R. L., 535–536
Bigelow, J. H., 329, 335, 503
Blamire, J., 105, 202
Blauberg, I. V., 182, 196
Bogdanov, A., 38–39, 196, 232, 237
Boltyanskii, V. G., 318, 322
Bolzano, B., 367
Booth, T. L., 37, 102, 196
Boulding, K. L., 33–34, 163, 181, 196, 211–212, 239, 344, 352, 363, 374, 621, 628–629
Bourbaki, N., 406, 410, 417, 510, 515, 519
Bowler, T. D., 196
Bradley, D. F., 177
Brandom, R., 187, 203, 376
Bremermann, H. J., 121–122, 124, 196, 226, 374, 508, 510, 587, 591
Broekstra, G., 347–348
Brown, G. S., 73, 196, 301, 374
Buck, R. C., 163–165, 196
Buckley, W., 167, 182, 197, 503, 505, 621, 628
Bunge, M., 182, 197, 212, 593, 597–598, 602–605
Butterfield, H., 260, 267

Cameron, R. H., 320, 323

Author Index

Cameron, S., 161
Cannon, W. C., 631
Carello, C., 589
Carlyle, J. W., 321, 323
Carnap, R., 374, 593–595, 604
Carroll, L., 285, 289
Carson, E. R., 71, 198
Cartwright, T. J., 105, 197
Casti, J. L., 101, 113, 187, 197
Caswell, H., 182, 197
Catalano, J. S., 360, 374
Cavallo, R. E., 182, 197, 226–227, 348
Chadwick, G., 182, 197
Checkland, P. B., 71, 197, 211, 259, 268
Cherry, C., 331, 335
Christensen, R., 350, 352, 403
Chuang Tzu, 87, 101
Churchman, C. W., 266, 268
Cellier, F. E., 348
Clark, W. A., 161, 198
Cleland, C. I., 182, 197
Coming, P. A., 182, 197
Conant, R. C., 36, 102, 161, 174, 197, 212, 227, 348, 352, 403, 419, 448, 511, 513, 519
Conrad, M., 161, 197, 587, 590, 591
Cook, N. D., 188, 197
Copernicus, N., 237
Crick, F. H. C., 178
Crombie, A. C., 260, 267

Dahl, R. A., 624, 628
Dalenoort, G. J., 161, 197
Dantzig, G. B., 22
Demko, S. G., 187, 195
Descartes, R., 230, 360, 374
Desoer, C. A., 205, 319, 322
Deutsch, K. W., 167, 182, 198, 501, 505
Devaney, R. L., 187, 198
De Zeeuw, G., 347
Dodd, S. C., 639, 648
Dorfler, M., 106, 187, 196
Downs, A., 624, 628
Dubois, D., 187, 198, 352
Duhl, F. J., 182, 199
Dutton, J. E., 646–647, 649

Easton, D., 167, 182, 198
Eddington, A., 251
Eigen, M., 588, 591
Eilenberg, S., 277

Einstein, A., 41, 271, 283, 366
Ellis, W. D., 25, 198
Empiricus, S., 359, 374
Euclid, 305–306, 472
Evans, C., 373–374

Farley, B. G., 161, 198
Farlow, S. J., 161, 198
Faucheux, C., 105, 227
Feder, J., 187, 198
Feigl, H., 594
Feisel, L. D., 343
Feyerabend, P., 233, 237, 364, 375
Fisher, R. A., 250, 415, 510, 522
Fishwick, P. A., 187, 198
Flagle, C. D., 38, 198
Fleischmann, M., 187, 198
Flood, R. L., 71, 198
Foerster, H. von, 161, 198, 298, 302, 397
Fogel, L. J., 587, 592
Folger, T. A., 117–119, 185, 187, 200, 351, 353, 403
Forrester, J. W., 167, 182, 198, 640–641, 648
Fox, S., 564, 566
Frechet, M., 320, 323

Gaines, B. R., 10, 49, 161, 198, 211, 223, 227, 301, 345, 347, 350, 352, 355, 375, 396
Galileo, G., 278–279, 472
Gallopin, G. C., 198, 227
Gamkrelidze, R. V., 318, 322
Gardner, M. R., 103–105, 198, 227, 564, 566
Garey, M. R., 89, 127, 198, 227
Garner, W. R., 425, 448, 522, 536
Gause, D., 342
Gauss, K. F., 306
Gelfand, A. E., 105, 198, 227
Gellner, E., 364, 375
Gerard, R., 33, 621
Gershuny, J., 218, 227
Gibbs, J. W., 450, 592
Gibson, J. J., 589
Gilbert, E. J., 321, 323
Gill, A., 102, 199
Glasersfeld, E. von, 12, 199, 211, 229, 232, 234, 238
Gleick, J., 187, 199
Glickberg, I., 319, 322
Goel, N., 106, 186–187, 199

Author Index

Goffman, E., 294, 301
Goguen, J. A., 11, 30–31, 199, 211, 293, 298, 301–302, 362
Goldstein, G. D., 161, 205
Gose, E., 587, 592
Gray, W., 182, 199
Greenspan, D., 378, 388, 395
Gross, O. A., 319, 322
Guetzkow, H., 637, 648

Hai, A., 348
Haken, H., 19, 157, 161, 199, 588, 592
Halfon, E., 182, 199
Hall, A. D., III, 199
Hamilton, W. R., 278, 614
Hanan, J., 202
Hanken, A. F. G., 182, 199
Harary, F., 187, 199
Harel, D., 127, 199
Harrison, M., 598, 604
Hartley, R. V. L., 116, 199, 350, 352
Hartmanis, J., 102, 199, 515, 519
Hartmann, G. W., 25, 199
Hartnett, W. E., 199
Hastings, H. M., 590–591
Hegel, G. W., 370, 375
Henderson, L. J., 631
Herman, G. T., 68, 187, 199, 648
Higashi, M., 348, 350, 353
Himmelblau, D. M., 71
Hockett, E., 583, 592
Hofstetter, E. M., 321, 323
Holden, A. V., 187, 199
Holland, J. H., 161, 199, 587, 592
Holling, C. S., 182, 199
Hoos, I. R., 165–168, 173, 199
Horn, W., 174, 204, 354
Hufford, K. D., 157, 200
Hume, D., 359–360, 375
Hutchins, R. M., 537–540, 557

Islam, S., 41, 46, 199
Ivakhnenko, A. G., 161, 199

Jacobi, G. T., 161, 205, 614
James, W., 232, 238, 375
Jantsch, E., 161, 199, 302
Jeffrey, R. C., 525, 536
Jennings, H. S., 534, 536
Johnson, D. S., 89, 127, 198, 227

Jones, B., 348

Kailath, T., 321, 323
Kalman, R. E., 178, 315, 319, 322–323, 396
Kandel, A., 187, 200
Kant, I., 230–232, 370, 375
Kaplan, M. M., 587
Kaplan, W., 314, 322
Karlqvist, A., 197
Katona, G., 625, 629
Katz, D., 25, 200
Kaufmann, S. A., 105, 200
Kelly, G., 234, 238
Kickert, W. J. M., 161, 200
Kierkegaard, S. A., 371
Kilmer, W. L., 599, 604
King, W. R., 182, 197
Klein, F., 306
Klir, G. J., 36–37, 48, 55, 75, 91, 94–95, 102, 106, 108, 117–120, 141, 157, 173–174, 185, 187, 200, 204, 211–212, 217, 226–227, 337, 339, 351, 353–354, 396–397, 403, 599, 604
Klopf, A. H., 587, 592
Koestler, A., 30, 200, 268
Koffka, K., 25, 200
Köhler, W., 25, 200
Kohout, L., 301
Kolmogorov, A. N., 302
Krinsky, V. I., 161, 200
Krippendorff, K., 348
Krohn, K. B., 200
Kugler, P., 589, 592
Kuhn, A., 621, 628
Kuhn, T. S., 360, 375

Landau, Y. D., 161, 200
Lange, O., 182, 201, 301
Langer, S. K., 327, 335
Laplace, P. S. de, 270, 502, 545
La Porte, T. R., 182, 201
Laszlo, E., 168–169, 172, 182, 201, 302
La Violette, P. A., 175, 201
Lavoisier, A. L., 170, 201
Leduc, S., 13, 157, 201
Lee, A. M., 182, 20
Leibniz, G. W., 35
Le Moigne, J.-L., 113
Lenin, V. I., 38, 375
Leontief, W. W., 622, 628, 643–644, 648

Lerner, 173, 201
Lewis, L. F., 644–645, 648
Lilienfeld, R., 165, 167–168, 173, 201
Lindblom, C. E., 624, 628
Lindenmayer, A., 187, 202
Locke, J., 230
Lofgren, L., 201
Losch, A., 624, 629
Lotka, A. J., 525, 536, 639, 648
Lowen, W., 182, 201, 342–343
Luhmann, N., 358, 375
Luker, P. A., 187, 198

Macko, D., 102, 142, 201
MacLane, S., 277
Madden, R. F., 347, 353, 402–403
Makridakis, S., 105, 201, 227
Mandelbrot, B. B., 187, 201
March, J. G., 521, 536
Marchal, J. H., 14, 164, 201
Marchetti, C., 645–646, 648
Marconi, G., 329
Margaleff, R., 182, 201
Margolus, N., 106, 187, 204
Mariano, M., 348, 351, 353
Martin, W. T., 320, 323
Maturana, H. R., 12–13, 159, 161, 201, 204, 212, 238, 302, 559, 561, 566
Maxwell, J. C., 170, 545
May, R. M., 391, 396
Mayr, O., 37, 201
McCulloch, W. A., 180, 201, 279
McGill, W. J., 425, 448, 522, 536
Meadows, D., 361, 366, 375
Mesarovic, M. D., 37, 42, 44–46, 101–102, 142, 177, 187, 201, 227, 301
Miller, G. A., 598, 604
Miller, J. G., 163–164, 181–182, 201, 212, 448, 631, 648
Miser, H. J., 38, 202
Monod, J., 215, 268, 280, 616
Montaigne, M. E. de, 230–231, 238
Moore, E. F., 314, 316, 322
Moore, R. E., 187 202
Moorehead, P. S., 587, 592
Moreno-Diaz, R., 86, 202
Morgenstern, O., 406, 417, 624, 628
Morris, C. W., 327, 335
Mountcastle, V. B., 640, 648
Mozart, W. A., 135

Nabokov, V., 261
Nagel, E., 580, 592
Naisbitt, J., 218, 227
Negoita, C. V., 187, 202
Neumann, J. von, 161, 202, 314, 322, 389, 396, 406, 417, 564, 566, 583, 587, 592, 624, 628
Newell, A., 588, 592
Newton, I., 19, 136, 270, 303, 495, 497, 499, 505, 507, 545
Nicolis, G., 161, 202, 492, 649
Noddings, N., 230, 233, 238

Odum, E. P., 182, 202
Odum, H. T., 639, 648–649
Orchard, R. A., 342, 345, 353

Padulo, L., 187, 202, 227, 380, 396, 599, 604
Pantin, C. F. A., 262, 268
Pareto, V., 170
Parsons, T., 170, 629, 631–632, 649
Parviz, B., 108, 200, 348, 353
Pask, G., 182, 202, 375
Pattee, H. H., 202, 212, 301, 481–482, 579, 582–584, 587, 591–592, 616
Patten, B. C., 182, 202
Pearl, J., 350, 353
Pedrycz, W., 187, 202
Peirce, C. S., 370, 376
Peterson, J. L., 187, 202
Phillips, P. C., 182, 202
Piaget, J., 12, 230–235, 238, 535
Pichler, F., 86, 202, 347
Pilette, R., 105, 202
Pittarelli, M., 348
Pitts, W., 180, 201, 279
Plato, 230, 238, 367, 375, 472
Poe, E. A., 304
Poincaré, H., 3, 276, 306, 487, 525
Polak, E., 205
Polanyi, M., 583, 592, 616
Pontryagin, L. S., 311, 317–319, 322, 362
Popper, K. R., 268, 360, 367–371, 376, 593, 604
Poston, T., 187, 202
Prade, H., 187, 198, 352
Pribram, K. H., 527, 536
Prigogine, I., 161, 202, 212, 273, 483, 492, 588, 592, 642–643, 648–649
Prusinkiewicz, P., 187, 202

Author Index

Pylyshyn, Z., 588, 592

Quade, E. S., 38, 202

Ralescu, D. A., 187, 202, 363, 376
Ramer, A., 351, 353
Rapoport, A., 33, 36, 202, 501, 505, 601, 604, 621, 639, 649
Rasband, S. N., 187, 202
Rashevsky, N., 540, 542
Reese, E. J., 345, 353
Rényi, A., 116–117, 202
Rescher, N., 108, 187, 202–203, 376
Reuver, H. A., 182, 199
Rhodes, J. L., 200
Riguet, J., 515, 519
Riker, W., 624, 628
Rizzo, N. D., 182, 199
Rogers, G., 174, 200, 353
Rosen, R., 4, 161, 179, 182, 203, 211–213, 269, 301, 303, 351, 353, 477–478, 480–482, 493, 537, 548–549, 551, 553–555, 557, 577
Rosenblueth, A., 36, 329, 335, 503, 607
Rothbart, H., 342
Rozehnal, I., 106, 186–187, 199
Rozenberg, G., 68, 187, 199
Rugh, W. J., 203
Russell, B., 359, 376, 496

Sadovsky, V. N., 182
Sage, A. P., 203
Samuelson, P., 622, 628
Sandelands, L. E., 646–647, 649
Sandquist, G. M., 203
Sartre, J. P., 360, 376
Sayre, K., 182, 203
Schrödinger, E., 278
Schuster, P., 588, 591
Shafer, G., 187, 203, 350, 353
Shannon, C. E., 36, 117, 167, 203, 243, 314, 322, 327, 335, 353, 405, 410–411, 413, 416–417, 448, 533–534, 536
Seidl, L. K., 339, 353
Sigal, R., 105, 202
Simmel, G., 232, 238
Simon, H. A., 85, 115, 135, 141, 203, 212, 448, 457, 521, 536, 623, 628
Singer, E. A., Jr., 329, 335
Singh, M. G., 354

Singleton, H. E., 314, 322
Skellam, J. G., 274
Skinner, B. F., 170
Skoglund, V., 33, 83, 203
Slonimsky, N., 283
Smuts, J. C., 28, 203, 302
Smythies, J. R., 30, 200, 268
Sommerhoff, G., 329, 335, 409, 417, 519, 526–528, 536, 577
Starr, T. B., 182, 195
Staw, B. M., 646–647, 649
Stearns, R. E., 102, 199, 515, 519
Stebakov, S. A., 314, 322
Steiglitz, K., 377, 384, 395
Steinbruner, J. D., 182, 203
Steiner, G. A., 637, 648
Stevens, S. S., 640, 649
Steward, I., 187, 202
Sugeno, M., 354
Suppes, P., 31, 115, 203, 604
Susiluoto, I., 39, 204
Sutherland, J. W., 378, 396
Svoboda, A., 338–339, 341–342, 354
Szilard, L., 271
Szucz, E., 33, 83, 204

Takahara, Y., 37, 42, 102–103, 142, 187, 201, 204, 227–228
Takai, T., 187, 204
Tausner, M. R., 345, 353
Taylor, R., 503
Teller, E., 135, 204
Thom, R., 187, 204, 301, 480, 482, 588, 592
Thorn, D. C., 32, 204
Thornthwaite, C. W., 332, 335
Toffler, A., 356, 376, 649
Toffoli, T., 106, 187, 204
Trappl, R., 174, 204, 347, 354, 640–641, 649
Turchin, V. F., 63, 204
Turing, A. M., 180, 313, 321, 587
Turvey, M., 589, 592

Uexküll, J. von, 234, 238
Uribe, R. B., 159, 161, 204, 212, 302, 559
Uyttenhove, H. J., 348

Vaihinger, H., 232, 238
Valach, M., 339, 353
Valkenburg, M. E. Van, 377
Van der Waerden, B. L., 214–215

Van Gigch, J. P., 204
Varela, F. J., 11–13, 30–31, 159, 161, 199, 201, 204, 211–212, 293, 301–302, 559, 561, 566
Vasspeg, K., 339–340, 344, 347, 354
Vico, G., 12–13, 229–232, 235, 238
Volterra, V., 320, 323

Waddington, C. H., 179, 588, 592
Walker, C. C., 105, 198, 204, 227–228
Walras, L., 622, 628
Wang, Z., 187, 204
Warfield, J. N., 182, 204
Watson, J. D., 178
Watts, J., 37
Way, E. C., 353, 403
Weaver, W., 20, 137, 204, 212, 335, 405, 417, 448–449, 536
Wedde, H., 355, 376
Weinberg, G. M., 9, 136, 204, 212, 501
Weir, M., 204
Weiss, P. A., 26, 204
Wertheimer, M., 25
Weyl, H., 545, 557
Whitson, T. G., 645, 649
Whitehead, A. N., 382–383, 396
Wiener, N., 34–36, 167, 180, 204, 269, 310–311, 321, 323, 327, 329, 335, 337, 376, 503

Wilder, R. L., 383, 396
Wiley, M. L., 360, 376
Windeknecht, J. G., 101, 204, 228
Wing, J., 319, 322
Wittgenstein, L., 365, 376, 607
Wolfram, S., 106, 187, 204
Wright, S., 623, 628
Wymore, A. W., 46, 204

Xenophanes, 230

Yager, R. R., 187, 351, 354
Yates, F. E., 161, 204
Yovitz, M. C., 161, 204–205
Yudin, E. G., 182

Zadeh, L. A., 37, 46, 120, 143, 205, 211, 309, 354, 376
Zeeman, E. C., 187, 205
Zeigler, B. P., 48–49, 72, 205, 228, 378, 380, 382, 392–393, 396
Zeleny, M., 157–158, 161, 200, 205
Ziegenhagen, E. A., 182, 205
Ziman, J., 260, 267
Zimmermann, H. J., 187, 205
Zopf, G. W., 161, 198

Subject Index

Abduction, 103
Abstraction, 16–17, 73, 382, 479–480, 553
Abstract system, 17
Acoustic system, 32
Adaptation, 63, 576
Adaptive system, 143, 149, 152, 154, 161
Algebra, 389
Algebraic equations, 33
Algebraic topology, 276–277
Algorithms, 127–129
    design of, 126
    exponential time, 129
    intractable, 131
    polynomial time, 129
    theory of, 126
    tractable, 131
Allopoietic systems, 561
Ampliative reasoning, 350, 399
Analog computer, 33, 81, 279, 338
Anticipatory behavior, 544–547, 556–557
Anticipatory system, 154, 161, 190, 281–282, 537, 546–547, 553, 555
Arithmetic, 389
Artificial intelligence, 37
Automata, 379–381, 388
Automata theory, 46, 598–599
Autonomy, 296, 298, 300, 561
Autonomous system, 298
Autopoiesis, 63
Autopoietic system, 158–161, 560–565

Basic variable, 49, 220
Behavior function, 56
Behavior probabilities, 56
Belousov–Zhabotinsky reaction, 486
Behavior system, 345

Bénard cell, 484–485
Bijective function, 192
Biological system, 298, 459, 484, 540, 608–609, 613, 615–616
Biology, 25, 29, 177–182, 269, 271–273, 282, 294, 409, 449, 474, 521, 537, 540, 546, 555, 559, 585, 607–612, 617
    molecular, 272, 584, 616
    organismic, 26
    relational, 540, 553
Bit, 117
Black box, 252–257, 315, 320, 516
Boundary, 564
Bremermann's limit, 121–126, 136, 226, 508

Calculus, 19
Cartesian product, 9–10, 192, 398
Catastrophe theory, 273
Categories of systems, 14
Category theory, 277
Causal explanation, 285
Chaos, 28
Church's thesis, 128
Circuit,
    electric, 32, 37, 338
    generalized, 32, 37, 338
    switching, 37
Circuit theory, 309, 316
Classical science, 6–7, 13
Classification,
    of science, 14
    of systems, 14–17
Coarsening of variables, 139
Cognitive sciences, 182
Combined machine, 95–97

Communication, 34, 36–37, 287, 289–290, 327, 419, 522
Communication theory, 36, 291, 312, 410, 412, 598
Complement of a set, 191
Complexity, 20, 113–116, 135, 185, 249–250, 263, 358, 363, 400–401, 449, 458, 475, 477, 483–484, 490–491, 503, 608–609
  computational, 88, 121, 126–127, 133–134, 173–174
  descriptive, 115–116, 119, 139
  disorganized, 20–23, 450–451
  organized, 21–24, 119, 137, 177, 185, 451–454
  uncertainty-based, 119, 139
Complex system, 281, 457–458, 464, 468, 470, 473, 475, 481, 483, 490, 499, 616–617
Computational complexity, 88, 121, 126–127, 133–134, 173–174
Computer, 24, 81, 83, 86, 102–103, 151, 368–369, 377–378, 388–389, 502
  analog, 33, 81, 279, 338
  as laboratory, 102–106, 172, 185
  digital, 37, 81, 377, 388, 395
  hybrid, 8
Computer technology, 19, 24, 185, 218, 225–226, 403
Conceptual frameworks, 41
  deductive approach to, 41–46
  inductive approach to, 41–42, 46–48
Conflict resolution, 365
Congruence, 306
Consistent system, 74
Constraint analysis, 397, 402
Constraints, 75, 77, 84–85, 234, 387, 398, 419, 522–523
Construction system, 13
Constructivism, 12–13, 229–237
Context, 287
Continuous model, 377–378, 388–392
Continuous system, 380, 388, 392
Continuous-time system, 385, 393–394
Continuous variable, 49, 57, 74
Control, 34, 36–37
  feedback, 37
Control theory, 36–37, 46, 296, 317–318
Crisp variable, 49, 74
Cryptography, 135
Cybernetic explanation, 285–286, 288

Cybernetics, 34–37, 241, 333, 337, 457
Cylindric extension, 402

Dataless system, 49
Data system, 51–53, 68–69, 221
Decision-making, 85, 156, 161, 537
Decomposition, 85, 420
Degrees of freedom, 255
Dempster–Shafer theory, 400
Descriptive complexity, 115–116, 119, 139
Deterministic system, 56, 74
Difference equation, 377, 380, 387, 390–394
Differential equations, 19, 33, 53, 57, 63, 82, 166, 295, 310, 314, 377–383, 388–394, 472, 503, 507, 525, 545, 610, 622
Digital computer, 37, 81, 377, 388, 395
Directed system, 49, 221
Discrete Fourier transform, 386
Discrete function, 383–387, 390–391, 393
Discrete model, 377–378, 380, 388–389, 391–392
Discrete-state system, 314
Discrete-time system, 379, 381–383, 388, 391–394
Discrete variable, 49, 74
Disorganized complexity, 20–23
Dissipative structure, 484
Dissipative system, 486–487, 491
Dissonance, 118–119, 351
Distinction, 10–14, 294–295
Dynamic programming, 317
Dynamic system, 45, 95, 97, 99, 523

Ecology, 182
Economics, 181–182, 621–628
Effector system, 550
Electric circuit, 32, 37, 338
Emergent properties, 135, 264
Empiricism, 269, 271
Empty set, 191
Energy, 287
Engineering, 19, 32, 37–38, 82, 84
Environment, 11, 48–49, 66, 242, 293–294, 297, 537, 549–550
Epistemological categories of systems, 48, 68–69, 74, 79
Epistemological hierarchy of systems, 48, 58, 68–69, 74, 83, 220, 222, 344
Equifinality, 272–273, 279
Equilibrium, 622

Subject Index

Equivalence classes
  of isomorphic systems, 16
  of systems, 15
Equivalence relation, 5, 15, 193
Equivalent states, 316
Equivalent systems, 316
Error, 481
Error correction, 152
Evolution, 28, 63, 135, 285, 303, 461–463, 587
Evolution theory, 285
Expansionism, 327
Experimental frame, 49, 344
Explanatory model, 83

Family of sets, 191
Feedback, 91, 93–94, 281, 288, 625
Feedback control, 37
Feedforward, 281
Finite-memory machine, 95–98
Finite-state automata, 37, 379
Finite-state machines, 53
  time-varying, 62
First world, 367, 370–373
Fourier transform, 386, 390
Function, 192
  discrete, 383–387, 390–391, 393
  real-valued, 384–386
Functional analysis, 379, 384
Fuzziness, 118–119
Fuzzy measure theory, 118–119, 142
Fuzzy variable, 49, 74
Fuzzy set theory, 118–120, 142

Game theory, 38, 624
Gedanken experiments, 81
Generalized circuit, 32, 37, 338
General system, 16–17, 33, 42, 222, 621–622
General systems problem solver (GSPS), 75–77, 86, 220, 328–329
General systems research, 33–34, 36
General systems theory, 10–11, 33–34, 38, 163, 166, 188–189, 239–243, 248–249, 251–252, 257, 269–270, 272–274, 277, 457, 525, 593–594, 597–604, 608
Generated variable, 54–46
Generating variable, 54–56
Generative system, 53, 56, 68–69, 90, 221
Genetic epistemology, 230
Genetic networks, 280

Geometry, 306–307
Gestalt, 25, 30
Gestalt psychology, 25
Gestalt theory, 25
Global inconsistency, 399
Global model, 356, 365
Goal, 143–145, 148–149
Goal-implementing element, 146, 149
Goal-oriented system, 143–147, 149–152
  paradigm of, 146–148
  feedback paradigm, 148
  feedforward paradigm, 148
  full-information paradigm, 148, 150
  informationless paradigm, 148
Goal-seeking element, 146, 149, 151
Goal-seeking system, 43–45, 148–151
Goal-seeking traits, 145
Goal-seeking variable, 146, 148
Growth theory, 243
GSPS framework, 42, 48, 71–72, 74

Hartley measure of uncertainty, 350
Hierarchically organized system, 141–142, 458, 465, 471, 615
Hierarchy, 30, 458–460, 462, 465, 468–470
  of epistemological systems categories, 48, 58, 68–69, 74, 83, 220, 222, 344
  of levels, 294–299
Holism, 24–26, 28–33, 173, 215, 263–264, 299, 301, 398, 402
Holon, 30
Homeostasis, 279
Homomorphic relation, 77–79, 516
Humanities, 333–334
Human mind, 12–13, 23
Hybrid computer, 8

Identification problem, 348, 398–399
Inconsistent system, 399
Inconsistency
  global, 399
  local, 399
Industrial society, 217, 225
Information, 26, 36–37, 119, 218, 254–257, 286–287, 289–292, 351, 368, 399–402, 470, 638
Information processing, 36
Information society, 7, 189–190, 218, 225
Information transmission, 118

Information theory, 241, 243, 312, 419–420, 426, 507, 534, 598–600
Input-output system, 43–45, 643–644
Input variable, 49, 221
Instrumentalism, 230
Intelligence, 413–414
Internal description, 610
Internal model, 282
Internal variable, 221
International Federation for Systems Research (IFSR), 39
International Society for the Systems Sciences (ISSS), 34
Interpretation, 16–17, 73
Interpreted system, 16–17, 78–80, 82
Interpretive structural modeling, 182
Intersection of sets, 192
Investigator, 48–49
Isomorphic relation, 15, 77, 515, 517
Isomorphic systems, 15, 222
Isomorphy, 32–33

Knowledge, 12, 14, 101, 231, 233, 237, 240, 357, 360
Knowledge structure, 101

Language, 579–580, 583–587, 590–591
Lattice, 194
Law of constraint loss, 442
Law of requisite hierarchy, 182–184, 219
Law of requisite variety, 155, 182–183, 219, 400, 405, 408, 413, 416, 533, 598
Laws of information, 102, 419–447
Laws of systems, 103–105, 173
Learning, 135, 491, 645
Linear systems, 74
Living systems, 632–637
Local inconsistency, 399
L-systems, 68

Machine, 494, 525
  combined, 95–97
  finite-memory, 95–98
  finite-state, 53
  Mealy, 95–98
  Moore, 95–97
  time-varying, 62
Magnetic system, 32
Management, 37–38, 182
Management science, 241, 248, 259, 333

Markov chain, 53
Mask, 55
Mathematical system, 80–83, 274, 525
Mathematical systems theory, 37, 101, 166, 219
Mathematical transformations, 83
Mathematics, 19, 24, 35, 83, 87–88, 166, 186–187, 224, 235–236, 239, 274–277, 311, 361, 372, 617
  continuous, 379
  discrete, 379
Maximal compatible, 193
Mealy machine, 95–98
Measurement, 271–272, 579–591
Mechanical system, 32
Medicine, 22, 449
Memory, 256
Metamethodological inquiry, 106, 134
Metasystem, 62–69, 222, 345
  of higher order, 62, 222
Method identification problem, 90
Methodological distinctions, 48–49, 74
Methodological paradigm, 89–91
  assumption free, 89, 91
Methods
  statistical, 19, 21, 23
  analytical, 23
  optimization, 38
Model, 77–86, 102, 382, 511–512, 514, 548, 556
  continuous, 377–378, 388–392
  discrete, 377–378, 380, 388–389, 391–392
  explanatory, 83
  global, 356, 365
  predictive, 83, 543, 546, 556
  prescriptive, 84–85
  retrodictive, 83
  internal, 282
Modeling relation, 79–80, 276–277, 281, 307
Modeling system, 77, 549
Molecular biology, 272
Moore machine, 95–97
Morphogenesis, 63
Multigoal-oriented system, 149—151

Natural law, 495–496
Natural selection, 285–286, 623
Nearly decomposable system, 465–468
Neural networks, 279–280
Neutral system, 49, 221
Newtonian mechanics, 495–497, 609

Subject Index

Nondeterministic system, 56, 74
Nonlinear system, 74
Nonspecificity, 118–119, 351
Novelty, 349
NP-complete problems, 132–133
NP-problems, 132
Numerical analysis, 380–381, 392

Object, 48
Objectives, 75, 77, 84–85
Open systems, 263, 272–273, 614
Operations research, 38, 313, 333, 416
Operon, 280
Order, 28
Organismic biology, 26
Organization, 521, 523, 525–528, 559–561, 571, 626, 638
Organization theory, 241, 523
Organized complexity, 21–24, 119, 137, 177, 185
Orthogonality, 320
Original system, 398–399, 401–402
Output variable, 49, 221
Overall system, 398–399, 401–402

Partial ordering, 194
Partition, 192
Partition law, 430
Parts, 24–27, 30–32, 250, 254, 264, 328, 397, 402, 523–525, 571
Performance, 143–145, 148
  of a method, 88
Performance modeling, 85
Perspective
  of classical science, 16–17
  of systems science, 16–17
Philosophy, 29, 182, 593, 602
Physics, 19, 32, 175–177, 272–273, 278–279, 332, 498–499, 545, 607, 609–612
Planning, 542–543, 548–551, 554, 557
Pluralism, 364
Policy-making 85
Political science, 182
Pontryagin maximum principle, 362
Possibility theory, 350, 400
Pragmatism, 230
Precision, 363
Predictive model, 83, 543, 546, 556
Pre-industrial society, 217, 225
Prescriptive model, 84–85

Primitives, 13–14
Primitive system, 49
Principle of incompatibility, 120
Principle of the least risk, 400
Principle of maximum uncertainty, 350, 400–402
Principle of minimum uncertainty, 350, 401
Probabilistic system, 314
Probability, 287
Probability theory, 117–119
Problem requirement, 75
Problem solvability, 128, 132–133
Projection, 402
Protocol, 253–255
Psychology, 25
  gestalt, 25

Q-analysis, 182
Quantum theory, 278

Randomness, 20–21, 489
Reactive system, 546
Realism, 231
Reality, 357–359, 370, 373, 399
Real-valued function, 384–386
Reconstructability analysis, 106, 141, 173–174, 348, 397–403
Reconstruction family, 399–400, 402
Reconstruction hypothesis, 401–402
Reconstruction principle of inductive reasoning, 106—107, 348
Reconstruction problem, 106–107, 110, 348–349, 398, 400–402
Reconstruction rule, 108–111
Reductionism, 24–26, 30–32, 173, 251, 261, 264, 269, 271, 278, 299, 301, 325, 327, 398, 402, 479, 608–609, 617
Redundancy, 288–290, 471
Refinement lattice, 401
Refinement relation, 401
Regulation, 155, 161, 400, 406–413, 511–515, 519
Relabeling, 16–17
Relation, 5, 9–10, 192, 219, 398
  compatibility, 193
  equivalence, 5, 15, 193
  homomorphic, 77–79
  isomorphic, 15, 77
  many-dimensional, 402
  modeling, 79–80, 276–277, 281, 307

Relation (*Cont.*)
  ordering, 194
  space-invariant, 57
  support-invariant, 53
  time-invariant, 55
Replacement procedure, 62–66
Reproduction, 561, 571, 575–576
Retrodictive model, 83

Sampling theorem, 385
Sampling variable, 54–55
Scepticism, 359–360
Science, 29, 31, 35, 73, 83, 101, 218, 237, 240, 249–250, 259–264, 331, 333, 449, 454–456, 472, 611
  classical, 6–7, 13, 491
  normal, 360
  two-dimensional, 217, 225
Sciences
  behavioral, 22
  cognitive, 182
  environmental, 22
  life, 22
  social, 22
Scientific ontology, 603–604
Scientific theory, 594–597
Second world, 367, 370–373
Self-organization, 28, 63
Self-organizing system, 156–158, 161, 528–530, 577, 642
Self-preserving system, 155
Self-reproducing system, 161, 571, 587
Self-reproduction, 135, 571–573, 576–577
Self-steering, 183–184
Semantic closure, 583, 585
Semilattice, 194
Set, 4–5, 9, 191, 213
Setness, 4, 213–215
Set theory, 214–215
Shannon cross-entropy, 402
Shannon entropy, 117—118, 155, 350–351, 400, 423–424, 426
Similitude, 33, 83
Simplicity, organized, 20–23
Simplification, 80, 135–142, 350, 504–505, 510, 616
Simplification principle, 401
Simulation, 102, 387, 392, 395
  computational, 387, 391
  symbolic, 387

Social sciences, 181–182, 459, 621
Social system, 298
Society for General Systems Research (SGSR), 33–34, 39, 621
Source system, 50–51, 68–69, 221, 344
Space complexity function, 134
Space-invariant relation, 57
Stability, 273, 279, 381
  structural, 480
State, 47, 313, 494
State-determined system, 494, 496, 499
State equation, 314
State set, 47, 398
State-space, 315, 610
State-transition system, 345
Statistical mechanics, 19
Steering, 183
Straight rule, 108–111
Strange attractor, 487, 489–490
Structuralism, 259
Structural stability, 480
Structure design, 90
Structure system, 58–62, 68–69, 90–91, 95, 221, 398
Subset, 191
Subsystem, 397–401, 420, 458–459, 465–469, 479, 616–618, 632–633
Support, 49
Supporting variable, 49, 220
Support-invariant relation, 53, 221
Support set, 49
Switching circuit, 37, 379
Switching circuit theory, 37, 379
Synergetics, 157
System
  abstract, 17
  acoustic, 32
  adaptive, 143, 149, 152, 154, 161
  allopoietic, 561
  anticipatory, 154, 161, 190, 281—282, 537, 546–547, 553, 555
  autonomous, 298
  autopoietic, 158–161, 560–565
  behavior, 345
  biological, 298, 459, 484, 540, 608–609, 613, 615–616
  complex, 281, 457–458, 464, 468, 470, 473, 475, 481, 483, 490, 499, 616–617
  consistent, 74
  construction of, 13

Subject Index

System (*Cont.*)
continuous, 380, 388, 392
continuous-time, 385, 393–394
data, 51–53, 68–69, 221
dataless, 49
definition of, 4–5, 9–14, 42, 163–164, 219, 310, 327, 398, 494
deterministic, 56, 74
directed, 49, 221
discrete-state, 314
discrete-time, 379, 381–383, 388, 391–394
dissipative, 486–487, 491
dynamic, 45, 95, 97, 99, 382, 523, 614
general, 16–17, 33, 42, 222, 621–622
generative, 53, 56, 68–69, 90, 221
goal-oriented, 143–147, 149–152
hierarchically organized, 141–142, 458, 465, 471, 615
inconsistent, 74, 399
input-output, 43–45, 643–644
interpreted, 16–17, 78–80, 82
isomorphic, 15, 222
Lindenmeyer (or L-), 68
linear, 74
living, 632–637
magnetic, 32
mathematical, 80–83, 274, 525
mechanical, 32
modeling, 77, 549
multigoal-oriented, 149–151
nearly decomposable, 465–468
neutral, 49, 221
nondeterministic, 56, 74
nonlinear, 74
open, 263, 272–273, 614
original, 77–81
overall, 398–399, 401–402
primitive, 49
probabilistic, 314
reactive, 546
self-organizing, 156–158, 161, 528–530, 577, 642
self-preserving, 155
self-reproducing, 161, 571, 587
social, 298
source, 50–51, 68–69, 221, 344
state-determined, 494, 496, 499
state-transition, 345
structure, 58–62, 68–69, 90–91, 95, 221, 398
thermal, 32

Systemhood, 4–7, 9–10, 13, 15–17, 32, 41, 50, 71–72, 87, 214–215
Systemology, 333
Systems analogy, 614
Systems analysis, 38, 165–167, 169
Systems classification, 320
Systems design, 77, 84
Systems engineering, 38, 313
Systems identification, 223, 320
Systems inquiry, 77
Systems knowledge, 101–102, 173, 219
Systems metamethodology, 87–88, 174, 224
Systems methodology, 71–75, 88, 173, 223–224, 265
hard, 71
soft, 71
Systems modelling, 77, 119
discovery approach to, 83–84
postulational approach to, 83–84
inductive, 348–350
Systems movement, 19, 24, 28, 30–34, 36–39, 263–265, 267, 346
Systems problem, 71–76, 87
Systems science, 13–14, 19, 23–24, 28, 41, 86–87, 101, 163, 169, 172–175, 177, 181–182, 185–190, 218–220, 225–226, 351–352, 378, 398
criticism of, 163–169
definition of, 3–7, 136
Systems theory, 309–313, 316, 319, 380, 507–508, 510, 632
Systems thinking, 19, 31

Taxonomy of systems, 220, 222, 340, 344
Technological primitives, 84–85
Tessellation automata, 62–63
Tektology, 38
Teleology, 327, 329, 495, 503, 546
Theory of similarity, 33, 83
Thermal system, 32
Thermodynamics, 273, 487
Thinghood, 4–6, 13, 16–17, 32, 50, 71–72, 213–215
Third world, 367–373
Time, 332, 556
Time complexity function, 129–130
Time-invariant relation, 55
Time-varying machine, 62
Topological boundary, 158–161
Topological invariants, 276

Topology, 276
  algebraic, 276–277
Total ordering, 194
Transcomputational problem, 132, 125
Translation rule, 53, 221
Transmission, 424–425
Turing machine, 127–128, 134, 313
  deterministic, 128
  nondeterministic, 128, 132
Two-dimensional science, 331, 333

Uncertainty, 116–119, 185, 350, 399–401, 424
Uncertainty-based complexity, 119, 139
Union of sets, 192
Universe of discourse, 191
Utility function, 156
U-uncertainty, 350–351

Variable, 47, 74, 220, 398, 498
  basic, 49, 220
  continuous, 49, 57, 74

Variable (*Cont.*)
  crisp, 49, 74
  discrete, 49, 74
  fuzzy, 49, 74
  generated, 54–56
  generating, 54–56
  goal-seeking, 146, 148
  input, 49, 221
  internal, 221
  output, 49, 221
  sampling, 54–55
  supporting, 49, 220
Variance analysis, 426
Variety, 405–406

Wholes, 24–32, 250, 264, 328, 397, 402, 523–525, 559, 571
World
  first, 367, 370–373
  second, 367, 370–373
  third, 367–373

CPSIA information can be obtained
at www.ICGtesting.com
Printed in the USA
BVHW010157100821
614077BV00002B/32